Computational Electromagnetics with MATLAB®
Fourth Edition

Computational Electromagnetics with MATLAB®

Fourth Edition

Matthew N.O. Sadiku

CRC Press
Taylor & Francis Group
Boca Raton London New York

CRC Press is an imprint of the
Taylor & Francis Group, an **informa** business

MATLAB® is a trademark of The MathWorks, Inc. and is used with permission. The MathWorks does not warrant the accuracy of the text or exercises in this book. This book's use or discussion of MATLAB® software or related products does not constitute endorsement or sponsorship by The MathWorks of a particular pedagogical approach or particular use of the MATLAB® software.

CRC Press
Taylor & Francis Group
6000 Broken Sound Parkway NW, Suite 300
Boca Raton, FL 33487-2742

First issued in paperback 2022

ISBN-13: 978-1-138-55815-1 (hbk)
ISBN-13: 978-1-03-233903-0 (pbk)
DOI: 10.1201/9781315151250

This book contains information obtained from authentic and highly regarded sources. Reasonable efforts have been made to publish reliable data and information, but the author and publisher cannot assume responsibility for the validity of all materials or the consequences of their use. The authors and publishers have attempted to trace the copyright holders of all material reproduced in this publication and apologize to copyright holders if permission to publish in this form has not been obtained. If any copyright material has not been acknowledged please write and let us know so we may rectify in any future reprint.

Publisher's Note

The publisher has gone to great lengths to ensure the quality of this reprint but points out that some imperfections in the original copies may be apparent.

**Visit the Taylor & Francis Web site at
http://www.taylorandfrancis.com**

**and the CRC Press Web site at
http://www.crcpress.com**

eResource material is available for this title at https://www.crcpress.com/9781138558151

To my teacher

Carl A. Ventrice

and my parents

Ayisat and Solomon Sadiku

Contents

Preface

Since the third edition of this book was published, there has been a noticeable increase of interest in computational electromagnetics (CEM), also known as numerical electromagnetics. This is evident by the amount of dissertations, theses, books, and articles on CEM appearing in journals and conferences each year. Along with this development is the rapid growth in commercial or free codes for designing complex EM problems. In spite of these cheap and powerful computational tools, there is a need to learn the fundamental analytical and numerical concepts behind the codes. It is beneficial to understand the inherent limitations of the commercial software. Experience shows that students learn more by developing their codes than just pushing buttons in a commercial software package. Also, a closer look at the newly published books reveals that they are not suitable for classroom use due to lack of examples and practice problems at the end of each chapter. There is still a need for a good introductory textbook for the CEM community. This book meets the need.

The book has the following features:

- It is comprehensive. Some CEM books cover just one numerical technique, while some cover only finite difference method (particularly FDTD), finite element method, and method of moments. In addition to these, this present book covers variational methods, transmission-line-modeling (TLM), method of lines, and Monte Carlo method.

- It presents several examples with MATLAB codes where applicable. I believe that CEM is best learned through direct programming. Commercially packaged programs can be useful, but they should not take priority over direct programming.

- It provides several end-of-chapter problems with answers to odd-numbered problems in Appendix E.

- Each chapter presents a clear, concise introduction to a numerical method in EM and provides up-to-date references to information on the method. The last section of the chapter is devoted to application(s) of the method.

When the first edition of this book was written, the term "Computational Electromagnetics" was not common. Now, it is the most common term used in describing the emerging field. Since it is expedient to use the latest term or development, the former name of the book, *Numerical Techniques in Electromagnetics*, has been changed to *Computational Electromagnetics with MATLAB®*. This is part of the process of making another edition—catching up with the trends in this exciting field.

Although the book can be covered in one semester, enough material is provided for two-semester coverage. For two-semesters, it is suggested that Chapters 1 through 5 be covered in one semester, while Chapters 6 through 9 is covered in the second semester. In addition to serving as an introductory text for students, the book will also serve as a concise, up-to-date reference for researchers and professionals in CEM.

The book provides a comprehensive bibliography that serves as the best resources for learning more about CEM. Appendix A is on vector analysis, while Appendix B provides programming in MATLAB. Appendix C covers briefly direct and iterative procedure for solving simultaneous equations. Appendix D provides a list of software packages that are either free or commercially available. Appendix E provides answers to odd-numbered problems.

Since the publication of the last edition, there has been increased awareness and utilization of computational tools. This edition adds noticeable changes in Section 5.2 on how moment methods can be used to solve differential equation, in section 310 on advanced applications of FDTD, in Section 6.12 on using a commercial solver to analyze microstrip lines, and in Section 8.7 on applying Monte Carlo Markov chain to Poisson equation.

MATLAB® is a registered trademark of The MathWorks, Inc. For product information, please contact:

The MathWorks, Inc.
3 Apple Hill Drive
Natick, MA 01760-2098 USA
Tel: 508 647 7000
Fax: 508-647-7001
E-mail: info@mathworks.com
Web: www.mathworks.com

Acknowledgment

I would like to thank Dr. Sarhan Musa of Prairie View A&M University, Dr. Andrew Peterson of Georgia Institute of Technology, Dr. Jian-Ming Jin of the University of Illinois at Urbana-Champaign, and Dr. David Davidson of University of Stellenbosch for allowing me to use their works. I would like to acknowledge the support of Dr. Shield Lin, dean of the College of Engineering, and Dr. Pamela Obiomon, head of the Department of Electrical and Computer Engineering, at Prairie View A&M University. I am also grateful to Nora Konopka, Kyra Lindholm, and other staff of CRC Press for their professional touch on the book. I express my profound gratitude to my wife, Kikelomo, for her sacrifices and prayer.

A Note to Students

Before you embark on writing your own computer program or using the ones in this text, you should try to understand all relevant theoretical backgrounds. A computer is no more than a tool used in the analysis of a problem. For this reason, you should be as clear as possible what the machine is really being asked to do before setting it off on several hours of expensive computations.

It has been well said by A. C. Doyle that, "It is a capital mistake to theorize before you have all the evidence. It biases the judgment." Therefore, you should never trust the results of numerical computation unless they are validated, as least in part. You validate the results by comparing them with those obtained by previous investigators or with similar results obtained using a different approach, which may be analytical or numerical. For this reason, it is advisable that you become familiar with as many numerical techniques as possible.

The references provided at the end of each chapter are by no means exhaustive but are meant to serve as the starting point for further reading.

Author

Matthew N.O. Sadiku earned his BSc in 1978 from Ahmadu Bello University, Zaria, Nigeria and his MSc and PhD from Tennessee Technological University, Cookeville, TN in 1982 and 1984, respectively. From 1984 to 1988, he was an assistant professor at Florida Atlantic University, Boca Raton, FL, where he worked as a graduate in computer science. From 1988 to 2000, he was at Temple University, Philadelphia, PA, where he became a full professor. From 2000 to 2002, he was with Lucent/Avaya, Holmdel, NJ as a system engineer and with Boeing Satellite Systems, Los Angeles, CA as a senior scientist. He is presently a professor of electrical and computer engineering at Prairie View A&M University, Prairie View, TX.

He is the author of over 450 professional papers and over 70 books including *Elements of Electromagnetics* (Oxford University Press, 7th ed., 2018), *Fundamentals of Electric Circuits* (McGraw-Hill, 6th ed., 2017, with C. Alexander), *Computational Electromagnetics with MATLAB®* (CRC, 4th ed., 2018), *Metropolitan Area Networks* (CRC Press, 1995), and *Principles of Modern Communication Systems* (Cambridge University Press, 2017, with S. O. Agbo). In addition to the engineering books, he has written Christian books including *Secrets of Successful Marriages, How to Discover God's Will for Your Life*, and commentaries on all the books of the New Testament Bible. Some of his books have been translated into French, Korean, Chinese (and Chinese Long Form in Taiwan), Italian, Portuguese, and Spanish.

He was the recipient of the 2000 McGraw-Hill/Jacob Millman Award for outstanding contributions in the field of electrical engineering. He was also the recipient of Regents Professor award for 2012–2013 by the Texas A&M University System. He is a registered professional engineer and a fellow of the Institute of Electrical and Electronics Engineers (IEEE) "for contributions to computational electromagnetics and engineering education." He was the IEEE Region 2 Student Activities Committee Chairman. He was an associate editor for *IEEE Transactions on Education*. He is also a member of Association for Computing Machinery (ACM) and American Society of Engineering Education (ASEE). His current research interests are in the areas of CEM, computer networks, and engineering education. His works can be found in his autobiography, *My Life and Work* (Trafford Publishing, 2017) or his website: www.matthewsadiku.com. He currently resides with his wife Kikelomo in Hockley, Texas. He can be reached via email at sadiku@ieee.org

1

Fundamental Concepts

We must keep innovating in order to stay relevant.

—Joel Comiskey

1.1 Introduction

Scientists and engineers use several techniques in solving continuum or field problems. Loosely speaking, these techniques can be classified as experimental, analytical, or numerical. The three are related as shown in Figure 1.1 [1]. Experiments are expensive, time consuming, sometimes hazardous, and usually do not allow much flexibility in parameter variation. However, every numerical method, as we shall see, involves an analytic simplification to the point where it is easy to apply the numerical method. Notwithstanding this fact, the following methods are among the most commonly used in electromagnetics (EM).

A. Analytical methods (exact solutions)
   1. Separation of variables
   2. Series expansion
   3. Conformal mapping
   4. Integral solutions, for example, Laplace and Fourier transforms
   5. Perturbation methods

B. Numerical methods (approximate solutions)
   1. Finite difference method
   2. Method of weighted residuals
   3. Moment method
   4. Finite element method
   5. Transmission-line modeling
   6. Monte Carlo method
   7. Method of lines

The numerical techniques mentioned above are usually known as low-frequency methods because they solve Maxwell's equation without making approximations and are limited to geometries of small electrical size. High-frequency methods include optical physics, geometrical theory of diffraction, and physical theory of diffraction. These techniques are too specialized and will not be covered in this book.

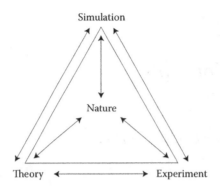

FIGURE 1.1
Relationship between experiment, theory, and simulation [1].

In the past 50 years, the electromagnetic (EM) community has witnessed a breathtaking evolution in the way we solve and apply EM concepts. With the ever-increasing power and memory of the digital computers, the art of computational electromagnetics (CEM) has gained momentum.

CEM deals with numerical methods applied in solving EM problems. It is based on computer implementation of mathematical models of EM systems using Maxwell equations. CEM tools are useful in analyzing and designing power systems, electrical machines, generators, transformers, microwave networks, waveguides, antennas, and aircraft. They are also used in predicting the electromagnetic compatibility (EMC) between complex electronic systems and their environment. For this reason, CEM is of increasing importance to the civil and defense sectors [2].

Application of these methods is not limited to EM-related problems; they find applications in other continuum problems such as in fluid, heat transfer, and acoustics.

As we shall see, some of the numerical methods are related and they all generally give approximate solutions of sufficient accuracy for engineering purposes. Since our objective is to study these methods in detail in the subsequent chapters, it may be premature to say more than this at this point.

The need for numerical solution of EM problems is best expressed in the words of Paris and Hurd: "Most problems that can be solved formally (analytically) have been solved." Until the 1940s, most EM problems were solved using the classical methods of separation of variables and integral equation solutions. Besides the fact that a high degree of ingenuity, experience, and effort were required to apply those methods, only a narrow range of practical problems could be investigated due to the complex geometries defining the problems.

EM started in the mid-1960s with the availability of modern high-speed digital computers. Since then, considerable effort has been expended on solving practical, complex EM-related problems for which closed-form analytical solutions are either intractable or do not exist. The numerical approach has the advantage of allowing the actual work to be carried out by operators without a knowledge of higher mathematics or physics, with a resulting economy of labor on the part of the highly trained personnel.

Before we set out to study the various techniques used in analyzing EM problems, it is expedient to remind ourselves of the physical laws governing EM phenomena in general. This will be done in Section 1.2. In Section 1.3, we shall become acquainted with different ways EM problems are categorized. The principle of superposition and the uniqueness theorem will be covered in Section 1.4.

1.2 Review of EM Theory

The whole subject of EM unfolds as a logical deduction from eight postulated equations, namely, Maxwell's four field equations and four medium-dependent equations [3–6]. Before we briefly review these equations, it may be helpful to state two important theorems commonly used in EM. These are the divergence (or Gauss's) theorem,

$$\oint_s \mathbf{F} \cdot d\mathbf{S} = \int_v \nabla \cdot \mathbf{F} \, dv \tag{1.1}$$

and Stokes's theorem

$$\oint_L \mathbf{F} \cdot d\mathbf{I} = \int_S \nabla \times \mathbf{F} \cdot d\mathbf{S} \tag{1.2}$$

Perhaps the best way to review EM theory is by using the fundamental concept of electric charge. EM theory can be regarded as the study of fields produced by electric charges at rest and in motion. Electrostatic fields are usually produced by static electric charges, whereas magnetostatic fields are due to motion of electric charges with uniform velocity (direct current). Dynamic or time-varying fields are usually due to accelerated charges or time-varying currents.

1.2.1 Electrostatic Fields

The two fundamental laws governing these electrostatic fields are Gauss's law,

$$\oint \mathbf{D} \cdot d\mathbf{S} = \int \rho_v dv \tag{1.3}$$

which is a direct consequence of Coulomb's force law, and the law describing electrostatic fields as conservative,

$$\oint \mathbf{E} \cdot d\mathbf{I} = 0 \tag{1.4}$$

In Equations 1.3 and 1.4,

$\mathbf{D}$ = the electric flux density (C/m^2)

ρ_v = the volume charge density (C/m^3)

$\mathbf{E}$ = the electric field intensity (V/m)

The integral form of the laws in Equations 1.3 and 1.4 can be expressed in the differential form by applying Equations 1.1 through 1.3 and Equations 1.2 through 1.4. We obtain

$$\nabla \cdot \mathbf{D} = \rho_v \tag{1.5}$$

and

$$\nabla \times \mathbf{E} = 0 \tag{1.6}$$

The vector fields **D** and **E** are related as

$$\mathbf{D} = \varepsilon \mathbf{E} \tag{1.7}$$

where ε is the dielectric permittivity (F/m) of the medium. In terms of the electric potential V (V), **E** is expressed as

$$\mathbf{E} = -\nabla V \tag{1.8}$$

or

$$V = -\int \mathbf{E} \cdot d\mathbf{I} \tag{1.9}$$

Combining Equations 1.5, 1.7, and 1.8 gives Poisson's equation:

$$\nabla \cdot \varepsilon \nabla V = -\rho_v \tag{1.10a}$$

or, if ε is constant,

$$\boxed{\nabla^2 V = -\frac{\rho_v}{\epsilon}} \tag{1.10b}$$

When $\rho_v = 0$, Equation 1.10 becomes Laplace's equation:

$$\nabla \cdot \varepsilon \nabla V = 0 \tag{1.11a}$$

or for constant ε

$$\boxed{\nabla^2 V = 0} \tag{1.11b}$$

1.2.2 Magnetostatic Fields

The basic laws of magnetostatic fields are Ampere's law

$$\oint_L \mathbf{H} \cdot d\mathbf{I} = \int_S \mathbf{J}_e \cdot d\mathbf{S} \tag{1.12}$$

which is related to Biot–Savart law, and the law of conservation of magnetic flux (also called Gauss's law for magnetostatics)

$$\oint_S \mathbf{B} \cdot d\mathbf{S} = 0 \tag{1.13}$$

where
  $\mathbf{H}$ = the magnetic field intensity (A/m)
  $\mathbf{J}_e$ = the electric current density (A/m^2)
  $\mathbf{B}$ = the magnetic flux density (T or Wb/m^2)

Applying Equations 1.2 to 1.12 and Equations 1.1 to 1.13 yields their differential forms as

$$\nabla \times \mathbf{H} = \mathbf{J}_e \tag{1.14}$$

and

$$\nabla \cdot \mathbf{B} = 0 \tag{1.15}$$

The vector fields $\mathbf{B}$ and $\mathbf{H}$ are related through the permeability μ (H/m) of the medium as

$$\mathbf{B} = \mu\mathbf{H} \tag{1.16}$$

Also, $\mathbf{J}_e$ is related to $\mathbf{E}$ through the conductivity σ (mhos/m) of the medium as

$$\mathbf{J}_e = \sigma\mathbf{E} \tag{1.17}$$

This is usually referred to as point form of Ohm's law. In terms of the magnetic vector potential $\mathbf{A}$ (Wb/m)

$$\mathbf{B} = \nabla \times \mathbf{A} \tag{1.18}$$

Applying the vector identity

$$\nabla \times (\nabla \times \mathbf{F}) = \nabla(\nabla \cdot \mathbf{F}) - \nabla^2\mathbf{F} \tag{1.19}$$

to Equations 1.14 and 1.18 and assuming Coulomb gauge condition ($\nabla \cdot \mathbf{A} = 0$) leads to Poisson's equation for magnetostatic fields:

$$\boxed{\nabla^2\mathbf{A} = -\mu\mathbf{J}_e} \tag{1.20}$$

When $\mathbf{J}_e = 0$, Equation 1.20 becomes Laplace's equation

$$\boxed{\nabla^2\mathbf{A} = 0} \tag{1.21}$$

1.2.3 Time-Varying Fields

In this case, electric and magnetic fields exist simultaneously. Equations 1.5 and 1.15 remain the same whereas Equations 1.6 and 1.14 require some modification for dynamic fields. Modification of Equation 1.6 is necessary to incorporate Faraday's law of induction, and that of Equation 1.14 is warranted to allow for displacement current. The time-varying EM fields are governed by physical laws expressed mathematically as

$$\nabla \cdot \mathbf{D} = \rho_v \tag{1.22a}$$

$$\nabla \cdot \mathbf{B} = 0 \tag{1.22b}$$

$$\nabla \times \mathbf{E} = -\frac{\partial \mathbf{B}}{\partial t} - \mathbf{J}_m \tag{1.22c}$$

$$\nabla \times \mathbf{H} = \mathbf{J}_e + \frac{\partial \mathbf{D}}{\partial t} \tag{1.22d}$$

where
  $\mathbf{J}_m = \sigma^*\mathbf{H}$ is the magnetic conductive current density (V/m²)
  $\sigma^* =$ the magnetic resistivity (Ω/m)

These equations are referred to as Maxwell's equations in the generalized form. They are first-order linear coupled differential equations relating the vector field quantities to each other. The equivalent integral form of Equation 1.22 is

$$\oint_S \mathbf{D} \cdot d\mathbf{S} = \int_v \rho_v \, dv \tag{1.23a}$$

$$\oint_S \mathbf{B} \cdot d\mathbf{S} = 0 \tag{1.23b}$$

$$\oint_L \mathbf{E} \cdot d\mathbf{l} = -\int_S \left(\frac{\partial \mathbf{B}}{\partial t} + \mathbf{J}_m \right) \cdot d\mathbf{S} \tag{1.23c}$$

$$\oint_L \mathbf{H} \cdot d\mathbf{l} = \int_S \left(\mathbf{J}_e + \frac{\partial \mathbf{D}}{\partial t} \right) \cdot d\mathbf{S} \tag{1.23d}$$

In addition to these four Maxwell's equations, there are four medium-dependent equations:

$$\mathbf{D} = \epsilon\mathbf{E} \tag{1.24a}$$

$$\mathbf{B} = \mu\mathbf{H} \tag{1.24b}$$

$$\mathbf{J}_e = \sigma\mathbf{E} \tag{1.24c}$$

$$\mathbf{J}_m = \sigma^*\mathbf{M} \tag{1.24d}$$

These are called *constitutive relations* for the medium in which the fields exist. Equations 1.22 and 1.24 form the eight postulated equations on which EM theory unfolds itself. We must note that in the region where Maxwellian fields exist, the fields are assumed to be

1. Single valued
2. Bounded and
3. Continuous functions of space and time with continuous derivatives.

It is worthwhile to mention two other fundamental equations that go hand-in-hand with Maxwell's equations. One is the Lorentz force equation

$$\mathbf{F} = Q(\mathbf{E} + \mathbf{u} \times \mathbf{B}) \tag{1.25}$$

where **F** is the force experienced by a particle with charge Q moving at velocity **u** in an EM field; the Lorentz force equation constitutes a link between EM and mechanics. The other is the continuity equation

$$\boxed{\nabla \cdot \mathbf{J} = -\frac{\partial \rho_v}{\partial t}} \tag{1.26}$$

which expresses the conservation (or indestructibility) of electric charge. The continuity equation is implicit in Maxwell's equations (see Example 1.2). Equation 1.26 is not peculiar to EM. In fluid mechanics, where **J** corresponds with velocity and ρ_v with mass, Equation 1.26 expresses the law of conservation of mass.

1.2.4 Boundary Conditions

The material medium in which an EM field exists is usually characterized by its constitutive parameters σ, ε, and μ. The medium is said to be *linear* if σ, ε, and μ are independent of **E** and **H** or nonlinear otherwise. It is *homogeneous* if σ, ε, and μ are not functions of space variables or inhomogeneous otherwise. It is *isotropic* if σ, ε, and μ are independent of direction (scalars) or anisotropic otherwise.

The boundary conditions at the interface separating two different media 1 and 2, with parameters $(\sigma_1, \varepsilon_1, \mu_1)$ and $(\sigma_2, \varepsilon_2, \mu_2)$ as shown in Figure 1.1, are easily derived from the integral form of Maxwell's equations. They are

$$E_{1t} = E_{2t} \quad \text{or} \quad (\mathbf{E}_1 - \mathbf{E}_2) \times \mathbf{a}_{n12} = 0 \tag{1.27a}$$

$$H_{1t} - H_{21} = K \quad \text{or} \quad (\mathbf{H}_1 - \mathbf{H}_2) \times \mathbf{a}_{n12} = \mathbf{K} \tag{1.27b}$$

$$D_{1n} - D_{2n} = \rho_S \quad \text{or} \quad (\mathbf{D}_1 - \mathbf{D}_2) \cdot \mathbf{a}_{n12} = \rho_S \tag{1.27c}$$

$$B_{1n} - B_{2n} = 0 \quad \text{or} \quad (\mathbf{B}_1 - \mathbf{B}_2) \cdot \mathbf{a}_{n12} = 0 \tag{1.27d}$$

where $\mathbf{a}_{n12}$ is a unit normal vector directed from medium 1 to medium 2, subscripts 1 and 2 denote fields in regions 1 and 2, and subscripts t and n, respectively, denote tangential and normal components of the fields. Equations 1.27a and 1.27d state that the tangential components of **E** and the normal components of **B** are continuous across the boundary. Equation 1.27b states that the tangential component of **H** is discontinuous by the surface current density **K** on the boundary. Equation 1.27c states that the discontinuity in the normal component of **D** is the same as the surface charge density ρ_s on the boundary.

In practice, only two of Maxwell's equations are used (Equations 1.22c and 1.22d) when a medium is source-free ($\mathbf{J} = 0$, $\rho_v = 0$) since the other two are implied (see Problem 1.4). Also, in practice, it is sufficient to make the tangential components of the fields satisfy the necessary boundary conditions since the normal components implicitly satisfy their corresponding boundary conditions.

1.2.5 Wave Equations

As mentioned earlier, Maxwell's equations are coupled first-order differential equations which are difficult to apply when solving boundary-value problems. The difficulty is overcome by decoupling the first-order equations, thereby obtaining the wave equation, a second-order differential equation which is useful for solving problems. To obtain the wave equation for a linear, isotropic, homogeneous, source-free medium ($\rho_v = 0$, $\mathbf{J} = 0$) from Equation 1.22, we take the curl of both sides of Equation 1.22c. This gives

$$\nabla \times \nabla \times \mathbf{E} = -\mu \frac{\partial}{\partial t} (\nabla \times \mathbf{H}) \tag{1.28}$$

From Equation 1.22d,

$$\nabla \times \mathbf{H} = \epsilon \frac{\partial \mathbf{E}}{\partial t}$$

since $\mathbf{J} = 0$, so that Equation 1.28 becomes

$$\nabla \times \nabla \times \mathbf{E} = -\mu\epsilon \frac{\partial^2 \mathbf{E}}{\partial t^2} \tag{1.29}$$

Applying the vector identity

$$\nabla \times \nabla \times \mathbf{F} = \nabla(\nabla \cdot \mathbf{F}) - \nabla^2 \mathbf{F} \tag{1.30}$$

in Equation 1.29,

$$\nabla(\nabla \cdot \mathbf{E}) - \nabla^2 \mathbf{E} = -\mu\epsilon \frac{\partial^2 \mathbf{E}}{\partial t^2}$$

Since $\rho_v = 0$, $\nabla \cdot \mathbf{E} = 0$ from Equation 1.22a, and hence we obtain

$$\boxed{\nabla^2 \mathbf{E} = -\mu\epsilon \frac{\partial^2 \mathbf{E}}{\partial t^2} = 0} \tag{1.31}$$

which is the time-dependent vector Helmholtz equation or simply wave equation. If we had started the derivation with Equation 1.22d, we would obtain the wave equation for **H** as

$$\boxed{\nabla^2 \mathbf{H} = -\mu\epsilon \frac{\partial^2 \mathbf{H}}{\partial t^2} = 0}$$

(1.32)

Equations 1.31 and 1.32 are the equations of motion of EM waves in the medium under consideration. The velocity (m/s) of wave propagation is

$$u = \frac{1}{\sqrt{\mu\epsilon}}$$

(1.33)

where $u = c \approx 3 \times 10^8$ m/s in free space. It should be noted that each of the vector equations in Equations 1.31 and 1.32 has three scalar components, so that altogether we have six scalar equations for E_x, E_y, E_z, H_x, H_y and H_z. Thus, each component of the wave equations has the form

$$\nabla^2 \Psi - \frac{1}{u^2} \frac{\partial^2 \Psi}{\partial t^2} = 0$$

(1.34)

which is the scalar wave equation.

1.2.6 Time-Varying Potentials

Although we are often interested in electric and magnetic field intensities (**E** and **H**), which are physically measurable quantities, it is often convenient to use auxiliary functions in analyzing an EM field. These auxiliary functions are the scalar electric potential V and vector magnetic potential **A**. Although these potential functions are arbitrary, they are required to satisfy Maxwell's equations. Their derivation is based on two fundamental vector identities (see Problem 1.1),

$$\nabla \times \nabla \Phi = 0$$

(1.35)

and

$$\nabla \nabla \times \mathbf{F} = 0$$

(1.36)

which an arbitrary scalar field Φ and vector field **F** must satisfy. Maxwell's equation 1.22b along with Equation 1.36 is satisfied if we define **A** such that

$$\boxed{\mathbf{B} = \nabla \times \mathbf{A}}$$

(1.37)

Substituting this into Equation 1.22c gives

$$-\nabla \times \left(\mathbf{E} + \frac{\partial \mathbf{A}}{\partial t} \right) = 0$$

Since this equation has to be compatible with Equation 1.35, we can choose the scalar field V such that

$$\mathbf{E} + \frac{\partial \mathbf{A}}{\partial t} = -\nabla V$$

or

$$\mathbf{E} = -\nabla V - \frac{\partial \mathbf{A}}{\partial t} \tag{1.38}$$

Thus, if we knew the potential functions V and $\mathbf{A}$, the fields $\mathbf{E}$ and $\mathbf{B}$ could be obtained from Equations 1.37 and 1.38. However, we still need to find the solution for the potential functions. Substituting Equations 1.37 and 1.38 into Equation 1.22d and assuming a linear, homogeneous medium,

$$\nabla \times \nabla \times \mathbf{A} = \mu \mathbf{J} + \epsilon \mu \frac{\partial}{\partial t} \left(-\nabla V - \frac{\partial \mathbf{A}}{\partial t} \right)$$

Applying the vector identity in Equation 1.30 leads to

$$\nabla^2 \mathbf{A} - \nabla(\nabla \cdot \mathbf{A}) = -\mu \mathbf{J} + \mu \epsilon \nabla \frac{\partial^2 \mathbf{A}}{\partial t^2} + \mu \epsilon \nabla \frac{\partial V}{\partial t} \tag{1.39}$$

Substituting Equation 1.38 into Equation 1.22a gives

$$\nabla \cdot \mathbf{E} = \frac{\rho}{\epsilon} = -\nabla^2 V - \frac{\partial(\nabla \cdot \mathbf{A})}{\partial t}$$

or

$$\nabla^2 V + \frac{\partial}{\partial t} \nabla \cdot \mathbf{A} = -\frac{\rho_v}{\epsilon} \tag{1.40}$$

According to the Helmholtz theorem of vector analysis, a vector is uniquely defined if and only if both its curl and divergence are specified. We have only specified the curl of $\mathbf{A}$ in Equation 1.37 we may choose the divergence of $\mathbf{A}$ so that the differential equations (1.39) and (1.40) have the simplest forms possible. We achieve this in the so-called *Lorentz condition*

$$\nabla \cdot \mathbf{A} = -\mu \epsilon \frac{\partial V}{\partial t} \tag{1.41}$$

Incorporating this condition into Equations 1.39 and 1.40 results in

$$\nabla^2 \mathbf{A} - \mu \epsilon \frac{\partial^2 \mathbf{A}}{\partial t^2} = -\mu \mathbf{J} \tag{1.42}$$

and

$$\nabla^2 V - \mu\epsilon \frac{\partial^2 V}{\partial t^2} = -\frac{\rho_v}{\epsilon} \qquad (1.43)$$

which are inhomogeneous wave equations. Thus, Maxwell's equations in terms of the potentials V and $\mathbf{A}$ reduce to the three Equations 1.41 through 1.43. In other words, the three equations are equivalent to the ordinary form of Maxwell's equations in that potentials satisfying these equations always lead to a solution of Maxwell's equations for $\mathbf{E}$ and $\mathbf{B}$ when used with Equations 1.37 and 1.38. Integral solutions to Equations 1.42 and 1.43 are the so-called *retarded* potentials

$$\mathbf{A} = \int \frac{\mu[\mathbf{J}]dv}{4\pi R} \qquad (1.44)$$

and

$$V = \int \frac{[\rho_v]dv}{4\pi\epsilon R} \qquad (1.45)$$

where R is the distance from the source point to the field point, and the square brackets denote ρ_v and $\mathbf{J}$ are specified at a time $R(\mu\epsilon)^{1/2}$ earlier than for which $\mathbf{A}$ or V is being determined.

1.2.7 Time-Harmonic Fields

Up to this point, we have considered the general case of arbitrary time variation of EM fields. In many practical situations, especially at low frequencies, it is sufficient to deal with only the steady-state (or equilibrium) solution of EM fields when produced by sinusoidal currents. Such fields are said to be sinusoidal time-varying or time-harmonic; that is, they vary at a sinusoidal frequency ω. An arbitrary time-dependent field $\mathbf{F}(x, y, z, t)$ or $\mathbf{F}(\mathbf{r}, t)$ can be expressed as

$$\mathbf{F}(\mathbf{r},t) = \mathrm{Re}\left[\mathbf{F}_s(\mathbf{r})e^{j\omega t}\right] \qquad (1.46)$$

where $\mathbf{F}_s(\mathbf{r}) = \mathbf{F}_s(x, y, z)$ is the phasor form of $\mathbf{F}(\mathbf{r}, t) = \mathbf{F}(x, y, z, t)$ and is in general complex, Re[] indicates "taking the real part of" quantity in brackets, and ω is the angular frequency (rad/s) of the sinusoidal excitation. The EM field quantities can be represented in phasor notation as

$$\begin{bmatrix} \mathbf{E}(\mathbf{r},t) \\ \mathbf{D}(\mathbf{r},t) \\ \mathbf{H}(\mathbf{r},t) \\ \mathbf{B}(\mathbf{r},t) \end{bmatrix} = \mathrm{Re}\left\{ \begin{bmatrix} \mathbf{E}_s(\mathbf{r}) \\ \mathbf{D}_s(\mathbf{r}) \\ \mathbf{H}_s(\mathbf{r}) \\ \mathbf{B}_s(\mathbf{r}) \end{bmatrix} e^{j\omega t} \right\} \qquad (1.47)$$

Using the phasor representation allows us to replace the time derivations $\partial/\partial t$ by $j\omega$ since

$$\frac{\partial e^{j\omega t}}{\partial t} = j\omega e^{j\omega t}$$

Thus, Maxwell's equations, in sinusoidal steady state, become

$$\boxed{\begin{aligned} \nabla \cdot \mathbf{D}_s &= \rho_{vs} \\ \nabla \cdot \mathbf{B}_s &= 0 \\ \nabla \times \mathbf{E}_s &= -j\omega \mathbf{B}_s - \mathbf{J}_{ms} \\ \nabla \times \mathbf{H}_s &= \mathbf{J}_{es} + j\omega \mathbf{D}_s \end{aligned}}$$

(1.48a)

(1.48b)

(1.48c)

(1.48d)

We should observe that the effect of the time-harmonic assumption is to eliminate the time dependence from Maxwell's equations, thereby reducing the time-space dependence to space dependence only. This simplification does not exclude more general time-varying fields if we consider ω to be one element of an entire frequency spectrum, with all the Fourier components superposed. In other words, a nonsinusoidal field can be represented as

$$\mathbf{F}(\mathbf{r},t) = \text{Re}\left[\int_{-\infty}^{\infty} \mathbf{F}_s(\mathbf{r},\omega)e^{j\omega t}\,d\omega\right]$$

(1.49)

Thus, the solutions to Maxwell's equations for a nonsinusoidal field can be obtained by summing all the Fourier components $\mathbf{F}_s(\mathbf{r},\omega)$ over ω. Henceforth, we drop the subscript s denoting phasor quantity when no confusion results.

Replacing the time derivative in Equation 1.34 by $(j\omega)^2$ yields the scalar wave equation in phasor representation as

$$\nabla^2 \Psi + k^2 \Psi = 0$$

(1.50)

where k is the propagation constant (rad/m), given by

$$k = \frac{\omega}{u} = \frac{2\pi f}{u} = \frac{2\pi}{\lambda}$$

(1.51)

We recall that Equations 1.31 through 1.34 were obtained assuming that $\rho_v = 0 = \mathbf{J}$. If $\rho_v \neq 0 \neq \mathbf{J}$, Equation 1.50 will have the general form (see Problem 1.5)

$$\boxed{\nabla^2 \Psi + k^2 \Psi = g}$$

(1.52)

We notice that this Helmholtz equation reduces to

1. Poisson's equation

$$\nabla^2 \Psi = g$$

(1.53)

when $k = 0$ (i.e., $\omega = 0$ for static case).

2. Laplace's equation

$$\nabla^2 \Psi = 0 \tag{1.54}$$

when $k = 0 = g$.

Thus, Poisson's and Laplace's equations are special cases of the Helmholtz equation. Note that function Ψ is said to be *harmonic* if it satisfies Laplace's equation.

EXAMPLE 1.1

From the divergence theorem, derive Green's theorem

$$\int_v (U\nabla^2 V - V\nabla^2 U)dv = \oint_S \left(U \frac{\partial V}{\partial n} - V \frac{\partial U}{\partial n} \right) \cdot d\mathbf{S}$$

where $(\partial\Phi/\partial n) = \nabla\Phi \cdot \mathbf{a}_n$ is the directional derivative of Φ along the outward normal to $\mathbf{S}$.

Solution

In Equation 1.1, let $\mathbf{F} = U\nabla V$, then

$$\int_v \nabla \cdot (U\nabla V)dv = \oint_S U\nabla V \cdot d\mathbf{S} \tag{1.55}$$

However,

$$\nabla \cdot (U\nabla V) = U\nabla \cdot \nabla V + \nabla V \cdot \nabla U$$
$$= U\nabla^2 V + \nabla U \cdot \nabla V$$

Substituting this into Equation 1.55 gives *Green's first identity*:

$$\int_v (U\nabla^2 V + \nabla U \cdot \nabla V)dv = \oint_S U\nabla V \cdot d\mathbf{S} \tag{1.56}$$

By interchanging U and V in Equation 1.56, we obtain

$$\int_v (V\nabla^2 U + \nabla V \cdot \nabla U)dv = \oint_S V\nabla U \cdot d\mathbf{S} \tag{1.57}$$

Subtracting Equation 1.57 from Equation 1.56 leads to *Green's second identity* or Green's theorem:

$$\int_v (U\nabla^2 V - V\nabla^2 U)dv = \oint_S (U\nabla V - V\nabla U) \cdot d\mathbf{S}$$

EXAMPLE 1.2

Show that the continuity equation is implicit (or incorporated) in Maxwell's equations.

Solution

According to Equation 1.36, the divergence of the curl of any vector field is zero. Hence, taking the divergence of Equation 1.22d gives

$$0 = \nabla \cdot \nabla \times \mathbf{H} = \nabla \cdot \mathbf{J} + \frac{\partial}{\partial t}\nabla \cdot \mathbf{D}$$

However, $\nabla \cdot \mathbf{D} = \rho_v$ is from Equation 1.22a. Thus,

$$0 = \nabla \cdot \mathbf{J} + \frac{\partial \rho_v}{\partial t}$$

which is the continuity equation.

EXAMPLE 1.3

Express

    a. $\mathbf{E} = 10\sin(\omega t - kz)\mathbf{a}_x + 20\cos(\omega t - kz)\mathbf{a}_y$ in phasor form.
    b. $\mathbf{H}_s = (4 - j3)\sin x\mathbf{a}_x + (e^{j10°}/x)\mathbf{a}_z$ in instantaneous form.

Solution

    a. We can express $\sin\theta$ as $\cos(\theta - \pi/2)$. Hence,

$$\mathbf{E} = 10\cos(\omega t - kz - \pi/2)\mathbf{a}_x + 20\cos(\omega t - kz)\mathbf{a}_y$$
$$= \mathrm{Re}\left[\left(10e^{-jkz}e^{-j\pi/2}\mathbf{a}_x + 20e^{-jkz}\mathbf{a}_y\right)e^{j\omega t}\right]$$
$$= \mathrm{Re}\left[\mathbf{E}_s e^{j\omega t}\right]$$

Thus,

$$\mathbf{E}_s = 10e^{-jkz}e^{-j\pi/2}\mathbf{a}_x + 20e^{-jkz}\mathbf{a}_y$$
$$= (-j10\mathbf{a}_x + 20\mathbf{a}_y)e^{-jkz}$$

    b. Since

$$\mathbf{H} = \mathrm{Re}\left[\mathbf{H}_s e^{j\omega t}\right]$$
$$= \mathrm{Re}\left[5\sin xe^{j(\omega t - 36.87°)}\mathbf{a}_x + \frac{1}{x}e^{j(\omega t + 10°)}\mathbf{a}_z\right]$$
$$= \left[5\sin x\cos(\omega t - 36.87°)\mathbf{a}_x + \frac{1}{x}\cos(\omega t + 10°)\mathbf{a}_z\right]$$

1.3 Classification of EM Problems

Classifying EM problems will help us later to answer the question of what method is best for solving a given problem. Continuum problems are categorized differently depending on the particular item of interest, which could be one of these:

1. The solution region of the problem
2. The nature of the equation describing the problem, or
3. The associated boundary conditions.

(In fact, the above three items define a problem uniquely.) It will soon become evident that these classifications are sometimes not independent of each other.

1.3.1 Classification of Solution Regions

In terms of the solution region or problem domain, the problem could be an interior problem, also variably called an inner, closed, or bounded problem, or an exterior problem, also variably called an outer, open, or unbounded problem.

Consider the solution region R with boundary S, as shown in Figure 1.2. If part or all of S is at infinity, R is exterior/open, otherwise R is interior/closed. For example, wave propagation in a waveguide is an interior problem, whereas wave propagations in free space scattering of EM waves by raindrops, and radiation from a dipole antenna are exterior problems.

A problem can also be classified in terms of the electrical, constitutive properties (σ, ϵ, μ) of the solution region. As mentioned in Section 1.2.4, the solution region could be linear (or nonlinear), homogeneous (or inhomogeneous), and isotropic (or anisotropic). We shall be concerned, for the most part, with linear, homogeneous, isotropic media in this chapter.

1.3.2 Classification of Differential Equations

EM problems are classified in terms of the equations describing them. The equations could be differential or integral or both. Most EM problems can be stated in terms of an operator equation

$$\boxed{L\Phi = g} \tag{1.58}$$

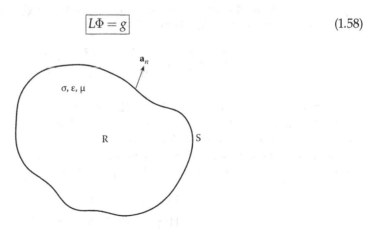

FIGURE 1.2
Solution region R with boundary S.

where L is an operator (differential, integral, or integro-differential), g is the known excitation or source, and Φ is the unknown function to be determined. A typical example is the electrostatic problem involving Poisson's equation. In differential form, Equation 1.58 becomes

$$-\nabla^2 V = \frac{\rho_v}{\epsilon} \qquad (1.59)$$

so that $L = -\nabla^2$ is the Laplacian operator, $g = \rho_v/\epsilon$ is the source term, and $\Phi = V$ is the electric potential. In integral form, Poisson's equation is of the form

$$\nabla = \int \frac{\rho_v dv}{4\pi\epsilon r} \qquad (1.60)$$

so that

$$L = \int \frac{dv}{4\pi r}, \quad g = V, \quad \text{and} \quad \Phi = \rho_v/\epsilon$$

In this section, we shall limit our discussion to differential equations; integral equations will be considered in detail in Chapter 5.

As observed in Equations 1.52 through 1.54, EM problems involve linear, second-order differential equations. In general, a second-order partial differential equation (PDE) is given by

$$a\frac{\partial^2 \Phi}{\partial x^2} + b\frac{\partial^2 \Phi}{\partial x\,\partial y} + c\frac{\partial^2 \Phi}{\partial y^2} + d\frac{\partial \Phi}{\partial x} + e\frac{\partial \Phi}{\partial y} + f\Phi = g$$

or simply

$$\boxed{a\Phi_{xx} + b\Phi_{xy} + c\Phi_{yy} + d\Phi_x + e\Phi_y + f\Phi = g} \qquad (1.61)$$

The coefficients, a, b, and c in general are functions of x and y; they may also depend on Φ itself, in which case the PDE is said to be *nonlinear*. A PDE in which $g(x, y)$ in Equation 1.61 equals zero is termed *homogeneous*; it is *inhomogeneous* if $g(x, y) \neq 0$. Notice that Equation 1.61 has the same form as Equation 1.58, where L is now a differential operator given by

$$L = a\frac{\partial^2}{\partial x^2} + b\frac{\partial^2}{\partial x\,\partial y} + c\frac{\partial^2}{\partial y^2} + d\frac{\partial}{\partial x} + e\frac{\partial}{\partial y} + f \qquad (1.62)$$

A PDE in general can be associated with both boundary values and initial values. PDEs whose boundary conditions are specified are called *steady-state equations*. If only initial values are specified, they are called *transient equations*.

Any linear second-order PDE can be classified as elliptic, hyperbolic, or parabolic depending on the coefficients a, b, and c. Equation 1.61 is said to be

$$
\begin{aligned}
&\text{Elliptic} && \text{if } b^2 - 4ac < 0 \\
&\text{Hyperbolic if } b^2 - 4ac > 0 \\
&\text{Parabolic} && \text{if } b^2 - 4ac = 0
\end{aligned} \qquad (1.63)
$$

The terms *hyperbolic, parabolic,* and *elliptic* are derived from the fact that the quadratic equation

$$ax^2 + bxy + cy^2 + dx + ey + f = 0$$

represents a hyperbola, parabola, or ellipse if $b^2 - 4ac$ is positive, zero, or negative, respectively. In each of these categories, there are PDEs that model certain physical phenomena. Such phenomena are not limited to EM but extend to almost all areas of science and engineering. Thus, the mathematical model specified in Equation 1.61 arises in problems involving heat transfer, boundary-layer flow, vibrations, elasticity, electrostatic, wave propagation, and so on.

Elliptic PDEs are associated with steady-state phenomena, that is, boundary-value problems. Typical examples of this type of PDE include Laplace's equation

$$\frac{\partial^2 \Phi}{\partial x^2} + \frac{\partial^2 \Phi}{\partial y^2} = 0 \tag{1.64}$$

and Poisson's equation

$$\frac{\partial^2 \Phi}{\partial x^2} + \frac{\partial^2 \Phi}{\partial y^2} = g(x, y) \tag{1.65}$$

where in both cases $a = c = 1$, $b = 0$. An elliptic PDE usually models an interior problem, and hence the solution region is usually closed or bounded as in Figure 1.3a.

Hyperbolic PDEs arise in propagation problems. They are usually posed as initial value problems. The solution region is usually open so that a solution advances outward indefinitely from initial conditions while always satisfying specified boundary conditions. A typical example of hyperbolic PDE is the wave equation in one dimension

$$\frac{\partial^2 \Phi}{\partial x^2} = \frac{1}{u^2} \frac{\partial^2 \Phi}{\partial t^2} \tag{1.66}$$

where $a = u^2, b = 0, c = -1$. Notice that the wave equation in Equation 1.50 is not hyperbolic but elliptic since the time-dependence has been suppressed and the equation is merely the steady-state solution of Equation 1.34.

Parabolic PDEs are generally associated with problems in which the quantity of interest varies slowly in comparison with the random motions which produce the variations. The most common parabolic PDE is the diffusion (or heat) equation in one dimension

$$\frac{\partial^2 \Phi}{\partial x^2} = k \frac{\partial \Phi}{\partial t} \tag{1.67}$$

where $a = 1, b = 0 = c$. Like hyperbolic PDE, the solution region for parabolic PDE is usually open, as in Figure 1.3b. The initial and boundary conditions typically associated with parabolic equations resemble those for hyperbolic problems except that only one initial

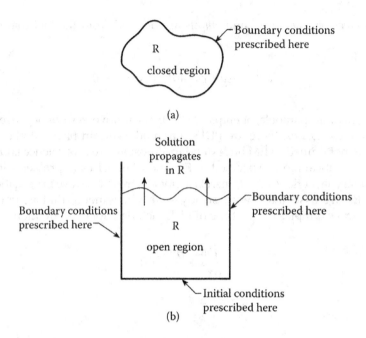

FIGURE 1.3
(a) Elliptic, (b) parabolic, or hyperbolic problem.

condition at $t = 0$ is necessary since Equation 1.67 is only first order in time. Also, parabolic and hyperbolic equations are solved using similar techniques, whereas elliptic equations require different techniques.

Note that (1) since the coefficients a, b, and c are in general functions of x and y, the classification of Equation 1.61 may change from point to point in the solution region, and (2) PDEs with more than two independent variables $(x, y, z, t, \ldots)$ may not fit as neatly into the classification above. A summary of our discussion so far in this section is shown in Table 1.1.

The type of problem represented by Equation 1.58 is said to be *deterministic* since the quantity of interest can be determined directly. Another type of problem where the quantity is found indirectly is called *nondeterministic* or *eigenvalue*. The *standard eigenproblem* is of the form

$$L\Phi = \lambda\Phi \tag{1.68}$$

TABLE 1.1

Classification of Partial Differential Equations

Type	Sign of $b^2 - 4ac$	Example	Solution Region
Elliptic	−	Laplace's equation: $\Phi_{xx} + \Phi_{yy} = 0$	Closed
Hyperbolic	+	Wave equation: $u^2\Phi_{xx} = \Phi_{tt}$	Open
Parabolic	0	Diffusion equation: $\Phi_{xx} = k\Phi_t$	Open

where the source term in Equation 1.58 has been replaced by $\lambda\Phi$. A more general version is the *generalized eigenproblem* having the form

$$\boxed{L\Phi = \lambda M\Phi} \tag{1.69}$$

where M, like L, is a linear operator for EM problems. In Equations 1.68 and 1.69, only some particular values of λ called *eigenvalues* are permissible; associated with these values are the corresponding solutions Φ called *eigenfunctions*. Eigenproblems are usually encountered in vibration and waveguide problems where the eigenvalues λ correspond to physical quantities such as resonance and cutoff frequencies, respectively.

1.3.3 Classification of Boundary Conditions

Our problem consists of finding the unknown function Φ of a partial differential equation. In addition to the fact that Φ satisfies Equation 1.58 within a prescribed solution region R, Φ must satisfy certain conditions on S, the boundary of R. Usually these boundary conditions are of the Dirichlet and Neumann types. Where a boundary has both, a mixed boundary condition is said to exist.

1. Dirichlet boundary condition:

$$\Phi(\mathbf{r}) = 0, \quad \mathbf{r} \text{ on S} \tag{1.70}$$

2. Neumann boundary condition:

$$\frac{\partial \Phi(\mathbf{r})}{\partial n} = 0, \quad \mathbf{r} \text{ on S}, \tag{1.71}$$

that is, the normal derivative of Φ vanishes on S.

3. Mixed boundary condition:

$$\frac{\partial \Phi(\mathbf{r})}{\partial n} + h(\mathbf{r})\Phi(\mathbf{r}) = 0, \quad \mathbf{r} \text{ on S}, \tag{1.72}$$

where $h(\mathbf{r})$ is a known function and $\partial\Phi/\partial n$ is the directional derivative of Φ along the outward normal to the boundary S, that is,

$$\frac{\partial \Phi}{\partial n} = \nabla\Phi \cdot \mathbf{a}_n \tag{1.73}$$

where $\mathbf{a}_n$ is a unit normal directed out of R, as shown in Figure 1.4. Note that the Neumann boundary condition is a special case of the mixed condition with $h(\mathbf{r}) = 0$.

The conditions in Equations 1.70 through 1.72 are called *homogeneous boundary conditions*. The more general ones are the inhomogeneous.

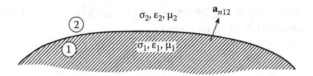

FIGURE 1.4
Interface between two media.

1. Dirichlet: $\Phi(\mathbf{r}) = p(\mathbf{r})$, $\mathbf{r}$ on S   (1.74)

2. Neumann: $\dfrac{\partial \Phi(\mathbf{r})}{\partial n} = q(\mathbf{r})$, $\mathbf{r}$ on S   (1.75)

3. Mixed: $\dfrac{\partial \Phi(\mathbf{r})}{\partial n} = h(\mathbf{r})\Phi(\mathbf{r}) = w(\mathbf{r})$, $\mathbf{r}$ on S   (1.76)

where $p(\mathbf{r})$, $q(\mathbf{r})$, and $w(\mathbf{r})$ are explicitly known functions on the boundary **S**. For example, $\Phi(0) = 1$ is an inhomogeneous Dirichlet boundary condition, and the associated homogeneous counterpart is $\Phi(0) = 0$. Also $\Phi'(1) = 2$ and $\Phi'(1) = 0$ are, respectively, inhomogeneous and homogeneous Neumann boundary conditions. In electrostatics, for example, if the value of electric potential is specified on **S**, we have Dirichlet boundary condition, whereas if the surface charge ($\rho_s = D_n = \epsilon(\partial V/\partial n)$) is specified, the boundary condition is Neumann. The problem of finding a function Φ that is harmonic in a region is called a *Dirichlet problem* (or *Neumann problem*) if Φ(or $(\partial \Phi/\partial n)$) is prescribed on the boundary of the region.

It is worth observing that the term *homogeneous* has been used to mean different things. The solution region could be homogeneous meaning that σ, ϵ, and μ are constant within R; the PDE could be homogeneous if $g = 0$ so that $L\Phi = 0$; and the boundary conditions are homogeneous when $p(\mathbf{r}) = q(\mathbf{r}) = w(\mathbf{r}) = 0$.

EXAMPLE 1.4

Classify these equations as elliptic, hyperbolic, or parabolic:

a. $4\Phi_{xx} + \partial 2\Phi_x + \Phi_y + x + y = 0$

b. $e^x \dfrac{\partial^2 V}{\partial x^2} + \cos y \dfrac{\partial^2 V}{\partial x \partial y} - \dfrac{\partial^2 V}{\partial y^2} = 0.$

State whether the equations are homogeneous or inhomogeneous.

Solution

a. In this PDE, $a = 4$, $b = 0 = c$. Hence,

$$b^2 - 4ac = 0,$$

that is, the PDE is parabolic. Since $g = -x - y$, the PDE is inhomogeneous.
b. For this PDE, $a = e^x$, $b = \cos y$, $c = -1$. Hence,

$$b^2 - 4ac = \cos^2 y + 4e^x > 0$$

and the PDE is hyperbolic. Since $g = 0$, the PDE is homogeneous.

1.4 Some Important Theorems

Two theorems are of fundamental importance in solving EM problems. These are the principle of superposition and the uniqueness theorem.

1.4.1 Superposition Principle

The principle of superposition is applied in several ways. We shall consider two of these.

If each member of a set of functions Φ_n, $n = 1, 2, ..., N$, is a solution to the PDE $L\Phi = 0$ with some prescribed boundary conditions, then a linear combination

$$\Phi_N = \Phi_0 + \sum_{n=1}^{N} a_n \Phi_n \tag{1.77}$$

also satisfies $L\Phi = g$.

Given a problem described by the PDE

$$L\Phi = g \tag{1.78}$$

subject to the boundary conditions

$$
\begin{aligned}
M_1(s) &= h_1 \\
M_2(s) &= h_2 \\
&\vdots \\
M_N(s) &= h_N,
\end{aligned}
\tag{1.79}
$$

as long as L is linear, we may divide the problem into a series of problems as follows:

$$
\begin{array}{cccc}
L\Phi_0 = g & L\Phi_1 = 0 & \cdots & L\Phi_N = 0 \\
M_1(s) = 0 & M_1(s) = h_1 & \cdots & M_1(s) = 0 \\
M_2(s) = 0 & M_2(s) = 0 & \cdots & M_2(s) = 0 \\
\vdots & \vdots & & \vdots \\
M_N(s) = 0 & M_N(s) = 0 & \cdots & M_N(s) = h_N
\end{array}
\tag{1.80}
$$

where $\Phi_0, \Phi_1, ..., \Phi_N$ are the solutions to the reduced problems, which are easier to solve than the original problem. The solution to the original problem is given by

$$\Phi = \sum_{n=0}^{N} \Phi_n \tag{1.81}$$

1.4.2 Uniqueness Theorem

This theorem guarantees that the solution obtained for a PDE with some prescribed boundary conditions is the only one possible. For EM problems, the theorem may be stated as follows: If in any way a set of fields $(\mathbf{E}, \mathbf{H})$ is found which satisfies simultaneously Maxwell's equations and the prescribed boundary conditions, this set is unique. Therefore, a field is uniquely specified by the sources $(\rho_v, \mathbf{J})$ within the medium plus the tangential components of $\mathbf{E}$ or $\mathbf{H}$ over the boundary.

To prove the uniqueness theorem, suppose there exist two solutions (with subscripts 1 and 2) that satisfy Maxwell's equations

$$\nabla \cdot \epsilon \mathbf{E}_{1,2} = \rho_v \tag{1.82a}$$

$$\nabla \cdot \mathbf{H}_{1,2} = 0 \tag{1.82b}$$

$$\nabla \times \mathbf{E}_{1,2} = -\mu \frac{\partial \mathbf{H}_{1,2}}{\partial t} \tag{1.82c}$$

$$\nabla \times \mathbf{H}_{1,2} = \mathbf{J} + \sigma \mathbf{E}_{1,2} + \epsilon \frac{\partial \mathbf{E}_{1,2}}{\partial t} \tag{1.82d}$$

If we denote the difference of the two fields as $\Delta \mathbf{E} = \mathbf{E}_2 - \mathbf{E}_1$ and $\Delta \mathbf{H} = \mathbf{H}_2 - \mathbf{H}_1$, $\Delta \mathbf{E}$ and $\Delta \mathbf{H}$ must satisfy the source-free Maxwell's equations, that is,

$$\nabla \cdot \epsilon \Delta \mathbf{E} = 0 \tag{1.83a}$$

$$\nabla \cdot \Delta \mathbf{H} = 0 \tag{1.83b}$$

$$\nabla \times \Delta \mathbf{E} = -\mu \frac{\partial \Delta \mathbf{H}}{\partial t} \tag{1.83c}$$

$$\nabla \times \Delta \mathbf{H} = \sigma \Delta \mathbf{E} + \epsilon \frac{\partial \Delta \mathbf{E}}{\partial t} \tag{1.83d}$$

Dotting both sides of Equation 1.83d with $\Delta \mathbf{E}$ gives

$$\Delta \mathbf{E} \cdot \nabla \times \Delta \mathbf{H} = \sigma \, |\Delta \mathbf{E}|^2 + \epsilon \nabla \mathbf{E} \cdot \frac{\partial \Delta \mathbf{E}}{\partial t} \tag{1.84}$$

Using the vector identity

$$\mathbf{A}(\nabla \times \mathbf{B}) = \mathbf{B} \cdot (\nabla \times \mathbf{A}) - \nabla \cdot (\mathbf{A} \times \mathbf{B})$$

Equation 1.84 becomes

$$\nabla \cdot (\Delta \mathbf{E} \times \Delta \mathbf{H}) = -\frac{1}{2} \frac{\partial}{\partial t} \left(\mu \, |\Delta \mathbf{H}|^2 + \epsilon \, |\Delta \mathbf{E}|^2 \right) - \sigma \, |\Delta \mathbf{E}|^2$$

Integrating over volume v bounded by surface $\mathbf{S}$ and applying divergence theorem to the left-hand side, we obtain

$$\oint_S (\Delta\mathbf{E}\times\Delta\mathbf{H})\cdot d\mathbf{S} = -\frac{\partial}{\partial t}\int_v \left[\frac{1}{2}\epsilon\,|\,\Delta\mathbf{E}\,|^2 + \frac{1}{2}\mu\,|\,\Delta\mathbf{H}\,|^2\right]dv$$
$$-\int_v \sigma\,|\,\Delta\mathbf{E}\,|\,dv \qquad (1.85)$$

showing that $\Delta\mathbf{E}$ and $\Delta\mathbf{H}$ satisfy the Poynting theorem just as $\mathbf{E}_{1,2}$ and $\mathbf{H}_{1,2}$. Only the tangential components of $\Delta\mathbf{E}$ and $\Delta\mathbf{H}$ contribute to the surface integral on the left-hand side of Equation 1.85. Therefore, if the tangential components of $\mathbf{E}_1$ and $\mathbf{E}_2$ or $\mathbf{H}_1$ and $\mathbf{H}_2$ are equal over $\mathbf{S}$ (thereby satisfying Equation 1.27), the tangential components of $\Delta\mathbf{E}$ and $\Delta\mathbf{H}$ vanish on $\mathbf{S}$. Consequently, the surface integral in Equation 1.85 is identically zero, and hence the right-hand side of the equation must vanish also. It follows that $\Delta\mathbf{E} = 0$ due to the second integral on the right-hand side, and hence also $\Delta\mathbf{H} = 0$ throughout the volume. Thus, $\mathbf{E}_1 = \mathbf{E}_2$ and $\mathbf{H}_1 = \mathbf{H}_2$, confirming that the solution is unique.

The theorem just proved for time-varying fields also holds for static fields as a special case. In terms of electrostatic potential V, the uniqueness theorem may be stated as follows: A solution to $\nabla^2 V = 0$ is uniquely determined by specifying either the value of V or the normal component of ∇V at each point on the boundary surface. For a magnetostatic field, the theorem becomes: A solution of $\nabla^2 \mathbf{A} = 0$ (and $\nabla\mathbf{A} = 0$) is uniquely determined by specifying the value of $\mathbf{A}$ or the tangential component of $\mathbf{B} = (\nabla \times \mathbf{A})$ at each point on the boundary surface.

PROBLEMS

1.1　In a coordinate system of your choice, prove that

　　a.　$\nabla \times \nabla\Phi = 0$,

　　b.　$\nabla \cdot \nabla \times \mathbf{F} = 0$,

　　c.　$\nabla \times \nabla \times \mathbf{F} = \nabla(\nabla \cdot \mathbf{F}) - \nabla^2\mathbf{F}$,

　　where Φ and $\mathbf{F}$ are scalar and vector fields, respectively.

1.2　If U and V are scalar fields, show that

$$\oint_L U\nabla V\cdot d\mathbf{l} = -\oint_L V\nabla U\cdot d\mathbf{l}$$

1.3　If $U(x, y, z)$ and $V(x, y, z)$ are two continuous functions with continuous derivatives over a smooth closed surface, show that

$$\int_S (U\nabla V)\cdot d\mathbf{S} = \int_v (U\nabla^2 V + \nabla U\cdot\nabla V)dv$$

1.4　Show that in a source-free region ($\mathbf{J} = 0$, $\rho_v = 0$), Maxwell's equations can be reduced to the two curl equations.

1.5   In deriving the wave equations (1.31) and (1.32), we assumed a source-free medium ($J = 0$, $\rho_v = 0$). Show that if $\rho_v \neq 0$, $J \neq 0$, the equations become

$$\nabla^2 \mathbf{E} - \frac{1}{c^2} \frac{\partial^2 \mathbf{E}}{\partial t^2} = \nabla(\rho_v / \epsilon) + \mu \frac{\partial \mathbf{J}}{\partial t},$$

$$\nabla^2 \mathbf{H} - \frac{1}{c^2} \frac{\partial^2 \mathbf{H}}{\partial t^2} = -\nabla \times \mathbf{J}$$

What assumptions have you made to arrive at these expressions?

1.6   Derive the continuity equation

$$\nabla \cdot \mathbf{J} = -\frac{\partial \rho_v}{\partial t}$$

from Maxwell's equations.

1.7   Starting with Maxwell's equations, derive the vector wave equation:

$$\varepsilon \frac{\partial^2 \mathbf{E}}{\partial t^2} + \nabla \times \frac{1}{\mu} \nabla \times \mathbf{E} = -\frac{\partial \mathbf{J}}{\partial t}$$

1.8   Starting with Maxwell's equations, show that

a. $\nabla \times \nabla \times \mathbf{H} + \mu\varepsilon \dfrac{\partial^2 \mathbf{H}}{\partial t^2} = \nabla \times \mathbf{J}$

b. $\nabla \times \nabla \times \mathbf{E} + \mu\varepsilon \dfrac{\partial^2 \mathbf{E}}{\partial t^2} = -\mu \dfrac{\partial \mathbf{J}}{\partial t}$

1.9   Given the total EM energy

$$W = \frac{1}{2} \int_v (\mathbf{E} \cdot \mathbf{D} + \mathbf{H} \cdot \mathbf{B}) dv$$

show from Maxwell's equations that

$$\frac{\partial W}{\partial t} = -\oint_S (\mathbf{E} \times \mathbf{H}) \cdot d\mathbf{S} - \int_v \mathbf{E} \cdot \mathbf{J} dv$$

1.10   Determine whether the fields

$$\mathbf{E} = 20 \sin(\omega t - kz)\mathbf{a}_x - 10 \cos(\omega t + kz)\mathbf{a}_y$$

$$\mathbf{H} = \frac{k}{\omega \mu_0}[-10 \cos(\omega t + kz)\mathbf{a}_x + 20 \sin(\omega t - kz)\mathbf{a}_y],$$

where $k = \omega\sqrt{\mu_0 \epsilon_0}$, satisfy Maxwell's equations.

1.11 In free space, the electric flux density is given by

$$\mathbf{D} = D_0\cos(\omega t + \beta z)\mathbf{a}_x$$

Use Maxwell's equation to find **H**.

1.12 In free space, a source radiates the magnetic field

$$\mathbf{H}_s = H_0 \frac{e^{-j\beta\rho}}{\sqrt{\rho}}\mathbf{a}_\phi$$

where $\beta = \omega\sqrt{\mu_0\epsilon_0}$. Determine $\mathbf{E}_s$.

1.13 *In a homogenous, lossless, source-free medium,*

$$\mathbf{E}_s = \mathbf{E}_o e^{-j\beta z}\mathbf{a}_x$$

$$\mathbf{H}_s = \frac{\mathbf{E}_0}{\eta} e^{-j\beta z}\mathbf{a}_x$$

Find β and η for which **E** and **H** satisfy Maxwell's equations.

1.14 The field of the TE_{12} mode of a rectangular waveguide for $0 < x < a$, $0 < y < b$ is

$$H_x = -\frac{\beta a}{\pi} H_o \sin\left(\frac{\pi x}{a}\right)\sin(\omega t - \beta z)$$

$$H_y = 0$$

$$H_z = H_o \cos\left(\frac{\pi x}{a}\right)\cos(\omega t - \beta z)$$

Determine the surface current densities that are required at $x = 0$, $x = a$, $y = 0$, $y = b$ to sustain the field.

1.15 In free space, the following electric field vector exists.

$$\mathbf{E} = \cos(\omega t - \beta z)\mathbf{a}_x$$

a. Does the field vector satisfy the wave equation

$$\left(\nabla^2 - \mu_0\varepsilon_o\frac{\partial^2}{\partial t^2}\right)\mathbf{E} = 0$$

b. What is the relationship between ω and β?

c. Find the magnetic field vector corresponding to the electric field vector.

1.16 An electric dipole of length L in free space has a field given in spherical system (r, θ, ϕ) as

$$\mathbf{H}_s = \frac{IL}{4\pi r}\sin\theta\left(\frac{1}{r}+j\beta\right)e^{-j\beta r}\mathbf{a}_\phi.$$

Find $\mathbf{E}_s$ using Maxwell's equations.

1.17 Show that the electric field

$$\mathbf{E}_s = 20\sin(k_x x)\cos(k_y y)\mathbf{a}_z$$

where $k_x^2 + k_y^2 = \omega^2\mu_0\varepsilon_0$ can be represented as the superposition of four propagating plane waves. Find the corresponding $\mathbf{H}_s$ field.

1.18 a. Express $I_s = e^{-jz}\sin\pi x\cos\pi y$ in instantaneous form.

b. Determine the phasor form of $V = 20\sin(\omega t - 2x) - 10\cos(\omega t - 4x)$

1.19 For each of the following phasors, determine the corresponding instantaneous form:

a. $\mathbf{A}_s = (\mathbf{a}_x + j\mathbf{a}_y)e^{-2jz}$

b. $\mathbf{B}_s = j\,10\sin x\mathbf{a}_x + 5e^{-j12z-\pi/4}\mathbf{a}_z$

c. $\mathbf{C}_s = \dfrac{2}{j}e^{-j3x}\cos 2x + e^{3x-j4x}$

1.20 Show that a time-harmonic EM field in a conducting medium ($\sigma \gg \omega\varepsilon$) satisfies the diffusion equation

$$\nabla^2\mathbf{E}_s - j\omega\mu\sigma\mathbf{E}_s = 0$$

1.21 Use Maxwell's equations to obtain a partial differential equation for A_z.

1.22 What is the relationship between the scalar and vector potentials used in defining Lorenz gauge?

1.23 Given two points $P(x, y, z)$ and $P'(x', y', z')$, let $\mathbf{R} = \mathbf{r} - \mathbf{r}'$ and $R = |\mathbf{R}|$.

$$\text{Show that}\quad \nabla\left(\frac{1}{R}\right) = -\nabla'\left(\frac{1}{R}\right) = -\frac{\mathbf{R}}{R^3}$$

1.24 Classify the following PDEs as elliptic, parabolic, or hyperbolic.

a. $\Phi_{xx} + 2\Phi_{xy} + 5\Phi_{yy} = 0$

b. $(y^2 + 1)\Phi_{xx} + (x^2 + 1)\Phi_{yy} = 0$

c. $\Phi_{xx} - 2\cos x\Phi_{xy} - (3 + \sin^2 x)\Phi_{yy} - y\Phi_y = 0$

d. $x^2\Phi_{xx} - 2xy\Phi_{xy} + y^2\Phi_{yy} + x\Phi_x + y\Phi_y = 0$

1.25 Repeat Problem 1.24 for the following PDEs.

a. $\alpha \dfrac{\partial^2 \Phi}{\partial x^2} = \beta \dfrac{\partial \Phi}{\partial x} + \dfrac{\partial \Phi}{\partial t}$ $(\alpha, \beta = \text{constant})$

which is called convective heat equation.

b. $\nabla^2 \phi + \lambda \Phi = 0$

which is the Helmholtz equation.

c. $\nabla^2 \Phi + [\lambda - \rho(x)]\Phi = 0$

which is the time-independent Schrodinger equation.

References

1. D. Landau and K. Binder, *A Guide to Monte Carlo Simulations in Statistical Physics*, 7th Edition, Cambridge, UK: Cambridge University Press, 2009, p. 5.
2. M. Lin and D.W. Walker, "GECEM: A portal-based grid application for computational electromagnetics," *Future Gener. Comput. Syst.*, vol. 24, Feb. 2008, pp. 66–72.
3. M.N.O. Sadiku, *Elements of Electromagnetics*, 7th Edition, New York, NY: Oxford University Press, 2018, chapter 9.
4. X.Q. Sheng and W. Song, *Essentials of Computational Electromagnetics*. Singapore: John Wiley & Sons, 2012, pp. 1–28.
5. J.P.A. Bastos and N. Sadowski, *Magnetic Materials and 3D Finite Element Modeling*. Boca Raton, FL: CRC Press, 2014, pp. 1–56.
6. W.C. Gibson, *The Method of Moments in Electromagnetics*. Boca Raton, FL: CRC Press, 2008, chapter 2, pp. 5–31.

2

Analytical Methods

Science never solves a problem without creating ten more.

—George B. Shaw

2.1 Introduction

The most satisfactory solution of a field problem is an exact mathematical one. Although in many practical cases such an analytical solution cannot be obtained and we must resort to numerical approximate solution, an analytical solution is useful in checking solutions obtained from numerical methods. Also, one would hardly appreciate the need for numerical methods without first seeing the limitations of the classical analytical methods. Hence, our objective in this chapter is to briefly examine the common analytical methods and thereby put numerical methods in proper perspective.

The most commonly used analytical methods in solving EM-related problems include [1]

1. Separation of variables
2. Series expansion method
3. Conformal mapping
4. Integral methods
5. Perturbation methods

Perhaps, the most powerful analytical method is the separation of variables; it is the method that will be emphasized in this chapter. Since the application of conformal mapping is restricted to certain EM problems, it will not be discussed here. The interested reader is referred to Gibbs [2]. The integral methods will be covered in Chapter 5.

2.2 Separation of Variables

The method of separation of variables (sometimes called the method of Fourier) is a convenient method for solving a partial differential equation (PDE). Basically, it entails seeking a solution which breaks up into a product of functions, each of which involves only one of the variables. For example, if we are seeking a solution $\Phi(x, y, z, t)$ to some PDE, we require that it has the product form

$$\Phi(x,y,z,t) = X(x)Y(y)Z(z)T(t) \tag{2.1}$$

A solution of the form in Equation 2.1 is said to be separable in x, y, z, and t. For example, consider the functions

1. $x^2 yz \sin 10t$,
2. $xy^2 + \dfrac{2}{t}$,
3. $(2x + y^2)z \cos 10t$

(1) Is completely separable, (2) is not separable, while (3) is separable only in z and t.

To determine whether the method of independent separation of variables can be applied to a given physical problem, we must consider the PDE describing the problem, the shape of the solution region, and the boundary conditions—the three elements that uniquely define a problem. For example, to apply the method to a problem involving two variables x and y (or ρ and ϕ, etc.), three things must be considered [3]:

1. The differential operator L must be separable, that is, it must be a function of $\Phi(x, y)$ such that

$$\frac{L\{X(x)Y(y)\}}{\Phi(x,y)X(x)Y(y)}$$

   is a sum of a function of x only and a function of y only.
2. All initial and boundary conditions must be on constant-coordinate surfaces, that is, $x = $ constant, $y = $ constant.
3. The linear operators defining the boundary conditions at $x = $ constant (or $y = $ constant) must involve no partial derivatives of Φ with respect to y (or x), and their coefficient must be independent of y (or x).

For example, the operator equation

$$L\Phi = \frac{\partial^2 \Phi}{\partial x^2} + \frac{\partial^2 \Phi}{\partial x \partial y} + \frac{\partial^2 \Phi}{\partial y^2}$$

violates (1). If the solution region R is not a rectangle with sides parallel to the x and y axes, (2) is violated. With a boundary condition $\Phi = 0$ on a part of $x = 0$ and $\partial \Phi / \partial x = 0$ on another part, (3) is violated.

With this preliminary discussion, we will now apply the method of separation of variables to PDEs in rectangular, circular cylindrical, and spherical coordinate systems. In each of these applications, we shall always take these three major steps:

1. Separate the (independent) variables
2. Find particular solutions of the separated equations, which satisfy some of the boundary conditions
3. Combine these solutions to satisfy the remaining boundary conditions.

We begin the application of separation of variables by finding the product solution of the homogeneous scalar wave equation

$$\nabla^2\Phi - \frac{1}{c^2}\frac{\partial^2\Phi}{\partial t^2} = 0 \tag{2.2}$$

Solution to Laplace's equation can be derived as a special case of the wave equation. Diffusion or heat equation can be handled in the same manner as we will treat the wave equation. To solve Equation 2.2, it is expedient that we first separate the time dependence. We let

$$\Phi(\mathbf{r},t) = U(\mathbf{r})T(t) \tag{2.3}$$

Substituting this in Equation 2.2,

$$T\nabla^2 U - \frac{1}{c^2}UT'' = 0$$

Dividing by UT gives

$$\frac{\nabla^2 U}{U} = \frac{T''}{c^2 T} \tag{2.4}$$

The left-hand side is independent of t, while the right-hand side is independent of $\mathbf{r}$; the equality can be true only if each side is independent of both variables. If we let an arbitrary constant $-k^2$ be the common value of the two sides, Equation 2.4 reduces to

$$T'' + c^2 k^2 T = 0, \tag{2.5a}$$

$$\nabla^2 U + k^2 U = 0 \tag{2.5b}$$

Thus, we have been able to separate the space variable $\mathbf{r}$ from the time variable t. The arbitrary constant $-k^2$ introduced in the course of the separation of variables is called the *separation constant*. We shall see that in general the total number of independent separation constants in a given problem is one less than the number of independent variables involved.

Equation 2.5a is an ordinary differential equation with solution

$$T(t) = a_1 e^{jckt} + a_2 e^{-jckt} \tag{2.6a}$$

or

$$T(t) = b_1 \cos(ckt) + b_2 \sin(ckt) \tag{2.6b}$$

Since the time dependence does not change with a coordinate system, the time dependence expressed in Equation 2.6a,b is the same for all coordinate systems. Therefore, we shall henceforth restrict our effort to seeking a solution to Equation 2.5b.

Notice that if $k = 0$, the time dependence disappears and Equation 2.5b becomes Laplace's equation.

2.3 Separation of Variables in Rectangular Coordinates

In order not to complicate things, we shall first consider Laplace's equation in two dimensions and later extend the idea to wave equations in three dimensions.

2.3.1 Laplace's Equation

Consider the Dirichlet problem of an infinitely long rectangular conducting trough whose cross section is shown in Figure 2.1. For simplicity, let three of its sides be maintained at zero potential while the fourth side is at a fixed potential V_o. This is a boundary value problem. The PDE to be solved is

$$\frac{\partial^2 V}{\partial x^2} + \frac{\partial^2 V}{\partial y^2} = 0$$

(2.7)

subject to (Dirichlet) boundary conditions

$$V(0, y) = 0$$ (2.8a)

$$V(a, y) = 0$$ (2.8b)

$$V(x, 0) = 0$$ (2.8c)

$$V(x, b) = V_o$$ (2.8d)

We let

$$V(x, y) = X(x)Y(y)$$ (2.9)

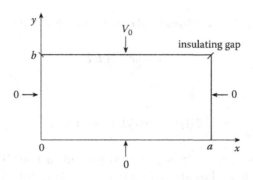

FIGURE 2.1
Cross section of the rectangular conducting trough.

Substitute this into Equation 2.7 and divide by XY. This leads to

$$\frac{X''}{X} + \frac{Y''}{Y} = 0$$

or

$$\frac{X''}{X} = -\frac{Y''}{Y} = \lambda \qquad (2.10)$$

where λ is the separation constant. Thus, the separated equations are

$$X'' - \lambda X = 0 \qquad (2.11)$$

$$Y'' + \lambda Y = 0 \qquad (2.12)$$

To solve the ordinary differential equations (2.11) and (2.12), we must impose the boundary conditions in Equation 2.8. However, these boundary conditions must be transformed so that they can be applied directly to the separated equations. Since $V = XY$,

$$V(0, y) = 0 \quad \rightarrow \quad X(0) = 0 \qquad (2.13a)$$

$$V(a, y) = 0 \quad \rightarrow \quad X(a) = 0 \qquad (2.13b)$$

$$V(x, 0) = 0 \quad \rightarrow \quad Y(0) = 0 \qquad (2.13c)$$

$$V(x, b) = V_o \quad \rightarrow \quad X(x)Y(b) = V_o \qquad (2.13d)$$

Notice that only the homogeneous conditions are separable. To solve Equation 2.11, we distinguish the three possible cases: $\lambda = 0$, $\lambda > 0$, and $\lambda < 0$.

Case 1: If $\lambda = 0$, Equation 2.11 reduces to

$$X'' = 0 \quad \text{or} \quad \frac{d^2 X}{dx^2} = 0 \qquad (2.14)$$

which has the solution

$$X(x) = a_1 x + a_2 \qquad (2.15)$$

where a_1 and a_2 are constants. Imposing the conditions in Equations 2.13a and 2.13b,

$$X(0) = 0 \quad \rightarrow \quad a_2 = 0$$
$$X(a) = 0 \quad \rightarrow \quad a_1 = 0$$

Hence, $X(x) = 0$, a trivial solution. This renders case $\lambda = 0$ as unacceptable.

Case 2: If $\lambda > 0$, say $\lambda = \alpha^2$, Equation 2.11 becomes

$$X'' - \alpha^2 X = 0 \qquad (2.16)$$

with the corresponding auxiliary equations $m^2 - \alpha^2 = 0$ or $m = \pm\alpha$. Hence, the general solution is

$$X = b_1 e^{-\alpha x} + b_2 e^{\alpha x} \tag{2.17}$$

or

$$X = b_3 \sinh\alpha x + b_4 \cosh\alpha x \tag{2.18}$$

The boundary conditions are applied to determine b_3 and b_4.

$$X(0) = 0 \quad \rightarrow \quad b_4 = 0$$
$$X(a) = 0 \quad \rightarrow \quad b_3 = 0$$

since $\sinh ax$ is never zero for $\alpha > 0$. Hence, $X(x) = 0$, a trivial solution, and we conclude that case $\lambda > 0$ is not valid.

Case 3: If $\lambda < 0$, say $\lambda = -\beta^2$, Equation 2.11 becomes

$$X'' + \beta^2 X = 0 \tag{2.19}$$

with the auxiliary equation $m^2 + \beta^2 = 0$ or $m = \pm j\beta$. The solution to Equation 2.19 is

$$X = A_1 e^{j\beta x} + A_2 e^{j\beta x} \tag{2.20a}$$

or

$$X = B_1 \sin\beta x + B_2 \cos\beta x \tag{2.20b}$$

Again,

$$X(0) = 0 \quad \rightarrow \quad B_2 = 0$$
$$X(a) = 0 \quad \rightarrow \quad \sin\beta a = 0 = \sin n\pi$$

or

$$\beta = \frac{n\pi}{a}, \quad n = 1, 2, 3, \dots \tag{2.21}$$

since B_1 cannot vanish for nontrivial solutions, whereas $\sin \beta a$ can vanish without its argument being zero. Thus, we have found an infinite set of discrete values of λ for which Equation 2.11 has nontrivial solutions, that is,

$$\lambda = -\beta^2 = \frac{-n^2\pi^2}{a^2}, \quad n = 1, 2, 3, \dots \tag{2.22}$$

These are the eigenvalues of the problem and the corresponding eigenfunctions are

$$X_n(x) = \sin \beta x = \sin \frac{n\pi x}{a} \tag{2.23}$$

From Equation 2.22 note that it is not necessary to include negative values of n since they lead to the same set of eigenvalues. Also we exclude $n = 0$ since it yields the trivial solution $X = 0$ as shown under Case 1 when $\lambda = 0$. Having determined λ, we can solve Equation 2.12 to find $Y_n(y)$ corresponding to $X_n(x)$. That is, we solve

$$Y'' - \beta^2 Y = 0, \tag{2.24}$$

which is similar to Equation 2.16, whose solution is in Equation 2.18. Hence, the solution to Equation 2.24 has the form

$$Y_n(y) = a_n \sinh \frac{n\pi y}{a} + b_n \cosh \frac{n\pi y}{a} \tag{2.25}$$

Imposing the boundary condition in Equation 2.13c,

$$Y(0) = 0 \quad \rightarrow \quad b_n = 0$$

so that

$$Y_n(y) = a_n \sinh \frac{n\pi y}{a} \tag{2.26}$$

Substituting Equations 2.23 and 2.26 into Equation 2.9, we obtain

$$V_n(x,y) = X_n(x)Y_n(y) = a_n \sin \frac{n\pi x}{a} \sinh \frac{n\pi y}{a}, \tag{2.27}$$

which satisfies Equation 2.7 and the three homogeneous boundary conditions in Equations 2.8a through 2.8c. By the superposition principle, a linear combination of the solutions V_n, each with different values of n and arbitrary coefficient a_n, is also a solution of Equation 2.7. Thus, we may represent the solution V of Equation 2.7 as an infinite series in the function V_n, that is,

$$V(x,y) = \sum_{n=1}^{\infty} a_n \sin \frac{n\pi x}{a} \sinh \frac{n\pi y}{a} \tag{2.28}$$

We now determine the coefficient a_n by imposing the inhomogeneous boundary condition in Equation 2.8d on Equation 2.28. We get

$$V(x,b) = V_o = \sum_{n=1}^{\infty} a_n \sin \frac{n\pi x}{a} \sinh \frac{n\pi b}{a}, \tag{2.29}$$

which is Fourier sine expansion of V_o. Hence,

$$a_n \sinh \frac{n\pi b}{a} = \frac{2}{b} \int_0^b V_o \sin \frac{n\pi x}{a} \, dx = \frac{2V_o}{n\pi} (1 - \cos n\pi)$$

or

$$a_n = \begin{cases} \dfrac{4V_o}{n\pi} \dfrac{1}{\sinh(n\pi b/a)}, & n = \text{odd}, \\ 0, & n = \text{even} \end{cases} \tag{2.30}$$

Substitution of Equation 2.30 into Equation 2.28 gives the complete solution as

$$V(x,y) = \frac{4V_o}{\pi} \sum_{n=\text{odd}}^{\infty} \frac{\sin \dfrac{n\pi x}{a} \sinh \dfrac{n\pi y}{a}}{n \sinh \dfrac{n\pi b}{a}} \tag{2.31a}$$

By replacing n by $2k-1$, Equation 2.31a may be written as

$$\boxed{V(x,y) = \frac{4V_o}{\pi} \sum_{k=1}^{\infty} \frac{\sin \dfrac{n\pi x}{a} \sinh \dfrac{n\pi y}{a}}{n \sinh \dfrac{n\pi b}{a}}, \quad n = 2k-1} \tag{2.31b}$$

2.3.2 Wave Equation

The time dependence has been taken care of in Section 2.2. We are left with solving the Helmholtz equation

$$\nabla^2 U + k^2 U = 0 \tag{2.5b}$$

In rectangular coordinates, Equation 2.5b becomes

$$\frac{\partial^2 U}{\partial x^2} + \frac{\partial^2 U}{\partial y^2} + \frac{\partial^2 U}{\partial z^2} + k^2 U = 0 \tag{2.32}$$

We let

$$U(x,y,z) = X(x)Y(y)Z(z) \tag{2.33}$$

Substituting Equation 2.33 into Equation 2.32 and dividing by XYZ, we obtain

$$\frac{X''}{X} + \frac{Y''}{Y} + \frac{Z''}{Z} + k^2 = 0 \tag{2.34}$$

Each term must be equal to a constant since each term depends only on the corresponding variable; X on x, etc. We conclude that

$$\frac{X''}{X} = -k_x^2, \quad \frac{Y''}{Y} = -k_y^2, \quad \frac{Z''}{Z} = -k_z^2 \tag{2.35}$$

so that Equation 2.34 reduces to

$$k_x^2 + k_y^2 + k_z^2 = k^2 \tag{2.36}$$

Notice that there are four separation constants k, k_x, k_y, and k_z since we have four variables t, x, y, and z. However from Equation 2.36, one is related to the other three so that only three separation constants are independent. As mentioned earlier, the number of independent separation constants is generally one less than the number of independent variables involved. The ordinary differential equations in Equation 2.35 have solutions

$$X = A_1 e^{jk_x x} + A_2 e^{-jk_x x} \tag{2.37a}$$

or

$$X = B_1 \sin k_x x + B_2 \cos k_x x, \tag{2.37b}$$

$$Y = A_3 e^{jk_y y} + A_4 e^{jk_y y} \tag{2.37c}$$

or

$$Y = B_3 \sin k_y y + B_4 \cos k_y y, \tag{2.37d}$$

$$Z = A_5 e^{jk_z z} + A_6 e^{-jk_z z} \tag{2.37e}$$

or

$$Z = B_5 \sin k_z z + B_6 \cos k_z z, \tag{2.37f}$$

Various combinations of X, Y, and Z will satisfy Equation 2.5b. Suppose we choose

$$X = A_1 e^{jk_x x}, \quad Y = A_3 e^{jk_y y}, \quad Z = A_5 e^{jk_z z}, \tag{2.38}$$

then

$$U(x, y, z) = A e^{j(k_x x + k_y y + k_z z)} \tag{2.39}$$

or

$$U(\mathbf{r}) = A e^{j\mathbf{k} \cdot \mathbf{r}} \tag{2.40}$$

Introducing the time dependence of Equation 2.6a gives

$$\boxed{\Phi(x,y,z,t) = Ae^{j(\mathbf{k}\cdot\mathbf{r}+\omega t)}}$$

(2.41)

where $\omega = kc$ is the angular frequency of the wave and k is given by Equation 2.36. The solution in Equation 2.41 represents a plane wave of amplitude A propagating in the direction of the wave vector $\mathbf{k} = k_x\mathbf{a}_x + k_y\mathbf{a}_y + k_z\mathbf{a}_z$ with velocity c.

EXAMPLE 2.1

In this example, we show that the method of separation of variables is not limited to a problem with only one inhomogeneous boundary condition as presented in Section 2.3.1. We reconsider the problem of Figure 2.1, but with four inhomogeneous boundary conditions as in Figure 2.2a.

Solution

The problem can be stated as solving Laplace's equation

$$\frac{\partial^2 V}{\partial x^2} + \frac{\partial^2 V}{\partial y^2} = 0$$

(2.42)

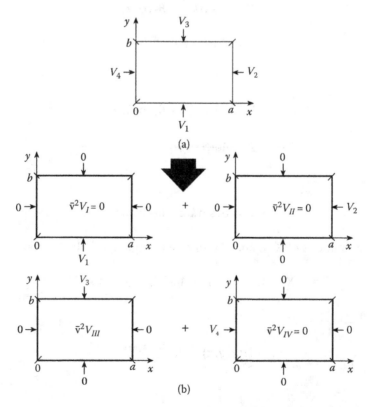

FIGURE 2.2
Applying the principle of superposition reduces the problem in (a) to those in (b).

subject to the following inhomogeneous Dirichlet conditions:

$$V(x, 0) = V_1$$

$$V(x, b) = V_3$$

$$V(0, y) = V_4$$

$$V(a, y) = V_2 \tag{2.43}$$

Since Laplace's equation is a linear homogeneous equation, the problem can be simplified by applying the superposition principle. If we let

$$V = V_I + V_{II} + V_{III} + V_{IV}, \tag{2.44}$$

we may reduce the problem to four simpler problems, each of which is associated with one of the inhomogeneous conditions. The reduced, simpler problems are illustrated in Figure 2.2b and stated as follows:

$$\frac{\partial^2 V_I}{\partial x^2} + \frac{\partial^2 V_I}{\partial y^2} = 0 \tag{2.45}$$

subject to

$$V_I(x, 0) = V_1$$
$$V_I(x, b) = 0$$
$$V_I(0, y) = 0$$
$$V_I(a, y) = 0; \tag{2.46}$$

$$\frac{\partial^2 V_{II}}{\partial x^2} + \frac{\partial^2 V_{II}}{\partial y^2} = 0 \tag{2.47}$$

subject to

$$V_{II}(x, 0) = 0$$
$$V_{II}(x, b) = 0$$
$$V_{II}(0, y) = 0$$
$$V_{II}(a, y) = V_2; \tag{2.48}$$

$$\frac{\partial^2 V_{III}}{\partial x^2} + \frac{\partial^2 V_{III}}{\partial y^2} = 0 \tag{2.49}$$

subject to

$$V_{III}(x, 0) = 0$$
$$V_{III}(x, b) = V_3$$
$$V_{III}(0, y) = 0$$
$$V_{III}(a, y) = 0; \tag{2.50}$$

and

$$\frac{\partial^2 V_{IV}}{\partial x^2} + \frac{\partial^2 V_{IV}}{\partial y^2} = 0 \tag{2.51}$$

subject to

$$\begin{aligned}
V_{IV}(x, 0) &= 0 \\
V_{IV}(x, b) &= 0 \\
V_{IV}(0, y) &= V_4 \\
V_{IV}(a, y) &= 0
\end{aligned} \tag{2.52}$$

It is obvious that the reduced problem in Equations 2.49 and 2.50 with solution V_{III} is the same as that in Figure 2.1. The other three reduced problems are quite similar. Hence, the solutions V_I, V_{II}, and V_{IV} can be obtained by taking the same steps as in Section 2.3.1 or by a proper exchange of variables in Equation 2.31. Thus,

$$V_I = \frac{4V_1}{\pi} \sum_{n=\text{odd}}^{\infty} \frac{\sin\dfrac{n\pi x}{a} \sinh\dfrac{n\pi(b-y)}{a}}{n \sinh\dfrac{n\pi b}{a}}, \tag{2.53}$$

$$V_{II} = \frac{4V_2}{\pi} \sum_{n=\text{odd}}^{\infty} \frac{\sin\dfrac{n\pi x}{b} \sinh\dfrac{n\pi y}{b}}{n \sinh\dfrac{n\pi a}{b}}, \tag{2.54}$$

$$V_{III} = \frac{4V_3}{\pi} \sum_{n=\text{odd}}^{\infty} \frac{\sin\dfrac{n\pi x}{a} \sinh\dfrac{n\pi y}{a}}{n \sinh\dfrac{n\pi b}{a}}, \tag{2.55}$$

$$V_{IV} = \frac{4V_4}{\pi} \sum_{n=\text{odd}}^{\infty} \frac{\sin\dfrac{n\pi(a-x)}{b} \sinh\dfrac{n\pi y}{b}}{n \sinh\dfrac{n\pi a}{b}} \tag{2.56}$$

We obtain the complete solution by substituting Equations 2.53 through 2.56 into Equation 2.44.

EXAMPLE 2.2

Find the product solution of the diffusion equation

$$\Phi_t + k\Phi_{xx}, \quad 0 < x < 1, \quad t > 0 \tag{2.57}$$

subject to the boundary conditions

$$\Phi(0, t) = 0 = \Phi(1, t), \quad t > 0 \tag{2.58}$$

and initial condition

$$\Phi(x, 0) = 5\sin 2\pi x, \quad 0 < x < 1 \tag{2.59}$$

Solution

Let

$$\Phi(x, t) = X(x)U(t) \tag{2.60}$$

Substitute this into Equation 2.57 and divide by kXT to obtain

$$\frac{U'}{kU} = \frac{X''}{X} = \lambda$$

where λ is the separation constant. Thus,

$$X'' - \lambda X = 0 \tag{2.61}$$

$$U' - \lambda kU = 0 \tag{2.62}$$

As usual, in order for the solution of Equation 2.61 to satisfy Equation 2.58, we must choose $\lambda = -\beta^2 = -n^2\pi^2$ so that $n = 1, 2, 3, \dots$ and

$$X_n(x) = \sin n\pi x \tag{2.63}$$

Equation 2.62 becomes

$$U' + kn^2\pi^2 U = 0,$$

which has solution

$$U_n(t) = e^{-kn^2\pi^2 t} \tag{2.64}$$

Substituting Equations 2.63 and 2.64 into Equation 2.60,

$$\Phi_n(x,t) = a_n \sin n\pi x \exp(-kn^2\pi^2 t)$$

where the coefficients a_n are to be determined from the initial condition in Equation 2.59. The complete solution is a linear combination of Φ_n, that is,

$$\Phi(x,t) = \sum_{n=1}^{\infty} a_n \sin n\pi x \exp(-kn^2\pi^2 t)$$

This satisfies Equation 2.59 if

$$\Phi(x,0) = \sum_{n=1}^{\infty} a_n \sin n\pi x = 5\sin 2\pi x \tag{2.65}$$

The coefficients a_n are determined as

$$a_n = \frac{2}{T} \int_0^1 5\sin 2\pi x \sin n\pi x \, dx = \begin{cases} 5, & n = 2 \\ 0, & n \neq 0 \end{cases}$$

Alternatively, by comparing the middle term in Equation 2.65 with the last term, the two are equal only when $n = 2$, $a_n = 5$, otherwise $a_n = 0$. Hence, the solution of the diffusion problem becomes

$$\Phi(x,t) = 5\sin 2\pi x \exp(-4k\pi^2 t)$$

2.4 Separation of Variables in Cylindrical Coordinates

Coordinate geometries other than rectangular Cartesian are used to describe many EM problems whenever it is necessary and convenient. For example, a problem having cylindrical symmetry is best solved in a cylindrical system where the coordinate variables (ρ, ϕ, z) are related as shown in Figure 2.3 and $0 \le \rho \le \infty$, $0 \le \phi \le 2n$, $-\infty \le z \le \infty$. In this system, the wave equation (2.5b) becomes

$$\nabla^2 U + k^2 U = \frac{1}{\rho^2}\frac{\partial}{\partial\rho}\left(\rho\frac{\partial U}{\partial\rho}\right) + \frac{1}{\rho^2}\frac{\partial^2 U}{\partial\phi^2} + \frac{\partial^2 U}{\partial z^2} + k^2 U = 0 \tag{2.66}$$

As we did in the previous section, we shall first solve Laplace's equation ($k = 0$) in two dimensions before we solve the wave equation.

Consider an infinitely long conducting cylinder of radius a with the cross section shown in Figure 2.4. Assume that the upper half of the cylinder is maintained at potential V_o while the lower half is maintained at potential $-V_o$. This is a Laplacian problem in two dimensions. Hence, we need to solve for $V(\rho, \phi)$ in Laplace's equation

$$\nabla^2 V = \frac{1}{\rho}\frac{\partial}{\partial\rho}\left(\rho\frac{\partial V}{\partial\rho}\right) + \frac{1}{\rho^2}\frac{\partial^2 V}{\partial\phi^2} = 0 \tag{2.67}$$

subject to the inhomogeneous Dirichlet boundary condition

$$V(a,\phi) = \begin{cases} V_o, & 0 < \phi < \pi \\ -V_o, & \pi < \phi < 2\pi \end{cases} \tag{2.68}$$

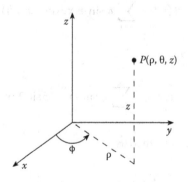

FIGURE 2.3
Coordinate relations in a cylindrical system.

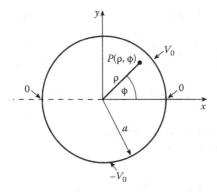

FIGURE 2.4
A two-dimensional Laplacian problem in cylindrical coordinates.

We let

$$V(\rho, \phi) = R(\rho)F(\phi) \tag{2.69}$$

Substituting Equation 2.69 into Equation 2.67 and dividing through by RF/ρ^2 result in

$$\frac{\rho}{R}\frac{d}{d\rho}\left(\rho\frac{dR}{d\rho}\right) + \frac{1}{F}\frac{d^2F}{d\phi^2} = 0$$

or

$$\frac{\rho^2}{R}\frac{d^2R}{d\rho^2} + \frac{\rho}{R}\frac{dR}{d\rho} = -\frac{1}{F}\frac{d^2F}{d\phi^2} = \lambda^2 \tag{2.70}$$

where λ is the separation constant. Thus, the separated equations are

$$F'' + \lambda^2 F = 0 \tag{2.71a}$$

$$\rho^2 R'' + \rho R' - \lambda^2 R = 0 \tag{2.71b}$$

It is evident that Equation 2.71a has the general solution of the form

$$F(\phi) = c_1 \cos(\lambda\phi) + c_2 \sin(\lambda\phi) \tag{2.72}$$

From the boundary conditions of Equation 2.68, we observe that $F(\phi)$ must be a periodic, odd function. Thus, $c_1 = 0$, $\lambda = n$, a real integer, and hence Equation 2.72 becomes

$$F_n(\phi) = c_2 \sin n\phi \tag{2.73}$$

Equation 2.71b, known as the *Cauchy–Euler equation*, can be solved by making a substitution $R = \rho^n$ and reducing it to an equation with constant coefficients. This leads to

$$R_n(\rho) = c_3 \rho^n + c_4 \rho^{-n}, \quad n = 1, 2, \dots \tag{2.74}$$

Note that case $n = 0$ is excluded; if $n = 0$, we obtain $R(\rho) = \ln \rho + \text{constant}$, which is not finite at $\rho = 0$. For the problem of a coaxial cable, $a < \rho < b$, $\rho \neq 0$ so that case $n = 0$ is the only solution. However, for the problem at hand, $n = 0$ is not acceptable.

Substitution of Equations 2.73 and 2.74 into Equation 2.69 yields

$$V_n(\rho, \phi) = \sin n\phi (A_n\rho^n + B_n\rho^{-n}) \tag{2.75}$$

where A_n and B_n are constants to be determined. As usual, it is possible by the superposition principle to form a complete series solution

$$V(\rho, \phi) = \sum_{n=1}^{\infty} (A_n\rho^n + B_n\rho^{-n})\sin n\phi \tag{2.76}$$

For $\rho < a$, inside the cylinder, V must be finite as $\rho \to 0$ so that $B_n = 0$. At $\rho = a$,

$$V(a, \phi) = \sum_{n=1}^{\infty} A_n a^n \sin n\phi = \begin{cases} V_o, & 0 < \phi < \pi \\ -V_o, & \pi < \phi < 2\pi \end{cases} \tag{2.77}$$

Multiplying both sides by $\sin m\phi$ and integrating over $0 < \phi < 2\pi$, we get

$$\int_0^\pi V_o \sin m\phi \, d\phi - \int_\pi^{2\pi} V_o \sin m\phi \, d\phi = \sum_{n=1}^{\infty} A_n a^n \int_0 \sin n\phi \sin m\phi \, d\phi$$

All terms on the right-hand side vanish except when $m = n$. Hence,

$$\frac{2V_o}{n}(1 - \cos n\pi) = A_n a^n \int_0^{2\pi} \sin^2 \phi \, d\phi = \pi A_n a^n$$

or

$$A_n = \begin{cases} \dfrac{4V_o}{n\pi a^n}, & n = \text{odd} \\ 0, & n = \text{even} \end{cases} \tag{2.78}$$

Thus,

$$V(\rho, \phi) = \frac{4V_o}{\pi} \sum_{n=\text{odd}}^{\infty} \frac{\rho^n \sin n\phi}{na^n}, \quad \rho < a \tag{2.79}$$

For $\rho > a$, outside the cylinder, V must be finite as $\rho \to \infty$ so that $A_n = 0$ in Equation 2.76 for this case. By imposing the boundary condition in Equation 2.68 and following the same steps as for case $\rho < a$, we obtain

$$B_n = \begin{cases} \dfrac{4V_o a^n}{n\pi}, & n = \text{odd} \\ 0, & n = \text{even} \end{cases} \tag{2.80}$$

Hence,

$$V(\rho, \phi) = \frac{4V_o}{\pi} \sum_{n=\text{odd}}^{\infty} \frac{a^n \sin n\phi}{na^n}, \quad \rho > a \tag{2.81}$$

2.4.1 Wave Equation

Having taken care of the time-dependence in Section 2.2, we now solve Helmholtz's equation (2.66), that is,

$$\frac{1}{\rho} \frac{\partial U}{\partial \rho} \left(\rho \frac{\partial U}{\partial \rho} \right) + \frac{1}{\rho^2} \frac{\partial^2 U}{\partial \phi^2} + \frac{\partial^2 U}{\partial z^2} + k^2 U = 0 \tag{2.66}$$

Let

$$U(\rho, \phi, z) = R(\rho)F(\phi)Z(z) \tag{2.82}$$

Substituting Equation 2.82 into Equation 2.66 and dividing by RFZ/ρ^2 yields

$$\frac{\rho}{R} \frac{d}{d\rho} \left(\rho \frac{dR}{d\rho} \right) + \frac{\rho^2}{z} \frac{d^2 Z}{dz^2} + k^2 \rho^2 = -\frac{1}{F} \frac{d^2 F}{d\phi^2} = n^2$$

where $n = 0, 1, 2, \ldots$ and n^2 is the separation constant. Thus,

$$F'' + n^2 F = 0 \tag{2.83}$$

and

$$\frac{\rho}{R} \frac{d}{d\rho} \left(\rho \frac{dR}{d\rho} \right) + \frac{\rho^2}{Z} \frac{d^2 Z}{dz^2} + k^2 \rho^2 = n^2 \tag{2.84}$$

Dividing both sides of Equation 2.84 by ρ^2 leads to

$$\frac{1}{\rho R} \frac{d}{d\rho} \left(\rho \frac{dR}{d\rho} \right) + \left(k^2 - \frac{n^2}{\rho^2} \right) = -\frac{1}{Z} \frac{d^2 Z}{dz^2} = \mu^2$$

where μ^2 is another separation constant. Hence,

$$-\frac{1}{Z} \frac{d^2 Z}{dz^2} = \mu^2 \tag{2.85}$$

and

$$\frac{1}{\rho R}\frac{d}{d\rho}\left(\rho\frac{dR}{d\rho}\right)+\left(k^2-\mu^2-\frac{n^2}{\rho^2}\right)=0 \tag{2.86}$$

If we let

$$\lambda^2=k^2-\mu^2, \tag{2.87}$$

the three separated Equations 2.83, 2.85, and 2.86 become

$$F''+n^2F=0, \tag{2.88}$$

$$Z''+\mu^2Z=0, \tag{2.89}$$

$$\rho^2R''+\rho R'+(\lambda^2\rho^2-n^2)R=0 \tag{2.90}$$

The solution to Equation 2.88 is given by

$$F(\phi)=c_1e^{jn\phi}+c_2e^{-jn\phi} \tag{2.91a}$$

or

$$F(\phi)=c_3\sin n\phi+c_4\cos n\phi \tag{2.91b}$$

Similarly, Equation 2.89 has the solution

$$Z(z)=c_5e^{jn\mu}+c_6e^{-jn\mu} \tag{2.92a}$$

or

$$Z(z)=c_7\sin n\mu+c_8\cos n\mu \tag{2.92b}$$

To solve Equation 2.90, we let $x=\lambda\rho$ and replace R by y; $R'=\lambda y'$ and $R''=\lambda^2y''$ and Equation 2.90 becomes

$$x^2y''+xy'+(x^2-n^2)y=0 \tag{2.93}$$

This is called *Bessel's equation*. It has a general solution of the form

$$y(x)=b_1J_n(x)+b_2Y_n(x) \tag{2.94}$$

where $J_n(x)$ and $Y_n(x)$ are, respectively, *Bessel functions* of the first and second kinds of order n and real argument x. Y_n is also called the *Neumann function*. If x in Equation 2.93 is imaginary so that we may replace x by jx, the equation becomes

$$x^2y''+xy'-(x^2+n^2)y=0 \tag{2.95}$$

which is called *modified Bessel's equation*. This equation has a solution of the form

$$y(x) = b_3 I_n(x) + b_4 K_n(x) \tag{2.96}$$

where $I_n(x)$ and $K_n(x)$ are, respectively, *modified Bessel functions* of the first and second kind of order n. For small values of x, Figure 2.5 shows the sketch of some typical Bessel functions (or cylindrical functions) $J_n(x)$, $Y_n(x)$, $I_n(x)$, and $K_n(x)$.

To obtain the Bessel functions from Equations 2.93 and 2.95, the method of Frobenius is applied. A detailed discussion is found in Kersten [4] and Myint-U [5]. For the Bessel function of the first kind,

$$y = J_n(x) = \sum_{m=0}^{\infty} \frac{(-1)^m (x/2)^{n+2m}}{m! \, \Gamma(n+m+1)} \tag{2.97}$$

where $\Gamma(k+1) = k!$ is the Gamma function. This is the most useful of all Bessel functions. Some of its important properties and identities are listed in Table 2.1. For the modified Bessel function of the first kind,

$$I_n(x) = j^{-n} J_n(jx) = \sum_{m=0}^{\infty} \frac{(x/2)^{n+2m}}{m! \, \Gamma(n+m+1)} \tag{2.98}$$

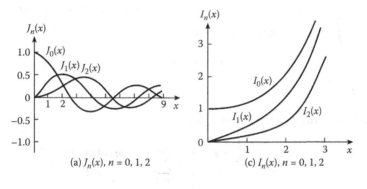

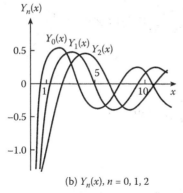

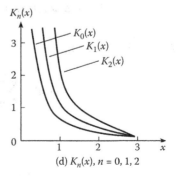

FIGURE 2.5
Bessel functions.

TABLE 2.1

Properties and Identities of Bessel Functions[a] $J_n(x)$

---

a. $J_{-n}(x) = (-1)^n J_n(x)$

b. $J_n(-x) = (-1)^n J_n(x)$

c. $J_{n+1}(x) = \dfrac{2n}{x} J_n(x) - J_{n-1}(x)$ (recurrence formula)

d. $\dfrac{d}{dx} J_n(x) = \dfrac{1}{2}[J_{n-1}(x) - J_{n+1}(x)]$

e. $\dfrac{d}{dx}[x^n J_n(x)] = x^n J_{n-1}(x)$

f. $\dfrac{d}{dx}[x^{-n} J_n(x)] = -x^{-n} J_{n-1}(x)$

g. $J_n(x) = \dfrac{1}{\pi}\displaystyle\int_0^\pi \cos(n\theta - x\sin\theta)d\theta, \quad n \geq 0$

h. Fourier–Bessel expansion of $f(x)$:

$$f(x) = \sum_{k=1}^{\infty} A_k J_n(\lambda_k x), \quad n \geq 0$$

$$A_k = \frac{2}{[a J_{n+1}(\lambda_i a)]^2}\int_0^a x f(x) J_n(\lambda_k x)dx, \quad 0 < x < a$$

where λ_k are the positive roots in ascending order of magnitude of $J_n(\lambda_k a) = 0$.

i. $\displaystyle\int_0^a \rho J_n(\lambda_i \rho) J_n(\lambda_j \rho)d\rho = \dfrac{a^2}{2}[J_{n+1}(\lambda_i a)]^2 \delta_{ij}$

where λ_i and λ_j are the positive roots of $J_n(\lambda a) = 0$.

---

[a] Properties (a) through (f) also hold for $Y_n(x)$.

For the Neumann function, when $n > 0$,

$$Y_n(x) = \frac{2}{\pi} J_n(x)\ln\frac{\gamma x}{2} - \frac{1}{\pi}\sum_{m=0}^{n-1}\frac{(n-m-1)!(x/2)^{2m-n}}{m!}$$
$$- \frac{1}{\pi}\sum_{m=0}^{\infty}\frac{(-1)^m (x/2)^{n+2m}}{m!\,\Gamma(n+m+1)}[p(m) + p(n+m)] \tag{2.99}$$

where $\gamma = 1.781$ is Euler's constant and

$$p(m) = \sum_{k=1}^{m}\frac{1}{k}, \quad p(0) = 0 \tag{2.100}$$

If $n = 0$,

$$Y_0(x) = \frac{2}{\pi} J_0(x)\ln\frac{\gamma x}{2} + \frac{2}{\pi}\sum_{m=0}^{\infty}\frac{(-1)^{m+1}(x/2)^{2m}}{(m!)^2}p(m) \tag{2.101}$$

For the modified Bessel function of the second kind,

$$K_n(x) = \frac{\pi}{2} j^{n+1}[J_n(jx) + jY_n(jx)] \tag{2.102}$$

If $n > 0$,

$$K_n(x) = \frac{1}{2} \sum_{m=0}^{n-1} \frac{(-1)^m (n-m-1)!(x/2)^{2m-n}}{m!}$$
$$+ (-1)^n \frac{1}{2} \sum_{m=0}^{\infty} \frac{(x/2)^{n+2m}}{m!(n+m)!} \left[p(m) + p(n+m) - 2\ln \frac{\gamma x}{2} \right] \tag{2.103}$$

and if $n = 0$,

$$K_0(x) = -I_0(x)\ln \frac{\gamma x}{2} + \sum_{m=0}^{\infty} \frac{(x/2)^{2m}}{(m!)^2} p(m) \tag{2.104}$$

Other functions closely related to Bessel functions are *Hankel functions* of the first and second kinds defined, respectively, by

$$H_n^{(1)}(x) = J_n(x) + jY_n(x) \tag{2.105a}$$

$$H_n^{(2)}(x) = J_n(x) - jY_n(x) \tag{2.105b}$$

Hankel functions are analogous to functions $\exp(\pm jx)$ just as J_n and Y_n are analogous to cosine and sine functions. This is evident from asymptotic expressions

$$J_n(x) \xrightarrow{x \to \infty} \sqrt{\frac{2}{\pi x}} \cos(x - n\pi/2 - \pi/4), \tag{2.106a}$$

$$Y_n(x) \xrightarrow{x \to \infty} \sqrt{\frac{2}{\pi x}} \sin(x - n\pi/2 - \pi/4), \tag{2.106b}$$

$$H_n^{(1)}(x) \xrightarrow{x \to \infty} \sqrt{\frac{2}{\pi x}} \exp[j(x - n\pi/2 - \pi/4)], \tag{2.106c}$$

$$H_n^{(2)}(x) \xrightarrow{x \to \infty} \sqrt{\frac{2}{\pi x}} \exp[-j(x - n\pi/2 - \pi/4)], \tag{2.106d}$$

$$I_n(x) \xrightarrow{x \to \infty} \frac{1}{\sqrt{2\pi x}} e^x, \tag{2.106e}$$

$$K_n(x) \xrightarrow{x \to \infty} \frac{1}{\sqrt{2\pi x}} e^{-x} \tag{2.106f}$$

With the time factor $e^{j\omega t}$, $H_n^{(1)}(x)$ and $H_n^{(2)}(x)$ represent inward and outward traveling waves, respectively, while $J_n(x)$ or $Y_n(x)$ represents a standing wave. With the time factor $e^{-j\omega t}$, the roles of $H_n^{(1)}(x)$ and $H_n^{(2)}(x)$ are reversed. For further treatment of Bessel and related functions, refer to the works of Watson [6] and Bell [7].

Any of the Bessel functions or related functions can be a solution to Equation 2.90 depending on the problem. If we choose $R(\rho) = J_n(x) = J_n(\lambda\rho)$ with Equations 2.91 and 2.92 and apply the superposition theorem, the solution to Equation 2.66 is

$$U(\rho,\phi,z) = \sum_n \sum_\mu A_{n\mu} J_n(\lambda\rho) \exp(\pm jn\phi \pm j\mu z) \tag{2.107}$$

Introducing the time dependence of Equation 2.6a, we finally get

$$\Phi(\rho,\phi,z,t) = \sum_m \sum_n \sum_\mu A_{mn\mu} J_n(\lambda\rho) \exp(\pm jn\phi \pm j\mu z \pm \omega t), \tag{2.108}$$

where $\omega = kc$.

EXAMPLE 2.3

Consider the skin effect on a solid cylindrical conductor. The current density distribution within a good conducting wire ($\sigma/\omega\varepsilon \gg 1$) obeys the diffusion equation

$$\nabla^2 J = \mu\sigma \frac{\partial J}{\partial t}$$

We want to solve this equation for a long conducting wire of radius a.

Solution

We may derive the diffusion equation directly from Maxwell's equation. We recall that

$$\nabla \times \mathbf{H} = \mathbf{J} + \mathbf{J}_d$$

where $\mathbf{J} = \sigma\mathbf{E}$ is the conduction current density and $\mathbf{J}_d = \dfrac{\partial \mathbf{D}}{\partial t}$ is the displacement current density. For $\sigma/\omega\varepsilon \gg 1$, $\mathbf{J}_d$ is negligibly small compared with $\mathbf{J}$. Hence,

$$\nabla \times \mathbf{H} \simeq \mathbf{J} \tag{2.109}$$

Also,

$$\Delta \times \mathbf{E} = -\mu \frac{\partial \mathbf{H}}{\partial t}$$

$$\nabla \times \nabla \times \mathbf{E} = \nabla\nabla \cdot \mathbf{E} - \nabla^2\mathbf{E} = -\mu \frac{\partial}{\partial t}\nabla \times \mathbf{H}$$

Since $\nabla\mathbf{E} = 0$, introducing Equation 2.109, we obtain

$$\nabla^2\mathbf{E} = \mu \frac{\partial \mathbf{J}}{\partial t} \tag{2.110}$$

Replacing $\mathbf{E}$ with $\mathbf{J}/\sigma$, Equation 2.110 becomes

$$\nabla^2 \mathbf{J} = \mu\sigma \frac{\partial \mathbf{J}}{\partial t}, \tag{2.111}$$

which is the diffusion equation.

Assuming time-harmonic field with time factor $e^{j\omega t}$,

$$\nabla^2 \mathbf{J} = j\omega\mu\sigma \mathbf{J} \tag{2.112}$$

For an infinitely long wire, Equation 2.112 reduces to a one-dimensional problem in cylindrical coordinates:

$$\frac{1}{\rho} \frac{\partial}{\partial \rho}\left(\rho \frac{\partial J_z}{\partial \rho} \right) = j\omega\mu\sigma J_z$$

or

$$\rho^2 J_z'' + \rho J_z' - j\omega\mu\sigma\rho^2 J_z = 0 \tag{2.113}$$

Comparing this with Equation 2.95 shows that Equation 2.113 is the modified Bessel equation of zero order. Hence, the solution of Equation 2.113 is

$$J_z(\rho) = c_1 I_0(\lambda\rho) + c_2 K_0(\lambda\rho) \tag{2.114}$$

where c_1 and c_2 are constants and

$$\lambda = \sqrt{j\omega\mu\sigma} = j^{1/2} \frac{\sqrt{2}}{\delta} \tag{2.115}$$

and $\delta = \sqrt{(2/\sigma\mu\omega)}$ is the skin depth. Constant c_2 must vanish if J_z is to be finite at $\rho = 0$. At $\rho = a$,

$$J_z(a) = c_1 I_0(\lambda a) \rightarrow c_1 = J_z(a)/I_0(\lambda a)$$

Thus,

$$J_z(\rho) = J_z(a) \frac{I_0(\lambda\rho)}{I_0(\lambda a)} \tag{2.116}$$

If we let $\lambda\rho = j^{1/2}(\sqrt{2}/\delta)\rho = j^{1/2}x$, it is convenient to replace

$$\begin{aligned} I_0(\lambda\rho) = I_0(j^{1/2}x) &= J_0(xe^{j3\pi/4}) \\ &= ber_0(x) + jbei_0(x) \end{aligned} \tag{2.117}$$

where ber_0 and bei_0 are *ber* and *bei* functions of zero order. Ber and bei functions are also known as *Kelvin functions*. For zero order, they are given by

$$ber_0(x) = \sum_{m=0}^{\infty} \frac{\cos(m\pi/2)(x/2)^{2m}}{(m!)^2}, \tag{2.118}$$

$$bei_0(x) = \sum_{m=0}^{\infty} \frac{\sin(m\pi/2)(x/2)^{2m}}{(m!)^2},$$

(2.119)

Using ber and bei functions, Equation 2.116 may be written as

$$J_z(\rho) = J_z(a) \frac{ber_0(x) + jbei_0(x)}{ber_0(y) + jbei_0(y)}$$

(2.120)

where $x = \sqrt{2}\rho/\delta, y = \sqrt{2}a/\delta$.

EXAMPLE 2.4

A semi-infinitely long cylinder ($z \geq 0$) of radius a has its end at $z = 0$ maintained at $V_o(a^2 - \rho^2)$, $0 \leq \rho \leq a$. Find the potential distribution within the cylinder.

Solution

The problem is that of finding a function $V(\rho, z)$ satisfying the PDE

$$\nabla^2 V = \frac{\partial^2 V}{\partial \rho^2} + \frac{1}{\rho} \frac{\partial V}{\partial \rho} + \frac{\partial^2 V}{\partial z^2} = 0$$

(2.121)

subject to the boundary conditions:

1. $V = V_o(a^2 - \rho^2)$, $z = 0$, $0 \leq \rho \leq a$,
2. $V \to 0$ as $z \to \infty$, that is, V is bounded,
3. $V = 0$ on $\rho = a$,
4. V is finite on $\rho = 0$.

Let $V = R(\rho)Z(z)$ and obtain the separated equations

$$Z'' - \lambda^2 Z = 0$$

(2.122a)

and

$$\rho^2 R'' + \rho R' + \lambda^2 \rho^2 R = 0$$

(2.122b)

where λ is the separated constant. The solution to Equation 2.122a is

$$Z_1 = c_1 e^{-\lambda z} + c_2 e^{\lambda z}$$

(2.123)

Comparing Equation 2.122b with Equation 2.93 shows that $n = 0$ so that Equation 2.122b is Bessel's equation with solution

$$R = c_3 J_0(\lambda\rho) + c_4 Y_0(\lambda\rho)$$

(2.124)

Condition (ii) forces $c_2 = 0$, while condition (iv) implies $c_4 = 0$, since $Y_0(\lambda\rho)$ blows up when $\rho = 0$. Hence, the solution to Equation 2.121 is

$$V(\rho, z) = \sum_{n=0}^{\infty} A_n e^{-\lambda_n z} J_0(\lambda_n \rho)$$

(2.125)

where A_n and λ_n are constants to be determined using conditions (i) and (iii). Imposing condition (iii) on Equation 2.125 yields the transcendent equation

$$J_0(\lambda_n a) = 0 \tag{2.126}$$

Thus, λ_n are the positive roots of $J_0(\lambda_n a)$. If we take λ_1 as the first root, λ_2 as the second root, etc., n must start from 1 in Equation 2.125. Imposing condition (i) on Equation 2.125, we obtain

$$V(\rho, 0) = V_o(a^2 - \rho^2) = \sum_{n=1}^{\infty} A_n J_0(\lambda_n \rho)$$

which is simply the Fourier–Bessel expansion of $V_o(a^2 - \rho^2)$. From Table 2.1, property (h),

$$A_n = \frac{2}{a^2[J_1(\lambda_n a)]^2} \int_0^a \rho V_o(a^2 - \rho^2) J_0(\lambda_n \rho) d\rho \tag{2.127}$$

To evaluate the integral, we utilize property (e) in Table 2.1:

$$\int_0^a x^n J_{n-1}(x) dx = x^n J_n(x)\Big|_0^a = a^n J_n(a), \quad n > 0$$

By changing variables, $x = \lambda \rho$,

$$\int_0^a \rho^n J_{n-1}(\lambda \rho) d\rho = \frac{a^n}{\lambda} J_n(\lambda a) \tag{2.128}$$

If $n = 1$,

$$\int_0^a \rho J_0(\lambda \rho) d\rho = \frac{a}{\lambda} J_1(\lambda a) \tag{2.129}$$

Similarly, using property (e) in Table 2.1, we may write

$$\int_0^a \rho^3 J_0(\lambda \rho) d\rho = \int_0^a \frac{\rho^2}{\lambda} \frac{\partial}{\partial \rho} [\rho J_1(\lambda \rho)] d\rho$$

Integrating the right-hand side by parts and applying Equation 2.128,

$$\int_0^a \rho^3 J_0(\lambda \rho) d\rho = \frac{a^3}{\lambda} J_1(\lambda a) - \frac{2}{\lambda} \int_0^a \rho^2 J_1(\lambda \rho) d\rho$$

$$= \frac{a^3}{\lambda} J_1(\lambda a) - \frac{2a^2}{\lambda^2} J_2(\lambda a)$$

$J_2(x)$ can be expressed in terms of $J_0(x)$ and $J_1(x)$ using the recurrence relations, that is, property (c) in Table 2.1:

$$J_2(x) = \frac{2}{x} J_1(x) - J_0(x)$$

Hence,

$$\int_0^a \rho^3 J_0(\lambda_n \rho) d\rho = \frac{2a^2}{\lambda_n^2} \left[J_0(\lambda_n a) + \left(\frac{a\lambda_n}{2} - \frac{2}{a\lambda_n} \right) J_1(\lambda_n a) \right] \qquad (2.130)$$

Substitution of Equations 2.129 and 2.130 into Equation 2.127 gives

$$\begin{aligned} A_n &= \frac{2V_o}{a^2 [J_1(\lambda_n a)]^2} \left[\frac{4a}{\lambda_n^3} J_1(\lambda_n a) - \frac{2a^2}{\lambda_n^2} J_0(\lambda_n a) \right] \\ &= \frac{8V_o}{a\lambda_n^3 J_1(\lambda_n a)} \end{aligned}$$

since $J_0(\lambda_n a) = 0$ from Equation 2.126. Thus, the potential distribution is given by

$$V(\rho, z) = \frac{8V_o}{a} \sum_{n=1}^{\infty} \frac{e^{-\lambda_n z} J_0(\lambda_n \rho)}{\lambda_n^3 J_1(\lambda_n a)}$$

EXAMPLE 2.5

A plane wave $\mathbf{E} = E_o e^{j(\omega t - kx)} \mathbf{a}_z$ is incident on an infinitely long conducting cylinder of radius a. Determine the scattered field.

Solution

Since the cylinder is infinitely long, the problem is two-dimensional as shown in Figure 2.6. We shall suppress the time factor $e^{j\omega t}$ throughout the analysis. For the sake of convenience, we need to express the plane wave in terms of cylindrical waves. We let

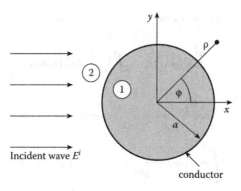

FIGURE 2.6
Scattering by a conducting cylinder.

$$e^{-jx} = e^{-j\rho\cos\phi} = \sum_{n=-\infty}^{\infty} a_n J_n(\rho) e^{jn\phi} \qquad (2.131)$$

where a_n are expansion coefficients to be determined. Since $e^{jn\phi}$ are orthogonal functions, multiplying both sides of Equation 2.131 by $e^{jm\phi}$ and integrating over $0 \leq \phi \leq 2\pi$ gives

$$\int_0^{2\pi} e^{-j\rho\cos\phi} e^{jm\phi} d\phi = 2\pi a_m J_m(\rho)$$

Taking the mth derivative of both sides with respect to ρ and evaluating at $\rho = 0$ leads to

$$2\pi \frac{j^{-m}}{2^m} = 2\pi a_m \frac{1}{2^m} \rightarrow a_m = j^{-m}$$

Substituting this into Equation 2.131, we obtain

$$e^{-jx} = \sum_{n=-\infty}^{\infty} j^{-n} J_n(\rho) e^{jn\phi}$$

(An alternative, easier way of obtaining this is using the generating function for $J_n(x)$ in Table 2.7.) Thus, the incident wave may be written as

$$E_z^i = E_o e^{-jkx} = E_o \sum_{n=-\infty}^{\infty} (-j)^n J_n(k\rho) e^{jn\phi} \qquad (2.132)$$

Since the scattered field E_z^s must consist of outgoing waves that vanish at infinity, it contains

$$J_n(k\rho) - jY_n(k\rho) = H_n^{(2)}(k\rho)$$

Hence,

$$E_z^s = \sum_{n=-\infty}^{\infty} A_n H_n^{(2)}(k\rho) e^{jn\phi} \qquad (2.133)$$

The total field in medium 2 is

$$E_2 = E_z^i + E_z^s$$

while the total field in medium 1 is $E_1 = 0$ since medium 1 is conducting. At $\rho = a$, the boundary condition requires that the tangential components of E_1 and E_2 be equal. Hence,

$$E_z^i(\rho = a) + E_z^s(\rho = a) = 0 \qquad (2.134)$$

Substituting Equations 2.132 and 2.133 into Equation 2.134,

$$\sum_{n=-\infty}^{\infty} \left[E_o(-j)^n J_n(ka) + A_n H_n^{(2)}(ka) \right] e^{jn\phi} = 0$$

From this, we obtain

$$A_n = -\frac{E_o(-j)^n J_n(ka)}{H_n^{(2)}(ka)}$$

Finally, substituting A_n into Equation 2.133 and introducing the time factor leads to the scattered wave as

$$\mathbf{E}_z^s = -E_o e^{j\omega t} \mathbf{a}_z \sum_{n=-\infty}^{\infty} (-j)^n \frac{J_n(ka) H_n^{(2)}(k\rho) e^{jn\phi}}{H_n^{(2)}(ka)}$$

2.5 Separation of Variables in Spherical Coordinates

Spherical coordinates (r, θ, ϕ) may be defined as in Figure 2.7, where $0 \leq r \leq \infty$, $0 \leq \theta \leq \pi$, $0 \leq \phi < 2\pi$. In this system, the wave equation (2.5b) becomes

$$\nabla^2 U + k^2 U = \frac{1}{r^2} \frac{\partial}{\partial r} \left(r^2 \frac{\partial U}{\partial r} \right) + \frac{1}{r^2 \sin \theta} \frac{\partial}{\partial \theta} \left(\sin \theta \frac{\partial U}{\partial \theta} \right)$$
$$+ \frac{1}{r^2 \sin^2 \theta} \frac{\partial^2 U}{\partial \phi^2} + k^2 U = 0 \tag{2.135}$$

As usual, we shall first solve Laplace's equation in two dimensions and later solve the wave equation in three dimensions.

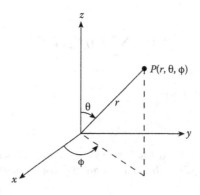

FIGURE 2.7
Coordinate relations in a spherical system.

2.5.1 Laplace's Equation

Consider the problem of finding the potential distribution due to an uncharged conducting sphere of radius a located in an external uniform electric field as in Figure 2.8. The external electric field can be described as

$$\mathbf{E} = E_o \mathbf{a}_z \tag{2.136}$$

while the corresponding electric potential can be described as

$$V = -\int \mathbf{E} \cdot d\mathbf{l} = -E_o z$$

or

$$V = -E_o r \cos\theta \tag{2.137}$$

where $V(\theta = \pi/2) = 0$ has been assumed. From Equation 2.137, it is evident that V is independent of ϕ, and hence our problem is solving Laplace's equation in two dimensions, namely,

$$\nabla^2 V = \frac{1}{r^2}\frac{\partial V}{\partial r}\left(r^2\frac{\partial V}{\partial r}\right) + \frac{1}{r^2\sin\theta}\frac{\partial}{\partial\theta}\left(\sin\theta\frac{\partial V}{\partial\theta}\right) = 0 \tag{2.138}$$

subject to the conditions

$$V(r, \theta) = -E_o r \cos\theta \quad \text{as} \quad r \to \infty, \tag{2.139a}$$

$$V(a, \theta) = 0 \tag{2.139b}$$

We let

$$V(r, \theta) = R(r)H(\theta) \tag{2.140}$$

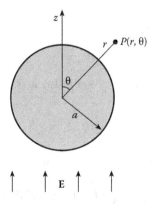

FIGURE 2.8
An uncharged conducting sphere in a uniform external electric field.

so that Equation 2.138 becomes

$$\frac{1}{R}\frac{d}{dr}(r^2R') = -\frac{1}{H\sin\theta}\frac{d}{d\theta}(\sin\theta H') = \lambda \tag{2.141}$$

where λ is the separation constant. Thus, the separated equations are

$$r^2R'' + 2rR' - \lambda R = 0 \tag{2.142}$$

and

$$\frac{d}{d\theta}(\sin\theta H') + \lambda\sin\theta H = 0 \tag{2.143}$$

Equation 2.142 is the *Cauchy–Euler equation*. It can be solved by making the substitution $R = r^k$. This leads to the solution

$$R_n(r) = A_n r^n + B_n r^{-(n+1)}, \quad n = 0,1,2, \ldots \tag{2.144}$$

with $\lambda = n(n+1)$. To solve Equation 2.143, we may replace H by y and $\cos\theta$ by x so that

$$\frac{d}{d\theta} = \frac{dx}{d\theta}\frac{d}{dx} = -\sin\theta\frac{d}{dx}$$

$$\frac{d}{d\theta}\left(\sin\theta\frac{dH}{d\theta}\right) = -\sin\theta\frac{d}{dx}\left(\sin\theta\frac{dx}{d\theta}\frac{dH}{dx}\right)$$

$$= \sin\theta\frac{d}{dx}\left(\sin^2\theta\frac{dy}{dx}\right)$$

$$= \sqrt{1-x^2}\frac{d}{dx}\left[(1-x^2)\frac{dy}{dx}\right]$$

Making these substitutions in Equation 2.143 yields

$$\frac{d}{dx}\left[(1-x^2)\frac{dy}{dx}\right] + n(n+1)y = 0$$

or

$$(1-x^2)y'' - 2xy' + n(n+1)y = 0 \tag{2.145}$$

which is the *Legendre differential equation*. Its solution is obtained by the method of Frobenius [5] as

$$y = c_n P_n(x) + d_n Q_n(x) \tag{2.146}$$

where $P_n(x)$ and $Q_n(x)$ are Legendre functions of the first and second kind, respectively.

$$\boxed{P_n(x) = \sum_{k=0}^{N} \frac{(-1)^k(2n-2k)!\,x^{n-2k}}{2^n k!(n-k)!(n-2k)!}} \tag{2.147}$$

where $N = n/2$ if n is even and $N = (n - 1)/2$ if n is odd. For example,

$$P_0(x) = 1$$
$$P_1(x) = x = \cos\theta$$
$$P_2(x) = \frac{1}{2}(3x^3 - 1) = \frac{1}{4}(3\cos 2\theta + 1)$$
$$P_3(x) = \frac{1}{2}(5x^3 - 3x) = \frac{1}{8}(5\cos 3\theta + 3\cos\theta)$$
$$P_4(x) = \frac{1}{8}(35x^4 - 30x^2 + 3) = \frac{1}{64}(35\cos 4\theta + 20\cos 2\theta + 9)$$
$$P_5(x) = \frac{1}{8}(65x^5 - 70x^3 + 15x) = \frac{1}{128}(30\cos\theta + 35\cos 3\theta + 63\cos 5\theta)$$

Some useful identities and properties [5] of Legendre functions are listed in Table 2.2. The Legendre functions of the second kind are given by

$$Q_n(x) = P_n(x)\left[\frac{1}{2}\ln\frac{1+x}{1-x} - p(n)\right]$$
$$+ \sum_{k=1}^{n} \frac{(-1)^k(n+k)!}{(k!)^2(n-k)!} p(k)\left[\frac{1-x}{2}\right]^k \tag{2.148}$$

where $p(k)$ is as defined in Equation 2.100. Typical graphs of $P_n(x)$ and $Q_n(x)$ are shown in Figure 2.9. Q_n are not as useful as P_n since they are singular at $x = \pm 1$ (or $\theta = 0, \pi$) due to the logarithmic term in Equation 2.148. We use Q_n only when $x \neq \pm 1$ (or $\theta \neq 0, \pi$), for example, in problems having conical boundaries that exclude the axis from the solution region. For the problem at hand, $\theta = 0, \pi$ is included so that the solution to Equation 2.143 is

$$H_n(\theta) = P_n(\cos\theta) \tag{2.149}$$

Substituting Equations 2.144 and 2.149 into Equation 2.140 gives

$$V_n(r, \theta) = [A_n r^n + B_n r^{-(n+1)}]P_n(\cos\theta) \tag{2.150}$$

To determine A_n and B_n, we apply the boundary conditions in Equation 2.139. Since as $r \to \infty$, $V = -E_o r \cos\theta$, it follows that $n = 1$ and $A_1 = -E_o$, that is,

$$V(r, \theta) = \left(-E_o r + \frac{B_1}{r^2}\right)\cos\theta$$

Also since $V = 0$ when $r = a$, $B_1 = E_o a^3$. Hence, the complete solution is

$$V(r, \theta) = -E_o\left(r - \frac{a^3}{r^2}\right)\cos\theta \tag{2.151}$$

TABLE 2.2

Properties and Identities of Legendre Functions[a]

a. For $n \geq 1, P_n(1) = 1, \quad P_n(-1) = (-1)^n,$

$$P_{2n+1} = 0, \quad P_{2n}(0) = (-1)^n \frac{(2n)!}{2^{2n}(n!)^2}$$

b. $P_n(-x) = (-1)^n P_n(x)$

c. $P_n(x) = \dfrac{1}{2^n n!} \dfrac{d^n}{dx^n}(x^2 - 1)^n, \quad n \geq 0$
(Rodriguez formula)

d. $(n+1)P_{n+1}(x) = (2n+1)xP_n(x) - nP_{n-1}(x), \quad n \geq 1$
(recurrence relation)

e. $P_n'(x) = xP_{n-1}'(x) + nP_{n-1}(x), \quad n \geq 1$

f. $P_n(x) = xP_{n-1}(x) + (x^2 - 1/n)P_{n-1}'(x), \quad n \geq 1$

g. $P_{n+1}'(x) - P_{n-1}'(x) = (2n+1)P_n(x), \quad n \geq 1$

or

$$\int P_n(x)\,dx = \frac{P_{n+1} - P_{n-1}}{2n+1}$$

h. Fourier–Legendre series expansion of $f(x)$:

$$f(x) = \sum_{n=0}^{\infty} A_n P_n(x), \quad -1 \leq x \leq 1$$

where

$$A_n = \frac{2n+1}{2} \int_{-1}^{1} f(x)P_n(x)\,dx, \quad n \geq 0$$

If $f(x)$ is odd,

$$A_n = (2n+1) \int_{0}^{1} f(x)P_n(x)\,dx, \quad n = 0, 2, 4 \ldots$$

and if $f(x)$ is even,

$$A_n = (2n+1) \int_{0}^{1} f(x)P_n(x)\,dx, \quad n = 1, 3, 5 \ldots$$

i. Orthogonality property

$$\int_{0}^{1} P_n(x)P_m(x)\,dx = \begin{cases} 0, & n \neq m \\ \dfrac{2}{2n+1} & n = m \end{cases}$$

[a] Properties (d) through (g) are also valid for $Q_n(x)$.

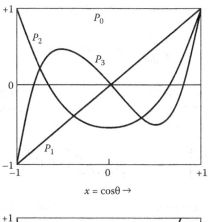

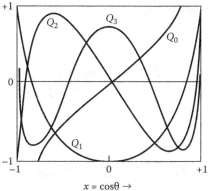

FIGURE 2.9
Typical Legendre functions of the first and second kinds.

The electric field intensity is given by

$$\mathbf{E} = -\nabla V = -\frac{\partial V}{\partial r}\mathbf{a}_r - \frac{1}{r}\frac{\partial V}{\partial \theta}\mathbf{a}_\theta$$

$$= E_o\left[1 + \frac{2a^3}{r^3}\right]\cos\theta\,\mathbf{a}_r + E_o\left[1 - \frac{a^3}{r^3}\right]\sin\theta\,\mathbf{a}_\theta \tag{2.152}$$

2.5.2 Wave Equation

To solve the wave equation (2.135), we substitute

$$U(r,\theta,\phi) = R(r)H(\theta)F(\phi) \tag{2.153}$$

into the equation. Multiplying the result by $r^2\sin^2\theta/RHF$ gives

$$\frac{\sin^2\theta}{R}\frac{d}{dr}\left(r^2\frac{dR}{dr}\right) + \frac{\sin\theta}{H}\frac{d}{d\theta}\left(\sin\theta\frac{dH}{d\theta}\right)$$

$$+ k^2 r^2 \sin^2\theta = -\frac{1}{F}\frac{d^2F}{d\phi^2} \tag{2.154}$$

Since the left-hand side of this equation is independent of ϕ, we let

$$-\frac{1}{F}\frac{d^2F}{d\phi^2} = m^2, \quad m = 0,1,2,\ldots$$

where m, the first separation constant, is chosen to be nonnegative integer such that U is periodic in ϕ. This requirement is necessary for physical reasons that will be evident later. Thus, Equation 2.154 reduces to

$$\frac{1}{R}\frac{d}{dr}\left(r^2\frac{dR}{dr}\right) + k^2r^2 = -\frac{1}{H\sin\theta}\frac{d}{d\theta}\left(\sin\theta\frac{dH}{d\theta}\right) + \frac{m^2}{\sin^2\theta} = \lambda$$

where λ is the second separation constant. As in Equations 2.141 through 2.144, $\lambda = n(n+1)$ so that the separated equations are now

$$F'' + m^2F = 0, \tag{2.155}$$

$$R'' + \frac{2}{r}R' + \left[k^2 - \frac{n(n+1)}{r^2}\right]R = 0, \tag{2.156}$$

and

$$\frac{1}{\sin\theta}\frac{d}{d\theta}(\sin\theta H') + \left[n(n+1) - \frac{m^2}{\sin^2\theta}\right]H = 0 \tag{2.157}$$

As usual, the solution to Equation 2.155 is

$$F(\phi) = c_1 e^{jm\phi} + c_2 e^{-jm\phi} \tag{2.158a}$$

or

$$F(\phi) = c_3\sin m\phi + c_4\cos m\phi \tag{2.158b}$$

If we let $R(r) = r^{-1/2}\tilde{R}(r)$, Equation 2.156 becomes

$$\tilde{R}'' + \frac{1}{r}\tilde{R}' + \left[k^2 - \frac{(n+1/2)^2}{r^2}\right]\tilde{R} = 0,$$

which has the solution

$$\tilde{R} = Ar^{1/2}z_n(kr) = BZ_{n+1/2}(kr) \tag{2.159}$$

Functions $z_n(x)$ are *spherical Bessel functions* and are related to ordinary Bessel functions $Z_{n+1/2}$ according to

$$z_n(x) = \sqrt{\frac{\pi}{2x}}Z_{n+1/2}(x) \tag{2.160}$$

In Equation 2.160, $Z_{n+1/2}(x)$ may be any of the ordinary Bessel functions of half-integer order, $J_{n+1/2}(x)$, $Y_{n+1/2}(x)$, $I_{n+1/2}(x)$, $K_{n+1/2}(x)$, $H^{(1)}_{n+1/2}(x)$, and $H^{(2)}_{n+1/2}(x)$, while $z_n(x)$ may be any of the corresponding spherical Bessel functions $j_n(x)$, $y_n(x)$, $i_n(x)$, $k_n(x)$, $h^{(1)}_n(x)$, and $h^{(2)}_n(x)$. Bessel functions of fractional order are, in general, given by

$$J_v(x) = \sum_{k=0}^{\infty} \frac{(-1)^k x^{2k+v}}{2^{2k+v} k! \Gamma(v+k+1)} \tag{2.161}$$

$$Y_v(x) = \frac{J_v(x)\cos(v\pi) - J_{-v}}{\sin(v\pi)} \tag{2.162}$$

$$I_v(x) = (-j)^v J_v(jx) \tag{2.163}$$

$$K_v(x) = \frac{\pi}{2}\left[\frac{I_{-v} - I_v}{\sin(v\pi)}\right] \tag{2.164}$$

where J_{-v} and I_{-v} are, respectively, obtained from Equations 2.161 and 2.163 by replacing v with $-v$. Although v in Equations 2.161 through 2.164 can assume any fractional value, in our specific problem, $v = n + 1/2$. Since Gamma function of half-integer order is needed in Equation 2.161, it is necessary to add that

$$\Gamma(n+1/2) = \begin{cases} \dfrac{2(n)!}{2^{2n}n!}\sqrt{\pi}, & n \ge 0 \\[2ex] \dfrac{(-1)^n 2^{2n} n!}{(2n)!}\sqrt{\pi}, & n < 0 \end{cases} \tag{2.165}$$

Thus, the lower-order spherical Bessel functions are as follows:

$$j_0(x) = \frac{\sin x}{x}, \qquad\qquad y_0(x) = -\frac{\cos x}{x},$$

$$h^{(1)}_0(x) = \frac{e^{jx}}{jx}, \qquad\qquad h^{(2)}_0(x) = \frac{e^{-jx}}{-jx},$$

$$i_0(x) = \frac{\sinh x}{x}, \qquad\qquad k_0(x) = \frac{e^{-x}}{x},$$

$$j_1(x) = \frac{\sin x}{x^2} - \frac{\cos x}{x}, \qquad y_1(x) = -\frac{\cos x}{x^2} - \frac{\sin x}{x},$$

$$h^{(1)}_1(x) = -\frac{(x+j)}{x^2}e^{jx}, \qquad h^{(2)}_1(x) = -\frac{(x-j)}{x^2}e^{-jx},$$

Other $z_n(x)$ can be obtained from the series expansion in Equations 2.161 and 2.162 or the recurrence relations and properties of $z_n(x)$ presented in Table 2.3.

By replacing H in Equation 2.157 with y, $\cos\theta$ by x, and making other substitutions as we did for Equation 2.143 we obtain

$$(1-x^2)y'' - 2xy' + \left[n(n+1) - \frac{m^2}{1-x^2}\right]y = 0, \tag{2.166}$$

TABLE 2.3

Properties and Identities of Spherical Bessel Functions

a. $z_{n+1} = \dfrac{(2n+1)}{x} z_n(x) - z_{n-1}(x)$ (recurrence relation)

b. $\dfrac{d}{dx} z_n(x) = \dfrac{1}{2n+1} [n z_{n-1} - (n+1) z_{n+1}(x)]$

c. $\dfrac{d}{dx}[x z_n(x)] = -n z_n(x) + x z_{n-1}(x)$

d. $\dfrac{d}{dx}[x^{n+1} z_n(x)] = -x^{n+1} z_{n-1}(x)$

e. $\dfrac{d}{dx}[x^{-n} z_n(x)] = -x^{-n} z_{n+1}(x)$

f. $\displaystyle\int x^{n+2} z_n(x)\,dx = x^{n+2} z_{n+1}(x)$

g. $\displaystyle\int x^{1-n} z_n(x)\,dx = -x^{1-n} z_{n-1}(x)$

h. $\displaystyle\int x^2 [z_n(x)]^2\,dx = \frac{1}{2} x^3 [z_n(x) - z_{n-1}(x) z_{n+1}(x)]$

which is Legendre's associated differential equation. Its general solution is of the form

$$y(x) = a_{mn} P_n^m(x) + d_{mn} Q_n^m(x) \tag{2.167}$$

where $P_n^m(x)$ and $Q_n^m(x)$ are called associated Legendre functions of the first and second kind, respectively. Equation 2.146 is a special case of Equation 2.167 when $m = 0$. $P_n^m(x)$ and $Q_n^m(x)$ can be obtained from ordinary Legendre functions $P_n(x)$ and $Q_n(x)$ using

$$P_n^m(x) = [1 - x^2]^{m/2} \frac{d^m}{dx^m} P_n(x) \tag{2.168}$$

and

$$Q_n^m(x) = [1 - x^2]^{m/2} \frac{d^m}{dx^m} Q_n(x) \tag{2.169}$$

where $-1 < x < 1$. We note that

$$\begin{aligned}
P_n^0(x) &= P_n(x), \\
Q_n^0(x) &= Q_n(x), \\
P_n^m(x) &= 0 \quad \text{for} \quad m > n
\end{aligned} \tag{2.170}$$

Typical associated Legendre functions are

$$P_1^1(x) = (1 - x^2)^{1/2} = \sin\theta$$
$$P_2^1(x) = 3x(1 - x^2)^{1/2} = 3\cos\theta\sin\theta,$$
$$P_2^2(x) = 3(1 - x^2) = 3\sin^2\theta,$$
$$P_3^1(x) = \frac{3}{2}(1 - x^2)^{1/2}(5x - 1) = \frac{3}{2}\sin\theta(5\cos\theta - 1),$$

$$Q_1^1(x) = (1-x^2)^{1/2}\left[\frac{1}{2}\ln\frac{1+x}{1-x} + \frac{x}{1-x^2}\right],$$

$$Q_2^1 = (1-x^2)^{1/2}\left[\frac{3x}{2}\ln\frac{1+x}{1-x} + \frac{3^2-2}{1-x^2}\right],$$

$$Q_2^2 = (1-x^2)^{1/2}\left[\frac{3}{2}\ln\frac{1+x}{1-x} + \frac{5x^2-3x^2}{[1-x^2]^2}\right]$$

Higher-order associated Legendre functions can be obtained using Equations 2.168 and 2.169 along with the properties in Table 2.4. As mentioned earlier, $Q_n^m(x)$ is unbounded at $x = \pm 1$, and hence it is only used when $x = \pm 1$ is excluded. Substituting Equations 2.158, 2.159, and 2.167 into Equation 2.153 and applying superposition theorem, we obtain

$$U(r,\theta,\phi,t) = \sum_{n=0}^{\infty}\sum_{m=0}^{n}\sum_{\ell=0}^{\infty} A_{mn\ell} z_n(k_{m\ell}r)P_n^m(\cos\theta)\exp(\pm jm\phi \pm j\omega t) \qquad (2.171)$$

Note that the products $H(\theta)F(\phi)$ are known as *spherical harmonics*.

EXAMPLE 2.6

A thin ring of radius a carries charge of density ρ. Find the potential at: (a) point $P(0, 0, z)$ on the axis of the ring, (b) point $P(r, \theta, \phi)$ in space.

Solution

Consider the thin ring as in Figure 2.10.

a. From elementary electrostatics, at $P(0, 0, z)$

$$V = \int \frac{\rho dl}{4\pi\varepsilon R}$$

where $dl = ad\phi$, $R = \sqrt{a^2 + z^2}$. Hence,

$$V = \int_{0}^{2\pi} \frac{\rho a d\phi}{4\pi\varepsilon[a^2 + z^2]^{1/2}} = \frac{a\rho}{2\varepsilon[a^2 + z^2]^{1/2}} \qquad (2.172)$$

b. To find the potential at $P(r, 0, \phi)$, we may evaluate the integral for the potential as we did in part (a). However, it turns out that the boundary-value solution is simpler. So we solve Laplace's equation $\nabla^2 V = 0$, where $V(0, 0, z)$ must conform with the result in part (a). From Figure 2.10, it is evident that V is invariant with ϕ. Hence, the solution to Laplace's equation is

$$V = \sum_{n=0}^{\infty}\left[A_n r^n + \frac{B_n}{r^{n+1}}\right]\left[A_n'P_n(u) + B_n'Q_n(u)\right]$$

where $u = \cos\theta$. Since Q_n is singular at $\theta = 0, \pi$, $B_n' = 0$. Thus,

$$V = \sum_{n=0}^{\infty}\left[C_n'r^n + \frac{D_n'}{r^{n+1}}\right]P_n(u) \qquad (2.173)$$

TABLE 2.4

Properties and Identities of Associated Legendre Functions[a]

a. $P_m(x) = 0, \quad m > n$

b. $P_n^m(x) = \dfrac{(2n-1)xP_{n-1}^m(x) - (n+m-1)P_{n-2}^m(x)}{n-m}$

(recurrence relations for fixed m)

c. $P_n^m(x) = \dfrac{2(m-1)x}{(1-x^2)^{1/2}} P_n^{m-1}(x) - (n-m+2)(n+m-1)P_n^{m-2}$

(recurrence relations for fixed n)

d. $P_n^m(x) = \dfrac{[1-x^2]^{m/2}}{2^n} \displaystyle\sum_{k=0}^{\left[\frac{m-n}{2}\right]} \dfrac{(-1)^k (2n-2k)! x^{n-2k-m}}{k!(n-k)!(n-2k-m)!}$

where $[t]$ is the bracket or greatest integer function, for example, $[3.54] = 3$.

e. $\dfrac{d}{dx} P_n^m(x) = \dfrac{(n+m)P_{n-1}^m(x) - nxP_n^m(x)}{1-x^2}$

f. $\dfrac{d}{d\theta} P_n^m(x) = \dfrac{1}{2}[(n-m+1)(n+m)P_n^{m-1}(x) - P_n^{m+1}(x)]$

g. $\dfrac{d}{dx} P_n^m(x) = -\dfrac{mxP_n^m(x)}{1-x^2} + \dfrac{(1-x^2)^{m/2}}{n} \displaystyle\sum_{k=0}^{\left[\frac{m-n}{2}\right]} \dfrac{(-1)^k (2n-2k)! x^{n-2k-m-1}}{k!(n-k)!(n-2k-m)!}$

h. $\dfrac{d}{d\theta} P_n^m(x) = -(1-x^2)^{1/2} \dfrac{d}{dx} P_n^m(x)$

i. The series expansion of $f(x)$:

$f(x) = \displaystyle\sum_{n=0}^{\infty} A_n P_n^m(x)$

where $A_n = \dfrac{(2n+1)(n-m)!}{2(n+m)!} \displaystyle\int_{-1}^{1} f(x)P_n^m(x)\,dx$

j. $\dfrac{d^m}{dx^m} P_n(x)\bigg|_{x=1} = \dfrac{(n+m)!}{2^m m!(n-m)!}$

  $\dfrac{d^m}{dx^m} P_n(x)\bigg|_{x=-1} = \dfrac{(-1)^{n+m}(n+m)!}{2^m m!(n-m)!}$

k. $P_n^{-m}(x) = (-1)^m \dfrac{(n-m)!}{(n+m)!} P_n^m(x), \quad m = 0,1,\dots,n$

l. $\displaystyle\int_{-1}^{1} P_n^m(x)P_n^m(x)\,dx = \dfrac{2}{2n+1} \dfrac{(n-m)!}{(n+m)!} \delta_{nk},$

where δ_{nk} is the Kronecker delta defined by $\delta_{nk} = \begin{cases} 0, & n \neq k \\ 1, & n = k \end{cases}$

---

[a] Properties (b) and (c) are also valid for $Q_n^m(x)$.

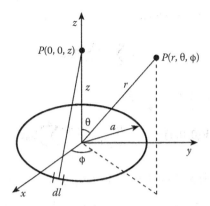

FIGURE 2.10
Charged ring of Example 2.6.

For $0 \leq r \leq a$, $D'_n = 0$ since V must be finite at $r = 0$.

$$V = \sum_{n=0}^{\infty} C'_n r^n P_n(u) \tag{2.174}$$

To determine the coefficients C'_n, we set $\theta = 0$ and equate V to the result in part (a). But when $\theta = 0$, $u = 1$, $P_n(1) = 1$, and $r = z$. Hence,

$$V(0, 0, z) = \frac{a\rho}{2\epsilon[a^2 + z^2]^{1/2}} = \frac{a\rho}{2\epsilon} \sum_{n=0}^{\infty} C_n z^n \tag{2.175}$$

Using the binomial expansion, the term $[a^2 + z^2]^{1/2}$ can be written as

$$\frac{1}{a}\left[1 + \frac{z^2}{a^2}\right]^{-1/2} = \frac{1}{a}\left[1 - \frac{1}{2}(z/a)^2 + \frac{1 \cdot 3}{2 \cdot 4}(z/a)^4 - \frac{1 \cdot 3 \cdot 5}{2 \cdot 4 \cdot 6}(z/a)^6 + \cdots\right]$$

Comparing this with the last term in Equation 2.175, we obtain

$$C_0 = 1, \quad C_1 = 0, \quad C_2 = -\frac{1}{2a^2}, \quad C_3 = 0,$$

$$C_4 = \frac{1 \cdot 3}{2 \cdot 4}\frac{1}{a^4}, \quad C_5 = 0, \quad C_6 = -\frac{1 \cdot 3 \cdot 5}{2 \cdot 4 \cdot 6}\frac{1}{a^6}, \cdots$$

or in general,

$$C_{2n} = (-1)^n \frac{(2n)!}{[n!2^n]^2 a^{2n}}$$

Substituting these into Equation (2.174) gives

$$V = \frac{a\rho}{2\epsilon} \sum_{n=0}^{\infty} \frac{(-1)^n (2n)!}{[n!2^n]^2}(r/a)^{2n} P_{2n}(\cos\theta), \quad 0 \leq r \leq a \tag{2.176}$$

For $r \geq a$, $C_n' = 0$ since V must be finite as $r \rightarrow \infty$, and

$$V = \sum_{n=0}^{\infty} \frac{D_n'}{r^{n+1}} P_n(u) \tag{2.177}$$

Again, when $\theta = 0$, $u = 1$, $P_n(1) = 1$, $r = z$,

$$V(0,0,z) = \frac{a\rho}{2\epsilon[a^2 + z^2]^{1/2}} = \frac{a\rho}{2\epsilon} \sum_{n=0}^{\infty} D_n z^{-(n+1)} \tag{2.178}$$

Using the binomial expansion, the middle term $[a^2 + z^2]^{-1/2}$ can be written as

$$\frac{1}{z}\left[1 + \frac{a^2}{z^2}\right]^{-1/2} = \frac{1}{z}\left[1 - \frac{1}{2}(a/z)^2 + \frac{1\cdot 3}{2\cdot 4}(a/z)^4 - \frac{1\cdot 3\cdot 5}{2\cdot 4\cdot 6}(a/z)^6 + \cdots\right]$$

Comparing this with the last term in Equation 2.178, we obtain

$$D_0 = 1, \quad D_1 = 0, \quad D_2 = -\frac{a^2}{2}, \quad D_3 = 0,$$

$$D_4 = \frac{1\cdot 3}{2\cdot 4}a^4, \quad D_5 = 0, \quad D_6 = -\frac{1\cdot 3\cdot 5}{2\cdot 4\cdot 6}a^6, \ldots$$

or in general,

$$D_{2n} = (-1)^n \frac{(2n)!}{[n!2^n]^2} a^{2n}$$

Substituting these into Equation 2.177 gives

$$V = \frac{a\rho}{2\epsilon r} \sum_{n=0}^{\infty} \frac{(-1)^n (2n)!}{[n!2^n]^2}(a/r)^{2n} P_{2n}(\cos\theta), \quad r \geq a \tag{2.179}$$

We may combine Equations 2.176 and 2.179 to get

$$V = \begin{cases} a \displaystyle\sum_{n=0}^{\infty} g_n (r/a)^{2n} P_{2n}(\cos\theta), & 0 \leq r \leq a \\[2ex] \displaystyle\sum_{n=0}^{\infty} g_n (a/r)^{2n+1} P_{2n}(\cos\theta), & r \geq a \end{cases}$$

where

$$g_n = (-1)^n \frac{\rho}{2\epsilon} \frac{2n!}{[n!2^n]^2}$$

EXAMPLE 2.7

A conducting spherical shell of radius a is maintained at potential $V_o \cos 2\phi$; determine the potential at any point inside the sphere.

Solution

The solution to this problem is somewhat similar to that of the previous problem except that V is a function of ϕ. Hence, the solution to Laplace's equation for $0 \leq r \leq a$ is of the form

$$V = \sum_{n=0}^{\infty} \sum_{m=0}^{\infty} (a_{mn} \cos m\phi + b_{mn} \sin m\phi)(r/a)^n P_n^m(\cos\theta)$$

Since $\cos m\phi$ and $\sin m\phi$ are orthogonal functions, $a_{mn} = 0 = b_{mn}$ except that $a_{n2} \neq 0$. Hence, at $r = a$

$$V_o \cos 2\phi = \cos 2\phi \sum_{n=2}^{\infty} a_{n2} P_n^2(\cos\theta)$$

or

$$V_o = \sum_{n=2}^{\infty} a_{n2} P_n^2(x), \quad x = \cos\theta$$

which is the Legendre expansion of V_o. Multiplying both sides by $P_m^2(x)$ gives

$$\frac{2}{2n+1} \frac{(n+2)!}{(n-2)!} a_{n2} = V_o \int_{-1}^{1} P_n^2(x)dx = V_o \int_{-1}^{1} (1-x^2) \frac{d}{dx^2} P_n(x)dx$$

Integrating by parts twice yields

$$a_{n2} = V_o \frac{2n+1}{2} \frac{(n-2)!}{(n+2)!} \left[2P_n(1) - 2P_n(-1) - 2\int_{-1}^{1} P_n(x)dx \right]$$

Using the generating functions for $P_n(x)$ (see Table 2.7 and Example 2.10), it is readily shown that

$$P_n(1) = 1, \quad P_n(-1) = (-1)^n$$

Also,

$$\int_{-1}^{1} P_n(x)dx = \int_{-1}^{1} P_0(x)P_n(x)dx = 0$$

by the orthogonality property of $P_n(x)$. Hence,

$$a_{n2} = V_o(2n+1) \frac{(n-2)!}{(n+2)!} [1 - (-1)^n]$$

and

$$V = V_o \cos 2\phi \sum_{n=2}^{\infty} (2n+1) \frac{(n-2)!}{(n+2)!} [1 - (-1)^n](r/a)^n P_n^2(\cos\theta)$$

EXAMPLE 2.8

Express (a) the plane wave e^{jz} and (b) the cylindrical wave $J_0(\rho)$ in terms of spherical wave functions.

Solution

  a. Since $e^{jz} = e^{jr\cos\theta}$ is independent of ϕ and finite at the origin, we let

$$e^{jz} = e^{jr\cos\theta} = \sum_{n=0}^{\infty} a_n j_n(r) P_n(\cos\theta) \tag{2.180}$$

where a_n are the expansion coefficients. To determine a_n, we multiply both sides of Equation 2.180 by $P_m(\cos\theta)\sin\theta$ and integrate over $0 < \theta < \pi$:

$$\int_0^\pi e^{jr\cos\theta} P_m(\cos\theta)\sin\theta\, d\theta = \sum_{n=0}^{\infty} a_n j_n(r) \int_{-1}^{1} P_n(x)P_m(x)\, dx$$

$$= \begin{cases} 0, & n \neq m \\ \dfrac{2}{2n+1} a_n j_n(r), & n = m \end{cases}$$

where the orthogonality property (i) of Table 2.2 has been utilized. Taking the nth derivative of both sides and evaluating at $r = 0$ gives

$$j^n \int_0^\pi \cos^n\theta P_n(\cos\theta)\sin\theta\, d\theta = \frac{2}{2n+1} a_n \frac{d^n}{dr^n} j_n(r)\bigg|_{r=0} \tag{2.181}$$

The left-hand side of Equation (2.181) yields

$$j^n \int_{-1}^{1} x^n P_n(x)\, dx = \frac{2^{n+1}(n!)^2}{(2n+1)!} j^n \tag{2.182}$$

To evaluate the right-hand side of Equation 2.181, we recall that

$$j_n(r) = \sqrt{\frac{\pi}{2x}} J_{n+1/2}(r) = \sqrt{\frac{\pi}{2}} \sum_{m=0}^{\infty} \frac{(-1)^m r^{2m+n}}{m!\Gamma(m+n+3/2)2^{2m+n+1/2}}$$

Hence,

$$\frac{d^n}{dr^n} j_n(r)\bigg|_{r=0} = \sqrt{\frac{\pi}{2}} \frac{n!}{\Gamma(n+3/2)2^{n+1/2}} = \frac{2^n(n!)^2}{(2n+1)!} \tag{2.183}$$

Substituting Equations 2.182 and 2.183 into Equation 2.181 gives

$$a_n = j^n(2n+1)$$

Thus,

$$e^{jz} = e^{jr\cos\theta} = \sum_{n=0}^{\infty} j^n (2n+1) j_n(r) P_n(\cos\theta) \tag{2.184}$$

b. Since $J_0(\rho) = J_0(r\sin\theta)$ is even, independent of ϕ, and finite at the origin,

$$J_0(\rho) = J_0(r\sin\theta) = \sum_{n=0}^{\infty} b_n j_{2n}(r) P_{2n}(\cos\theta) \tag{2.185}$$

To determine the coefficients of expansion b_n, we multiply both sides by $P_m(\cos\theta)\sin\theta$ and integrate over $0 < \theta < \pi$. We obtain

$$\int_0^\pi J_0(r\sin\theta) P_m(\cos\theta)\sin\theta\, d\theta = \begin{cases} 0, & m \neq 2n \\ \dfrac{2b_n}{4n+1} j_{2n}(r), & m = 2n \end{cases}$$

Differentiating both sides $2n$ times with respect to r and setting $r = 0$ gives

$$b_n = \frac{(-1)^n (4n+1)(2n-1)!}{2^{2n-1} n!(n-1)!}$$

Hence,

$$J_0(\rho) = \sum_{n=0}^{\infty} \frac{(-1)^n (4n+1)(2n-1)!}{2^{2n-1} n!(n-1)!} j_{2n}(r) P_{2n}(\cos\theta)$$

2.6 Some Useful Orthogonal Functions

Orthogonal functions are of great importance in mathematical physics and engineering. A system of real functions $\Phi_n(n = 0, 1, 2, \ldots)$ is said to be *orthogonal with weight* $w(x)$ on the interval (a, b) if

$$\int_a^b w(x)\Phi_m(x)\Phi_n(x)\,dx = 0 \tag{2.186}$$

for every $m \neq n$. For example, the system of functions $\cos(nx)$ is orthogonal with weight 1 on the interval $(0, \pi)$ since

$$\int_0^\pi \cos mx \cos nx\, dx = 0, \quad m \neq n$$

Orthogonal functions usually arise in the solution of PDEs governing the behavior of certain physical phenomena. These include Bessel, Legendre, Hermite, Laguerre, and Chebyshev functions. In addition to the orthogonality properties in Equation 2.186, these functions have many other general properties, which will be discussed briefly in this section. They are very useful in series expansion of functions belonging to very general classes, for example, Fourier–Bessel series, Legendre series, etc. Although Hermite, Laguerre, and Chebyshev functions are of less importance in EM problems than Bessel and Legendre functions, they are sometimes useful and therefore deserve some attention.

An arbitrary function $f(x)$, defined over interval (a, b), can be expressed in terms of any complete, orthogonal set of functions:

$$f(x) = \sum_{n=0}^{\infty} A_n \Phi_n(x) \tag{2.187}$$

where the expansion coefficients are given by

$$A_n = \frac{1}{N_n} \int_a^b w(x) f(x) \Phi_n(x) dx \tag{2.188}$$

and the (weighted) norm N_n is defined as

$$N_n = \int_a^b w(x) \Phi_n^2(x) dx \tag{2.189}$$

Simple orthogonality results when $w(x) = 1$ in Equations 2.186 through 2.189.

Perhaps the best way to briefly describe the orthogonal functions is in table form. This is done in Tables 2.5 through 2.7. The differential equations giving rise to each function are provided in Table 2.5. The orthogonality relations in Table 2.6 are necessary for expanding a given arbitrary function $f(x)$ in terms of the orthogonal functions as in Equations 2.187 through 2.189. Most of the properties of the orthogonal functions can be proved using the generating functions of Table 2.7. To the properties in Tables 2.5 through 2.7 we may add the recurrence relations and series expansion formulas for calculating the functions for specific argument x and order n. These have been provided for $J_n(x)$ and $Y_n(x)$ in Table 2.1 and Equations 2.97 and 2.99, for $P_n(x)$ and $Q_n(x)$ in Table 2.2 and Equations 2.147 and 2.148, for $j_n(x)$ and $y_n(x)$ in Table 2.3 and Equation 2.160, and for $P_n^m(x)$ and $Q_n^m(x)$ in Table 2.4 and Equations 2.168 and 2.169. For Hermite polynomials, the series expansion formula is

$$H_n(x) = \sum_{k=0}^{[n/2]} \frac{(-1)^k n! (2x)^{n-2k}}{k!(n-2k)!} \tag{2.190}$$

where $[n/2] = N$ is the largest even integer $\leq n/2$ or simply the greatest integer function. Thus,

$$H_0(x) = 1, \ H_1(x) = 2x, \ H_2(x) = 4x^2 - 2, \text{ etc.}$$

TABLE 2.5

Differential Equations with Solutions

Equations		Solutions
$x^2 y'' + xy' + (x^2 - n^2)y = 0$	$J_n(x)$	Bessel functions of the first kind
	$Y_n(x)$	Bessel functions of the second kind
	$H_n^{(1)}(x)$	Hankel functions of the first kind
	$H_n^{(2)}(x)$	Hankel functions of the second kind
$x^2 y'' + xy' - (x^2 + n^2)y = 0$	$I_n(x)$	Modified Bessel functions of the first kind
	$K_n(x)$	Modified Bessel functions of the second kind
$x^2 y'' + 2xy' - [x^2 - n(n+1)]y = 0$	$j_n(x)$	Spherical Bessel functions of the first kind
	$y_n(x)$	Spherical Bessel functions of the second kind
$(1 - x^2)y'' - 2xy + n(n+1)y = 0$	$P_n(x)$	Legendre polynomials
	$Q_n(x)$	Legendre functions of the second kind
$(1 - x^2)y'' - 2xy' + \left[n(n+1) - \dfrac{m^2}{1-x^2} \right]y = 0$	$P_n^m(x)$	Associated Legendre polynomials
	$Q_n^m(x)$	Associated Legendre functions of the second kind
$y'' - 2xy' + 2ny = 0$	$H_n(x)$	Hermite polynomials
$xy'' + (1-x)y' + ny = 0$	$L_n(x)$	Laguerre polynomials
$xy'' + (m+1-x)y' + ny = 0$	$L_n^m(x)$	Associated Laguerre polynomials
$(1 - x^2)y'' - xy' + n^2 y = 0$	$T_n(x)$	Chebyshev polynomials of the first kind
	$U_n(x)$	Chebyshev polynomials of the second kind

The recurrence relations are

$$H_{n+1}(x) = 2xH_n(x) - 2nH_{n-1}(x) \tag{2.191a}$$

and

$$H_n'(x) = 2nH_{n-1}(x) \tag{2.191b}$$

For Laguerre polynomials,

$$L_n(x) = \sum_{k=0}^{n} \frac{n!(-x)^k}{(k!)^2(n-k)!} \tag{2.192}$$

so that

$$L_0(x) = 1, \quad L_1(x) = -x+1, \quad L_2(x) = \frac{1}{2!}(x^2 - 4x + 2), \text{ etc.}$$

The recurrence relations are

$$L_{n+1}(x) = (2n+1-x)L_n(x) - n^2 L_{n-1}(x) \tag{2.193a}$$

and

$$\frac{d}{dx}L_n(x) = \frac{1}{x}\left[nL_n(x) - n^2 L_{n+1}(x) \right] \tag{2.193b}$$

TABLE 2.6

Orthogonality Relations

Functions	Relations
Bessel functions	$\int_0^a x J_n(\lambda_i x) J_n(\lambda_j x)\,dx = \dfrac{a^2}{2}[J_{n+1}(\lambda_i a)]^2 \delta_{ij}$ where λ_i and λ_j are the roots of $J_n(\lambda a) = 0$
Spherical Bessel functions	$\int_\infty^\infty j_n(x) j_m(x)\,dx = \dfrac{\pi}{2n+1}\delta_{mn}$
Legendre polynomials	$\int_{-1}^1 P_n(x) P_m(x)\,dx = \dfrac{2}{2n+1}\delta_{mn}$
Associated Legendre polynomials	$\int_{-1}^1 P_n^k(x) P_m^k(x)\,dx = \dfrac{2(n+k)!}{(2n+1)(n-k)}\delta_{mn}$ $\int_{-1}^1 \dfrac{P_n^m(x) P_n^k(x)}{1-x^2}\,dx = \dfrac{(n+m)!}{m(n-m)!}\delta_{mk}$
Hermite polynomials	$\int_{-\infty}^\infty e^{-x^2} H_n(x) H_m(x)\,dx = 2^n n!(\sqrt{\pi})\delta_{mn}$
Laguerre polynomials	$\int_0^\infty e^{-x} L_n(x) L_m(x)\,dx = \delta_{mn}$
Associated Laguerre polynomials	$\int_0^\infty e^{-x} x^k L_n^k(x) L_m^k(x)\,dx = \dfrac{(n+k)!}{n!}\delta_{mn}$
Chebyshev polynomials	$\int_{-1}^1 \dfrac{T_n(x) T_m(x)}{(1-x^2)^{1/2}}\,dx = \begin{cases} 0, & m \neq n \\ \pi/2, & m = n \neq 0 \\ \pi, & m = n = 0 \end{cases}$ $\int_{-1}^1 \dfrac{U_n(x) U_m(x)}{(1-x^2)^{1/2}}\,dx = \begin{cases} 0, & m \neq n \\ \pi/2, & m = n \neq 0 \\ \pi, & m = n = 0 \end{cases}$

For the associated Laguerre polynomials,

$$L_n^m(x) = (-1)^m \frac{d^m}{dx^m} L_{n+m}(x) = \sum_{k=0}^n \frac{(m+n)!(-x)^k}{k!(n-k)!(m+k)!} \tag{2.194}$$

so that

$$L_1^1(x) = -x+2, \quad L_2^1(x) = \frac{x^2}{2} - 3x + 3, \quad L_2^2(x) = \frac{x^2}{2} - 4x + 6, \text{ etc.}$$

TABLE 2.7

Generating Functions

Functions	Generating Function
	$R = [1 - 2xt + t^2]^{1/2}$
Bessel function	$\exp\left[\frac{x}{2}\left(t - \frac{1}{t}\right)\right] = \sum_{n=-\infty}^{\infty} t^n J_n(x)$
Legendre polynomial	$\frac{1}{R} = \sum_{n=0}^{\infty} t^n P_n(x)$
Associated Legendre polynomial	$\frac{(2m)!(1-x^2)^{m/2}}{2^m m! R^{m+1}} = \sum_{n=0}^{\infty} t^n P_{n+m}^m(x)$
Hermite polynomial	$\exp(2tx - t^2) = \sum_{n=0}^{\infty} \frac{t^n}{n!} H_n(x)$
Laguerre polynomial	$\frac{\exp[-xt/(1-t)]}{1-t} = \sum_{n=0}^{\infty} t^n L_n(x)$
Associated Laguerre polynomial	$\frac{\exp[-xt/(1-t)]}{(1-t)^{m+1}} = \sum_{n=0}^{\infty} t^n L_n^m(x)$
Chebyshev polynomial	$\frac{1-t^2}{R^2} = T_0(x) + 2\sum_{n=1}^{\infty} t^n T_n(x)$
	$\frac{\sqrt{1-x^2}}{R^2} = \sum_{n=0}^{\infty} t^n U_{n+1}(x)$

Note that $L_n^m(x) = 0$, $m > n$. The recurrence relations are

$$L_{n+1}^m(x) = \frac{1}{n+1}\left[(2n+m+1-x)L_n^m(x) - (n+m)L_{n-1}^m(x)\right] \qquad (2.195)$$

For Chebyshev polynomials of the first kind,

$$T_n(x) = \sum_{k=0}^{[n/2]} \frac{(-1)^k n! x^{n-2k}(1-x^2)^k}{(2k)!(n-2k)!}, \quad -1 \le x \le 1 \qquad (2.196)$$

so that

$$T_0(x) = 1, \quad T_1(x) = x, \quad T_2(x) = 2x^2 - 1, \text{ etc.}$$

The recurrence relation is

$$T_{n+1}(x) = 2xT_n(x) - T_{n-1}(x) \qquad (2.197)$$

For Chebyshev polynomials of the second kind,

$$U_n(x) = \sum_{k=0}^{N} \frac{(-1)^{k-1}(n+1)! x^{n-2k+2}(1-x^2)^{k-1}}{(2k+1)!(n-2k+2)!}, \quad -1 \le x \le 1 \tag{2.198}$$

where $N = [n + 1/2]$ so that

$$U_0(x) = 1, \quad U_1(x) = 2x, \quad U_2(x) = 4x^2 - 1, \text{ etc.}$$

The recurrence relation is the same as that in Equation 2.197.

For example, if a function $f(x)$ is to be expanded on the interval $(0, \infty)$, Laguerre functions can be used as the orthogonal functions with an exponential weighting function, that is, $w(x) = e^{-x}$. If $f(x)$ is to be expanded on the interval $(-\infty, \infty)$, we may use Hermite functions with $w(x) = e^{-x^2}$. As we have noticed earlier, if $f(x)$ is defined on the interval $(-1, 1)$, we may choose Legendre functions with $w(x) = 1$. For more detailed treatment of these functions, see Andrews et al. [7] or Johnson and Johnson [8].

EXAMPLE 2.9

Expand the function

$$f(x) = |x|, -1 \le x \le 1$$

in a series of Chebyshev polynomials.

Solution

The given function can be written as

$$f(x) = \begin{cases} -x, & -1 \le x < 0 \\ x, & 0 < x \le 1 \end{cases}$$

Let

$$f(x) = \sum_{n=0}^{\infty} A_n T_n(x)$$

where A_n are expansion coefficients to be determined. Since $f(x)$ is an even function, the odd terms in the expansion vanish. Hence,

$$f(x) = A_0 + \sum_{n=1}^{\infty} A_{2n} T_{2n}(x)$$

If we multiply both sides by $w(x) = \left(T_{2m}/\sqrt{1-x^2}\right)$ and integrate over $-1 \le x \le 1$, all terms in the summation vanish except when $m = n$. That is, from Table 2.6, the orthogonality property of $T_n(x)$ requires that

$$\int_{-1}^{1} \frac{T_m(x)T_n(x)}{(1-x^2)^{1/2}} dx = \begin{cases} 0, & m \ne n \\ \pi/2, & m = n \ne 0 \\ \pi, & m = n = 0 \end{cases}$$

Hence,

$$A_0 = \frac{1}{\pi} \int_{-1}^{1} \frac{f(x)T_0(x)}{(1-x^2)^{1/2}} dx = \frac{2}{\pi} \int_{0}^{1} \frac{x}{\left(1-x^2\right)^{1/2}} dx = \frac{2}{\pi},$$

$$A_{2n} = \frac{2}{\pi} \int_{-1}^{1} \frac{f(x)T_{2n}(x)}{(1-x^2)^{1/2}} dx = \frac{4}{\pi} \int_{0}^{1} \frac{xT_{2n}}{\left(1-x^2\right)^{1/2}} dx$$

Since $T_n(x) = \cos(n \cos^{-1} x)$, it is convenient to let $x = \cos\theta$ so that

$$A_{2n} = \frac{4}{\pi} \int_{\pi/2}^{0} \frac{\cos\theta \cos 2n\theta}{\sin\theta} (-\sin\theta \, d\theta) = \frac{4}{\pi} \int_{0}^{\pi/2} \cos\theta \cos 2n\theta \, d\theta$$

$$= \frac{4}{\pi} \int_{0}^{\pi/2} \frac{1}{2} [\cos(2n+1)\theta + \cos(2n-1)\theta] d\theta = \frac{4}{\pi} \frac{(-1)^{n+1}}{4n^2 - 1}$$

Hence,

$$f(x) = \frac{2}{\pi} + \frac{4}{\pi} \sum_{n=1}^{\infty} \frac{(-1)^{n+1}}{4n^2 - 1} T_{2n}(x)$$

EXAMPLE 2.10

Evaluate $\dfrac{P_n^1(x)}{\sin\theta}$ at $x = 1$ and $x = -1$.

Solution

This example serves to illustrate how the generating functions are useful in deriving some properties of the corresponding orthogonal functions. Since

$$\frac{P_n^1(x)}{\sin\theta} = \frac{P_n^1(x)}{\sqrt{1-x^2}},$$

direct substitution of $x = 1$ or $x = -1$ gives 0/0, which is indeterminate. But $P_n^1(x) = (1-x^2)^{1/2}(d/dx)P_n$ by definition. Hence,

$$\frac{P_n^1(x)}{\sin\theta} = \frac{d}{dx} P_n,$$

that is, the problem is reduced to evaluating dP_n/dx at $x = \pm 1$. We use the generating function for P_n, namely,

$$(1 - 2xt + t^2)^{-1/2} = \sum_{n=0}^{\infty} t^n P_n(x)$$

Differentiating both sides with respect to x,

$$\frac{t}{(1-2xt+t^2)^{3/2}} = \sum_{n=0}^{\infty} t^n \frac{d}{dx} P_n \tag{2.199}$$

When $x = 1$,

$$\frac{1}{(1-t)^3} = \sum_{n=0}^{\infty} t^{n-1} \frac{d}{dx} P_n \bigg|_{x=1} \tag{2.200}$$

But

$$(1-t)^{-3} = 1 + 3t + 6t^2 + 10t^3 + 15t^4 + \cdots = \sum_{n=1}^{\infty} \frac{n}{2}(n+1)t^{n-1} \tag{2.201}$$

Comparing this with Equation 2.200 clearly shows that

$$\frac{d}{dx} P_n \bigg|_{x=1} = n(n+1)/2$$

Similarly, when $x = -1$, Equation 2.199 becomes

$$\frac{1}{(1+t)^3} = \sum_{n=0}^{\infty} t^{n-1} \frac{d}{dx} P_n \bigg|_{x=-1} \tag{2.202}$$

But

$$(1+t)^{-3} = 1 - 3t + 6t^2 - 10t^3 + 15t^4 - \cdots = \sum_{n=1}^{\infty} (-1)^{n+1} \frac{n}{2}(n+1)t^{n-1}$$

Hence,

$$\frac{d}{dx} P_n \bigg|_{x=-1} = (-1)^{n+1} n(n+1)/2$$

EXAMPLE 2.11

Write a program to generate Hermite functions $H_n(x)$ for any argument x and order n. Use the series expansion and recurrence formulas and compare your results. Take $x = 0.5$, $0 \leq n \leq 15$.

Solution

The program is shown in Figure 2.11. Equation 2.190 is used for the series expansion method, while Equation 2.191a with $H_0(x) = 1$ and $H_1(x) = 2x$ is used for the recurrence formula. Note that in the program, we have replaced n by $n-1$ in Equation 2.191a so that

$$H_n(x) = 2xH_{n-1}(x) - 2(n-1)H_{n-2}(x)$$

```
clear all; format compact;format short g; tic

%%%%%%%%%%%%%%%%%%%%%%%%%%%%%%%%%%%%%%%%%%%%%%%%%%%%%%%%%%%%%%%%%%%%%%%%%%%%%%
% THIS PROGRAM GENERATES HERMITE'S FUNCTIONS HN(X) IN TWO WAYS USING:
%    1) SERIES EXPANSION
%    2) RECURRENCE RELATION
% THE TWO METHODS ARE COMPARED
% X = ARGUMENT (FIXED IN THIS PROGRAM)
% N = ORDER OF THE FUNCTION
%%%%%%%%%%%%%%%%%%%%%%%%%%%%%%%%%%%%%%%%%%%%%%%%%%%%%%%%%%%%%%%%%%%%%%%%%%%%%%


    X = 0.5;    %Argument
    NMAX = 15;  %Order of Function

    % METHOD 1:  SERIES EXPANSION FORMULA Equation (2.190)
    for N = 0:NMAX
        SUM = 0;
        FN = factorial(N);
        I = floor(N/2);     %Greatest Integer Function
        for K = 0:I
            M = N - 2*K;
            FM = factorial(M);
            FK = factorial(K);
            SUM = SUM + ( ((-1)^K)*FN*((2*X)^M) )/( FK*FM );
        end
        HS(N+1) = SUM;
    end

    % METHOD 2:  RECURRENCE FORMULA Equation (2.191a)
    HR(1) = 1;
    HR(2) = 2*X;
    for k = 2:N
        n = k-1;    %MATLAB HR vector starts at 1 while equation
                    % subscript starts at 0
        HR(k+1) = 2*X*HR(k) - 2*(n)*HR(k-1);
    end

    Difference = HS-HR;

    hdr = [{'N'},{'Series Expansion'},{'Recurrence'},{'Difference'}];
    output = [hdr; num2cell([(0:N)',HS',HR', Difference'])];
    disp(['Values of Hn(x) for x = ',num2str(X),', 0<=n<=',num2str(N)])
    disp(output)
```

FIGURE 2.11
Program for Hermite function *Hn(x)*.

The result of the computation is in Table 2.8. In this case, the two methods give identical results. In general, the series expansion method gives results of greater accuracy since error in one computation is not propagated to the next as is the case when using recurrence relations.

Generating functions such as this is sometimes needed in numerical computations. This example has served to illustrate how this can be done in two ways. Special techniques may be required for very large or very small values of *x* or *n*.

TABLE 2.8

Results of the Program in Figure 2.11

	Values of $H_n(x)$ for $x = 0.5, 0 \leq n \leq 15$		
n	Series Expansion	Recurrence	Difference
0	1.00	1.00	0.00
1	1.00	1.00	0.00
2	−1.00	−1.00	0.00
3	−5.00	−5.00	0.00
4	1.00	1.00	0.00
5	44.00	44.00	0.00
6	31.00	31.00	0.00
7	−461.00	−461.00	0.00
8	−895.00	−895.00	0.00
9	6181.00	6181.00	0.00
10	22591.00	22591.00	0.00
11	−107029.00	−107029.00	0.00
12	−604031.00	−604031.00	0.00
13	1964665.00	1964665.00	0.00
14	17669472.00	17669472.00	0.00
15	−37341152.00	−37341148.00	−4.00

2.7 Series Expansion

As we have noticed in earlier sections, PDEs can be solved with the aid of infinite series and, more generally, with the aid of series of orthogonal functions. In this section we apply the idea of infinite series expansion to those PDEs in which the independent variables are not separable or, if they are separable, the boundary conditions are not satisfied by the particular solutions. We will illustrate the technique in the following three examples.

2.7.1 Poisson's Equation in a Cube

Consider the problem

$$\nabla^2 V = \frac{\partial^2 V}{\partial x^2} = \frac{\partial^2 V}{\partial y^2} + \frac{\partial^2 V}{\partial z^2} = -f(x, y, z) \tag{2.203}$$

subject to the boundary conditions

$$V(0, y, z) = V(a, y, z) = V(x, 0, z) = 0$$

$$V(x, b, z) = V(x, y, 0) = V(x, y, c) = 0 \tag{2.204}$$

where $f(x, y, z)$, the source term, is given. We should note that the independent variables in Equation 2.203 are not separable. However, in Laplace's equation, $f(x, y, z) = 0$, and the variables are separable. Although the problem defined by Equations 2.203 and 2.204 can be solved in several ways, we stress the use of series expansion in this section.

Let the solution be of the form

$$V(x, y, z) = \sum_{m=1}^{\infty} \sum_{n=1}^{\infty} \sum_{p=1}^{\infty} A_{mnp} \sin \frac{m\pi x}{a} \sin \frac{n\pi y}{b} \sin \frac{p\pi z}{c} \qquad (2.205)$$

where the triple sine series is chosen so that the individual terms and the entire series would satisfy the boundary conditions of Equation 2.204. However, the individual terms do not satisfy either Poisson's or Laplace's equation. Since the expansion coefficients A_{mnp} are arbitrary, they can be chosen such that Equation 2.205 satisfies Equation 2.203. We achieve this by substituting Equation 2.205 into Equation 2.203. We obtain

$$-\sum\sum\sum A_{mnp}(m\pi/a)^2 \sin\frac{m\pi x}{a} \sin\frac{n\pi y}{b} \sin\frac{p\pi z}{c}$$

$$-\sum\sum\sum A_{mnp}(n\pi/b)^2 \sin\frac{m\pi x}{a} \sin\frac{n\pi y}{b} \sin\frac{p\pi z}{c}$$

$$-\sum\sum\sum A_{mnp}(p\pi/c)^2 \sin\frac{m\pi x}{a} \sin\frac{n\pi y}{b} \sin\frac{p\pi z}{c} = -f(x, y, z)$$

Multiplying both sides by $\sin(i\pi x/a)$, $\sin(j\pi y/b)$, $\sin(k\pi z/c)$, and integrating over $0 < x < a$, $0 < y < b$, $0 < z < c$ gives

$$\sum\sum\sum A_{mnp}\left[(m\pi/a)^2 + (n\pi/b)^2 + (p\pi/c)^2\right].$$

$$\int_0^a \sin\frac{mnx}{a} \sin\frac{i\pi x}{a} dx \int_0^b \sin\frac{n\pi y}{b} \sin\frac{j\pi y}{b} dy \int_0^c \sin\frac{p\pi z}{c} \sin\frac{k\pi z}{c} dz$$

$$= \int_0^a \int_0^b \int_0^c f(x, y, z) \sin\frac{i\pi x}{a} \sin\frac{j\pi y}{b} \sin\frac{k\pi z}{c} dx\,dy\,dz$$

Each of the integrals on the left-hand side vanishes except when $m = i$, $n = j$, and $p = k$. Hence,

$$A_{mnp}\left[(m\pi/a)^2 + (n\pi/b)^2 + (p\pi/c)^2\right]\frac{a}{2}\cdot\frac{b}{2}\cdot\frac{c}{2}$$

$$= \int_0^a \int_0^b \int_0^c f(x, y, z) \sin\frac{i\pi x}{a} \sin\frac{j\pi y}{b} \sin\frac{k\pi z}{c} dx\,dy\,dz$$

or

$$A_{mnp} = \frac{8}{abc}\left[(m\pi/a)^2 + (n\pi/b)^2 + (p\pi/c)^2\right]^{-1}.$$

$$\int_0^a \int_0^b \int_0^c f(x, y, z) \sin\frac{i\pi x}{a} \sin\frac{j\pi y}{b} \sin\frac{k\pi z}{c} dx\,dy\,dz \qquad (2.206)$$

Thus, the series expansion solution to the problem is in Equation 2.205 with A_{mnp} given by Equation 2.206.

2.7.2 Poisson's Equation in a Cylinder

The problem to be solved is shown in Figure 2.12, which illustrates a cylindrical metal tank partially filled with charged liquid [9]. To find the potential distribution V in the tank, we let V_ℓ and V_g be the potential in the liquid and gas portions, respectively, that is,

$$V = \begin{cases} V_\ell, & 0 < z < b & \text{(liquid)} \\ V_g, & b < z < b+c & \text{(gas)} \end{cases}$$

Thus, we need to solve a two-dimensional problem:

$$\frac{1}{\rho}\frac{\partial}{\partial\rho}\left(\rho\frac{\partial V_\ell}{\partial\rho}\right) + \frac{\partial^2 V_\ell}{\partial z^2} = -\frac{\rho_v}{\epsilon}, \quad \text{for liquid space} \tag{2.207a}$$

$$\frac{1}{\rho}\frac{\partial}{\partial\rho}\left(\rho\frac{\partial V_g}{\partial\rho}\right) + \frac{\partial^2 V_g}{\partial z^2} = 0, \quad \text{for gas space} \tag{2.207b}$$

subject to

$$V = 0, \rho = a \qquad \text{(at the wall)}$$
$$V_g = V_\ell, z = b \qquad \text{(at the gas-liquid interface)}$$
$$\frac{\partial V_g}{\partial_z} = \epsilon_r\frac{\partial V_\ell}{\partial z}, z = b \quad \text{(at the gas-liquid interface)}$$

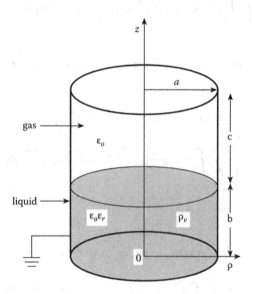

FIGURE 2.12
A cylindrical metal tank partially filled with charged liquid.

Applying the series expansion techniques, we let

$$V_\ell = \sum_{n=1}^{\infty} J_0(\lambda_n \rho) F_n(z) \tag{2.208a}$$

$$V_g = \sum_{n=1}^{\infty} J_0(\lambda_n \rho)[A_n \sinh[\lambda_n(b+c-z)] + B_n \cosh[\lambda_n(b+c-z)]] \tag{2.208b}$$

where $F_n(z)$, A_n, and B_n are to be determined.

At $z = b + c$, $V_g = 0$, which implies that $B_n = 0$. Hence, Equation 2.208b becomes

$$V_g = \sum_{n=1}^{\infty} A_n J_0(\lambda_n \rho) \sinh[\lambda_n(b+c-z)] \tag{2.209}$$

Substituting Equation 2.208a into 2.207a yields

$$\sum_{n=1}^{\infty} J_0(\lambda_n \rho)\left[F_n'' - \lambda_n^2 F_n\right] = -\frac{\rho_v}{\epsilon}$$

If we let $F_n'' - \lambda_n^2 F_n = G_n$, then

$$\sum_{n=1}^{\infty} G_n J_0(\lambda_n \rho) = -\frac{\rho_v}{\epsilon} \tag{2.210}$$

At $\rho = a$, $V_g = V_\ell = 0$, which makes

$$J_0(\lambda_n a) = 0$$

indicating that λ_n are the roots of J_0 divided by a. Multiplying Equation 2.210 by $\rho J_0(\lambda_m \rho)$ and integrating over the interval $0 < \rho < a$ gives

$$\sum_{n=1}^{\infty} G_n \int_0^a \rho J_0(\lambda_m \rho) J_0(\lambda_n \rho) d\rho = -\frac{\rho_v}{\epsilon} \int_0^a \rho J_0(\lambda_m \rho) d\rho$$

The left-hand side is zero except when $m = n$.

$$\int_0^a \rho J_0^2(\lambda_m \rho) d\rho = \frac{1}{2} a^2 [J_0^2(\lambda_n a) + J_1^2(\lambda_n a)] = \frac{a^2}{2} J_1^2(\lambda_n a)$$

since $J_0(\lambda_n a) = 0$. Also,

$$\int_0^a \rho J_0(\lambda_m \rho) d\rho = \frac{a}{\lambda_n} J_1(\lambda_n a)$$

Hence,

$$G_n \frac{a^2}{2} J_1^2(\lambda_n a) = -\frac{\rho_v}{\epsilon} \frac{a}{\lambda_n} J_1(\lambda_n a)$$

or

$$G_n = -\frac{2\rho_v}{\epsilon a \lambda_n J_1(\lambda_n a)}$$

showing that G_n is a constant. Thus,

$$F_n'' - \lambda_n^2 F_n = G_n$$

which is an inhomogeneous ordinary differential equation. Its solution is

$$F_n(z) = C_n \sinh(\lambda_n z) + D_n \cosh(\lambda_n z) - \frac{G_n}{\lambda_n^2}$$

However,

$$F_n(0) = 0 \quad \rightarrow \quad D_n = \frac{G_n}{\lambda_n^2}$$

Thus,

$$V_\ell = \sum_{n=1}^{\infty} J_0(\lambda_n \rho) \left[C_n \sinh(\lambda_n z) + \frac{G_n}{\lambda_n^2} [\cosh(\lambda_n z) - 1] \right] \tag{2.211}$$

Imposing the conditions at $z = b$, that is,

$$V_\ell(\rho, b) = V_g(\rho, b)$$

we obtain

$$A_n \sinh(\lambda_n c) = C_n \sinh(\lambda_n b) + \frac{G_n}{\lambda_n^2} [\cosh(\lambda_n b) - 1] \tag{2.212}$$

Also,

$$\left. \frac{\partial V_g}{\partial z} \right|_{z=b} = \epsilon_r \left. \frac{\partial V_\ell}{\partial z} \right|_{z=b}$$

gives

$$\lambda_n A_n \cosh(\lambda_n c) = -\epsilon_r \lambda_n C_n \cosh(\lambda_n b) - \frac{\epsilon_r G_n}{\lambda_n} \sinh(\lambda_n b) \tag{2.213}$$

Solving Equations 2.212 and 2.213, we get

$$A_n = \frac{2\rho_v}{R_n K_n}[\cosh(\lambda_n b) - 1]$$

$$C_n = \frac{2\rho_v}{R_n K_n}[\cosh(\lambda_n b)\cosh(\lambda_n c) + \epsilon_r \sinh(\lambda_n b)\sinh(\lambda_n c) - \cosh(\lambda_n c)]$$

where

$$K_n = \sinh(\lambda_n b)\cosh(\lambda_n c) + \epsilon_r \cosh(\lambda_n b)\sinh(\lambda_n c)$$

$$R_n = \epsilon_0 a\lambda_n^3 J_1(\lambda_n a)$$

Substituting A_n and C_n in Equations 2.209 and 2.211, we obtain the complete solution as

$$V_\ell = \sum_{n=1}^{\infty} \frac{2\rho_v}{R_n \epsilon_r} J_0(\lambda_n \rho)\left[\frac{\sinh(\lambda_n z)}{K_n}[\cosh(\lambda_n b)\cosh(\lambda_n c)\right.$$
$$\left. + \epsilon_r \sinh(\lambda_n b)\sinh(\lambda_n c) - \cosh(\lambda_n c)] - \cosh(\lambda_n z) + 1\right] \quad (2.214a)$$

$$V_g = \sum_{n=1}^{\infty} \frac{2\rho_v}{R_n K_n} J_0(\lambda_n \rho)[\cosh(\lambda_n b) - 1]\sinh[\lambda_n(b + c - z)] \quad (2.214b)$$

2.7.3 Strip Transmission Line

Consider a strip conductor enclosed in a shielded box containing homogeneous medium as shown in Figure 2.13a. If TEM mode of propagation is assumed, our problem is reduced to finding V satisfying Laplace's equation $\nabla^2 V = 0$. Due to symmetry, we need only consider one quarter-section of the line as in Figure 2.13b. This quadrant can be subdivided into regions 1 and 2, where region 1 is under the center conductor and region 2 is not. We now seek solutions V_1 and V_2 for regions 1 and 2, respectively.

If $w \gg b$, region 1 is similar to a parallel-plate problem. Thus, we have a one-dimensional problem similar to Equation 2.14 with solution

$$V_1 = a_1 y + a_2$$

Since $V_1(y = 0) = V_0$ and $V_1(y = -b/2) = V_0$, $a_2 = 0$, $a_1 = -2V_0/b$. Hence,

$$V_1(x, y) = \frac{-2V_0}{b} y \quad (2.215)$$

For region 2, the series expansion solution is of the form

$$V_2(x, y) = \sum_{n=1,3,5}^{\infty} A_n \sin\frac{n\pi y}{b} \sinh\frac{n\pi}{b}(a/2 - x), \quad (2.216)$$

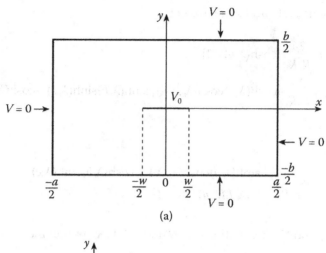

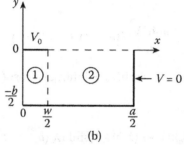

FIGURE 2.13
Strip line example.

which satisfies Laplace's equation and the boundary condition along the box. Notice that the even-numbered terms could not be included because they do not satisfy the boundary condition requirements about line $y = 0$, that is, $E_y(y = 0) = -\partial V_2/\partial y|_{y=0} \neq 0$. To determine the expansion coefficients A_n in Equation 2.216, we utilize the fact that V must be continuous at the interface $x = w/2$ between regions 1 and 2, that is,

$$V_1(x = w/2, y) = V_2(x = w/2, y)$$

or

$$-\frac{2V_o y}{b} = \sum_{n=odd}^{\infty} A_n \sin\frac{n\pi y}{b} \sinh\frac{n\pi}{2b}(a-w),$$

which is Fourier series. Thus,

$$A_n \sinh\frac{n\pi}{2b}(a-w) = -\frac{2}{b}\int_{-b/2}^{b/2}\frac{2V_o y}{b}\sin\frac{n\pi y}{b}\,dy = -\frac{8V_o \sin\left(\frac{n\pi}{2}\right)}{n^2\pi^2}$$

Hence,

$$A_n = -\frac{8V_o \sin\frac{n\pi}{2}}{n^2\pi^2 \sinh\frac{n\pi}{b}(a-w)} \tag{2.217}$$

It is instructive to find the capacitance per unit length C of the strip line using the fact that the energy stored per length is related to C according to

$$W = \frac{1}{2}CV_o^2 \tag{2.218}$$

where

$$W = \frac{1}{2}\int \mathbf{D}\cdot\mathbf{E}\,dv = \frac{1}{2}\epsilon\int |\mathbf{E}|^2\,dv \tag{2.219}$$

For region 1,

$$\mathbf{E} = -\nabla V = -\frac{\partial V}{\partial y}\mathbf{a}_x - \frac{\partial V}{\partial y}\mathbf{a}_y = \frac{2V_o}{b}\mathbf{a}_y$$

Hence,

$$W_1 = \frac{1}{2}\epsilon\int_{x=0}^{w/2}\int_{y=-b/2}^{0}\frac{4V_o^2}{b^2}\,dy\,dx = \frac{\epsilon V_o^2 w}{2b} \tag{2.220}$$

For region 2,

$$E_x = -\frac{\partial V_2}{\partial x} = \sum\frac{n\pi}{b}A_n\cosh\frac{n\pi}{b}(a/2-x)\sin\frac{n\pi y}{b}$$

$$E_y = -\frac{\partial V_2}{\partial y} = -\sum\frac{n\pi}{b}A_n\sinh\frac{n\pi}{b}(a/2-x)\cos\frac{n\pi y}{b}$$

and

$$W_2 = \frac{1}{2}\epsilon\iint (E_x^2 + E_y^2)\,dx\,dy$$

$$= \frac{1}{2}\epsilon\int_{y=-b/2}^{0}\int_{x=w/2}^{a/2}\sum_n\sum_m\frac{mn\pi^2}{b^2}A_nA_m.$$

$$\left[\sinh\frac{m\pi}{b}(a/2-x)\sinh^2\frac{n\pi}{b}(a/2-x)\cos\frac{m\pi y}{b}\cos\frac{n\pi y}{b}\right.$$

$$\left.+\cosh\frac{m\pi}{b}(a/2-x)\cosh^2\frac{n\pi}{b}(a/2-x)\sin\frac{m\pi y}{b}\sin\frac{n\pi y}{b}\right]dx\,dy$$

where the double summation is used to show that we are multiplying two series which may have different indices m and n. Due to the orthogonality properties of sine and cosine functions, all terms vanish except when $m = n$. Thus,

$$W_2 = \frac{1}{2}\epsilon \sum_{n=\text{odd}}^{\infty} \frac{n^2\pi^2 A_n^2}{b^2} \cdot \frac{b/2}{2} \int_{w/2}^{a/2} \left[\sinh^2 \frac{n\pi}{b}(a/2 - x) + \cosh^2 \frac{n\pi}{b}(a/2 - x) \right] dx$$

$$= \frac{1}{2}\epsilon \sum_{n=\text{odd}}^{\infty} \frac{n^2\pi^2 A_n^2}{4b} \frac{b}{n\pi} \cosh \frac{m\pi}{2b}(a - w)\sinh \frac{n\pi}{2b}(a - w)$$

Substituting for A_n gives

$$W_2 = \sum_{n=1,3,5}^{\infty} \frac{8\epsilon V_o^2}{n^3\pi^3} \coth \frac{n\pi}{2b}(a - w) \tag{2.221}$$

The total energy in the four quadrants is

$$W = 4(W_1 + W_2)$$

Thus,

$$C = \frac{2W}{V_o^2} = \frac{8}{V_o^2}(W_1 + W_2)$$

$$= \epsilon \left[\frac{4w}{b} + \frac{64}{\pi^3} \sum_{n=1,3,5}^{\infty} \frac{1}{n^3} \coth \frac{n\pi}{2b}(a - w) \right] \tag{2.222}$$

The characteristic impedance of the lossless line is given by

$$Z_o = \frac{\sqrt{\mu\epsilon}}{C} = \frac{\sqrt{\mu_r\epsilon_r}}{cC} = \sqrt{\frac{\mu}{\epsilon}}\frac{1}{C/\epsilon}$$

or

$$Z_o = \frac{120\pi}{\sqrt{\epsilon_r}\left[\frac{4w}{b} + \frac{64}{\pi^3} \sum_{n=1,3,5}^{\infty} \frac{1}{n^3} \coth \frac{n\pi}{2b}(a - w) \right]} \tag{2.223}$$

where $c = 3 \times 10^8$ m/s, the speed of light in a vacuum, and $\mu_r = 1$ is assumed.

EXAMPLE 2.12

Solve the two-dimensional problem

$$\nabla^2 V = -\frac{\rho_s}{\epsilon_o}$$

where

$$\rho_s = x(y-1)\text{nC/m}^2$$

subject to

$$eV(x, 0) = 0, \quad V(x, b) = V_o, \quad V(0, y) = 0 = V(a, y)$$

Solution

If we let

$$\nabla^2 V_1 = 0, \tag{2.224a}$$

subject to

$$V_1(x, 0) = 0, \quad V_1(x, b) = V_o, \quad V_1(0, y) = 0 = V(a, y) \tag{2.224b}$$

and

$$\nabla^2 V_2 = -\frac{\rho_s}{\epsilon_o}, \tag{2.225a}$$

subject to

$$V_2(x, 0) = 0, \quad V_2(x, b) = 0, \quad V_2(0, y) = 0 = V(a, y) \tag{2.225b}$$

By the superposition principle, the solution to the given problem is

$$V = V_1 + V_2 \tag{2.226}$$

The solution to Equation 2.224 is already found in Section 2.3.1, that is,

$$V_1(x, y) = \frac{4V_o}{\pi} \sum_{n=1,3,5}^{\infty} \frac{\sin\dfrac{m\pi x}{a} \sinh\dfrac{n\pi y}{a}}{n\sinh\dfrac{n\pi b}{a}} \tag{2.227}$$

The solution to Equation 2.225 is a special case of that of Equation 2.205. The only difference between this problem and that of Equations 2.203 and 2.204 is that this problem is two-dimensional while that of Equations 2.203 and 2.204 is three-dimensional. Hence,

$$V_2(x, y) = \sum_{m=1}^{\infty}\sum_{n=1}^{\infty} A_{mn} \sin\frac{m\pi x}{a} \sin\frac{n\pi y}{b} \tag{2.228}$$

where, according to Equation 2.206, A_{mn} is given by

$$A_{mn} = \frac{4}{ab}\left[(m\pi/a)^2 + (n\pi/b)^2\right]^{-1}$$

$$\times \int_0^b \int_0^a f(x, y)\sin\frac{m\pi x}{a}\sin\frac{n\pi x}{a}\,dx\,dy \tag{2.229}$$

But $f(x, y) = x(y-1)/\varepsilon_0 \, \text{nC/m}^2$,

$$\int_0^b \int_0^a f(x,y)\sin\frac{m\pi x}{a}\sin\frac{n\pi y}{a}\,dx\,dy = \frac{10^{-9}}{\epsilon_0}\int_0^a x\sin\frac{m\pi x}{a}\,dx\int_0^b (y-1)\sin\frac{n\pi y}{b}\,dy$$

$$= \frac{10^{-9}}{10^{-9}/36\pi}\left(-\frac{a^2\cos m\pi}{m\pi}\right)\left(-\frac{b^2\cos n\pi}{n\pi}+\frac{b}{n\pi}[\cos n\pi - 1]\right)$$

$$= \frac{36\pi(-1)^{m+n}a^2b^2}{mn\pi^2}\left(1-\frac{1}{b}[1-(-1)^n]\right) \qquad (2.230)$$

since $\cos n\pi = (-1)^n$. Substitution of Equation 2.230 into Equation 2.229 leads to

$$A_{mn} = \left[(m\pi/a)^2 + (n\pi/b)^2\right]^{-1}\frac{(-1)^{m+n}144ab}{mn\pi}\left(1-\frac{1}{b}[1-(-1)^n]\right) \qquad (2.231)$$

Substituting Equations 2.227 and 2.228 into Equation 2.226 gives the complete solution as

$$V(x,y) = \frac{4V_o}{\pi}\sum_{n=1,3,5}^\infty \frac{\sin\dfrac{m\pi x}{a}\sinh\dfrac{n\pi y}{a}}{n\sinh\dfrac{n\pi b}{a}} + \sum_{m=1}^\infty\sum_{n=1}^\infty A_{mn}\sin\frac{m\pi x}{a}\sin\frac{n\pi y}{b} \qquad (2.232)$$

where A_{mn} is in Equation 2.231.

2.8 Practical Applications

The scattering of EM waves by a dielectric sphere, known as the Mie scattering problem due to its first investigator in 1908, is an important problem whose analytic solution is usually referred to in assessing some numerical computations. Though the analysis of the problem is more rigorous, the procedure is similar to that of Example 2.5, where scattering due to a conducting cylinder was treated. Our treatment here will be brief; for an in-depth treatment, consult Stratton [10].

2.8.1 Scattering by Dielectric Sphere

Consider a dielectric sphere illuminated by a plane wave propagating in the z direction and **E** polarized in the x direction as shown in Figure 2.14. The incident wave is described by

$$\mathbf{E}^i = E_o e^{j(\omega t - kz)}\mathbf{a}_x \qquad (2.233a)$$

$$\mathbf{H}^i = \frac{E_o}{\eta}e^{j(\omega t - kz)}\mathbf{a}_y \qquad (2.233b)$$

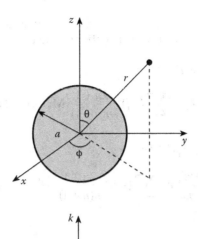

FIGURE 2.14
Incident EM plane wave on a dielectric sphere.

The first step is to express this incident wave in terms of spherical wave functions as in Example 2.8. Since

$$\mathbf{a}_x = \sin\theta\cos\phi\,\mathbf{a}_r + \cos\theta\cos\phi\,\mathbf{a}_\theta - \sin\phi\,\mathbf{a}_\phi,$$

the r-component of $\mathbf{E}^i$, for example, is

$$E_r^i = \cos\phi\sin\theta E_x^i = E_o e^{j\omega t}\frac{\cos\phi}{jkr}\frac{\partial}{\partial\theta}\left(e^{-jkr\cos\theta}\right)$$

Introducing Equation 2.184,

$$E_r^i = E_o e^{j\omega t}\frac{\cos\phi}{jkr}\sum_{n=0}^{\infty}(-j)^n(2n+1)j_n(kr)\frac{\partial}{\partial\theta}P_n(\cos\theta)$$

However,

$$\frac{\partial P_n}{\partial\theta} = P_n^1$$

hence

$$E_r^i = E_o e^{j\omega t}\frac{\cos\phi}{jkr}\sum_{n=1}^{\infty}(-j)^n(2n+1)j_n(kr)P_n^1(\cos\theta) \qquad (2.234)$$

where the $n = 0$ term has been dropped since $P_0^1 = 0$. The same steps can be taken to express E_θ^i and E_ϕ^i in terms of the spherical wave functions. The result is

$$\mathbf{E}^i = \mathbf{a}_x E_o e^{j(\omega t - kz)}$$

$$= E_o e^{j\omega t} \sum_{n=1}^{\infty} (-j)^n \frac{2n+1}{n(n+1)} \left[\mathbf{M}_n^{(1)}(k) + j\mathbf{N}_n^{(1)}(k) \right] \tag{2.235a}$$

$$\mathbf{H}^i = \mathbf{a}_y H_o e^{j(\omega t - kz)}$$

$$= -\frac{kE_o}{\mu\omega} e^{j\omega t} \sum_{n=1}^{\infty} (-j)^n \frac{2n+1}{n(n+1)} \left[\mathbf{M}_n^{(1)}(k) - j\mathbf{N}_n^{(1)}(k) \right] \tag{2.235b}$$

where

$$\mathbf{M}_n(k) = \frac{1}{\sin\theta} z_n(kr) P_n^1(\cos\theta) \cos\phi\, \mathbf{a}_\theta - z_n(kr) \frac{\partial P_n^1(\cos\theta)}{\partial\theta} \sin\phi\, \mathbf{a}_\phi \tag{2.236}$$

$$\mathbf{N}_n(k) = \frac{n(n+1)}{kr} z_n(kr) P_n^1(\cos\theta) \cos\phi\, \mathbf{a}_r + \frac{1}{kr} \frac{\partial}{\partial r} [z_n(kr)] \frac{\partial P_n^1(\cos\theta)}{\partial\theta} \cos\phi\, \mathbf{a}_\theta \tag{2.237}$$

$$+ \frac{1}{kr\sin\theta} \frac{\partial}{\partial r} [z_n(kr)] P_n^1(\cos\theta) \sin\phi\, \mathbf{a}_\phi$$

The superscript (1) on the spherical vector functions $\mathbf{M}$ and $\mathbf{N}$ in Equation 2.235 indicates that these functions are constructed with spherical Bessel function of the first kind; that is, $z_n(kr)$ in Equations 2.236 and 2.237 is replaced by $j_n(kr)$ when $\mathbf{M}$ and $\mathbf{N}$ are substituted in Equation 2.235.

The induced secondary field consists of two parts. One part applies to the interior of the sphere and is referred to as the transmitted field, while the other applies to the exterior of the sphere and is called the scattered field. Thus, the total field outside the sphere is the sum of the incident and scattered fields. We now construct these fields in a fashion similar to that of the incident field. For the scattered field, we let

$$\mathbf{E}^s = E_o e^{j\omega t} \sum_{n=1}^{\infty} (-j)^n \frac{2n+1}{n(n+1)} \left[a_n \mathbf{M}_n^{(4)}(k) + jb_n \mathbf{N}_n^{(4)}(k) \right] \tag{2.238a}$$

$$\mathbf{H}^s = -\frac{kE_o}{\mu\omega} e^{j\omega t} \sum_{n=1}^{\infty} (-j)^n \frac{2n+1}{n(n+1)} \left[a_n \mathbf{M}_n^{(4)}(k) - jb_n \mathbf{N}_n^{(4)}(k) \right] \tag{2.238b}$$

where a_n and b_n are expansion coefficients and the superscript (4) on $\mathbf{M}$ and $\mathbf{N}$ shows that these functions are constructed with spherical Bessel function of the fourth kind (or Hankel function of the second kind); that is, $z_n(kr)$ in Equations 2.236 and 2.237 is replaced by $h_n^{(2)}(kr)$ when $\mathbf{M}$ and $\mathbf{N}$ are substituted into Equation 2.238. The spherical Hankel function has been chosen to satisfy the radiation condition. In other words, the asymptotic behavior of $h_n^{(2)}(kr)$, namely,

$$h_n^{(2)}(kr) \sim j^{n+1} \frac{e^{-kr}}{kr}, \tag{2.239}$$

when combined with the time factor $e^{j\omega t}$, represents an outgoing spherical wave (see Equation 2.106d). Similarly, the transmitted field inside the sphere can be constructed as

$$\mathbf{E}^t = E_o e^{j\omega t} \sum_{n=1}^{\infty} (-j)^n \frac{2n+1}{n(n+1)} \left[c_n \mathbf{M}_n^{(1)}(k_1) + j d_n \mathbf{N}_n^{(1)}(k_1) \right] \tag{2.240a}$$

$$\mathbf{H}^t = -\frac{kE_o}{\mu\omega} e^{j\omega t} \sum_{n=1}^{\infty} (-j)^n \frac{2n+1}{n(n+1)} \left[c_n \mathbf{M}_n^{(1)}(k_1) - j d_n \mathbf{N}_n^{(1)}(k_1) \right] \tag{2.240b}$$

where c_n and d_n are expansion coefficients, k_1 is the propagation constant in the sphere. The functions $\mathbf{M}_n^{(1)}$ and $\mathbf{N}_n^{(1)}$ in Equation 2.240 are obtained by replacing $z_n(kr)$ in Equations 2.236 and 2.237 by $j_n(k_1 r)$; j_n is the only solution in this case since the field must be finite at the origin, the center of the sphere.

The unknown expansion coefficients a_n, b_n, c_n, and d_n are determined by letting the fields satisfy the boundary conditions, namely, the continuity of the tangential components of the total electric and magnetic fields at the surface of the sphere. Thus at $r = a$,

$$\mathbf{a}_r \times (\mathbf{E}^i + \mathbf{E}^s - \mathbf{E}^t) = 0 \tag{2.241a}$$

$$\mathbf{a}_r \times (\mathbf{H}^i + \mathbf{H}^s - \mathbf{H}^t) = 0 \tag{2.241b}$$

This is equivalent to

$$E_\theta^i + E_\theta^s = E_\theta^t, \quad r = a \tag{2.242a}$$

$$E_\phi^i + E_\phi^s = E_\phi^t, \quad r = a \tag{2.242b}$$

$$H_\theta^i + H_\theta^s = H_\theta^t, \quad r = a \tag{2.242c}$$

$$H_\phi^i + H_\phi^s = H_\phi^t, \quad r = a \tag{2.242d}$$

Substituting Equations 2.235, 2.238, and 2.240 into Equation 2.242, multiplying the resulting equations by $\cos\phi$ or $\sin\phi$ and integrating over $0 \le \phi < 2\pi$, and then multiplying by $(d\rho_m^1 / d\theta)$ and integrating over $0 \le \theta \le \pi$, we obtain

$$j_n(ka) + a_n h_n^{(2)}(ka) = c_n j_n(k_1 a) \tag{2.243a}$$

$$\mu_1 \left[ka j_n(ka) \right]' + a_n \mu_1 \left[ka h_n^{(2)}(ka) \right]' = c_n \mu \left[k_1 a j_n(k_1 a) \right]' \tag{2.243b}$$

$$\mu_1 j_n(ka) + b_n \mu_1 h_n^{(2)}(ka) = d_n \mu j_n(k_1 a) \tag{2.243c}$$

$$k \left[ka j_n(ka) \right]' + b_n k \left[ka h_n^{(2)}(ka) \right]' = d_n k_1 \left[k_1 a j_n(k_1 a) \right]' \tag{2.243d}$$

Solving Equations 2.243a and 2.243b gives a_n and c_n, while solving Equations 2.243c and 2.243d gives b_n and d_n. Thus, for $\mu = \mu_o = \mu_1$,

$$a_n = \frac{j_n(m\alpha)[\alpha j_n(\alpha)]' - j_n(\alpha)[m\alpha j_n(m\alpha)]'}{j_n(m\alpha)[\alpha h_n^{(2)}(\alpha)]' - h_m^{(2)}(\alpha)[m\alpha j_n(m\alpha)]'} \tag{2.244a}$$

$$b_n = \frac{j_n(\alpha)[m\alpha j_n(m\alpha)]' - m^2 j_n(m\alpha)[(\alpha j_n(\alpha)]'}{h_n^{(2)}(\alpha)[m\alpha j_n(m\alpha)]' - m^2 j_n(m\alpha)[\alpha h_n^{(2)}(\alpha)]'} \tag{2.244b}$$

$$c_n = \frac{j/\alpha}{h_n^{(2)}(\alpha)[m\alpha j_n(m\alpha)]' - j_{nn}(m\alpha)[\alpha h_n^{(2)}(\alpha)]'} \tag{2.244c}$$

$$d_n = \frac{j/\alpha}{h_n^{(2)}(\alpha)[m\alpha j_n(m\alpha)]' - m^2 j_n(m\alpha)[\alpha h_n^{(2)}(\alpha)]'} \tag{2.244d}$$

where $\alpha = ka = 2\pi a/\lambda$ and $m = k_1/k$ is the refractive index of the dielectric, which may be real or complex depending on whether the dielectric is lossless or lossy. The primes at the square brackets indicate differentiation with respect to the argument of the Bessel function inside the brackets, that is, $[xz_n(x)]' = (\partial/\partial x)[xz_n(x)]$. To obtain Equations 2.244c and 2.244d, we have made use of the Wronskian relationship

$$j_n(x)\left[xh_n^{(2)}(x)\right]' - h_n^{(2)}(x)[xj_n(x)]' = -j/x \tag{2.245}$$

If the dielectric is lossy and its surrounding medium is free space,

$$k_1^2 = \omega\mu_o(\omega\epsilon_1 - j\sigma), \quad k^2 = \omega^2\mu_o\epsilon_o \tag{2.246}$$

so that the (complex) refractive index m becomes

$$m = \frac{k_1}{k} = \sqrt{\epsilon_c} = \sqrt{\epsilon_{r1} - j\frac{\sigma_1}{\omega\epsilon_o}} = m' - jm'' \tag{2.247}$$

The problem of scattering by a conducting sphere can be obtained as a special case of the problem considered above. Since the EM fields must vanish inside the conducting sphere, the right-hand sides of Equations 2.242a, 2.242b, 2.243a, and 2.243d must be equal to zero so that ($c_n = 0 = d_n$)

$$a_n = -\frac{j_n(\alpha)}{h_n^{(2)}(\alpha)} \tag{2.248a}$$

$$b_n = -\frac{[\alpha j_n(\alpha)]'}{[\alpha h_n^{(2)}(\alpha)]'} \tag{2.248b}$$

Thus we have completed the Mie solution; the field at any point inside or outside the sphere can now be determined. We will now apply the solution to problems of practical interest.

2.8.2 Scattering Cross Sections

Often scattered radiation is most conveniently measured by the scattering cross section Q_{sca} (in meter^2) which may be defined as the ratio of the total energy scattered per second W_s to the energy density P of the incident wave, that is,

$$Q_{sca} = \frac{W_s}{P} \tag{2.249}$$

The energy density of the incident wave is given by

$$P = \frac{E_o^2}{2\eta} = \frac{1}{2} E_o^2 \sqrt{\frac{\epsilon}{\mu}} \tag{2.250}$$

The scattered energy from the sphere is

$$W_s = \frac{1}{2} \text{Re} \int\limits_0^{2\pi} \int\limits_0^{\pi} \left[E_\theta H_\phi^* - E_\phi H_\theta^* \right] r^2 \sin\theta \, d\theta \, d\phi$$

where the star sign denotes complex conjugation and field components are evaluated at far field ($r \gg a$). By using the asymptotic expressions for spherical Bessel functions, we can write the resulting field components as

$$E_\theta^s = \eta H_\phi^s = -\frac{j}{kr} E_o e^{j(\omega t - kr)} \cos\phi S_2(\theta) \tag{2.251a}$$

$$-E_\phi^s = \eta H_\theta^s = -\frac{j}{kr} E_o e^{j(\omega t - kr)} \cos\phi S_1(\theta) \tag{2.251b}$$

where the amplitude functions $S_1(\theta)$ and $S_2(\theta)$ are given by [11]

$$S_1(\theta) = \sum_{n=1}^{\infty} \frac{2n+1}{n(n+1)} \left(\frac{a_n}{\sin\theta} P_n^1(\cos\theta) + b_n \frac{d P_n^1(\cos\theta)}{d\theta} \right) \tag{2.252a}$$

$$S_2(\theta) = \sum_{n=1}^{\infty} \frac{2n+1}{n(n+1)} \left(\frac{b_n}{\sin\theta} P_n^1(\cos\theta) + a_n \frac{d P_n^1(\cos\theta)}{d\theta} \right) \tag{2.252b}$$

Thus,

$$W_s = \frac{\pi E_o^2}{2k^2\eta} \text{Re} \int\limits_0^{\pi} \left(|S_1(\theta)|^2 + |S_2(\theta)|^2 \right) \sin\theta \, d\theta$$

This is evaluated with the help of the identities [10]

$$\int_0^\pi \left(\frac{dP_n^1}{d\theta} \frac{dP_m^1}{d\theta} + \frac{1}{\sin^2\theta} P_n^1 P_m^1 \right) \sin\theta\, d\theta = \begin{cases} 0, & n \neq m \\ \dfrac{2}{2n+1} \dfrac{(n+1)!}{(n-1)} n(n+1), & n = m \end{cases}$$

and

$$\int_0^\pi \left(\frac{dP_m^1}{\sin\theta} \frac{dP_n^1}{d\theta} + \frac{P_n^1}{\sin\theta} \frac{P_m^1}{d\theta} \right) \sin\theta\, d\theta = 0$$

We obtain

$$W_s = \frac{\pi E_o^2}{k^2 \eta} \sum_{n+1}^\infty (2n+1)\left(|a_n|^2 + |b_n|^2 \right) \tag{2.253}$$

Substituting Equations 2.250 and 2.253 into Equation 2.249, the scattering cross section is found to be

$$Q_{sca} = \frac{2\pi}{k^2} \sum_{n=1}^\infty (2n+1)\left(|a_n|^2 + |b_n|^2 \right) \tag{2.254}$$

Similarly, the *cross section for extinction* Q_{ext} (in meter^2) is obtained [11] from the amplitude functions for $\theta = 0$, that is,

$$Q_{ext} = \frac{4\pi}{k^2} \operatorname{Re} S(0)$$

or

$$Q_{ext} = \frac{2\pi}{k^2} \operatorname{Re} \sum_{n=1}^\infty (2n+1)(a_n + b_n) \tag{2.255}$$

where

$$S(0) = S_1(0°) = S_2(0°) = \frac{1}{2} \sum_{n=1}^\infty (2n+1)(a_n + b_n) \tag{2.256}$$

In obtaining Equation 2.256, we have made use of

$$\left. \frac{P_n^1}{\sin\theta} \right|_{\theta=0} = \left. \frac{dP_n^1}{d\theta} \right|_{\theta=0} = n(n+1)/2$$

If the sphere is absorbing, the *absorption cross section* Q_{abs} (in meter^2) is obtained from

$$Q_{abs} = Q_{ext} - Q_{sca} \qquad (2.257)$$

since the energy removed is partly scattered and partly absorbed.

A useful, measurable quantity in radar communications is the *radar cross section* or *back-scattering cross section* σ_b of a scattering obstacle. It is a lump measure of the efficiency of the obstacle in scattering radiation back to the source ($\theta = 180°$). It is defined in terms of the far zone scattered field as

$$\sigma_b = 4\pi r^2 \frac{|\mathbf{E}^s|^2}{E_o^2}, \quad \theta = \pi \qquad (2.258)$$

From Equation 2.251,

$$\sigma_b = \frac{2\pi}{k^2} \left[|S_1(\pi)|^2 + |S_2(\pi)|^2 \right]$$

But

$$-S(\pi) = S_2(\pi) = \frac{1}{2} \sum_{n=1}^{\infty} (-1)^n (2n+1)(a_n - b_n)$$

where we have used

$$-\frac{P_n^1}{\sin\theta}\bigg|_{\theta=\pi} = \frac{dP_n^1}{d\theta}\bigg|_{\theta=\pi} = (-1)^n n(n+1)/2$$

Thus,

$$\sigma_b = \frac{\pi}{k^2} \left| \sum_{n=1}^{\infty} (-1)^n (2n+1)(a_n - b_n) \right|^2 \qquad (2.259)$$

Similarly, we may determine the *forward-scattering cross section* ($\theta = 0°$) as

$$\sigma_f = \frac{2\pi}{k^2} \left[|S_1(0)|^2 + |S_2(0)|^2 \right]$$

Substituting Equation 2.256 into this yields

$$\sigma_f = \frac{\pi}{k^2} \left| \sum_{n=1}^{\infty} (2n+1)(a_n + b_n) \right|^2 \qquad (2.260)$$

2.9 Attenuation due to Raindrops

The rapid growth in demand for additional communication capacity has put pressure on engineers to develop microwave systems operating at higher frequencies. It turns out, however, that at frequencies above 10 GHz attenuation caused by atmospheric particles can reduce the reliability and performance of radar and space communication links. Such particles include oxygen, ice crystals, rain, fog, and snow. Prediction of the effect of these precipitates on the performance of a system becomes important. In this final subsection, we will examine attenuation and phase shift of an EM wave propagating through rain drops. We will assume that raindrops are spherical so that Mie rigorous solution can be applied. This assumption is valid if the rate intensity is low. For high rain intensity, an oblate spheroidal model would be more realistic [12].

The magnitude of an EM wave traveling through a homogeneous medium (with N identical spherical particles per unit volume) in a distance ℓ is given by $e^{-\gamma \ell}$, where γ is the attenuation coefficient given by [11]

$$\gamma = N Q_{\text{ext}}$$

or

$$\gamma = \frac{N \lambda^2}{\pi} \operatorname{Re} S(0) \tag{2.261}$$

Thus, the wave is attenuated by

$$A = 10 \log_{10} \frac{1}{e^{-\gamma \ell}} = \gamma \ell 10 \log_{10} e$$

or

$$A = 4.343 \gamma \ell \quad (\text{in dB})$$

The attenuation per length (in dB/m) is

$$A = 4.343 \gamma$$

or

$$A = 4.343 \frac{\lambda^2 N}{\pi} \operatorname{Re} S(0) \tag{2.262}$$

Similarly, it can be shown [11] that the phase shift of the EM wave caused by the medium is

$$\Phi = -\frac{\lambda^2 N}{2\pi} \operatorname{Im} S(0) \quad (\text{in rad/unit length})$$

or

$$\Phi = -\frac{\lambda^2 N}{2\pi} \operatorname{Im} S(0) \frac{180}{\pi} \quad (\text{in deg}/\text{m}) \tag{2.263}$$

To relate attenuation and phase shift to a realistic rainfall rather than identical drops assumed so far, it is necessary to know the drop-size distribution for a given rate intensity. Representative distributions were obtained by Laws and Parsons [13] as shown in Table 2.9. To evaluate the effect of the drop-size distribution, suppose for a particular rain rate R, p is the percent of the total volume of water reaching the ground (as in Table 2.9), which consists of drops whose diameters fall in the interval centered in D cm ($D = 2a$), the number of drops in that interval is given by

$$N_c = pN(D) \tag{2.264}$$

The total attenuation and phase shift over the entire volume become

$$A = 0.4343 \frac{\lambda^2}{\pi} \cdot 10^6 \sum pN(D) \operatorname{Re} S(0) \quad (\text{dB}/\text{km}) \tag{2.265}$$

$$\Phi = -\frac{9\lambda^2}{\pi^2} \cdot 10^6 \sum pN(D) \operatorname{Im} S(0) \quad (\text{deg}/\text{km}) \tag{2.266}$$

where λ is the wavelength in cm and $N(D)$ is the number of raindrops with equivolumic diameter D per cm^3. The summations are taken over all drop sizes. In order to relate the attenuation and phase shift to the rain intensity measured in rain rate R (in mm/h), it is

TABLE 2.9

Laws and Parsons' Drop-Size Distributions for Various Rain Rates

Drop	Rain Rate (mm/h)								
	0.25	1.25	2.5	5	12.5	25	50	100	150
Diameter (cm)	Percent of Total Volume								
0.05	28.0	10.9	7.3	4.7	2.6	1.7	1.2	1.0	1.0
0.1	50.1	37.1	27.8	20.3	11.5	7.6	5.4	4.6	4.1
0.15	18.2	31.3	32.8	31.0	24.5	18.4	12.5	8.8	7.6
0.2	3.0	13.5	19.0	22.2	25.4	23.9	19.9	13.9	11.7
0.25	0.7	4.9	7.9	11.8	17.3	19.9	20.9	17.1	13.9
0.3	–	1.5	3.3	5.7	10.1	12.8	15.6	18.4	17.7
0.35	–	0.6	1.1	2.5	4.3	8.2	10.9	15.0	16.1
0.4	–	0.2	0.6	1.0	2.3	3.5	6.7	9.0	11.9
0.45	–	–	0.2	0.5	1.2	2.1	3.3	5.8	7.7
0.5	–	–	–	0.3	0.6	1.1	1.8	3.0	3.6
0.55	–	–	–	–	0.2	0.5	1.1	1.7	2.2
0.6	–	–	–	–	–	0.3	0.5	1.0	1.2
0.65	–	–	–	–	–	–	0.2	0.7	1.0
0.7	–	–	–	–	–	–	–	–	0.3

necessary to have a relationship between N and R. The relationship obtained by Setzer [13], shown in Table 2.10, involves the terminal velocity u (in m/s) of the rain drops, that is,

$$R = u \cdot N \cdot \text{ (volume of a drop)}$$

$$= uN \frac{4\pi a^3}{3} \quad \text{(in m/s)}$$

or

$$R = 6\pi N u D^3 \cdot 10^5 \quad \text{(mm/h)}$$

Thus,

$$N(D) = \frac{R}{6\pi u D^3} 10^{-5} \tag{2.267}$$

Substituting this into Equations 2.265 and 2.266 leads to

$$A = 4.343 \frac{\lambda^2}{\pi^2} R \sum \frac{p}{6uD^3} \operatorname{Re} S(0) \quad \text{(dB/km)} \tag{2.268}$$

$$\Phi = -90 \frac{\lambda^2}{\pi^3} R \sum \frac{P}{6uD^3} \operatorname{Im} S(0) \quad \text{(deg/km)}, \tag{2.269}$$

where $N(D)$ is in per cm^3, D and λ are in cm, u is in m/s, p is in percent, and $S(0)$ is the complex forward-scattering amplitude defined in Equation 2.256. The complex refractive index of raindrops [14] at 20°C required in calculating attenuation and phase shift is shown in Table 2.11.

TABLE 2.10

Raindrop Terminal Velocity

Radius (cm)	Velocity (m/s)
0.025	2.1
0.05	3.9
0.075	5.3
0.10	6.4
0.125	7.3
0.15	7.9
0.175	8.35
0.20	8.7
0.225	9.0
0.25	9.2
0.275	9.35
0.30	9.5
0.325	9.6

TABLE 2.11

Refractive Index of Water at 20°C

Frequency (GHz)	Refractive Index ($m = m' - jm''$)
0.6	8.960 – j0.1713
0.8	8.956 – j0.2172
1.0	8.952 – j0.2648
1.6	8.933 – j0.4105
2.0	8.915 – j0.5078
3.0	8.858 – j0.7471
4.0	8.780 – j0.9771
6.0	8.574 – j1.399
11	7.884 – j2.184
16	7.148 – j2.614
20	6.614 – j2.780
30	5.581 – j2.848
40	4.886 – j2.725
60	4.052 – j2.393
80	3.581 – j2.100
100	3.282 – j1.864
160	2.820 – j1.382
200	2.668 – j1.174
300	2.481 – j0.8466

EXAMPLE 2.13

For ice spheres, plot the normalized back-scattering cross section, $\sigma_b/\pi a^2$, as a function of the normalized circumference, $\alpha = 2\pi a/\lambda$. Assume that the refractive index of ice is independent of wavelength, making the normalized cross section for ice applicable over the entire microwave region. Take $m = 1.78 - j2.4 \times 10^{-3}$ at 0°C.

Solution

From Equation 2.259,

$$\sigma_b = \frac{\pi}{k^2}\left|\sum_{n=1}^{\infty}(-1)^n(2n+1)(a_n - b_n)\right|^2$$

Since $\alpha = ka$, the normalized back-scattering cross section is

$$\frac{\sigma_b}{\pi a^2} = \frac{1}{\alpha^2}\left|\sum_{n=1}^{\infty}(-1)^n(2n+1)(a_n - b_n)\right|^2 \tag{2.270}$$

Using this expression in conjunction with Equation 2.244, the section SCATTERING in the MATLAB code of Figure 2.15 was used as the main program to determine $\sigma_b/\pi a^2$ for $0.2 < \alpha < 4$. Details on the program will be explained in the next example. It suffices to mention that the maximum number of terms of the infinite series in Equation 2.270 was 10. It has been found that truncating the series at $n = 2\alpha$ provides sufficient accuracy. The plot of the normalized radar cross section versus a is shown in Figure 2.16. From the plot, we note that back-scattering oscillates between very large and small values. If α is increased

```
%%%%%%%%%%%%%%%%%%%%%%%%%%%%%%%%%%%%%%%%%%%%%%%%%%%%%%%%%%%%%%%%%%%%%%%
% MAIN PROGRAM
%
% FOR SPHERICAL RAIN DROPS, THIS PROGRAM CALCULATES ATTENUATION IN dB/KM
% AND PHASE SHIFT IN DEG/KM FOR A GIVEN RAIN RATE
%
% R = RAIN RATE IN MM/HR
% D = DROP DIAMETER IN CM
% F = FREQUENCY IN GHZ
% AT = ATTENUATION IN dB/KM
% PH = PHASE SHIFT IN DEG/KM
% V = TERMINAL VELOCITY OF RAIN DROPS
% P = PERCENT OF TOTAL VOLUME AS MEASURED
% M = COMPLEX REFRACTIVE INDEX OF WATER AT T = 20 C
% X = ALPHA = K*A

clear all; format compact; tic

%Inputs:
F = 11.0; %Frequency (GHz)
NMAX = 10; %Ideally inf but 10 is sufficient eqn (2.256)

LAM = 30.0/F;   % WAVELENGTH IN CM

% Raindrop Terminal Velocity Data
V=[2.1,3.9,5.3,6.4,7.3,7.9,8.35,8.7,9.0,9.2,9.35,9.5,9.6,9.6];

% Rain rate (mm/hr)
R=[0.25,1.25,2.5,5.0,12.5,25.0,50.0,100.0,150.0];

% Laws and Parsons drop-size distribution for various rain rate (% of to-
tal volume)
P =1/100*[28.0, 50.1, 18.2, 3.0,  0.7,  0    0 0 0 0 0 0 0;...
          10.9, 37.1, 31.3, 13.5, 4.9,  1.5,  0.6,  0.2, 0 0 0 0 0 0;...
          7.3,  27.8, 32.8, 19.0, 7.9,  3.3,  1.1,  0.6,  0.2, 0 0 0 0 0;...
          4.7,  20.3, 31.0, 22.2, 11.8, 5.7,  2.5,  1.0,  0.5, 0.3, 0 0 0 0;...
          2.6,  11.5, 24.5, 25.4, 17.3, 10.1, 4.3,  2.3,  1.2, 0.6, 0.2 0 0 0;...
          1.7,  7.6,  18.4, 23.9, 19.9, 12.8, 8.2,  3.5,  2.1, 1.1, 0.5, 0.3 0 0;...
          1.2,  5.4,  12.5, 19.9, 20.9, 15.6, 10.9, 6.7,  3.3, 1.8, 1.1, 0.5, 0.2, 0;...
          1.0,  4.6,  8.8,  13.9, 17.1, 18.4, 15.0, 9.0,  5.8, 3.0, 1.7, 1.0, 0.7, 0;...
          1.0,  4.1,  7.6,  11.7, 13.9, 17.7, 16.1, 11.9, 7.7, 3.6, 2.2, 1.2, 1.0, 0.3]';

f = [0.6 0.8 1 1.6 2 3 4 6 11 16 20 30 40 60 80 100 160 200 300];
% Refractive Index of Water at 20 degrees C, real and complex portions
mr= [8.96 8.956 8.952 8.933 8.915 8.858 8.78 8.574 7.884 7.148 6.614 5.581 4.886
4.052 3.581 3.282 2.82 2.668 2.481];
mi= [0.1713 0.2172 0.2648 0.4105 0.5078 0.7471 0.9771 1.399 2.184 2.614 2.78 2.848
2.725 2.393 2.100 1.864 1.382 1.174 0.8466];

% Interpolate refractive index of water to target frequency. F must be
% within the range of f.
mr_i = interp1(f,mr,F);
mi_i = interp1(f,mi,F);
M    = mr_i - mi_i*i;
```

FIGURE 2.15
MATLAB program for Examples 2.13 and 2.14.   (Continued)

further, the normalized radar cross section increases rapidly. The unexpectedly large cross sections have been attributed to a lens effect; the ice sphere acts like a lens which focuses the incoming wave on the back side from which it is reflected backward in a concentrated beam. This is recognized as a crude description, but it at least permits visualization of a physical process which may have some reality.

```
N = 1:NMAX;
D = 0.05*(1:length(V));
SO = zeros(1,length(V));
SB = zeros(1,length(V));
for I = 1:length(V) %I sweeps rain drop diameter

    X = pi*D(I)/LAM;

    % CALCULATE THE SCATTERING COEFFICIENTS an and bn

    % Spherical Bessel function of the first kind of order n
    J  = sqrt(pi./(2*X)).*besselj(N+0.5,X);
    JD = -N.*sqrt(pi./(2*X)).*besselj(N+0.5,X)+ X.*sqrt(pi./(2*X)).*besselj(N+0.5-
1,X);
    JM = sqrt(pi./(2*M*X)).*besselj(N+0.5,M*X);
    JMD= -N.*sqrt(pi./(2*M*X)).*besselj(N+0.5,M*X)+ M*X.*sqrt(pi./(2*M*X)).*besselj(N+0.5-
1,M*X);
    H  = sqrt(pi./(2*X)).*besselh(N+0.5,2,X);
    HD = -N.*sqrt(pi./(2*X)).*besselh(N+0.5,2,X)+X.*sqrt(pi./(2*X)).*besselh(N+0.5-
1,2,X);


    A1 = JM.*JD - J.*JMD;
    A2 = JM.*HD - H.*JMD;
    B1 = J.*JMD - (M^2)*JM.*JD;
    B2 = H.*JMD - (M^2)*JM.*HD;
    an =  A1./A2;
    bn =  B1./B2;
    cn = i./(X*A2);
    dn = -i*M./(X*B2);

    % FORWARD SCATTERING AMPLITUDE FUNCTION S(O)
    SO(I) = sum((2*N+1).*(an+bn))/2;
    % NORMALIZED RADAR CROSS-SECTION
    SB(I)=(abs( sum((2*N+1).*((-1).^N).*( an - bn )) )/X).^2;
end

AT = zeros(1,size(P,2));
PH = zeros(1,size(P,2));
for n = 1:size(P,2) % N sweeps rain rate
    % CALCULATE ATTENUATION AND PHASE SHIFT
    ATO = real(SO).*P(:,n)'./(6*D.^3.* V);
    AT(n) = 4.343*R(n)*(LAM/pi)^2*sum(ATO);

    PHO = imag(SO).*P(:,n)'./(6*D.^3.*V);
    PH(n) = 90.0*R(n)*(LAM/pi)^2/pi*sum(PHO);
end

disp(['Attenuation and Phase Shift at ',num2str(F),' GHz'])
hdr = [{'Rain rate (mm/hr)'},{'Attenuation (dB/km)'},{'Phase shift (deg/km)'}];
out = [hdr; num2cell([R', AT',  PH'])];
disp(out)

figure(1),
    semilogy(R,AT,'--'),
        xlabel('rain rate (mm/hr)'),ylabel('Attenuation (dB/km)')
        legend([num2str(F),' GHz'],'location','southeast')
```

FIGURE 2.15 (Continued)
MATLAB program for Examples 2.13 and 2.14.

EXAMPLE 2.14

Assuming the Laws and Parsons' rain drop-size distribution, calculate the attenuation in dB/km for rain rates of 0.25, 1.25, 2.5, 5.0, 12.5, 50.0, 100.0, and 150.0 mm/h. Consider the incident microwave frequencies of 6, 11, and 30 GHz.

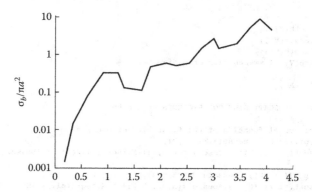

FIGURE 2.16
Normalized back-scattering (radar) cross sections $\alpha = 2\pi a/\lambda$ for ice at 0°C.

Solution

The MATLAB code developed for calculating attenuation and phase shift of microwaves due to rain is shown in Figure 2.15. The main program calculates attenuation and phase shift for given values of frequency and rain rate by employing Equations 2.268 and 2.269. For each frequency, the corresponding value of the refractive index of water at 20°C is taken from Table 2.11. The data in Tables 2.9 and 2.10 on the drop-size distributions and terminal velocity are incorporated in the main program.

MATLAB provides commands for calculating Bessel functions. The derivative of Bessel–Riccati function $[xz_n(x)]$ is obtained from (see Problem 2.14)

$$[xz_n(x)]' = -nz_n(x) + xz_{n-1}(x)$$

where z_n is j_n, j_{-n}, y_n or $h_n(x)$. Subroutine GAMMA calculates $\Gamma(n + 1/2)$ using Equation 2.165, while subroutine FACTORIAL determines $n!$. All computations were done in double precision arithmetic, although it was observed that using single precision would only alter results slightly.

Typical results for 11 GHz are tabulated in Table 2.12. A graph of attenuation vs. rain rate is portrayed in Figure 2.17. The plot shows that attenuation increases with rain rate and conforms with the common rule of thumb. We must note that the underlying assumption of spherical raindrops renders the result as only a first-order approximation of the practical rainfall situation.

TABLE 2.12

Attenuation and Phase Shift at 11 GHz

Rain Rate (mm/h)	Attenuation (dB/km)	Phase Shift (deg/km)
0.25	2.56×10^{-3}	0.4119
1.25	1.702×10^{-3}	1.655
2.5	4.072×10^{-3}	3.040
5.0	9.878×10^{-3}	5.601
12.5	0.3155	12.58
25	0.7513	23.19
50	1.740	42.74
100	3.947	78.59
150	6.189	112.16

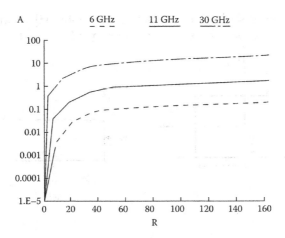

FIGURE 2.17
Attenuation vs. rain rate.

2.10 Concluding Remarks

We have reviewed analytic methods for solving PDEs. Analytic solutions are of major interest as test models for comparison with numerical techniques. The emphasis has been on the method of separation of variables, the most powerful analytic method. For an excellent, more in-depth exposition of this method, consult Myint-U [5]. In the course of applying the method of separation of variables, we have encountered some mathematical functions such as Bessel functions and Legendre polynomials. For a thorough treatment of these functions and their properties, Johnson and Johnson [8] is recommended. The mathematical handbook by Abramowitz and Stegun [15] provides tabulated values of these functions for specific orders and arguments. A few useful texts on the topics covered in this chapter are also listed in the references.

As an example of real-life problems, we have applied the analytical techniques developed in this chapter to the problem of attenuation of microwaves due to spherical raindrops. Spherical models have also been used to assess the absorption characteristics of the human skull exposed to EM plane waves [16–20] (see Problems 2.58 through 2.61).

We conclude this chapter by remarking that the most satisfactory solution of a field problem is an exact analytical one. In many practical situations, no solution can be obtained by the analytical methods, and one must therefore resort to numerical approximation or graphical or experimental solutions. (Experimental solutions are usually very expensive, while graphical solutions are not so accurate.) More information on analytical modeling in EM can be found in References 1, 21–24. The remainder of this book will be devoted to a study of the numerical methods commonly used in EM.

PROBLEMS

2.1 Consider the PDE

$$a\Phi_{xx} + b\Phi_{xy} + c\Phi_{yy} + d\Phi_x + e\Phi_y + f\Phi = 0$$

where the coefficients a, b, c, d, e, and f are in general functions of x and y. Under what conditions is the PDE separable?

Computational Electromagnetics with MATLAB®

2.2   Determine the distribution of electrostatic potential inside the conducting rectangular boxes with cross sections shown in Figure 2.18.

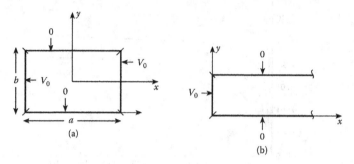

FIGURE 2.18
For Problem 2.2.

2.3   The cross sections of the cylindrical systems that extend to infinity in the z-direction are shown in Figure 2.19. The potentials on the boundaries are as shown. For each system, find the potential distribution.

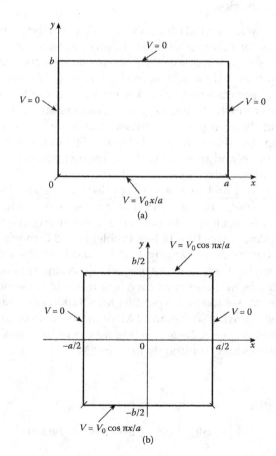

FIGURE 2.19
For Problem 2.3.

2.4 Find the solution U of

a. Laplace equation

$$\nabla^2 U = 0, \qquad\qquad 0 < x, y < \pi$$
$$U_x(0,y) = 0 = U_x(\pi,y), \qquad U(x,0) = 0,$$
$$U(x,\pi) = -x, \qquad\qquad 0 < x < \pi$$

b. Heat equation

$$kU_{xx} = U_t, \qquad\qquad 0 \le x \le 1, t > 0$$
$$U(0,t) = 0, t > 0, \qquad U(1,t) = 1, t > 0$$
$$U(x,0) = 0, \qquad\qquad 0 \le x \le 1$$

c. Wave equation

$$a^2 U_{xx} = U_{tt}, \qquad\qquad 0 \le x \le 1, t > 0$$
$$U(0,t) = 0 = U(1,t), \qquad t > 0$$
$$U(x,0) = 0, \qquad\qquad U_t(x,0) = x$$

2.5 Find the solution Φ of

a. Laplace equation

$$\nabla^2 \Phi = 0, \qquad \rho \ge 1, 0 < \phi < \pi$$
$$\Phi(1,\phi) = \sin\phi, \qquad \Phi(\rho,0) = \Phi(\rho,\pi) = 0$$

b. Laplace equation

$$\nabla^2 \Phi = 0, \qquad 0 < \rho < q, 0 < z < L$$
$$\Phi(\rho,0) = 0 = \Phi(\rho,L), \qquad \Phi(a,z) = 1$$

c. Heat equation

$$\Phi_t = k\nabla^2 \Phi, \qquad 0 \le \rho \le 1, -\infty < z < \infty, t > 0$$
$$\Phi(a,\phi,t) = 0, \, t > 0, \qquad \Phi(\rho,\phi,0) = \rho^2 \cos 2\phi, 0 \le \phi < 2\pi$$

2.6 The one-dimensional heat equation is given by

$$U_{xx} = U_t, \quad 0 < x < 1, \, t > 0$$

subject to Neumann boundary condition

$$U_x(0,t) = 0 = U_x(1,t)$$

and an inhomogeneous initial condition

$$U(x,0) = f(x) = \cos(2\pi x)$$

Obtain the solution for $U(x, t)$.

2.7   Solve the diffusion equation

$$U_{xx} = U_t, \quad 0 < x < 1, \, t > 0$$

subject to boundary conditions

$$U(0,t) = 0 = U(1,t)$$

and the initial condition

$$U(x,0) = x(1-x)$$

2.8   Solve the two-dimensional wave equation

$$U_{xx} + U_{yy} = U_{tt}, \quad 0 < x < \pi, \, 0 < y < \pi, \, t > 0$$

subject to the following conditions:

$$U(0,y,t) = 0, \quad U(\pi,y,t) = 1$$
$$U(x,0,t) = 0 = U(x,\pi,t)$$
$$U(x,y,0) = \sin(3\pi x)\sin(4\pi y) + \frac{x}{\pi}$$
$$U_t(x,y,0) = 0$$

2.9   Solve the two-dimensional heat equation

$$U_{xx} + U_{yy} = U_t, \quad 0 < x < 1, \, 0 < y < 1, \, t > 0$$

Boundary conditions:

$$U(0,y,t) = 0 = U(1,y,t), \quad 0 < y < 1, \, t > 0$$
$$U(x,0,t) = 0 = U(x,1,t), \quad 0 < x < 1, \, t > 0$$

Initial condition:

$$U(x,y,0) = 10xy, \quad 0 < x < 1, \, 0 < y < 1$$

2.10 Solve the PDE

$$4\frac{\partial^4 \Phi}{\partial x^4} + \frac{\partial^2 \Phi}{\partial t^2} = 0, \quad 0 < x < 1, \, t > 0$$

subject to the boundary conditions:

$$\Phi(0,t) = 0 = \Phi(1,t) = \Phi_{xx}(0,t) = \Phi_{xx}(1,t)$$

and initial conditions:

$$\Phi_t(x,0) = 0, \quad \Phi(x,0) = x$$

2.11 Consider the following Laplace's equation with Neumann boundary conditions:

$$\nabla^2 V(x,y) = 0, \quad 0 < x < \pi, 0 < y < \pi$$
$$V_x(0,y) = 0 = V_x(\pi,y)$$
$$V_y(x,0) = \cos x, \quad V_y(x,\pi) = 0$$

Show that the analytical solution is

$$V(x,y) = \left[\sinh y - \frac{\cosh y}{\tanh \pi}\right]\cos x$$

2.12 Solve Laplace's equation

$$\nabla^2 V = 0, \quad 0 < y < 1, 0 < x < \infty$$

subject to:

$$V(x,0) = 0 = V(x,1)$$
$$V(0,y) = 10, \quad V(\infty,y) = 0$$

2.13 Determine the solution to the wave equation

$$U_{tt}(x,t) = a^2 U_{xx}(x,t), \quad 0 < x < 1, \, t > 0$$

subject to

$$U(0,t) = 0 = U(1,t), \quad t > 0$$
$$U(x,0) = x(1-x), \quad 0 < x < 1$$
$$U_t(x,0) = 0, \quad 0 < x < 1$$

2.14 Find the solution to Laplace's equation

$$\nabla^2 U = 0, \quad 0 < x < 1, 0 < y < 1$$

subject to the boundary conditions

$$U_x(0,y) = 0 = U_x(1,y)$$
$$U(x,0) = 0, \quad U(x,1) = x$$

2.15 A cylinder similar to the one in Figure 2.20 has its ends $z = 0$ and $z = L$ held at zero potential. If

$$V(a,z) = \begin{cases} V_o z/L, & 0 < z < L/2 \\ V_o(1-z/L), & L/2 < z < L \end{cases}$$

find $V(\rho, z)$. Calculate the potential at $(\rho, z) = (0.8a, 0.3L)$.

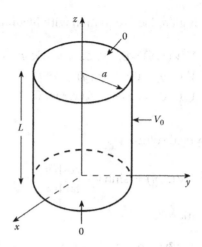

FIGURE 2.20
For Problem 2.9.

2.16 Determine the potential distribution in a hollow cylinder of radius a and length L with ends held at zero potential while the lateral surface is held at potential V_o as in Figure 2.20. Calculate the potential along the axis of the cylinder when $L = 2a$.

2.17 The conductor whose cross section is shown in Figure 2.21 is maintained at $V = 0$ everywhere except on the curved electrode where it is held at $V = V_o$. Find the potential distribution $V(\rho, \phi)$.

2.18 Solve the PDE

$$\frac{\partial^2 \Phi}{\partial \rho^2} + \frac{1}{\rho}\frac{\partial \Phi}{\partial \rho} = \frac{\partial^2 \Phi}{\partial t^2}, \quad 0 \le \rho \le a, t \ge 0$$

under the conditions

$$\Phi(0,t) \quad \text{is bounded}, \qquad \Phi(a,t) = 0, t \ge 0,$$

$$\Phi(\rho,0) = (1 - \rho^2/a^2), \qquad \left.\frac{\partial \Phi}{\partial t}\right|_{t=0} = 0, 0 \le \rho \le a$$

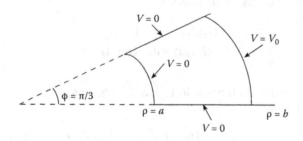

FIGURE 2.21
For Problem 2.11.

2.19 Find the solution of

$$\nabla^2 U = \frac{\partial U}{\partial t} \rightarrow U_{\rho\rho} + \frac{1}{\rho} U_\rho = U_t, \quad 0 < \rho < 1, t > 0$$

subject to

$$U(1,t) = 0, \quad U(\rho,0) = V_o(\text{constant})$$

2.20 Determine the solution to

$$U_{\rho\rho} + \frac{1}{\rho} U_\rho + U_{zz} = U_t, \quad 0 < \rho < 1, 0 < z < 1, t > 0$$

subject to the boundary conditions

$$U(\rho,0,t) = 0 = U(\rho,1,t) = U(1,z,t)$$

and initial condition

$$U(\rho,z,0) = V_o(\text{constant})$$

2.21 a. Prove that

$$e^{\pm j\rho\sin\phi} = \sum_{n=-\infty}^{\infty} (\pm 1)^n J_n(\rho) e^{jn\phi}$$

b. Derive the *Bessel's integral formula*

$$J_n(\rho) = \frac{1}{\pi} \int_0^\pi \cos(n\theta - \rho\sin\theta)\, d\theta$$

2.22 Show that

$$\cos x = J_0(x) + 2\sum_{n=1}^{\infty} (-1)^n J_{2n}(x)$$

and

$$\sin x = 2\sum_{n=1}^{\infty} (-1)^{n+1} J_{2n+1}(x)$$

2.23 Show that

a. $J_{1/2}(x) = \sqrt{\dfrac{2}{\pi x}} \sin x,$

b. $J_{-1/2}(x) = \sqrt{\dfrac{2}{\pi x}} \cos x,$

c. $\dfrac{d}{dx}[x^{-n}J_n(x)] = -x^n J_{n+1}(x).$

d. $\dfrac{d^n}{dx^n} J_n(x)\Big|_{x=0} = \dfrac{1}{2^n},$

e. $\dfrac{d}{dx}[xz_n(x)] = -nz_n(x) + xz_{n-1}(x) = (n+1)z_n(x) + xz_{n+1}(x)$

2.24 Prove that

$$J_3(x) = -\frac{4}{x}J_0(x) + \left(\frac{8}{x^2} - 1\right)J_1(x)$$

2.25 Evaluate the following integrals:

a. $\displaystyle\int x^n J_{n-1}(x)\,dx,$

b. $\displaystyle\int \frac{1}{x^n} J_{n+1}(x)\,dx$

2.26 Reduce $\int x^2 J_2(x)\,dx$ to an integral involving only $J_0(x)$.

2.27 Write a computer program that will evaluate the first roots λ_{nm} of Bessel function $J_n(x)$ for $n = 1, 2, \ldots, 5$, that is, $J_n(\lambda_{nm}) = 0$.

2.28 Evaluate

a. $\displaystyle\int_{-1}^{1} P_1(x)P_2(x)\,dx,$

b. $\displaystyle\int_{-1}^{1} [P_4(x)]^2\,dx,$

c. $\displaystyle\int_{0}^{1} x^2 P_3(x)\,dx$

2.29 In Legendre series of the form $\displaystyle\sum_{n=0}^{\infty} A_n P_n(x)$, expand

a. $f(x) = \begin{cases} 0, & -1 < x < 0, \\ 1, & 0 < x < 1 \end{cases}$

b. $f(x) = x^3, \quad -1 < x < 1,$

c. $f(x) = \begin{cases} 0, & -1 < x < 0, \\ x, & 0 < x < 1 \end{cases}$

d. $f(x) = \begin{cases} 1+x, & -1 < x < 0, \\ 1-x, & 0 < x < 1 \end{cases}$

2.30 Derive the following associated Legendre functions:

a. $P_3^2(x)$,

b. $P_2^3(x)$,

c. $P_3^3(x)$

2.31 Determine the following polynomials:

a. $L_3^1(x)$,

b. $T_3(x)$, $T_4(x)$

c. $U_3(x)$, $U_4(x)$

2.32 Use MATLAB to determine:

a. $J_3(0.5)$,

b. $P_3(0.5)$,

c. $T_3(0.5)$,

d. $U_3(0.5)$

2.33 Evaluate

a. $H_3(x)$ and $H_4(x)$

b. $L_3(x)$ and $L_4(x)$

2.34 Laguerre polynomials can be defined as

$$L_n(t) = \frac{e^t}{n!} \frac{d^n}{dt^n} \left(t^n e^{-t} \right), \quad n = 0,1,2,\dots$$

Use this definition to obtain $L_0(t)$, $L_1(t)$, $L_2(t)$, and $L_3(t)$.

2.35 Hermite polynomials may be given by Rodrigue's formula:

$$H_n(x) = (-1)^n e^{x^2} \frac{d^n}{dx^n} \left(e^{-x^2} \right), \quad n = 0,1,2,\dots$$

Use this formula to determine $H_0(x)$, $H_1(x)$, $H_2(x)$, and $H_3(x)$.

2.36 Solve Laplace's equation:

a. $\nabla^2 U = 0$, $0 \le r \le a$, $U(a,\theta) = \begin{cases} 1, & 0 < \theta < \pi/2 \\ 0, & \text{otherwise} \end{cases}$

b. $\nabla^2 U = 0$, $r > a$, $\left. \dfrac{\partial U}{\partial r} \right|_{r=a} = \cos\theta + 3\cos^3\theta$, $0 < \theta < \pi$,

c. $\nabla^2 U = 0$, $r < a$, $0 < \theta < \pi$, $0 < \phi < 2\pi$,

$$U(a,\theta,\phi) = \sin^2\theta$$

2.37 A hollow conducting sphere of radius a has its upper half charged to potential V_o while its lower half is grounded. Find the potential distribution inside and outside the sphere.

2.38 A circular disk of radius a carries charge of surface charge density ρ_o. Show that the potential at point $(0, 0, z)$ on its axis $\theta = 0$ is

$$V = \frac{\rho_o}{2\epsilon}\left[(z^2 + a^2)^{1/2} - z\right]$$

From this deduce the potential at any point (r, θ, ϕ).

2.39 a. Verify the three-term recurrence relation

$$(2n+1)xP_n(x) = (n+1)P_{n+1}(x) + nP_{n-1}(x)$$

b. Use the recurrence relation to find $P_6(x)$ and $P_7(x)$.

2.40 Establish the formula

$$P_n(-x) = (-1)^n P_n(x)$$

2.41 Verify the following identities:

a. $\displaystyle\int_{-1}^{1} P_n(x)P_m(x)\,dx = \frac{2}{2n+1}\delta_{nm},$

b. $\displaystyle\int_{-1}^{1} P_n^m(x)P_k^m(x)\,dx = \frac{2}{2n+1}\frac{(n+m)!}{(n-m)!}\delta_{nk}$

2.42 Rework the problem in Figure 2.8 if the boundary conditions are now

$$V(r=a) = V_o, \quad V(r \to \infty) = E_o r \cos\theta + V_o$$

Find V and E everywhere. Determine the maximum value of the field strength.

2.43 In a sphere of radius a, obtain the solution $V(r, \theta)$ of Laplace's equation

$$\nabla^2 V(r,\theta) = 0, \quad r \le a$$

subject to

$$V(a,\theta) = 3\cos^2\theta + 3\cos\theta + 1$$

2.44 Determine the solution to Laplace's equation

$$\nabla^2 V = 0$$

outside a sphere $r > a$ subject to the boundary condition

$$\frac{\partial}{\partial r}V(a,\theta) = \cos\theta + 3\cos^3\theta$$

2.45 Find the potential distribution inside and outside a dielectric sphere of radius a placed in a uniform electric field E_0.

Hint: The problem to be solved is $\nabla^2 V = 0$ subject to

$$\epsilon_r \frac{\partial V_1}{\partial r} = \frac{\partial V_2}{\partial r} \quad \text{on } r = a, \quad V_1 = V_2 \quad \text{on } r = a,$$
$$V_2 = -E_0 r \cos\theta \quad \text{as } r \to \infty$$

2.46 a. Derive the recurrence relation of the associated Legendre polynomials

$$P_n^{m+1}(x) = \frac{2mx}{(1-x^2)^{1/2}} P_n^m(x) - [n(n+1) - m(m-1)]P_n^{m-1}(x)$$

b. Using the recurrence relation on the formula for P_n^m, find P_3^2, P_3^3, P_4^1, and P_4^2.

2.47 Expand $V = \cos 2\phi \sin^2\phi$ in terms of the spherical harmonics $P_n^m(\cos\theta)\sin m\phi$ and $P_n^m(\cos\theta)\cos m\phi$.

2.48 In the prolate spheroidal coordinates (ξ, η, ϕ), the equation

$$\nabla^2 \Phi + k^2 \Phi = 0$$

assumes the form

$$\frac{\partial}{\partial \xi}\left[(\xi^2 - 1)\frac{\partial \Phi}{\partial \xi}\right] + \frac{\partial}{\partial \eta}\left[(1-\eta^2)\frac{\partial \Phi}{\partial \eta}\right] + \left|\frac{1}{\xi^2 - 1}\right.$$
$$\left. + \frac{1}{1-\eta^2}\right|\frac{\partial^2 \Phi}{\partial \phi^2} + k^2 d^2(\xi^2 - \eta^2)\Phi = 0$$

Show that the separated equations are

$$\frac{d}{d\xi}\left[(\xi^2 + 1)\frac{d\Psi_1}{d\xi}\right] + \left[k^2 d^2 \xi^2 - \frac{m^2}{\xi^2 - 1} - c\right]\Psi_1 = 0$$
$$\frac{d}{d\eta}\left[(1-\eta^2)\frac{d\Psi_2}{d\eta}\right] - \left[k^2 d^2 \eta^2 + \frac{m^2}{1-\eta^2} - c\right]\Psi_2 = 0$$
$$\frac{d^2\Psi_3}{d\phi^2} + m^2\Psi_3 = 0$$

where m and c are separation constants.

2.49 Solve Equation 2.203 if $a = b = c = \pi$ and

a. $f(x, y, z) = e^{-x}$,

b. $f(x, y, z) = \sin^2 x$.

2.50 Solve the inhomogeneous potential problem

$$U_{xx} + U_{yy} = -xy, \quad 0 < x < \pi, \quad 0 < y < \pi$$

Subject to the following boundary conditions

$$U(0,y) = 0, \quad U(\pi,y) = 0$$
$$U(x,0) = 0, \quad U(x,\pi) = U_o(\text{constant})$$

2.51 Solve the inhomogeneous PDE

$$\frac{\partial^2 \Phi}{\partial \rho^2} + \frac{1}{\rho}\frac{\partial \Phi}{\partial \rho} - \frac{\partial^2 \Phi}{\partial t^2} = -\Phi_o \sin \omega t, \quad 0 \le \rho \le a, t \ge 0$$

subject to the conditions $\Phi(a, t) = 0$, $\Phi(\rho, 0) = 0$, $\Phi_t(\rho, 0) = 0$, Φ is finite for all $0 \le \rho \le a$. Take Φ_o as a constant and $a\omega$ not being a zero of $J_0(x)$.

2.52 Infinitely long metal box has a rectangular cross section shown in Figure 2.22. If the box is filled with charge $\rho_v = \rho_o x/a$, find V inside the box.

2.53 In Section 2.7.2, find $\mathbf{E}_g$ and $\mathbf{E}_\ell$, the electric field intensities in gas and liquid, respectively.

2.54 Consider the potential problem shown in Figure 2.23. The potentials at $x = 0$, $x = a$, and $y = 0$ sides are zero while the potential at $y = b$ side is V_o. Using the series expansion technique similar to that used in Section 2.7.2, find the potential distribution in the solution region.

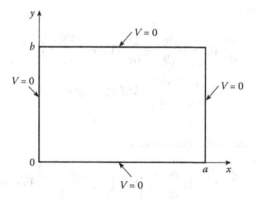

FIGURE 2.22
For Problem 2.52.

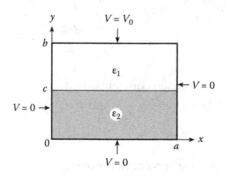

FIGURE 2.23
Potential system for Problem 2.54.

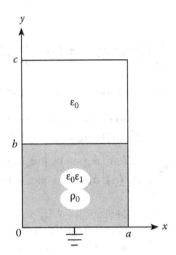

FIGURE 2.24
Earthed rectangular pipe partially filled with charged liquid—for Problem 2.36.

2.55 Consider a grounded rectangular pipe with the cross section shown in Figure 2.24. Assuming that the pipe is partially filled with hydrocarbons with charge density ρ_o, apply the same series expansion technique used in Section 2.7.2 to find the potential distribution in the pipe.

2.56 Write a program to generate associated Legendre polynomial, with $x = \cos\theta = 0.5$. You may use either series expansion or recurrence relations. Take $0 \le n \le 15, 0 \le m \le n$. Compare your results with those tabulated in standard tables.

2.57 Show that

$$\int T_0(x)dx = T_1(x)$$

$$\int T_1(x)dx = \frac{1}{4}T_2(x) + \frac{1}{4}$$

$$\int T_n(x)dx = \frac{1}{2}\left(\frac{T_{n+1}(x)}{n+1} - \frac{T_{n-1}(x)}{n-1}\right), \quad n > 1$$

2.58 A function is defined by

$$f(x) = \begin{cases} 1, & -1 \le x \le 1 \\ 0, & \text{otherwise} \end{cases}$$

a. Expand $f(x)$ in a series of Hermite functions,

b. Expand $f(x)$ in a series of Laguerre functions.

2.59 By expressing E_θ^i and E_ϕ^i in terms of the spherical wave functions, show that Equation 2.235 is valid.

2.60 By defining

$$\rho_n(x) = \frac{d}{dx}\ln[xh_n^{(2)}(x)], \quad \sigma_n(x) = \frac{d}{dx}\ln[xj_n(x)],$$

show that the scattering amplitude coefficients can be written as

$$a_n = \frac{j_n(\alpha)}{h_n^{(2)}(\alpha)}\left[\frac{\sigma_n(\alpha) - m\sigma_n(m\alpha)}{\rho_n(\alpha) - m\sigma_n(m\alpha)}\right]$$

$$b_n = \frac{j_n(\alpha)}{h_n^{(2)}(\alpha)}\left[\frac{\sigma_n(m\alpha) - m\sigma_n(\alpha)}{\sigma_n(m\alpha) - m\sigma_n(\alpha)}\right]$$

2.61 For the problem in Figure 2.14, plot $|E_z^t|/|E_x^i|$ for $-a < z < a$ along the axis of the dielectric sphere of radius $a = 9$ cm in the $x - z$ plane. Take $E_o = 1$, $\omega = 2\pi \times 5 \times 10^9$ rad/s, $\varepsilon_1 = 4\varepsilon_o$, $\mu_1 = \mu_o$, $\sigma_1 = 0$. You may modify the program in Figure 2.15 or write your own.

2.62 In analytical treatment of the radio-frequency radiation effect on the human body, the human skull is frequently modeled as a lossy sphere. Of major concern is the calculation of the normalized heating potential

$$\Phi(r) = \frac{1}{2}\sigma\frac{|E^t(r)|^2}{|E_o|^2} \quad (\Omega \cdot m)^{-1},$$

where E_t is the internal electric field strength and E_o is the peak incident field strength. If the human skull can be represented by a homogeneous sphere of radius $a = 10$ cm, plot $\Phi(r)$ against the radial distance $-10 \le r = z \le 10$ cm. Assume an incident field as in Figure 2.14 with $f = 1$ GHz, $\mu_r = 1$, $\varepsilon_r = 60$, $\sigma = 0.9$ mhos/m, $E_o = 1$.

2.63 Instead of the homogeneous spherical model assumed in the previous problem, consider the multilayered spherical model shown in Figure 2.25 with each region labeled by an integer p, such that $p = 1$ represents the central core region and $p = 4$ represents air. At $f = 2.45$ GHz, plot the heating potential along the x axis, y axis, and z axis. Assume the data given below.

Region p	Tissue	Radius (mm)	ε_r	σ (mho/m)
1	Muscle	18.5	46	2.5
2	Fat	19	6.95	0.29
3	Skin	20	43	2.5
4	Air		1	0

$\mu_r = 1$

Note that for each region p, the resultant field consists of the transmitted and scattered fields and in general given by

$$E_p(r,\theta,\phi) = E_o e^{j\omega t}\sum_{n=1}^{\infty}(-j)^n\frac{2n+1}{n(n+1)}[a_{np}\mathbf{M}_{np}^{(4)}(k)]$$
$$+ jb_{np}\mathbf{N}_{np}^{(4)}(k) + c_{np}\mathbf{M}_{np}^{(1)}(k_1) + jd_{np}\mathbf{N}_{np}^{(1)}(k1)$$

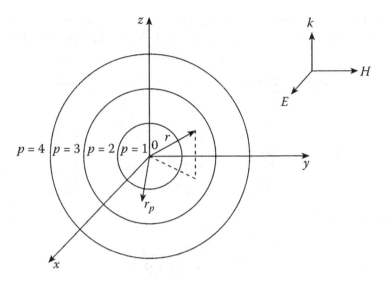

FIGURE 2.25
For Problem 2.63, a multilayered spherical model of the human skull.

2.64 The absorption characteristic of biological bodies is determined in terms of the specific absorption rate (SAR) defined as the total power absorbed divided as the power incident on the geometrical cross section. For an incident power density of 1 mW/cm^2 in a spherical model of the human head,

$$\text{SAR} = 2\frac{Q_{\text{abs}}}{\pi a} \quad \text{mW/cm}^3$$

where a is in centimeters. Using the above relation, plot SAR against frequency for $0.1 < f < 3$ GHz, $a = 10$ cm assuming frequency-dependent and dielectric properties of head as

$$\epsilon_r = 5\left[\frac{12 + (f/f_o)^2}{1 + (f/f_o)^2}\right]$$

$$\sigma = 6\left[\frac{1 + 62(f/f_o)^2}{1 + (f/f_o)^2}\right]$$

where f is in GHz and $f_o = 20$ GHz.

2.65 For the previous problem, repeat the calculations of SAR assuming a six-layered spherical model of the human skull (similar to that of Figure 2.25) of outer radius $a = 10$ cm. Plot P_a/P_i vs. frequency for $0.1 < f < 3$ GHz where

$$\frac{P_a}{P_i} = \frac{2}{\alpha^2}\sum(2n+1)\left[\text{Re}(a_n + b_n) - \left(|a_n|^2 + |b_n|^2\right)\right],$$

P_a = absorbed power, P_i = incident power, $\alpha = 2\pi a/\lambda$, λ is the wavelength in the external medium. Use the dimensions and electrical properties shown below.

Layer p	Tissue	Radius (mm)	ε_r	σ_o (mho/m)
1	Brain	9	$5\nabla(f)$	$6\nabla(f)$
2	CSF	12	$7\nabla(f)$	$8\nabla(f)$
3	Dura	13	$4\nabla(f)$	$8\nabla(f)$
4	Bone	17.3	5	62
5	Fat	18.5	6.95	0.29
6	Skin	20	43	2.5

where $\mu_r = 1$,

$$\nabla(f) = \frac{1 + 12(f/f_o)^2}{1 + (f/f_o)^2},$$

$$\Delta(f) = \frac{1 + 62(f/f_o)^2}{1 + (f/f_o)^2},$$

f is in GHz, and $f_o = 20$ GHz. Compare your result with that from the previous problem.

References

1. M.N.O. Sadiku and S. R. Nelatury, *Analytical Techniques in Electromagnetics*. Boca Raton, FL: CRC Press, 2016.
2. N. Morita et al., *Integral Equation Methods for Electromagnetics*. Boston, MA: Artech House, 1990.
3. H.F. Weinberger, *A First Course in Partial Differential Equations*. New York: John Wiley, 1965, Chap. 5, pp. 63–116.
4. R.D. Kersten, *Engineering Differential Systems*. New York: McGraw-Hill, 1969, Chap. 5, pp. 66–106.
5. T. Myint-U, *Partial Differential Equations of Mathematical Physics*, 2nd Edition. New York: North-Holland, 1980, pp. 186–188, 198–1204.
6. G.N. Watson, *Theory of Bessel Functions*. London: Cambridge University Press, 1966.
7. G. E. Andrews, R.A. Askey, and R. Roy, *Special Functions*. Cambridge, UK: Cambridge University Press, 1999.
8. D.E. Johnson and J.R. Johnson, *Mathematical Methods in Engineering and Physics*. Englewood Cliffs, NJ: Prentice-Hall, 1982.
9. K. Asano, "Electrostatic potential and field in a cylindrical tank containing charged liquid," *Proc. IEEE*, vol. 124, no. 12, December 1977, pp. 1277–1281.
10. J.A. Stratton, *Electromagnetic Theory*. New York: McGraw-Hill, 1941, pp. 394–421, 563–573.
11. H.C. Van de Hulst, *Light Scattering of Small Particles*. New York: John Wiley, 1957, pp. 28–37, 114–136, 284.
12. J. Morrison and M.J. Cross, "Scattering of a plane electromagnetic wave by axisymmetric raindrops," *Bell Syst. Tech. J.*, vol. 53, no. 6, July–August 1974, pp. 955–1019.
13. D.E. Setzer, "Computed transmission through rain at microwave and visible frequencies," *Bell Syst. Tech. J.*, vol. 49, no. 8, October 1970, pp. 1873–1892.

14. M.N.O. Sadiku, "Refractive index of snow at microwave frequencies," *Appl. Optics*, vol. 24, no. 4, February 1985, pp. 572–575.
15. M. Abramowitz and I.A. Stegun, *Handbook of Mathematical Functions*. New York: Dover, 1965.
16. C.H. Durney, "Electromagnetic dosimetry for models of humans and animals: A review of theoretical and numerical techniques," *Proc. IEEE.*, vol. 68, no. 1, January 1980, pp. 33–40.
17. M.A. Morgan, "Finite element calculation of microwave absorption by the cranial structure," *IEEE Trans. Bio. Engr.*, vol. BME-28, no. 10, October 1981, pp. 687–695.
18. J.W. Hand, "Microwave heating patterns in simple tissue models," *Phys. Med. Biol.*, vol. 22, no. 5, 1977, pp. 981–987.
19. W.T. Joines and R.J. Spiegel, "Resonance absorption of microwaves by the human skull," *IEEE Trans. Bio. Engr.*, vol. BME-21, January 1974, pp. 46–48.
20. C.M. Weil, "Absorption characteristics of multilayered sphere models exposed to UHF/microwave radiation," *IEEE Trans. Bio. Engr.*, vol. BME-22, no. 6, November 1975, pp. 468–476.
21. S. Tretyakov, *Analytical Modeling in Applied Electromagnetics*. Norwood, MA: Artech House, 2003.
22. K.J. Binns, P.J. Lawrence, and C.W. Trowbridge, *The Analytical and Numerical Solution of Electric and Magnetic Fields*. Chichester, UK: John Wiley & Sons, 1992.
23. R. Garg and R. Mittra, *Analytical and Computational Methods in Electromagnetics*. Norwood, MA: Artech House, 2008.
24. H.J. Eom, *Electromagnetic Wave Theory for Boundary-Value Problems: An Advanced Course on Analytical Methods*. Berlin: Springer-Verlag, 2004.

3

Finite Difference Methods

The reason why worry kills more people than hard work is that more people worry than work.

—Robert Frost

3.1 Introduction

It is rare for real-life EM problems to fall neatly into a class that can be solved by the analytical methods presented in the preceding chapter. Classical approaches may fail if [1]

- PDE is not linear and cannot be linearized without seriously affecting the result
- The solution region is complex
- The boundary conditions are of mixed types
- The boundary conditions are time-dependent
- The medium is inhomogeneous or anisotropic

Whenever a problem with such complexity arises, numerical solutions must be employed. Of the numerical methods available for solving PDEs, those employing finite differences are more easily understood, more frequently used, and more universally applicable than any other.

The finite difference method (FDM) was first developed by A. Thom [2] in the 1920s under the title "the method of squares" to solve nonlinear hydrodynamic equations. Since then, the method has found applications in solving different field problems. The finite difference techniques are based upon approximations which permit replacing differential equations by finite difference equations. These finite difference approximations are algebraic in form; they relate the value of the dependent variable at a point in the solution region to the values at some neighboring points. Thus a finite difference solution basically involves three steps:

1. Dividing the solution region into a grid of nodes
2. Approximating the given differential equation by finite difference equivalent that relates the dependent variable at a point in the solution region to its values at the neighboring points
3. Solving the difference equations subject to the prescribed boundary conditions and/or initial conditions

The course of action taken in three steps is dictated by the nature of the problem being solved, the solution region, and the boundary conditions. The most commonly used grid

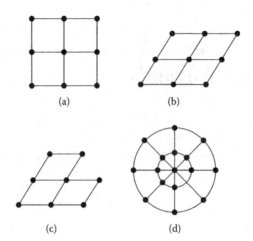

FIGURE 3.1
Common grid patterns: (a) rectangular grid, (b) skew grid, (c) triangular grid, (d) circular grid.

patterns for two-dimensional problems are shown in Figure 3.1. A three-dimensional grid pattern will be considered later in the chapter.

3.2 Finite Difference Schemes

Before finding the finite difference solutions to specific PDEs, we will look at how one constructs finite difference approximations from a given differential equation. This essentially involves estimating derivatives numerically.

Given a function $f(x)$ shown in Figure 3.2, we can approximate its derivative, slope or the tangent at P by the slope of the arc PB, giving the *forward-difference* formula

$$f'(x_o) \simeq \frac{f(x_o + \Delta x) - f(x_o)}{\Delta x} \tag{3.1}$$

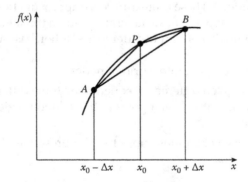

FIGURE 3.2
Estimates for the derivative of $f(x)$ at P using forward, backward, and central differences.

or the slope of the arc AP, yielding the *backward-difference* formula

$$\boxed{f'(x_o) \simeq \frac{f(x_o) - f(x_o - \Delta x)}{\Delta x}} \tag{3.2}$$

or the slope of the arc AB, resulting in the *central-difference* formula

$$\boxed{f'(x_o) \simeq \frac{f(x_o + \Delta x) - f(x_o - \Delta x)}{2\Delta x}} \tag{3.3}$$

We can also estimate the second derivative of $f(x)$ at P by applying Equation 3.3 twice

$$f''(x_o) \simeq \frac{f'(x_o + \Delta x/2) - f'(x_o - \Delta x/2)}{\Delta x}$$

$$= \frac{1}{\Delta x}\left[\frac{f(x_o + \Delta x) - f(x_o)}{\Delta x} - \frac{f(x_o) - f(x_o - \Delta x)}{\Delta x}\right]$$

or

$$f''(x_o) \simeq \frac{f(x_o + \Delta x) - 2f(x_o) + f(x_o - \Delta x)}{(\Delta x)^2} \tag{3.4}$$

Any approximation of a derivative in terms of values at a discrete set of points is called *finite difference* approximation.

The approach used above in obtaining finite difference approximations is rather intuitive. A more general approach is using Taylor's series. According to the well-known expansion,

$$f(x_o + \Delta x) = f(x_o) + \Delta x f'(x_o) + \frac{1}{2!}(\Delta x)^2 f''(x_o) + \frac{1}{3!}(\Delta x)^3 f'''(x_o) + \cdots \tag{3.5}$$

and

$$f(x_o - \Delta x) = f(x_o) - \Delta x f'(x_o) + \frac{1}{2!}(\Delta x)^2 f''(x_o) - \frac{1}{3!}(\Delta x)^3 f'''(x_o) + \cdots \tag{3.6}$$

Upon adding these expansions,

$$f(x_o + \Delta x) + f(x_o - \Delta x) = 2f(x_o) + (\Delta x)^2 f''(x_o) + O(\Delta x)^4 \tag{3.7}$$

where $O(\Delta x)^4$ is the error introduced by truncating the series. We say that this error is of the order $(\Delta x)^4$ or simply $O(\Delta x)^4$. Therefore, $O(\Delta x)^4$ represents terms that are not greater than $(\Delta x)^4$. Assuming that these terms are negligible,

$$f''(x_o) \simeq \frac{f(x_o + \Delta x) - 2f(x_o) + f(x_o - \Delta x)}{(\Delta x)^2}$$

which is Equation 3.4. Subtracting Equation 3.6 from Equation 3.5 and neglecting terms of the order $(\Delta x)^3$ yields

$$f'(x_o) \simeq \frac{f(x_o + \Delta x) - f(x_o - \Delta x)}{2\Delta x}$$

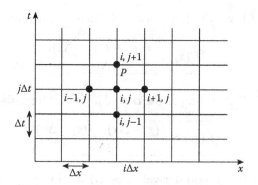

FIGURE 3.3
Finite difference mesh for two independent variables x and t.

which is Equation 3.3. This shows that the leading errors in Equations 3.3 and 3.4 are of the order $(\Delta x)^2$. Similarly, the difference formula in Equations 3.1 and 3.2 have truncation errors of $O(\Delta x)$. Higher-order finite difference approximations can be obtained by taking more terms in Taylor-series expansion. If the infinite Taylor series were retained, an exact solution would be realized for the problem. However, for practical reasons, the infinite series is usually truncated after the second-order term. This imposes an error which exists in all finite difference solutions.

To apply the difference method to find the solution of a function $\Phi(x, t)$, we divide the solution region in the $x - t$ plane into equal rectangles or meshes of sides Δx and Δt as in Figure 3.3. We let the coordinates (x, t) of a typical grid point or node be

$$x = i\Delta x, \quad i = 0, 1, 2, \ldots$$
$$t = j\Delta t, \quad j = 0, 1, 2, \ldots \tag{3.8a}$$

and the value of Φ at P be

$$\Phi_P = \Phi(i\,\Delta x, j\,\Delta t) = \Phi(i, j) \tag{3.8b}$$

With this notation, the central difference approximations of the derivatives of Φ at the (i, j)th node are

$$\Phi_x|_{i,j} \simeq \frac{\Phi(i+1, j) - \Phi(i-1, j)}{2\Delta x}, \tag{3.9a}$$

$$\Phi_t|_{i,j} \simeq \frac{\Phi(i, j+1) - \Phi(i, j-1)}{2\Delta t}, \tag{3.9b}$$

$$\Phi_{xx}|_{i,j} \simeq \frac{\Phi(i+1, j) - 2\Phi(i, j) + \Phi(i-1, j)}{(\Delta x)^2}, \tag{3.9c}$$

$$\Phi_{tt}|_{i,j} \simeq \frac{\Phi(i, j+1) - 2\Phi(i, j) + \Phi(i, j-1)}{(\Delta t)^2} \tag{3.9d}$$

Table 3.1 gives some useful finite difference approximations for Φ_x and Φ_{xx}.

TABLE 3.1

Finite Difference Approximations for Φ_x and Φ_{xx}

Derivative	Finite Difference Approximation	Type	Error
Φ_x	$\dfrac{\Phi_{i+1} - \Phi_i}{\Delta x}$	FD	$O(\Delta x)$
	$\dfrac{\Phi_i - \Phi_{i-1}}{\Delta x}$	BD	$O(\Delta x)$
	$\dfrac{\Phi_{i+1} - \Phi_{i-1}}{2\Delta x}$	CD	$O(\Delta x)^2$
	$\dfrac{-\Phi_{i+2} + 4\Phi_{i+1} - 3\Phi_i}{2\Delta x}$	FD	$O(\Delta x)^2$
	$\dfrac{3\Phi_i - 4\Phi_{i-1} + \Phi_{i-2}}{2\Delta x}$	BD	$O(\Delta x)^2$
Φ_{xx}	$\dfrac{-\Phi_{i+2} + 8\Phi_{i+1} - 8\Phi_{i-1} + \Phi_{i-2}}{12\Delta x}$	CD	$O(\Delta x)^4$
	$\dfrac{\Phi_{i+2} - 2\Phi_{i+1} + \Phi_i}{(\Delta x)^2}$	FD	$O(\Delta x)^2$
	$\dfrac{\Phi_i - 2\Phi_{i-1} + \Phi_{i-2}}{(\Delta x)^2}$	BD	$O(\Delta x)^2$
	$\dfrac{\Phi_{i+1} - 2\Phi_i + \Phi_{i-1}}{(\Delta x)^2}$	CD	$O(\Delta x)^2$
	$\dfrac{-\Phi_{i+2} + 16\Phi_{i+1} - 30\Phi_i + 16\Phi_{i-1} - \Phi_{i-2}}{12(\Delta x)^2}$	CD	$O(\Delta x)^4$

where FD = Forward Difference, BD = Backward Difference, and CD = Central Difference.

3.3 Finite Differencing of Parabolic PDEs

Consider a simple example of a parabolic (or diffusion) partial differential equation with one spatial independent variable

$$k\frac{\partial \Phi}{\partial t} = \frac{\partial^2 \Phi}{\partial x^2} \tag{3.10}$$

where k is a constant. The equivalent finite difference approximation is

$$k\frac{\Phi(i, j+1) - \Phi(i, j)}{\Delta t} = \frac{\Phi(i+1, j) - 2\Phi(i, j) + \Phi(i-1, j)}{(\Delta x)^2} \tag{3.11}$$

where $x = i\,\Delta x$, $i = 0, 1, 2, \ldots, n$, $t = j\,\Delta t$, $j = 0, 1, 2, \ldots$. In Equation 3.11, we have used the forward difference formula for the derivative with respect to t and central difference formula for that with respect to x. If we let

$$r = \frac{\Delta t}{k(\Delta x)^2}, \tag{3.12}$$

Equation 3.11 can be written as

$$\boxed{\Phi(i, j+1) = r\Phi(i+1, j) + (1 - 2r)\Phi(i, j) + r\Phi(i-1, j)} \tag{3.13}$$

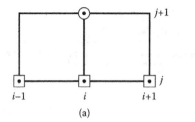 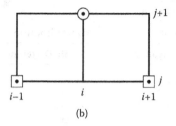

FIGURE 3.4
Computational molecule for parabolic PDE: (a) for $0 < r \le 1/2$, (b) for $r = 1/2$.

This *explicit formula* can be used to compute $\Phi(x, t + \Delta t)$ explicitly in terms of $\Phi(x, t)$. Thus, the values of Φ along the first time row (see Figure 3.3), $t = \Delta t$, can be calculated in terms of the boundary and initial conditions, then the values of Φ along the second time row, $t = 2\Delta t$, are calculated in terms of the first time row, and so on.

A graphic way of describing the difference formula of Equation 3.13 is through the *computational molecule* of Figure 3.4a, where the square is used to represent the grid point where Φ is presumed known and a circle where Φ is unknown.

In order to ensure a stable solution or reduce errors, care must be exercised in selecting the value of r in Equations 3.12 and 3.13. It will be shown in Section 3.6 that Equation 3.13 is valid only if the coefficient $(1 - 2r)$ in Equation 3.13 is nonnegative or $0 < r \le 1/2$. If we choose $r = 1/2$, Equation 3.13 becomes

$$\Phi(i, j+1) = \frac{1}{2}[\Phi(i+1, j) + \Phi(i-1, j)] \tag{3.14}$$

so that the computational molecule becomes that shown in Figure 3.4b.

The fact that obtaining stable solutions depends on r or the size of the time step Δt renders the explicit formula of Equation 3.13 inefficient. Although the formula is simple to implement, its computation is slow. An *implicit formula*, proposed by Crank and Nicholson in 1974, is valid for all finite values of r. We replace $\partial^2\Phi/\partial x^2$ in Equation 3.10 by the average of the central difference formulas on the jth and $(j + 1)$th time rows so that

$$k\frac{\Phi(i, j+1) - \Phi(i, j)}{\Delta t} = \frac{1}{2}\left[\frac{\Phi(i+1, j) - 2\Phi(i, j) + \Phi(i-1, j)}{(\Delta x)^2} \right.$$
$$\left. + \frac{\Phi(i+1, j+1) - 2\Phi(i, j+1) + \Phi(i-1, j+1)}{(\Delta x)^2} \right]$$

This can be rewritten as

$$-r\Phi(i-1, j+1) + 2(1+r)\Phi(i, j+1) - r\Phi(i+1, j+1)$$
$$= r\Phi(i-1, j) + 2(1-r)\Phi(i, j) + r\Phi(i+1, j) \tag{3.15}$$

where r is given by Equation 3.12. The right-hand side of Equation 3.15 consists of three known values, while the left-hand side has the three unknown values of Φ. This is illustrated in the computational molecule of Figure 3.5a. Thus if there are n free nodes along each time row, then for $j = 0$, applying Equation 3.15 to nodes $i = 1, 2, \ldots, n$ results in n simultaneous equations with n unknown values of Φ and known initial and boundary values of Φ. Similarly, for $j = 1$, we obtain n simultaneous equations for n unknown values of Φ in terms of the known values $j = 0$, and so on. The combination of accuracy and unconditional

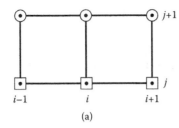
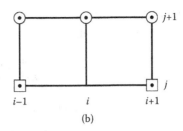

FIGURE 3.5
Computational molecule for Crank–Nicholson method: (a) for finite values of r, (b) for $r = 1$.

stability allows the use of a much larger time step with the Crank–Nicholson method than is possible with the explicit formula. Although the method is valid for all finite values of r, a convenient choice of $r = 1$ reduces Equation 3.15 to

$$-\Phi(i-1, j+1) + 4\Phi(i, j+1) - \Phi(i+1, j+1) = \Phi(i-1, j) + \Phi(i+1, j) \qquad (3.16)$$

with the computational molecule of Figure 3.5b.

More complex finite difference schemes can be developed by applying the same principles discussed above. Two of such schemes are the Leapfrog method and the Dufort–Frankel method [3,4]. These and those discussed earlier are summarized in Table 3.2. Notice that the last two methods are two-step finite difference schemes in that finding Φ at time $j+1$ requires knowing Φ at two previous time steps j and $j-1$, whereas the first two methods are one-step schemes. For further treatment on the finite difference solution of parabolic PDEs, see Smith [5] and Ferziger [6].

TABLE 3.2

Finite Difference Approximation to the Parabolic Equation: $\dfrac{\partial \Phi}{\partial t} = \dfrac{1}{k}\dfrac{\partial^2 \Phi}{\partial x^2}, k > 0$

Method	Algorithm	Molecule
1. First order (Euler)	$\dfrac{\Phi_i^{j+1} - \Phi_i^j}{\Delta t} = \dfrac{\Phi_{i+1}^j - 2\Phi_i^j + \Phi_{i-1}^j}{k(\Delta x)^2}$ explicit, stable for $r = \Delta t/k(\Delta x)^2 \le 0.5$	
2. Crank–Nicholson	$\dfrac{\Phi_i^{j+1} - \Phi_i^j}{\Delta t} = \dfrac{\Phi_{i+1}^{j+1} - 2\Phi_i^{j+1} + \Phi_{i-1}^{j+1}}{2k(\Delta x)^2}$ $\qquad + \dfrac{\Phi_{i+1}^j - 2\Phi_i^j + \Phi_{i-1}^j}{2k(\Delta x)^2}$ implicit, always stable	
3. Leapfrog	$\dfrac{\Phi_i^{j+1} - \Phi_i^{j-1}}{2\Delta t} = \dfrac{\Phi_{i+1}^j - 2\Phi_i^j + \Phi_{i-1}^j}{k(\Delta x)^2}$ explicit, always unstable	
4. Dufort–Frankel	$\dfrac{\Phi_i^{j+1} - \Phi_i^{j-1}}{2\Delta t} = \dfrac{\Phi_{i+1}^j - \Phi_i^{j+1} - \Phi_i^{j-1} + \Phi_{i-1}^j}{k(\Delta x)^2}$ explicit, unconditionally stable	

EXAMPLE 3.1

Solve the diffusion equation

$$\frac{\partial^2 \Phi}{\partial x^2} = \frac{\partial \Phi}{\partial t}, \quad 0 \leq x \leq 1 \tag{3.17}$$

subject to the boundary conditions

$$\Phi(0, t) = 0 = \Phi(1, t) = 0, \quad t > 0 \tag{3.18a}$$

and initial condition

$$\Phi(x, 0) = 100 \tag{3.18b}$$

Solution

This problem may be regarded as a mathematical model of the temperature distribution in a rod of length $L = 1$ m with its end in contact with ice blocks (or held at 0°C) and the rod initially at 100°C. With that physical interpretation, our problem is finding the internal temperature Φ as a function of position and time. We will solve this problem using both explicit and implicit methods.

a. *Explicit Method*

For easy hand calculations, let us choose $\Delta x = 0.1$, $r = 1/2$ so that

$$\Delta t = kr(\Delta x)^2 = 0.005$$

since $k = 1$. We need the solution for only $0 \leq x \leq 0.5$ because the problem is symmetric with respect to $x = 0.5$. First, we calculate the initial and boundary values using Equation 3.18. These values of Φ at the fixed nodes are shown in Table 3.3 for $x = 0$, $x = 1$, and $t = 0$. Notice that the values of $\Phi(0, 0)$ and $\Phi(1, 0)$ are taken as the average of 0 and 100. We now calculate Φ at the free nodes using Equation 3.14 or the molecule of Figure 3.4b. The result is shown in Table 3.3. The analytic solution to Equation 3.17 subject to Equation 3.18 is

$$\Phi(x,t) = \frac{400}{\pi} \sum_{k=0}^{\infty} \frac{1}{n} \sin n\pi \, x \exp(-n^2\pi^2 t), \quad n = 2k + 1$$

TABLE 3.3

Results for Example 3.1

x	0	0.1	0.2	0.3	0.4	0.5	0.6	...	1.0
t									
0	50	100	100	100	100	100	100		50
0.005	0	75.0	100	100	100	100	100		0
0.01	0	50	87.5	100	100	100	100		0
0.015	0	43.75	75	93.75	100	100	100		0
0.02	0	37.5	68.75	87.5	96.87	100	96.87		0
0.025	0	34.37	62.5	82.81	93.75	96.87	93.75		0
0.03	0	31.25	58.59	78.21	89.84	93.75	89.84		0
⋮									
0.1	0	14.66	27.92	38.39	45.18	47.44	45.18		0

TABLE 3.4

Comparison of Explicit Finite Difference Solution with Analytic Solution; for Example 3.1

t	Finite Difference Solution at $x = 0.4$	Analytic Solution at $x = 0.4$	Percentage Error
0.005	100	99.99	0.01
0.01	100	99.53	0.47
0.015	100	97.85	2.2
0.02	96.87	95.18	1.8
0.025	93.75	91.91	2.0
0.03	89.84	88.32	1.7
0.035	85.94	84.61	1.6
0.04	82.03	80.88	1.4
⋮			
0.10	45.18	45.13	0.11

Comparison of the explicit finite difference solution with the analytic solution at $x = 0.4$ is shown in Table 3.4. The table shows that the finite difference solution is reasonably accurate. Greater accuracy can be achieved by choosing smaller values of Δx and Δt.

b. *Implicit Method*

Let us choose $\Delta x = 0.2$, $r = 1$ so that $\Delta t = 0.04$. The values of Φ at the fixed nodes are calculated as in part (a) (see Table 3.3). For the free nodes, we apply Equation 3.16 or the molecule of Figure 3.5b. If we denote $\Phi(i, j + 1)$ by $\Phi_i (i = 1, 2, 3, 4)$, the values of Φ for the first time step (Figure 3.6) can be obtained by solving the following simultaneous equations:

$$-0 + 4\Phi_1 - \Phi_2 = 50 + 100$$
$$-\Phi_1 + 4\Phi_2 + \Phi_3 = 100 + 100$$
$$-\Phi_2 + 4\Phi_3 - \Phi_4 = 100 + 100$$
$$-\Phi_3 + 4\Phi_4 - 0 = 100 + 50$$

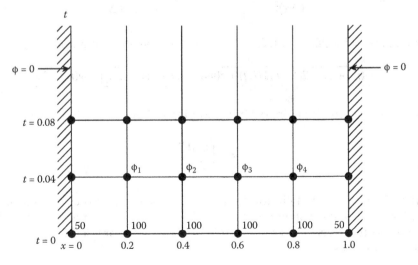

FIGURE 3.6
For Example 3.1, part (b).

We obtain

$$\Phi_1 = 58.13, \quad \Phi_2 = 82.54, \quad \Phi_3 = 72, \quad \Phi_4 = 55.5$$

at $t = 0.04$. Using these values of Φ, we apply Equation 3.16 to obtain another set of simultaneous equations for $t = 0.08$ as

$$-0 + 4\Phi_1 - \Phi_2 = 0 + 82.54$$
$$-\Phi_1 + 4\Phi_2 - \Phi_3 = 58.13 + 72$$
$$-\Phi_2 + 4\Phi_3 - \Phi_4 = 82.54 + 55.5$$
$$-\Phi_3 + 4\Phi_4 - 0 = 72 + 0$$

which results in

$$\Phi_1 = 34.44, \quad \Phi_2 = 55.23, \quad \Phi_3 = 56.33, \quad \Phi_4 = 32.08$$

This procedure can be programmed and accuracy can be increased by choosing more points for each time step.

3.4 Finite Differencing of Hyperbolic PDEs

The simplest hyperbolic PDE is the wave equation of the form

$$u^2 \frac{\partial^2 \Phi}{\partial x^2} = \frac{\partial^2 \Phi}{\partial t^2} \tag{3.19}$$

where u is the speed of the wave. An equivalent finite difference formula is

$$u^2 \frac{\Phi(i+1, j) - 2\Phi(i, j) + \Phi(i-1, j)}{(\Delta x)^2} = \frac{\Phi(i, j+1) - 2\Phi(i, j) + \Phi(i, j-1)}{(\Delta t)^2}$$

where $x = i\,\Delta x$, $t = j\,\Delta t$, $i, j = 0, 1, 2, \ldots$ This equation can be written as

$$\boxed{\Phi(i, j+1) = 2(1-r)\Phi(i, j) + r[\Phi(i+1, j) + \Phi(i-1, j)] - \Phi(i, j-1)} \tag{3.20}$$

where $\Phi(i, j)$ is an approximation to $\Phi(x, t)$ and r is the "aspect ratio" given by

$$r = \left(\frac{u\Delta t}{\Delta x} \right)^2 \tag{3.21}$$

Equation 3.20 is an explicit formula for the wave equation. The corresponding computational molecule is shown in Figure 3.7a. For the solution algorithm in Equation 3.20 to be stable, the aspect ratio $r \leq 1$, as will be shown in Example 3.5. If we choose $r = 1$, Equation 3.20 becomes

$$\Phi(i, j+1) = \Phi(i+1, j) + \Phi(i-1, j) - \Phi(i, j-1) \tag{3.22}$$

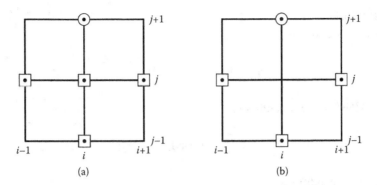

FIGURE 3.7
Computational molecule for wave equation: (a) for arbitrary $r \leq 1$, (b) for $r = 1$.

with the computational molecule in Figure 3.7b. Unlike the single-step schemes of Equations 3.13 and 3.15, the two-step schemes of Equations 3.20 and 3.22 require that the values of Φ at times j and $j - 1$ be known to get Φ at time $j + 1$. Thus, we must derive a separate algorithm to "start" the solution of Equation 3.20 or 3.22; that is, we must compute $\Phi(i, 1)$ and $\Phi(i, 2)$. To do this, we utilize the prescribed initial condition. For example, suppose the initial condition on the PDE in Equation 3.19 is

$$\frac{\partial \Phi}{\partial t}\bigg|_{t=0} = 0$$

We use the backward-difference formula

$$\frac{\partial \Phi(x,0)}{\partial t} \simeq \frac{\Phi(i,1) - \Phi(i,-1)}{2\Delta t} = 0$$

or

$$\Phi(i, 1) = \Phi(i, -1) \tag{3.23}$$

Substituting Equation 3.23 into Equation 3.20 and taking $j = 0$ (i.e., at $t = 0$), we get

$$\Phi(i, 1) = 2(1 - r)\Phi(i, 0) + r[\Phi(i - 1, 0) + \Phi(i + 1, 0)] - \Phi(i, 1)$$

or

$$\Phi(i,1) = (1-r)\Phi(i,0) + \frac{r}{2}[\Phi(i-1,0) + \Phi(i+1,0)] \tag{3.24}$$

Using the starting formula in Equation 3.24 together with the prescribed boundary and initial conditions, the value of Φ at any grid point (i, j) can be obtained directly from Equation 3.20.

There are implicit methods for solving hyperbolic PDEs just as we have implicit methods for parabolic PDEs. However, for hyperbolic PDEs, implicit methods result in an infinite number of simultaneous equations to be solved and therefore cannot be used without making some simplifying assumptions. Interested readers are referred to Smith [5] or Ferziger [6].

EXAMPLE 3.2

Solve the wave equation

$$\Phi_{tt} = \Phi_{xx}, \quad 0 < x < 1, \quad t \geq 0$$

subject to the boundary conditions

$$\Phi(0, t) = 0 = \Phi(1, t), \quad t \geq 0$$

and the initial conditions

$$\Phi(x, 0) = \sin\pi x, \quad 0 < x < 1,$$

$$\Phi_t(x, 0) = 0, \quad 0 < x < 1$$

Solution

The analytical solution is easily obtained as

$$\Phi(x, t) = \sin \pi x \cos \pi t \tag{3.25}$$

Using the explicit finite difference scheme of Equation 3.20 with $r = 1$, we obtain

$$\Phi(i, j + 1) = \Phi(i - 1, j) + \Phi(i + 1, j) - \Phi(i, j - 1), \quad j \geq 1 \tag{3.26}$$

For $j = 0$, substituting

$$\Phi_t = \frac{\Phi(i,1) - \Phi(i,-1)}{2\Delta t} = 0$$

or

$$\Phi(i, 1) = \Phi(i, -1)$$

into Equation 3.26 gives the starting formula

$$\Phi(i,1) = \frac{1}{2}[\Phi(i-1,0) + \Phi(i+1,0)] \tag{3.27}$$

Since $u = 1$, and $r = 1$, $\Delta t = \Delta x$. Also, since the problem is symmetric with respect to $x = 0.5$, we solve for Φ using Equations 3.26 and 3.27 within $0 < x < 0.5$, $t \geq 0$. We can either calculate the values by hand or write a simple computer program. With the MATLAB code in Figure 3.8, the result shown in Table 3.5 is obtained for $\Delta t = \Delta x = 0.1$. The finite difference solution agrees with the exact solution in Equation 3.25 to six decimal places. The accuracy of the FD solution can be increased by choosing a smaller spatial increment Δx and a smaller time increment Δt.

```
% ********************************************************************
% MATLAB code for example 3.2 on  one-dimensional wave equation solved
% using an explicit finite difference scheme
% ********************************************************************

clear all; format compact; tic

%Explicit Method
    delx = 0.1;          % resolution size
    r = 1;               % 'aspect ratio'
    u = 1;               % Constant of given wave equation
    delt = r^2*delx/u;   % time step size
    Tsteps = round(1/delt); % Number of time steps

    % X1 is the potential grid of the simulation, due to symetry only half
    % of the field is calculated.
    X1 = zeros(Tsteps,1/(2*delx)+2);    % Initilize X1

    %Initial conditions and reflection line defined
    x = 0:delx:.5+delx;
    X1(1,:) = sin(pi*x);
    X1(2,2:end-1) = .5*(X1(1,1:end-2)+X1(1,3:end));
    X1(2,end) = X1(2,end-2); %reflection line

    for row = 3:size(X1,1)
        for col = 2:size(X1,2)-1
            X1(row,col) = X1(row-1,col-1)+X1(row-1,col+1)-X1(row-2,col); % eqn. (3.26)
        end
        X1(row,end) = X1(row,end-2);     %reflected line
    end

    %Use symetry condition to create entire field
    X2 = [X1,fliplr(X1(:,1:end-3))];

    figure(1),imagesc(0:delx:1,(0:delt:Tsteps*delt),X2),colorbar
        ylabel('\leftarrow time (sec)')
        xlabel('x')
        title('Hyperbolic PDE')

    if (delx==.1)
        dispmat = [X1(1:8,1:7)];
        disp(sprintf('\nCompare to Table 3.5, Solution of the Wave Equation in Exam-
ple 3.2'))
        disp(num2str(dispmat))
    end
```

FIGURE 3.8
MATLAB code for Example 3.2.

TABLE 3.5

Solution of the Wave Equation in Example 3.2

x / t	0	0.1	0.2	0.3	0.4	0.5	0.6	...
0.0	0	0.3090	0.5879	0.8990	0.9511	1.0	0.9511	
0.1	0	0.2939	0.5590	0.7694	0.9045	0.9511	0.9045	
0.2	0	0.2500	0.4755	0.6545	0.7694	0.8090	0.7694	
0.3	0	0.1816	0.3455	0.4755	0.5590	0.5878	0.5590	
0.4	0	0.0955	0.1816	0.2500	0.2939	0.3090	0.2939	
0.5	0	0	0	0	0	0	0	
0.6	0	−0.0955	−0.1816	−0.2500	−0.2939	−0.3090	−0.2939	
0.7	0	−0.1816	−0.3455	−0.4755	−0.5590	−0.5878	−0.5590	
⋮	⋮	⋮	⋮	⋮	⋮	⋮	⋮	

3.5 Finite Differencing of Elliptic PDEs

A typical elliptic PDE is Poisson's equation, which in two dimensions is given by

$$\nabla^2 \Phi = \frac{\partial^2 \Phi}{\partial x^2} + \frac{\partial^2 \Phi}{\partial y^2} = g(x, y) \tag{3.28}$$

We can use the central difference approximation for the partial derivatives of which the simplest forms are

$$\frac{\partial^2 \Phi}{\partial x^2} = \frac{\Phi(i+1, j) - 2\Phi(i, j) + \Phi(i-1, j)}{(\Delta x)^2} + O(\Delta x)^2 \tag{3.29a}$$

$$\frac{\partial^2 \Phi}{\partial y^2} = \frac{\Phi(i, j+1) - 2\Phi(i, j) + \Phi(i, j-1)}{(\Delta y)^2} + O(\Delta y)^2 \tag{3.29b}$$

where $x = i\,\Delta x$, $y = j\,\Delta y$, and $i, j = 0, 1, 2, \ldots$. If we assume that $\Delta x = \Delta y = h$, to simplify calculations, substituting Equation 3.29 into Equation 3.28 gives

$$\big[\Phi(i+1, j) + \Phi(i-1, j) + \Phi(i, j) + \Phi(i, j-1)\big] - 4\Phi(i, j) = h^2 g(i, j)$$

or

$$\Phi(i, j) = \frac{1}{4}\big[\Phi(i+1, j) + \Phi(i-1, j) + \Phi(i, j+1) + \Phi(i, j-1) - h^2 g(i, j)\big] \tag{3.30}$$

at every point (i, j) in the mesh for Poisson's equation. The spatial increment h is called the *mesh size*. A special case of Equation 3.28 is when the source term vanishes, that is, $g(x, y) = 0$. This leads to Laplace's equation. Thus for Laplace's equation, Equation 3.30 becomes

$$\Phi(i, j) = \frac{1}{4}[\Phi(i+1, j) + \Phi(i-1, j) + \Phi(i, j+1) + \Phi(i, j-1)] \tag{3.31}$$

It is worth noting that Equation 3.31 states that the value of Φ for each point is the average of those at the four surrounding points. The five-point computational molecule for the difference scheme in Equation 3.31 is illustrated in Figure 3.9a, where values of the

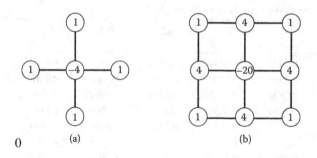

(a)   (b)

FIGURE 3.9
Computational molecules for Laplace's equation based on: (a) second-order approximation, (b) fourth-order approximation.

coefficients are shown. This is a convenient way of displaying finite difference algorithms for elliptic PDEs. The molecule in Figure 3.9a is the second-order approximation of Laplace's equation. This is obviously not the only way to approximate Laplace's equation, but it is the most popular choice. An alternative fourth-order difference is

$$\begin{aligned}
-20\Phi(i,j) &+ 4[\Phi(i+1,j) + \Phi(i-1,j) + \Phi(i,j+1) + \Phi(i,j-1)] \\
&+ \Phi(i+1,j-1) + \Phi(i-1,j-1) + \Phi(i-1,j+1) \\
&+ \Phi(i+1,j+1) = 0
\end{aligned}$$
(3.32)

The corresponding computational molecule is shown in Figure 3.9b.

The application of the finite difference method to elliptic PDEs often leads to a large system of algebraic equations, and their solution is a major problem in itself. Two commonly used methods of solving the system of equations are band matrix and iterative methods.

3.5.1 Band Matrix Method

From Equations 3.30 through 3.32, we notice that only nearest-neighboring nodes affect the value of Φ at each node. Hence, application of any of Equations 3.30 through 3.32 to all free nodes in the solution region results in a set of simultaneous equations of the form

$$[A][X] = [B]$$
(3.33)

where $[A]$ is a *sparse* matrix (it has many zero elements), $[X]$ is a column matrix consisting of the unknown values of Φ at the free nodes, and $[B]$ is a column matrix containing the known values of Φ at fixed nodes. Matrix $[A]$ is also banded in that its nonzero terms appear clustered near the main diagonal. Matrix $[X]$, containing the unknown elements, can be obtained from

$$[X] = [A]^{-1}[B]$$
(3.34)

or by solving Equation 3.33 using the Gauss elimination discussed in Appendix C.1.

3.5.2 Iterative Methods

The iterative methods are generally used to solve a large system of simultaneous equations. An iterative method for solving equations is one in which a first approximation is used to calculate a second approximation, which in turn is used to calculate a third approximation, and so on. The three common iterative methods (Jacobi, Gauss–Seidel, and successive over-relaxation [SOR]) are discussed in Appendix C.2. We will apply only SOR here.

To apply the method of SOR to Equation 3.30, for example, we first define the *residual* $R(i,j)$ at node (i,j) as the amount by which the value of $\Phi(i,j)$ does not satisfy Equation 3.30, that is,

$$\begin{aligned}
R(i,j) = \Phi(i+1,j) &+ \Phi(i-1,j) + \Phi(i,j+1) \\
&+ \Phi(i,j-1) - 4\Phi(i,j) - h^2 g(i,j)
\end{aligned}$$
(3.35)

The value of the residual at kth iteration, denoted by $R^k(i,j)$, may be regarded as a correction which must be added to $\Phi(i,j)$ to make it nearer to the correct value. As convergence to the correct value is approached, $R^k(i,j)$ tends to zero. Hence to improve the rate of convergence, we multiply the residual by a number ω and add that to $\Phi(i, j)$ at the kth iteration to get Φ (i, j) at $(k + 1)$th iteration. Thus,

$$\Phi^{k+1}(i, j) = \Phi^k(i, j) + \frac{\omega}{4} R^k(i, j)$$

or

$$\Phi^{k+1}(i, j) = \Phi^k(i, j) + \frac{\omega}{4} \Big[\Phi^k(i+1, j) + \Phi^k(i-1, j) + \Phi^k(i, j-1)$$
$$+ \Phi^k(i, j+1) - 4\Phi^k(i, j) - h^2 g(i, j) \Big] \tag{3.36}$$

The parameter ω is called the *relaxation factor* while the technique is known as the method of *successive over-relaxation*. The value of ω lies between 1 and 2. (When $\omega = 1$, the method is simply called successive relaxation.) Its optimum value ω_{opt} must be found by trial and error. In order to start Equation 3.36, an initial guess, $\Phi^0(i, j)$, is made at every free node. Typically, we may choose $\Phi^0(i, j) = 0$ or the average of Φ at the fixed nodes.

EXAMPLE 3.3

Solve Laplace's equation

$$\nabla^2 V = 0, \quad 0 \le x \le 1, \quad 0 \le y \le 1$$

with $V(x, 1) = 45x(1 - x)$, $V(x, 0) = 0 = V(0, y) = V(1, y)$.

Solution

Let $h = 1/3$ so that the solution region is as in Figure 3.10. Applying Equation 3.31 to each of the four points leads to

$$4V_1 - V_2 - V_3 - 0 = 10$$

$$-V_1 + 4V_2 - 0 - V_4 = 10$$

$$-V_1 - 0 + 4V_3 - V_4 = 0$$

$$-0 - V_2 - V_3 + 4V_4 = 0$$

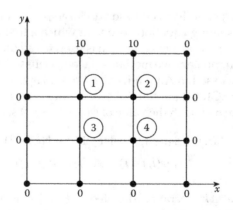

FIGURE 3.10
Finite difference grid for Example 3.3.

This can be written as

$$\begin{bmatrix} 4 & -1 & -1 & 0 \\ -1 & 4 & 0 & -1 \\ -1 & 0 & 4 & -1 \\ 0 & -1 & -1 & 4 \end{bmatrix} \begin{bmatrix} V_1 \\ V_2 \\ V_3 \\ V_4 \end{bmatrix} = \begin{bmatrix} 10 \\ 10 \\ 0 \\ 0 \end{bmatrix}$$

or

$$[A][V] = [B]$$

where [A] is the band matrix, [V] is the column matrix containing the unknown potentials at the free nodes, and [B] is the column matrix of potentials at the fixed nodes. Solving the equations either by matrix inversion or by Gauss elimination, we obtain

$$V_1 = 3.75, \quad V_2 = 3.75, \quad V_3 = 1.25, \quad V_4 = 1.25$$

with MATLAB, $[V] = \text{inv}[A]|[B]$.

EXAMPLE 3.4

Solve Poisson's equation

$$\nabla^2 V = -\frac{\rho_S}{\varepsilon}, \quad 0 \le x, \quad y \le 1$$

and obtain the potential at the grid points shown in Figure 3.11. Assume $\rho_S = x(y - 1)$ nC/m³ and $\epsilon_r = 1.0$. Use the method of successive over-relaxation.

Solution

This problem has an exact analytical solution and is deliberately chosen so that we can verify the numerical results with exact ones, and we can also see how a problem with a complicated analytical solution is easily solved using finite difference method. For the exact solution, we use the superposition theorem and let

$$V = V_1 + V_2$$

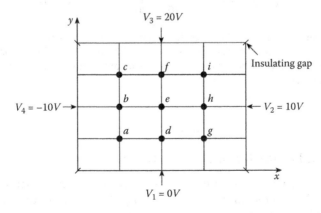

FIGURE 3.11
Solution region for the problem in Example 3.4.

where V_1 is the solution to Laplace's equation $\nabla^2 V_1 = 0$ with the inhomogeneous boundary conditions shown in Figure 3.11 and V_2 is the solution to Poisson's equation $\nabla^2 V_2 = g = -\rho_s/\varepsilon$ subject to homogeneous boundary conditions. From Example 2.1, it is evident that

$$V_1 = V_I + V_{II} + V_{III} + V_{IV}$$

where V_I to V_{IV} are defined by Equations 2.53 through 2.56. V_2 can be obtained by the series expansion method of Section 2.7. From Example 2.12,

$$V_2 = \sum_{m=1}^{\infty} \sum_{n=1}^{\infty} A_{mn} \sin \frac{m\pi x}{a} \sin \frac{n\pi y}{b}$$

where

$$A_{mn} = \int_0^a \int_0^b g(x,y) \sin \frac{m\pi x}{a} \sin \frac{n\pi y}{b} \, dx \, dy$$

$$= \frac{\left[1.0 - \frac{1}{b}[1-(-1)^n]\right]}{[(m\pi/a)^2 + (n\pi/b)^2]} \cdot \frac{(-1)^{m+n} 144ab}{mn\pi},$$

$a = b = 1$, and $g(x,y) = -x(y-1) \cdot 10^{-9}/\varepsilon_o$.

For the finite difference solution, it can be shown that in a rectangular region, the optimum over-relaxation factor is given by the smaller root of the quadratic equation [7]

$$t^2 \omega^2 - 16\omega + 16 = 0$$

where $t = \cos(\pi/N_x) + \cos(\pi/N_y)$ and N_x and N_y are the number of intervals along x and y axes, respectively. Hence,

$$\omega = \frac{8 - \sqrt{64 - 16t^2}}{t^2}$$

We try three cases of $N_x = N_y = 4$, 12, and 20 so that $\Delta x = \Delta y = h = 1/4$, 1/12, and 1/20, respectively. Also, we set

$$g(x,y) = -\frac{\rho_s}{\varepsilon} = -\frac{x(y-1)\times 10^{-9}}{10^{-9}/36\pi} = -36\pi x(y-1)$$

Figure 3.12 presents the MATLAB code for the solution of this problem. The potentials at the free nodes for the different cases of h are shown in Table 3.6. Notice that as the mesh size h reduces, the solution becomes more accurate, but it takes more iterations for the same tolerance.

3.6 Accuracy and Stability of FD Solutions

The question of accuracy and stability of numerical methods is extremely important if our solution is to be reliable and useful. Accuracy has to do with the closeness of the approximate solution to exact solutions (assuming they exist). Stability is the requirement that the scheme does not increase the magnitude of the solution with increase in time.

The analysis of errors in numerical schemes is important because it indicates where errors come from and how to minimize them. There are three sources of errors that are nearly unavoidable in numerical solution of physical problems [8]:

1. Modeling errors,
2. Truncation (or discretization) errors, and
3. Roundoff errors.

```
%%%%%%%%%%%%%%%%%%%%%%%%%%%%%%%%%%%%%%%%%%%%%%%%%%%%%%%
%     FINITE DIFFERENCE SOLUTION OF POISSON'S EQUATION:
%                    Vxx + Vyy = G
%     USING THE METHOD OF SUCCESSIVE OVER-RELAXATION
%
%     NX    :  NO. OF INTERVALS ALONG X-AXIS
%     NY    :  NO. OF INTERVALS ALONG Y-AXIS
%     A X B :  DIMENSION OF THE SOLUTION REGION
%     V(I,J) :  POTENTIAL AT GRID POINT (X,Y) = H*(I,J)
%                 WHERE I = 0,1,...,NX, J = 0,1,....,NY
%     H     :  MESH SIZE
%     ****************************************************

      % SPECIFY BOUNDARY VALUES AND NECESSARY PARAMETERS
      A=1;B=1;
      V1=0;V2=10;V3=20;V4=-10;
      NX= 20; %4 12 20
      NY= NX;
      H = A/NX;
      % SET INITIAL GUESS EQUAL TO ZEROS OR TO AVERAGE OF FIXED VALUES

      for I=1:NX-1
          for J=1:NY-1
              V(I+1,J+1)=(V1 + V2 + V3 + V4)/4.0;
          end
      end
      % SET POTENTIALS AT FIXED NODES
      for I = 1:NX-1
          V(I+1,1)=V1;
          V(I+1,NY+1)=V3;
      end
      for J=1:NY-1
          V(1,J+1)=V4;
          V(NX+1,J+1)=V2;
      end
      V(1,1)=(V1 + V4)/2.0;
      V(NX+1,1)=(V1 + V2)/2.0;
      V(1,NY+1)=(V3 + V4)/2.0;
      V(NX+1,NY+1)=(V2 + V3)/2.0;
      % FIND THE OPTIMUM OVER-RELAXATION FACTOR
      T = cos(pi/NX) + cos(pi/NY);
      W = ( 8 - sqrt(64 - 16*T^2))/(T^2);
      disp(['SOR Factor Omega = ',num2str(W)])
      W4 = W/4;
      % ITERATION BEGINS
      NCOUNT = 0;

      loop = 1;
      while loop == 1;
          RMIN = 0;
          for I =1:NX-1
              X = H*I;
              for J = 1:NY-1
                  Y = H*J;
                  G = -36.0*pi*X*(Y - 1.0);
                  R = W4*( V(I+2,J+1) + V(I,J+1) + V(I+1,J+2) + V(I+1,J)-
4.0*V(I+1,J+1) - G*H*H  );
```

FIGURE 3.12
MATLAB code for Example 3.4.

(Continued)

```
                        RMIN = RMIN + abs(R);
                        V(I+1,J+1) =  V(I+1,J+1) + R;
                    end
            end

            RMIN = RMIN/(NX*NY);
            if(RMIN>=0.0001)
                NCOUNT = NCOUNT + 1;
                if(NCOUNT>100)
                    loop = 0;
                    disp('SOLUTION DOES NOT CONVERGE IN 100 ITERATIONS')
                end
            else
                %Then RMIN is less than .0001 and then solution has converged
                loop = 0;
                disp(['Solution Converges in ',num2str(NCOUNT),' iterations'])
                disp(['h = ', num2str(H)])
            end
        end

Vnum = V;

%Grab original points a through i
        abc = zeros(1,9);
        a_tic = 1;
        vec = [0:H:1];
        for ii = .25:.25:.75
            for jj = .25:.25:.75
                xind = find(vec==ii);
                yind = find(vec==jj);
                %disp([xind,yind])
                abc(a_tic) = Vnum(xind,yind);
                a_tic = a_tic + 1;
            end
        end


%     OUTPUT THE FINITIE DIFFERENCE APPROX. RESULTS


%     ------------------------------------------------------------
%     CALCULATE THE EXACT SOLUTION
%
%     POISSON'S EQUATION WITH HOMOGENEOUS BOUNDARY CONDITIONS
%     SOLVED BY SERIES EXPANSION
%
    for I =1:NX-1
        X = H*I;
        for J = 1:NY-1
            Y = H*J;
            SUM = 0;
            for M = 1:10    % TAKE ONLY 10 TERMS OF THE SERIES
                FM = M;
                for N = 1:10
                    FN = N;
                    FACTOR1 = (FM*pi/A)^2  +  (FN*pi/B)^2;
                    FACTOR2 = ( (-1)^(M+N) )*144*A*B/(pi*FM*FN);
                    FACTOR3 = 1 - (1 - (-1)^N)/B;
                    FACTOR = FACTOR2*FACTOR3/FACTOR1;
                    SUM = SUM + FACTOR*sin(FM*pi*X/A)*sin(FN*pi*Y/B);
                end
            end
            VH = SUM;
```

FIGURE 3.12 (Continued)
MATLAB code for Example 3.4.   *(Continued)*

```
%       LAPLACE'S EQUATION WITH INHOMOGENEOUS BOUNDARY CONDITIONS
%       SOLVED USING THE METHOD OF SEPARATION OF VARIABLES

                C1=4*V1/pi;
                C2=4*V2/pi;
                C3=4*V3/pi;
                C4=4*V4/pi;
                SUM=0;
                for K =1:10   % TAKE ONLY 10 TERMS OF THE SERIES
                    N=2*K-1;
                    AN=N;
                    A1=sin(AN*pi*X/B);
                    A2=sinh(AN*pi*(A-Y)/B);
                    A3=AN*sinh(AN*pi*A/B);
                    TERM1=C1*A1*A2/A3;
                    B1=sinh(AN*pi*X/A);
                    B2=sin(AN*pi*Y/A);
                    B3=AN*sinh(AN*pi*B/A);
                    TERM2=C2*B1*B2/B3;
                    D1=sin(AN*pi*X/B);
                    D2=sinh(AN*pi*Y/B);
                    D3=AN*sinh(AN*pi*A/B);
                    TERM3=C3*D1*D2/D3;
                    E1=sinh(AN*pi*(B-X)/A);
                    E2=sin(AN*pi*Y/A);
                    E3=AN*sinh(AN*pi*B/A);
                    TERM4=C4*E1*E2/E3;
                    TERM = TERM1 + TERM2 + TERM3 + TERM4;
                    SUM=SUM + TERM;
                end
                VI = SUM;
                Vexact(I+1,J+1) = VH + VI;
        end
    end

%Grab original points a through i
        abc2 = zeros(1,9);
        a_tic = 1;
        vec = [0:H:1];
        for ii = .25:.25:.75
            for jj = .25:.25:.75
                xind = find(vec==ii);
                yind = find(vec==jj);
                %disp([xind,yind])
                abc2(a_tic) = Vexact(xind,yind);
                a_tic = a_tic + 1;
            end
        end

figure(1),
        imagesc(flipud(Vnum')),
        colorbar
        ylabel('y'),       xlabel('x')
        title('Example 3.4: Poisson PDE')


format short g
disp('    numerical     exact')
disp([abc' abc2'])
```

FIGURE 3.12 (Continued)
MATLAB code for Example 3.4.

Each of these error types will affect accuracy and therefore degrade the solution.

The modeling errors are due to several assumptions made in arriving at the mathematical model. For example, a nonlinear system may be represented by a linear PDE. Truncation errors arise from the fact that in numerical analysis, we can deal only with a finite number of terms from processes which are usually described by infinite series. For example, in

TABLE 3.6

Successive Over-Relaxation Solution of Example 3.4

Node	$h = 1/4$ $\omega_{opt} = 1.171$ 8 iterations	$h = 1/12$ $\omega_{opt} = 1.729$ 26 iterations	$h = 1/20$ $\omega_{opt} = 1.729$ 43 iterations	Exact Solution
a	−3.247	−3.409	−3.424	−3.429
b	−1.703	−1.982	−2.012	−2.029
c	4.305	4.279	4.277	4.277
d	−0.0393	−0.0961	−0.1087	−0.1182
e	3.012	2.928	2.921	2.913
f	9.368	9.556	9.578	9.593
g	3.044	2.921	2.909	2.902
h	6.111	6.072	6.069	6.065
i	11.04	11.12	11.23	11.13

deriving finite difference schemes, some higher-order terms in the Taylor series expansion were neglected, thereby introducing truncation error. Truncation errors may be reduced by using finer meshes, that is, by reducing the mesh size h and time increment Δt. Alternatively, truncation errors may be reduced by using a large number of terms in the series expansion of derivatives, that is, by using higher-order approximations. However, care must be exercised in applying higher-order approximations. Instability may result if we apply a difference equation of an order higher than the PDE being examined. These higher-order difference equations may introduce "spurious solutions."

Roundoff errors reflect the fact that computations can be done only with a finite precision on a computer. This unavoidable source of errors is due to the limited size of registers in the arithmetic unit of the computer. Roundoff errors can be minimized by the use of double-precision arithmetic. The only way to avoid roundoff errors completely is to code all operations using integer arithmetic. This is hardly possible in most practical situations.

Although it has been noted that reducing the mesh size h will increase accuracy, it is not possible to indefinitely reduce h. Decreasing the truncation error by using a finer mesh may result in increasing the roundoff error due to the increased number of arithmetic operations. A point is reached where the minimum total error occurs for any particular algorithm using any given word length [9]. This is illustrated in Figure 3.13. The concern about accuracy leads us to question whether the finite difference solution can grow unbounded, a property termed the instability of the difference scheme. A numerical algorithm is said to be stable if a small error at any stage produces a smaller cumulative error. It is unstable otherwise. The consequence of instability (producing unbonded solution) is disastrous. To determine whether a finite difference scheme is stable, we define an error, ϵ^n, which occurs at time step n, assuming that there is one independent variable. We define the amplification of this error at time step $n + 1$ as

$$\epsilon^{n+1} = g\epsilon^n \tag{3.37}$$

where g is known as the *amplification factor*. In more complex situations, we have two or more independent variables, and Equation 3.37 becomes

$$[\epsilon]^{n+1} = [G][\epsilon]^n \tag{3.38}$$

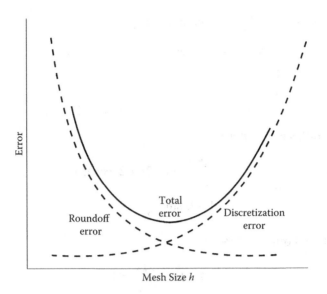

FIGURE 3.13
Error as a function of the mesh size.

where $[G]$ is the amplification matrix. For the stability of the difference scheme, it is required that Equation 3.37 satisfies

$$|\epsilon^{n+1}| \leq |\epsilon^n|$$

or

$$|g| \leq 1 \tag{3.39a}$$

For the case in Equation 3.38,

$$\|G\| \leq 1 \tag{3.39b}$$

One useful and simple method of finding a stability criterion for a difference scheme is to construct a Fourier analysis of the difference equation and thereby derive the amplification factor. We illustrate this technique, known as *von Neumann's method* [4,5,7,10], by considering the explicit scheme of Equation 3.13:

$$\Phi_i^{n+1} = (1-2r)\Phi_i^n + r(\Phi_{i+1}^n + \Phi_{i-1}^n) \tag{3.40}$$

where $r = \Delta t/k(\Delta x)^2$. We have changed our usual notation so that we can use $j = \sqrt{-1}$ in the Fourier series. Let the solution be

$$\Phi_i^n = \sum A^n(t)e^{jkix}, \quad 0 \leq x \leq 1 \tag{3.41a}$$

where k is the wave number. Since the differential equation (3.10) approximated by Equation 3.13 is linear, we need consider only a Fourier mode, that is,

$$\Phi_i^n = A^n(t)e^{jkix} \tag{3.41b}$$

Substituting Equation 3.41b into Equation 3.40 gives

$$A^{n+1}e^{jkix} = (1 - 2r)A^n e^{jkix} + r\left(e^{jkx} + e^{-jkx}\right) A^n e^{jkix}$$

or

$$A^{n+1} = A^n[1 - 2r + 2r \cos kx] \tag{3.42}$$

Hence, the amplification factor is

$$g = \frac{A^{n+1}}{A^n} = 1 - 2r + 2r \cos kx$$

$$= 1 - 4r \sin^2 \frac{kx}{2} \tag{3.43}$$

In order to satisfy Equation 3.39a,

$$\left| 1 - 4r \sin^2 \frac{kx}{2} \right| \leq 1$$

Since this condition must hold for every wave number k, we take the maximum value of the sine function so that

$$1 - 4r \geq -1 \quad \text{and} \quad r \geq 0$$

or

$$r \geq \frac{1}{2} \quad \text{and} \quad r \geq 0$$

Of course, $r = 0$ implies $\Delta t = 0$, which is impractical. Thus,

$$0 < r \leq \frac{1}{2} \tag{3.44}$$

EXAMPLE 3.5

For the finite difference scheme of Equation 3.20, use the von Neumann approach to determine the stability condition.

Solution

We assume a trial solution of the form

$$\Phi_i^n = A^n e^{jkix}$$

Substituting this into Equation 3.20 results in

$$A^{n+1}e^{jkix} = 2(1 - r) A^n e^{jkix} + r\left(e^{jkx} + e^{-jkx}\right) A^n e^{jkix} - A^{n-1}e^{jkix}$$

or

$$A^{n+1} = A^n \left[2(1 - r) + 2r \cos kx\right] - A^{n-1} \tag{3.45}$$

In terms of $g = A^{n+1}/A^n$, Equation 3.45 becomes

$$g^2 - 2pg + 1 = 0 \tag{3.46}$$

where $p = 1 - 2r \sin^2 kx/2$. The quadratic equation 3.46 has solutions

$$g_1 = p + [p^2 - 1]^{1/2}, \quad g_2 = p - [p^2 - 1]^{1/2}$$

For $|g_i| \leq 1$, where $i = 1, 2$, p must lie between 1 and -1, that is, $-1 \leq p \leq 1$ or

$$-1 \leq 1 - 2r \sin^2 \frac{kx}{2} \leq 1$$

which implies that $r \leq 1$ or $u\,\Delta t \leq \Delta x$ for stability. This idea can be extended to show that the stability condition for two-dimensional wave equation is $u\Delta t/h < \dfrac{1}{\sqrt{2}}$, where $h = \Delta x = \Delta y$.

3.7 Practical Applications I: Guided Structures

The finite difference method has been applied successfully to solve many EM-related problems. Besides those simple examples we have considered earlier in this chapter, the method has been applied to diverse problems [11] including

- Transmission-line problems [12–21],
- Waveguides [21–26],
- Microwave circuit [27–30],
- EM penetration and scattering problems [31,32],
- EM pulse (EMP) problems [33],
- EM exploration of minerals [34], and
- EM energy deposition in human bodies [35,36].

It is practically impossible to cover all those applications within the limited scope of this chapter. In this section, we consider the relatively easier problems of transmission lines and waveguides while the problems of penetration and scattering of EM waves will be treated in the next section. Other applications utilize basically similar techniques.

3.7.1 Transmission Lines

The finite difference techniques are suited for computing the characteristic impedance, phase velocity, and attenuation of several transmission lines—polygonal lines, shielded strip lines, coupled strip lines, microstrip lines, coaxial lines, and rectangular lines [12–19]. The knowledge of the basic parameters of these lines is of paramount importance in the design of microwave circuits.

For concreteness, consider the microstrip line shown in Figure 3.14a. The geometry in Figure 3.14a is deliberately selected to be able to illustrate how one accounts for discrete inhomogeneities (i.e., homogeneous media separated by interfaces) and lines of symmetry using a finite difference technique. The techniques presented are equally applicable to other

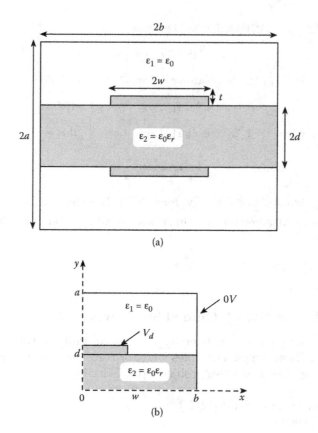

(a)

(b)

FIGURE 3.14
(a) Shielded double-strip line with partial dielectric support; (b) problem in (a) simplified by making full use of symmetry.

lines. Assuming that the mode is TEM, having components of neither **E** nor **H** fields in the direction of propagation, the fields obey Laplace's equation over the line cross section. The TEM mode assumption provides good approximations if the line dimensions are much smaller than half a wavelength, which means that the operating frequency is far below cutoff frequency for all higher-order modes [16]. Also due to biaxial symmetry about the two axes only one quarter of the cross section needs to be considered as shown in Figure 3.14b.

The finite difference approximation of Laplace's equation, $\nabla^2 V = 0$, was derived in Equation 3.31, namely,

$$V(i,j) = \frac{1}{4}[V(i+1,j) + V(i-1,j) + V(i,j+1) + V(i,j-1)] \tag{3.47}$$

For the sake of conciseness, let us denote

$$\begin{aligned}
V_o &= V(i,\ j) \\
V_1 &= V(i,\ j+1) \\
V_2 &= V(i-1,\ j) \\
V_3 &= V(i,\ j-1) \\
V_4 &= V(i+1,\ j)
\end{aligned} \tag{3.48}$$

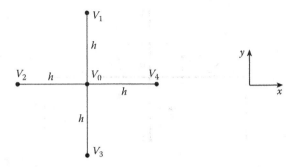

FIGURE 3.15
Computation molecule for Laplace's equation.

so that Equation 3.47 becomes

$$\boxed{V_o = \frac{1}{4}[V_1 + V_2 + V_3 + V_4]}$$ (3.49)

with the computation molecule shown in Figure 3.15. Equation 3.49 is the general formula to be applied to all free nodes in the free space and dielectric region of Figure 3.14b.

On the dielectric boundary, the boundary condition,

$$D_{1n} = D_{2n},$$ (3.50)

must be imposed. We recall that this condition is based on Gauss's law for the electric field, that is,

$$\int_\ell \mathbf{D} \cdot d\mathbf{l} = \oint_\ell \varepsilon \mathbf{E} \cdot d\mathbf{l} = Q_{\text{enc}} = 0$$ (3.51)

since no free charge is deliberately placed on the dielectric boundary. Substituting $\mathbf{E} = -\nabla V$ in Equation 3.51 gives

$$0 = \oint_\ell \varepsilon \nabla V \cdot d\mathbf{l} = \oint_\ell \varepsilon \frac{\partial V}{\partial n} dl$$ (3.52)

where $\partial V/\partial n$ denotes the derivative of V normal to the contour ℓ. Applying Equation 3.52 to the interface in Figure 3.16 yields

$$0 = \varepsilon_1 \frac{(V_1 - V_0)}{h} h + \varepsilon_1 \frac{(V_2 - V_0)}{h} \frac{h}{2} + \varepsilon_2 \frac{(V_2 - V_0)}{h} \frac{h}{2}$$
$$+ \varepsilon_2 \frac{(V_3 - V_0)}{h} h + \varepsilon_2 \frac{(V_4 - V_0)}{h} \frac{h}{2} + \varepsilon_1 \frac{(V_4 - V_0)}{h} \frac{h}{2}$$

Rearranging the terms,

$$2(\varepsilon_1 + \varepsilon_2)V_0 = \varepsilon_1 V_1 + \varepsilon_2 V_3 + \frac{(\varepsilon_1 + \varepsilon_2)}{2}(V_2 + V_4)$$

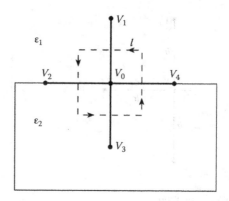

FIGURE 3.16
Interface between media of dielectric permittivities ϵ_1 and ϵ_2.

or

$$V_0 = \frac{\varepsilon_1}{2(\varepsilon_1 + \varepsilon_2)} V_1 + \frac{\varepsilon_2}{2(\varepsilon_1 + \varepsilon_2)} V_3 + \frac{1}{4} V_2 + \frac{1}{4} V_4 \qquad (3.53)$$

This is the finite difference equivalent of the boundary condition in Equation 3.50. Notice that the discrete inhomogeneity does not affect points 2 and 4 on the boundary but affects points 1 and 3 in proportion to their corresponding permittivities. Also note that when $\epsilon_1 = \epsilon_2$, Equation 3.53 reduces to Equation 3.49.

On the line of symmetry, we impose the condition

$$\frac{\partial V}{\partial n} = 0 \qquad (3.54)$$

This implies that on the line of symmetry along the y-axis ($x = 0$ or $i = 0$), $\frac{\partial V}{\partial x} = (V_4 - V_2)/2h = 0$ or $V_2 = V_4$ so that Equation 3.49 becomes

$$V_o = \frac{1}{4}[V_1 + V_3 + 2V_4] \qquad (3.55a)$$

or

$$V(0, j) = \frac{1}{4}[V(0, j+1) + V(0, j-1) + 2V(1, j)] \qquad (3.55b)$$

On the line of symmetry along the x-axis ($y = 0$ or $j = 0$), $\frac{\partial V}{\partial y} = (V_1 - V_3)/2h = 0$ or $V_3 = V_1$ so that

$$V_o = \frac{1}{4}[2V_1 + V_2 + V_4] \qquad (3.56a)$$

or

$$V(i, 0) = \frac{1}{4}[2V(i, 1) + V(i-1, 0) + V(i+1, 0)] \qquad (3.56b)$$

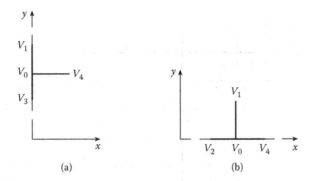

FIGURE 3.17
Computation molecule used for satisfying symmetry conditions: (a) $\partial V/\partial x = 0$, (b) $\partial V/\partial y = 0$.

The computation molecules for Equations 3.55 and 3.56 are displayed in Figure 3.17.

By setting the potential at the fixed nodes equal to their prescribed values and applying Equations 3.49, 3.53, 3.55, and 3.56 to the free nodes according to the band matrix or iterative methods discussed in Section 3.5, the potential at the free nodes can be determined. Once this is accomplished, the quantities of interest can be calculated.

The characteristic impedance Z_o and phase velocity u of the line are defined as

$$Z_o = \sqrt{\frac{L}{C}} \tag{3.57a}$$

$$u = \frac{1}{\sqrt{LC}} \tag{3.57b}$$

where L and C are the inductance and capacitance per unit length, respectively. If the dielectric medium is nonmagnetic ($\mu = \mu_o$), the characteristic impedance Z_{oo} and phase velocity u_o with the dielectric removed (i.e., the line is air-filled) are given by

$$Z_{oo} = \sqrt{\frac{L}{C_o}} \tag{3.58a}$$

$$u_o = \frac{1}{\sqrt{LC_o}} \tag{3.58b}$$

where C_o is the capacitance per unit length without the dielectric. Combining Equations 3.57 and 3.58 yields

$$Z_o = \frac{1}{u_o\sqrt{CC_o}} = \frac{1}{uC} \tag{3.59a}$$

$$u = u_o\sqrt{\frac{C_o}{C}} = \frac{u_o}{\sqrt{\varepsilon_{\text{eff}}}} \tag{3.59b}$$

$$\varepsilon_{\text{eff}} = \frac{C}{C_o} \tag{3.59c}$$

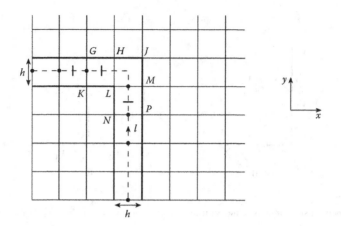

FIGURE 3.18
Rectangular path ℓ used in calculating charge enclosed.

where $u_o = c = 3 \times 10^8$ m/s, the speed of light in free space, and ϵ_{eff} is the effective dielectric constant. Thus to find Z_o and u for an inhomogeneous medium require calculating the capacitance per unit length of the structure, with and without the dielectric substrate.

If V_d is the potential difference between the inner and the outer conductors,

$$C = \frac{4Q}{V_d},$$
(3.60)

so that the problem is reduced to finding the charge per unit length Q. (The factor 4 is needed since we are working on only one quarter of the cross section.) To find Q, we apply Gauss's law to a closed path ℓ enclosing the inner conductor. We may select ℓ as the rectangular path between two adjacent rectangles as shown in Figure 3.18.

$$Q = \oint_\ell \mathbf{D} \cdot d\mathbf{l} = \oint_\ell \varepsilon \frac{\partial V}{\partial n} d\ell$$

$$= \varepsilon \left(\frac{V_P - V_N}{\Delta x} \right) \Delta y + \varepsilon \left(\frac{V_M - V_L}{\Delta x} \right) \Delta y + \varepsilon \left(\frac{V_H - V_L}{\Delta y} \right) \Delta x$$

$$+ \varepsilon \left(\frac{V_G - V_K}{\Delta y} \right) \Delta x + \cdots$$
(3.61)

Since $\Delta x = \Delta y = h$,

$$Q = (\epsilon V_P + \epsilon V_M + \epsilon V_H + \epsilon V_G + \cdots) - (\epsilon V_N + 2\epsilon V_L + \epsilon V_K + \cdots)$$

or

$$Q = \varepsilon_o \left[\sum \varepsilon_{ri} V_i \quad \text{for nodes } i \text{ on outer rectangle GHJMP} \right.$$
$$\left. \text{with corners (such as J) not counted} \right]$$
$$- \varepsilon_o \left[\sum \varepsilon_{ri} V_i \quad \text{for nodes } i \text{ on inner rectangle KLN} \right.$$
$$\left. \text{with corners (such as L) counted twice} \right],$$
(3.62)

where V_i and ϵ_{ri} are the potential and dielectric constant at the *i*th node. If *i* is on the dielectric interface, $\epsilon_{ri} = (\epsilon_{r1} + \epsilon_{r2})/2$. Also if *i* is on the line of symmetry, we use $V_i/2$ instead of V_i to avoid including V_i twice in Equation 3.60, where factor 4 is applied. We also find

$$C_o = 4Q_o/V_d \tag{3.63}$$

where Q_o is obtained by removing the dielectric, finding V_i at the free nodes and then using Equation 3.62 with $\epsilon_{ri} = 1$ at all nodes. Once Q and Q_o are calculated, we obtain C and C_o from Equations 3.60 and 3.63 and Z_o and u from Equation 3.59.

An outline of the procedure is given below.

1. Calculate V (with the dielectric space replaced by free space) using Equations 3.49, 3.53, 3.55, and 3.56.

2. Determine Q using Equation 3.62.

3. Find $C_o = \dfrac{4Q_o}{V_d}$.

4. Repeat steps 1 and 2 (with the dielectric space) and find $C = \dfrac{4Q}{V_d}$.

5. Finally, calculate $Z_o = \dfrac{1}{c\sqrt{CC_o}}, c = 3 \times 10^8 \text{m/s}$.

The attenuation of the line can be calculated by following similar procedure outlined in References 14,20,21. The procedure for handling boundaries at infinity and that for boundary singularities in finite difference analysis are discussed in References 37,38.

3.7.2 Waveguides

The solution of waveguide problems is well suited for finite difference schemes because the solution region is closed. This amounts to solving the Helmholtz or wave equation

$$\nabla^2 \Phi + k^2 \Phi = 0 \tag{3.64}$$

where $\Phi = E_z$ for TM modes or $\Phi = H_z$ for TE modes, while k is the wave number given by

$$k^2 = \omega^2 \mu \epsilon - \beta^2 \tag{3.65}$$

The permittivity ϵ of the dielectric medium can be real for a lossless medium or complex for a lossy medium. We consider all fields to vary with time and axial distance as $\exp j(\omega t - \beta z)$. In the eigenvalue problem of Equation 3.64, both k and Φ are to be determined. The cutoff wavelength is $\lambda_c = 2\pi/k_c$. For each value of the cutoff wave number k_c, there is a solution for the eigenfunction Φ_i, which represents the field configuration of a propagating mode.

To apply the finite difference method, we discretize the cross section of the waveguide by a suitable square mesh. Applying Equation 3.29 to Equation 3.64 gives

$$\boxed{\Phi(i+1,j) + \Phi(i-1,j) + \Phi(i,j+1) + \Phi(i,j-1) - (4 - h^2 k^2)\Phi(i,j) = 0} \tag{3.66}$$

where $\Delta x = \Delta y = h$ is the mesh size. Equation 3.66 applies to all the free or interior nodes. At the boundary points, we apply Dirichlet condition ($\Phi = 0$) for the TM modes and

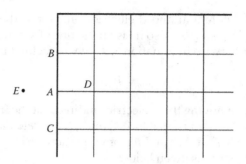

FIGURE 3.19
Finite difference mesh for a waveguide.

Neumann condition $(\partial\Phi/\partial n = 0)$ for the TE modes. This implies that at point A in Figure 3.19, for example,

$$\Phi_A = 0 \tag{3.67}$$

for TM modes. At point A, $\partial\Phi/\partial n = 0$ implies that $\Phi_D = \Phi_E$ so that Equation 3.64 becomes

$$\Phi_B + \Phi_C + 2\Phi_D - (4 - h^2k^2)\,\Phi_A = 0 \tag{3.68}$$

for TE modes. By applying Equation 3.66 and either Equation 3.67 or Equation 3.68 to all mesh points in the waveguide cross section, we obtain m simultaneous equations involving the m unknowns $(\Phi_1, \Phi_2, ..., \Phi_m)$. These simultaneous equations may be conveniently cast into the matrix equation

$$(A - \lambda I)\Phi = 0 \tag{3.69a}$$

or

$$A\Phi = \lambda\Phi \tag{3.69b}$$

where A is an $m \times m$ band matrix of known integer elements, I is an identity matrix, $\Phi = (\Phi_1, \Phi_2, ..., \Phi_m)$ is the eigenvector, and

$$\lambda = (kh^2) = \left(\frac{2\pi h}{\lambda_c}\right)^2 \tag{3.70}$$

is the eigenvalue. There are several ways of determining λ and the corresponding Φ. We consider two of these options.

The first option is the *direct method*. Equation 3.69 can be satisfied only if the determinant of $(A - \lambda I)$ vanishes, that is,

$$|A - \lambda I| = 0 \tag{3.71}$$

This results in a polynomial in λ, which can be solved [39] for the various eigenvalues λ. For each λ, we obtain the corresponding Φ from Equation 3.66. This method requires storing the relevant matrix elements and does not take advantage of the fact that matrix A is sparse. In favor of the method is the fact that a computer subroutine usually exists (see Reference 40 or Appendix C.4) that solves the eigenvalue problem in Equation 3.71 and that determines all the eigenvalues of the matrix. These eigenvalues give the dominant and higher modes of the waveguide, although accuracy deteriorates rapidly with mode number.

The second option is the *iterative method*. In this case, the matrix elements are usually generated rather than stored. We begin with $\Phi_1 = \Phi_2 = \cdots = \Phi_m = 1$ and a guessed value for k. The field Φ_{ij}^{k+1} at the (i, j)th node in the $(k + 1)$th iteration is obtained from its known value in the kth iteration using

$$\Phi^{k+1}(i, j) = \Phi^k(i, j) + \frac{\omega R_{ij}}{(4 - h^2 k^2)} \tag{3.72}$$

where ω is the acceleration factor, $1 < \omega < 2$, and R_{ij} is the residual at the (i, j)th node given by

$$R_{ij} = \Phi(i, j+1) + \Phi(i, j-1) + \Phi(i+1, j)$$
$$+ \Phi(i-1, j) - (4 - h^2 k^2)\Phi(i, j) \tag{3.73}$$

After three or four scans of the complete mesh using Equation 3.73, the value of $\lambda = h^2 k^2$ should be updated using Raleigh formula

$$k^2 = \frac{-\displaystyle\int_S \Phi \nabla^2 \Phi \, dS}{\displaystyle\int_S \Phi^2 \, dS} \tag{3.74}$$

The finite difference equivalent of Equation 3.74 is

$$k^2 = \frac{-\sum_{i=1} \sum_{j=1} \Phi(i, j)[\Phi(i+1, j) + \Phi(i-1, j) + \Phi(i, j+1) + \Phi(i, j-1) - 4\Phi(i, j)]}{h^2 \sum_{i=1} \sum_{j=1} \Phi^2(i, j)} \tag{3.75}$$

where Φ_s are the latest field values after three or four scans of the mesh and the summation is carried out over all points in the mesh. The new value of k obtained from Equation 3.75 is now used in applying Equation 3.72 over the mesh for another three or four times to give more accurate field values, which are again substituted into Equation 3.75 to update k. This process is continued until the difference between consecutive values of k is within a specified acceptable tolerance.

If the first option is to be applied, matrix A must first be found. To obtain matrix A is not easy. Assuming TM modes, one way of calculating A is to number the free nodes from left to right, bottom to top, starting from the left-hand corner as shown typically in Figure 3.20. If there are n_x and n_y divisions along the x and y directions, the number of free nodes is

$$n_f = (n_x - 1)(n_y - 1) \tag{3.76}$$

Each free node must be assigned two sets of numbers, one to correspond to m in Φ_m and the other to correspond to (i, j) in $\Phi(i, j)$. An array $NL(i, j) = m$, $i = 1, 2, ..., n_x - 1$, $j = 1, 2, ..., n_y - 1$ is easily developed to relate the two numbering schemes. To determine the

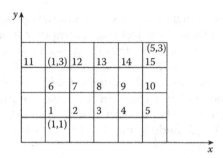

FIGURE 3.20
Relating node numbering schemes for $n_x = 6$, $n_y = 4$.

value of element A_{mn}, we search $NL(i, j)$ to find (i_m, j_m) and (i_n, j_n), which are the values of (i, j) corresponding to nodes m and n, respectively. With these ideas, we obtain

$$A_{mn} = \begin{cases} 4, & m = n \\ -1, & i_m = i_n, \quad j_m = j_n + 1 \\ -1, & i_m = i_n, \quad j_m = j_n - 1 \\ -1, & i_m = i_n + 1, \quad j_m = j_n \\ -1, & i_m = i_n - 1, \quad j_m = j_n \\ 0, & \text{otherwise} \end{cases} \tag{3.77}$$

EXAMPLE 3.6

Calculate Z_o for the microstrip transmission line in Figure 3.14 with

$$a = b = 2.5 \text{ cm}, \quad d = 0.5 \text{ cm}, \quad w = 1 \text{ cm}$$
$$t = 0.001 \text{ cm}, \quad \epsilon_1 = \epsilon_o, \quad \epsilon_2 = 2.35\epsilon_o$$

Solution

This problem is representative of the various types of problems that can be solved using the concepts developed in Section 3.7.1. The computer program in Figure 3.21 was developed based on the five-step procedure outlined above. By specifying the step size h and the number of iterations, the program first sets the potential at all nodes equal to zero. The potential on the outer conductor is set equal to zero, while that on the inner conductor is set to 100 V so that $V_d = 100$. The program finds C_o when the dielectric slab is removed and C when the slab is in place and finally determines Z_o. For a selected h, the number of iterations must be large enough and greater than the number of divisions along x or y direction. Table 3.7 shows some typical results.

3.8 Practical Applications II: Wave Scattering (FDTD)

The finite-difference time-domain (FDTD) formulation of EM field problems is a convenient tool for solving scattering problems. The FDTD method, first introduced by Yee [42] in 1966 and later developed by Taflove and others [31,32,35,43–46], is a direct solution of Maxwell's two time-dependent curl equations. The scheme treats the irradiation of the scatterer as an initial value problem. Our discussion on the FD TD method will cover

- Yee's finite difference algorithm,
- Accuracy and stability,
- Lattice truncation conditions,
- Initial fields, and
- Programming aspects.

Some model examples with MATLAB codes will be provided to illustrate the method.

```
%****************************************************************
%  Using the finite difference method
%  This program calculates the characteristic impedance of the transmission
%  line shown in Figure 3.14.
%****************************************************************

clear all; format compact;

% Output:
%
%    H          NT          Zo
% --------------------------------
%  0.25        700         69.77
%   0.1        500         65.75
%  0.05        500         70.53
%  0.05        700         67.36
%  0.05       1000         65.50

    H = 0.05;
    NT = 1000;

    A = 2.5; B = 2.5; D = 0.5; W = 1.0;

    ER=2.35;
    EO=8.81E-12;
    U=3.0E+8;


    NX = A/H;
    NY = B/H;
    ND = D/H;
    NW = W/H;
    VD = 100.0;

% CALCULATE CHARGE WITH AND WITHOUT DIELECTRIC
    ERR = 1.0;
    for L=1:2
        E1 = EO;
        E2 = EO*ERR;

% INITIALIZATION
        V = zeros(NX+2,NY+2);

% SET POTENTIAL ON INNER CONDUCTOR (FIXED NODES) EQUAL TO VD
        V(2:NW+1,ND+2) = VD;

% CALCULATE POTENTIAL AT FREE NODES
        P1 = E1/(2*(E1 + E2));
        P2 = E2/(2*(E1 + E2));
        for K=1:NT
            for I=0:NX-1
                for J=0:NY-1
                    if( (J==ND)&(I<=NW) )
                        %do nothing
                    elseif (J==ND)
                        % IMPOSE BOUNDARY CONDITION AT THE INTERFACE
                        V(I+2,J+2) = 0.25*(V(I+3,J+2) + V(I+1,J+2)) + ...
                            P1*V(I+2,J+3) + P2*V(I+2,J+1);
```

FIGURE 3.21
MATLAB code for Example 3.6.   (*Continued*)

```
                    elseif(I==0)
                        % IMPOSE SYMMETRY CONDITION ALONG Y-AXIS
                        V(I+2,J+2) = (2*V(I+3,J+2) + V(I+2,J+3) + V(I+2,J+1))/4.0;
                    elseif(J==0)
                        % IMPOSE SYMMETRY CONDITION ALONG X-AXIS
                        V(I+2,J+2) = (V(I+3,J+2) + V(I+1,J+2) + 2*V(I+2,J+3))/4.0;
                    else
                        V(I+2,J+2) = (V(I+3,J+2)+V(I+1,J+2)+V(I+2,J+3)+V(I+2,J+1))/4.0;
                    end
                end
            end
        % Animation of calculation
        % figure(1),imagesc(V),colorbar,title([num2str(K),'/',num2str(NT)])
        % drawnow
        end

% NOW, CALCULATE THE TOTAL CHARGE ENCLOSED IN A
% RECTANGULAR PATH SURROUNDING THE INNER CONDUCTOR
            IOUT = round((NX + NW)/2);
            JOUT = round((NY + ND)/2);
% SUM POTENTIAL ON INNER AND OUTER LOOPS
            for K=1:2
                SUM = E1*sum(V(3:IOUT+1,JOUT+2)) ...
                    + E1*V(2,JOUT+2)/2 + E2*V(IOUT+2,2)/2;
                for J=1:JOUT-1
                    if(J<ND)
                        SUM = SUM + E2*V(IOUT+2,J+2);
                    elseif(J==ND)
                        SUM = SUM + (E1+E2)*V(IOUT+2,J+2)/2;
                    else
                        SUM = SUM + E1*V(IOUT+2,J+2);
                    end
                end
                if K==1
                    SV(1) = SUM;
                end
                IOUT = IOUT - 1;
                JOUT = JOUT - 1;
            end
            SUM = SUM + 2.0*E1*V(IOUT+2,JOUT+2);
            SV(2) = SUM;
            Q(L) = abs( SV(1) - SV(2) );
            ERR = ER;
        end

% FINALLY, CALCULATE Zo
        C0 = 4.0*Q(1)/VD;
        C1 = 4.0*Q(2)/VD;
        Z0 = 1.0/( U*sqrt(C0*C1) );

        disp([H,NT,Z0])
```

FIGURE 3.21 (Continued)
MATLAB code for Example 3.6.

TABLE 3.7

Characteristic Impedance of a Microstrip Line for Example 3.6

h	Number of Iterations	Z_o
0.25	700	69.77
0.1	500	65.75
0.05	500	70.53
0.05	700	67.36
0.05	1000	65.50

Other method [41]: $Z_o = 62.50$.

3.8.1 Yee's Finite Difference Algorithm

In an isotropic medium, Maxwell's equations can be written as

$$\nabla \times \mathbf{E} = -\mu \frac{\partial \mathbf{H}}{\partial t} \tag{3.78a}$$

$$\nabla \times \mathbf{H} = \sigma \mathbf{E} + \varepsilon \frac{\partial \mathbf{E}}{\partial t} \tag{3.78b}$$

The vector Equation 3.78 represents a system of six scalar equations, which can be expressed in rectangular coordinate system as

$$\frac{\partial H_x}{\partial t} = \frac{1}{\mu} \left(\frac{\partial E_y}{\partial z} - \frac{\partial E_z}{\partial y} \right), \tag{3.79a}$$

$$\frac{\partial H_y}{\partial t} = \frac{1}{\mu} \left(\frac{\partial E_z}{\partial x} - \frac{\partial E_x}{\partial z} \right), \tag{3.79b}$$

$$\frac{\partial H_z}{\partial t} = \frac{1}{\mu} \left(\frac{\partial E_x}{\partial y} - \frac{\partial E_y}{\partial x} \right), \tag{3.79c}$$

$$\frac{\partial E_x}{\partial t} = \frac{1}{\varepsilon} \left(\frac{\partial H_z}{\partial y} - \frac{\partial H_y}{\partial z} - \sigma E_x \right), \tag{3.79d}$$

$$\frac{\partial E_y}{\partial t} = \frac{1}{\varepsilon} \left(\frac{\partial H_x}{\partial z} - \frac{\partial H_z}{\partial x} - \sigma E_y \right), \tag{3.79e}$$

$$\frac{\partial E_z}{\partial t} = \frac{1}{\varepsilon} \left(\frac{\partial H_y}{\partial x} - \frac{\partial H_x}{\partial y} - \sigma E_z \right) \tag{3.79f}$$

Following Yee's notation, we define a grid point in the solution region as

$$(i, j, k) \equiv (i\, \Delta x, j\, \Delta y, k\, \Delta z) \tag{3.80}$$

and any function of space and time as

$$F^n(i, j, k) \equiv F(i\delta, j\delta, k\delta, n\, \Delta t) \tag{3.81}$$

where $\delta = \Delta x = \Delta y = \Delta z$ is the space increment, and Δt is the time increment, while i, j, k, and n are integers. Using central finite difference approximation for space and time derivatives that are second-order accurate,

$$\frac{\partial F^n(i, j, k)}{\partial x} = \frac{F^n(i+1/2, j, k) - F^n(i-1/2, j, k)}{\delta} + O(\delta^2) \tag{3.82}$$

$$\frac{\partial F^n(i, j, k)}{\partial t} = \frac{F^{n+1/2}(i, j, k) - F^{n-1/2}(i, j, k)}{\Delta t} + O(\Delta t^2) \tag{3.83}$$

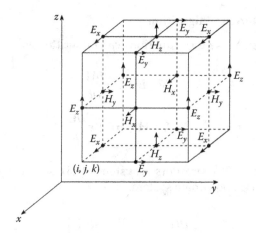

FIGURE 3.22
Positions of the field components in a unit cell of the Yee's lattice.

In applying Equation 3.82 to all the space derivatives in Equation 3.79, Yee positions the components of **E** and **H** about a unit cell of the lattice as shown in Figure 3.22. To incorporate Equation 3.83, the components of **E** and **H** are evaluated at alternate half-time steps. Thus, we obtain the explicit finite difference approximation of Equation 3.79 as

$$
\begin{aligned}
H_x^{n+1/2}(i,j+1/2,k+1/2) = {}& H_x^{n-1/2}(i,j+1/2,k+1/2) + \frac{\delta t}{\mu(i,j+1/2,k+1/2)\delta} \\
& \times \big[E_y^n(i,j+1/2,k+1) - E_y^n(i,j+1/2,k) \\
& + E_z^n(i,j,k+1/2) - E_z^n(i,j+1,k+1/2) \big],
\end{aligned}
\tag{3.84a}
$$

$$
\begin{aligned}
H_y^{n+1/2}(i+1/2,j,k+1/2) = {}& H_y^{n-1/2}(i+1/2,j,k+1/2) + \frac{\delta t}{\mu(i+1/2,j,k+1/2)\delta} \\
& \times \big[E_z^n(i+1,j,k+1/2) - E_z^n(i,j,k+1/2) \\
& + E_x^n(i+1/2,j,k) - E_x^n(i+1/2,j,k+1) \big],
\end{aligned}
\tag{3.84b}
$$

$$
\begin{aligned}
H_z^{n+1/2}(i+1/2,j+1/2,k) = {}& H_z^{n-1/2}(i+1/2,j+1/2,k) + \frac{\delta t}{\mu(i+1/2,j+1/2,k)\delta} \\
& \times \big[E_x^n(i+1/2,j+1,k) - E_x^n(i+1/2,j,k) \\
& + E_y^n(i,j+1/2,k) - E_y^n(i+1,j+1/2,k) \big],
\end{aligned}
\tag{3.84c}
$$

$$
\begin{aligned}
E_x^{n+1}(i+1/2,j,k) = {}& \left(1 - \frac{\sigma(i+1/2,j,k)\delta t}{\varepsilon(i+1/2,j,k)} \right) \cdot E_x^n(i+1/2,j,k) + \frac{\delta t}{\varepsilon(i+1/2,j,k)\delta} \\
& \times \big[H_z^{n+1/2}(i+1/2,j+1/2,k) - H_z^{n+1/2}(i+1/2,j-1/2,k) \\
& + H_y^{n+1/2}(i+1/2,j,k-1/2) - H_y^{n+1/2}(i+1/2,j,k+1/2) \big],
\end{aligned}
\tag{3.84d}
$$

$$E_y^{n+1}(i,j+1/2,k) = \left(1 - \frac{\sigma(i,j+1/2,k)\delta t}{\varepsilon(i,j+1/2,k)}\right) \cdot E_y^n(i,j+1/2,k) + \frac{\delta t}{\epsilon(i,j+1/2,k)\delta}$$
$$\times \left[H_x^{n+1/2}(i,j+1/2,k+1/2) - H_x^{n+1/2}(i,j+1/2,k-1/2)\right.$$
$$\left. + H_z^{n+1/2}(i-1/2,j+1/2,k) - H_z^{n+1/2}(i+1/2,j+1/2,k)\right], \tag{3.84e}$$

$$E_z^{n+1}(i,j,k+1/2) = \left(1 - \frac{\sigma(i,j,k+1/2)\delta t}{\varepsilon(i,j,k+1/2)}\right) \cdot E_z^n(i,j,k+1/2) + \frac{\delta t}{\epsilon(i,j,k+1/2)\delta}$$
$$\times \left[H_y^{n+1/2}(i+1/2,j,k+1/2) - H_y^{n+1/2}(i-1/2,j,k+1/2)\right.$$
$$\left. + H_x^{n+1/2}(i,j-1/2,k+1/2) - H_x^{n+1/2}(i,j+1/2,k+1/2)\right] \tag{3.84f}$$

Notice from Equations 3.84a through 3.84f and Figure 3.22 that the components of **E** and **H** are interlaced within the unit cell and are evaluated at alternate half-time steps. All the field components are present in a quarter of a unit cell as shown typically in Figure 3.23a. Figure 3.23b illustrates typical relations between field components on a plane; this is particularly useful when incorporating boundary conditions. The figure can be inferred from Equation 3.79d or Equation 3.84d. In translating the hyperbolic system of Equations 3.84a through 3.84f into a computer code, one must make sure that, within the same time loop, one type of field component is calculated first and the results obtained are used in calculating another type.

3.8.2 Accuracy and Stability

One factor that dictates the accuracy of FDTD technique is the number of points per wavelength for any given frequency. To ensure the accuracy of the computed results, the spatial increment δ must be small compared to the wavelength (usually $\leq \lambda/10$) or minimum dimension of the scatterer. This amounts to having 10 or more cells per wavelength. To ensure the stability of the finite difference scheme of Equations 3.84a through 3.84f, the time increment Δt must satisfy the following stability condition [43,47]:

$$u_{max}\Delta t \leq \left[\frac{1}{\Delta x^2} + \frac{1}{\Delta y^2} + \frac{1}{\Delta z^2}\right]^{-1/2} \tag{3.85}$$

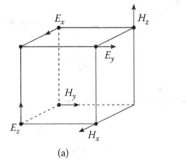
(a)

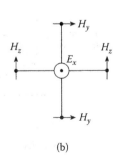
(b)

FIGURE 3.23
Typical relations between field components: (a) within a quarter of a unit cell, (b) in a plane.

where u_{max} is the maximum wave phase velocity within the model. Since we are using a cubic cell with $\Delta x = \Delta y = \Delta z = \delta$, Equation 3.85 becomes

$$\boxed{\frac{u_{max}\Delta t}{\delta} \leq \frac{1}{\sqrt{n}}} \tag{3.86}$$

where n is the number of space dimensions. (n here should not be confused with n in Equation 3.84. The former n refers to the number of dimensions, whereas the latter refers to time.) For practical reasons, it is best to choose the ratio of the time increment to spatial increment as large as possible yet satisfying Equation 3.86.

3.8.3 Lattice Truncation Conditions

A basic difficulty encountered in applying the FDTD method to scattering problems is that the domain in which the field is to be computed is open or unbounded (see Figure 1.3). Since no computer can store an unlimited amount of data, a finite difference scheme over the whole domain is impractical. We must limit the extent of our solution region. In other words, an artificial boundary must be enforced, as in Figure 3.24, to create the numerical illusion of an infinite space. The solution region must be large enough to enclose the scatterer, and suitable boundary conditions on the artificial boundary must be used to simulate the extension of the solution region to infinity. Outer boundary conditions of this type have been called *radiation conditions, absorbing boundary conditions,* or *lattice truncation conditions.* Although several types of boundary conditions have been proposed [48,49], we will only consider those developed by Taflove et al. [43,44].

The lattice truncation conditions developed by Taflove et al. allow excellent overall accuracy and numerical stability even when the lattice truncation planes are positioned

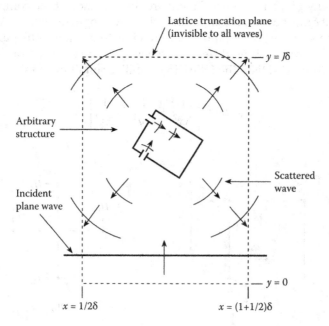

FIGURE 3.24
Solution region with lattice truncation.

no more than 5δ from the surface of the scatterer. The conditions relate in a simple way the values of the field components at the truncation planes to the field components at points one or more δ within the lattice (or solution region).

For simplicity, we first consider one-dimensional wave propagation. Assume waves have only E_z and H_x components and propagate in the $\pm y$ directions. Also assume a time step of $\delta t = \delta y/c$, the maximum allowed by the stability condition of Equation 3.86. If the lattice extends from $y = 0$ to $y = J\,\Delta y$, with E_z component at the end points, the truncation conditions are

$$E_z^n(0) = E_z^{n-1}(1) \tag{3.87a}$$

$$E_z^n(J) = E_z^{n-1}(J-1) \tag{3.87b}$$

With these lattice conditions, all possible $\pm y$-directed waves are absorbed at $y = 0$ and $J\,\Delta y$ without reflection. Equation 3.87 assumes free-space propagation. If we wish to simulate the lattice truncation in a dielectric medium of refractive index m, Equation 3.87 is modified to

$$E_z^n(0) = E_z^{n-m}(1) \tag{3.88a}$$

$$E_z^n(J) = E_z^{n-m}(J-1) \tag{3.88b}$$

For the three-dimensional case, we consider scattered waves having all six field components and propagating in all possible directions. Assume a time step of $\delta t = \delta/2c$, a value which is about 13% lower than the maximum allowed ($\delta t = \delta/\sqrt{3}c$) by Equation 3.86. If the lattice occupies $1/2\delta < x < (I_{max} + 1/2)\delta, 0 < y < J_{max}\delta, 0 < z < K_{max}\delta$, the truncation conditions are [36,44]:

a. Plane $i = 1/2$

$$H_y^n(1/2, j, k+1/2) = \frac{1}{3}\Big[H_y^{n-2}(3/2, j, k-1/2) + H_y^{n-2}(3/2, j, k+1/2)$$
$$+ H_y^{n-2}(3/2, j, k+3/2)\Big], \tag{3.89a}$$

$$H_z^n(1/2, j+1/2, k) = \frac{1}{3}\Big[H_z^{n-2}(3/2, j+1/2, k-1) + H_z^{n-2}(3/2, j+1/2, k)$$
$$+ H_z^{n-2}(3/2, j+1/2, k+1)\Big], \tag{3.89b}$$

b. Plane $i = I_{max} + 1/2$

$$H_y^n(I_{max} + 1/2, j, k+1/2)$$
$$= \frac{1}{3}\Big[H_y^{n-2}(I_{max} - 1/2, j, k-1/2) + H_y^{n-2}(I_{max} - 1/2, j, k+1/2)$$
$$+ H_y^{n-2}(I_{max} - 1/2, j, k+3/2)\Big], \tag{3.89c}$$

$$H_z^n(I_{max} + 1/2, j+1/2, k)$$
$$= \frac{1}{3}\Big[H_z^{n-2}(I_{max} - 1/2, j+1/2, k-1) + H_z^{n-2}(I_{max} - 1/2, j+1/2, k)$$
$$+ H_z^{n-2}(I_{max} - 1/2, j+1/2, k+1)\Big], \tag{3.89d}$$

c. Plane $j = 0$,

$$E_x^n(i+1/2,0,k) = E_x^{n-2}(i+1/2,1,k), \tag{3.89e}$$

$$E_z^n(i,0,k+1/2) = E_z^{n-2}(i,1,k+1/2), \tag{3.89f}$$

d. Plane $j = J_{max}$

$$E_x^n(i+1/2,J_{max},k) = E_x^{n-2}(i+1/2,J_{max}-1,k) \tag{3.89g}$$

$$E_z^n(i,J_{max},k+1/2) = E_z^{n-2}(i,J_{max}-1,k+1/2), \tag{3.89h}$$

e. Plane $k = 0$,

$$E_x^n(i+1/2,j,0) = \frac{1}{3}\left[E_x^{n-2}(i-1/2,j,1) + E_x^{n-2}(i+1/2,j,1) + E_x^{n-2}(i+3/2,j,1)\right], \tag{3.89i}$$

$$H_y^n(i,j+1/2,0) = \frac{1}{3}\left[E_y^{n-2}(i-1,j+1/2,1) + E_y^{n-2}(i,j+1/2,1) + E_y^{n-2}(i+1,j+1/2,1)\right], \tag{3.89j}$$

f. Plane $k = K_{max}$,

$$E_x^n(i+1/2,j,K_{max}) = \frac{1}{3}\left[E_x^{n-2}(i-1/2,j,K_{max}-1) + E_x^{n-2}(i+1/2,j,K_{max}-1)\right.$$
$$\left. + E_x^{n-2}(i+3/2,j,K_{max}-1)\right], \tag{3.89k}$$

$$E_y^n(i,j+1/2,K_{max}) = \frac{1}{3}\left[E_y^{n-2}(i-1,j+1/2,K_{max}-1) + E_y^{n-2}(i,j+1/2,K_{max}-1)\right.$$
$$\left. + E_y^{n-2}(i+1,j+1/2,K_{max}-1)\right], \tag{3.89l}$$

These boundary conditions minimize the reflection of any outgoing waves by simulating the propagation of the wave from the lattice plane adjacent to the lattice truncation plane in a number of time steps corresponding to the propagation delay. The averaging process is used to take into account all possible local angles of incidence of the outgoing wave at the lattice boundary and possible multiple incidences [43]. If the solution region is a dielectric medium of refractive index m rather than free space, we replace the superscript $n - 2$ in Equations 3.89a through 3.89l by $n - m$.

3.8.4 Initial Fields

The initial field components are obtained by simulating either an incident plane wave pulse or single-frequency plane wave. The simulation should not take excessive storage nor cause spurious wave reflections. A desirable plane wave source condition takes into account the scattered fields at the source plane. For the three-dimensional case, a typical wave source condition at plane $y = j_s$ (near $y = 0$) is

$$E_z^n(i,j_s,k+1/2) \leftarrow 1000\sin(2\pi f n\, \delta t) + E_z^n(i,j_s,k+1/2) \tag{3.90}$$

where f is the irradiation frequency. Equation 3.90 is a modification of the algorithm for all points on plane $y = j_s$; the value of the sinusoid is added to the value of E_z^n obtained from Equations 3.84a through 3.84f.

At $t = 0$, the plane wave source of frequency f is assumed to be turned on. The propagation of waves from this source is simulated by time stepping, that is, repeatedly implementing Yee's finite difference algorithm on a lattice of points. The incident wave is tracked as it first propagates to the scatterer and then interacts with it via surface–current excitation, diffusion, penetration, and diffraction. Time stepping is continued until the sinusoidal steady state is achieved at each point. The field envelope, or maximum absolute value, during the final half-wave cycle of time stepping is taken as the magnitude of the phasor of the steady-state field [32,43].

From experience, the number of time steps needed to reach the sinusoidal steady state can be greatly reduced by introducing a small isotropic conductivity σ_{ext} within the solution region exterior to the scatterer. This causes the fields to converge more rapidly to the expected steady-state condition.

3.8.5 Programming Aspects

Since most EM scattering problems involve nonmagnetic media ($\mu_r = 1$), the quantity $\delta t/\mu(i, j, k)\delta$ can be assumed constant for all (i, j, k). The nine multiplications per unit cell per time required by Yee's algorithm of Equations 3.84a through 3.84f can be reduced to six multiplications, thereby reducing computer time. Following Taflove et al. [31,35,44], we define the following constants:

$$R = \delta t/2\epsilon_o, \tag{3.91a}$$

$$R_a = (c\delta t/\delta)^2, \tag{3.91b}$$

$$R_b = \delta t/\mu_o\delta, \tag{3.91c}$$

$$C_a = \frac{1 - R\sigma(m)/\varepsilon_r(m)}{1 + R\sigma(m)/\varepsilon_r(m)}, \tag{3.91d}$$

$$C_b = \frac{R_a}{\varepsilon_r(m) + R\sigma(m)} \tag{3.91e}$$

where $m = \text{MEDIA}(i, j, k)$ is an integer referring to the dielectric or conducting medium type at location (i, j, k). For example, for a solution region comprising three different homogeneous media shown in Figure 3.25, m is assumed to be 1–3. (This m should not be confused with

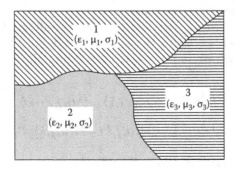

FIGURE 3.25
A typical inhomogeneous solution region with integer m assigned to each medium.

the refractive index of the medium, mentioned earlier.) In addition to the constants in Equation 3.91, we define proportional electric field

$$\tilde{\mathbf{E}} = R_b \mathbf{E} \tag{3.92}$$

Thus, Yee's algorithm is modified and simplified for easy programming as [50,51]:

$$H_x^n(i,j,k) = H_x^{n-1}(i,j,k) + \tilde{E}_y^{n-1}(i,j,k+1)$$
$$- \tilde{E}_y^{n-1}(i,j,k) - \tilde{E}_z^{n-1}(i,j+1,k) + \tilde{E}_z^{n-1}(i,j,k), \tag{3.93a}$$

$$H_y^n(i,j,k) = H_y^{n-1}(i,j,k) + \tilde{E}_z^{n-1}(i+1,j,k) - \tilde{E}_z^{n-1}(i,j,k)$$
$$- \tilde{E}_x^{n-1}(i,j,k+1) - \tilde{E}_x^{n-1}(i,j,k), \tag{3.93b}$$

$$H_z^n(i,j,k) = H_z^{n-1}(i,j,k) + \tilde{E}_x^{n-1}(i,j+1,k) - \tilde{E}_x^{n-1}(i,j,k)$$
$$- \tilde{E}_y^{n-1}(i+1,j,k) + \tilde{E}_y^{n-1}(i,j,k), \tag{3.93c}$$

$$\tilde{E}_x^n(i,j,k) = C_a(m)\tilde{E}_x^{n-1}(i,j,k)$$
$$+ C_b(m)\big[H_z^{n-1}(i,j,k) - H_z^{n-1}(i,j-1,k)$$
$$- H_y^{n-1}(i,j,k) + H_y^{n-1}(i,j,k-1) \big], \tag{3.93d}$$

$$\tilde{E}_y^n(i,j,k) = C_a(m)\tilde{E}_y^{n-1}(i,j,k)$$
$$+ C_b(m)\big[H_x^{n-1}(i,j,k) - H_x^{n-1}(i,j,k-1)$$
$$- H_z^{n-1}(i,j,k) + H_z^{n-1}(i-1,j,k) \big], \tag{3.93e}$$

$$\tilde{E}_z^n(i,j,k) = C_a(m)\tilde{E}_z^{n-1}(i,j,k)$$
$$+ C_b(m)\big[H_y^{n-1}(i,j,k) - H_y^{n-1}(i-1,j,k)$$
$$- H_x^{n-1}(i,j,k) + H_x^{n-1}(i,j-1,k) \big] \tag{3.93f}$$

The relationship between the original and modified algorithms is illustrated in Figure 3.26 and shown in Table 3.8. Needless to say, the truncation conditions in Equations 3.89a through 3.89l must be modified accordingly. This modification eliminates the need for computer storage of separate ϵ and σ arrays; only a MEDIA array which specifies the type-integer of the dielectric or conducting medium at the location of each electric field component in the lattice need be stored. Also the programming problem of handling half integral values of i, j, k has been eliminated.

With the modified algorithm, we determine the scattered fields as follows. Let the solution region, completely enclosing the scatterer, be defined by $0 < i < I_{max}$, $0 < j < J_{max}$, $0 < k < K_{max}$. At $t \leq 0$, the program is started by setting all field components at the grip points equal to zero:

$$\tilde{E}_x^0(i,j,k) = \tilde{E}_y^0(i,j,k) = \tilde{E}_z^0(i,j,k) = 0 \tag{3.94a}$$

$$H_x^0(i,j,k) = H_y^0(i,j,k) = H_z^0(i,j,k) = 0 \tag{3.94b}$$

for $0 < i < I_{max}$, $0 < j < J_{max}$, $0 < k < K_{max}$. If we know

$$H_x^{n-1}(i,j,k), \quad E_z^{n-1}(i,j,k),$$

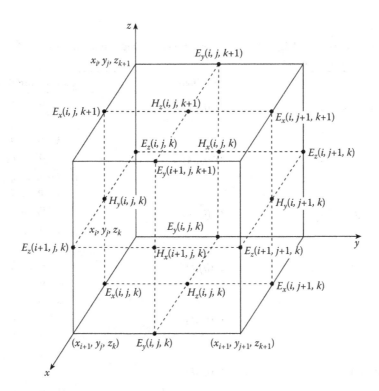

FIGURE 3.26
Modified node numbering.

and

$$E_y^{n-1}(i,j,k)$$

at all grid points in the solution region, we can determine new $H_x^n(i,j,k)$ everywhere from Equation 3.93a. The same applies for finding other field components except that the lattice truncation conditions of Equations 3.89a through 3.89l must be applied when necessary. The plane wave source is activated at $t = \delta t$, the first time step, and left on during the entire run. The field components are advanced by Yee's finite difference formulas in Equations 3.93a through 3.93f and by the lattice truncation condition in Equations 3.89a through 3.89l. The time stepping is continued for $t = N_{max}\delta t$, where N_{max} is chosen large enough that the sinusoidal steady state is achieved. In obtaining the steady-state solutions, the program must not be left for too long (i.e., N_{max} should not be too large), otherwise the imperfection of the boundary conditions causes the model to become unstable.

The FDTD method has the following inherent advantages over other modeling techniques, such as the moment method and transmission-line modeling:

- It is conceptually simple, general, and robust.
- The algorithm does not require the formulation of integral equations, and relatively complex scatterers can be treated without the inversion of large matrices.
- It is simple to implement for complicated, inhomogeneous conducting or dielectric structures because constitutive parameters (σ, μ, ϵ) can be assigned to each lattice point.

TABLE 3.8

Relationship between Original and Modified Field Components (lattice size $= I_{max}\delta \times J_{max}\delta \times K_{max}\delta$)

Original	Modified	Limits on Modified (i, j, k)
$H_x^{n+1/2}(x_i, y_{j+1/2}, z_{k+1/2})$	$H_x^n(i, j, k)$	$i = 0, \ldots, I_{max}$
		$j = 0, \ldots, J_{max} - 1$
		$k = 0, \ldots, K_{max} - 1$
$H_y^{n+1/2}(x_{i+1/2}, y_j, z_{k+1/2})$	$H_y^n(i, j, k)$	$i = 0, \ldots, I_{max} - 1$
		$j = 0, \ldots, J_{max}$
		$k = 0, \ldots, K_{max} - 1$
$H_z^{n+1/2}(x_{i+1/2}, y_{j+1/2}, z_k)$	$H_z^n(i, j, k)$	$i = 0, \ldots, I_{max} - 1$
		$j = 0, \ldots, J_{max} - 1$
		$k = 0, \ldots, K_{max}$
$E_x^n(x_{i+1/2}, y_j, z_k)$	$E_x^n(i, j, k)$	$i = 0, \ldots, I_{max} - 1$
		$j = 0, \ldots, J_{max}$
		$k = 0, \ldots, K_{max}$
$E_y^n(x_i, y_{j+1/2}, z_k)$	$E_y^n(i, j, k)$	$i = 0, \ldots, I_{max}$
		$j = 0, \ldots, J_{max} - 1$
		$k = 0, \ldots, K_{max}$
$E_z^n(x_i, y_j, z_{k+1/2})$	$E_z^n(i, j, k)$	$i = 0, \ldots, I_{max}$
		$j = 0, \ldots, J_{max}$
		$k = 0, \ldots, K_{max} - 1$

- Its computer memory requirement is not prohibitive for many complex structures of interest.
- The algorithm makes use of the memory in a simple sequential order.
- It is much easier to obtain frequency domain data from time-domain results than the converse. Thus, it is more convenient to obtain frequency domain results via time domain when many frequencies are involved.

The method has the following disadvantages:

- Its implementation necessitates modeling an object as well as its surroundings. Thus, FDTD requires a lot of memory space and the required program execution time may be excessive.
- Its accuracy is at least one order of magnitude worse than that of the method of moments, for example.
- FDTD employs a low-order approximation in space that requires at least 10 cells per wavelength to achieve acceptable accuracy.
- Since the computational meshes are rectangular in shape, they do not conform to scatterers with curved surfaces, as is the case of the cylindrical or spherical boundary.
- As in all finite difference algorithms, the field quantities are only known at grid nodes.

Time-domain modeling in three dimensions involves a number of issues which are yet to be resolved even for frequency-domain modeling. Among these are whether it is best to reduce Maxwell's equations to a second-order equation for the electric (or magnetic) field or to work directly with the coupled first-order equation. The former approach is used in Reference 35, for example, for solving the problem of EM exploration for minerals. The latter approach has been used with great success in computing EM scattering from objects as demonstrated in this section. In spite of these unresolved issues, the FDTD algorithm has been applied to solve scattering and other problems including the following:

1. Aperture penetration [44,52,53],
2. Antenna/radiation problems [54–60],
3. Microwave circuits [61–66],
4. Eigenvalue problems [67],
5. EM absorption in human tissues (bioelectromagnetics) [35,36,68–72], and
6. Other areas [73–77].

The following two examples are taken from the work of Taflove et al. [32,43,44]. The problems whose exact solutions are known will be used to illustrate the applications and accuracy of FDTD algorithm.

EXAMPLE 3.7

Consider the scattering of $a + y$-directed plane wave of frequency 2.5 GHz by a uniform, circular, dielectric cylinder of radius 6 cm. We assume that the cylinder is infinite in the z direction and that the incident fields do not vary along z. Thus, $\partial/\partial z = 0$ and the problem is reduced to the two-dimensional scattering of the incident wave with only E_z, H_x, and H_y components. Our objective is to compute one of the components, say E_z, at points within the cylinder.

Solution

Assuming a lossless dielectric with

$$\epsilon_d = 4\epsilon_o, \quad \mu_d = \mu_o, \quad \sigma_d = 0, \tag{3.95}$$

the speed of the wave in the cylinder is

$$u_d = \frac{c}{\sqrt{\varepsilon_r}} = 1.5 \times 10^8 \, \text{m/s} \tag{3.96}$$

Hence, $\lambda_d = u_d/f = 6$ cm. We may select $\delta = \Delta x = \Delta y = \Delta z = \lambda_d/20 = 0.3$ cm and $\delta t = \delta/2c = 5$ ps. Thus, we use the two-dimensional grid of Figure 3.27 as the solution domain. Due to the symmetry of the scatterer, the domain can be reduced relative to Figure 3.27 to the 25×49 subdomain of Figure 3.28. Choosing the cylinder axis as passing through point $(i, j) = (25.5, 24.5)$ allows the *symmetry condition* to be imposed at line $i = 26$, that is,

$$\tilde{E}_z^n(26, j) = \tilde{E}_z^n(25, j) \tag{3.97}$$

Soft-grid truncation conditions are applied at $j = 0, 49$ and $i = 1/2$, that is,

$$\tilde{E}_z^n(i, 0) = \frac{1}{3}\left[\tilde{E}_z^{n-2}(i-1, 1) + \tilde{E}_z^{n-2}(i, 1) + \tilde{E}_z^{n-2}(i+1, 1)\right], \tag{3.98}$$

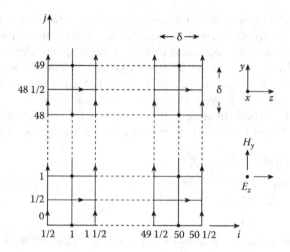

FIGURE 3.27
Two-dimensional lattice for Example 3.7.

$$\tilde{E}_z^n(i,49) = \frac{1}{3}\left[\tilde{E}_z^{n-2}(i-1,48) + \tilde{E}_z^{n-2}(i,48) + \tilde{E}_z^{n-2}(i+1,48)\right],$$ (3.99)

$$H_y^n(0.5,49) = \frac{1}{3}\left[H_y^{n-2}(1.5,j) + H_y^{n-2}(1.5,j-1) + H_y^{n-2}(1.5,j+1)\right]$$ (3.100)

Assumptions:

$$E_x = E_y = 0; \quad H_z = 0$$

$$\frac{\partial}{\partial z} = 0$$

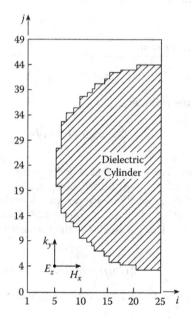

FIGURE 3.28
Finite difference model of cylindrical dielectric scatterer relative to the grid of Figure 3.27.

Maxwell's equations:

$$\frac{\partial H_x}{\partial t} = -\frac{1}{\mu}\frac{\partial E_z}{\partial y}$$

$$\frac{\partial H_y}{\partial t} = \frac{1}{\mu}\frac{\partial E_z}{\partial x}$$

$$\frac{\partial E_z}{\partial t} = \frac{1}{\varepsilon}\left(\frac{\partial H_y}{\partial x} - \frac{\partial H_x}{\partial y} - \sigma E_x\right)$$

where $n-2$ is due to the fact that $\delta = 2c\delta t$ is selected. Notice that the actual values of (i, j, k) are used here, while the modified values for easy programming are used in the program; the relationship between the two types of values is in Table 3.8.

Grid points (i, j) internal to the cylinder, determined by

$$[(i - 25.5)^2 + (j - 24.5)^2]^{1/2} \leq 20, \tag{3.101}$$

are assigned the constitutive parameters ϵ_d, μ_d, and ϵ_d, while grid points external to the cylinder are assigned parameters of free space ($\epsilon = \epsilon_o$, $\mu = \mu_o$, $\sigma = 0$).

A FORTRAN program has been developed by Bemmel [78] based on the ideas expounded above. A similar but more general code is THREDE developed by Holland [50]. The program starts by setting all field components at grid points equal to zero. A plane wave source

$$\tilde{E}_z^n(i,2) \leftarrow 1000\sin(2\pi f\, n\delta t) + \tilde{E}_z^n(i,2) \tag{3.102}$$

is used to generate the incident wave at $j = 2$ and $n = 1$, the first time step, and left on during the entire run. The program is time stepped to $t = N_{max}\delta t$, where N_{max} is large enough that sinusoidal steady state is achieved. Since $f = 2.5$ GHz, the wave period $T = 1/f = 400$ ps $= 80\delta t$. Hence, $N_{max} = 500 = 6.25\ T/\delta t$ is sufficient to reach steady state. Thus, the process is terminated after 500 timesteps. Typical results are portrayed in Figure 3.29 for the envelope of $E_z^n(15, j)$ for $460 \leq n \leq 500$. Figure 3.29 also shows the exact solution using series expansion [79]. Bemmel's code has both the numerical and exact solutions. By simply changing the constitutive parameters of the media and specifying the boundary of the scatterer (through a look-up table for complex objects), the program can be applied to almost any two-dimensional scattering or penetration problem.

EXAMPLE 3.8

Consider the penetration of $a + y$-directed plane wave of frequency 2.5 GHz by a uniform, dielectric sphere of radius 4.5 cm. The problem is similar to the previous example except that it is three dimensional and more general. We assume that the incident wave has only E_z and H_x components.

Solution

As in the previous example, we assume that internal to the lossless dielectric sphere,

$$\epsilon_d = 4\epsilon_o, \quad \mu_d = \mu_o, \quad \sigma_d = 0 \tag{3.103}$$

We select

$$\delta = \lambda_d/20 = 0.3 \text{ cm} \tag{3.104}$$

and

$$\delta t = \delta/2c = 5 \text{ ps} \tag{3.105}$$

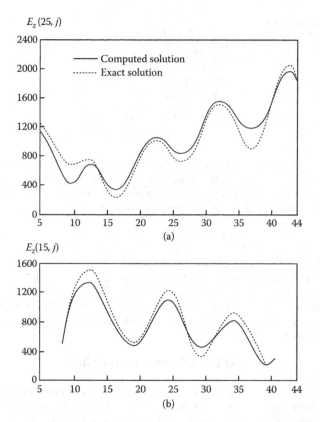

FIGURE 3.29
Computed internal E_z on line: (a) $i = 25$, (b) $i = 15$.

This choice of the grid size implies that the radius of the sphere is 4.5/0.3 = 15 units. The sphere model centered at grid point (19.5, 20, 19) in a 19 × 39 × 19 lattice is portrayed in Figure 3.30 at two lattice symmetry planes $k = 19$ and $i = 19.5$. Grid points (i, j, k) internal to the sphere are determined by

$$[(i - 19.5)^2 + (j - 20)^2 + (k - 19)^2]^{1/2} \leq 15 \qquad (3.106)$$

Rather than assigning $\sigma = 0$ to points external to the sphere, a value $\sigma = 0.1$ mho/m is assumed to reduce spurious wave reflections. The MATLAB code shown in Figure 3.31, a modified version of Bemmel's [78], is used to generate field components E_y and E_z near the sphere irradiation axis. With the dimensions and constitutive parameters of the sphere specified as input data, the program is developed based on the following steps:

1. Compute the parameters of each medium using Equation 3.91 where $m = 1, 2$.
2. Initialize field components.
3. Use the FDTD algorithm in Equations 3.93a through 3.93f to generate field components. This is the heart of the program. It entails taking the following steps:
   a. Calculate actual values of grid point (x, y, z) using the relationship in Table 3.8. This will be needed later to identify the constitutive parameters of the medium at that point using subroutine MEDIA.
   b. Apply soft lattice truncation conditions in Equations 3.89a through 3.89l at appropriate boundaries, that is, at $x = \delta/2, y = 0, y_{max}$, and $z = 0$. Notice that

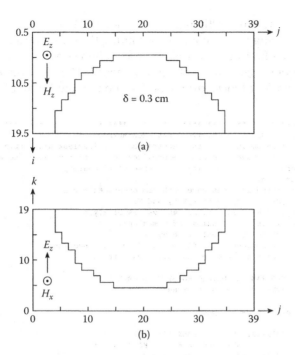

FIGURE 3.30
FDTD model of dielectric sphere.

some of the conditions in Equations 3.89a through 3.89l are not necessary in this case because we restrict the solution to one fourth of the sphere due to geometrical symmetry. At other boundaries ($x = x_{max}$ and $z = z_{max}$), the symmetry conditions are imposed. For example, at $k = 19$,

$$\tilde{E}_x^n(i, j, 20) = \tilde{E}_x^n(i, j, 18)$$

   c. Apply FDTD algorithm in Equations 3.93a through 3.93f.
   d. Activate the plane wave source, that is,

$$\tilde{E}_z^n(i, j, k) \leftarrow \sin(2\pi f n \,\delta t) + \tilde{E}_z^n(i, j_s, k)$$

where $j_s = 3$ or any plane near $y = 0$.
   e. Time step until steady state is reached.
 4. Obtain the maximum absolute values (envelopes) of field components in the last half-wave and output the results.

Figure 3.32 illustrates the results of the program. The values of $|E_y|$ and $|E_z|$ near the sphere axis are plotted against j for observation period $460 \le n \le 500$. The computed results are compared with Mie's exact solution [80] covered in Section 2.8. The code for calculating the exact solution is also found in Bemmel's work [78].

3.9 Absorbing Boundary Conditions for FDTD

The FDTD method is a robust, flexible (adaptable to complex geometries), efficient, versatile, easy-to-understand, easy-to-implement, and user-friendly technique to solve Maxwell's equations in the time domain. Although the method did not receive as much attention as it deserved when it was suggested, it is now becoming the most popular method of choice in computational EM. It is finding widespread use for solving open-region scattering, radiation, penetration/absorption, electromagnetic interference (EMI), electromagnetic compatibility (EMC), diffusion, transient, bioelectromagnetics, and microwave circuit modeling problems.

```
%*********************************************************************
% APPLICATION OF THE FINITE DIFFERENCE METHOD
% This program involves the penetration of a lossless dielectric SPHERE
% by a plane wave. The program provides in the maximum absolute value of
% Ey and Ez during the final half-wave of time-stepping
% Assumption:
% +y-directed incident wave with components Ez and Hx.
% I,J,K,NN correspond to X,Y,Z, and Time.
% IMAX,JMAX,KMAX are the maximum values of x,y,z
% NNMAX is the total number of timesteps.
% NHW represents one half-wave cycle.
% MED is the number of different uniform media sections.
% JS is the j-position of the plane wave front.

% THIS PROGRAM WAS DEVELOPED BY V. BEMMEL [80]
% AND LATER IMPROVED BY D. TERRY

clear all; format compact; tic

        IMAX=19; JMAX=39; KMAX=19;
        NMAX=2; NNMAX=500; NHW=40; MED=2; JS=3;
        DELTA=3E-3; CL=3.0E8; F=2.5E9;

% Define scatterer dimensions
    OI=19.5; OJ=20.0; OK=19.0; RADIUS=15.0;

    ER=[1.0,4.0];    % CONSTITUTIVE PARAMETERS
    SIG=[0.1,0.0];

% Statement function to compute position w.r.t. center of the sphere

    E0=(1E-9)/(36*pi);
    U0=(1E-7)*4*pi;
    DT=DELTA/(2*CL);
    R=DT/E0;
    RA=(DT^2)/(U0*E0*(DELTA^2));
    RB=DT/(U0*DELTA);
    TPIFDT = 2.0*pi*F*DT;

%*******************************************************
% STEP # 1 - COMPUTE MEDIA PARAMETERS
%*******************************************************

CA = 1-R*SIG./ER;
   CB = RA./ER;
   CBMRB = CB/RB;

%  (i) CALCULATE THE REAL/ACTUAL GRID POINTS

% Initialize the media arrays.Index (M) determines which
% medium each point is actually located in and is used to
% index into arrays which determine the constitutive
% parameters of the medium.There are separate M determining
% arrays for EX, EY, and EZ.  These arrays correlate the
% integer values of I,J,K to the actual position within
% the lattice.  Computing these values now and storing them  in these
% arrays as opposed to computing them each time they are
% needed saves a large amount of computation time.
```

FIGURE 3.31
Computer program for FDTD three-dimensional scattering problem.

(Continued)

```
x = 0:(IMAX+1); y = 0:(JMAX+1); z = 0:(KMAX+1);
[Mx,My,Mz]=ndgrid(x,y,z);
IXMED = (sqrt((Mx-OI+.5).^2+(My-OJ).^2+(Mz-OK).^2)<=RADIUS)+1;
IYMED = (sqrt((Mx-OI).^2+(My-OJ+.5).^2+(Mz-OK).^2)<=RADIUS)+1;
IZMED = (sqrt((Mx-OI).^2+(My-OJ).^2+(Mz-OK+.5).^2)<=RADIUS)+1;

% ************************************************
%   STEP    # 2 - INITIALIZE FIELD COMPONENTS
% ************************************************
%  components for output

    EY1 = zeros(1,JMAX+2);
    EZ1 = zeros(1,JMAX+2);

    EX=zeros(IMAX+2,JMAX+2,KMAX+2,NMAX+1);
    EY=zeros(IMAX+2,JMAX+2,KMAX+2,NMAX+1);
    EZ=zeros(IMAX+2,JMAX+2,KMAX+2,NMAX+1);
    HX=zeros(IMAX+2,JMAX+2,KMAX+2,NMAX+1);
    HY=zeros(IMAX+2,JMAX+2,KMAX+2,NMAX+1);
    HZ=zeros(IMAX+2,JMAX+2,KMAX+2,NMAX+1);


% ********************************************************
% STEP # 3 - USE FD/TD ALGORITHM TO GENERATE
%  FIELD COMPONENTS
% ********************************************************
%   SINCE ONLY FIELD COMPONENTS AT CURRENT TIME (t) AND   PREVIOUS
%   TWO TIME STEPS ( t-1 AND t-2) ARE REQUIRED FOR  COMPUTATION,
%   WE SAVE MEMORY SPACE BY USING THE FOLLOWING INDICES
%       NCUR is index in for time t
%       NPR1 is index in for t-1
%       NPR2 is index in for t-2


% NOTES %%%%%%%%%%%%%%%%%%%%%%%%%%%%%%%%%%%%%%%%%%%%%%%%%%%%%%%%%%%
% *ind03c.m I incremented the time so it goes 1 2 3, instead of 0 1 2

%%%%%%%%%%%%%%%%%%%%%%%%%%%%%%%%%%%%%%%%%%%%%%%%%%%%%%%%%%%%%%%%%%%

    NCUR = 3;
    NPR1 = 2;
    NPR2 = 1;
    for NN = 1: NNMAX  % TIME LOOP
        if mod(NN,10)==0
            disp(['NN = ',num2str(NN)]) % DISPLAY PROGRESS
        end
        % Next time step - move indices up a notch.
        NPR2 = NPR1;
        NPR1 = NCUR;
        NCUR = mod( NCUR, 3)+1;
        for K=0:KMAX  % Z LOOP
            for J=0:JMAX  % Y LOOP
                for I=0:IMAX  % X LOOP
                    % (ii)-APPLY SOFT LATTICE TRUNCATION CONDITIONS
                    %---x=delta/2
                    if (I==0)
                        if ((K~=KMAX)&&(K~=0))

                            HY(0+1,J+1,K+1,NCUR) = (HY(1+1,J+1,K-
1+1,NPR2) + HY(1+1,J+1,K+1,NPR2)+ HY(1+1,J+1,K+1+1,NPR2))/3;
```

FIGURE 3.31 (Continued)
Computer program for FDTD three-dimensional scattering problem.   (*Continued*)

```
                        HZ(0+1,J+1,K+1,NCUR) = (HZ(1+1,J+1,K-
1+1,NPR2) + HZ(1+1,J+1,K+1,NPR2)+ HZ(1+1,J+1,K+1+1,NPR2))/3;

                   else
                      if (K==KMAX)
                               HY(0+1,J+1,KMAX+1,NCUR) = (HY(1+1,J+1,KMAX-
1+1,NPR2)+ HY(1+1,J+1,KMAX+1,NPR2))/2;

                      HZ(0+1,J+1,K+1,NCUR)=( HZ(1+1,J+1,K-
1+1,NPR2)+ HZ(1+1,J+1,K+1,NPR2) )/2;
                          else
                               HY(0+1,J+1,K+1,NCUR) = ( HY(1+1,J+1,K+1,NPR2)
+ HY(1+1,J+1,K+1+1,NPR2))/2;
                               HZ(0+1,J+1,K+1,NCUR)=(HZ(1+1,J+1,K+1,NPR2)+
HZ(1+1,J+1,K+1,1+1,NPR2))/2;
                          end
                      end
                  end
                  % ---y=0
                  if (J==0)
                      EX(I+1,0+1,K+1,NCUR)=EX(I+1,1+1,K+1,NPR2);
                      EZ(I+1,0+1,K+1,NCUR)=EZ(I+1,1+1,K+1,NPR2);
                  else
                      %---y=ymax
                      if (J==JMAX)
                          EX(I+1,JMAX+1,K+1,NCUR)=EX(I+1,JMAX-1+1,K+1,NPR2);
                          EZ(I+1,JMAX+1,K+1,NCUR)=EZ(I+1,JMAX-1+1,K+1,NPR2);
                      end
                  end
                  %---z=0
                  if(K==0)
                      if ((I~=0)&&(I~=IMAX))
                          EX(I+1,J+1,0+1,NCUR) = (EX(I-
1+1,J+1,1+1,NPR2) + EX(I+1,J+1,1+1,NPR2)+EX(I+1+1,J+1,1+1,NPR2))/3;
                          EY(I+1,J+1,0+1,NCUR) = (EY(I-
1+1,J+1,1+1,NPR2) + EY(I+1,J+1,1+1,NPR2)+EY(I+1+1,J+1,1+1,NPR2))/3;
                      else
                          if (I==0)
                              EX(0+1,J+1,0+1,NCUR)=(EX(0+1,J+1,1+1,NPR2)+
EX(1+1,J+1,1+1,NPR2))/2;
                              EY(I+1,J+1,0+1,NCUR)=(EY(I+1,J+1,1+1,NPR2)+
EY(I+1+1,J+1,1+1,NPR2))/2;
                          else
                              EX(I+1,J+1,0+1,NCUR)=(EX(I-
1+1,J+1,1+1,NPR2)+EX(I+1,J+1,1+1,NPR2))/2;
                              EY(I+1,J+1,0+1,NCUR)=(EY(I-
1+1,J+1,1+1,NPR2)+EY(I+1,J+1,1+1,NPR2))/2;
                          end
                      end
                  end

                  % (iii)  APPLY  FD/TD ALGORITHM

                  %-----a. HX  generation:
                  HX(I+1,J+1,K+1,NCUR)=HX(I+1,J+1,K+1,NPR1)+RB*(EY(I+1,J+1,K+1+1,NPR1)-
EY(I+1,J+1,K+1,NPR1)+EZ(I+1,J+1,K+1,NPR1)-EZ(I+1,J+1+1,K+1,NPR1));
                  %-----b. HY  generation:
                  HY(I+1,J+1,K+1,NCUR)=HY(I+1,J+1,K+1,NPR1)+RB*(EZ(I+1+1,J+1,K+1,NPR1)-
EZ(I+1,J+1,K+1,NPR1)+EX(I+1,J+1,K+1,NPR1)-EX(I+1,J+1,K+1+1,NPR1));
```

FIGURE 3.31 (Continued)
Computer program for FDTD three-dimensional scattering problem.   *(Continued)*

```
                %-----c.  HZ   generation:
                HZ(I+1,J+1,K+1,NCUR)=HZ(I+1,J+1,K+1,NPR1)+RB*(EX(I+1,J+1+1,K+1,NPR1)-
EX(I+1,J+1,K+1,NPR1)+EY(I+1,J+1,K+1,NPR1)-EY(I+1+1,J+1,K+1,NPR1));
                %---k=kmax   ! SYMMETRY
                if (K==KMAX)
                    HX(I+1,J+1,KMAX+1,NCUR)=HX(I+1,J+1,KMAX-1+1,NCUR);
                    HY(I+1,J+1,KMAX+1,NCUR)=HY(I+1,J+1,KMAX-1+1,NCUR);
                end
                % -----d.  EX   generation:
                if ((J~=0)&&(J~=JMAX)&&(K-=0))
                    M = IXMED( I+1, J+1, K+1 );
EX(I+1,J+1,K+1,NCUR) = CA(M)*EX(I+1,J+1,K+1,NPR1) + CBMRB(M)*(HZ(I+1,J+1,K+1,NCUR)-
HZ(I+1,J-1+1,K+1,NCUR)+HY(I+1,J+1,K-1+1,NCUR)-HY(I+1,J+1,K+1,NCUR));
                end
                %-----e.  EY   generation:
                if(K~=0)
                    M = IYMED( I+1, J+1, K+1 );
                    if I~=0
EY(I+1,J+1,K+1,NCUR)=CA(M)*EY(I+1,J+1,K+1,NPR1) + CBMRB(M)*(HX(I+1,J+1,K+1,NCUR)-
HX(I+1,J+1,K-1+1,NCUR)+HZ(I-1+1,J+1,K+1,NCUR)-HZ(I+1,J+1,K+1,NCUR));
                    else
EY(I+1,J+1,K+1,NCUR)=CA(M)*EY(I+1,J+1,K+1,NPR1) + CBMRB(M)*(HX(I+1,J+1,K+1,NCUR)-
HX(I+1,J+1,K-1+1,NCUR)+    0    -HZ(I+1,J+1,K+1,NCUR));
                    end
                end
                %-----f.  EZ   generation:
                if ((J~=0)&&(J~=JMAX))

                    M = IZMED( I+1, J+1, K+1 );
                    %  sig(ext)=0 for Ez only from Taflove[32]
                    if(M==1)
                        CAM=1;
                    else
                        CAM=CA(M);
                    end

                    if I~=0
EZ(I+1,J+1,K+1,NCUR)=CAM*EZ(I+1,J+1,K+1,NPR1)+CBMRB(M)*(HY(I+1,J+1,K+1,NCUR)-HY(I-
1+1,J+1,K+1,NCUR)+HX(I+1,J-1+1,K+1,NCUR)-HX(I+1,J+1,K+1,NCUR));
                    else
EZ(I+1,J+1,K+1,NCUR)=CAM*EZ(I+1,J+1,K+1,NPR1)+CBMRB(M)*(HY(I+1,J+1,K+1,NCUR)-
0              +HX(I+1,J-1+1,K+1,NCUR)-HX(I+1,J+1,K+1,NCUR));
                    end

                    % (iv)   APPLY THE PLANE-WAVE SOURCE
                    if (J==JS)
                        EZ(I+1,JS+1,K+1,NCUR) = EZ(I+1,JS+1,K+1,NCUR)+sin( TPIFDT*NN );
                    end
                end
                %---i=imax+1/2   ! SYMMETRY
                if(I==IMAX)
                    EY(IMAX+1+1,J+1,K+1,NCUR)=EY(IMAX+1,J+1,K+1,NCUR);
                    EZ(IMAX+1+1,J+1,K+1,NCUR)=EZ(IMAX+1,J+1,K+1,NCUR);
                end
                %---k=kmax
                if(K==KMAX)
                    EX(I+1,J+1,KMAX+1+1,NCUR)=EX(I+1,J+1,KMAX-1+1,NCUR);
                    EY(I+1,J+1,KMAX+1+1,NCUR)=EY(I+1,J+1,KMAX-1+1,NCUR);
                end
            end % X LOOP
```

FIGURE 3.31 (Continued)
Computer program for FDTD three-dimensional scattering problem.   (*Continued*)

```
% ***********************************************************
%   STEP # 4 - RETAIN THE MAXIMUM ABSOLUTE VALUES DURING
%                 THE LAST HALF-WAVE
% ***********************************************************
              if ( (K==KMAX)&&(NN>(NNMAX-NHW)) )
                  TEMP = abs( EY(IMAX+1,J+1,KMAX-1+1,NCUR) );
                  if (TEMP > EY1(J+1) )
                      EY1(J+1) = TEMP;
                  end
                  TEMP = abs( EZ(IMAX+1,J+1,KMAX+1,NCUR) );
                  if (TEMP > EZ1(J+1) )
                      EZ1(J+1) = TEMP;
                  end
              end
          end % Y LOOP
      end % Z LOOP
    end % TIME LOOP
% ***********************************************************
toc

figure(3),plot(6:34,EY1(6:34),'.-')
    ylabel('Computed |E_y|/|E_i_n_c|')
    xlabel('j')
    grid on
figure(4),plot(5:34,EZ1(5:34),'.-')
    ylabel('Computed |E_z|/|E_i_n_c|')
    xlabel('j')
    grid on
```

FIGURE 3.31 (Continued)
Computer program for FDTD three-dimensional scattering problem.

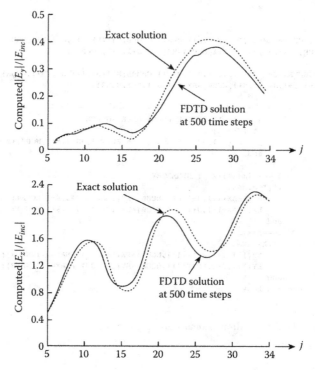

FIGURE 3.32
Computed $E_y(19.5, j, 18)$ and $E_z(19, j, 18.5)$ within the lossless dielectric sphere.

However, the method exhibits some problems such as slow convergence for solving resonant structures, requirement of large memory for inhomogeneous waveguide structures due to the necessity of a full-wave analysis, inability to properly handle curved boundaries due to its orthogonal nature, low stability, and low accuracy unless fine mesh is used, to mention a few. These problems prohibit the application of the standard FDTD technique and have led to various forms of its modifications [81–91] and hybrid FDTD methods [92–94]. Although these new FDTD methods have enhanced the standard FDTD (increase accuracy and stability, etc.), some researchers still prefer the standard FDTD.

One of the major problems inherent in the standard FDTD, however, is the requirement for artificial mesh truncation (boundary) condition. The artificial termination truncates the solution region electrically close to the radiating/scattering object but effectively simulates the solution to infinity. These artificial termination conditions are known as *absorbing boundary conditions* (ABCs) as they theoretically absorb incident and scattered fields. The accuracy of the ABC dictates the accuracy of the FDTD method. The need for accurate ABCs has resulted in various types of ABCs [95–105], which are fully discussed in Reference 102. Due to space limitation, we will consider only Berenger's *perfectly matched layer* (PML) type of ABC [98–102] since PML has been the most widely accepted and is set to revolutionize the FDTD method. It acts in the same way as absorbing material in anechoic rooms.

In the PML truncation technique, an artificial layer of absorbing material is placed around the outer boundary of the computational domain. The goal is to ensure that a plane wave that is incident from FDTD free space to the PML region at an arbitrary angle is completely absorbed there without reflection. This is the same as saying that there is complete transmission of the incident plane wave at the interface between free space and the PML region (see Figure 3.33). Thus, the FDTD and the PML region are said to be *perfectly matched.*

To illustrate the PML technique, consider Maxwell's equation in two dimensions for transverse electric (TE) case with field components E_x, E_y and H_z and no variation with z. Expanding Equations 1.22c and 1.22d in Cartesian coordinates and setting $E_z = 0 = \partial/\partial z$, we obtain

$$\varepsilon_o \frac{\partial E_x}{\partial t} + \sigma E_x = \frac{\partial H_z}{\partial y} \tag{3.107a}$$

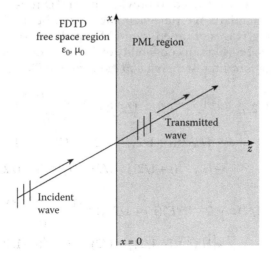

FIGURE 3.33
Reflectionless transmission of a plane wave at a PML/free-space interface.

$$\varepsilon_0 \frac{\partial E_y}{\partial t} + \sigma E_y = -\frac{\partial H_z}{\partial x} \tag{3.107b}$$

$$\mu_0 \frac{\partial H_z}{\partial t} + \sigma^* H_z = \frac{\partial E_x}{\partial y} - \frac{\partial E_y}{\partial x} \tag{3.107c}$$

where the PML, as a lossy medium, is characterized by an electrical conductivity σ and a magnetic conductivity σ^*. The conductivities are related as

$$\frac{\sigma}{\varepsilon_0} = \frac{\sigma^*}{\mu_0} \tag{3.108}$$

This relationship ensures a required level of attenuation and forces the wave impedance of the PML to be equal to that of the free space. Thus, a reflectionless transmission of a plane wave propagation across the interface is permitted. For oblique incident angles, the conductivity of the PML must have a certain anisotropy characteristic to ensure reflectionless transmission. To achieve this, the H_z component must be split into two subcomponents, H_{zx} and H_{zy}, with the possibility of assigning losses to the individual split field components. This is the cornerstone of the PML technique. It leads to four components E_x, E_y, H_{zx}, and H_{zy} and four (rather than the usual three) coupled field equations.

$$\varepsilon_0 \frac{\partial E_x}{\partial t} + \sigma_y E_x = \frac{\partial (H_{zx} + H_{zy})}{\partial y} \tag{3.109a}$$

$$\varepsilon_0 \frac{\partial E_y}{\partial t} + \sigma_x E_y = -\frac{\partial (H_{zx} + H_{zy})}{\partial x} \tag{3.109b}$$

$$\mu_0 \frac{\partial H_{zx}}{\partial t} + \sigma_x^* H_{zx} = -\frac{\partial E_y}{\partial x} \tag{3.109c}$$

$$\mu_0 \frac{\partial H_{zy}}{\partial t} + \sigma_y^* H_{zy} = \frac{\partial E_x}{\partial y} \tag{3.109d}$$

These equations can be discretized to provide the FDTD time-stepping equations for the PML region. The standard Yee time-stepping cannot be used because of the rapid attenuation to outgoing waves afforded by a PML medium. We use the exponentially differenced equations to preclude any possibility of diffusion instability. In the usual FDTD notations, the resulting four time-stepping equations for the PML region are [99]

$$E_x^{n+1}(i+1/2,j) = e^{-\sigma_y(j)\delta t/\varepsilon_0} E_x^n(i+1/2,j) + \frac{(1-e^{-\sigma_y(j)\delta t/\varepsilon_0})}{\sigma_y(j)\delta}$$
$$\times \left[H_{zx}^{n+1/2}(i+1/2,j+1/2) + H_{zy}^{n+1/2}(i+1/2,j+1/2) \right.$$
$$\left. - H_{zx}^{n+1/2}(i+1/2,j-1/2) - H_{zy}^{n+1/2}(i+1/2,j-1/2) \right] \tag{3.110a}$$

$$E_y^{n+1}(i,j+1/2) = e^{-\sigma_x(i)\delta t/\varepsilon_0} E_y^n(i,j+1/2,k) + \frac{(1-e^{-\sigma_x(i)\delta t/\varepsilon_0})}{\sigma_x(i)\delta}$$
$$\times \left[H_{zx}^{n+1/2}(i-1/2,j+1/2) + H_{zy}^{n+1/2}(i-1/2,j+1/2) \right.$$
$$\left. - H_{zx}^{n+1/2}(i+1/2,j+1/2) - H_{zy}^{n+1/2}(i+1/2,j+1/2) \right] \tag{3.110b}$$

$$H_{zx}^{n+1/2}(i+1/2, j+1/2) = e^{-\sigma_x^*(i+1/2)\delta t/\mu_0} H_{zx}^{n-1/2}(i+1/2, j+1/2)$$

$$+ \frac{(1 - e^{-\sigma_x^*(i+1/2)\delta t/\mu_0})}{\sigma_x^*(i+1/2)\delta} \Big[E_y^n(i, j+1/2) - E_y^n(i+1, j+1/2) \Big] \qquad (3.110c)$$

$$H_{zy}^{n+1/2}(i+1/2, j+1/2) = e^{-\sigma_y^*(i+1/2)\delta t/\mu_0} H_{zy}^{n-1/2}(i+1/2, j+1/2)$$

$$+ \frac{(1 - e^{-\sigma_y^*(i+1/2)\delta t/\mu_0})}{\sigma_y^*(i+1/2)\delta} \Big[E_x^n(i+1/2, j+1) - E_x^n(i+1/2, j) \Big] \qquad (3.110d)$$

These equations can be directly implemented in an FDTD simulation to model PML medium. All that is required is to select the depth of the PML and its conductivity. In theory, the PML could be δ deep and have near-infinite conductivity. It has been shown, however, that increasing the conductivity gradually with depth minimizes reflections; hence, the "layering" of the medium and the dependence of σ on i and j.

The TM case can be obtained by duality, with E_z split so that $E_z = E_{zx} + E_{zy}$. In three dimensions, all six Cartesian field components are split so that the resulting PML modification of Maxwell's equations yields 12 equations [102].

3.10 Advanced Applications of FDTD

As a versatile, powerful technique, FDTD is being used in several areas including computing, communications, and medicine. The simple technique has been improved upon in the following ways:

- Perfectly matched layers (PML), discussed earlier.
- The segmented FDTD (SFDTD), which divides the problem space into segments so that the computational redundancy is reduced [106].
- Fast and memory-efficient algorithms of the high-order difference equations [107].
- Pseudospectral time-domain (PSTD) methods for broadband electromagnetic [108,109].

These changes coupled with the advances in computer hardware have expanded the popularity, accuracy, and speed of FDTD modeling.

In this section, we consider some topics that are useful in a variety of FDTD applications [110]. Since the topics are advanced and there is space limitation, only an introductory treatment is provided; the reader is encouraged to get more information from the references provided [102,111].

3.10.1 Periodic Structures

Periodic structures, such as frequency selective surfaces and volumes, are useful in practice because of their spatial filtering characteristics. They are found in photonic bandgap structures and antenna arrays. Due to the geometric nature of the structure, this kind of problem can be handled by modeling a single period of the structure [112,113]. Since the structure consists of replicas of the basic element, one can model the basic element using FDTD algorithm and apply boundary conditions to simulate the periodic replication.

3.10.2 Antennas

FDTD is lagging behind the method of moments (MOM) in antenna modeling because MOM can handle antennas with less time and memory. Originally, FDTD was basically used to model the scattering of EM waves from objects. The technique has now been developed to model radiating structures (transmitting and receiving antennas) with realistic complexity. It has been accurately used to model antennas and obtain their farfield radiation and reception properties such as farfield radiation patterns, receiving apertures, gain, efficiency, and driving-point impedances.

3.10.3 PSTD Techniques

FDTD modeling of electrically large-scale problems is very challenging for scattering and propagation problems. To address this challenge, various hybrid methods have been proposed recently. These include the PSTD techniques and the general vector auxiliary differential equation (GVADE). We will consider only PSTD here. There are two common types of PSTD techniques for solving Maxwell's equations: the Fourier and Chebyshev methods. PSTD applications for large-scale problems include wave propagation, wave scattering, nonlinear optics, and photonics. The emerging PSTD techniques have expanded the scope and applicability of the original FDTD scheme. The major advantage of the PSTD algorithm over the standard FDTD algorithm lies in its higher accuracy achieved by less sampling density. Its low sampling rate, as low as two cells per wavelength, makes it much more efficient than the conventional FDTD [114].

3.10.4 Photonics

The application of FDTD has been expanded to embrace a closely related PSTD. FDTD together with PSTD can be used to discretize Maxwell's equations on a spatial mesh and apply a leapfrog time-matching schemes. This can be applied to wide variety of photonic technologies such as optical waveguides, microcavity rings, photonic crystals, biophotonics, nanotechnology, and biomedical applications of light. It requires simulating the behavior of the materials involved at optical wavelengths. FDTD and PSTD can put Maxwell's equations to work in the analysis and simulation of a wide range of biophotonic devices. This application will grow with the continuing improvement in computer capabilities.

3.10.5 Metamaterials

Until recently, research by CEM community has focused on fields and waves in vacuum and in other simple mediums. Interest in complex-mediums electromagnetics (CME) has been demonstrated and given impetus by the emergence of nanoscience and nanotechnologies [115]. There is the potential ability to engineer the exotic properties of metamaterials for a variety of engineering applications. Two of those properties are negative refractive index and decrease in size and weight of components and devices. The FDTD technique is regarded as the most effective numerical method for studying metamaterial-based structures. The incorporation of material nonlinearity and handling of metamaterials is an emerging area in FDTD modeling [116,117].

3.10.6 MEEP

This is a free, open-source implementation of the FDTD simulation. The software package was developed at MIT. MEEP is an acronym for MIT electromagnetic equation propagation. Such a free FDTD package will greatly aid research in EM. It was first released in 2006 and it can be downloaded from http://ab-initio.mit.edu/meep

MEEP can handle: arbitrary anisotropic, nonlinear media; a variety of boundary conditions including symmetries and perfectly matched layers (PML); distributed-memory parallelism; and Cartesian and cylindrical coordinates [118].

3.11 Finite Differencing for Nonrectangular Systems

So far in this chapter, we have considered only rectangular solution regions within which a rectangular grid can be readily placed. Although we can always replace a nonrectangular solution region by an approximate rectangular one, our discussion in this chapter would be incomplete if we failed to apply the method to nonrectangular coordinates since it is sometimes preferable to use these coordinates. We will demonstrate the finite differencing technique in cylindrical coordinates (ρ, ϕ, z) and spherical coordinates (r, θ, ϕ) by solving Laplace's equation $\nabla^2 V = 0$. The idea is readily extended to other PDEs.

3.11.1 Cylindrical Coordinates

Laplace's equation in cylindrical coordinates can be written as

$$\nabla^2 V = \frac{\partial^2 V}{\partial \rho^2} + \frac{1}{\rho} \frac{\partial V}{\partial \rho} + \frac{1}{\rho^2} \frac{\partial^2 V}{\partial \phi^2} + \frac{\partial^2 V}{\partial z^2}, \tag{3.111}$$

Refer to the cylindrical system and finite difference molecule shown in Figure 3.34. At point $O(\rho_o, \phi_o, z_o)$, the equivalent finite difference approximation is

$$\frac{V_1 - 2V_0 + V_2}{(\Delta \rho)^2} + \frac{1}{\rho_o} \frac{V_1 - V_2}{2 \Delta \rho} + \frac{V_3 - 2V_0 + V_4}{(\rho_o \Delta \phi)^2} + \frac{V_5 - 2V_0 + V_6}{(\Delta z)^2} = 0 \tag{3.112}$$

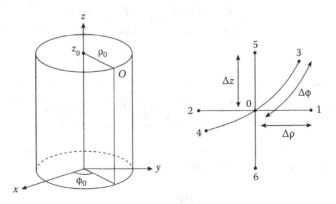

FIGURE 3.34
Typical node in cylindrical coordinate.

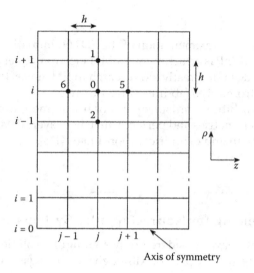

FIGURE 3.35
Finite difference grid for an axisymmetric system.

where $\Delta\rho$, $\Delta\phi$, and Δz are the step sizes along ρ, ϕ, and z, respectively, and

$$
\begin{aligned}
&V_o = V(\rho_o, \phi_o, z_o), \quad V_1 = V(\rho_o + \Delta\rho, \phi_o, z_o), \quad V_2 = V(\rho_o - \Delta\rho, \phi_o, z_o), \\
&V_3 = V(\rho_o, \phi_o + \rho_o\Delta\phi, z_o), \quad V_4 = V(\rho_o, \phi_o - \rho_o\Delta\phi, z_o), \\
&V_5 = V(\rho_o, \phi_o, z_o + \Delta z), \quad V_6 = V(\rho_o, \phi_o, z_o - \Delta z)
\end{aligned} \tag{3.113}
$$

We now consider a special case of Equation 3.112 for an axisymmetric system [119]. In this case, there is no dependence on ϕ so that $V = V(\rho, z)$. If we assume square nets so that $\Delta\rho = \Delta z = h$, the solution region is discretized as in Figure 3.35 and Equation 3.112 becomes

$$
\left(1 + \frac{h}{2\rho_o}\right)V_1 + \left(1 - \frac{h}{2\rho_o}\right)V_2 + V_5 + V_6 - 4V_o = 0 \tag{3.114}
$$

If point O is at $(\rho_o, z_o) = (ih, jh)$, then

$$
1 + \frac{h}{2\rho_o} = \frac{2i+1}{2i}, \quad 1 - \frac{h}{2\rho_o} = \frac{2i-1}{2i}
$$

so that Equation 3.114 becomes

$$
\boxed{V(i,j) = \frac{1}{4}\left[V(i,j-1) + V(i,j+1) + \left(\frac{2i-1}{2i}\right)V(i-1,j) + \left(\frac{2i+1}{2i}\right)V(i+1,j)\right]} \tag{3.115}
$$

Notice that in Equation 3.114, it appears we have a singularity for $\rho_o = 0$. However, by symmetry, all odd order derivatives must be zero. Hence,

$$
\left.\frac{\partial V}{\partial \rho}\right|_{\rho=0} = 0 \tag{3.116}
$$

since

$$V(\Delta\rho, z_o) = V(-\Delta\rho, z_o) \tag{3.117}$$

Therefore by L'Hopital's rule,

$$\lim_{\rho_o \to 0} \frac{1}{\rho_o} \frac{\partial V}{\partial \rho}\bigg|_{\rho_o} = \frac{\partial^2 V}{\partial \rho^2}\bigg|_{\rho_o} \tag{3.118}$$

Thus, at $\rho = 0$, Laplace's equation becomes

$$2\frac{\partial^2 V}{\partial \rho^2} + \frac{\partial^2 V}{\partial z^2} = 0 \tag{3.119}$$

The finite difference equivalent to Equation 3.119 is

$$V_o = \frac{1}{6}(4V_1 + V_5 + V_6)$$

or

$$\boxed{V(0,j) = \frac{1}{6}[V(0,j-1) + V(0,j+1) + 4V(1,j)]} \tag{3.120}$$

which is used at $\rho = 0$.

To solve Poisson's equation $\nabla^2 V = -\rho_v/\epsilon$ in cylindrical coordinates, we obtain the finite difference form by replacing zero on the right-hand side of Equation 3.112 with $g = -\rho_v/\epsilon$. We obtain

$$\boxed{V(i,j) = \frac{1}{4}\left[V(i,j+1) + V(i,j-1) + \frac{2i-1}{2i}V(i-1,j) + \frac{2i+1}{2i}V(i+1,j) + gh^2\right]} \tag{3.121}$$

where h is the step size.

As in Section 3.7.1, the boundary condition $D_{1n} = D_{2n}$ must be imposed at the interface between two media. As an alternative to applying Gauss's law as in Section 3.7.1, we will apply Taylor-series expansion [120]. Applying the series expansion to points 1, 2, 5 in medium 1 in Figure 3.36, we obtain

$$V_1 = V_o + \frac{\partial V_o^{(1)}}{\partial \rho}h + \frac{\partial^2 V_o^{(1)}}{\partial \rho^2}\frac{h^2}{2} + \cdots$$

$$V_2 = V_o - \frac{\partial V_o^{(1)}}{\partial \rho}h + \frac{\partial^2 V_o^{(1)}}{\partial \rho^2}\frac{h^2}{2} - \cdots \tag{3.122}$$

$$V_5 = V_o + \frac{\partial V_o^{(1)}}{\partial z}h + \frac{\partial^2 V_o^{(1)}}{\partial z^2}\frac{h^2}{2} + \cdots$$

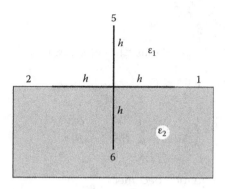

FIGURE 3.36
Interface between two dielectric media.

where superscript (1) denotes medium 1. Combining Equations 3.111 and 3.122 results in

$$h^2 \nabla^2 V = V_1 + V_2 + 2V_5 - 4V_o - 2h \frac{\partial V_o^{(1)}}{\partial z} + \frac{h(V_1 - V_2)}{2\rho_o} = 0$$

or

$$\frac{\partial V_o^{(1)}}{\partial z} = \frac{V_1 + V_2 + 2V_5 - 4V_o + \dfrac{h(V_1 - V_2)}{2\rho_o}}{2h} \qquad (3.123)$$

Similarly, applying Taylor series to points 1, 2, and 6 in medium 2, we get

$$V_1 = V_o + \frac{\partial V_o^{(1)}}{\partial \rho} h + \frac{\partial^2 V_o^{(1)}}{\partial \rho^2} \frac{h^2}{2} + \cdots$$

$$V_2 = V_o - \frac{\partial V_o^{(1)}}{\partial \rho} h + \frac{\partial^2 V_o^{(1)}}{\partial \rho^2} \frac{h^2}{2} - \cdots \qquad (3.124)$$

$$V_6 = V_o - \frac{\partial V_o^{(2)}}{\partial z} h + \frac{\partial^2 V_o^{(2)}}{\partial z^2} \frac{h^2}{2} - \cdots$$

Combining Equations 3.111 and 3.124 leads to

$$h^2 \nabla^2 V = V_1 + V_2 + 2V_6 - 4V_o - 2h \frac{\partial V_o^{(2)}}{\partial z} + \frac{h(V_1 - V_2)}{2\rho_o} = 0$$

or

$$-\frac{\partial V_o^{(2)}}{\partial z} = \frac{V_1 + V_2 + 2V_6 - 4V_o + \dfrac{h(V_1 - V_2)}{2\rho_o}}{2h} \qquad (3.125)$$

But $D_{1n} = D_{2n}$ or

$$\varepsilon_1 \frac{\partial V_o^{(1)}}{\partial z} = \varepsilon_2 \frac{\partial V_o^{(2)}}{\partial z} \qquad (3.126)$$

Substituting Equations 3.123 and 3.125 into Equation 3.126 and solving for V_o yields

$$V_o = \frac{1}{4}\left(1 + \frac{h}{2\rho_o}\right)V_1 + \frac{1}{4}\left(1 - \frac{h}{2\rho_o}\right)V_2 + \frac{\varepsilon_1}{2(\varepsilon_1 + \varepsilon_2)}V_5 + \frac{\varepsilon_2}{2(\varepsilon_1 + \varepsilon_2)}V_6 \qquad (3.127)$$

Equation 3.127 is only applicable to interface points. Notice that Equation 3.127 becomes Equation 3.114 if $\epsilon_1 = \epsilon_2$.

Typical examples of finite difference approximations for boundary points, written for square nets in rectangular and cylindrical systems, are tabulated in Table 3.9. For more examples, see References 12,121. The FDTD has also been applied in solving time-varying axisymmetric problems [91,122].

3.11.2 Spherical Coordinates

In spherical coordinates, Laplace's equation can be written as

$$\nabla^2 V = \frac{\partial^2 V}{\partial r^2} + \frac{2}{r}\frac{\partial V}{\partial r} + \frac{1}{r^2}\frac{\partial^2 V}{\partial \theta^2} + \frac{\cot\theta}{r^2}\frac{\partial V}{\partial \theta} + \frac{1}{r^2\sin^2\theta}\frac{\partial^2 V}{\partial \phi^2} = 0 \qquad (3.128)$$

TABLE 3.9

Finite Difference Approximations at Boundary Points

Description	Figure	Cartesian Equation	Cylindrical Equation
1. Bottom edge		$4V_0 = V_1 + V_2 + 2V_3$	$4V_0 = V_1 + V_2 + 4V_3$
2. Top edge		$4V_0 = V_1 + V_2 + 2V_4$	$4V_0 = V_1 + V_2 + 2V_3$
3. Left edge		$4V_0 = 2V_2 + V_3 + V_4$	$8V_0 = 4V_2 + \left(\frac{2i+1}{i}\right)V_3$ $+ \left(\frac{2i-1}{i}\right)V_4$
4. Right edge		$4V_0 = 2V_2 + V_3 + V_4$	$8V_0 = 4V_2 + \left(\frac{2i+1}{i}\right)V_3$ $+ \left(\frac{2i-1}{i}\right)V_4$
5. Bottom left corner point		$2V_0 = V_1 + V_3$	$3V_0 = V_1 + 2V_3$
6. Bottom right corner point		$2V_0 = V_2 + V_3$	$3V_0 = V_2 + 2V_3$
7. Top left corner point		$2V_0 = V_1 + V_4$	$3V_0 = V_1 + 2V_4$
8. Top right corner point		$2V_0 = V_2 + V_4$	$3V_0 = V_2 + 2V_4$

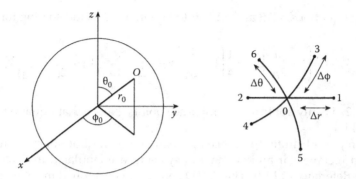

FIGURE 3.37
Typical node in spherical coordinates.

At a grid point $O(r_o, \theta_o, \phi_o)$ shown in Figure 3.37, the finite difference approximation to Equation 3.128 is

$$\frac{V_1 - 2V_o + V_2}{(\Delta r)^2} + \frac{2}{r_o}\left(\frac{V_1 - V_2}{2\Delta r}\right) + \frac{V_6 - 2V_o + V_5}{(r_o\Delta\theta)^2}$$
$$+ \frac{\cot\theta_o}{r_o^2}\left(\frac{V_5 - V_6}{2\Delta\theta}\right) + \frac{V_3 - 2V_o + V_4}{(r_o\Delta\phi\sin\theta_o)^2} = 0 \qquad (3.129)$$

Note that θ increases from node 6 to 5, and hence we have $V_5 - V_6$ and not $V_6 - V_5$ in Equation 3.129.

EXAMPLE 3.9

Consider an earthed metal cylindrical tank partly filled with a charge liquid, such as hydrocarbons, as illustrated in Figure 3.38a. Using the finite difference method, determine the potential distribution in the entire domain. Plot the potential along $\rho = 0.5, 0 < z < 2$ m and on the surface of the liquid. Take

$a = b = c = 1.0$ m,
$\epsilon_r = 2.0$ (hydrocarbons),
$\rho_v = 10^{-5}$ C/m³

Solution

The exact analytic solution to this problem was given in Section 2.7.2.

It is apparent from Figure 3.38a and from the fact that ρ_v is uniform that $V = V(\rho, z)$ (i.e., the problem is two-dimensional) and the domain of the problem is symmetrical about the z-axis. Therefore, it is only necessary to investigate the solution region in Figure 3.38b and impose the condition that the z-axis is a flux line, that is, $\partial V/\partial n = \partial V/\partial \rho = 0$.

The finite difference grid of Figure 3.35 is used with $0 \le i \le I_{max}$ and $0 \le j \le J_{max}$. Choosing $\Delta\rho = \Delta z = h = 0.05$ m makes $I_{max} = 20$ and $J_{max} = 40$. Equation 3.115 is applied for gas space, and Equation 3.121 for liquid space. Along the z-axis, that is, $i = 0$, we impose the Neumann condition in Equation 3.120. To account for the fact that the gas has dielectric constant ϵ_{r1} while the liquid has ϵ_{r2}, we impose the boundary condition in Equation 3.127 on the liquid–gas interface.

Based on these ideas, the computer program shown in Figure 3.39 was developed to determine the potential distribution in the entire domain. The values of the potential along $\rho = 0.5, 0 < z < 2$ and along the gas–liquid interface are plotted in Figure 3.40.

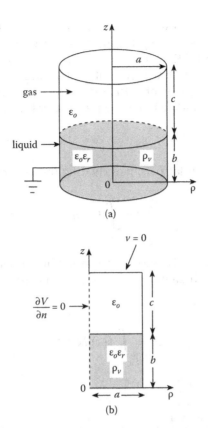

FIGURE 3.38
For Example 3.9: (a) earthed cylindrical tank, (b) solution region.

It is evident from the figure that the finite difference solution compares well with the exact solution in Section 2.7.2. It is the simplicity in concept and ease of programming finite difference schemes that make them very attractive for solving problems such as this.

3.12 Numerical Integration

Numerical integration (also called *numerical quadrature*) is used in science and engineering whenever a function cannot easily be integrated in closed form or when the function is described in the form of discrete data. Integration is a more stable and reliable process than differentiation. The term *quadrature* or *integration rule* will be used to indicate any formula that yields an integral approximation. Several integration rules have been developed over the years. The common ones include

1. Euler's rule,
2. Trapezoidal rule,
3. Simpson's rule,
4. Newton–Cotes rules, and
5. Gaussian (quadrature) rules.

The first three are simple and will be considered first to help build up background for other rules which are more general and accurate. A discussion on the subject of numerical integration with diverse FORTRAN codes can be found in Davis and Rabinowitz [123]. A program package called QUADPACK for automatic integration covering a wide variety of problems and various degrees of difficulty is presented in Piessens et al. [124]. Our discussion will be brief but sufficient for the purpose of this text.

3.12.1 Euler's Rule

To apply the Euler or rectangular rule in evaluating the integral

$$I = \int_a^b f(x)dx, \tag{3.130}$$

where $f(x)$ is shown in Figure 3.41, we seek an approximation for the area under the curve. We divide the curve into n equal intervals as shown in Figure 3.41. The subarea under the curve within $x_{i-1} < x < x_i$ is

$$A_i = \int_{x_{i-1}}^{x_i} f(x)dx \simeq h\,f_i \tag{3.131}$$

```
%%%%%%%%%%%%%%%%%%%%%%%%%%%%%%%%
% MATLAB CODE FOR EXAMPLE 3.9:
% AN AXISYMMETRIC PROBLEM OF AN EARTHED CYLINDER PARTIALLY FILLED WITH
% CHARGED LIQUID SOLVED USING FINITE DIFFERENCE SCHEME
%%%%%%%%%%%%%%%%%%%%%%%%%%%%%%%%

A=1; % Radius of cylindrical tank (meters)
B=1; % Height of liquid in tank
C=1; % Height of gas in tank

ER1=1;ER2=2;EO=8.854E-12; %dielectric parameters

H = 0.05;    %spacial step size
NA = A/H;    %number of points along A
NB = B/H;    %number of points along B
NC = C/H;    %number of points along C
NBC = NB + NC;  %%number of points along B & C
NMAX = 500;    %number of iterations
RHOV = 1E-5;
G = -RHOV/(ER2*EO);
GH2 = G*H^2;

%INITIALIZE - THIS ALSO TAKES CARE OF DIRICHLET CONDITIONS
V = zeros(NA+1,NBC+1);   %V(radius, height)

%NOW, APPLY FINITE DIFFERENCE SCHEME

for N = 1:NMAX
    for I = 2:NA %step through radius
        FM = (2*(I-1) - 1)/(2*(I-1)); %note that indicie starts at 1 instead of 0
        FP = (2*(I-1) + 1)/(2*(I-1));
```

FIGURE 3.39
MATLAB code for Example 3.9.   *(Continued)*

```
%step through liquid (z+ direction)
for J = 2:NB
    V(I,J) = 0.25*( V(I,J-1) + V(I,J+1)...
        + FM*V(I-1,J) + FP*V(I+1,J) - GH2 );
end

%step through gas (z+ direction)
for J = NB+2:NBC
    V(I,J) = 0.25*( V(I,J-1) + V(I,J+1)...
        + FM*V(I-1,J) + FP*V(I+1,J) );
end
%ALONG THE GASS-LIQUID INTERFACE
V(I,NB+1) = 0.5*( V(I,NB+2)*ER1/(ER1+ER2)+...
    V(I,NB)*ER2/(ER1+ER2) )...
    + 0.25*(FM*V(I-1,NB+1)+FP*V(I+1,NB+1) );
end
%IMPOSE NEUMANN CONDITION ALONG THE Z-AXIS
for J = 2:NBC
    V(1,J) = ( 4.0*V(2,J) + V(1,J-1)+V(1,J+1))/6.0;
end
end

% OUTPUT THE POTENTIAL ALONG RHO = 0.5, 0 < Z < 1.0
NR5 = 0.5/H;

radius = 0:H:A;
height = 0:H:(B+C);

figure(1),
    subplot(211),
        plot(height,V(NR5+1,(0:NBC)+1)/1000.0)
            ylabel('V(kV)')
            title('Potential distribution in the tank of Figure 3.38 along \rho=0.5m')
    subplot(212)
        plot(radius ,V((0:NA)+1,NB+1)/1000.0)
            title('Potential distribution in the tank of Figure 3.38 along gas-
liquid interface')
            ylabel('V(kV)')

    figure(2),surf(height,radius,V/1000)
        ylabel('radius in tank')
        xlabel('height in tank')
        zlabel('Potential (kV)')
```

FIGURE 3.39 (Continued)
MATLAB code for Example 3.9.

where $f_i = f(x_i)$. The total area under the curve is

$$I = \int_a^b f(x)dx \simeq \sum_{i=1}^n A_i$$
$$= h[f_1 + f_2 + \cdots + f_n]$$

or

$$I = h \sum_{i=1}^n f_i \tag{3.132}$$

It is clear from Figure 3.41 that this quadrature method gives an inaccurate result since each A_i is less or greater than the true area introducing negative or positive error, respectively.

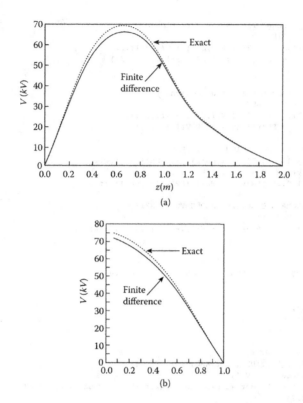

(a)

(b)

FIGURE 3.40
Potential distribution in the tank of Figure 3.38: (a) along $\rho = 0.5$ m, $0 \leq z \leq 2$ m; (b) along the gas–liquid interface.

3.12.2 Trapezoidal Rule

To evaluate the same integral in Equation 3.130 using the trapezoidal rule, the subareas are chosen as shown in Figure 3.42. For the interval $x_{i-1} < x < x_i$,

$$A_i = \int_{x_i-1}^{x_i} f(x)dx \simeq \left(\frac{f_{i-1} + f_i}{2}\right)h \tag{3.133}$$

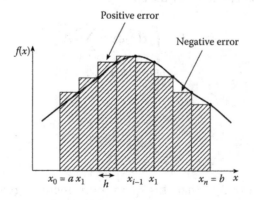

FIGURE 3.41
Integration using Euler's rule.

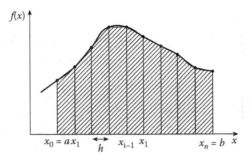

FIGURE 3.42
Integration using the trapezoidal rule.

Hence,

$$I = \int_a^b f(x)dx \simeq \sum_{i=1}^n A_i$$

$$= h\left[\frac{f_0 + f_1}{2} + \frac{f_1 + f_2}{2} + \cdots + \frac{f_{n-2} + f_{n-1}}{2} + \frac{f_{n-1} + f_n}{2}\right]$$

$$= \frac{h}{2}[f_0 + 2f_1 + 2f_2 + \cdots + 2f_{n-1} + f_n]$$

or

$$I = h\sum_{i=1}^{n-1} f_i + \frac{h}{2}(f_0 + f_n) \qquad (3.134)$$

3.12.3 Simpson's Rule

Simpson's rule gives a still more accurate result than the trapezoidal rule. While the trapezoidal rule approximates the curve by connecting successive points on the curve by straight lines, Simpson's rule connects successive groups of three points on the curve by a second-degree polynomial (i.e., a parabola). Thus,

$$A_i = \int_{x_i-1}^{x_i} f(x)dx \simeq \frac{h}{3}(f_{i-1} + f_i + f_{i-1}) \qquad (3.135)$$

Therefore,

$$I = \int_a^b f(x)dx \simeq \sum_{i=1}^n A_i \qquad (3.136)$$

$$I = \frac{h}{3}[f_0 + 4f_1 + 2f_2 + 4f_3 + \cdots + 2f_{n-2} + 4f_{n-1} + f_n]$$

where n is even.

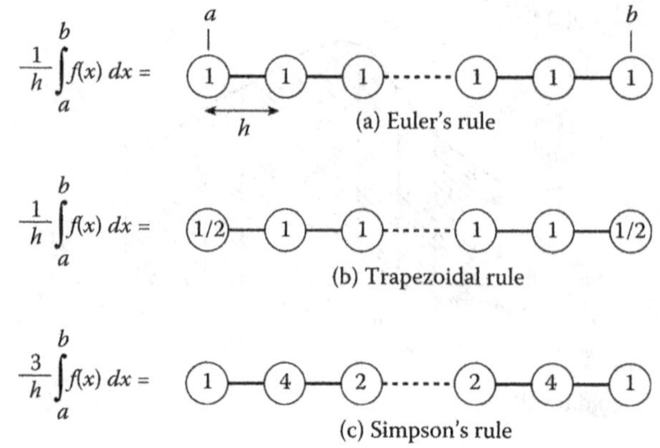

FIGURE 3.43
Computational molecules for integration.

The computational molecules for Euler's, trapezoidal, and Simpson's rules are shown in Figure 3.43. Now that we have considered simple quadrature rules to help build up background, we now consider more general, accurate methods.

3.12.4 Newton–Cotes Rules

To apply a Newton–Cotes rule to evaluate the integral in Equation 3.130, we divide the interval $a < x < b$ into m equal intervals so that

$$h = \frac{b-a}{m} \tag{3.137}$$

where m is a multiple of n, and n is the number of intervals covered at a time or the order of the approximating polynomial. The subarea in the interval $x_{n(i-1)} < x < x_{ni}$ is

$$A_i = \int_{x_n(i-1)}^{x_{ni}} f(x)dx \simeq \frac{nh}{N} \sum_{k=0}^{n} C_k^n f(x_{n(i-1)+k}) \tag{3.138}$$

The coefficients $C_k^n, 0 \leq k \leq n$, are called Newton–Cotes numbers and tabulated in Table 3.10. The numbers are obtained from

$$C_k^n = \frac{1}{n} \int_0^N L_k(s)ds \tag{3.139}$$

where

$$L_k(s) = \prod_{j=0,\neq k}^{n} \frac{s-j}{k-j} \tag{3.140}$$

It is easily shown that the coefficients are symmetric, that is,

$$C_k^n = C_{n-k}^n \tag{3.141a}$$

TABLE 3.10

Newton–Cotes Numbers

n	N	NC_0^n	NC_1^n	NC_2^n	NC_3^n	NC_4^n	NC_5^n	NC_6^n	NC_7^n	NC_8^n
1	2	1	1							
2	6	1	4	1						
3	8	1	3	3	1					
4	90	7	32	12	32	7				
5	288	19	75	50	50	75	19			
6	840	41	216	27	272	27	216	41		
7	17280	751	3577	1323	2989	2989	1323	3577	751	
8	28350	989	5888	–928	10496	–4540	10946	–928	5888	989

and they sum up to unity, that is,

$$\sum_{k=0}^{n} C_k^n = 1 \tag{3.141b}$$

For example, for $n = 2$,

$$C_0^2 = \frac{1}{2} \int_0^6 \frac{(s-1)(s-2)}{(-1)(-2)} ds = \frac{1}{6},$$

$$C_1^2 = \frac{1}{2} \int_0^6 \frac{s(s-2)}{1(-1)} ds = \frac{4}{6},$$

$$C_2^2 = \frac{1}{2} \int_0^6 \frac{(s-1)}{2(1)} ds = \frac{1}{6}$$

Once the subareas are found using Equation 3.138, then

$$\boxed{I = \int_a^b f(x)dx \simeq \sum_{i=1}^{m/n} A_i} \tag{3.142}$$

The most widely known Newton–Cotes formulas are

$n = 1$ (2-point; trapezoidal rule)

$$A_i \simeq \frac{h}{2}(f_{i+1} + f_i), \tag{3.143}$$

$n = 2$ (3-point; Simpson's 1/3 rule)

$$A_i \simeq \frac{h}{3}(f_{i-1} + 4f_i + f_{i+1}), \tag{3.144}$$

$n = 3$ (4-point; Newton's rule)

$$A_i \simeq \frac{3h}{8}(f_i + 3f_{i+1} + 3f_{i+2} + f_{i+3}) \tag{3.145}$$

3.12.5 Gaussian Rules

The integration rules considered so far involve the use of equally spaced abscissa points. The idea of integration rules using unequally spaced abscissa points stems from Gauss. The Gaussian rules are more complicated but more accurate than the Newton–Cotes rules. A Gaussian rule has the general form

$$\boxed{\int_a^b f(x)dx \simeq \sum_{i=1}^n w_i f(x_i)} \tag{3.146}$$

where (a, b) is the interval for which a sequence of orthogonal polynomials $\{w_i(x)\}$ exists, x_i are the zeros of $w_i(x)$, and the weights w_i are such that Equation 3.146 is of degree of precision $2n - 1$. Any of the orthogonal polynomials discussed in Chapter 2 can be used to give a particular Gaussian rule. Commonly used rules are Gauss–Legendre, Gauss–Chebyshev, etc., since the sample points x_i are the roots of the Legendre, Chebyshev, etc., of degree n. For the Legendre ($n = 1$–16) and Laguerre ($n = 1$–16) polynomials, the zeros x_i and weights w_i have been tabulated in Reference 112.

Using Gauss–Legendre rule,

$$\int_a^b f(x)dx \simeq \frac{b-a}{2} \sum_{i=1}^n w_i f(u_i) \tag{3.147}$$

where $u_i = [(b-a)/2]x_i + [(b+a)/2]$ are the transformation of the roots x_i of Legendre polynomials from limits $(-1, 1)$ to finite limits (a, b). The values of the abscissas x_i and weights w_i for n up to 7 are presented in Table 3.11; for higher values of n, the interested reader is referred to [125,126]. Note that $-1 < x_i < 1$ and $\sum_{i=1}^n w_i = 2$.

The Gauss–Chebyshev rule is similar to the Gauss–Legendre rule. We use Equation 3.147 except that the sample points x_i, the roots of Chebyshev polynomial $T_n(x)$, are

$$x_i = \cos\frac{(2i-1)}{2n}, \quad i = 1, 2, \ldots, n \tag{3.148}$$

and the weights are all equal [127], that is,

$$w_i = \frac{\pi}{n} \tag{3.149}$$

When either of the limits of integration a or b or both are $\pm\infty$, we use Gauss–Laguerre or Gauss–Hermite rule. For the Gauss–Laguerre rule,

$$\int_0^\infty f(x)dx \simeq \sum_{i=1}^n w_i f(x_i) \tag{3.150}$$

where the appropriate abscissas x_i, the roots of Laguerre polynomials, and weights w_i are listed for n up to 7 in Table 3.12. For the Gauss–Hermite rule,

$$\int_{-\infty}^\infty f(x)dx \simeq \sum_{i=1}^n w_i f(x_i) \tag{3.151}$$

TABLE 3.11

Abscissas (Roots of Legendre Polynomials) and
Weights for Gauss–Legendre Integration

$\pm x_i$	w_i
$n = 2$	
0.57735 02691 89626	1.00000 00000 00000
$n = 3$	
0.00000 00000 00000	0.88888 88888 88889
0.77459 66692 41483	0.55555 55555 55556
$n = 4$	
0.33998 10435 84856	0.65214 51548 62546
0.86113 63115 94053	0.34785 48451 37454
$n = 5$	
0.00000 00000 00000	0.56888 88888 88889
0.53846 93101 05683	0.47862 86704 99366
0.90617 98459 38664	0.23692 68850 56189
$n = 6$	
0.23861 91860 83197	0.46791 39345 72691
0.66120 93864 66265	0.36076 15730 48139
0.93246 95142 03152	0.17132 44923 79170
$n = 7$	
0.00000 00000 00000	0.41795 91836 73469
0.40584 51513 77397	0.38183 00505 05119
0.74153 11855 99394	0.27970 53914 89277
0.94910 79123 42759	0.12948 49661 68870

where the abscissas x_i, the roots of the Hermite polynomials, and weights w_i are listed for n up to 7 in Table 3.13. An integral over (a, ∞) is taken care of by a change of variable so that

$$\int_a^\infty f(x)\,dx = \int_0^\infty f(y+a)\,dy \tag{3.152}$$

We apply Equation 3.146 with $f(x)$ evaluated at points $x_i + a$, $i = 1, 2, \ldots, n$ and x_is are tabulated in Table 3.12.

A major drawback with Gaussian rules is that if one wishes to improve the accuracy, one must increase n which means that the values of w_i and x_i must be included in the program for each value of n. Another disadvantage is that the function $f(x)$ must be explicit since the sample points x_i are unassigned.

3.12.6 Multiple Integration

This is an extension of one-dimensional (1D) integration discussed so far. A double integral is evaluated by means of two successive applications of the rules presented above for single integral [128]. To evaluate the integral using the Newton–Cotes or Simpson's 1/3 rule ($n = 2$), for example,

$$I = \int_a^b \int_c^d f(x,y)\,dx\,dy \tag{3.153}$$

TABLE 3.12

Abscissas (Roots of Laguerre Polynomials) and
Weights for Gauss–Laguerre Integration

$\pm x_i$	w_i
$n = 2$	
0.58578 64376 27	1.53332 603312
3.41421 35623 73	4.45095 733505
$n = 3$	
0.41577 45567 83	1.07769 285927
2.29428 03602 79	2.76214 296190
6.28994 50829 37	5.60109 462543
$n = 4$	
0.32254 76896 19	0.83273 912383
1.74576 11011 58	2.04810 243845
4.53662 02969 21	3.63114 630582
9.39507 09123 01	6.48714 508441
$n = 5$	
0.26356 03197 18	0.67909 404220
1.41340 30591 07	1.63848 787360
3.59642 57710 41	2.76944 324237
12.64080 08442 76	7.21918 635435
$n = 6$	
0.22284 66041 79	0.57353 550742
1.18893 21016 73	1.36925 259071
2.99273 63260 59	2.26068 459338
5.77514 35691 05	3.35052 458236
9.83746 74183 83	4.88682 680021
15.98287 39806 02	7.84901 594560
$n = 7$	
0.19304 36765 60	0.49647 759754
1.02666 48953 39	1.17764 306086
2.56787 67449 51	1.91824 978166
4.90035 30845 26	2.77184 863623
8.18215 34445 63	3.84124 912249
12.73418 02917 98	5.38067 820792
19.39572 78622 63	8.40543 248683

over a rectangular region $a < x < b$, $c < y < d$, we divide the region into $m \cdot l$ smaller rectangles with sides

$$h_x = \frac{b-a}{m} \tag{3.154a}$$

$$h_y = \frac{d-c}{l} \tag{3.154b}$$

where m and l are multiples of $n = 2$. The subarea

$$A_{ij} = \int_{y_{n(j-1)}}^{y_{n(j+1)}} dy \int_{x_{n(i-1)}}^{x_{n(i+1)}} f(x,y)dx \tag{3.155}$$

TABLE 3.13

Abscissas (Roots of Hermite Polynomials) and Weights for
Gauss–Hermite Integration

$\pm x_i$	w_i
$n = 2$	
0.70710 67811 86548	1.46114 11826 611
$n = 3$	
0.00000 00000 00000	1.18163 59006 037
1.22474 48713 91589	1.32393 11752 136
$n = 4$	
0.52464 76232 75290	1.05996 44828 950
1.65068 01238 85785	1.24022 58176 958
$n = 5$	
0.00000 00000 00000	0.94530 87204 829
0.95857 24646 13819	0.98658 09967 514
2.02018 28704 56086	1.18148 86255 360
$n = 6$	
0.43607 74119 27617	0.87640 13344 362
1.33584 90740 13697	0.93558 05576 312
2.35060 49736 74492	1.13690 83326 745
$n = 7$	
0.00000 00000 00000	0.81026 46175 568
0.81628 78828 58965	0.82868 73032 836
1.67355 16287 67471	0.89718 46002 252
2.65196 13568 35233	1.10133 07296 103

is evaluated by integrating along x and then along y according to Equation 3.144:

$$A_{ij} \simeq \frac{h_x}{3}(g_{j-1} + 4g_j + g_{j+1}) \tag{3.156}$$

where

$$g_j \simeq \frac{h_y}{3}(f_{i-1,j} + 4f_{i,j} + f_{i+1,j}) \tag{3.157}$$

Substitution of Equation 3.157 into Equation 3.156 yields

$$A_{ij} = \frac{h_x h_y}{9}\Big[(f_{i+1,j+1} + f_{i+1,j-1} + f_{i-1,j+1} + f_{i-1,j-1})$$
$$+ 4(f_{i,j+1} + f_{i,j-1} + f_{i+1,j} + f_{i-1,j}) + 16f_{i,j}\Big] \tag{3.158}$$

The corresponding schematic or integration molecule is shown in Figure 3.44. Summing the value of A_{ij} for all subareas yields

$$I = \sum_{i=1}^{m/n} \sum_{j=1}^{l/n} A_{ij} \tag{3.159}$$

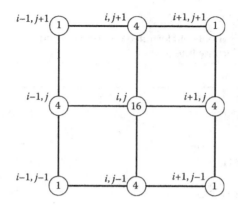

FIGURE 3.44
Double integration molecule for Simpson's 1/3 rule.

The procedure applied in the 2D integral can be extended to a 3D integral. To evaluate

$$I = \int\limits_{a}^{b} \int\limits_{c}^{d} \int\limits_{e}^{f} f(x,y,z)\,dx\,dy\,dz \tag{3.160}$$

using the $n = 2$ rule, the cuboid $a < x < b, c < y < d, e < z < f$ is divided into $m \cdot l \cdot p$ smaller cuboids of sides

$$h_x = \frac{b-a}{m}$$

$$h_y = \frac{d-c}{l} \tag{3.161}$$

$$h_z = \frac{f-e}{p}$$

where m, l, and p are multiples of $n = 2$. The subvolume A_{ijk} is evaluated by integrating along x according to Equation 3.144 to obtain

$$g_{j,k} = \frac{h_x}{3}(f_{i+1,j,k} + 4f_{i,j,k} + f_{i-1,j,k}), \tag{3.162}$$

then along y to obtain

$$g_k = \frac{h_y}{3}(g_{j+1,k} + 4g_{j,k} + g_{j-1,k}), \tag{3.163}$$

and finally along z to obtain

$$A_{ijk} = \frac{h_z}{3}(g_{k+1} + 4g_k + g_{k-1}) \tag{3.164}$$

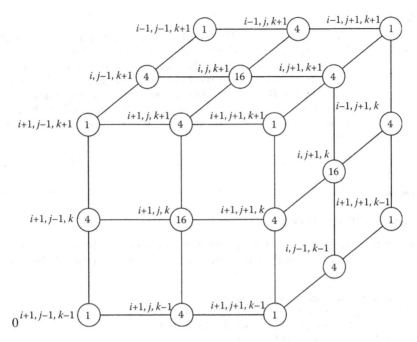

FIGURE 3.45
Triple integration molecule for Simpson's 1/3 rule.

Substituting Equations 3.162 and 3.163 into Equation 3.164 results in [115]

$$
\begin{aligned}
A_{ijk} = \frac{h_x h_y h_z}{27} \big[& (f_{i-1,j-1,k+1} + 4f_{i-1,j,k+1} + f_{i-1,j+1,k+1}) \\
& + (4f_{i,j-1,k+1} + 16f_{i,j,k+1} + 4f_{i,j+1,k+1}) \\
& + (f_{i+1,j-1,k+1} + 4f_{i+1,j,k+1} + f_{i+1,j+1,k+1}) \\
& + (4f_{i-1,j-1,k} + 16f_{i-1,j,k} + 4f_{i-1,j+1,k}) \\
& + (16f_{i,j-1,k} + 64f_{i,j,k} + 16f_{i,j+1,k}) \\
& + (4f_{i+1,j-1,k} + 16f_{i+1,j,k} + 4f_{i+1,j+1,k}) \\
& + (f_{i-1,j-1,k-1} + 4f_{i-1,j,k-1} + f_{i-1,j+1,k-1}) \\
& + (4f_{i,j-1,k-1} + 16f_{i,j,k-1} + 4f_{i,j+1,k-1}) \\
& + (f_{i+1,j-1,k-1} + 4f_{i+1,j,k-1} + f_{i+1,j+1,k-1}) \big]
\end{aligned}
$$

(3.165)

The integration molecule is portrayed in Figure 3.45. Observe that the molecule is symmetric with respect to all planes that cut the molecule in half.

EXAMPLE 3.10

Write a program that uses the Newton–Cotes rule ($n = 6$) to evaluate Bessel function of order m, that is,

$$
J_m(x) = \frac{1}{\pi} \int_0^\pi \cos(x\sin\theta - m\theta)d\theta
$$

Run the program for $m = 0$ and $x = 0.1, 0.2, \ldots, 2.0$.

Solution

The computer program is shown in Figure 3.46. The program is based on Equations 3.138 and 3.142. It evaluates the integral within a subinterval $\theta_{n(i-1)} < \theta < \theta_{ni}$. The summation over all the subintervals gives the required integral. The result for $m = 0$ and $0.1 < x < 2.0$ is shown in Table 3.14; the values agree up to six significant figures with those in standard tables [126, p. 390]. The program is intentionally made general so that n, the corresponding Newton–Cotes numbers, and the integrand can be changed easily. Although the integrand in Figure 3.46 is real, the program can be modified for complex integrand.

3.13 Concluding Remarks

Only a brief treatment of the finite difference analysis of PDEs is given here. There are many valuable references on the subject which answer many of the questions left

```
% Create custom Bessel function
lwr = 0;    %lower limit of integration
uppr= pi;   %upper limit of integration
theta = linspace(0,pi,N+1);

for ii = 1:length(x)
    % Create function to be integrated
    A = cos(x(ii)*sin(theta)-m*theta)/pi;

    %Integrate with Newton-Cotes method
    J_m(ii) = nc_method(A,lwr,uppr,n);
end

% Difference between MATLAB and custom Bessel
compare = J-J_m;

% eps is the Spacing of floating point numbers in MATLAB.
% Any number smaller than in magnitude than eps is essen-
tially zero
% Try 1 + eps, this is equal to 1.
v_eps = [eps, eps];
x_eps = [x(1),x(end)];

figure(1),
    subplot(211),plot(x,J,x,J_m,':')
        title('Comparison between MATLAB and Custom Bessel func-
tion')
        legend('MATLAB','Custom')
    subplot(212),plot(x,compare,'-o',x_eps,v_eps,x_eps,-v_eps)
        title('Difference between MATLAB and Custom Bessel')
        legend('difference','+eps','-eps')

%figure(2),plot(theta,A)

(a)
```

FIGURE 3.46
(a) Main program for Example 3.10. (b) Function for the main program in (a); for Example 3.10.   *(Continued)*

```
function I = nc_method(f,a,b,n)

% I = nc_method_fun(f,a,b,n)
%
% Calculates Newton-Cotes Method Quadrature
%
% Inputs:
% f : data vector from function to be integrated
% a : lower integration limit
% b : upper integration limit
% n : order of approximating polynomial
%
% Output:
% I : Value of Integral
%
% NOTE that m (the number of sections) must be a multiple of the or-
der of
% the approximation polynomial

m = length(f)-1;     %Divide the data into m sections

%Below checks if the requirement that m is a multiple of n
if rem(m,n) ~= 0
    disp('Error: length(f)-1 must be a multiple of n')
    I = NaN; return
end

h = (b-a)/m;         %step size.
C = nc_weight (n);   %Get Newton-Cotes Number.
I = 0;               %intitialize I
for ii = 1:n:m,
    ind2 = (ii:ii+n);   %select indicies of function
    Ai = n*h*f(ind2)*C;       %Equation (3.138)
    I = I+Ai;                 %Equation (3.142)
end

function q = nc_weight(n)
% returns in q the weights for the N-point Newton-Cotes quadra-
ture rule.
% N must be between 1 and 8.

if (n<1)
    disp('error: newton-cotes order must be at least 1!')
elseif (n==1) %  Trapezoidal rule:
    q = [1 1]'/2;
elseif (n==2) %  Simpson's rule:
    q = [1 4 1]'/6;
elseif (n==3) %  Simpson's 3/8 rule:
    q = [1 3 3 1]'/8;
elseif (n==4)%  Boole's rule:
    q = [7 32 12 32 7]'/90;
```

FIGURE 3.46 (Continued)
(a) Main program for Example 3.10. (b) Function for the main program in (a); for Example 3.10.   (*Continued*)

```
elseif (n==5)
    q = [19 75 50 50 75 19]'/288;
elseif (n==6)
    q = [41 216 27 272 27 216 41]'/840;
elseif (n==7)
    q = [751 3577 1323 2989 2989 1323 3577 751]'/17280;
elseif (n==8)
    q = [989 5888 -928 10496 -4540 10496 -928 5888 989]'/28350;
else
    disp('error: N must be no more than 8!')
end
```

(b)

FIGURE 3.46 (Continued)
(a) Main program for Example 3.10. (b) Function for the main program in (a); for Example 3.10.

unanswered here [3–8,10, 102, 103]. The book by Smith [5] gives an excellent exposition with numerous examples. The problems of stability and convergence of finite difference solutions are further discussed in References 129,130, while the error estimates are discussed in Reference 131.

As noted in Section 3.8, the finite difference method has some inherent advantages and disadvantages. It is conceptually simple and easy to program. The finite difference approximation to a given PDE is by no means unique; more accurate expressions can be obtained by employing more elaborate and complicated formulas. However, the relatively simple approximations may be employed to yield solutions of any specified accuracy simply by reducing the mesh size provided that the criteria for stability and convergence are met.

A very important difficulty in finite differencing of PDEs, especially parabolic and hyperbolic types, is that if one value of Φ is not calculated and therefore is set equal to zero by mistake, the solution may become unstable. For example, in finding the difference between $\Phi_i = 1000$ and $\Phi_{i+1} = 1002$, if Φ_{i+1} is set equal to zero by mistake, the difference of 1000 instead of 2 may cause instability. To guard against such error, care must be taken to ensure that Φ is calculated at every point, particularly at boundary points.

TABLE 3.14

Result of the Program
in Figure 3.46 for $m = 0$

x	$J_0(x)$
0.1	0.9975015
0.2	0.9900251
0.3	0.9776263
0.4	0.9603984
0.5	0.9384694
⋮	⋮
1.5	0.5118274
1.6	0.4554018
1.7	0.3979859
1.8	0.3399859
1.9	0.2818182
2.0	0.2238902

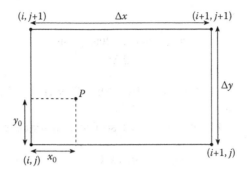

FIGURE 3.47
Evaluating Φ at a point P not on the grid.

The applications of finite difference method in general and FDTD method in particular have been limited due to memory requirement in spite of the fact that they are simple in concept and implementation. Several memory-efficient algorithms and hybrid methods have been proposed to increase the efficiency of FDTD method [132].

A serious limitation of the finite difference method is that interpolation of some kind must be used to determine solutions at points not on the grid. Suppose we want to find Φ at a point P which is not on the grid, as in Figure 3.47. Assuming Φ is known at the four grid points surrounding P, at a distance x_o along the bottom edge of the rectangle in Figure 3.47,

$$\Phi_b = \frac{x_o}{\Delta x}[\Phi(i+1,j) - \Phi(i,j)] + \Phi(i,j) \tag{3.166}$$

At a distance x_o along the top edge,

$$\Phi_t = \frac{x_o}{\Delta x}[\Phi(i+1,j+1) - \Phi(i,j+1)] + \Phi(i,j+1) \tag{3.167}$$

The value of Φ at P is estimated by combining Equations 3.166 and 3.167, that is,

$$\Phi_P = \frac{y_o}{\Delta y}(\Phi_t - \Phi_b) + \Phi_b \tag{3.168}$$

One obvious way to avoid interpolation is to use a finer grid if possible.

PROBLEMS

3.1 Show that the following finite difference approximations for Φ_x are valid:

a. forward difference,

$$\frac{-\Phi_{i+2} + 4\Phi_{i+1} - 3\Phi_i}{2\Delta x}$$

b. backward difference,

$$\frac{3\Phi_i - 4\Phi_{i-1} + \Phi_{i-2}}{2\Delta x}$$

    c. central difference

$$\frac{-\Phi_{i+2} + 8\Phi_{i+1} - 8\Phi_{i-1} + \Phi_{i-2}}{12\Delta x}$$

3.2 Solve $y'' - y = -1$, $0 < x < 1$ with $y'(0) = 0$, $y(1) = 2$. You should use finite difference method and take $\Delta x = 0.25$.

3.3 Solve the equation $\Phi_t = \Phi_{xx}$, $0 \le x \le 1$, subject to initial and boundary conditions

$$\Phi(x, 0) = \sin\pi x, \, 0 \le x \le 1,$$

$$\Phi(0, t) = 0 = \Phi(1, t) \, t > 0$$

Obtain the solution by hand calculation and use $\Delta x = 0.25$ and $r = 0.5$.

3.4 Obtain the finite difference formula for the differential equation:

$$U_x + U_t = 0$$

around the grid point (i, n), where $x = i\Delta x$, $t = n\Delta t$.

3.5 A one-dimensional wave equation is given by

$$\frac{\partial^2 V}{\partial x^2} = \frac{\partial^2 V}{\partial t^2}$$

Let $V(x, y) = V(i, n)$, $x = i\Delta x, t = n\Delta t$. Obtain the finite difference equivalent.

3.6 Derive the finite difference scheme for the one-dimension heat equation in cylindrical coordinates.

$$U_{\rho\rho} + \frac{1}{\rho}U_\rho = U_t, \quad 0 < \rho < 1, \quad t > 0$$

around the grid point (i, n), where $\rho = i\Delta\rho = ih$, $t = n\Delta t$.

3.7 Repeat the previous problem for two-dimensional heat equation

$$U_{\rho\rho} + \frac{1}{\rho}U_\rho + U_{zz} = U_t, \quad 0 < \rho < 1, \quad 0 < z < 1, \quad t > 0$$

Let $U(\rho, z, t) = U(i, j, n)$, where $\Delta\rho = \Delta z = h$, $\rho = ih, z = ih, t = n\Delta t$.

3.8 Derive the Crank–Nicholson implicit algorithm for the hyperbolic equation $\Phi_{xx} = a^2\Phi_{yy}$, $a^2 = $ constant. Let $\Delta x = \Delta y = \Delta$.

3.9 Given a boundary-value problem defined by

$$\frac{d^2\Phi}{dx^2} = x + 1, \quad 0 < x < 1$$

subject to $\Phi(0) = 0$ and $\Phi(1) = 1$, use the finite difference method to find $\Phi(0.5)$. You may take $\Delta = 0.25$ and perform five iterations. Compare your result with the exact solution.

3.10 Prove that the fourth-order approximation of Laplace's equation $\Phi_{xx} + \Phi_{yy} = 0$ is

$$60\Phi(i,j) - 16[\Phi(i+1,j) + \Phi(i-1,j) + \Phi(i,j+1) + \Phi(i,j-1)]$$
$$+ \Phi(i+2,j) + \Phi(i-2,j) + \Phi(i,j+2) + \Phi(i,j-2) = 0$$

Draw the computational molecule for the finite difference scheme.

3.11 a. If $\Delta x \neq \Delta y$, show that for the computational molecule in Figure 3.48a, Equation 3.49 becomes

$$V_o = \frac{V_1}{2(1+\alpha)} + \frac{V_2}{2(1+\alpha)} + \frac{V_3}{2(1+1/\alpha)} + \frac{V_4}{2(1+1/\alpha)}$$

where $\alpha = (\Delta x/\Delta y)^2$.

b. Show that for the molecule in Figure 3.48b, Equation 3.49 becomes

$$V_o = \frac{V_1}{(1+\Delta x_1/\Delta x_2)(1+\Delta x_1\Delta x_2/\Delta y_3\Delta y_4)}$$
$$+ \frac{V_2}{(1+\Delta x_2/\Delta x_1)(1+\Delta x_1\Delta x_2/\Delta y_3\Delta y_4)}$$
$$+ \frac{V_3}{(1+\Delta y_3/\Delta y_4)(1+\Delta y_3\Delta y_4/\Delta x_1\Delta x_2)}$$
$$+ \frac{V_4}{(1+\Delta y_4/\Delta y_3)(1+\Delta y_3\Delta y_4/\Delta x_1\Delta x_2)}$$

The molecule in Figure 3.48b is useful in treating irregular boundaries.

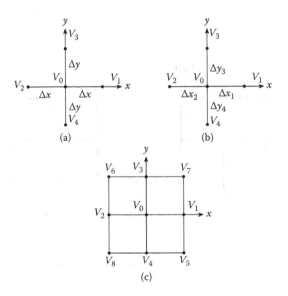

FIGURE 3.48
For Problem 3.11.

c. For the nine-point molecule in Figure 3.48c, show that

$$V_o = \frac{1}{8}\sum_{i=1}^{8} V_i$$

This is a more accurate difference equation than Equation 3.49.

3.12 A Dirichlet problem is characterized by

$$U_{xx} + U_{yy} = 0, \quad 0 < x < 1, \quad 0 < y < 1$$
$$U(0,y) = 0, \quad U(x,0) = 0$$
$$U(1,y) = 100y, \quad U(x,1) = 100x$$

By selecting $\Delta x = \Delta y = 0.25$, we have the square grid shown in Figure 3.49. Determine the potential at the nine free nodes.

3.13 For a long hollow conductor with a uniform U-shape cross section shown in Figure 3.50, find the potential at points A, B, C, D, and E.

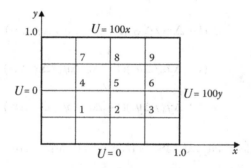

FIGURE 3.49
For Problem 3.12.

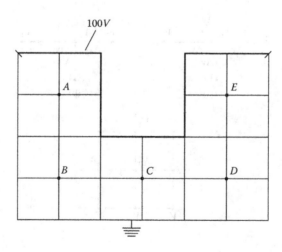

FIGURE 3.50
For Problem 3.13.

3.14 It is desired to solve

$$\frac{\partial^2 \Phi}{\partial x^2} + \frac{\partial^2 \Phi}{\partial y^2} + 50 = 0$$

in the square region $0 \le x \le 1$, $0 \le y \le 1$ subject to the boundary conditions $\Phi = 10$ at $x = 0, 1$, $\Phi_y = 40$ at $y = 0$, $\Phi_y = -20$ at $y = 1$.

a. Set up a system of finite difference equations which will allow the solution to be found at $x = y = 0.25$ using $\Delta x = \Delta y = h = 0.25$. Perform three iterations.

b. Develop a program to solve the same problem using $h = 0.05$, 0.1, and 0.2.

3.15 A potential problem is characterized by Poisson's equation

$$U_{xx} + U_{yy} = -2, \quad 0 < x < 6, \quad 0 < y < 8$$

with zero potential $U = 0$ on the boundaries. By selecting $\Delta x = \Delta y = h = 2$, we realize that there are six free nodes as shown in Figure 3.51. Use finite difference to determine the potential at the nodes.

3.16 Modify the code of Figure 3.12 to solve the following three-dimensional problem:

$$\nabla^2 V = -\rho_v / \varepsilon, \quad 0 \le x \le 1, \quad 0 \le y \le 1, \quad 0 \le z \le 1 \quad \text{meter},$$

where $\rho_v = xyz^2 nC/m^2$ and $\varepsilon = 2\varepsilon_o$ subject to the boundary conditions

$$V(0, y, z) = 0 = V(1, y, z)$$
$$V(x, 0, z) = 0 = V(x, 1, z)$$
$$V(x, y, 0) = 0 = V(x, y, 1) = V_o$$

Find the potential at the center of the cube and compare your result with the analytic solution. Take $V_o = 100$ V.

3.17 Show that the leapfrog method applied to the parabolic equation (3.10) is unstable, whereas applying the DuFort–Frankel scheme yields an unconditionally stable solution.

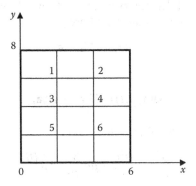

FIGURE 3.51
For Problem 3.15.

3.18 The advective equation

$$\frac{\partial \Phi}{\partial t} + u \frac{\partial \Phi}{\partial x} = 0, \quad u > 0$$

can be discretized as

$$\Phi_i^{n+1} = \Phi_i^n - r(\Phi_{i+1}^n - \Phi_{i-1}^n),$$

where $r = u\Delta t / 2\Delta x$. Show that the difference scheme is unstable. An alternative scheme is

$$\Phi_{n+1} = \frac{1}{2}(\Phi_{i+1}^n + \Phi_{i-1}^n) - r(\Phi_{i+1}^n - \Phi_{i-1}^n)$$

Find the condition on r for which this scheme is stable.

3.19 The two-dimensional parabolic equation

$$\frac{\partial U}{\partial t} = \frac{\partial^2 U}{\partial x^2} + \frac{\partial^2 U}{\partial y^2}, \quad 0 \le x \le 1, \quad 0 \le y \le 1, \quad t > 0$$

is approximated by the finite difference methods:

i. $U_{i,j}^{n+1} = [1 + r(\delta_x^2 + \delta_y^2)]U_{ij}^n$

ii. $U_{i,j}^{n+1} = (1 + r\delta_x^2)(1 + r\delta_y^2)]U_{ij}^n$

where
$r = \Delta t / h^2, h = \Delta x = \Delta y$

and

$$\delta_x^2 U_{i,j}^n = U_{i-1,j}^n - 2U_{i,j}^n + U_{i+1,j}^n$$
$$\delta_y^2 U_{i,j}^n = U_{i,j-1}^n - 2U_{i,j}^n + U_{i,j+1}^n$$

Show that (i) is stable for $r \le 1/4$ and (ii) is stable for $r \le 1/2$.

3.20 a. The constitutive parameters of the earth allow the displacement currents to be negligibly small. In this type of medium, show that Maxwell's equation for two-dimensional TM mode, where

$$\mathbf{E}(x, y, t) = E_z \mathbf{a}_z$$

and

$$\mathbf{H}(x, y, t) = H_x \mathbf{a}_x + H_y \mathbf{a}_y,$$

reduce to the diffusion equation

$$\frac{\partial^2 E}{\partial x^2} + \frac{\partial^2 E}{\partial y^2} - \mu\sigma \frac{\partial E}{\partial t} = \mu \frac{\partial J_s}{\partial t}$$

where $E = E_z$ and J_s is the source current density in the z direction.

b. Taking $J_s = 0$, $\Delta x = \Delta y = \Delta$, and

$$\sum E_{i,j} = E_{i+1,j}^n + E_{i-1,j}^n + E_{i,j+1}^n + E_{i,j-1}^n,$$

show that applying Euler, leapfrog, and DuFort–Frankel difference methods to the diffusion equation gives

Euler:

$$E_{i,j}^{n+1} = (1 - 4r)E_{i,j}^n + r\sum E_{i,j}^n,$$

Leapfrog:

$$E_{i,j}^{n+1} = E_{i,j}^{n-1} + 2r\left(\sum E_{i,j}^n - 4E_{i,j}^n\right),$$

DuFort–Frankel:

$$E_{i,j}^{n+1} = \frac{1-4r}{1+4r} E_{i,j}^{n-1} + \frac{2r}{1+4r}\sum E_{i,j}^n$$

where $r = \Delta t/(\sigma\mu\Delta^2)$.

c. Analyze the stability of these finite difference schemes by substituting for $E_{i,j}^n$ a Fourier mode of the form

$$E_{i,j}^n = E(x = i\Delta, y = j\Delta, t = n\Delta t) = A_n \cos(k_x i\Delta)\cos(k_y J\Delta)$$

3.21 Yee's FDTD algorithm for one-dimensional wave problems is given by

$$H_y^{n+1/2}(k+1/2) = H_y^{n-1/2}(k+1/2) + \frac{\delta t}{\mu\delta}[E_x^n(k) - E_x^n(k+1)]$$

Determine the stability criterion for the scheme by letting

$$E_x^n(k) = A^n e^{j\beta k\delta}, \quad H_y^n(k) = \frac{A^n}{\eta} e^{j\beta k\delta},$$

where $\eta = (\mu/\epsilon)^{1/2}$ is the intrinsic impedance of the medium.

3.22 a. The potential system in Figure 3.52a is symmetric about the y-axis. Set the initial values at free nodes equal to zero and calculate (by hand) the potential at nodes 1–5 for five or more iterations.

b. Consider the square mesh in Figure 3.52b. By setting initial values at the free nodes equals to zero, find (by hand calculation) the potential at nodes 1–4 for five or more iterations.

3.23 The potential system shown in Figure 3.53 is a quarter section of a transmission line. Using hand calculation, find the potential at nodes 1, 2, 3, 4, and 5 after five iterations.

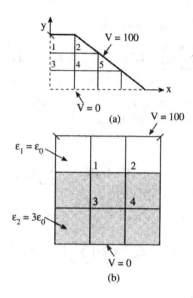

FIGURE 3.52
For Problem 3.22.

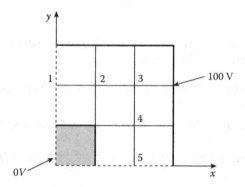

FIGURE 3.53
For Problem 3.23.

3.24 Modify the program in Figure 3.21 or write your own program to calculate Z_0 for the microstrip line shown in Figure 3.54. Take $a = 2.02$, $b = 7.0$, $h = 1.0 = w$, $t = 0.01$, $\epsilon_1 = \epsilon_0$, $\epsilon_2 = 9.6\epsilon_0$.

3.25 Use the FDM to calculate the characteristic impedance of the high-frequency, air-filled rectangular transmission line shown in Figure 3.55. Take advantage of the symmetry of the problem and consider cases for which

  a. $B/A = 1.0$, $a/A = 1/2$, $b/B = 1/2$, $a = 1$,

  b. $B/A = 1/2$, $a/A = 1/3$, $b/B = 1/3$, $a = 1$.

3.26 Figure 3.56 shows a shield microstrip line. Write a program to calculate the potential distribution within the cross section of the line. Take $\epsilon_1 = \epsilon_0$, $\epsilon_2 = 3.5\epsilon_0$ and $h = 0.5$ mm. Find the potential at the middle of the conducting plates.

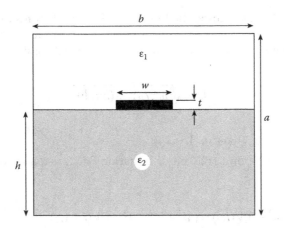

FIGURE 3.54
For Problem 3.24.

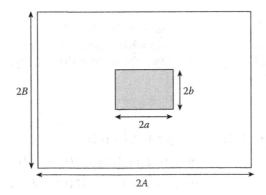

FIGURE 3.55
For Problem 3.25.

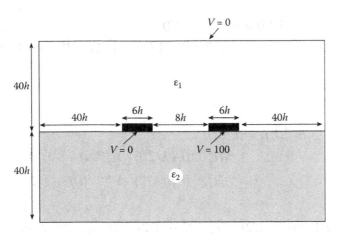

FIGURE 3.56
For Problem 3.26.

3.27 Use the FDM to determine the lowest (or dominant) cut-off wave-number k_c of the TM_{11} mode in waveguides with square $(a \times a)$ and rectangular $(a \times b, b = 2a)$ cross sections. Compare your results with the exact solution

$$k_c = \sqrt{(m\pi/a)^2 + (n\pi/b)^2}$$

where $m = n = 1$. Take $a = 1$.

3.28 Instead of the 5-point schema of Equation 3.121, use a more accurate 5-point formula

$$2i(8i^2 - 5)V(i,j) = (4i^3 + 2i^2 - 4i + 1)V(i+1,j)$$
$$+ (4i^3 - 2i^2 - 4i - 1)V(i-1,j)$$
$$i(4i^2 - 1)V(i,j+1) + i(4i^2 - 1)V(i,j-1)$$

in Example 3.9 while other things remain the same.

3.29 For two-dimensional problems in which the field components do not vary with z coordinate $(\partial/\partial z = 0)$, show that Yee's algorithm of Equations 3.84a through 3.84f becomes

a. for TE waves $(E_z = 0)$

$$H_z^{n+1/2}(i+1/2, j+1/2) = H_z^{n-1/2}(i+1/2, j+1/2)$$
$$- \alpha[E_y^n(i+1, j+1/2) - E_y^n(i, j+1/2)]$$
$$+ \alpha[E_x^n(i+1/2, j+1) - E_x^n(i+1/2, j)],$$
$$E_x^{n+1}(i+1/2, j) = E_x^n(i+1/2, j)$$
$$+ \beta[H_z^{n+1/2}(i+1/2, j+1/2) - H_z^{n+1/2}(i+1/2, j-1/2)],$$

$$E_z^{n+1}(i, j+1/2) = \gamma E_y^n(i, j+1/2)$$
$$- \beta[H_z^{n+1/2}(i+1/2, j+1/2) - H_z^{n+1/2}(i-1/2, j+1/2)];$$

b. for TM waves $(H_z = 0)$

$$E_z^{n+1}(i,j) = \gamma E_z^n(i,j)$$
$$+ \beta[H_y^{n+1/2}(i+1/2, j) - H_y^{n+1/2}(i-1/2, j)]$$
$$- \beta\left[H_x^{n+1/2}(i, j+1/2) - H_x^{n+1/2}(i, j-1/2)\right],$$

$$H_x^{n+1/2}(i, j+1/2) = H_x^{n-1/2}(i, j+1/2) - \alpha[E_z^n(i, j+1) - E_z^n(i, j)],$$
$$H_y^{n+1/2}(i+1/2, j) = H_y^{n-1/2}(i+1/2, j) + \alpha[E_z^n(i+1, j) - E_z^n(i, j)],$$

where

$$\alpha = \frac{\delta t}{\mu \delta}, \quad \beta = \frac{\delta t}{\varepsilon \delta}, \quad \gamma = 1 - \frac{\sigma \delta t}{\varepsilon},$$

and $\delta = \Delta x = \Delta y$.

3.30 Consider the diffraction/scattering of an incident TM wave by a perfectly conducting square of side $4a$. The conducting obstacle occupies $17 < i < 49$, $33 < j < 65$, while artificial boundaries are placed at $i = 1, 81, j = 0.5, 97.5$ as shown in Figure 3.57. Assume an incident wave with only E_z and H_y components given by

$$E_z = \begin{cases} \sin \pi \theta, & 0 < \theta < 1 \\ 0, & \text{otherwise} \end{cases}$$

$$H_y = \frac{1}{\eta_o} E_z$$

where $\eta_o = 120\pi \Omega, \theta = ((x - 50a + ct)/8a), \Delta x = \Delta y = a/8, \Delta t = c\Delta x = a/16$. Write a program that applies the algorithm in Problem 3.25(b). Assume "hard lattice truncation conditions" at the artificial boundaries shown in Figure 3.57 and reproduce Yee's result [42] in his figure 3.

3.31 Repeat the previous problem but assume "soft lattice truncation condition" of Equations 3.87 through 3.89a to 3.891 at the artificial boundaries.

3.32 For axisymmetric problems (no variation with respect to ϕ), show that Yees algorithm for TM waves can be written as

$$H_\phi^{n+1}(i,j) = H_\phi^n(i,j) + \alpha[E_z^{n+1/2}(i,j+1/2) - E_z^{n+1/2}(i,j-1/2)]$$
$$- \alpha[E_\rho^{n+1/2}(i+1/2,j) - E_\rho^{n+1/2}(i-1/2,j)]$$

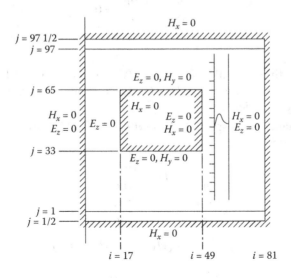

FIGURE 3.57
For Problem 3.30.

$$H_z^{n+3/2}(i,j+1/2) = \gamma E_z^{n+1/2}(i,j+1/2)$$

$$+ \beta \left[\frac{1}{j} H_\phi^{n+1}(i,j+1/2) + H_\phi^{n+1}(i,j+1) - H_\phi^{n+1}(i,j) \right]$$

$$E_\rho^{n+3/2}(i+1/2,j) = \gamma E_\rho^{n+1/2}(i+1/2,j) - \beta[H_\phi^{n+1}(i+1,j) - H_\phi^{n+1}(i,j)],$$

where

$$\alpha = \frac{\delta t}{\mu \delta}, \quad \beta = \frac{\delta t}{\varepsilon \delta}, \quad \gamma = 1 - \frac{\sigma \delta t}{\varepsilon}, \quad \delta = \Delta\rho = \Delta z,$$

and $H_\phi(z,\rho,t) = H_\phi(z = i\Delta z, \rho = (j-1/2)\Delta\rho, t = n\delta t) = H_\phi^n(i,j).$

3.33 a. Show that the finite difference discretization of Mur's ABC for two-dimensional problem

$$\frac{\partial E_z}{\partial x} - \frac{1}{c_o} \frac{\partial E_z}{\partial t} - \frac{c_o\mu_o}{2} \frac{\partial H_x}{\partial y} = 0$$

at the boundary $x = 0$ is

$$E_z^{n+1}(0,j) = E_z^n(i,j) + \frac{c_o\delta t - \delta}{c_o\delta t + \delta}[E_z^{n+1}(i,j) - E_z^n(0,j)]$$

$$- \frac{\mu_o c_o}{2(c_o\delta t + \delta)}[H_x^{n+1/2}(0,j+1/2) - H_x^{n+1/2}(0,j-1/2)]$$

$$+ H_x^{n+1/2}(1,j+1/2) - H_x^{n+1/2}(1,j-1/2)$$

where c_o is the velocity of wave propagation.

b. Discretize the first-order boundary condition

$$\frac{\partial E_z}{\partial x} - \frac{1}{c_o} \frac{\partial E_z}{\partial t} = 0$$

at $x = 0$

3.34 For a three-dimensional problem, the PML modification of Maxwell's equations yields 12 equations because all six Cartesian field components split. Obtain the 12 resulting equations.

3.35 In a PML region, E_z is split into E_{zx} and E_{zy} for the TM case. Show that Maxwell's equation becomes

$$\varepsilon_o \frac{\partial E_{zx}}{\partial t} + \sigma_x E_{zx} = \frac{\partial H_y}{\partial x}$$

$$\varepsilon_o \frac{\partial E_{zy}}{\partial t} + \sigma_y E_{zy} = \frac{\partial H_x}{\partial y}$$

$$\mu_o \frac{\partial H_x}{\partial t} + \sigma_y^* H_x = -\frac{\partial}{\partial y}(E_{zx} + E_{zy})$$

$$\mu_o \frac{\partial H_y}{\partial t} + \sigma_x^* H_y = \frac{\partial}{\partial x}(E_{zx} + E_{zy})$$

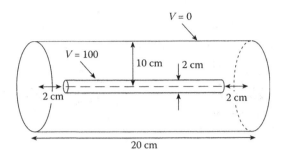

FIGURE 3.58
For Problem 3.37.

3.36 An FDTD equation for a PML region is given by

$$H_z^{n+1/2}(i+1/2,k) = H_z^{n-1/2}(i+1/2,k)$$

$$- \frac{\delta t}{\mu \delta}[E_{yx}^n(i+1,k) + E_{yz}^n(i+1,k) - E_{yx}^n(i,k) - E_{yz}^n(i,k)]$$

where δ, δt, n, i, and k have their usual FDTD meanings. By substituting the harmonic dependence $e^{jwt}e^{-jk_z z}$, show that the impedance of the PML region is

$$Z_z = \frac{E_y}{H_z} = \frac{\mu_o \delta}{\delta t} \frac{\sin(\omega \delta t/2)}{\sin(k_o \delta/2)}$$

3.37 Consider the finite cylindrical conductor held at $V = 100$ V and enclosed in a larger grounded cylinder as in Figure 3.58. Such a deceptively simple looking problem is beyond closed-form solution, but by employing finite difference techniques, the problem can be solved without much effort. Using the finite difference method, write a program that determines the potential distribution in the axisymmetric solution region. Output the potential at $(\rho, z) = (2, 10)$, $(5, 10)$, $(8, 10)$, $(5, 2)$, and $(5, 18)$.

3.38 The problem in Figure 3.59 is a prototype of an electrostatic particle focusing system which is employed in a recoil-mass time-of-flight spectrometer. Write a program to determine the potential distribution in the system. The problem is similar to the previous problem except that the outer conductor abruptly expands radius by a factor of 2. Output the potential at $(\rho, z) = (5, 18)$, $(5, 10)$, $(5, 2)$, $(10, 2)$, and $(15, 2)$.

3.39 The conventional 3-D FDTD lattice in cylindrical coordinates is shown is Figure 3.60a while its projection on the $\rho - z$ plane is in Figure 3.60b. Show that by discretizing Maxwell's equation,

$$E_\rho^{n+1}(i,j) = \frac{\left(1 - \dfrac{\sigma\delta}{2\varepsilon}\right)}{\left(1 + \dfrac{\sigma\delta}{2\varepsilon}\right)} E_\rho^n(i,j) - \frac{\delta t}{\varepsilon\delta} \frac{1}{\left(1 + \dfrac{\sigma\delta}{2\varepsilon}\right)} \cdot$$

$$\left[H_\phi^{n+1/2}(i,j) - H_\phi^{n+1/2}(i,j-1)\right]$$

where $\delta = \Delta z = \Delta \rho$. Obtain the FDTD equations for H_ρ and H_ϕ.

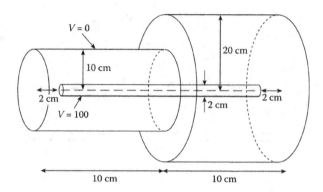

FIGURE 3.59
For Problem 3.37.

3.40 Consider the one-dimensional parabolic equation in cylindrical coordinates

$$\nabla^2 U = \frac{\partial U}{\partial t}$$

or

$$U_{\rho\rho} + \frac{1}{\rho} U_\rho = U_t, \quad 0 < \rho < 1, \quad t > 0$$

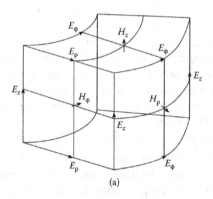

(a)

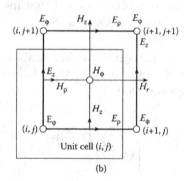

(b)

FIGURE 3.60
For Problem 3.39: (a) A conventional 3-D FDTD lattice in cylindrical coordinates, (b) projection of 3-D FDTD cell at $\rho - z$ plane.

subject to $U(1, t) = 0, t > 0$

$$U(\rho, 0) = T_o (\text{constant})$$

By selecting $\Delta\rho - h = 0.1$, and $T_o = 10$, calculate U at $\rho = 0.5$, $t = 0.1, 0.2, 0.3$, 0.4, 0.5, 1.0 using finite difference. Compare your result with the exact solution

$$U(\rho, t) = 2T_o \sum_{n=1}^{\infty} \frac{J_0(\lambda_n \rho)}{\lambda_n J_1(\lambda_n)} \exp(-\lambda_n^2 t)$$

where J_0 and J_1 are Bessel function of orders 0 and 1, respectively, and λ_n are the positive roots of J_0.

3.41 For a two-dimensional heat equation in cylindrical system, consider

$$\nabla^2 U = \frac{\partial U}{\partial t}$$

or

$$U_{\rho\rho} + \frac{1}{\rho} U_\rho + U_{zz} = U_t, \ 0 < \rho < 1, 0 < z < 1, t > 0$$

with the boundary conditions

$$U(\rho, 0, t) = 0 = U(\rho, 1, t), \quad 0 < \rho < 1, \quad t > 0$$
$$U(1, z, t) = 0, \quad 0 < z < 1, \quad t > 0$$

and initial condition

$$U(\rho, z, 0) = T_o, \quad 0 < \rho < 1, \quad 0 < z < 1$$

By selecting $\Delta z = \Delta\rho = h = 0.1$, and $T_o = 10$, use the finite difference method to determine U at $\rho = 0.5$, $z = 0.5$, $t = 0.05, 0.1, 0.15, 0.2, 0.25, 0.3$.

3.42 Given the tabulated values of $y = \sin x$ for $x = 0.4$ to 0.52 rad in intervals of $\Delta x = 0.02$, find (a) $\frac{dy}{dx}$ at $x = 0.44$, (b) $\int_{0.4}^{0.52} y \, dx$ using Simpson's rule.

x	Sin x
0.40	0.38942
0.42	0.40776
0.44	0.42594
0.46	0.44395
0.48	0.46178
0.50	0.47943
0.52	0.49688

3.43 a. Use a pocket calculator to determine the approximate area under the curve $f(x) = 4 - x^2, 0 < x < 1$ by the trapezoidal rule with $h = 0.2$.

   b. Repeat part (a) using the Newton–Cotes rules with $n = 3$.

3.44 For a half-wave dipole, evaluation the integral

$$\int_0^{1/2} \frac{\cos^2\left(\frac{\pi}{2}\cos\theta\right)}{\sin\theta}\, d\theta$$

is usually required. Evaluate this integral numerically using any quadrature rule of your choice.

3.45 Compute

$$\int_0^1 e^{-x}dx$$

using the Newton–Cotes rule for cases $n = 2$, 4, and 6. Compare your results with exact values.

3.46 Given that

$$J_n(z) = \frac{1}{\pi}\int_0^\pi \cos(z\cos\theta + n\theta)\,d\theta$$

Let $n = 1, z = 4$ so that

$$I = \frac{1}{\pi}\int_0^\pi \cos(4\cos\theta + \theta)\,d\theta$$

Evaluate I,

   a. using the trapezoidal rule with $\Delta\theta = \pi/10$

   b. using Simpson's 1/3- rule with $\Delta\theta = \pi/10$

   c. using Gaussian quadrature.

   Compare numerical results with the exact value of $J_1(4)$.

3.47 The criterion for accuracy of the numerical approximation of an integral

$$I = \int_a^b f(x)\,dx \simeq \sum_{i=0}^\infty a_i f(x_i)$$

is that the formula is exact for all polynomials of degree less than or equals to n. If $a = 0$, $b = 4$, and the values of $f(x)$ are available at points $x_0 = 0$, $x_1 = 1$, $x_2 = 3$, $x_4 = 4$, find the values of the coefficients a_i for which the above requirement of accuracy is met.

3.48 The elliptic integral of the first type

$$F(k,\phi) = \int\limits_{0}^{\phi} (1 - k^2 \sin^2 \theta)^{-1/2} d\theta$$

cannot be evaluated in a closed form. Write a program using Simpson's rule to determine $F(k, \phi)$ for $k = 2$ and $\phi = \pi/2$.

3.49 The following integral represents radiation from a circular aperture antenna with a constant current amplitude and phase distribution

$$I = \int\limits_{0}^{1} \int\limits_{0}^{2\pi} e^{j\alpha\rho\cos\phi} \rho \, d\phi \, d\rho$$

Find I numerically for $\alpha = 5$ and compare your result with the exact result

$$I(\alpha) = \frac{2\pi J_1(\alpha)}{\alpha}$$

3.50 Evaluate the following integrals numerically:

a. $I = \int_0^1 \int_0^1 e^{x+y} dx \, dy$

b. $J = \int_{y=-2}^{2} \int_{x=0}^{2} (2x + 3y^2) dx \, dy$

References

1. L.D. Kovach, *Boundary-Value Problems*. Reading, MA: Addison-Wesley, 1984, pp. 355–379.
2. A. Thom and C.J. Apelt, *Field Computations in Engineering and Physics*. London: D. Van Nostrand, 1961, p. v.
3. R.D. Richtmyer and K.W. Morton, *Difference Methods for Initial-Value Problems*, 2nd Edition. New York: Interscience Publ., 1976, pp. 185–193.
4. D. Potter, *Computational Physics*. London: John Wiley, 1973, pp. 40–79.
5. G.D. Smith, *Numerical Solution of Partial Differential Equations: Finite Difference Methods*, 3rd Edition. Oxford Univ. Press, New York, 1985.
6. J.H. Ferziger, *Numerical Methods for Engineering Application*. New York: John Wiley, 1981.
7. G. de Vahl Davis, *Numerical Methods in Engineering and Science*. London: Allen & Unwin, 1986.
8. V. Vemuri and W.J. Karplus, *Digital Computer Treatment of Partial Differential Equations*. Englewood Cliffs, NJ: Prentice-Hall, 1981, pp. 88–92.
9. A. Wexler, "Computation of electromagnetic fields," *IEEE Trans. Micro. Theo. Tech.*, vol. MTT-17, no. 8, Aug. 1969, pp. 416–439.
10. L. Lapidus and G.F. Pinder, *Numerical Solution of Partial Differential Equations in Science and Engineering*. New York: John Wiley, 1982, pp. 166–185.
11. Computer-oriented microwave practices covers various applications of finite difference methods to EM problems, *Special issue of IEEE Transactions on Microwave Theory and Techniques*, vol. MTT-17, no. 8, Aug. 1969 on.

12. H.E. Green, "The numerical solution of some important transmission-line problems," *IEEE Trans. Micro. Theo. Tech.*, vol. MTT-13, no. 5, Sept. 1965, pp. 676–692.

13. M.N.O. Sadiku, "Finite difference solution of electrodynamic problems," *Int. Elect. Engr. Educ.*, vol. 28, April 1991, pp. 107–122.

14. M.F. Iskander, "A new course on computational methods in electromagnetics," *IEEE Trans. Educ.*, vol. 31, no. 2, May 1988, pp. 101–115.

15. W.S. Metcalf, "Characteristic impedance of rectangular transmission lines," *Proc. IEE*, vol. 112, no. 11, Nov. 1965, pp. 2033–2039.

16. M.V. Schneider, "Computation of impedance and attenuation of TEM-lines by," *IEEE Micro. Theo. Tech.*, vol. MTT-13, no. 6, Nov. 1965, pp. 793–800.

17. M. Sendaula, M. Sadiku, and R. Heiman, "Crosstalk computation in coupled transmission lines," *Proc. IEEE Southeastcon*, April 1991, pp. 790–795.

18. A.R. Djordjevic et al. "Time-domain response of multiconductor transmission lines," *Proc. IEEE*, vol. 75, no. 6, June 1987, pp. 743–764.

19. R.R. Gupta, "Accurate impedance determination of coupled TEM conductors," *IEEE Trans. Micro. Theo. Tech.*, vol. MTT-17, no. 12, Aug. 1969, pp. 479–489.

20. E. Yamashita et al. "Characterization method and simple design formulas of MDS lines proposed for MMIC's," *IEEE Trans. Micro. Theo. Tech.*, vol. MTT-35, no. 12, Dec. 1987, pp. 1355–1362.

21. J.R. Molberg and D.K. Reynolds, "Iterative solutions of the scalar Helmholtz equations in lossy regions," *IEEE Trans. Micro. Theo. Tech.*, vol. MTT-17, no. 8, Aug. 1969, pp. 460–477.

22. J.B. Davies and C.A. Muilwyk, "Numerical solution of uniform hollow waveguides with boundaries of arbitrary shape," *Proc. IEEE*, vol. 113, no. 2, Feb. 1966, pp. 277–284.

23. J.S. Hornsby and A. Gopinath, "Numerical analysis of a dielectric-loaded waveguide with a microstrip line—finite-difference methods," *IEEE Trans. Micro. Theo. Tech.*, vol. MTT-17, no. 9, Sept. 1969, pp. 684–690.

24. M.J. Beubien and A. Wexler, "An accurate finite-difference method for higher-order waveguide modes," *IEEE Trans. Micro. Theo. Tech.*, vol. MTT-16, no. 12, Dec. 1968, pp. 1007–1017.

25. C.A. Muilwyk and J.B. Davies, "The numerical solution of rectangular waveguide junctions and discontinuities of arbitrary cross section," *IEEE Trans. Micro. Theo. Tech.*, vol. MTT-15, no. 8, Aug. 1967, pp. 450–455.

26. J.H. Collins and P. Daly, "Calculations for guided electromagnetic waves using finite-difference methods," *J. Electron. Control*, vol. 14, 1963, pp. 361–380.

27. T. Itoh (ed.), *Numerical Techniques for Microwaves and Millimeterwave Passive Structures*. New York: John Wiley, 1989.

28. D.H. Sinnott et al. "The finite difference solution of microwave circuit problems," *IEEE Trans. Micro. Theo. Tech.*, vol. MTT-17, no. 8, Aug. 1969, pp. 464–478.

29. W.K. Gwarek, "Analysis of an arbitrarily-shaped planar circuit—a time-domain approach," *IEEE Trans. Micro. Theo. Tech.*, vol. MTT-33, no. 10, Oct. 1985, pp. 1067–1072.

30. M. De Pourceq, "Field and power-density calculations in closed microwave systems by three-dimensional finite differences," *IEEE Proc.*, vol. 132, Pt. H, no. 6, Oct. 1985, pp. 360–368.

31. A. Taflove and K.R. Umashankar, "Solution of complex electromagnetic penetration and scattering problems in unbounded regions," in A.J. Kalinowski (ed.), *Computational Methods for Infinite Domain Media-Structure Interaction*. Washington, DC: ASME, vol. 46, 1981, pp. 83–113.

32. A. Taflove, "Application of the finite-difference time-domain method to sinusoidal steady-state electromagnetic-penetration problems," *IEEE Trans. EM Comp.*, vol. EMC-22, no. 3, Aug. 1980, pp. 191–202.

33. K.S. Kunz and K.M. Lee, "A three-dimensional finite-difference solution of the external response of an aircraft to a complex transient EM environment" (2 parts), *IEEE Trans. EM Comp.*, vol. EMC-20, no. 2, May 1978, pp. 328–341.

34. M.L. Oristaglio and G.W. Hohman, "Diffusion of electromagnetic fields into a two-dimensional earth: A finite-difference approach," *Geophysics*, vol. 49, no. 7, July 1984, pp. 870–894.

35. K. Umashankar and A. Taflove, "A novel method to analyze electromagnetic scattering of complex objects," *IEEE Trans. EM Comp.*, vol. EMC-24, no. 4, Nov. 1982, pp. 397–405.

36. R.W.M. Lau and R.J. Sheppard, "The modelling of biological systems in three dimensions using the time domain finite-difference method" (2 parts), *Phys. Med. Biol.*, vol. 31, no. 11, 1986, pp. 1247–1266.

37. F. Sandy and J. Sage, "Use of finite difference approximations to partial differential equations for problems having boundaries at infinity," *IEEE Trans. Micro. Theo. Tech.*, May 1971, pp. 484–486.

38. K.B. Whiting, "A treatment for boundary singularities in finite difference solutions of Laplace's equation," *IEEE Trans. Micro. Theo. Tech.*, vol. MTT-16, no. 10, Oct. 1968, pp. 889–891.

39. G.E. Forsythe and W.R. Wasow, *Finite Difference Methods for Partial Differential Equations.* New York: John Wiley, 1960.

40. M.L. James et al. *Applied Numerical Methods for Digital Computation*, 3rd Edition. New York: Harper & Row, 1985, pp. 203–274.

41. Y. Naiheng and R.F. Harrington, "Characteristic impedance of transmission lines with arbitrary dielectrics under the TEM approximation," *IEEE Trans. Micro. Theo. Tech.*, vol. MTT-34, no. 4, April 1986, pp. 472–475.

42. K.S. Yee, "Numerical solution of initial boundary-value problems involving Maxwell's equations in isotropic media," *IEEE Trans. Ant. Prop.*, vol. AP-14, May 1966, pp. 302–307.

43. A. Taflove and M.E. Brodwin, "Numerical solution of steady-state electromagnetic scattering problems using the time-dependent Maxwell's equations," *IEEE Micro. Theo. Tech.*, vol. MTT-23, no. 8, Aug. 1975, pp. 623–630.

44. A. Taflove and K. Umashankar, "A hybrid moment method/finite-difference time-domain approach to electromagnetic coupling and aperture penetration into complex geometries," *IEEE Trans. Ant. Prop.*, vol. AP-30, no. 4, July 1982, pp. 617–627. Also in B. J. Strait (ed.), *Applications of the Method of Moments to Electromagnetic Fields.* Orlando, FL: SCEE Press, Feb. 1980, pp. 361–426.

45. M. Okoniewski, "Vector wave equation 2D-FDTD method for guided wave equation," *IEEE Micro. Guided Wave Lett.*, vol. 3, no. 9, Sept. 1993, pp. 307–309.

46. A. Taflove and K.R. Umashankar, "The finite-difference time-domain method for numerical modeling of electromagnetic wave interactions," *Electromagnetics*, vol. 10, 1990, pp. 105–126.

47. M.N.O. Sadiku, V. Bemmel, and S. Agbo, "Stability criteria for finite-difference time-domain algorithm," *Proc. IEEE Southeastcon*, April 1990, pp. 48–50.

48. J.G. Blaschak and G.A. Kriegsmann, "A comparative study of absorbing boundary conditions," *J. Comp. Phys.*, vol. 77, 1988, pp. 109–139.

49. G. Mux, "Absorbing boundary conditions for the finite-difference approximation of the time-domain electromagnetic-field equations," *IEEE Trans. EM Comp.*, vol. EMC-23, no. 4, Nov. 1981, pp. 377–382.

50. R. Holland, "THREDE: A free-field EMP coupling and scattering code," *IEEE Trans. Nucl. Sci.*, vol. NS-24, no. 6, Dec. 1977, pp. 2416–2421.

51. R.W. Ziolkowski et al. "Three-dimensional computer modeling of electromagnetic fields: A global lookback lattice truncation scheme," *J. Comp. Phy.*, vol. 50, 1983, pp. 360–408.

52. E.R. Demarest, "A finite difference—time domain technique for modeling narrow apertures in conducting scatterers," *IEEE Trans. Ant. Prog.*, vol. AP-35, no. 7, July 1987, pp. 826–831.

53. A. Taflove et al. "Detailed FDTD analysis of electromagnetic fields penetrating narrow slots and lapped joints in thick conducting screens," *IEEE Trans. Ant. Prog.*, vol. 36, no. 2, Feb. 1988, pp. 24–257.

54. H. Meskanen and O. Pekonen, "FDTD analysis of field distribution in an elevator car by using various antenna positions and orientations," *Elect. Lett.*, vol. 34, no. 6, March 1998, pp. 534–535.

55. J.G. Maloney, G.S. Smith, and W.R. Scott, "Accurate computation of the radiation from simple antennas using the finite-difference time-domain method," *IEEE Trans. Ant. Prog.*, vol. 38, no. 7, July 1990, pp. 1059–1068.

56. J.G. Maloney and G.D. Smith, "The efficient modeling of thin material sheets in the finite-difference time-domain (FDTD) method," *IEEE Trans. Ant. Prog.*, vol. 40, no. 3, March 1992, pp. 323–330.

57. P.A. Tirkas and C.A. Balanis, "Finite-difference time-domain method for antenna radiation," *IEEE Trans. Ant. Prog.*, vol. 40, no. 4, March 1992, pp. 334–340.

58. E. Thiele and A. Taflove, "FDTD analysis of vivaldi flared horn antennas and arrays," *IEEE Trans. Ant. Prog.*, vol. 42, no. 5, May 1994, pp. 633–641.

59. J.S. Colburn and Y. Rahmat-Samii, "Human proximity effects on circular polarized handset antennas in personal satellite communications," *IEEE Trans. Ant. Prog.*, vol. 46, no. 6, June 1998, pp. 813–820.

60. K. Uehara and K. Kagoshima, "Rigorous analysis of microstrip phased array antennas using a new FDTD method," *Elect. Lett.*, vol. 30, no. 2, Jan. 1994, pp. 100–101.

61. T. Shibata et al. "Analysis of microstrip circuits using three-dimensional full-wave electromagnetic field analysis in the time-domain," *IEEE Trans. Micro. Theo. Tech.*, vol. 36, no. 6, June 1988, pp. 1064–1070.

62. W.K. Gwarek, "Analysis of arbitrarily shaped two-dimensional microwave circuits by finite-difference time-domain method," *IEEE Trans. Micro. Theo. Tech.*, vol. 36, no. 4, April 1988, pp. 738–744.

63. X. Zhang et al. "Calculation of the dispersive characteristics of microstrips by the time-domain finite-difference method," *IEEE Trans. Micro. Theo. Tech.*, vol. 36, no. 2, Feb. 1988, pp. 263–267.

64. X. Zhang and K.K. Mei, "Time-domain finite-difference approach to the calculation of the frequency-dependent characteristics of microstrip discontinuities," *IEEE Trans. Micro. Theo. Tech.*, vol. 36, no. 12, Dec. 1988, pp. 1775–1787.

65. R.W. Larson, "Special purpose computers for the time domain advance of Maxwell's equations," *IEEE Trans. Magnetics*, vol. 25, no. 4, July 1989, pp. 2913–2915.

66. K.K. Mei et al. "Conformal time domain finite difference method," *Radio Sci.*, vol. 19, no. 5, Sept./Oct. 1984, pp. 1145–1147.

67. D.H. Choi and W.J.R. Hoefer, "The finite-difference time-domain method and its application to eigenvalue problems," *IEEE Trans. Micro. Theo. Tech.*, vol. MTT-36, no. 12, Dec. 1986, pp. 1464–1470.

68. D.M. Sullivan et al. "Use of the finite-difference time-domain method in calculating EM absorption in human tissues," *IEEE Trans. Biomed. Engr.*, vol. BME-34, no. 2, Feb. 1987, pp. 148–157.

69. A.D. Tinniswood, C.M. Furse, and O.P. Gandhi, "Computations of SAR distributions for two anatomically based models of the human head using CAD files of commercial telephone and the parallelized FDTD code," *IEEE Trans. Ant. Prog.*, vol. 46, no. 6, June 1998, pp. 829–833.

70. D. Dunn, C.M. Rappaport, and A.J. Terzuoli, "FDTD verification of deep-set brain tumor hyperthermia using a spherical microwave source distribution," *IEEE Trans. Micro. Theo. Tech.*, vol. 44, no. 10, Oct. 1996, pp. 1769–1777.

71. V. Hombach et al. "The dependence of EM energy absorption upon human head modeling at 900 MHz," *IEEE Trans. Micro. Theo. Tech.*, vol. 44, no. 10, Oct. 1996, pp. 1865–1873.

72. O. Fujiwara and A. Kato, "Computation of SAR inside eyeball for 1.5-GHz microwave exposure using finite-difference time-domain technique," *IEICE Trans. Comm.*, vol. E77-B, no. 6, June 1994, pp. 732–737.

73. A. Christ and H.L. Hartnagel, "Three-dimensional finite-difference method for the analysis of microwave-device embedding," *IEEE Trans. Micro. Theo. Tech.*, vol. MTT-35, no. 8, Aug. 1987, pp. 688–696.

74. J.H. Whealton, "A 3D analysis of Maxwell's equations for cavities of arbitrary shape," *J. Comp. Phys.*, vol. 75, 1988, pp. 168–189.

75. R. Luebbers et al. "A frequency-dependence finite-difference time-domain formulation for dispersive materials," *IEEE Trans. EMC*, vol. 32, no. 3, Aug. 1990, pp. 222–227.

76. R. Holland, "Finite-difference time-domain (FDTD) analysis of magnetic diffusion," *IEEE Trans. EMC*, vol. 36, no. 1, Feb. 1994, pp. 32–39.

77. A. Taflove and K.R. Umashankar, "Review of FDTD numerical modeling of electromagnetic wave scattering and radar cross section," *Proc. IEEE*, vol. 77, no. 5, May 1989, pp. 682–699.

78. V. Bemmel, "Time-domain finite-difference analysis of electromagnetic scattering and penetration problems," M.Sc. Thesis, Dept. of Electrical and Computer Engr., Florida Atlantic University, Boca Raton, FL, Dec. 1987.

79. D.S. Jones, *The Theory of Electromagnetism*. New York: Macmillan, 1964, pp. 450–452.

80. J.A. Stratton, *Electromagnetic Theory*. New York: McGraw-Hill, 1941, pp. 563–573.

81. G.A. Kriegsmann, A. Taflove, and K.R. Umashankar, "A new formulation of electromagnetic wave scattering using an on-surface radiation boundary condition approach," *IEEE Trans. Ant. Prop.*, vol. 35, no. 2, Feb. 1987, pp. 153–161.

82. B. Zhiqiang et al. "A new finite-difference time-domain algorithm for solving Maxwell's equation," *IEEE Micro. Guided Wave Lett.*, vol. 1, no. 12, Dec. 1991, pp. 382–384.

83. S.X.R. Vahldieck and H. Jin, "Full-wave analysis of guided wave structures using a novel 2-D FDTD," *IEEE Micro. Guided Wave Lett.*, vol. 2, no. 5, May 1992, pp. 165–167.

84. R. Mittra and P.H. Harms, "A new finite-difference time-domain (FDTD) algorithm for efficient field computation in resonator narrow-band structures," *IEEE Micro. Guided Wave Lett.*, vol. 3, no. 9, Sept. 1993, pp. 336–318.

85. J.B. Cole, "A high accuracy realization of the Yee algorithm using non-standard finite differences," *IEEE Trans. Micro. Theo. Tech.*, vol. 45, no. 6, June 1997, pp. 991–996.

86. J.B. Cole, "A high accuracy FDTD algorithm to solve microwave propagation and scattering problems on a coarse grid," *IEEE Trans. Micro. Theo. Tech.*, vol. 43, no. 9, Sept. 1995, pp. 2053–2058.

87. U. Oguz, L. Gurel, and O. Arkan, "An efficient and accurate technique for the incident-wave excitations in the FDTD method," *IEEE Trans. Micro. Theo. Tech.*, vol. 46, no. 6, June 1998, pp. 869–882.

88. I.J. Craddock and C.J. Railton, "A new technique for the stable incorporation of static field solutions in the FDTD method for the analysis of thin wires and narrow strips," *IEEE Trans. Micro. Theo. Tech.*, vol. 46, no. 8, Aug. 1998, pp. 1091–1096.

89. J.B. Cole et al. "Finite-difference time-domain simulations of wave propagation and scattering as a research and educational tool," *Computer in Physics*, vol. 9, no. 2, March/April 1995, pp. 235–239.

90. P.H. Harms, J.F. Lee, and R. Mittra, "A study of the nonorthogonal FDTD method versus the conventional FDTD technique for computing resonant frequency of cylindrical cavities," *IEEE Trans. Micro. Theo. Tech.*, vol. 40, no. 4, April 1992, pp. 741–746.

91. Y. Chen, R. Mittra, and P. Harms, "Finite-difference time-domain algorithm for solving Maxwell's equations in rotationally symmetric geometries," *IEEE Trans. Micro. Theo. Tech.*, vol. 44, no. 6, June 1996, pp. 832–839.

92. C. Wang, B.Q. Gao, and C.P. Deng, "Q factor of a resonator by the finite difference time-domain method incorporating perturbation techniques," *Elect. Lett.*, vol. 29, no. 21, Oct. 1993, pp. 1866–1867.

93. G. Cerri et al. "MoM-FDTD hybrid technique for analysing scattering problems," *Elect. Lett.*, vol. 34, no. 5, March 1998, pp. 438–440.

94. A.R. Bretones, R. Mittra, and R.G. Martin, "A hybrid technique combining the method of moments in the time domain and FDTD," *IEEE Micro. Guided Wave Lett.*, vol. 8, no. 8, Aug. 1998, pp. 281–283.

95. C.J. Railton, E.M. Daniel, and J.P. McGeehan, "Use of second order absorbing boundary conditions for the termination of planar waveguides in the FDTD method," *Elect. Lett.*, vol. 29, no. 10, May 1993, pp. 900–902.

96. P.Y. Wang et al. "Higher order formulation of absorbing boundary conditions for finite-difference time-domain method," *Elect. Lett.*, vol. 29, no. 23, Nov. 1993, pp. 2018–2019.

97. J.C. Olivier, "On the synthesis of exact free space absorbing boundary conditions for the finite-difference time-domain method," *IEEE Trans. Ant. Prog.*, vol. 40, no. 4, April 1992, pp. 456–460.

98. D.S. Katz, E.T. Thiele, and A. Taflove, "Validation and extension to three dimensions of the Berenger PML absorbing boundary conditions for FDTD meshes," *IEEE Micro. Guided Wave Lett.*, vol. 4, no. 6, Aug. 1994, pp. 268–270.

99. J.P. Berenger, "A perfectly matched layer for the absorption of electromagnetic waves," *J. Comp. Phys.*, vol. 114, Aug. 1994, pp. 185–200.

100. J.P. Berenger, "Perfectly matched layer for the FDTD solution of wave-structure interaction problems," *IEEE Trans. Ant. Prop.*, vol. 44, no. 1, Jan. 1996, pp. 110–117.

101. D.T. Prescott and N.V. Shuley, "Reflection analysis of FDTD boundary conditions – Part I: Time-space absorbing boundaries," *IEEE Trans. Micro. Theo. Tech.*, vol. 45, no. 8, Aug. 1997, pp. 1162–1170. Part II, pp. 1171–1178.

102. A. Taflove and S. C. Hagness, *Computational Electrodynamics: the Finite-Difference Time-Domain Method*. Boston, MA: Artech House, 3rd edition, 2005.

103. K.S. Kunz and R.J. Luebbers, *The Finite-Difference Time-Domain Method for Electromagnetic*. Boca Raton, FL: CRC Press, 1993, pp. 347–358.

104. A. Taflove and K.R. Umashankar, "The finite-difference time-domain method for numerical modeling of electromagnetic wave interactions with arbitrary structures," in M.A. Morgan (ed.), *Finite Element and Difference Methods in Electromagnetic Scattering*. New York: Elsevier, 1990, pp. 287–373.

105. K.K. Mei and J. Fang, "Superabsorption — a method to improve absorbing boundary conditions," *IEEE Trans. Ant. Prop.*, vol. 40, no. 9, Sept. 1992, pp. 1001–1010.

106. Y. Wu and I. Wassell, "Introduction to the segmented finite-difference time-domain method," *IEEE Transactions on Magnetics*, vol. 45, no. 3, March 2009, pp. 1364–1367.

107. W. C. Chew et al. *Fast and Efficient Algorithms in Computational Electromagnetics*. Norwood, MA: Artech House, 2001.

108. Q. H. Liu and G. Zhao, "Review of PSTD methods for transient electromagnetic," *Int. J. Numer. Modell.: Electron. Networks, Devices and Fields*, vol. 17, 2007, pp. 299–323.

109. Q. H. Liu, "The PSTD algorithm: A time-domian method requiring only two cells per wavelength," *Microwave and Optical Technology Letters*, vol. 15, no. 3, June 1997, pp. 158–165.

110. U. S. Inan and R. A. Marshall, *Numerical Electromagnetics: the FDTD Method*. Cambridge, UK: Cambridge University Press, chapter 12, 2011, pp. 291–326.

111. A. Taflove, A. Oskooi, and S. G. Johnson (eds.), *Advances in FDTD Computational Electrodynamics: Photonics and Nanoteechnology*. Boston, MA: Artech House, 2013.

112. E. A. Navarro, B. Gimeno, and J. L. Cruz, "Modeling of periodic structures using the finite difference time domain method combined with the Floquet theorem," *Electronic Letters*, vol. 29, no. 5, March 1993, pp. 446–447.

113. Q. Li, Y. Chen, and D. Ge, "Comparison study of the PSTD and FDTD methods for scattering analysis," *Microwave and Optical Technology Letters*, vol. 25, no. 3, May 2000, pp. 220–226.

114. P. Samineni, T. Khan, and A. De, "Modeling of electromagnetic band gap structures: A Review," *International Journal of RF and Microwave Computer-Aided Engineering*, vol. 27, Septemebr 2017, pp. 1–19.

115. A. Lakhtakia, "Special issue on exotic electromagnetics: Guest Editorial," *Electromagnetics*, vol. 25, no. 5, 2005, pp. 363–364.

116. Y. Hao and R. Mittra, *FDTD Modeling of Metamaterials: Theory and Applications*. Norwood, MA: Artech House, 2009.

117. N. Engheta and R. W. Ziolkowski, *Metamaterials: Physics and Engineering Explorations*. Hoboken, NJ: John Wiley & Sons, 2006.

118. A. F. Oskooi et al. "Meep: A flexible free-software package for electromagnetic simulations by the FDTD method," *Computer Physics Communications*, vol. 181, no. 3, 2010, pp. 687–702.

119. M. DiStasio and W.C. McHarris, "Electrostatic problems? Relax!" *Am. J. Phys.*, vol. 47, no. 5, May 1979, pp. 440–444.

120. M.N.O. Sadiku, "Finite difference solution of axisymmetric potential problems," *Int. J. Appl. Engr. Educ.*, vol. 6, no. 4, 1990, pp. 479–485.

121. H.E. Green, "The numerical solution of transmission line problems," in L. Young (ed.), *Advances in Microwaves*, vol. 2. New York: Academic Press, 1967, pp. 327–393.

122. C.D. Taylor et al. "Electromagnetic pulse scattering in time varying inhomogeneous media," *IEEE Trans. Ant. Prog.*, vol. AP-17, no. 5, Sept. 1969, pp. 585–589.

123. P.J. Davis and P. Rabinowitz, *Methods of Numerical Integration*. New York: Academic Press, 1975.
124. R. Piessens et al. *QUADPACK: A Subroutine Package for Automatic Integration*. Berlin: Springer-Verlag, 1980.
125. *Tables of Functions and Zeros of Functions*. Washington, DC: National Bureau of Standards, Applied Mathematical Series, no. 37, 1954.
126. M. Abramwitz and I.A. Stegun (eds.), *Handbook of Mathematical Functions*. Washington, DC: National Bureau of Standards, Applied Mathematical Series, no. 55, 1964.
127. L.G. Kelly, *Handbook of Numerical Methods and Applications*. Reading, MA: Addison-Wesley, 1967, pp. 57–61.
128. M.N.O. Sadiku and R. Jongakiem, "Newton–Cotes rules for triple integrals," *Proc. IEEE Southeastcon*, April 1990, pp. 471–475.
129. B.P. Rynne, "Instabilities in time marching methods for scattering problems," *Electromagnetics*, vol. 6, no. 2, 1986, pp. 129–144.
130. J.I. Steger and R.F. Warming, "On the convergence of certain finite-difference schemes by an inverse-matrix method," *J. Comp. Phys.*, vol. 17, 1975, pp. 103–121.
131. D.W. Kelly et al. "A posteriori error estimates in finite difference techniques," *J. Comp. Phys.*, vol. 74, 1988, pp. 214–232.
132. N. Marais and D. B. Davidson, "Efficient high-order time domain hybrid implicit/explicit finite element methods for microwave electromagnetic," *Electromagnetics*, vol. 30, 2010, pp. 127–148.

4

Variational Methods

Faith ends where worry begins, and worry ends where faith begins.

—George Mueller

4.1 Introduction

In solving problems arising from mathematical physics and engineering, we find that it is often possible to replace the problem of integrating a differential equation by the equivalent problem of seeking a function that gives a minimum value of some integral. Problems of this type are called *variational problems*. The methods that allow us to reduce the problem of integrating a differential equation to the equivalent variational problem are usually called *variational methods* [1]. The variational methods form a common base for both the method of moments (MoM) and finite element method (FEM). Therefore, it is appropriate that we study the variational methods before MoM and FEM. Besides, it is relatively easy to formulate the solution of certain differential and integral equations in variational terms. Also, variational methods give accurate results without making excessive demands on computer storage and time.

Variational methods can be classified into two groups: direct and indirect methods. The direct method is the classical Rayleigh–Ritz method, while the indirect methods are collectively referred to as the method of weighted residuals: collocation (or point-matching), subdomain, Galerkin, and least-squares methods. The variational solution of a given PDE using an indirect method usually involves two basic steps [2]:

- Casting the PDE into variational form, and
- Determining the approximate solution using one of the methods.

The literature on the theory and applications of variational methods to EM problems is quite extensive, and no attempt will be made to provide an exhaustive list of references. Numerous additional references may be found in those cited in this chapter. Owing to a lack of space, we can only hint at some of the topics usually covered in an introduction to this subject.

4.2 Operators in Linear Spaces

In this section, we will review some principles of operators in linear spaces and establish notation [2–5]. We define the *inner (dot or scalar) product* of functions u and v as

$$\langle u,v \rangle = \int_\Omega uv^* \, d\Omega \qquad (4.1)$$

where * denotes the complex conjugate and the integration is performed over Ω, which may be one-, two-, or three-dimensional physical space depending on the problem. In a sense, the inner product $\langle u, v \rangle$ gives the component or projection of function u in the direction of v. If **u** and **v** are vector fields, we modify Equation 4.1 slightly to include a dot between them, that is,

$$\langle \mathbf{u},\mathbf{v} \rangle = \int_\Omega \mathbf{u} \cdot \mathbf{v}^* \, d\Omega \qquad (4.2)$$

However, we shall consider u and v to be complex-valued scalar functions. For each pair of u and v belonging to the linear space, a number $\langle u, v \rangle$ is obtained that satisfies

$$
\begin{aligned}
&(1)\quad \langle u,v \rangle = \langle v,u \rangle^*, &&(4.3a)\\
&(2)\quad \langle \alpha u_1 + \beta u_2, v \rangle = \alpha\langle u_1,v \rangle + \beta\langle u_2,v \rangle, &&(4.3b)\\
&(3)\quad \langle u,v \rangle > 0 \quad \text{if } u \neq 0, &&(4.3c)\\
&(4)\quad \langle u,v \rangle = 0 \quad \text{if } u = 0 &&(4.3d)
\end{aligned}
$$

If $\langle u, v \rangle = 0$, u and v are said to be *orthogonal*. Notice that these properties mimic familiar properties of the dot product in three-dimensional space. Equation 4.3 is easily derived from Equation 4.1. Note that from Equations 4.3a and 4.3b,

$$\langle u,\alpha v \rangle = \alpha^* \langle v,u \rangle^* = \alpha^* \langle u,v \rangle$$

where α is a complex scalar.

Equation 4.1 is called an *unweighted* or *standard inner product*. A *weighted inner product* is given by

$$\langle u,v \rangle = \int_\Omega uv^* w \, d\Omega \qquad (4.4)$$

where w is a suitable weight function.

We define the norm of the function u as

$$\|u\| = \sqrt{\langle u,u \rangle} \qquad (4.5)$$

The norm is a measure of the "length" or "magnitude" of the function. (As far as a field is concerned, the norm is its rms value.) A vector is said to be *normal* if its norm is 1. Since the *Schwarz inequality*

$$|\langle u,v \rangle| \leq \|u\| \, \|v\| \qquad (4.6)$$

holds for any inner product space, the angle θ between two nonzero vectors $\mathbf{u}$ and $\mathbf{v}$ can be obtained as

$$\theta = \cos^{-1} \frac{\langle \mathbf{u}, \mathbf{v} \rangle}{\|\mathbf{u}\| \, \|\mathbf{v}\|} \tag{4.7}$$

We now consider the operator equation

$$\boxed{L\Phi = g} \tag{4.8}$$

where L is any linear operator, Φ is the unknown function, and g is the source function. The space spanned by all functions resulting from the operator L is

$$\langle L\Phi, g \rangle = \langle \Phi, L^a g \rangle \tag{4.9}$$

The operator L is said to be

1. Self-adjoint if $L = L^a$, i.e., $\langle L\Phi, g \rangle = \langle \Phi, Lg \rangle$,
2. Positive definite if $\langle L\Phi, \Phi \rangle > 0$ for any $\Phi \neq 0$ in the domain of L,
3. Negative definite if $\langle L\Phi, \Phi \rangle < 0$ for any $\Phi \neq 0$ in the domain of L.

The properties of the solution of Equation 4.8 depend strongly on the properties of the operator L. If, for example, L is positive definite, we can easily show that the solution of Φ in Equation 4.8 is unique, that is, Equation 4.8 cannot have more than one solution. To do this, suppose that Φ and Ψ are two solutions to Equation 4.8 such that $L\Phi = g$ and $L\Psi = g$. Then, by virtue of linearity of L, $f = \Phi - \Psi$ is also a solution. Therefore, $Lf = 0$. Since L is positive definite, $f = 0$ implies that $\Phi = \Psi$ and confirms the uniqueness of the solution Φ.

EXAMPLE 4.1

Find the inner product of $u(x) = 1 - x$ and $v(x) = 2x$ in the interval $(0, 1)$.

Solution

In this case, both u and v are real functions. Hence,

$$\langle u, v \rangle = \langle v, u \rangle = \int_0^1 (1 - x) 2x \, dx$$

$$= 2 \left(\frac{x^2}{2} - \frac{x^3}{3} \right) \Big|_0^1 = 0.333$$

EXAMPLE 4.2

Show that the operator

$$L = -\nabla^2 = -\frac{\partial^2}{\partial x^2} - \frac{\partial^2}{\partial y^2}$$

is self-adjoint.

Solution

$$\langle Lu, v \rangle = -\int_S v \nabla^2 u \, dS$$

Taking u and v to be real functions (for convenience) and applying Green's identity

$$\oint_\ell v \frac{\partial u}{\partial n} dl = \int_S \nabla u \cdot \nabla v \, dS + \int_S v \nabla^2 u \, dS$$

yields

$$\langle Lu, v \rangle = \int_S \nabla u \cdot \nabla v \, dS - \oint_\ell v \frac{\partial u}{\partial n} dl \qquad (4.10)$$

where S is bounded by ℓ and $\mathbf{n}$ is the outward normal. Similarly,

$$\langle u, Lv \rangle = \int_S \nabla u \cdot \nabla v \, dS - \oint_\ell u \frac{\partial v}{\partial n} dl \qquad (4.11)$$

The line integrals in Equations 4.10 and 4.11 vanish under either the homogeneous Dirichlet or Neumann boundary conditions. Under the homogeneous mixed boundary conditions, they become equal. Thus, L is self-adjoint under any one of these boundary conditions. L is also positive definite.

4.3 Calculus of Variations

The calculus of variations, an extension of ordinary calculus, is a discipline that is concerned primarily with the theory of maxima and minima. Here we are concerned with seeking the extrema (minima or maxima) of an integral expression involving a function of functions or *functionals*. Whereas a function produces a number as a result of giving values to one or more independent variables, a functional produces a number that depends on the entire form of one or more functions between prescribed limits. In a sense, a functional is a measure of the function. A simple example is the inner product $\langle u, v \rangle$. Variational formulation refers to the construction of a functional that is equivalent to the governing equation of the given problem.

In the calculus of variation, we are interested in the necessary condition for a functional to achieve a stationary value. This necessary condition on the functional is generally in the form of a differential equation with boundary conditions on the required function.

Consider the problem of finding a function $y(x)$ such that the function

$$I(y) = \int_a^b F(x, y, y') dx, \qquad (4.12a)$$

subject to the boundary conditions

$$y(a) = A, \quad y(b) = B, \tag{4.12b}$$

is rendered stationary. In other words, we want to find $y(x)$ from which the integral in Equation 4.12a is an extremum. The integrand $F(x, y, y')$ is a given function of x, y, and $y' = dy/dx$. In Equation 4.12a, $I(y)$ is called a *functional* or *variational* (or *stationary*) *principle*. The problem here is finding an extremizing function $y(x)$ for which the functional $I(y)$ has an extremum. Before attacking this problem, it is necessary that we introduce the operator δ, called the *variational symbol*.

The variation δy of a function $y(x)$ is an infinitesimal change in y for a fixed value of the independent variable x, that is, for $\delta x = 0$. The variation δy of y vanishes at points where y is prescribed (since the prescribed value cannot be varied) and it is arbitrary elsewhere (see Figure 4.1). Due to the change in y (i.e., $y \rightarrow y + \delta y$), there is a corresponding change in F. The first variation of F at y is defined by

$$\delta F = \frac{\partial F}{\partial y} \delta y + \frac{\partial F}{\partial y'} \delta y' \tag{4.13}$$

This is analogous to the total differential of F,

$$dF = \frac{\partial F}{\partial x} dx + \frac{\partial F}{\partial y} dy + \frac{\partial F}{\partial y'} dy' \tag{4.14}$$

where $\delta x = 0$ since x does not change as y changes to $y + \delta y$. Thus, we note that the operator δ is similar to the differential operator. Therefore, if $F_1 = F_1(y)$ and $F_2 = F_2(y)$, then

$$\text{(i)} \quad \delta(F_1 \pm F_2) = \delta F_1 \pm \delta F_2, \tag{4.15a}$$

$$\text{(ii)} \quad \delta(F_1 F_2) = F_2 \delta F_1 + F_1 \delta F_2, \tag{4.15b}$$

$$\text{(iii)} \quad \delta\left(\frac{F_1}{F_2}\right) = \frac{F_2 \delta F_1 - F_1 \delta F_2}{F_2^2}, \tag{4.15c}$$

$$\text{(iv)} \quad \delta(F_1)^n = n(F_1)^{n-1} \delta F_1, \tag{4.15d}$$

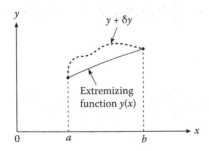

FIGURE 4.1
Variation of extremizing function with fixed ends.

(v) $\dfrac{d}{dx}(\delta y) = \delta\left(\dfrac{dy}{dx}\right),$ (4.15e)

(vi) $\delta\displaystyle\int_a^b y(x)dx = \int_a^b \delta y(x)dx$ (4.15f)

A necessary condition for the function $I(y)$ in Equation 4.12a to have an extremum is that the variation vanishes, that is,

$$\boxed{\delta I = 0}$$ (4.16)

To apply this condition, we must be able to find the variation δI of I in Equation 4.12a. To this end, let $h(x)$ be an increment in $y(x)$. For Equation 4.12b to be satisfied by $y(x) + h(x)$,

$$h(a) = h(b) = 0$$ (4.17)

The corresponding increment in I in Equation 4.12a is

$$\Delta I = I(y + h) - I(y)$$
$$= \int_a^b [F(x, y + h, y' + h') - F(x, y, y')]dx$$

On applying Taylor's expansion,

$$\Delta I = \int_a^b [F_y(x, y, y')h - F_{y'}(x, y, y')h']dx$$
$$+ \text{higher-order terms}$$
$$= \delta I + O(h^2)$$

where

$$\delta I = \int_a^b [F_y(x, y, y')h - F_{y'}(x, y, y')h']dx$$

Integration by parts leads to

$$\delta I = \int_a^b \left[\frac{\partial F}{\partial y} - \frac{d}{dx}\left(\frac{\partial F}{\partial y'}\right)\right]h\,dx + \frac{\partial F}{\partial y'}h\bigg|_{x=0}^{x=b}$$

The last term vanishes since $h(b) = h(a) = 0$ according to Equation 4.17. In order that $\delta I = 0$, the integrand must vanish, that is,

$$\frac{\partial F}{\partial y} - \frac{d}{dx}\left(\frac{\partial F}{\partial y'}\right) = 0$$

or

$$\boxed{F_y - \frac{d}{dx}F_{y'} = 0} \tag{4.18}$$

This is called *Euler's* (or *Euler–Lagrange*) *equation*. Thus, a necessary condition for $I(y)$ to have an extremum for a given function $y(x)$ is that $y(x)$ satisfies Euler's equation.

This idea can be extended to more general cases. In the case considered so far, we have one dependent variable y and one independent variable x, that is, $y = y(x)$. If we have one dependent variable u and two independent variables x and y, that is, $u = u(x, y)$, then

$$I(u) = \int_S F(x, y, u, u_x, u_y)\,dS \tag{4.19}$$

where $u_x = \partial u/\partial x$, $u_y = \partial u/\partial y$, and $dS = dx\,dy$. The functional in Equation 4.19 is stationary when $\delta I = 0$, and it is easily shown that the corresponding Euler's equation is [6]

$$\boxed{\frac{\partial F}{\partial u} - \frac{\partial}{\partial x}\left(\frac{\partial F}{\partial u_x}\right) - \frac{\partial}{\partial y}\left(\frac{\partial F}{\partial u_y}\right) = 0} \tag{4.20}$$

Next, we consider the case of two independent variables x and y and two dependent variables $u(x, y)$ and $v(x, y)$. The functional to be minimized is

$$I(u, v) = \int_S F(x, y, u, v, u_x, u_y, v_x, v_y)\,dS \tag{4.21}$$

The corresponding Euler's equation is

$$\boxed{\frac{\partial F}{\partial u} - \frac{\partial}{\partial x}\left(\frac{\partial F}{\partial u_x}\right) - \frac{\partial}{\partial y}\left(\frac{\partial F}{\partial u_y}\right) = 0} \tag{4.22a}$$

$$\frac{\partial F}{\partial v} - \frac{\partial}{\partial x}\left(\frac{\partial F}{\partial v_x}\right) - \frac{\partial}{\partial y}\left(\frac{\partial F}{\partial v_y}\right) = 0 \tag{4.22b}$$

Another case is when the functional depends on second- or higher-order derivatives. For example,

$$I(y) = \int_a^b F(x, y, y', y'', \ldots, y^{(n)})\,dx \tag{4.23}$$

In this case, the corresponding Euler's equation is

$$F_y - \frac{d}{dx}F_{y'} + \frac{d^2}{dx^2}F_{y''} - \frac{d^3}{dx^3}F_{y'''} + \cdots + (-1)^n \frac{d^n}{dx^n}F_{y(n)} = 0 \tag{4.24}$$

Note that each of Euler's equations (4.18), (4.20), (4.22), and (4.24) is a differential equation.

EXAMPLE 4.3

Given the functional

$$I(\Phi) = \int_S \left[\frac{1}{2}\left(\Phi_x^2 + \Phi_y^2\right) - f(x,y)\Phi \right] dx\, dy,$$

obtain the relevant Euler's equation.

Solution

Let

$$F(x,y,\Phi,\Phi_x,\Phi_y) = \frac{1}{2}\left(\Phi_x^2 + \Phi_y^2\right) - f(x,y)\Phi$$

showing that we have two independent variables x and y and one dependent variable Φ. Hence, Euler's equation (4.20) becomes

$$-f(x,y) - \frac{\partial}{\partial x}\Phi_x - \frac{\partial}{\partial y}\Phi_y = 0$$

or

$$\Phi_{xx} + \Phi_{yy} = -f(x,y),$$

that is,

$$\nabla^2\Phi = -f(x,y)$$

which is Poisson's equation. Thus, solving Poisson's equation is equivalent to finding Φ that extremizes the given functional $I(\Phi)$.

4.4 Construction of Functionals from PDEs

In the previous section, we noticed that Euler's equation produces the governing differential equation corresponding to a given functional or variational principle. Here, we seek the inverse procedure of constructing a variational principle for a given differential equation. The procedure for finding the functional associated with the differential equation involves four basic steps [2,7]:

- Multiply the operator equation $L\Phi = g$ (Euler's equation) with the variational $\delta\Phi$ of the dependent variable Φ and integrate over the domain of the problem.
- Use the divergence theorem or integration by parts to transfer the derivatives to variation $\delta\Phi$.
- Express the boundary integrals in terms of the specified boundary conditions.
- Bring the variational operator δ outside the integrals.

The procedure is best illustrated with an example. Suppose we are interested in finding the variational principle associated with the Poisson's equation

$$\nabla^2\Phi = -f(x, y) \tag{4.25}$$

which is the converse of what we did in Example 4.3. After taking step 1, we have

$$\delta I = \int\int [-\nabla^2\Phi - f]\delta\Phi \, dx \, dy = 0$$

$$= -\int\int \nabla^2\Phi \delta\Phi \, dx \, dy - \int\int f\delta\Phi \, dx \, dy$$

This can be evaluated by applying divergence theorem or integrating by parts. To integrate by parts, let $u = \delta\Phi$, $dv = (\partial/\partial x)(\partial\Phi/\partial x)dx$ so that $du = (\partial/\partial x)\delta\Phi dx$, $v = (\partial\Phi/\partial x)$ and

$$-\int\left[\int \frac{\partial}{\partial x}\left(\frac{\partial\Phi}{\partial x}\right)\delta\Phi \, dx\right]dy = -\int\left[\delta\Phi\frac{\partial\Phi}{\partial x} - \int \frac{\partial\Phi}{\partial x}\frac{\partial}{\partial x}\delta\Phi \, dx\right]dy$$

Thus,

$$\delta I = \int\int\left[\frac{\partial\Phi}{\partial x}\frac{\partial}{\partial x}\delta\Phi + \frac{\partial\Phi}{\partial y}\frac{\partial}{\partial y}\delta\Phi - \delta f\Phi\right]dx \, dy$$

$$- \int \delta\Phi\frac{\partial\Phi}{\partial x}dy - \int \delta\Phi\frac{\partial\Phi}{\partial y}dx$$

$$\delta I = \frac{\delta}{2}\int\int\left[\left(\frac{\partial\Phi}{\partial x}\right)^2 + \left(\frac{\partial\Phi}{\partial y}\right)^2 - 2f\Phi\right]dx \, dy$$

$$- \delta\int \Phi\frac{\partial\Phi}{\partial x}dy - \delta\int \Phi\frac{\partial\Phi}{\partial y}dx \tag{4.26}$$

The last two terms vanish if we assume either the homogeneous Dirichlet or Neumann conditions at the boundaries. Hence,

$$\delta I = \delta\int\int \frac{1}{2}[\Phi_x^2 + \Phi_y^2 - 2\Phi f]dx \, dy,$$

that is,

$$I(\Phi) = \frac{1}{2}\int\int [\Phi_x^2 + \Phi_y^2 - 2\Phi f]dx \, dy \tag{4.27}$$

as expected.

Rather than following the four steps listed above to find the function $I(\Phi)$ corresponding to the operator equation (4.8), an alternative approach is provided by Mikhlin [1 pp. 74–78]. According to Mikhlin, if L in Equation 4.8 is real, self-adjoint, and positive definite, the solution of Equation 4.8 minimizes the functional

$$\boxed{I(\Phi) = \langle L\Phi, \Phi \rangle - 2\langle \Phi, g \rangle} \tag{4.28}$$

(See Problem 4.6 for a proof.) Thus, Equation 4.27, for example, can be obtained from Equation 4.25 by applying Equation 4.28. This approach has been applied to derive variational solutions of integral equations [8].

Other systematic approaches for the derivation of variational principles for EM problems include Hamilton's principle or the principle of least action [9,10], Lagrange multipliers [10–14], and a technique described as variational electromagnetics [15,16]. The method of Lagrange undetermined multipliers is particularly useful for deriving a functional for a PDE whose arguments are constrained. Table 4.1 provides the variational principles for some differential equations commonly found in EM-related problems.

EXAMPLE 4.4

Find the functional for the ordinary differential equation

$$y'' + y + x = 0, \quad 0 < x < 1$$

subject to $y(0) = y(1) = 0$.

TABLE 4.1

Variational Principle Associated with Common PDEs in EM[a]

Name of Equation	PDE	Variational Principle
Inhomogeneous wave equation	$\nabla^2 \Phi + k^2 \Phi = g$	$I(\Phi) = \dfrac{1}{2} \displaystyle\int_v [|\nabla\Phi|^2 - k^2\Phi^2 + 2g\Phi]dv$
Homogeneous wave equation	$\nabla^2 \Phi + k^2 \Phi = 0$	$I(\Phi) = \dfrac{1}{2} \displaystyle\int_v [|\nabla\Phi|^2 - k^2\Phi^2]dv$
	or	
	$\nabla^2 \Phi - \dfrac{1}{u^2}\Phi_{tt} = 0$	$I(\Phi) = \dfrac{1}{2} \displaystyle\int^{t_o}\!\!\int_v [|\nabla\Phi|^2 - \tfrac{1}{u^2}\Phi_t^2]dv\,dt$
Diffusion equation	$\nabla^2 \Phi - k\Phi_t = 0$	$I(\Phi) = \dfrac{1}{2} \displaystyle\int^{t_o}\!\!\int_v [|\nabla\Phi|^2 - k\Phi\Phi_t]dv\,dt$
Poisson's equation	$\nabla^2 \Phi = g$	$I(\Phi) = \dfrac{1}{2} \displaystyle\int_v [|\nabla\Phi|^2 + 2g\Phi]dv$
Laplace's equation	$\nabla^2 \Phi = 0$	$I(\Phi) = \dfrac{1}{2} \displaystyle\int_v [|\nabla\Phi|^2]dv$

[a] Note that $|\nabla\Phi|^2 = \nabla\Phi \cdot \nabla\Phi = \Phi_x^2 + \Phi_y^2 + \Phi_z^2$.

Solution

Given that

$$\frac{d^2y}{dx^2} + y + x = 0, \quad 0 < x < 1,$$

we obtain

$$\delta I = \int_0^1 \left(\frac{d^2y}{dx^2} + y + x \right) \delta y\, dx = 0$$

$$= \int_0^1 \frac{d^2y}{dx^2} \delta y\, dx + \int_0^1 y \delta y\, dx + \int_0^1 x \delta y\, dx$$

Integrating the first term by parts,

$$\delta I = \delta y \frac{dy}{dx}\Big|_{x=0}^{x=1} - \int_0^1 \frac{dy}{dx} \frac{d}{dx} \delta y + \int_0^1 \frac{1}{2} \delta(y^2)\, dx + \delta \int_0^1 xy\, dx$$

Since y is fixed at $x = 0, 1$, $\delta y(1) = \delta y(0) = 0$. Hence,

$$\delta I = -\delta \int_0^1 \frac{1}{2}\left(\frac{dy}{dx} \right)^2 dx + \frac{1}{2}\delta \int_0^1 y^2 dx + \delta \int_0^1 xy\, dx$$

$$= \frac{\delta}{2} \int_0^1 [-y'^2 + y^2 + 2xy]\, dx$$

or

$$I(y) = \frac{1}{2} \int_0^1 [-y'^2 + y^2 + 2xy]\, dx$$

Check: Taking $F(x, y, y') = y'^2 - y^2 - 2xy$, Euler's equation $F_y - (d/dx)F_{y'} = 0$ gives the differential equation

$$y'' + y + x = 0$$

4.5 Rayleigh–Ritz Method

The Rayleigh–Ritz method is the direct variational method for minimizing a given functional. It is direct in that it yields a solution to the variational problem without recourse to the associated differential equation [17]. In other words, it is the direct application of variational principles discussed in the previous sections. The method was first presented by Rayleigh in 1877 and extended by Ritz in 1909. Without loss of generality, let the associated variational principle be

$$I(\Phi) = \int_S F(x, y, \Phi, \Phi_x, \Phi_y) \, dS \qquad (4.29)$$

Our objective is to minimize this integral. In the Rayleigh–Ritz method, we select a linearly independent set of functions called *expansion functions* (or *basis functions*) u_n and construct an approximate solution to Equation 4.29, satisfying some prescribed boundary conditions. The solution is in the form of a finite series

$$\tilde{\Phi} \simeq \sum_{n=1}^{N} a_n u_n + u_o \qquad (4.30)$$

where u_o meets the nonhomogeneous boundary conditions, and u_n satisfies homogeneous boundary conditions. a_n are expansion coefficients to be determined and $\tilde{\Phi}$ is an approximate solution to Φ (the exact solution). We substitute Equation 4.30 into Equation 4.29 and convert the integral $I(\Phi)$ into a function of N coefficients $a_1, a_2, ..., a_N$, that is,

$$I(\Phi) = I(a_1, a_2, ..., a_N)$$

The minimum of this function is obtained when its partial derivatives with respect to each coefficient are zero:

$$\frac{\partial I}{\partial a_1} = 0, \quad \frac{\partial I}{\partial a_2} = 0, ..., \frac{\partial I}{\partial a_N} = 0$$

or

$$\boxed{\frac{\partial I}{\partial a_n} = 0, \quad n = 1, 2, ..., N} \qquad (4.31)$$

Thus, we obtain a set of N simultaneous equations. The system of linear algebraic equations obtained is solved to get a_n, which are finally substituted into the approximate solution of Equation 4.30. In the approximate solution of Equation 4.30, if $\tilde{\Phi} \to \Phi$ as $N \to \infty$ in some sense, then the procedure is said to *converge* to the exact solution.

An alternative, perhaps easier, procedure to determine the expansion coefficients a_n is by solving a system of simultaneous equations obtained as follows [4,18]. Substituting Equation 4.30 (ignoring u_o since it can be lumped with the right-hand side of the equation) into Equation 4.28 yields

$$I = \left\langle \sum_{m=1}^{N} a_m L u_m, \sum_{n=1}^{N} a_n u_n \right\rangle - 2 \left\langle \sum_{m=1}^{N} a_m u_m, g \right\rangle$$

$$= \sum_{m=1}^{N} \sum_{n=1}^{N} \langle L u_m, u_n \rangle a_n a_m - 2 \sum_{m=1}^{N} \langle u_m, g \rangle a_m$$

Expanding this into powers of a_m results in

$$I = \langle Lu_m, u_m \rangle a_m^2 + \sum_{n \neq m}^{N} \langle Lu_m, u_n \rangle a_m a_n + \sum_{k \neq m}^{N} \langle Lu_k, u_m \rangle a_k a_m$$
$$- 2\langle g, u_m \rangle a_m + \text{terms not containing } a_m \qquad (4.32)$$

Assuming L is self-adjoint and replacing k with n in the second summation,

$$I = \langle Lu_m, u_m \rangle a_m^2 + 2\sum_{n \neq m}^{N} \langle Lu_m, u_n \rangle a_n a_m - 2\langle g, u_m \rangle a_m + \cdots \qquad (4.33)$$

Since we are interested in selecting a_m such that I is minimized, Equation 4.33 must satisfy Equation 4.31. Thus, differentiating Equation 4.33 with respect to a_m and setting the result equal to zero leads to

$$\sum_{n=1}^{N} \langle Lu_m, u_n \rangle a_n = \langle g, u_m \rangle, \quad m = 1, 2, \ldots, N \qquad (4.34)$$

which can be put in matrix form as

$$\begin{bmatrix} \langle Lu_1, u_1 \rangle & \langle Lu_1, u_2 \rangle & \cdots & \langle Lu_1, u_N \rangle \\ \vdots & & & \vdots \\ \langle Lu_N, u_1 \rangle & \langle Lu_N, u_2 \rangle & \cdots & \langle Lu_N, u_n \rangle \end{bmatrix} \begin{bmatrix} a_1 \\ \vdots \\ a_N \end{bmatrix} = \begin{bmatrix} \langle g, u_1 \rangle \\ \vdots \\ \langle g, u_N \rangle \end{bmatrix} \qquad (4.35a)$$

or

$$[A][X] = [B] \qquad (4.35b)$$

where $A_{mn} = \langle Lu_m, u_n \rangle$, $B_m = \langle g, u_m \rangle$, $X_n = a_n$. Solving for $[X]$ in Equation 4.35 and substituting a_m in Equation 4.30 gives the approximate solution $\tilde{\Phi}$. Equation 4.35 is called the *Rayleigh–Ritz system*.

We are yet to know how the expansion functions are selected. They are selected to satisfy the prescribed boundary conditions of the problem. u_o is chosen to satisfy the inhomogeneous boundary conditions, while $u_n (n = 1, 2, \ldots, N)$ are selected to satisfy the homogeneous boundary conditions. If the prescribed boundary conditions are all homogeneous (Dirichlet conditions), $u_o = 0$. The next section will discuss more on the selection of the expansion functions.

The Rayleigh–Ritz method has two major limitations. First, the variational principle in Equation 4.29 may not exist in some problems such as in nonself-adjoint equations (odd order derivatives). Second, it is difficult, if not impossible, to find the functions u_o satisfying the global boundary conditions for the domains with complicated geometries [19].

EXAMPLE 4.5

Use the Rayleigh–Ritz method to solve the ordinary differential equation:

$$\Phi'' + 4\Phi - x^2 = 0, \quad 0 < x < 1$$

subject to $\Phi(0) = 0 = \Phi(1)$.

Solution

The exact solution is

$$\Phi(x) = \frac{\sin 2(1-x) - \sin 2x}{8\sin 2} + \frac{x^2}{4} - \frac{1}{8}$$

The variational principle associated with $\Phi'' + 4\Phi - x^2 = 0$ is

$$I(\Phi) = \int_0^1 [(\Phi')^2 - 4\Phi^2 + 2x^2\Phi]dx \tag{4.36}$$

This is readily verified using Euler's equation. We let the approximate solution be

$$\tilde{\Phi} = u_o + \sum_{n=1}^{N} a_n u_n \tag{4.37}$$

where $u_o = 0$, $u_n = x^n(1 - x)$ since $\Phi(0) = 0 = \Phi(1)$ must be satisfied. (This choice of u_n is not unique. Other possible choices are $u_n = x(1 - x^n)$ and $u_n = \sin n\pi x$. Note that each choice satisfies the prescribed boundary conditions.) Let us try different values of N, the number of expansion coefficients. We can find the expansion coefficients a_n in two ways: using the functional directly as in Equation 4.31 or using the Rayleigh–Ritz system of Equation 4.35.

Method 1

For $N = 1$, $\tilde{\Phi} = a_1 u_1 = a_1 x(1 - x)$. Substituting this into Equation 4.36 gives

$$I(a_1) = \int_0^1 [a_1^2(1-2x)^2 - 4a_1^2(x-x^2)^2 + 2a_1 x^3(1-x)]dx$$

$$= \frac{1}{5}a_1^2 + \frac{1}{10}a_1$$

$I(a_1)$ is minimum when

$$\frac{\partial I}{\partial a_1} = 0 \rightarrow \frac{2}{5}a_1 + \frac{1}{10} = 0 \quad \text{or} \quad a_1 = -\frac{1}{4}$$

Hence, the quadratic approximate solution is

$$\tilde{\Phi} = -\frac{1}{4}x(1-x) \tag{4.38}$$

For $N = 2$, $\tilde{\Phi} = a_1 u_1 + a_2 u_2 = a_1 x(1-x) + a_2 x^2(1-x)$. Substituting $\tilde{\Phi}$ into Equation 4.36,

$$I(a_1, a_2) = \int_0^1 [[a_1(1-2x) + a_2(2x-3x^2)]^2 - 4[a_1(x-x^2) + a_2(x^2-x^3)]^2$$

$$+ 2a_1x^2(x-x^2) + 2a_1x^2(x^2-x^3)]dx$$

$$= \frac{1}{5}a_1^2 + \frac{2}{21}a_2^2 + \frac{1}{5}a_1a_2 + \frac{1}{10}a_1 + \frac{1}{15}a_2$$

$$\frac{\partial I}{\partial a_1} = 0 \rightarrow \frac{2}{5}a_1 + \frac{1}{5}a_2 + \frac{1}{10} = 0$$

or

$$4a_1 + 2a_2 = -1 \qquad (4.39a)$$

$$\frac{\partial I}{\partial a_2} = 0 \rightarrow \frac{4}{21}a_2 + \frac{1}{5}a_1 + \frac{1}{15} = 0$$

or

$$21a_1 + 20a_2 = -7 \qquad (4.39b)$$

Solving Equation 4.39 gives

$$a_1 = -\frac{6}{38}, \quad a_2 = -\frac{7}{38}$$

and hence the cubic approximate solution is

$$\tilde{\Phi} = -\frac{6}{38}x(1-x) - \frac{7}{38}x^2(1-x)$$

or

$$\tilde{\Phi} = \frac{x}{38}(7x^2 - x - 6)$$

Method 2

We now determine a_m using Equation 4.35. From the given differential equation,

$$L = \frac{d^2}{dx^2} + 4, \quad g = x^2$$

Hence,

$$A_{mn} = \langle Lu_m, u_n \rangle = \langle u_m, Lu_n \rangle$$

$$= \int_0^1 x^m(1-x)\left[\left(\frac{d^2}{dx^2} + 4\right)x^n(1-x)\right]dx,$$

$$A_{mn} = \frac{n(n-1)}{m+n-1} - \frac{2n^2}{m+n} + \frac{n(n+1)+4}{m+n+1} - \frac{8}{m+n+2} + \frac{4}{m+n+3},$$

$$B_n = \langle g, u_n \rangle = \int_0^1 x^2 n^n(1-x)dx = \frac{1}{n+3} - \frac{1}{n+4}$$

When $N = 1$, $A_{11} = -\dfrac{1}{5}$, $B_1 = \dfrac{1}{20}$, that is,

$$-\frac{1}{5}a_1 = \frac{1}{20} \rightarrow a_1 = -\frac{1}{4}$$

as before. When $N = 2$,

$$A_{11} = -\frac{1}{5},\quad A_{12} = A_{21} = -\frac{1}{10},\quad A_{22} = -\frac{2}{21},\quad B_1 = \frac{1}{20},\quad B_2 = \frac{1}{30}$$

Hence,

$$\begin{bmatrix} -\dfrac{1}{5} & -\dfrac{1}{10} \\[2mm] -\dfrac{1}{10} & -\dfrac{2}{21} \end{bmatrix}\begin{bmatrix} a_1 \\[2mm] a_2 \end{bmatrix} = \begin{bmatrix} \dfrac{1}{20} \\[2mm] \dfrac{1}{30} \end{bmatrix}$$

which gives $a_1 = -6/38$, $a_2 = -7/38$ as obtained previously. When $N = 3$,

$$A_{13} = A_{31} = -\frac{13}{210},\quad A_{23} = A_{32} = -\frac{28}{105},\quad A_{33} = -\frac{22}{315},\quad B_3 = \frac{1}{42},$$

that is,

$$\begin{bmatrix} -\dfrac{1}{5} & -\dfrac{1}{10} & -\dfrac{13}{210} \\[2mm] -\dfrac{1}{10} & -\dfrac{2}{21} & -\dfrac{28}{105} \\[2mm] -\dfrac{13}{210} & -\dfrac{28}{105} & -\dfrac{22}{315} \end{bmatrix}\begin{bmatrix} a_1 \\[2mm] a_2 \\[2mm] a_3 \end{bmatrix} = \begin{bmatrix} \dfrac{1}{20} \\[2mm] \dfrac{1}{30} \\[2mm] \dfrac{1}{42} \end{bmatrix}$$

from which we obtain

$$a_1 = -\frac{6}{38},\quad a_2 = -\frac{7}{38},\quad a_3 = 0$$

showing that we obtain the same solution as for the case $N = 2$. For different values of $x, 0 < x < 1$, the Rayleigh–Ritz solution is compared with the exact solution in Table 4.2.

EXAMPLE 4.6

Using the Rayleigh–Ritz method, solve Poisson's equation:

$$\nabla^2 V = -\rho_o, \quad \rho_o = \text{constant}$$

in a square $-1 \le x \le 1$, $-1 \le y \le 1$, subject to the homogeneous boundary conditions $V(x, \pm 1) = 0 = V(\pm 1, y)$.

Solution

Due to the symmetry of the problem, we choose the basis functions as

$$u_{mn} = (1 - x^2)(1 - y^2)(x^{2m}y^{2n} + x^{2n}y^{2m}), \quad m, n = 0, 1, 2, \ldots$$

TABLE 4.2

Comparison of Exact Solution with the Rayleigh–Ritz Solution
of $\Phi'' + 4\Phi - x^2 = 0$, $\Phi(0) = 0 = \Phi(1)$

x	Exact Solution	Rayleigh–Ritz $N = 1$	Solution $N = 2$
0.0	0.0	0.0	0.0
0.2	−0.0301	−0.0400	−0.0312
0.4	−0.0555	−0.0600	−0.0556
0.6	−0.0625	−0.0625	−0.0644
0.8	−0.0489	−0.0400	−0.0488
1.0	0.0	0.0	0.0

Hence,

$$\tilde{\Phi} = (1 - x^2)(1 - y^2)[a_1 + a_2(x^2 + y^2) + a_3 x^2 y^2 + a_4(x^4 + y^4) + \cdots]$$

Case 1: When $m = n = 0$, we obtain the first approximation ($N = 1$) as

$$\tilde{\Phi} = a_1 u_1$$

where $u_1 = (1 - x^2)(1 - y^2)$.

$$A_{11} = \langle Lu_1, u_1 \rangle = \int_{-1}^{1} \int_{-1}^{1} \left(\frac{\partial^2 u_1}{\partial x^2} + \frac{\partial^2 u_1}{\partial y^2} \right) u_1 \, dx \, dy$$

$$= -8 \int_{0}^{1} \int_{0}^{1} (2 - x^2 - y^2)(1 - x^2)(1 - y^2) \, dx \, dy$$

$$= -\frac{256}{45},$$

$$B_1 = \langle g, u_1 \rangle = -\int_{-1}^{1} \int_{-1}^{1} (1 - x^2)(1 - y^2) \rho_o \, dx \, dy = -\frac{16\rho_o}{9}$$

Hence,

$$-\frac{256}{45} a_1 = -\frac{16}{9} \rho_o \rightarrow a_1 = \frac{5}{16} \rho_o$$

and

$$\tilde{\Phi} = \frac{5}{16} \rho_o (1 - x^2)(1 - y^2)$$

Case 2: When $m = n = 1$, we obtain the second-order approximation ($N = 2$) as

$$\tilde{\Phi} = a_1 u_1 + a_2 u_2$$

where $u_1 = (1 - x^2)(1 - y^2)$, $u_2 = (1 - x^2)(1 - y^2)(x^2 + y^2)$. A_{11} and B_1 are the same as in case 1.

$$A_{12} = A_{21} = \langle Lu_1, u_2 \rangle = -\frac{1024}{525},$$

$$A_{22} = \langle Lu_2, u_2 \rangle = -\frac{11,264}{4725},$$

$$B_2 = \langle g, u_2 \rangle = -\frac{32}{45}\rho_o$$

Hence,

$$\begin{bmatrix} -\dfrac{256}{45} & -\dfrac{1024}{525} \\ -\dfrac{1024}{525} & -\dfrac{11,264}{4725} \end{bmatrix} \begin{bmatrix} a_1 \\ a_2 \end{bmatrix} = \begin{bmatrix} -\dfrac{16}{9}\rho_o \\ -\dfrac{32}{45}\rho_o \end{bmatrix}$$

Solving this yields

$$a_1 = \frac{1295}{4432}\rho_o = 0.2922\rho_o, \quad a_2 = \frac{525}{8864}\rho_o = 0.0592\rho_o$$

and

$$\tilde{\Phi} = (1 - x^2)(1 - y^2)(0.2922 + 0.0592(x^2 + y^2))\rho_0$$

4.6 Weighted Residual Method

As noted earlier, the Rayleigh–Ritz method is applicable when a suitable functional exists. In cases where such a functional cannot be found, we apply one of the techniques collectively referred to as the *method of weighted residuals*. The method is more general and has wider application than the Rayleigh–Ritz method because it is not limited to a class of variational problems. It does not require that the governing equation be stationary.

Consider the operator equation

$$L\Phi = g \tag{4.40}$$

In the weighted residual method, the solution to Equation 4.40 is approximated, in the same manner as in the Rayleigh–Ritz method, using the expansion functions, u_n, that is,

$$\tilde{\Phi} = \sum_{n=1}^{N} a_n u_n \tag{4.41a}$$

where a_n are the expansion coefficients. We seek to make

$$L\tilde{\Phi} \approx g \tag{4.41b}$$

Substitution of the approximate solution in the operator equation results in a *residual R* (an error in the equation), that is,

$$R = L(\tilde{\Phi} - \Phi) = L\tilde{\Phi} - g \tag{4.42}$$

In the weighted residual method, the weighting functions w_m (which, in general, are not the same as the expansion functions u_n) are chosen such that the integral of a weighted residual of the approximation is zero, that is,

$$\int w_m R \, dv = 0$$

or

$$\langle w_m, R \rangle = 0 \tag{4.43}$$

If a set of *weighting functions* $\{w_m\}$ (also known as *testing functions*) are chosen and the inner product of Equation 4.41b is taken for each w_m, we obtain

$$\sum_{n=1}^{N} a_n \langle w_m, L u_n \rangle = \langle w_m, g \rangle, \quad m = 1, 2, \ldots, N \tag{4.44}$$

The system of linear equations (4.44) can be cast into matrix form as

$$[A][X] = [B] \tag{4.45}$$

where $A_{mn} = \langle w_m, L u_n \rangle$, $B_m = \langle w_m, g \rangle$, $X_n = a_n$. Solving for $[X]$ in Equation 4.45 and substituting for a_n in Equation 4.41a gives the approximate solution to Equation 4.40. However, there are different ways of choosing the weighting functions w_m leading to

- Collocation (or point-matching) method,
- Subdomain method,
- Galerkin method,
- Least-squares method.

4.7 Collocation Method

We select the Dirac delta function as the weighting function, that is,

$$w_m(\mathbf{r}) = \delta(\mathbf{r} - \mathbf{r}_m) = \begin{cases} 1, & \mathbf{r} = \mathbf{r}_m \\ 0, & \mathbf{r} \neq \mathbf{r}_m \end{cases} \tag{4.46}$$

Substituting Equation 4.46 into Equation 4.43 results in

$$R(\mathbf{r}) = 0 \tag{4.47}$$

Thus, we select as many collocation (or matching) points in the interval as there are unknown coefficients a_n and make the residual zero at those points. This is equivalent to enforcing

$$\sum_{n=1}^{N} La_n u_n = g \tag{4.48}$$

at discrete points in the region of interest, generally where boundary conditions must be met. Although the point-matching method is the simplest specialization for computation, it is not possible to determine in advance for a particular operator equation what collocation points would be suitable. An accurate result is ensured only if judicious choice of the match points is taken. (This will be illustrated in Example 4.7.) It is important to note that the finite difference method is a particular case of collocation with locally defined expansion functions [20]. The validity and legitimacy of the point-matching technique are discussed in References 21,22.

4.7.1 Subdomain Method

We select weighting functions w_m, each of which exists only over subdomains of the domain of Φ. Typical examples of such functions for one-dimensional problems are illustrated in Figure 4.2 and defined as follows:

1. Piecewise uniform (or pulse) function:

$$w_m(x) = \begin{cases} 1, & x_{m-1} < x < x_{m+1} \\ 0, & \text{otherwise} \end{cases} \tag{4.49a}$$

2. Piecewise linear (or triangular) function:

$$w_m(x) = \begin{cases} \dfrac{\Delta - |x - x_m|}{\Delta}, & x_{m-1} < x < x_{m+1} \\ 0, & \text{otherwise} \end{cases} \tag{4.49b}$$

3. Piecewise sinusoidal function:

$$w_m(x) = \begin{cases} \dfrac{\sin k(x - |x - x_m|)}{\Delta}, & x_{m-1} < x < x_{m+1} \\ 0, & \text{otherwise} \end{cases} \tag{4.49c}$$

Using the unit pulse function, for example, is equivalent to dividing the domain of Φ into as many subdomains as there are unknown terms and letting the average value of R over such subdomains vanish.

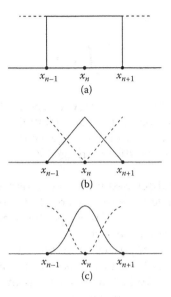

FIGURE 4.2
Typical subdomain weighting functions: (a) piecewise uniform function, (b) piecewise linear function, (c) piecewise sinusoidal function.

4.7.2 Galerkin Method

We select basis functions as the weighting function, that is, $w_m = u_m$. When the operator is a linear differential operator of even order, the Galerkin method[*] reduces to the Rayleigh–Ritz method. This is due to the fact that the differentiation can be transferred to the weighting functions and the resulting coefficient matrix $[A]$ will be symmetric [7]. In order for the Galerkin method to be applicable, the operator must be of a certain type. Also, the expansion function u_n must span both the domain and the range of the operator; that is, the Galerkin method is a global approach (or entire domain basis function). It is commonly used in FEM and in moment methods.

4.7.3 Least Squares Method

This involves minimizing the integral of the square of the residual, that is,

$$\frac{\partial}{\partial a_m} \int R^2 dv = 0$$

or

$$\int \frac{\partial R}{\partial a_m} R \, dv = 0 \tag{4.50}$$

Comparing Equation 4.50 with Equation 4.43 shows that we must choose

$$w_m = \frac{\partial R}{\partial a_m} = L u_m \tag{4.51}$$

---

[*] The Galerkin method was developed by the Russian engineer B.G. Galerkin in 1915.

This may be viewed as requiring that

$$\frac{1}{2} \int R^2 \, dv$$

be minimum. In other words, the choice of w_m corresponds to minimizing the mean square residual. It should be noted that the least-squares method involves higher-order derivatives which will, in general, lead to a better convergence than the Rayleigh–Ritz method or Galerkin method, but it has the disadvantage of requiring higher-order weighting functions [19].

The concept of convergence discussed in the previous section applies here as well. That is, if the approximate solution $\tilde{\Phi}$ were to converge to the exact solution Φ as $N \to \infty$, the residual must approach zero as $N \to \infty$. Otherwise, the sequence of approximate solutions may not converge to any meaningful result.

The inner product involved in applying a weighted residual method can sometimes be evaluated analytically, but in most practical situations it is evaluated numerically. Due to a careless evaluation of the inner product, one may think that the least-squares technique is being used when the resulting solution is identical to a point-matching solution. To avoid such erroneous results or conclusions, one must be certain that the overall number of points involved in the numerical integration is not smaller than the number of unknowns, N, involved in the weighted residual method [23].

The accuracy and efficiency of a weighted residual method are largely dependent on the selection of expansion functions. The solution may be exact or approximate depending on how we select the expansion and weighting functions [17]. The criteria for selecting expansion and weighting functions in a weighted residual method are provided in the work of Sarkar and others [24–27]. We summarize their results here. The expansion functions u_n are selected to satisfy the following requirements [27]:

1. The expansion functions should be in the domain of the operator L in some sense, that is, they should satisfy the differentiability criterion and they must satisfy the boundary conditions for the operator. It is not necessary for each expansion function to satisfy exactly the boundary conditions. What is required is that the total solution must satisfy the boundary conditions at least in some distributional sense. The same holds for the differentiability conditions.

2. The expansion functions must be such that $L u_n$ form a complete set for the range of the operator. It really does not matter whether the expansion functions are complete in the domain of the operator. What is important is that u_n must be chosen in such a way that $L u_n$ is complete. This will be illustrated in Example 4.8.

From a mathematical point of view, the choice of expansion functions does not depend on the choice of weighting functions. It is required that the weighting functions w_n must make the difference $\Phi - \tilde{\Phi}$ small. For the Galerkin method to be applicable, the expansion functions u_n must span both the domain and the range of the operator. For the least-squares method, the weighting functions are already presented and defined by $L u_n$. It is necessary that $L u_n$ form a complete set. The least-squares technique mathematically and numerically is one of the safest techniques to utilize when very little is known about the nature of the operator and the exact solution.

EXAMPLE 4.7

Find an approximate solution to

$$\Phi'' + 4\Phi - x^2 = 0, \quad 0 < x < 1,$$

with $\Phi(0) = 0$, $\Phi'(1) = 1$, using the method of weighted residuals.

Solution

The exact solution is

$$\Phi(x) = \frac{\cos 2(x-1) + 2\sin 2x}{8\cos 2} + \frac{x^2}{4} - \frac{1}{8} \tag{4.52}$$

Let the approximate solution be

$$\tilde{\Phi} = u_0 + \sum_{n=1}^{N} a_n u_n \tag{4.53}$$

The boundary conditions can be decomposed into two parts:

1. Homogeneous part $\rightarrow \Phi(0) = 0$, $\Phi'(0) = 0$,
2. Inhomogeneous part $\rightarrow \Phi'(1) = 1$.

We choose u_0 to satisfy the inhomogeneous boundary condition. A reasonable choice is

$$u_0 = x \tag{4.54a}$$

We choose $u_n (n = 1, 2, \ldots, N)$ to satisfy the homogeneous boundary condition. Suppose we select

$$u_n(x) = x^n \left(x - \frac{n+1}{n} \right) \tag{4.54b}$$

Thus, if we take $N = 2$, the approximate solution is

$$\begin{aligned}
\tilde{\Phi} &= u_0 + a_1 u_1 + a_2 u_2 \\
&= x + a_1 x(x-2) + a_2 x^2 (x - 3/2)
\end{aligned} \tag{4.55}$$

where the expansion coefficients, a_1 and a_2, are to be determined. We find the residual R using Equation 4.42, namely,

$$\begin{aligned}
R &= L\tilde{\Phi} - g \\
&= \left(\frac{d^2}{dx^2} + 4 \right) \tilde{\Phi} - x^2 \\
&= a_1(4x^2 - 8x + 2) + a_2(4x^3 - 6x^2 + 6x - 3) - x^2 + 4x
\end{aligned} \tag{4.56}$$

We now apply each of the four techniques discussed and compare the solutions.

Method 1: Collocation or point-matching method

Since we have two unknowns a_1 and a_2, we select two match points, at $x = 1/3$ and $x = 2/3$, and set the residual equal to zero at those points, that is,

$$R\left(\frac{1}{3}\right) = 0 \;\rightarrow\; 6a_1 + 41a_2 = 33$$

$$R\left(\frac{2}{3}\right) = 0 \;\rightarrow\; 42a_1 + 13a_2 = 60$$

Solving these equations,

$$a_1 = \frac{677}{548}, \quad a_2 = \frac{342}{548}$$

Substituting a_1 and a_2 into Equation 4.55 gives

$$\tilde{\Phi}(x) = -1.471x + 0.2993x^2 + 0.6241x^3 \tag{4.57}$$

To illustrate the dependence of the solution on the match points, suppose we select $x = 1/4$ and $x = 3/4$ as the match points. Then,

$$R = \left(\frac{1}{4}\right) = 0 \;\rightarrow\; -4a_1 + 29a_2 = 15$$

$$R = \left(\frac{3}{4}\right) = 0 \;\rightarrow\; 28a_1 + 3a_2 = 39$$

Solving for a_1 and a_2, we obtain

$$a_1 = \frac{543}{412}, \quad a_2 = \frac{288}{412}$$

with the approximate solution

$$\tilde{\Phi}(x) = -1.636x + 0.2694x^2 + 0.699x^3 \tag{4.58}$$

We will refer to the solutions in Equations 4.57 and 4.58 as collocation 1 and collocation 2, respectively. It is evident from Table 4.3 that collocation 2 is more accurate than collocation 1.

Method 2: Subdomain method

Divide the interval $0 < x < 1$ into two segments since we have two unknowns a_1 and a_2. We select pulse functions as weighting functions:

$$w_1 = 1, \quad 0 < x < \frac{1}{2},$$

$$w_2 = 1, \quad \frac{1}{2} < x < 1$$

so that

$$\int_0^{1/2} w_1 R\, dx = 0 \;\rightarrow\; -8a_1 + 45a_2 = 22$$

$$\int_{1/2}^1 w_2 R\, dx = 0 \;\rightarrow\; 40a_1 + 3a_2 = 58$$

TABLE 4.3

Comparison of the Weighted Residual Solutions of the Problem in Example 4.7 with the Exact Solution in Equation 4.52

x	Exact Solution	Collocation 1	Collocation 2	Subdomain	Galerkin	Least Squares
0.0	0.0000	0.0000	0.0000	0.0000	0.0000	0.0000
0.1	−0.1736	−0.1435	−0.1602	−0.1753	−0.1794	−0.1657
0.2	−0.3402	−0.2772	−0.3108	−0.3403	−0.3451	−0.3217
0.3	−0.4928	−0.3975	−0.4477	−0.4907	−0.4935	−0.4635
0.4	−0.6248	−0.5006	−0.5666	−0.6221	−0.6208	−0.5869
0.5	−0.7303	−0.5827	−0.6633	−0.7300	−0.7233	−0.6877
0.6	−0.8042	−0.6400	−0.7336	−0.8100	−0.7972	−0.7615
0.7	−0.8424	−0.6690	−0.7734	−0.8577	−0.8390	−0.8041
0.8	−0.8422	−0.6657	−0.7785	−0.8687	−0.8449	−0.8113
0.9	−0.8019	−0.6264	−0.7446	−0.8385	−0.8111	−0.7788
1.0	−0.7216	−0.5476	−0.6676	−0.7627	−0.7340	−0.7022

Solving the two equations gives

$$a_1 = \frac{53}{38}, \quad a_2 = \frac{28}{38}$$

and hence Equation 4.55 becomes

$$\tilde{\Phi}(x) = -1.789x + 0.2895x^2 + 0.7368x^3 \tag{4.59}$$

Method 3: Galerkin method

In this case, we select $w_m = u_m$, that is,

$$w_1 = x(x-2), \quad w_2 = x^2(x-3/2)$$

We now apply Equation 4.43, namely, $\int w_m R\, dx = 0$. We obtain

$$\int_0^1 (x^2 - 2x) R\, dx = 0 \;\rightarrow\; 24a_1 + 11a_2 = 41$$

$$\int_0^1 \left(x^3 - \frac{3}{2}x^2\right) R\, dx = 0 \;\rightarrow\; 77a_1 + 15a_2 = 119$$

Solving these leads to

$$a_1 = \frac{694}{487}, \quad a_2 = \frac{301}{487}$$

Substituting a_1 and a_2 into Equation 4.55 gives

$$\tilde{\Phi}(x) = -1.85x + 0.4979x^2 + 0.6181x^3 \tag{4.60}$$

Method 4: Least-squares method

For this method, we select $w_m = \partial R/\partial a_m$, that is,

$$w_1 = 4x^2 - 8x + 2, \quad w_2 = 4x^3 - 6x^2 + 6x - 3$$

Applying Equation 4.43

$$\int_0^1 w_1 R\, dx = 0 \;\rightarrow\; 7a_1 - 2a_2 = 8$$

$$\int_0^1 w_2 R\, dx = 0 \;\rightarrow\; -112a_1 - 438a_2 = 161$$

Thus,

$$a_1 = \frac{3826}{2842}, \quad a_2 = \frac{2023}{2842}$$

and Equation 4.55 becomes

$$\tilde{\Phi}(x) = -1.6925x + 0.2785x^2 + 0.7118x^3 \tag{4.61}$$

Notice that the approximate solutions in Equations 4.57 through 4.61 all satisfy the boundary conditions $\Phi(0) = 0$ and $\Phi'(1) = 1$. The five approximate solutions are compared in Table 4.3.

EXAMPLE 4.8

This example illustrates the fact that expansion functions u_n must be selected such that $L\, u_n$ form a complete set for the range of the operator L. Consider the differential equation

$$-\Phi'' = 2 + \sin x, \quad 0 \le x \le 2\pi \tag{4.62}$$

subject to

$$\Phi(0) = \Phi(2\pi) = 0 \tag{4.63}$$

Suppose we carelessly select

$$u_n = \sin nx, \quad n = 1, 2, \ldots \tag{4.64}$$

as the expansion functions, the approximate solution is

$$\tilde{\Phi} = \sum_{n=1}^{N} a_n \sin nx \tag{4.65}$$

If we apply the Galerkin method, we obtain

$$\tilde{\Phi} = \sin x \tag{4.66}$$

Although u_n satisfies both the differentiability and boundary conditions, Equation 4.66 does not satisfy Equation 4.62. Hence, Equation 4.66 is an incorrect solution. The problem is that the set {sin nx} does not form a complete set. If we add constant and cosine terms to the expansion functions in Equation 4.65, then

$$\tilde{\Phi} = a_0 + \sum_{n=1}^{N} [a_n \sin nx + b_n \cos nx] \tag{4.67}$$

As $N \to \infty$, Equation 4.67 is the classical Fourier series solution. Applying the Galerkin method leads to

$$\tilde{\Phi} = \sin nx \tag{4.68}$$

which is again an incorrect solution. The problem is that even though u_n forms a complete set, $L\,u_n$ do not. For example, nonzero constants cannot be approximated by $L\,u_n$. In order for $L\,u_n$ to form a complete set, $\tilde{\Phi}$ must be of the form

$$\tilde{\Phi} = \sum_{n=1}^{n} [a_n \sin nx + b_n \cos nx] + a_0 + cx + dx^2 \tag{4.69}$$

Notice that the expansion functions {1, x, x^2, sin nx, cos nx} in the interval $[0, 2\pi]$ form a linearly dependent set. This is because any function such as x or x^2 can be represented in the interval $[0, 2\pi]$ by the set {sin nx, cos nx}. Applying the Galerkin method, Equation 4.69 leads to

$$\tilde{\Phi} = \sin x + x(2\pi - x) \tag{4.70}$$

which is the exact solution Φ.

4.8 Eigenvalue Problems

As mentioned in Section 1.3.2, eigenvalue (nondeterministic) problems are described by equations of the type

$$L\Phi = \lambda M\Phi \tag{4.71}$$

where L and M are differential or integral, scalar or vector operators. The problem here is the determination of the eigenvalues λ and the corresponding eigenfunctions Φ. It can be shown [11] that the variational principle for λ takes the form

$$\lambda = \min \frac{\langle \Phi, L\Phi \rangle}{\langle \Phi, M\Phi \rangle} = \min \frac{\int \Phi L\Phi \, dv}{\int \Phi M\Phi \, dv} \tag{4.72}$$

We may apply Equation 4.72 to the Helmholtz's equation for scalar waves, for example,

$$\nabla^2 \Phi + k^2 \Phi = 0 \tag{4.73}$$

Comparing Equation 4.73 with Equation 4.71, we obtain $L = -\nabla^2$, $M = 1$ (the identity operator), $\lambda = k^2$ so that

$$k^2 = \min \frac{\int \Phi \nabla^2 \Phi \, dv}{\int \Phi^2 \, dv} \tag{4.74}$$

Applying Green's identity (see Example 1.1),

$$\int_v (U \nabla^2 V + \nabla U \cdot \nabla V) \, dv = \oint U \frac{\partial V}{\partial n} \, dS,$$

to Equation 4.74 yields

$$k^2 = \min \frac{\int_v |\nabla \Phi|^2 \, dv - \oint \Phi \frac{\partial \Phi}{\partial n} \, dS}{\oint_v \Phi^2 \, dv} \tag{4.75}$$

Consider the following special cases.

a. For homogeneous boundary conditions of the Dirichlet type ($\Phi = 0$) or Neumann type ($\partial \Phi / \partial n = 0$). Equation 4.75 reduces to

$$k^2 = \min \frac{\int_v |\nabla \Phi|^2 \, dv}{\int_v \Phi^2 \, dv} \tag{4.76}$$

b. For mixed boundary conditions (($\partial \Phi / \partial n) + h\Phi = 0$) Equation 4.75 becomes

$$k^2 = \min \frac{\int_v |\nabla \Phi|^2 \, dv + \oint h\Phi^2 \, dS}{\int_v \Phi^2 \, dv} \tag{4.77}$$

It is usually possible to solve Equation 4.71 in a different way. We choose the basis functions $u_1, u_2, \ldots, u_N$ which satisfy the boundary conditions and assume the approximate solution

$$\tilde{\Phi} = a_1 u_1 + a_2 u_2 + \cdots + a_N u_N$$

or

$$\tilde{\Phi} = \sum_{n=1}^{N} a_n u_n \tag{4.78}$$

Substituting this into Equation 4.71 gives

$$\sum_{n=1}^{N} a_n Lu_n = \lambda \sum_{n=1}^{N} a_n Mu_n \qquad (4.79)$$

Choosing the weighting functions w_m and taking the inner product of Equation 4.79 with each w_m, we obtain

$$\sum_{n=1}^{N} [\langle w_m, Lu_n \rangle - \lambda \langle w_m, Mu_n \rangle] a_n = 0, \quad m = 1, 2, \dots, N \qquad (4.80)$$

This can be cast into matrix form as

$$\sum_{n=1}^{N} (A_{mn} - \lambda B_{mn}) X_n = 0 \qquad (4.81)$$

where $A_{mn} = \langle w_m, Lu_n \rangle$, $B_{mn} = \langle w_m, Mu_n \rangle$, $X_n = a_n$. Thus, we have a set of homogeneous equations. In order for $\tilde{\Phi}$ in Equation 4.78 not to vanish, it is necessary that the expansion coefficients a_n not all be equal to zero. This implies that the determinant of simultaneous equations (4.81) vanishes, that is,

$$\begin{vmatrix} A_{11} - \lambda B_{11} & A_{12} - \lambda B_{12} & \cdots & A_{1N} - \lambda B_{1N} \\ \vdots & & & \vdots \\ A_{N1} - \lambda B_{N1} & A_{N2} - \lambda B_{N2} & \cdots & A_{NN} - \lambda B_{NN} \end{vmatrix} = 0$$

or

$$|[A] - \lambda[B]| = 0 \qquad (4.82)$$

Solving this gives N approximate eigenvalues $\lambda_1, \dots, \lambda_N$. The various ways of choosing w_m leads to different weighted residual techniques as discussed in the previous section.

Examples of eigenvalue problems for which variational methods have been applied include [28–37]:

- The cutoff frequency of a waveguide,
- The propagation constant of a waveguide, and
- The resonant frequency of a resonator.

EXAMPLE 4.9

Solve the eigenvalue problem

$$\Phi'' + \lambda \Phi = 0, \quad 0 < x < 1$$

with boundary conditions $\Phi(0) = 0 = \Phi(1)$.

Solution

The exact eigenvalues are

$$\lambda_n = (n\pi)^2, \quad n = 1, 2, 3, \dots \qquad (4.83)$$

and the corresponding (normalized) eigenfunctions are

$$\Phi_n = \sqrt{2}\sin(n\pi x) \tag{4.84}$$

where Φ_n has been normalized to unity, that is, $\langle \Phi_n, \Phi_n \rangle = 1$.

The approximate eigenvalues and eigenfunctions can be obtained by using either Equation 4.72 or Equation 4.82. Let the approximate solution be

$$\tilde{\Phi}(x) = \sum_{k=0}^{N} a_k u_k, \quad u_k = x(1 - x^k) \tag{4.85}$$

From the given problem, $L = -(d^2/dx^2)$, $M = 1$ (identity operator). Using the Galerkin method, $w_m = u_m$.

$$A_{mn} = \langle u_m, L u_n \rangle = \int_0^1 (x - x^{m+1})\left[-\frac{d^2}{dx^2}(x - x^{n+1}) \right] dx$$

$$= \frac{mn}{m + n + 1}, \tag{4.86}$$

$$B_{mn} = \langle u_m, M u_n \rangle = \int_0^1 (x - x^{m+1})(x - x^{n+1}) dx$$

$$= \frac{mn(m + n + 6)}{3(m + 3)(n + 3)(m + n + 3)} \tag{4.87}$$

The eigenvalues are obtained from

$$|[A] - \lambda[B]| = 0 \tag{4.88}$$

For $N = 1$,

$$A_{11} = \frac{1}{3}, \quad B_1 = \frac{1}{30},$$

giving

$$\frac{1}{3} - \lambda\frac{1}{30} = 0 \ \rightarrow \ \lambda = 10$$

The first approximate eigenvalue is $\lambda = 10$, a good approximation to the exact value of $\pi^2 = 9.8696$. The corresponding eigenfunction $\tilde{\Phi} = a_1(x - x^2)$ can be normalized to unity so that

$$\tilde{\Phi} = \sqrt{30}(x - x^2)$$

For $N = 2$, evaluating Equations 4.86 and 4.87, we obtain

$$\begin{bmatrix} \frac{1}{3} & \frac{1}{2} \\ \frac{1}{2} & \frac{4}{5} \end{bmatrix}\begin{bmatrix} a_1 \\ a_2 \end{bmatrix} = \begin{bmatrix} \frac{1}{30} & \frac{1}{20} \\ \frac{1}{20} & \frac{8}{105} \end{bmatrix}\begin{bmatrix} a_1 \\ a_2 \end{bmatrix}$$

or

$$\begin{vmatrix} 10 - \lambda & 0 \\ 0 & 42 - \lambda \end{vmatrix} = 0$$

TABLE 4.4

Comparison between Approximate and Exact Eigenvalues
for Example 4.9

Exact	Approximate			
	$N = 1$	$N = 2$	$N = 3$	$N = 4$
9.870	10.0	10.0	9.8697	9.8697
39.478	–	42.0	39.497	39.478
88.826	–	–	102.133	102.133
157.914	–	–	–	200.583

giving eigenvalues $\lambda_1 = 10$, $\lambda_2 = 42$, compared with the exact values $\lambda_1 = \pi^2 = 9.8696$, $\lambda_2 = 4\pi^2 = 39.4784$, and the corresponding normalized eigenfunctions are

$$\tilde{\Phi}_1 = \sqrt{30}(x - x^2)$$
$$\tilde{\Phi}_2 = 2\sqrt{210}(x - x^2) - 2\sqrt{210}(x - x^3)$$

Continuing this way for higher N, the approximate eigenvalues shown in Table 4.4 are obtained. Unfortunately, the labor of computation increases as more u_k are included in $\tilde{\Phi}$. Notice from Table 4.4 that the approximate eigenvalues are always greater than the exact values. This is always true for a self-adjoint, positive-definite operator [17]. Figure 4.3 shows the comparison between the approximate and exact eigenfunctions.

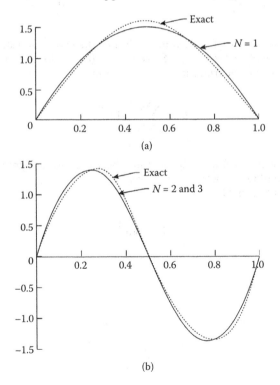

FIGURE 4.3

Comparison of approximate eigenfunctions with the exact solutions: (a) first eigenfunction, (b) second eigenfunction. (After R.F. Harrington, *Field Computation by Moment Methods*. Malabar, FL: R.E. Krieger, 1968, pp. 19, 126–131; with permission of Krieger Publishing Co.)

EXAMPLE 4.10

Calculate the cutoff frequency of the inhomogeneous rectangular waveguide shown in Figure 4.4. Take $\epsilon = 4\epsilon_o$ and $s = a/3$.

Solution

We will find the lowest mode having $\partial/\partial y \equiv 0$. It is this dominant mode that is of most practical value. Since the dielectric constant varies from one region to another, it is reasonable to choose Φ to be an electric field, that is, $\Phi = E_y$. Also, since $k^2 = \omega^2/u^2 = \omega^2\mu\epsilon$, Equation 4.74 becomes

$$\omega^2\mu_0\epsilon_0 \int_0^s E_y^2\, dx + \omega^2\mu_0\epsilon_0\epsilon_r \int_s^{a-s} E_y^2\, dx + \omega^2\mu_0\epsilon_0 \int_{a-s}^a E_y^2\, dx$$

$$= -\int_0^a E_y \frac{d^2 E_y}{dx^2}\, dx \tag{4.89}$$

Notice that in this implementation of Equation 4.74, there are no coefficients so that there is nothing to minimize. We simply take k^2 as a ratio. Equation 4.89 can be written as

$$\omega^2\mu_0\epsilon_0 \int_0^a E_y^2\, dx + \omega^2\mu_0\epsilon_0(\epsilon_r - 1) \int_s^{a-s} E_y^2\, dx = -\int_0^a E_y \frac{d^2 E_y}{dx^2}\, dx \tag{4.90}$$

We now choose the trial function for E_y. It must be chosen to satisfy the boundary conditions, namely, $E_y = 0$ at $x = 0, a$. Since $E_y \sim \sin(n\pi x/a)$ for the empty waveguide, it makes sense to choose the trial function of the form

$$E_y = \sum_{n=1,3,5}^{\infty} c_n \sin \frac{n\pi x}{a} \tag{4.91}$$

We choose the odd values of n because the dielectric is symmetrically placed; otherwise, we would have both odd and even terms.

Let us consider the trial function

$$E_y = \sin \frac{\pi x}{a} \tag{4.92}$$

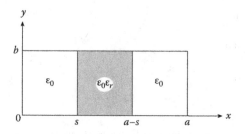

FIGURE 4.4
Symmetrically loaded rectangular waveguide.

Substituting Equation 4.92 into Equation 4.90 yields

$$
\omega^2 \mu_0 \epsilon_0 \int_0^a \sin^2 \frac{\pi x}{a} \, dx + \omega^2 \mu_0 \epsilon_0 (\epsilon_r - 1) \int_s^{a-s} \sin^2 \frac{\pi x}{a} \, dx
$$

$$
= \frac{\pi^2}{a^2} \int_0^a \sin^2 \frac{\pi x}{a} \, dx \tag{4.93}
$$

which leads to

$$
\omega^2 \mu_0 \epsilon_0 \left\{ 1 + (\epsilon_r - 1) \left[\left(1 - \frac{2s}{a} \right) + \frac{1}{\pi} \sin \frac{2\pi s}{a} \right] \right\} = \frac{\pi^2}{a^2}
$$

But $k_o^2 = \omega^2 \mu_0 \epsilon_0 = (4\pi^2/\lambda_c^2)$, where λ_c is the cutoff wavelength of the waveguide filled with vacuum. Hence,

$$
\frac{4\pi^2}{\lambda_c^2} = \frac{(\pi/a)^2}{1 + (\epsilon_r - 1) \left[\left(1 - \frac{2s}{a} \right) + \frac{1}{\pi} \sin \frac{2\pi s}{a} \right]}
$$

Taking $\epsilon_r = 4$ and $s = a/3$ gives

$$
\frac{4\pi^2}{\lambda_c^2} = \frac{(\pi/a)^2}{2 + \dfrac{3\sqrt{3}}{2\pi}}
$$

or

$$
\frac{a}{\lambda_c} = 0.2974
$$

This is a considerable reduction in a/λ_c compared with the value of $a/\lambda_c = 0.5$ for the empty guide. The accuracy of the result may be improved by choosing more terms in Equation 4.91.

4.9 Practical Applications

The various techniques discussed in this chapter have been applied to solve a considerable number of EM problems. We select a simple example for illustration [38,39]. This example illustrates the conventional use of the least-squares method.

Consider a strip transmission line enclosed in a box containing a homogeneous medium as shown in Figure 4.5. If a TEM mode of propagation is assumed, Laplace's equation

$$
\nabla^2 V = 0 \tag{4.94}
$$

is obeyed. Due to symmetry, we will consider only one quarter section of the line as in Figure 4.6 and adopt a boundary condition $\partial V/\partial x = 0$ at $x = -W$. We allow for the

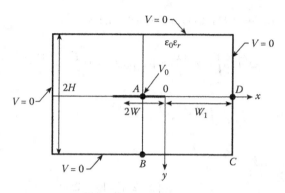

FIGURE 4.5
Strip line enclosed in a shielded box.

singularity at the edge of the strip. The variation of the potential in the vicinity of such a singularity is approximated, in terms of trigonometric basis functions, as

$$V = V_o + \sum_{k=1,3,5}^{\infty} c_k \rho^{k/2} \cos\frac{k\phi}{2}, \tag{4.95}$$

where V_o is the potential on the trip conductor and the expansion coefficients c_k are to be determined. If we truncate the infinite series in Equation 4.95 so that we have N unknown coefficients, we determine the coefficients by requiring that Equation 4.95 be satisfied at $M(\geq N)$ points on the boundary. If $M = N$, we are applying the collocation method. If $M > N$, we obtain an overdetermined system of equations which can be solved by the method of least squares. Enforcing Equation 4.95 at M boundary points, we obtain M simultaneous equations

$$\begin{bmatrix} V_1 \\ V_2 \\ \vdots \\ V_M \end{bmatrix} = \begin{bmatrix} A_{11} & A_{12} & \cdots & A_{1N} \\ A_{21} & A_{22} & \cdots & A_{2N} \\ \vdots & & & \vdots \\ A_{M1} & A_{M2} & \cdots & A_{MN} \end{bmatrix} \begin{bmatrix} c_1 \\ c_2 \\ \vdots \\ c_N \end{bmatrix}$$

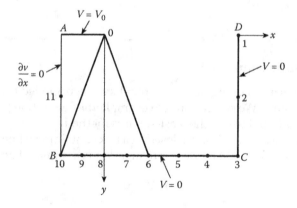

FIGURE 4.6
Quarter-section of the strip line.

that is,

$$[V] = [A][X] \tag{4.96}$$

where $[X]$ is an $N \times 1$ matrix containing the unknown expansion coefficients, $[V]$ is an $M \times 1$ column matrix containing the boundary conditions, and $[A]$ is the $M \times N$ coefficient matrix. Due to redundancy, $[X]$ cannot be uniquely determined from Equation 4.96 if $M > N$. To solve this redundant system of equations by the method of least squares, we define the residual matrix $[R]$ as

$$[R] = [A][X] - [V] \tag{4.97}$$

We seek for $[X]$, which minimizes $[R]^2$. Consider

$$[I] = [R]^t[R] = [[A][X] - [V]]^t[[A][X] - [V]]$$

$$\frac{\partial[I]}{\partial[X]} = 0 \;\rightarrow\; [A]^t[A][X] - [A]^t[V] = 0$$

or

$$[X] = \left[[A]^t[A]\right]^{-1}[A]^t[V] \tag{4.98}$$

where the superscript t denotes the transposition of the relevant matrix. Thus, we have reduced the original redundant system of equations to a determinate set of N simultaneous equations in N unknown coefficients $c_1, c_2, \ldots, c_N$.

Once $[X] = [c_1, c_2, \ldots, c_N]$ is determined from Equation 4.98, the approximate solution in Equation 4.95 is completely determined. We can now determine the capacitance and consequently the characteristic impedance of the line for a given value of width-to-height ratio. The capacitance is determined from

$$C = \frac{Q}{V_o} = Q \tag{4.99}$$

if we let $V_o = 1$ V. The characteristic impedance is found from [40]

$$Z_o = \frac{\sqrt{\epsilon_r}}{cC} \tag{4.100}$$

where $c = 3 \times 10^8$ m/s, the speed of light in vacuum. The major problem here is finding Q in Equation 4.99. If we divide the boundary BCD into segments,

$$Q = \int_{BCD} \rho_L \, dl = 4 \sum \rho_L \Delta l$$

where the charge density $\rho_L = \mathbf{D} \cdot \mathbf{a}_n = \epsilon \mathbf{E} \cdot \mathbf{a}_n$, $\mathbf{E} = -\nabla V$, and the factor 4 is due to the fact that we consider only one quarter section of the line. However,

$$\nabla V = \frac{\partial V}{\partial \rho} \mathbf{a}_\rho + \frac{1}{\rho} \frac{\partial V}{\partial \phi} \mathbf{a}_\phi,$$

$$\mathbf{E} = -\sum_{k=\text{odd}} \frac{k}{2} c_k \rho^{k/2-1} \left(\cos\frac{k\phi}{2} \mathbf{a}_\rho - \sin\frac{k\phi}{2} \mathbf{a}_\phi \right)$$

Since $\mathbf{a}_x = \cos\phi\mathbf{a}_\rho - \sin\phi\mathbf{a}_\phi$ and $\mathbf{a}_y = \sin\phi\mathbf{a}_\rho + \cos\phi\mathbf{a}_\phi$,

$$\rho_L|_{CD} = \epsilon\mathbf{E}\cdot\mathbf{a}_x$$

$$= -\epsilon\sum_{k=\text{odd}}\frac{k}{2}c_k\rho^{k/2-1}\left(\cos\frac{k\phi}{2}\cos\phi + \sin\frac{k\phi}{2}\sin\phi\right) \tag{4.101a}$$

and

$$\rho_L|_{BC} = \epsilon\mathbf{E}\cdot\mathbf{a}_y$$

$$= -\epsilon\sum_{k=\text{odd}}\frac{k}{2}c_k\rho^{k/2-1}\left(\cos\frac{k\phi}{2}\sin\phi - \sin\frac{k\phi}{2}\cos\phi\right) \tag{4.101b}$$

EXAMPLE 4.11

Using the collocation (or point matching) method, write a computer program to calculate the characteristics impedance of the line shown in Figure 4.5. Take

(a) $W = H = 1.0$ m, $W_1 = 5.0$ m, $\epsilon_r = 1$, $V_0 = 1$V,
(b) $W = H = 0.5$ m, $W_1 = 5.0$ m, $\epsilon_r = 1$, $V_0 = 1$V.

Solution

The computer program is presented in Figure 4.7. For the first run, we take the number of matching points $N = 11$; the points are selected as illustrated in Figure 4.6. The selection of the points is based on our prior knowledge of the fact that the flux lines are concentrated on the side of the strip line numbered 6–10; hence, more points are chosen on that side.

The first step is to determine the potential distribution within the strip line using Equation 4.95. In order to determine the expansion coefficients c_k in Equation 4.95, we let Equation 4.95 be satisfied at the matching points. On points 1–10 in Figure 4.6, $V = 0$ so that Equation 4.95 can be written as

$$-V_o = \sum_{k=1,3,5}^{\infty} c_k\rho^{k/2}\cos\frac{k\phi}{2} \tag{4.102}$$

The infinite series is terminated at $k = 19$ so that 10 points are selected on the sides of the strip line. The 11th point is selected such that $\partial V/\partial x = 0$ is satisfied at the point. Hence at point 11,

$$0 = \frac{\partial V}{\partial x} = \cos\phi\frac{\partial V}{\partial\rho} - \frac{\sin\phi}{\rho}\frac{\partial V}{\partial\phi}$$

or

$$0 = \sum_{k=1,3,5}\frac{k}{2}c_k\rho^{k/2-1}\left(\cos\frac{k\phi}{2}\cos\phi + \sin\frac{k\phi}{2}\sin\phi\right) \tag{4.103}$$

With Equations 4.102 and 4.103, we set up a matrix equation of the form

$$[B] = [F][A] \tag{4.104}$$

```
clear all; format compact; tic

% Output of this program
% W = H          N           Nx          Ny          Zo
% 1              7           5           2           67.735
% 1              11          8           3           64.963
% 0.5            7           5           2           96.758
% 0.5            11          8           3           99.098


% ***************************************************************
% THIS PROGRAM CALCULATES THE CHARACTERISTIC IMPEDANCE
% Zo OF A BOXED MICROSTRIP LINE USING COLLOCATION
% POINT-MATCHING METHOD
% ***************************************************************

    X=[5 5 5, 4.0, 2.0, 1.0, 0.5, 0.0, -0.5, -1.0 -1]; Y=[0.0,
0.5,1 1 1 1 1 1 1 ,0.5];
    EO=8.8541E-12;VL=3.0E+8;

%    INPUT DATA
% W    : Width of conductor
% H    : Vertical separation of conductor and ground planes
% V0   : Potential at the conductor
% W1   : Horizontal separation of conductor and ground plane
% NMAX : Number of matching points
% NBC  : NO. OF POINTS ON BC OR X-AXIS
% ER   : Dielectric parameter

for qq = 1:4
    switch qq
        case 1
            W = 1; NMAX = 7;  NBC = 5;
            str = sprintf('\n\nW=H     N    Nx    Ny    Zo ');
            disp(str)
        case 2
            W = 1; NMAX = 11; NBC = 8;
        case 3
            W =.5; NMAX = 7;  NBC = 5;
        case 4
            W =.5; NMAX = 11; NBC = 8;
    end

    H = W;
    W1 = 5.0;
    ER = 1.0;
    V0 = 1.0;
    NCD = NMAX - NBC;    % NO. OF POINTS ON DC OR Y-AXIS
    DX = (W + W1)/NBC;
    DY = H/NCD;
    EPS = EO*ER;
```

FIGURE 4.7
Computer program for Example 4.11.

(Continued)

```
%      CALCULATE R & PHI FOR EACH POINT (X,Y)

       for N = 1:NMAX
           R(N) = sqrt( X(N)^2 + Y(N)^2 );
           if (X(N))<0
               PHI(N) = pi - atan( Y(N)/abs(X(N)) );
           elseif X(N)==0
               PHI(N) = pi/2;
           elseif X(N)>0
               PHI(N) = atan( Y(N)/X(N) );
           end
       end

%      CALCULATE MATRICES F(I,J) AND B(I)

       for I = 1:NMAX
           B(I) = -V0;
           M = 0;
           for J = 1:NMAX
               M = 2*J - 1;
               FM = (M)/2.0;
               if (I==NMAX)
                   CC = cos(PHI(I))*cos(PHI(I)*FM);
                   SS = sin(PHI(I))*sin(PHI(I)*FM);
                   F(I,J) = ( R(I)^(FM-1.) )*(CC + SS)*FM;
                   B(I) = 0.0;
               else
                   F(I,J) = ( R(I)^(FM) )*cos(FM*PHI(I));
               end
           end
       end

%      DETERMINE THE EXPANSION COEFFICIENTS A(I)

       IDM = 30;
       F = inv(F);
       for I = 1:NMAX
           A(I) = 0.0;
           for J = 1:NMAX
               A(I) = A(I) + F(I,J)*B(J);
           end
       end

%      NOW, CALCULATE CHARGE ON THE X-SIDE, i.e. BC

       RHO = 0.0;
       YC = H;
       XC = -W - DX/2.0;
       for I = 1:NBC
           XC = XC + DX;
           RC = sqrt(XC^2 + YC^2);
```

FIGURE 4.7 (Continued)
Computer program for Example 4.11.   (*Continued*)

```
        if(XC)<=0
            PC = pi - atan(YC/abs(XC));
        else
            PC = atan(YC/XC);
        end

        for K = 1:NMAX
            FK = (2*K - 1)/2.0;
            RRO = sin(PC)*cos(FK*PC) - cos(PC)*sin(FK*PC);
            RHO = RHO - A(K)*FK*(RC^(FK-1))*RRO*DX*EPS;
        end
    end

%   NEXT, CALCULATE THE CHARGE ON THE Y-SIDE, i.e. DC

    XC = W1;
    YC = -DY/2;
    for I = 1:NCD
        YC = YC + DY;
        RC = sqrt(XC^2 + YC^2);
        PC = atan(YC/XC);
        for K = 1:NMAX
            FK = (2*K - 1)/2;
            RRO = cos(PC)*cos(FK*PC) + sin(PC)*sin(FK*PC);
            RHO = RHO - A(K)*FK*(RC^(FK-1))*RRO*DY*EPS;
        end
    end

%   CALCULATE THE CHARACTERISTIC IMPEDANCE Zo

    Q = 4.0*RHO;
    C = abs(Q)/V0;
    Z0 = sqrt(ER)/( C*VL );
    %disp([C,Z0])
    %disp([W N NBC NCD C Z0])
    str = sprintf('%3g  %3g  %3g  %3g    %g',W, N, NBC, NCD, Z0);
    disp(str)
end
```

FIGURE 4.7 (Continued)
Computer program for Example 4.11.

where

$$
B_k = \begin{cases} -V_o, & k \neq N \\ 0, & k = N, \end{cases}
$$

$$
F_{ki} = \begin{cases} \rho_i^{k/2} \cos k\phi_i / 2 & \begin{array}{l} i = 1, \ldots, N-1, \\ k = 1, \ldots, N \end{array} \\ \dfrac{k}{2} \rho_i^{k/2-1} (\cos(k\phi_i/2)\cos\phi_i + \sin(k\phi_i/2\sin\phi_i), & i = N, k = 1, \ldots, N \end{cases}
$$

where (ρ_i, ϕ_i) are the cylindrical coordinates of the ith point. Matrix $[A]$ consists of the unknown expansion coefficients c_k. By matrix inversion, we obtain $[A]$ as

$$
[A] = [F]^{-1}[B] \tag{4.105}
$$

TABLE 4.5

Characteristic Impedance of the Strip Transmission Line of Figure 4.5; for Example 4.11 with $W_1 = 5.0$

$W = H$	N	N_x	N_y	c_1	Calculated $Z_o(\Omega)$	Exact [39] $Z_o(\Omega)$
1.0	7	5	2	−1.1549	67.735	65.16
	11	8	3	−1.1266	64.963	
0.5	7	5	2	−1.1549	96.758	100.57
	11	8	3	−1.1266	99.098	

Once the expansion coefficients are determined, we now calculate the total charge on the sides of the strip line using Equation 4.101 and

$$Q = 4 \sum_{BDC} \rho_L \Delta l$$

Finally, we obtain Z_o from Equations 4.99 and 4.100. Table 4.5 shows the results obtained using the program in Figure 4.7 for different cases. In Table 4.5, $N = N_x + N_y$ where N_x and N_y are the number of matching points selected along the x and y axes, respectively. By comparing Figure 4.5 with Figure 2.13, one may be tempted to apply Equation 2.223 to obtain the exact solution of part (a) as 61.1 Ω. But we must recall that Equation 2.223 was derived based on the assumption that $w \gg b$ in Figure 2.12 or $W \gg H$ in Figure 4.5. The assumption is not valid in this case, the exact solution given in Reference 39 is more appropriate.

4.10 Concluding Remarks

This chapter has provided an elementary introduction to the variational techniques. The variational methods provide simple, elegant, and powerful solutions to physical problems provided we can find approximate basis functions. A prominent feature of the variational method lies in the ability to achieve high accuracy with few terms in the approximate solution. A major drawback is the difficulty encountered in selecting the basis functions. In spite of the drawback, the variational methods have been very useful and provide basis for both the method of moments and FEM to be discussed in the forthcoming chapters.

Needless to say, our discussion on variational techniques in this chapter has been only introductory. An exhaustive treatment of the subject can be found in References 1,6,10,11,41–43. Various applications of variational methods to EM-related problems include

- Waveguides and resonators [28–37]
- Transmission lines [38,39,44–47]
- Acoustic radiation [48]
- Wave propagation [49–51]
- Transient problems [52]
- Scattering problems [53–59].
- Optical fibers [60]

The problem of variational principles for EM waves in inhomogeneous media is discussed in Reference 60. Basic information on calculus of variations can be found in References 61–63.

PROBLEMS

4.1 Find $\langle u, v \rangle$ if
     a. $u = x^2$, $v = 2 - x$ in the interval $-1 < x < 1$,
     b. $u = 1$, $v = x^2 - 2y^2$ in the rectangular region $0 < x < 1$, $1 < y < 2$,
     c. $u = x + y$, $v = xz$ in the cylindrical region $x^2 + y^2 \leq 4$, $0 < z < 5$.

4.2 Show that
     a. $\langle h(x), f(x) \rangle = \langle h(x), f(-x) \rangle$,

     b. $\langle h(ax), f(x) \rangle = \left\langle h(x), \dfrac{1}{a} f\left(\dfrac{x}{a}\right) \right\rangle$,

     c. $\left\langle \dfrac{df}{dx}, h(x) \right\rangle = -\left\langle f(x), \dfrac{dh}{dx} \right\rangle$,

     d. $\left\langle \dfrac{d^n f}{dx^n}, h(x) \right\rangle = (-1)^n - \left\langle f(x), \dfrac{d^n h}{dx^n} \right\rangle$

Note from (d) that $L = (d/dx)$, (d^3/dx^3), etc., are not self-adjoint, whereas $L = (d^2/dx^2)$, (d^4/dx^4), etc., are.

4.3 Given the functions
$$f(x) = x - 1, \qquad 0 < x < 2$$
$$g(x) = \begin{cases} x, & 0 < x < 1 \\ 2 - x, & 1 < x < 2 \\ 0, & \text{otherwise} \end{cases}$$

Calculate their inner product.

4.4 Show that the functional
$$I(y) = \int_0^1 \left[\left(\frac{dy}{dx}\right)^2 + y^2 \right] dx$$

with boundary conditions $y(0) = 0$, $y(1) = 1$ is made stationary by the solution of the differential equation $(d^2y/dx^2) - y = 0$.

4.5 Obtain an admissible extremals to minimize:

     a. $I[y] = \int_0^1 (y'^2 + 2ye^x)dx, \qquad y(0) = 0, y(1) = 1$

     b. $J[y] = \int_0^1 (y^2 + y'^2 + 2y)dx, \qquad y(0) = 0, y(1) = 0$

     c. $K[y] = \int_0^1 (y^2 + y'^2 + 2ye^x)dx, \quad y(0) = 0, y(1) = 0$

4.6   Repeat Problem 4.3 for the following functionals:

   a. $\displaystyle\int_a^b (y'^2 - y^2)\,dx$

   b. $\displaystyle\int_a^b [5y^2 - (y'')^2 + 10x]\,dx$

   c. $\displaystyle\int_a^b (3uv - u^2 + u'^2 - v'^2)\,dx$

4.7   Determine the extremal $y(x)$ for each of the following variational problems:

   a. $\displaystyle\int_0^1 (2y'^2 + yy' + y' + y)\,dx,\, y(0) = 0,\, y(1) = 1$

   b. $\displaystyle\int (y'^2 - y^2)\,dx,\, y(0) = 1,\, y(\pi/2) = 0$

4.8   If L is a positive definite, self-adjoint operator and $L\Phi = g$ has a solution Φ_o, show that the function

$$I = \langle L\Phi, \Phi \rangle - 2\langle \Phi, g \rangle,$$

where Φ and g are real functions, is minimized by the solution Φ_o.

4.9   Show that Euler's equation

$$\frac{d}{dx}\left(\frac{dF}{dy'}\right) - \frac{dF}{dy} = 0$$

can be written as

$$\frac{d}{dx}\left[F - y'\frac{dF}{dy'}\right] - \frac{dF}{dx} = 0$$

4.10 Obtain Euler's equation (or first variation) corresponding to the integral

$$I = \int \left[\left(\frac{dV}{dx}\right)^2 - 2g(x)V\right]dx$$

4.11 Show that a function that minimizes the functional

$$I(\Phi) = \frac{1}{2}\int_S [|\nabla\Phi|^2 - k^2\Phi^2 + 2g\Phi]\,dS$$

is the solution to the inhomogeneous Helmholtz equation

$$\nabla^2\Phi + k^2\Phi = g$$

4.12 Using Euler's equation, obtain the differential equation corresponding to the electrostatic field functional

$$I = \int_v \left[\frac{1}{2} \epsilon E^2 - \rho_v V \right] dv$$

where $E = |\mathbf{E}|$ and ρ_v is the volume charge density.

4.13 Repeat Problem 4.12 for the energy function for steady-state currents

$$I = \int_v \frac{1}{2} \mathbf{J} \cdot \mathbf{E} \, dv$$

where $\mathbf{J} = \sigma \mathbf{E}$.

4.14 Poisson's equation in an anisotropic medium is

$$\frac{\partial}{\partial x}\left(\epsilon_x \frac{\partial V}{\partial x} \right) + \frac{\partial}{\partial y}\left(\epsilon_y \frac{\partial V}{\partial y} \right) + \frac{\partial}{\partial z}\left(\epsilon_z \frac{\partial V}{\partial z} \right) = -\rho_v$$

in three dimensions. Derive the functional for the boundary-value problem. Assume ϵ_x, ϵ_y and ϵ_z are constants.

4.15 Show that the variational principle for the boundary-value problem

$$\nabla^2 \Phi = f(x, y, z)$$

subject to the mixed boundary condition

$$\frac{\partial \Phi}{\partial n} + g\Phi = h \text{ on } S$$

is

$$I(\Phi) = \int_v [|\nabla \Phi|^2 - 2fg] dv + \oint [g\Phi^2 - 2h\Phi] dS$$

4.16 Obtain the variational principle for the differential equation

$$-\frac{d^2 y}{dx^2} + y = \sin \pi x, \quad 0 < x < 1$$

subject to $y(0) = 0 = y(1)$.

4.17 Determine the variational principle for

$$\Phi'' = \Phi - 4xe^x, \quad 0 < x < 1$$

subject to $\Phi'(0) = \Phi(0) + 1$, $\Phi'(1) = \Phi(1) - e$.

4.18 For the boundary-value problem

$$-\Phi'' = x, \quad 0 < x < 1$$
$$\Phi(0) = 0, \quad \Phi(1) = 2$$

determine the approximate solution using the Rayleigh–Ritz method with basis functions

$$uk = x^k(x - 1), \quad k = 0,1,2, \ldots, M$$

Try cases when $M = 1, 2$, and 3.

4.19 Use Rayleigh–Ritz method to solve

$$\frac{d^2U}{dx^2} + U - x^2 = 0, \quad 0 < x < 1$$

Consider the following sets of boundary conditions:

$$U(0) = 0 = U(1)$$
$$U(0) = 0, \quad \underline{U}'(1) = 1.$$

4.20 Using Rayleigh–Ritz method, find the solution to the differential equation

$$\frac{d^2\Phi}{dx^2} - \Phi + 4x = 0, \quad 0 < x < 2$$

subject to $\Phi(0) = 0 = \Phi(2)$. Assume the trial function $\tilde{\Phi}(x) = x^2(4 - x^2)$.

4.21 Use the Rayleigh method and the least-squares method to solve the differential equation:

$$\frac{d^2\Phi}{dx^2} + \cos\pi x = 0, \quad 0 < x < 1$$

subject to:
a. $\Phi(0) = 0 = \Phi(1)$
b. $\Phi(0) = 0 = \Phi'(1)$

Consider a two-term solution.

4.22 Given the functional

$$I(y) = \int_0^1 \left[\frac{1}{2}\left(\frac{dy}{dt}\right)^2 - y \right] dx, \quad y(0) = 0 = y(1)$$

a. Show that the corresponding Euler's equation is

$$\frac{d^2y}{dx^2} + 1 = 0$$

b. Use Rayleigh–Ritz method to obtain the approximate solution. Use the two-parameter trial function

$$y(x) = a_1 \sin(\pi x) + a_2 \sin(2\pi x)$$

c. Compare the solution in part (b) with the exact solution

$$y(x) = \frac{1}{2}x(1-x)$$

4.23 Rework Example 4.5 using

a. $u_m = x(1 - x^m)$,

b. $u_m = \sin m\pi x$, $m = 1, 2, 3, \dots, M$. Try cases when $M = 1, 2,$ and 3.

4.24 Solve the differential equation

$$-\Phi''(x) + 0.1\,\Phi(x) = 1, \quad 0 \le x \le 10$$

subject to the boundary conditions $\Phi'(0) = 0 = \Phi(0)$ using the trial function

$$\tilde{\Phi}(x) = a_1 \cos\frac{\pi x}{20} + a_2 \cos\frac{3\pi x}{20} + a_3 \cos\frac{5\pi x}{20}$$

Determine the expansion coefficients using (a) collocation method, (b) subdomain method, (c) Galerkin method, and (d) least-squares method.

4.25 For the boundary-value problem

$$\Phi'' + \Phi + x = 0, \quad 0 < x < 1$$

with homogeneous boundary conditions $\Phi = 0$ at $x = 0$ and $x = 1$, determine the coefficients of the approximate solution function

$$\tilde{\Phi}(x) = x(1-x)(a_1 + a_2 x)$$

using (a) collocation method (choose $x = 1/4$, $x = 1/2$ as collocation points), (b) Galerkin method, and (c) least-squares method.

4.26 Use the Ritz method to solve the ordinary differential equation

$$\frac{d^2 y}{dx^2} + 100x^2 = 0, \quad 0 \le x \le 1$$

subject to $y(0) = 0 = y(1)$. Use the trial function $u_1 = x(1 - x^2)$.

4.27 Rework Problem 4.18 using the trial function $u_1 = x(1 - x^3)$.

4.28 Given the boundary-value problem

$$y'' + (1 + x^2)y + 1 = 0, \quad -1 < x < 1 \quad y(-1) = 0 = y(1),$$

solve for y assuming the approximate solution

$$\tilde{y} = a_1(1 - x^2)(1 - 4x^2) + a_2 x^2(1 - x^2)$$

Use the Galerkin and the least-squares methods to determine a_1 and a_2.

4.29 Consider the problem

$$\Phi'' + x\Phi' + \Phi = 2x, \quad 0 < x < 1$$

subject to $\Phi(0) = 1$, $\Phi(1) = 0$. Find the approximate solution using the Galerkin method. Use $u_k = x^k(1 - x)$, $k = 0, 1, \ldots, N$. Try $N = 3$.

4.30 Solve the differential equation

$$\frac{d^2\Phi}{dx^2} - \Phi + 100x = 0, \quad 0 \le x \le 10$$

subject to $\Phi(0) = 0 = \Phi(10)$. Assume the two trial functions

$$u_1(x) = x(1 - x^2), \quad u_2(x) = x(1 - x^4)$$

Take the two collocation points at $x = 1/3$ and $x = 2/3$.

4.31 Determine the first three eigenvalues of the equation

$$y'' + \lambda y = 0, \quad 0 < x < 1,$$

$y(0) = 0 = y(1)$ using collocation at $x = 1/4, 1/2, 3/4$.

4.32 The differential equation governing the vibration of a string is given by

$$\frac{d^2\Phi}{dx^2} + \lambda\Phi = 0, \quad 0 \le x \le 1$$

with the boundary conditions

$$\Phi(0) = 0 = \Phi(1)$$

where λ is the eigenvalue. Using the trial solution

$$\Phi(x) = a_1 x(1 - x) + a_2 x^2(1 - x)$$

where a_1 and a_2 are constants, use Garlekin's method to determine the eigenvalues of the string.

4.33 Determine the fundamental eigenvalue of the problem

$$\Phi''(x) + 0.1\Phi(x) = \lambda\Phi(x), \quad 0 < x < 10$$

subject to $\Phi(0) = 0 = \Phi(10)$. Use the trial function

$$\tilde{\Phi}(x) = x(x - 10)$$

4.34 Obtain the lowest eigenvalue of the problem

$$\nabla^2\Phi + \lambda\Phi = 0, \quad 0 < \rho < 1$$

with $\Phi = 0$ at $\rho = 1$.

4.35 Calculate the smallest eigenvalue of

$$y'' + \lambda y = 0, \quad 0 < x < 1$$

with $y(0) = 0 = y(1)$.

Hint: Consider $\lambda = \left(\int_0^1 (y')^2 \, dx \Big/ \int_0^1 y^2 \, dx \right)$

and assume the approximate solution

$$\tilde{y} = a_0 + a_1 x + a_2 x^2 + a_3 x^3$$

4.36 Determine the first two eigenvalues of the differential equation

$$\frac{d^2 U}{dx^2} + \lambda U = 0, \quad 0 < x < 1$$

subject to:

$$U(0) = 0, \quad U(1) + U'(1) = 0$$

4.37 The two-dimensional Helmholtz wave equation is

$$LU = -\nabla^2 U - k^2 U = f(x, y), \quad 0 < x < a, \quad 0 < y < b$$

subject to a homogeneous Dirichlet boundary conditions. Prove that the eigenvalues of the operator L is

$$\lambda_{mn} = \left(\frac{m\pi}{a} \right)^2 + \left(\frac{n\pi}{a} \right)^2 - k^2$$

4.38 Rework Example 4.10 using the trial function

$$E_y = \sin \frac{\pi x}{a} + c_1 \sin \frac{3\pi x}{a}$$

where c_1 is a coefficient to be chosen such that $\omega^2 \epsilon_o \mu_o$ is minimized.

4.39 Consider the waveguide in Figure 4.4 as homogeneous. To determine the cutoff frequency, we may use the polynomial trial function

$$H_z = Ax^3 + Bx^2 + Cx + D$$

By applying the conditions

$$H_z = 1 \quad \text{at } x = 0, \quad H_z = -1 \quad \text{at } x = a,$$

$$\frac{\partial H_z}{\partial x} = 0 \quad \text{at } x = 0, a,$$

determine $A, B, C,$ and D. Using the trial function, calculate the cutoff frequency.

References

1. S.G. Mikhlin, *Variational Methods in Mathematical Physics*. New York: Macmillan, 1964, pp. xv, 4–78.
2. J.N. Reddy, *An Introduction to the Finite Element Method*, 2nd Edition. New York: McGraw-Hill, 1993, pp. 18–64.
3. R.B. Guenther and J.W. Lee, *Partial Differential Equations of Mathematical Physics and Integral Equations*. Englewood Cliffs, NJ: Prentice-Hall, 1988, pp. 434–485.
4. A. Wexler, "Computation of electromagnetic fields," *IEEE Transactions on Microwave Theory and Techniques*, vol. MTT-17, no. 8, August 1969, pp. 416–439.
5. M.M. Ney, "Method of moments as applied to electromagnetic problems," *IEEE Trans. Micro. Theo. Tech.*, vol. MTT-33, no. 10, October 1985, pp. 972–980.
6. I.M. Gelfand and S.V. Fomin (translated by R. A. Silvermans), *Calculus of Variations*. Mineola, NY: Dover Publications, 2000.
7. J.N. Reddy and M.L. Rasmussen, *Advanced Engineering Analysis*. New York: John Wiley, 1982, pp. 377–386.
8. B.H. McDonald et al., "Variational solution of integral equations," *IEEE Trans. Micro. Theo. Tech.*, vol. MTT-22, no. 3, Mar. 1974, pp. 237–248. See also vol. MTT-23, no. 2, Feb. 1975, pp. 265–266 for correction to the paper.
9. K. Morishita and N. Kumagai, "Unified approach to the derivation of variational expression for electromagnetic fields," *IEEE Trans. Micro. Theo. Tech.*, vol. MTT-25, no. 1, Jan. 1977, pp. 34–39.
10. B.L. Moiseiwitch, *Variational Principles*. London: Interscience Pub., 1966.
11. L. Cairo and T. Kahan, *Variational Technique in Electromagnetics*. New York: Gordon & Breach, 1965, pp. 48–65.
12. K. Kalikstein, "Formulation of variational principles via Lagrange multipliers," *J. Math. Phys.*, vol. 22, no. 7, July 1981, pp. 1433–1437.
13. K. Kalikstein and A. Sepulveda, "Variational principles and variational functions in electromagnetic scattering," *IEEE Trans. Ant. Prog.*, vol. AP-29, no. 5, Sept. 1981, pp. 811–815.
14. P. Hammond, "Equilibrium and duality in electromagnetic field problems," *J. Frank. Inst.*, vol. 306, no. 1, July 1978, pp. 133–157.
15. S.K. Jeng and C.H. Chen, "On variational electromagnetics," *IEEE Trans. Ant. Prog.*, vol. AP-32, no. 9, Sept. 1984, pp. 902–907.
16. S.J. Chung and C.H. Chen, "Partial variational principle for electromagnetic field problems: Theory and applications," *IEEE Trans. Micro. Theo. Tech.*, vol. 36, no. 3, Mar. 1988, pp. 473–479.
17. R.F. Harrington, *Field Computation by Moment Methods*. Malabar, FL: R.E. Krieger, 1968, pp. 19, 126–131.
18. S.G. Mikhlin and K.I. Smolitskiy, *Approximate Methods for Solution of Differential and Integral Equations*. New York: Elsevier, 1967, pp. 147–270.
19. T.J. Chung, *Finite Element Analysis in Fluid Dynamics*. New York: McGraw-Hill, 1978, pp. 36–43.
20. O.C. Zienkiewicz and R.L. Taylor, *The Finite Element Method*, vol. 1, 4th Edition. London: McGraw-Hill, 1989, pp. 206–259.
21. L. Lewin, "On the restricted validity of point-matching techniques," *IEEE Trans. Micro. Theo. Tech.*, vol. MTT-18, no. 12, Dec. 1970, pp. 1041–1047.
22. R.F. Muller, "On the legitimacy of an assumption underlying the point-matching method," *IEEE Trans. Micro. Theo. Tech.*, vol. MTT-18, June 1970, pp. 325–327.
23. A.R. Djordjevic and T.K. Sarkar, "A theorem on the moment methods," *IEEE Trans. Ant. Prog.*, vol. AP-35, no. 3, Mar. 1987, pp. 353–355.
24. T.K. Sarkar, "A study of the various methods for computing electromagnetic field utilizing thin wire integral equations," *Radio Sci.*, vol. 18, no. 1, Jan./Feb. 1983, pp. 29–38.
25. T.K. Sarkar, "A note on the variational method (Rayleigh–Ritz), Galerkin's method, and the method of least squares," *Radio Sci.*, vol. 18, no. 6, Nov./Dec. 1983, pp. 1207–1224.

26. T.K. Sarkar, "A note on the choice of weighting functions in the method of moments," *IEEE Trans. Ant. Prog.,* vol. AP-33, no. 4, April 1985, pp. 436–441.

27. T.K. Sarkar et al., "On the choice of expansion and weighting functions in the numerical solution of operator equations," *IEEE Trans. Ant. Prog.,* vol. AP-33, no. 9, Sept. 1985, pp. 988–996.

28. A.D. Berk, "Variational principles for electromagnetic resonators and waveguides," *IRE Trans. Ant. Prog.,* vol. AP-4, April 1956, pp. 104–111.

29. G.J. Gabriel and M.E. Brodwin, "The solution of guided waves in inhomogeneous anisotropic media by perturbation and variation methods," *IEEE Trans. Micro. Theo. Tech.,* vol. MTT-13, May 1965, pp. 364–370.

30. W. English and F. Young, "An E vector variational formulation of the Maxwell equations for cylindrical waveguide problems," *IEEE Trans. Micro. Theo. Tech.,* vol. MTT-19, Jan. 1971, pp. 40–46.

31. J.R. James, "Point-matched solutions for propagating modes on arbitrarily-shaped dielectric rods," *Radio Electron. Eng.,* vol. 42, no. 3, March 1972, pp. 103–113.

32. J.A. Fuller and N.F. Audeh, "The point-matching solution of uniform nonsymmetric waveguides," *IEEE Trans. Micro. Theo. Tech.,* vol. MTT-17, no. 2, Feb. 1969, pp. 114–115.

33. R.B. Wu and C.H. Chen, "On the variational reaction theory for dielectric waveguides," *IEEE Trans. Micro. Theo. Tech.,* vol. M-33, no. 6, June 1985, pp. 477–483.

34. T.E. Rozzi, "The variational treatment of thick interacting inductive irises," *IEEE Trans. Micro. Theo. Tech.,* vol. MTT-21, no. 2, Feb. 1973, pp. 82–88.

35. A.D. McAulay, "Variational finite-element solution of dissipative waveguide and transportation application," *IEEE Trans. Micro. Theo. Tech.,* vol. MTT-25, no. 5, May 1977, pp. 382–392.

36. L.V. Lindell, "A variational method for nonstandard eigenvalue problems in waveguides and resonator analysis," *IEEE Trans. Micro. Theo. Tech.,* vol. MTT-30, no. 8, Aug. 1982, pp. 1194–1204. See comment on this paper in vol. MTT-31, no. 9, Sept. 1983, pp. 786–789.

37. K. Chang, "Variational solutions on two opposite narrow resonant strips in waveguide," *IEEE Trans. Micro. Theo. Tech.,* vol. MTT-35, no. 2, Feb. 1987, pp. 151–158.

38. T.K. Seshadri et al., "Application of "corner function approach' to strip line problems," *Int. J. Electron.,* vol. 44, no. 5, May 1978, pp. 525–528.

39. T.K. Seshadri et al., "Least squares collocation as applied to the analysis of strip transmission lines," *Proc. IEEE,* vol. 67, no. 2, Feb. 1979, pp. 314–315.

40. M.N.O. Sadiku, *Elements of Electromagnetics,* 7th Edition. New York: Oxford University Press, 2018, Chap. 11.

41. P.M. Morse and H. Feshback, *Methods of Theoretical Physics.* New York: McGraw-Hill, 2 volumes, 1953.

42. R.E. Collin, *Field Theory of Guided Waves.* New York: McGraw-Hill, 1960, pp. 148–164, 314–367.

43. D.G. Bodner and D.T. Paris, "New variational principle in electromagnetics," *IEEE Trans. Ant. Prog.,* vol. AP-18, no. 2, March 1970, pp. 216–223.

44. T.D. Tsiboukis, "Estimation of the characteristic impedance of a transmission line by variational methods," *IEEE Proc.,* vol. 132, Pt. H, no. 3, June 1985, pp. 171–175.

45. E. Yamashita and R. Mittra, "Variational method for the analysis of microstrip lines," *IEEE Trans. Micro. Theo. Tech.,* vol. MTT-16, no. 4, Apr. 1968, pp. 251–256.

46. E. Yamashita, "Variational method for the analysis of microstrip-like transmission lines," *IEEE Trans. Micro. Theo. Tech.,* vol. MTT-16, no. 8, Aug. 1968, pp. 529–535.

47. F. Medina and M. Horno, "Capacitance and inductance matrices for multistrip structures in multilayered anisotropic dielectrics," *IEEE Trans. Micro. Theo. Tech.,* vol. MTT-35, no. 11, Nov. 1987, pp. 1002–1008.

48. F.H. Fenlon, "Calculation of the acoustic radiation of field at the surface of a finite cylinder by the method of weighted residuals," *Proc. IEEE,* vol. 57, no. 3, March 1969, pp. 291–306.

49. C.H. Chen and Y.W. Kiang, "A variational theory for wave propagation in a one-dimensional inhomogeneous medium," *IEEE Trans. Ant. Prog.,* vol. AP-28, no. 6, Nov. 1980, pp. 762–769.

50. S.K. Jeng and C.H. Chen, "Variational finite element solution of electromagnetic wave propagation in a one-dimensional inhomogeneous anisotropic medium," *J. Appl. Phys.*, vol. 55, no. 3, Feb. 1984, pp. 630–636.

51. J.A. Bennett, "On the application of variation techniques to the ray theory of radio propagation," *Radio Sci.*, vol. 4, no. 8, Aug. 1969, pp. 667–678.

52. J.T. Kuo and D.H. Cho, "Transient time-domain electromagnetics," *Geophysics*, vol. 45, no. 2, Feb. 1980, pp. 271–291.

53. R.D. Kodis, "An introduction to variational methods in electromagnetic scattering," *J. Soc. Industr. Appl. Math.*, vol. 2, no. 2, June 1954, pp. 89–112.

54. D.S. Jones, "A critique of the variational method in scattering problems," *IRE Trans.*, vol. AP-4, no. 3, 1965, pp. 297–301.

55. R.J. Wagner, "Variational principles for electromagnetic potential scattering," *Phys. Rev.*, vol. 131, no. 1, July 1963, pp. 423–434.

56. J.A. Krill and R.A. Farrell, "Comparison between variational, perturbational, and exact solutions for scattering from a random rough-surface model," *J. Opt. Soc. Am.*, vol. 68, June 1978, pp. 768–774.

57. R.B. Wu and C.H. Chen, "Variational reaction formulation of scattering problem for anisotropic dielectric cylinders," *IEEE Trans. Ant. Prog.*, vol. 34, no. 5, May 1986, pp. 640–645.

58. J.A. Krill and R.H. Andreo, "Vector stochastic variational principles for electromagnetic wave scattering," *IEEE Trans. Ant. Prog.*, vol. AP-28, no. 6, Nov. 1980, pp. 770–776.

59. R.W. Hart and R.A. Farrell, "A variational principle for scattering from rough surfaces," *IEEE Trans. Ant. Prog.*, vol. AP-25, no. 5, Sept. 1977, pp. 708–713.

60. R. Kamala and V. Kudalkar, "Variational method in the study of optical fibres," *J. Mod. Opt.*, vol. 38, no. 4, 1991, pp. 755–760.

61. J.R. Willis, "Variational principles and operator equations for electromagnetic waves in inhomogeneous media," *Wave Motion*, vol. 6, no. 2, 1984, pp. 127–139.

62. T. Mura and T. Koya, *Variational Methods in Mechanics*. New York: Oxford University Press, 1992.

63. M. Mesterton-Gibbons, *A Primer on the Calculus of Variations and Optimal Control*, Volume 50. Providence, RI: American Mathematical Society, 2000.

5

Moment Methods

There are no shortcuts to any place worth going.

—**Beverly Sills**

5.1 Introduction

In Section 1.3.2, it was mentioned that most EM problems can be stated in terms of an inhomogeneous equation

$$L\Phi = g \tag{5.1}$$

where L is an operator which may be differential, integral, or integro-differential, g is the known excitation or source function, and Φ is the unknown function to be determined. So far, we have limited our discussion to cases for which L is differential. In this chapter, we will treat L as an integral or integro-differential operator.

The *method of moments* (MoM) is a general procedure for solving Equation 5.1. The method owes its name to the process of taking moments by multiplying with appropriate weighing functions and integrating, as discussed in Section 4.6. The name "method of moments" has its origin in Russian literature [1,2]. In western literature, the first use of the name is usually attributed to Harrington [3]. The origin and development of the moment method are fully documented by Harrington [4,5].

The MoM is essentially the method of weighted residuals discussed in Section 4.6. Therefore, the method is applicable for solving both differential and integral equations (IEs). The method is also known as the boundary element method (BEM), the Galerkin method, or the surface integral method (SIE).

The use of MoM in EM has become popular since the work of Richmond [6] in 1965 and Harrington [7] in 1967. The method has been successfully applied to a wide variety of EM problems of practical interest such as radiation due to thin-wire elements and arrays, scattering problems, analysis of microstrips and lossy structures, propagation over an inhomogeneous earth, and antenna beam pattern, to mention a few. An updated review of the method is found in Ney [8]. The literature on MoM is already so large as to prohibit a comprehensive bibliography. A partial bibliography is provided by Adams [9].

The procedure for applying MoM to solve Equation 5.1 usually involves four steps:

1. Derivation of the appropriate IE,
2. Conversion (discretization) of the IE into a matrix equation using basis (or expansions) functions and weighting (or testing) functions,

3. Evaluation of the matrix elements, and

4. Solving the matrix equation and obtaining the parameters of interest.

The basic tools for step 2 have already been mastered in Section 4.6; in this chapter we will apply them to IEs rather than PDEs.

5.2 Differential Equations

We first consider the case in which the operator L in Equation 5.1 is differential. The following is a list of examples L may assume:

$$L[\Phi(x)] = \Phi(x) \tag{5.2a}$$

$$L[\Phi(x)] = \frac{d\Phi}{dx} + 3\frac{d^2\Phi}{dx^2} \tag{5.2b}$$

$$L[\Phi(x,y)] = \nabla^2\Phi(x,y) \tag{5.2c}$$

Our objective in this section to present the procedure for applying the MoM to a differential equation [9]. This basically involves utilizing the variational techniques covered in the previous chapter.

Given the linear-operator equation of Equation 5.1, the MoM starts by expanding the unknown function Φ into a series of known *expansion functions*, $u_1, u_2, ..., u_N$ in the domain of L, as follows:

$$\Phi(x) \simeq \sum_{n=1}^{N} a_n u_n(x) \tag{5.3}$$

where a_n are expansion coefficients to be determined. Substituting Equation 5.3 into Equation 5.1 gives

$$g(x) \simeq L[\Phi(x)] = L\left[\sum_{n=1}^{N} a_n u_n(x)\right] = a_1 L[u_1(x)] + a_2 L[u_2(x)] + \cdots + a_N L[u_N(x)] \tag{5.4}$$

Taking the inner product of this equation with each of the weighting or testing functions, $w_1, w_2, ..., w_N$, produces

$$\left\langle w_m(x), L\left[\sum_{n=1}^{N} a_n u_n(x)\right]\right\rangle$$

$$= a_1\langle w_m(x), L[u_1(x)]\rangle + a_2\langle w_m(x), L[u_2(x)]\rangle + \cdots + a_N\langle w_m(x), L[u_N(x)]\rangle, \ m = 1, 2, ..., N \tag{5.5}$$

This can be written in matrix form as

$$\begin{bmatrix} A_{11} & A_{12} & \cdots & A_{1N} \\ A_{21} & A_{22} & \cdots & A_{2N} \\ \vdots & \vdots & \cdots & \vdots \\ A_{N1} & A_{N2} & \cdots & A_{NN} \end{bmatrix} \begin{bmatrix} a_1 \\ a_2 \\ \vdots \\ a_N \end{bmatrix} = \begin{bmatrix} \langle w_1(x), g(x) \rangle \\ \langle w_2(x), g(x) \rangle \\ \vdots \\ \langle w_N(x), g(x) \rangle \end{bmatrix} \tag{5.6a}$$

Or

$$[A][a] = [B] \tag{5.6b}$$

where $A_{mn} = \langle w_m(x), L[u_n(x)] \rangle$, $B_m = \langle w_m(x), g(x) \rangle$, and $[a]$ consists of unknown coefficients.

Solving the simultaneous equations in Equation 5.6 yields the set of unknown expansion coefficients, a_n. This completes the MoM analysis. The choice of expansion and testing functions determines the accuracy of the solution to Equation 5.6.

EXAMPLE 5.1

Use the MoM to solve the differential equation [10]

$$\frac{d^2 U}{dx^2} = -x^2, \quad 0 < x < 1$$

subject to $U(0) = 0 = U(1)$.

Solution

In this example,

$$L = \frac{d^2}{dx^2}, \quad g(x) = -x^2.$$

We may select the basis function

$$u_n(x) = x - x^{n+1}, \quad n = 1, 2, \ldots, N$$

If we choose Galerkin approach, the weighting function $w_n = u_n$. For example, let $N = 2$ so that

$$U(x) = a_1 u_1(x) + a_2 u_2(x) = a_1(x - x^2) + a_2(x - x^3) \tag{5.1.1}$$

We determine a_1 and a_2 by solving Equation 5.6. The coefficients are obtained from Equation 5.6, namely

$$A_{11} = \langle w_1(x), L[u_1(x)] \rangle = \left\langle (x - x^2), \frac{d^2}{dx^2}(x - x^2) \right\rangle = \langle (x - x^2), -2 \rangle = \int\limits_1^0 (x - x^2)(-2)dx = -\frac{1}{3}$$

Similarly,

$$A_{12} = \int_1^0 (x - x^2)(-6x)dx = -\frac{1}{2}$$

$$A_{21} = \int_1^0 (x - x^2)(-2)dx = -\frac{1}{2}$$

$$A_{22} = \int_1^0 (x - x^3)(-6x)dx = -\frac{4}{5}$$

$$B_1 = \int_1^0 (x - x^2)(-x^2)dx = -\frac{1}{20}$$

$$B_2 = \int_1^0 (x - x^3)(-x^2)dx = -\frac{1}{12}$$

Putting these in matrix form,

$$\begin{bmatrix} -1/3 & -1/2 \\ -1/2 & -4/5 \end{bmatrix}\begin{bmatrix} a_1 \\ a_2 \end{bmatrix} = \begin{bmatrix} -1/20 \\ -1/12 \end{bmatrix} \quad \text{or} \quad \begin{bmatrix} 1/3 & 1/2 \\ 1/2 & 4/5 \end{bmatrix}\begin{bmatrix} a_1 \\ a_2 \end{bmatrix} = \begin{bmatrix} 1/20 \\ 1/12 \end{bmatrix}$$

We can solve this in many ways. Using Cramer's rule, we obtain the determinants as

$$\Delta = \begin{vmatrix} 1/3 & 1/2 \\ 1/2 & 4/5 \end{vmatrix} = \frac{4}{15} - \frac{1}{4} = \frac{1}{60}$$

$$\Delta_1 = \begin{vmatrix} 1/20 & 1/2 \\ 1/12 & 4/5 \end{vmatrix} = \frac{1}{25} - \frac{1}{24} = -\frac{1}{600}$$

$$\Delta_2 = \begin{vmatrix} 1/3 & 1/20 \\ 1/2 & 1/12 \end{vmatrix} = \frac{1}{36} - \frac{1}{40} = -\frac{1}{360}$$

$$a_1 = \frac{\Delta_1}{\Delta} = -\frac{1}{10}, \quad a_2 = \frac{\Delta_2}{\Delta} = \frac{1}{6}$$

Thus, Equation 5.1.1 becomes

$$U(x) \simeq -\frac{1}{10}(x - x^2) + \frac{1}{6}(x - x^3)$$

This approximate solution may be compared with the exact solution:

$$U(x) = \frac{1}{12}(x - x^4)$$

5.3 Integral Equations

An IE is any equation involving unknown function Φ under the integral sign. Simple examples of IEs are Fourier, Laplace, and Hankel transforms.

5.3.1 Classification of IEs

Linear IEs that are most frequently studied fall into two categories named after Fredholm and Volterra. One class is the Fredholm equations of the first, second, and third kind, namely,

$$f(x) = \int_a^b K(x,t)\Phi(t)\,dt, \tag{5.7a}$$

$$f(x) = \Phi(x) - \lambda \int_a^b K(x,t)\Phi(t)\,dt, \tag{5.7b}$$

$$f(x) = a(x)\Phi(x) - \lambda \int_a^b K(x,t)\Phi(t)\,dt, \tag{5.7c}$$

where λ is a scalar (or possibly complex) parameter. Functions $K(x, t)$ and $f(x)$ and the limits a and b are known, while $\Phi(x)$ is unknown. The function $K(x, t)$ is known as the *kernel* of the IE. The parameter λ is sometimes equal to unity.

The second class of IEs is the Volterra equations of the first, second, and third kind, namely,

$$f(x) = \int_a^x K(x,t)\Phi(t)\,dt, \tag{5.8a}$$

$$f(x) = \Phi(x) - \lambda \int_a^x K(x,t)\Phi(t)\,dt, \tag{5.8b}$$

$$f(x) = a(x)\Phi(x) - \lambda \int_a^x K(x,t)\Phi(t)\,dt, \tag{5.8c}$$

with a variable upper limit of integration. If $f(x) = 0$, the IEs (5.7a) through (5.8c) become homogeneous. Note that Equations 5.7a through 5.8c are all linear equations in that Φ enters the equations in a linear manner. An IE is nonlinear if Φ appears in the power of $n > 1$ under the integral sign. For example, the IE

$$f(x) = \Phi(x) - \int_a^b K(x,t)\Phi^2(t)\,dt \tag{5.8}$$

is nonlinear. Also, if limit a or b or the kernel $K(x, t)$ becomes infinite, an IE is said to be singular. Finally, a kernel $K(x, t)$ is said to be symmetric if $K(x, t) = K(t, x)$.

5.3.2 Connection between Differential and IEs

The above classification of one-dimensional IEs arises naturally from the theory of differential equations, thereby showing a close connection between the integral and differential formulation of a given problem. Most ordinary differential equations can be expressed as IEs, but the reverse is not true. While boundary conditions are imposed externally in differential equations, they are incorporated within an IE.

For example, consider the first-order ordinary differential equation

$$\frac{d\Phi}{dx} = F(x, \Phi), \quad a \leq x \leq b \tag{5.9}$$

subject to $\Phi(a) = $ constant. This can be written as the Volterra integral of the second kind. Integrating Equation 5.9 gives

$$\Phi(x) = \int_a^x F(t, \Phi(t)) dt + c_1$$

where $c_1 = \Phi(a)$. Hence, Equation 5.9 is the same as

$$\Phi(x) = \Phi(a) + \int_a^x F(t, \Phi) dt \tag{5.10}$$

Any solution of Equation 5.10 satisfies both Equation 5.9 and the boundary conditions. Thus, an IE formulation incorporates both the differential equation and the boundary conditions.

Similarly, consider the second-order ordinary differential equation

$$\frac{d^2\Phi}{dx^2} = F(x, \Phi), \quad a \leq x \leq b \tag{5.11}$$

Integrating once yields

$$\frac{d\Phi}{dx} = \int_a^x F(x, \Phi(t)) dt + c_1 \tag{5.12}$$

where $c_1 = \Phi'(a)$. Integrating Equation 5.12 by parts,

$$\Phi(x) = c_2 + c_1 x + \int_a^x (x - t) F(x, \Phi(t)) dt$$

where $c_2 = \Phi(a) - \Phi'(a)a$. Hence,

$$\Phi(x) = \Phi(a) + (x - a)\Phi'(a) + \int_a^x (x - t) F(x, \Phi) dt \tag{5.13}$$

Again, we notice that IE (5.13) represents both the differential equation (5.11) and the boundary conditions. We have considered only one-dimensional IEs. IEs involving unknown functions in two or more space dimensions will be discussed later.

EXAMPLE 5.2

Solve the Volterra IE

$$\Phi(x) = 1 + \int_0^x \Phi(t)\,dt$$

Solution

This can be solved directly or indirectly by finding the solution of the corresponding differential equation. To solve it directly, we differentiate both sides of the given IE. In general, given an integral

$$g(x) = \int_{\alpha(x)}^{\beta(x)} f(x,t)\,dt \tag{5.14}$$

with variable limits, we differentiate this using the Leibnitz rule, namely,

$$g'(x) = \int_{\alpha(x)}^{\beta(x)} \frac{\partial f(x,t)}{\partial x}\,dt + f(x,\beta)\beta' - f(x,\alpha)\alpha' \tag{5.15}$$

Differentiating the given IE, we obtain

$$\frac{d\Phi}{dx} = \Phi(x) \tag{5.16a}$$

or

$$\frac{d\Phi}{\Phi(x)} = dx \tag{5.16b}$$

Integrating gives

$$\ln \Phi = x + \ln c_o$$

or

$$\Phi = c_o e^x$$

where $\ln c_o$ is the integration constant. From the given IE

$$\Phi(0) = 1 = c_o$$

Hence,

$$\Phi(x) = e^x \tag{5.17}$$

is the required solution. This can be checked by substituting it into the given IE.
An indirect way of solving the IE is by comparing it with Equation 5.10 and noting that

$$a = 0, \quad \Phi(a) = \Phi(0) = 1$$

and that $F(x, \Phi) = \Phi(x)$. Hence, the corresponding first-order differential equation is

$$\frac{d\Phi}{dx} = \Phi, \quad \Phi(0) = 1$$

which is the same as Equation 5.16, and the solution in Equation 5.17 follows.

EXAMPLE 5.3
Find the IE corresponding to the differential equation

$$\Phi''' - 3\Phi'' - 6\Phi' + 8\Phi = 0$$

subject to $\Phi''(0) = \Phi'(0) = \Phi(0) = 1$.

Solution
Let $\Phi''' = F(\Phi, \Phi, \phi, x) = 3\Phi'' + 6\Phi' - 8\Phi$. Integrating both sides,

$$\Phi'' = 3\Phi' + 6\Phi - 8 \int_0^x \Phi(t)\,dt + c_1 \tag{5.18}$$

where c_1 is determined from the initial values, that is,

$$1 = 3 + 6 + c_1 \rightarrow c_1 = -8$$

Integrating both sides of Equation 5.18 gives

$$\Phi' = 3\Phi + 6 \int_0^x \Phi(t)\,dt - 8 \int_0^x (x - t)\Phi(t)\,dt - 8x + c_2 \tag{5.19}$$

where

$$1 = 3 + c_2 \rightarrow c_2 = -2$$

Finally, we integrate both sides of Equation 5.19 to get

$$\Phi = 3 \int_0^x \Phi(t)\,dt + 6 \int_0^x (x - 1)\Phi(t)\,dt - 8 \int_0^x (x - t)^2 \Phi(t)\,dt - 4x^2 - 2x + c_3$$

where $1 = c_3$. Thus, the IE equivalent to the given differential equation is

$$\Phi(x) = 1 - 2x - 4x^2 + \int_0^x [3 + 6(x-t) - 4(x-t)^2]\Phi(t)\,dt$$

5.4 Green's Functions

A more systematic means of obtaining an IE from a PDE is by constructing an auxiliary function known as *Green's function** for that problem [10–13]. Green's function, also known as the *source function* or *influence function,* is the kernel function obtained from a linear boundary value problem and forms the essential link between the differential and integral formulations. Green's function also provides a method of dealing with the source term (g in $L\Phi = g$) in a PDE. In other words, it provides an alternative approach to the series expansion method of Section 2.7 for solving inhomogeneous boundary-value problems by reducing the inhomogeneous problem to a homogeneous one.

To obtain the field caused by a distributed source by Green's function technique, we find the effects of each elementary portion of source and sum them. If $G(\mathbf{r}, \mathbf{r}')$ is the field at the observation point (or field point) $\mathbf{r}$ caused by a unit point source at the source point $\mathbf{r}'$, then the field at $\mathbf{r}$ by a source distribution $g(\mathbf{r}')$ is the integral of $g(\mathbf{r}')G(\mathbf{r}, \mathbf{r}')$ over the range of $\mathbf{r}'$ occupied by the source. The function G is the Green function. Thus, physically, Green's function $G(\mathbf{r}, \mathbf{r}')$ represents the potential at $\mathbf{r}$ due to a unit point charge at $\mathbf{r}'$. For example, the solution to the Dirichlet problem

$$\nabla^2 \Phi = g \quad \text{in } R$$
$$\Phi = f \quad \text{on } B \tag{5.20}$$

is given by

$$\Phi = \int_R g(\mathbf{r}')G(\mathbf{r},\mathbf{r}')\,dv' + \oint_B f\frac{\partial G}{\partial n}\,dS \tag{5.21}$$

where n denotes the outward normal to the boundary B of the solution region R. It is obvious from Equation 5.21 that the solution Φ can be determined provided Green's function G is known. So the real problem is not that of finding the solution but that of constructing Green's function for the problem.

Consider the linear second-order PDE

$$L\Phi = g \tag{5.22}$$

---

* Named after George Green (1793–1841), an English mathematician.

We define Green's function corresponding to the differential operator L as a solution of the point source inhomogeneous equation

$$LG(\mathbf{r}, \mathbf{r}') = \delta(\mathbf{r}, \mathbf{r}') \tag{5.23}$$

where $\mathbf{r}$ and $\mathbf{r}'$ are the position vectors of the field point (x, y, z) and source point (x', y', z'), respectively, and $\delta(\mathbf{r}, \mathbf{r}')$ is the Dirac delta function, which vanishes for $\mathbf{r} \neq \mathbf{r}'$ and satisfies

$$\int \delta(\mathbf{r}, \mathbf{r}')g(\mathbf{r}')dv' = g(\mathbf{r}) \tag{5.24}$$

From Equation 5.23, we notice that the Green function $G(\mathbf{r}, \mathbf{r}')$ can be interpreted as the solution to the given boundary value problem with the source term g replaced by the unit impulse function. Thus, $G(\mathbf{r}, \mathbf{r}')$ physically represents the response of the linear system to a unit impulse applied at the point $\mathbf{r} = \mathbf{r}'$.

Green's function has the following properties [13]:

a. G satisfies the equation $LG = 0$ except at the source point $\mathbf{r}'$, that is,

$$LG(\mathbf{r}, \mathbf{r}') = \delta(\mathbf{r}, \mathbf{r}') \tag{5.23}$$

b. G is symmetric in the sense that

$$G(\mathbf{r}, \mathbf{r}') = G(\mathbf{r}', \mathbf{r}) \tag{5.25}$$

c. G satisfies that prescribed boundary value f on B, that is,

$$G = f \text{ on } B \tag{5.26}$$

d. The directional derivative $\partial G/\partial n$ has a discontinuity at $\mathbf{r}'$ which is specified by the equation

$$\lim_{\epsilon \to 0} \oint_s \frac{\partial G}{\partial n} ds = 1 \tag{5.27}$$

where n is the outward normal to the sphere of radius ϵ as shown in Figure 5.1, that is,

$$|\mathbf{r} - \mathbf{r}'| = \epsilon^2$$

Property (b) expresses the *principle of reciprocity*; it implies that an exchange of source and observer does not affect G. The property is proved by Myint-U [13] by applying Green's second identity in conjunction with Equation 5.23 while property (d) is proved by applying divergence theorem along with Equation 5.23.

5.4.1 For Free Space

We now illustrate how to construct the free space Green's function G corresponding to a PDE. It is usually convenient to let G be the sum of a particular integral of the inhomogeneous

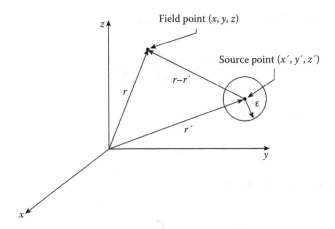

FIGURE 5.1
Illustration of the field point (x, y, z) and source point (x', y', z').

equation $LG = g$ and the solution of the associated homogeneous equation $LG = 0$. In other words, we let

$$G(\mathbf{r}, \mathbf{r}') = F(\mathbf{r}, \mathbf{r}') + U(\mathbf{r}, \mathbf{r}') \tag{5.28}$$

where F, known as the free-space Green's function or fundamental solution, satisfies

$$LF = \delta(\mathbf{r}, \mathbf{r}') \text{ in } R \tag{5.29}$$

and U satisfies

$$LU = 0 \text{ in } R \tag{5.30}$$

so that by superposition $G = F + U$ satisfies Equation 5.23. Also $G = f$ on the boundary B requires that

$$U = -F + f \text{ on } B \tag{5.31}$$

Notice that F need not satisfy the boundary condition.

We apply this to two specific examples. First, consider the two-dimensional problem for which

$$L = \frac{\partial^2}{\partial x^2} + \frac{\partial^2}{\partial y^2} = \nabla^2 \tag{5.32}$$

The corresponding Green's function $G(x, y; x', y')$ satisfies

$$\nabla^2 G(x, y; x', y') = \delta(x - x')\delta(y - y') \tag{5.33}$$

Hence, F must satisfy

$$\nabla^2 F = \delta(x - x')\delta(y - y')$$

For $\rho = [(x - x')^2 + (y - y')^2]^{1/2} > ,0$, i.e., for $x \neq x', y \neq y'$,

$$\nabla^2 F = \frac{1}{\rho} \frac{\partial}{\partial \rho} \left(\rho \frac{\partial F}{\partial \rho} \right) = 0 \tag{5.34}$$

which is integrated twice to give

$$F = A \ln \rho + B \tag{5.35}$$

Applying the property in Equation 5.27

$$\lim_{\epsilon \to 0} \oint \frac{dF}{d\rho} dl = \lim_{\epsilon \to 0} \int_0^{2\pi} \frac{A}{\rho} \rho d\phi = 2\pi A = 1$$

or $A = (1/2\pi)$. Since B is arbitrary, we may choose $B = 0$. Thus,

$$F = \frac{1}{2\pi} \ln \rho$$

and

$$G = F + U = \frac{1}{2\pi} \ln \rho + U \tag{5.36}$$

We choose U so that G satisfies prescribed boundary conditions. For the three-dimensional problem,

$$L = \nabla^2 = \frac{\partial^2}{\partial x^2} + \frac{\partial^2}{\partial y^2} + \frac{\partial^2}{\partial z^2} \tag{5.37}$$

and the corresponding Green's function $G(x, y, z; x', y', z')$ satisfies

$$LG(x, y, z; x', y', z') = \delta(x - x')\delta(y - y')\delta(z - z') \tag{5.38}$$

Hence, F must satisfy

$$\nabla^2 F = \delta(x - x')\delta(y - y')\delta(z - z')$$
$$= \delta(\mathbf{r} - \mathbf{r}')$$

For $\mathbf{r} \neq \mathbf{r}'$,

$$\nabla^2 F = \frac{1}{r^2} \frac{d}{dr} \left(r^2 \frac{dF}{dr} \right) = 0 \tag{5.39}$$

which is integrated twice to yield

$$F = -\frac{A}{r} + B \tag{5.40}$$

Applying Equation 5.27,

$$1 = \lim_{\epsilon \to 0} \oint \frac{dF}{dr} dS = \lim_{\epsilon \to 0} \int_0^{2\pi} \int_0^\pi \frac{A}{r^2} r^2 \sin\phi \, d\theta \, d\phi = 4\pi A$$

or $A = (1/4\pi)$. Choosing $B = 0$ leads to

$$F = -\frac{1}{4\pi r}$$

and

$$G = F + U = -\frac{1}{4\pi r} + U \tag{5.41}$$

where U is chosen so that G satisfies prescribed boundary conditions.

Table 5.1 lists some Green functions that are commonly used in the solution of EM-related problems. It should be observed from Table 5.1 that the form of the three-dimensional Green's function for the steady-state wave equation tends to Green's function for Laplace's equation as the wave number k approaches zero. It is also worthy of remark that each of Green's functions in closed form as in Table 5.1 can be expressed in series form. For example, Green's function

$$F = -\frac{j}{4} H_0^{(1)}(k \, | \rho - \rho' |)$$

$$= -\frac{j}{4} H_0^{(1)}(k[\rho^2 + \rho'^2 - 2\rho\rho' \cos(\phi - \phi')]^{1/2}) \tag{5.42}$$

TABLE 5.1

Free-Space Green's Functions

Operator Equation	Laplace's Equation	Steady-State Helmholtz's (or wave) Equation[a]	Modified Steady-State Helmholtz's (or Wave) Equation
Solution Region	$\nabla^2 G = \delta(\mathbf{r}, \mathbf{r}')$	$\nabla^2 G + k^2 G = \delta(\mathbf{r}, \mathbf{r}')$	$\nabla^2 G - k^2 G = \delta(\mathbf{r}, \mathbf{r}')$
1-Dimensional	No solution for $(-\infty, \infty)$	$-\dfrac{j}{2k} \exp(jk \, |x - x'|)$	$-\dfrac{1}{2k} \exp(-k \, |x - x'|)$
2-Dimensional	$\dfrac{1}{2\pi} \ln|\rho - \rho'|$	$-\dfrac{j}{4} H_0^{(1)}(k \, |\rho - \rho'|)$	$-\dfrac{1}{2\pi} K_0(k \, |\rho - \rho'|)$
3-Dimensional	$-\dfrac{1}{4\pi(\mathbf{r} - \mathbf{r}')}$	$-\dfrac{\exp(jk \, |\mathbf{r} - \mathbf{r}'|)}{4\pi \, |\mathbf{r} - \mathbf{r}'|}$	$\dfrac{\exp(-k \, |\mathbf{r} - \mathbf{r}'|)}{4\pi \, |\mathbf{r} - \mathbf{r}'|}$

[a] The wave equation has the time factor $e^{j\omega t}$ so that $k = \omega \sqrt{\mu\epsilon}$.

can be written in series form as

$$
F = \begin{cases}
-\dfrac{j}{4} \displaystyle\sum_{n=-\infty}^{\infty} H_n^{(1)}(k\rho')J_n(k\rho)e^{-jn(\phi-\phi')}, & \rho < \rho' \\[4mm]
-\dfrac{j}{4} \displaystyle\sum_{n=-\infty}^{\infty} H_n^{(1)}(k\rho)J_n(k\rho')e^{-jn(\phi-\phi')}, & \rho > \rho'
\end{cases}
\tag{5.43}
$$

This is obtained from addition theorem for Hankel functions [14]. It should be noted that Green's functions are very difficult to construct in an explicit form except for the simplest shapes of domain.

With the aid of Green's function, we can construct the IE corresponding to Poisson's equation in three dimensions

$$
\nabla^2 V = -\frac{\rho_v}{\epsilon}
\tag{5.44}
$$

as

$$
V = \int \frac{\rho_v}{\epsilon} G(\mathbf{r},\mathbf{r}')dv'
$$

or

$$
V = \int \frac{\rho_v dv'}{4\pi\epsilon r}
\tag{5.45}
$$

Similarly, the IE corresponding to Helmholtz's equation in three dimensions

$$
\nabla^2 \Phi + k^2 \Phi = g
\tag{5.46}
$$

as

$$
\Phi = \int g G(\mathbf{r},\mathbf{r}')dv'
$$

or

$$
\Phi = \int \frac{g e^{jkr} dv'}{4\pi r}
\tag{5.47}
$$

where an outgoing wave is assumed.

5.4.2 For Domain with Conducting Boundaries

Green's functions derived so far are useful if the domain is free space. When the domain is bounded by one or more grounded planes, there are two ways to obtain Green's function:

a. The method of images [12,15–22] and
b. The eigenfunction expansion [12,16,17,22–30].

5.4.2.1 Method of Images

The method of images is a powerful technique for obtaining the field due to one or more sources with conducting boundary planes. If a point charge q is at some distance h from a grounded conducting plane, the boundary condition imposed by the plane on the resulting potential field may be satisfied by replacing the plane with an "image charge" $-q$ located at a position which is the mirror location of q. Using this idea to obtain Green's function is perhaps best illustrated with an example.

Consider the region between the ground planes at $y = 0$ and $y = h$ as shown in Figure 5.2. Green's function $G(x, y; x', y')$ is the potential at the point (x, y), which results when a unit line charge of 1 C/m is placed at the point (x', y'). If no ground planes were present, the potential at distance ρ from a unit line charge would be

$$V(\rho) = \frac{1}{4\pi\epsilon} \ln \rho^2 \tag{5.48}$$

In order to satisfy the boundary conditions on the ground planes, an infinite set of images is derived as shown in Figure 5.2. The potential due to such a sequence of line charges (including the original) within the strip is the superposition of an infinite series of images:

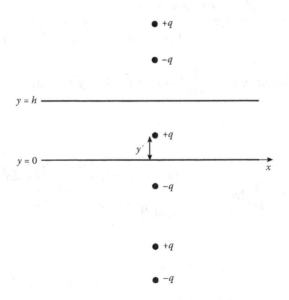

FIGURE 5.2
A single charge placed between two conducting planes produces the same potential as does the system of image charges when no conducting planes are present.

$$G(x,y;x',y') = \frac{1}{4\pi\epsilon}\left[\ln[(x-x')^2+(y+y')^2]-\ln[(x-x')^2+(y-y')^2]\right.$$

$$+\sum_{n=1}^{\infty}(-1)^n\{\ln[(x-x')^2+(y+y'-2nh)^2]$$

$$-\ln[(x-x')^2+(y-y'-2nh)^2]$$

$$+\ln[(x-x')^2+(y+y'-2nh)^2]$$

$$\left.-\ln[(x-x')^2+(y-y'-2nh)^2]\}\right]$$

$$= \frac{1}{4\pi\epsilon}\sum_{n=-\infty}^{\infty}\ln\left[\frac{(x-x')^2+(y+y'-2nh)^2}{(x-x')^2+(y-y'-2nh)^2}\right] \tag{5.49}$$

This series converges slowly and is awkward for numerical computation. It can be summed to give [15]

$$G(x,y;x',y') = \frac{1}{4\pi\epsilon}\ln\left[\frac{\sinh^2\left(\frac{\pi(x-x')}{2h}\right)+\sin^2\left(\frac{\pi(y+y')}{2h}\right)}{\sinh^2\left(\frac{\pi(x-x')}{2h}\right)+\sin^2\left(\frac{\pi(y-y')}{2h}\right)}\right] \tag{5.50}$$

This expression can be shown to satisfy the appropriate boundary conditions along the ground plane, that is, $G(x, y; x', y') = 0$ at $y = 0$ or $y = h$. Note that G has exactly one singularity at $x = x'$, $y = y'$ in the region $0 \leq y \leq h$.

In order to evaluate an integral involving $G(x, y; x', y')$ in Equation 5.50, it is convenient to take out the singular portion of the unit source function. We rewrite Equation 5.50 as

$$G(x,y;x',y') = -\frac{1}{4\pi\epsilon}\ln[(x-x')^2+(y+y')^2]+g(x,y;x',y') \tag{5.51}$$

where

$$g(x,y;x',y') = \frac{1}{4\pi\epsilon}\ln\left[\frac{[(x-x')^2(y-y')^2]\left[\sinh^2\left(\frac{\pi(x-x')}{2h}\right)+\sin^2\left(\frac{\pi(y-y')}{2h}\right)\right]}{\sinh^2\left(\frac{\pi(x-x')}{2h}\right)+\sin^2\left(\frac{\pi(y-y')}{2h}\right)}\right] \tag{5.52}$$

Note that $g(x, y; x', y')$ is finite everywhere in $0 \leq y \leq h$. The integral involving g is evaluated numerically, while the one involving the singular logarithmic term is evaluated analytically with the aid of integral tables.

The method of images has been applied in deriving Green's functions for multiconductor transmission lines [18–20] and planar microwave circuits [16,17,21]. The method is restricted to the shapes enclosed by boundaries that are straight conductors.

5.4.2.2 Eigenfunction Expansion

This method is suitable for deriving Green's function for differential equations whose homogeneous solution is known. Green's function is represented in terms of a series of orthonormal functions that satisfy the boundary conditions associated with the differential equation. To illustrate the eigenfunction expansion procedure, suppose we are interested in Green's function for the wave equation

$$\frac{\partial^2 \Psi}{\partial x^2} + \frac{\partial^2 \Psi}{\partial y^2} + k^2 \Psi = 0 \tag{5.53}$$

subject to

$$\frac{\partial \Psi}{\partial n} = 0 \quad \text{or} \quad \Psi = 0 \tag{5.54}$$

Let the eigenfunctions and eigenvalues of Equation 5.53 that satisfy Equation 5.54 be Ψ_j and k_j, respectively, that is,

$$\nabla^2 \Psi_j + k_j^2 \Psi_j = 0 \tag{5.55}$$

Assuming that Ψ_j form a complete set of orthonormal functions,

$$\int_S \Psi_j^* \Psi_i \, dx \, dy = \begin{cases} 1, & j = i \\ 0, & j \neq i \end{cases} \tag{5.56}$$

where the asterisk (*) denotes complex conjugation. $G(x, y; x', y')$ can be expanded in terms of Ψ_j, that is,

$$G(x, y; x', y') = \sum_{j=1}^{\infty} a_j \Psi_j(x, y) \tag{5.57}$$

Since Green's function must satisfy

$$(\nabla^2 + k^2) G(x, y; x', y') = \delta(x - x') \delta(y - y'), \tag{5.58}$$

substituting Equations 5.55 and 5.57 into Equation 5.58, we obtain

$$\sum_{j=1}^{\infty} a_j (k^2 - k_j^2) \Psi_j = \delta(x - x') \delta(y - y') \tag{5.59}$$

Multiplying both sides by Ψ_i^* and integrating over the region S gives

$$\sum_{j=1}^{\infty} a_j (k^2 - k_j^2) \int_S \Psi_j \Psi_i^* \, dx \, dy = \Psi_i^*(x', y') \tag{5.60}$$

Imposing the orthonormal property in Equation 5.56 leads to

$$a_i(k^2 - k_i^2) = \Psi_i^*(x', y')$$

or

$$a_i = \frac{\Psi_i^*(x', y')}{(k^2 - k_i^2)} \tag{5.61}$$

Thus,

$$G(x, y; x', y') = \sum_{j=1}^{\infty} \frac{\Psi_j(x, y)\Psi_j^*(x', y')}{(k^2 - k_j^2)} \tag{5.62}$$

The eigenfunction expansion approach has been applied to derive Green's functions for plane conducting boundaries such as rectangular box and prism [22], planar microwave circuits [16,17,25], multilayered dielectric structures [23,24], waveguides [28], and surfaces of revolution [27]. The approach is limited to separable coordinate systems since the requisite eigenfunctions can be determined for only these cases.

EXAMPLE 5.4

Construct a Green's function for

$$\nabla^2 V = 0$$

subject to $V(a, \phi) = f(\phi)$ within a circular disk $\rho \leq a$.

Solution

Since $g = 0$, the solution is obtained from Equation 5.21 as

$$V = \oint_C f \frac{\partial G}{\partial n} dl \tag{5.63}$$

where the circle C is the boundary of the disk as shown in Figure 5.3. Let

$$G = F + U,$$

where F is already found to be

$$F = \frac{1}{2\pi} \ln|\rho - \rho'|,$$

that is,

$$F(\rho^2, \phi; \rho', \phi') = \frac{1}{4\pi} \ln\left[\rho^2 + \rho'^2 - 2\rho\rho' \cos(\phi - \phi')\right] \tag{5.64}$$

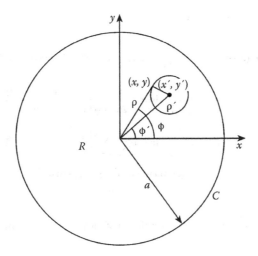

FIGURE 5.3
A disk of radius *a*.

The major problem is finding U. But

$$\nabla^2 U = 0 \text{ in } R \tag{5.65a}$$

with

$$U = -F \text{ on } C$$

or

$$U(a, \phi; \rho', \phi') = -\frac{1}{4\pi} \ln[a^2 + \rho'^2 - 2a\rho' \cos(\phi - \phi')] \tag{5.65b}$$

Thus, U can be found by solving the PDE in Equation 5.65a subject to the condition in Equation 5.65b. Applying the separation of variables method,

$$U = \frac{A_0}{2} + \sum_{n=1}^{\infty} \rho^n [A_n \cos n\phi + B_n \sin n\phi] \tag{5.66}$$

The term ρ^{-n} is not included since U must be bounded at $\rho = 0$. To impose the boundary condition in Equation 5.65b on the solution in Equation 5.66, we first express Equation 5.65b in Fourier series using the identity

$$\sum_{n=1}^{\infty} \frac{z^n}{n} \cos n\theta = \int_0^z \frac{\cos\theta - \lambda}{1 + \lambda^2 - 2\lambda \cos} d\lambda = -\frac{1}{2} \ln[1 + z^2 - 2z \cos\theta] \tag{5.67}$$

Hence, Equation 5.65b becomes

$$U(a,\phi;\rho',\phi') = -\frac{1}{4\pi}\ln a^2\left[1+(\rho'/a)^2-\frac{2\rho'}{a}\cos(\phi-\phi')\right]$$

$$= -\frac{1}{2\pi}\ln a + \frac{1}{2\pi}\sum_{n=1}^{\infty}\left[\frac{\rho'}{a}\right]^n\frac{\cos n(\phi-\phi')}{n}$$

$$= -\frac{1}{2\pi}\ln a + \frac{1}{2\pi}\sum_{n=1}^{\infty}\left[\frac{\rho'}{a}\right]^n \cdot$$

$$\frac{(\cos n\phi\cos n\phi'+\sin n\phi\sin n\phi'}{n} \tag{5.68}$$

Comparing Equation 5.66 with Equation 5.68 at $\rho=a$, we obtain the coefficients A_n and B_n as

$$\frac{A_0}{2}=-\frac{1}{2\pi}\ln a$$

$$a^n A_n = \frac{1}{2\pi n}\left[\frac{\rho'}{a}\right]^n\cos n\phi'$$

$$a^n B_n = \frac{1}{2\pi n}\left[\frac{\rho'}{a}\right]^n\sin n\phi'$$

Thus, Equation 5.66 becomes

$$U(\rho,\phi;\rho',\phi') = -\frac{1}{2\pi}\ln a + \frac{1}{2\pi}\sum_{n=1}^{\infty}\left[\frac{\rho'}{a}\right]^n\left[\frac{\rho}{a}\right]^n\frac{\cos n(\phi-\phi')}{n}$$

$$= -\frac{1}{2\pi}\ln a - \frac{1}{4\pi}\ln\left[1+\left[\frac{\rho\rho'}{a^2}\right]^2-\frac{2\rho\rho'}{a^2}\cos(\phi-\phi')\right] \tag{5.69}$$

From Equations 5.64 and 5.69, we obtain Green's function as

$$G = \frac{1}{4\pi}\ln[\rho^2+\rho'^2-2\rho\rho'\cos(\phi-\phi')]$$

$$-\frac{1}{4\pi}\ln\left[a^2+\frac{\rho'^2\rho^2}{a^2}-2\rho\rho'\cos(\phi-\phi')\right] \tag{5.70}$$

An alternative means of constructing Green's function is the method of images. Let us obtain Equation 5.70 using the method of images. Let

$$G(P,P') = \frac{1}{2\pi}\ln r + U$$

The problem reduces to finding the induced field U, which is harmonic within the disk and is equal to $-(1/2\pi)\ln r$ on C. Let P' be the singular point of Green's function and let P_o be the image of P' with respect to the circle C as shown in Figure 5.4. The triangles

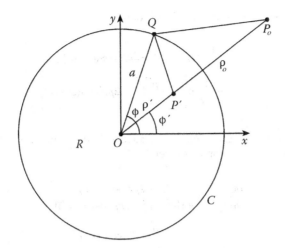

FIGURE 5.4
Image point P_o of P' with respect to circle C so that $O\,P' \times O\,P_o = O\,Q = a^2$ and $O\,Q\,P'$ and $O\,P_o\,Q$ are similar triangles.

$O\,Q\,P'$ and $O\,Q\,P_o$ are similar because the angle at O is common and the sides adjacent to it are proportional. Thus,

$$\frac{\rho'}{a} = \frac{a}{\rho_o} \rightarrow \rho'\rho_o = a^2 \tag{5.71}$$

That is, the product of $O\,P'$ and $O\,P_o$ is equal to the square of the radius $O\,Q$. At a point Q on C, it is evident from Figure 5.4 that

$$r_{QP'} = \frac{\rho'}{a} r_{QP_o}$$

Therefore,

$$U = -\frac{1}{2\pi} \ln \frac{\rho' r_{PP_o}}{a} \tag{5.72}$$

and

$$G = \frac{1}{2\pi} \ln r_{PP'} - \frac{1}{2\pi} \ln \frac{\rho'}{a} r_{PP_o} \tag{5.73}$$

Since $r_{PP'}$ is the distance between $P(\rho, \phi)$ and $P'(\rho', \phi')$ while r_{PP_o} is the distance between $P(\rho, \phi)$ and $P_o(\rho'_o, \phi) = P_o(a^2/\rho', \phi)$,

$$r_{PP'}^2 = \rho^2 + \rho'^2 - 2\rho\rho' \cos(\phi - \phi'),$$

$$r_{PP_o}^2 = \rho^2 + \frac{a^4}{\rho'^2} - 2\rho \frac{a^2}{\rho'} \cos(\phi - \phi')$$

Substituting these into Equation 5.73, we obtain

$$G = \frac{1}{4\pi} \ln[\rho^2 + \rho'^2 - 2\rho\rho'\cos(\phi - \phi')]$$
$$- \frac{1}{4\pi} \ln\left[a^2 + \frac{\rho'^2\rho^2}{a^2} - 2\rho\rho'\cos(\phi - \phi')\right] \qquad (5.74)$$

which is the same as Equation 5.70. From Equation 5.70 or Equation 5.74, the directional derivative $\partial G/\partial n = (\nabla G \cdot \mathbf{a}_n)$ on C is given by

$$\frac{\partial G}{\partial \rho'}\bigg|_{\rho'=a} = \frac{2a - 2\rho\cos(\phi - \phi')}{4\pi[a^2 + \rho^2 - 2a\rho\cos(\phi - \phi')]}$$
$$\frac{\dfrac{2\rho^2}{a} - 2\rho\cos(\phi - \phi')}{4\pi[a^2 + \rho^2 - 2a\rho\cos(\phi - \phi')]},$$
$$= \frac{a^2 - \rho^2}{2\pi a[a^2 + \rho^2 - 2a\rho\cos(\phi - \phi')]}$$

Hence, the solution in Equation 5.63 becomes (with $dl = ad\phi'$)

$$V(\rho, \phi) = \frac{1}{2\pi} \int\limits_{0}^{2\pi} \frac{(a^2 - \rho^2)f(\phi')d\phi'}{[a^2 + \rho^2 - 2a\rho\cos(\phi - \phi')]} \qquad (5.75)$$

which is known as Poisson's integral formula.

EXAMPLE 5.5

Obtain the solution for the Laplace operator on unbounded half-space, $z \leq 0$, with the condition $V(z = 0) = f$.

Solution

Again the solution is

$$V = \oint\limits_{S} f \frac{\partial G}{\partial n} dS$$

where S is the plane $z = 0$. We let

$$G = \frac{1}{4\pi |\mathbf{r} - \mathbf{r}'|} + U,$$

so that the major problem is reduced to finding U. Using the method of images, it is easy to see that the image point of $P'(x', y', z')$ is $P_o(x', y', -z')$ as shown in Figure 5.5. Hence,

$$U = -\frac{1}{4\pi |\mathbf{r} - \mathbf{r}_o|}$$

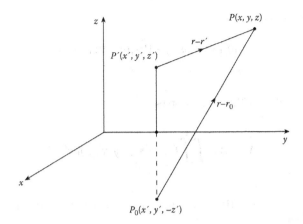

FIGURE 5.5
Half-space problem of Example 5.5.

and

$$G = \frac{1}{4\pi |\mathbf{r} - \mathbf{r}'|} - \frac{1}{4\pi |\mathbf{r} - \mathbf{r}_0|},$$

where

$$|\mathbf{r} - \mathbf{r}'| = [(x - x')^2 + (y - y')^2 + (z - z')^2]^{1/2}$$
$$|\mathbf{r} - \mathbf{r}_0| = [(x - x')^2 + (y - y')^2 + (z + z')^2]^{1/2}$$

Notice that G reduces to zero at $z = 0$ and has the required singularity at $P'(x', y', z')$. The directional derivative $\partial G/\partial n$ on plane $z = 0$ is

$$\frac{\partial G}{\partial z'}\bigg|_{z'=0} = \frac{1}{4\pi}\left[\frac{(z - z')}{|\mathbf{r} - \mathbf{r}'|^3} + \frac{(z + z')}{|\mathbf{r} - \mathbf{r}_0|^3}\right]\bigg|_{z'=0}$$

$$= \frac{z}{2\pi[(x - x')^2 + (y - y')^2 + z^2]^{3/2}}$$

Hence,

$$V(x, y, z) = \frac{1}{2\pi}\int_{-\infty}^{\infty}\int_{-\infty}^{\infty}\frac{zf(x', y')\,dx'\,dy'}{[(x - x')^2 + (y - y')^2 + z^2]^{3/2}}$$

EXAMPLE 5.6

Using Green's function, construct the solution for Poisson's equation

$$\frac{\partial^2 V}{\partial x^2} + \frac{\partial^2 V}{\partial y^2} = f(x, y)$$

subject to the boundary conditions

$$V(0,y) = V(a,y) = V(x,0) = V(x,b) = 0$$

Solution

According to Equation 5.21, the solution is

$$V(x,y) = \int_0^b \int_0^a f(x',y')G(x,y;x',y')dx'\,dy' \tag{5.76}$$

so that our problem is essentially that of obtaining Green's function $G(x, y; x', y')$. Green's function satisfies

$$\frac{\partial^2 G}{\partial x^2} + \frac{\partial^2 G}{\partial y^2} = \delta(x - x')\delta(y - y') \tag{5.77}$$

To apply the series expansion method of finding G, we must first determine eigenfunctions $\Psi(x, y)$ of Laplace's equation, that is,

$$\nabla^2 \Psi = \lambda \Psi$$

where Ψ satisfies the boundary conditions. It is evident that the normalized eigenfunctions are

$$\Psi_{mn} = \frac{2}{\sqrt{ab}} \sin\frac{m\pi x}{a} \sin\frac{n\pi y}{b}$$

with the corresponding eigenvalues

$$\lambda_{mn} = -\left(\frac{m^2\pi^2}{a^2} + \frac{n^2\pi^2}{b^2}\right)$$

Thus,

$$G(x,y;x',y') = \frac{2}{\sqrt{ab}} \sum_{m=1}^{\infty} \sum_{n=1}^{\infty} A_{mn}(x',y')\sin\frac{m\pi x}{a} \sin\frac{n\pi y}{b} \tag{5.78}$$

The expansion coefficients, A_{mn}, are determined by substituting Equation 5.78 into Equation 5.77, multiplying both sides by $\sin(m\pi x/a)\sin(n\pi y/b)$, and integrating over $0 < x < a$, $0 < y < b$. Using the orthonormality property of the eigenfunctions and the shifting property of the delta function results in

$$-\left(\frac{m^2\pi^2}{a^2} + \frac{n^2\pi^2}{b^2}\right)A_{mn} = \frac{2}{\sqrt{ab}} \sin\frac{m\pi x'}{a} \sin\frac{n\pi y'}{b}$$

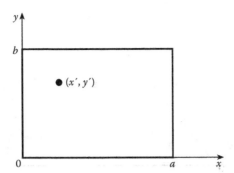

FIGURE 5.6
Line source in a rectangular region.

Obtaining A_{mn} from this and substituting into Equation 5.78 gives

$$G(x,y;x',y') = -\frac{4}{ab}\sum_{m=1}^{\infty}\sum_{n=1}^{\infty}\frac{\sin\frac{m\pi x}{a}\sin\frac{m\pi x'}{a}\sin\frac{n\pi y}{b}\sin\frac{n\pi y'}{b}}{m^2\pi^2/a^2 + n^2\pi^2/b^2} \qquad (5.79)$$

Another way of obtaining Green's function is by means of a single series rather than a double summation in Equation 5.79. It can be shown that [28,29]

$$G(x,y;x',y') = \begin{cases} -\dfrac{2}{\pi}\displaystyle\sum_{n=1}^{\infty}\dfrac{\sin\dfrac{n\pi x}{b}\sinh\dfrac{n\pi(a-x')}{b}\sin\dfrac{n\pi y}{b}\sinh\dfrac{n\pi y'}{b}}{n\sinh\dfrac{n\pi a}{b}}, & x<x' \\[2em] -\dfrac{2}{\pi}\displaystyle\sum_{n=1}^{\infty}\dfrac{\sinh\dfrac{n\pi x'}{b}\sinh\dfrac{n\pi(a-x)}{b}\sin\dfrac{n\pi y}{b}\sinh\dfrac{n\pi y'}{b}}{n\sinh\dfrac{n\pi a}{b}}, & x>x' \end{cases} \qquad (5.80)$$

By Fourier series expansion, it can be verified that the expressions in Equations 5.79 and 5.80 are identical. Besides the factor $1/\varepsilon$, Green's function in Equation 5.79 or Equation 5.80 gives the potential V due to a unit line source at (x', y') in the region $0 < x < a$, $0 < y < b$ as shown in Figure 5.6.

EXAMPLE 5.7

An infinite line source I_z is located at (ρ', ϕ') in a wedge waveguide shown in Figure 5.7. Derive the electric field due to the line.

Solution

Assuming the time factor $e^{j\omega t}$, the z-component of **E** for the TE mode satisfies the wave equation

$$\nabla^2 E_z + k^2 E_z = j\omega\mu I_z \qquad (5.81)$$

with

$$\frac{\partial E_z}{\partial n} = 0$$

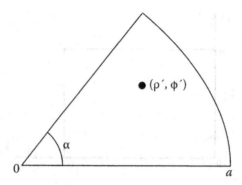

FIGURE 5.7
Line source in a waveguide.

where $k = \omega\sqrt{\mu\epsilon}$ and n is the outward unit normal at any point on the periphery of the cross section. Green's function for this problem satisfies

$$\nabla^2 G + k^2 G = j\omega\mu\delta(\rho - \rho') \tag{5.82}$$

with

$$\frac{\partial G}{\partial n} = 0$$

so that the solution to Equation 5.81 is

$$E_z = j\omega\mu \int_S I_z(\rho',\phi')G(\rho,\phi;\rho',\phi')dS \tag{5.83}$$

To determine Green's function $G(\rho, \phi; \rho', \phi')$, we find Ψ_i so that Equation 5.62 can be applied. The boundary condition $\partial G/\partial n = 0$ implies that

$$\frac{1}{\rho}\frac{\partial G}{\partial \phi}\bigg|_{\phi=0} = 0 = -\frac{1}{\rho}\frac{\partial G}{\partial \phi}\bigg|_{\phi=\alpha} = \frac{\partial G}{\partial \rho}\bigg|_{\rho=a} \tag{5.84}$$

The set of functions which satisfy the boundary conditions are

$$\Psi_{mv}(\rho,\phi) = J_v(k_{mv}\rho)\cos v\phi \tag{5.85}$$

where

$$v = n\pi/\alpha, \quad n = 0, 1, 2, \ldots, \tag{5.86a}$$

k_{mv} are chosen to satisfy

$$\frac{\partial}{\partial \rho} J_v(k_{mv}\rho)\bigg|_{\rho=a} = 0, \tag{5.86b}$$

and the subscript m is used to denote the mth root of Equation 5.86b; m can take the value zero for $n = 0$. The functions Ψ_{mv} are orthogonal if and only if v is an integer which implies that v is an integral multiple of α. Let $\alpha = n/\ell$, where ℓ is a positive integer, so that Φ_{mv} are mutually orthogonal. To obtain Green's function using Equation 5.62, these eigenfunctions must be normalized over the region, that is,

$$\int_0^a J_v^2(k_{mv}\rho)d\rho = \begin{cases} a^2/2, & m = v \\ \dfrac{1}{2}[a^2 - (v^2/k_{mv}^2)]J_v'^2(k_{mv}a), & \text{otherwise} \end{cases} \tag{5.87a}$$

$$\int_0^\alpha \cos^2 v\phi\, d\phi = \begin{cases} \dfrac{\pi}{\ell}, & v = 0 \\ \dfrac{\pi}{2\ell}, & \text{otherwise} \end{cases} \tag{5.87b}$$

where $v = n\ell$. Using the normalized eigenfunctions, we obtain

$$G(\rho,\phi;\rho',\phi') = \frac{j2\ell}{\omega\epsilon\pi a^2}$$
$$-4j\ell\omega\mu\sum_{n=1}^\infty\sum_{m=1}^\infty \frac{J_v(k_{mv}\rho)J_v(k_{mv}\rho')\cos v\phi\cos v\phi'}{\epsilon_v\pi\left(a^2 - \dfrac{v^2}{k_{mv}^2}\right)J_v^2(k_{mv}a)(k^2 - k_{mv}^2)} \tag{5.88}$$

where

$$\epsilon_v = \begin{cases} 2, & v = 0 \\ 1, & v \neq 0 \end{cases} \tag{5.89}$$

We have employed the fact that $\omega\mu/k^2 = 1/\omega\varepsilon$ to obtain the first term on the right-hand side of Equation 5.88.

5.5 Applications I: Quasi-Static Problems

The MoM has been applied to so many EM problems that covering all of them is practically impossible. We will consider only the relatively easy ones to illustrate the techniques involved. Once the basic approach has been mastered, it will be easy for the reader to extend the idea to attack more complicated problems.

We will apply MoM to a static problem in this section; more involved application will be considered in the sections to follow. We will consider the problem of determining the characteristic impedance Z_o of a strip transmission line [31].

Consider the strip transmission of Figure 5.8a. If the line is assumed to be infinitely long, the problem is reduced to a two-dimensional TEM problem of line sources in a plane as in Figure 5.8b. Let the potential difference of the strips be $V_d = 2V$ so that strip 1 is maintained

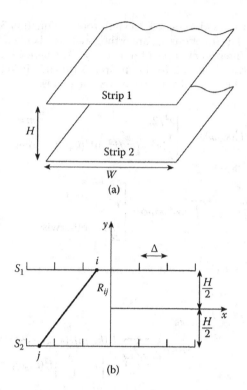

FIGURE 5.8
(a) Strip transmission line. (b) The two-dimensional view.

at $+1V$ while strip 2 is at $-1V$. Our objective is to find the surface charge density $\rho(x, y)$ on the strips so that the total charge per unit length on one strip can be found as

$$Q_\ell = \int \rho \, dl \tag{5.90}$$

(Q_ℓ is charge per unit length as distinct from the total charge on the strip because we are treating a three-dimensional problem as a two-dimensional one.) Once Q is known, the capacitance per unit length C_ℓ can be found from

$$C_\ell = \frac{Q_\ell}{V_d} \tag{5.91}$$

Finally, the line characteristic impedance is obtained:

$$Z_o = \frac{(\mu\epsilon)^{1/2}}{C_\ell} = \frac{1}{uC_\ell} \tag{5.92}$$

where $u = 1/\sqrt{\mu\epsilon}$ is the speed of the wave in the (lossless) dielectric medium between the strips. Everything is straightforward once the charge density $\rho(x, y)$ in Equation 5.90 is known. To find ρ using MoM, we divide each strip into n subareas of equal width Δ so that subareas in strip 1 are numbered 1, 2, ..., n, while those in strip 2 are numbered $n + 1$, $n + 2, n + 3, ..., 2n$. The potential at an arbitrary field point is

$$V(x,y) = \frac{1}{2\pi\epsilon} \int \rho(x',y') \ln \frac{R}{r_o} dx' dy' \tag{5.93}$$

where R is the distance between source and field points, that is,

$$R = [(x-x')^2 + (y-y')^2]^{1/2} \tag{5.94}$$

Since the integral in Equation 5.93 may be regarded as rectangular subareas in a numerical sense, the potential at the center of a typical subarea S_i is

$$V_i = \frac{1}{2\pi\epsilon} \sum_{j=1}^{2n} \rho_j \int_{S_i} \ln \frac{R_{ij}}{r_o} dx'$$

or

$$V_i = \sum_{j=1}^{2n} A_{ij}\rho_j \tag{5.95}$$

where

$$A_{ij} = \frac{1}{2\pi\epsilon} \int_{S_i} \ln \frac{R_{ij}}{r_o} dx', \tag{5.96}$$

R_{ij} is the distance between ith and jth subareas, and $A_{ij}\, \rho_j$ represents the potential at point i due to subarea j. In Equation 5.95, we have assumed that the charge density ρ is constant within each subarea. For all the subareas $S_i, i = 1, 2, ..., 2n$ we have

$$V_1 = \sum_{j=1}^{2n} \rho_j A_{1j} = 1$$

$$V_2 = \sum_{j=1}^{2n} \rho_j A_{2j} = 1$$

$$\vdots$$

$$V_n = \sum_{j=1}^{2n} \rho_j A_{nj} = 1$$

$$V_{n+1} = \sum_{j=1}^{2n} \rho_j A_{n+1,j} = -1$$

$$\vdots$$

$$V_{2n} = \sum_{j=1}^{2n} \rho_j A_{2n,j} = -1$$

Thus, we obtain a set of $2n$ simultaneous equations with $2n$ unknown charge densities ρ_i. In matrix form,

$$
\begin{bmatrix}
A_{11} & A_{12} & \cdots & A_{1,2n} \\
A_{21} & A_{22} & \cdots & A_{2,2n} \\
\vdots & & & \\
A_{2n,1} & A_{2n,2} & \cdots & A_{2n,2n}
\end{bmatrix}
\begin{bmatrix}
\rho_1 \\
\rho_2 \\
\vdots \\
\rho_{2n}
\end{bmatrix}
=
\begin{bmatrix}
1 \\
1 \\
\vdots \\
-1 \\
-1
\end{bmatrix}
$$

or simply

$$[A][\rho] = [B] \tag{5.97}$$

It can be shown that [32] the elements of matrix $[A]$ expressed in Equation 5.96 can be reduced to

$$
A_{ij} =
\begin{cases}
\dfrac{\Delta}{2\pi\epsilon} \ln \dfrac{R_{ij}}{r_o}, & i \neq j \\[2ex]
\dfrac{\Delta}{2\pi\epsilon}\left[\ln \dfrac{\Delta}{r_o} - 1.5\right], & i = j
\end{cases}
\tag{5.98}
$$

where r_o is a constant scale factor (commonly taken as unity). From Equation 5.97, we obtain $[\rho]$ either by solving the simultaneous equation or by matrix inversion, that is,

$$[\rho] = [A]^{-1}[B] \tag{5.99}$$

Once $[\rho]$ is known, we determine C_ℓ from Equations 5.90 and 5.91 as

$$C_\ell = \sum_{j=1}^{n} \rho_j \Delta / V_d \tag{5.100}$$

where $V_d = 2\ V$. Obtaining Z_o follows from Equations 5.92 and 5.100.

EXAMPLE 5.8

Write a program to find the characteristic impedance Z_o of a strip line with $H = 2$ m, $W = 5$ m, $\epsilon = \epsilon_o$, $\mu_o = \mu_o$ and $V_d = 2\ V$.

Solution

The MATLAB program is shown in Figure 5.9. With the given data, the program calculates the elements of matrices $[A]$ and $[B]$ and determines $[\rho]$ by matrix inversion. With the computed charge densities the capacitance per unit length is calculated using Equation 5.100 and the characteristic impedance from Equation 5.92. Table 5.2 presents the computed values of Z_o for a different number of segments per strip, n. The results agree well with $Z_o = 50\ \Omega$ from Wheeler's curve [33].

```
%    ****************************************************************
%    USING MOMENTS METHOD,
%    THIS PROGRAM DETERMINES THE CHARACTERISTIC IMPEDANCE
%    OF A STRIP TRANSMISSION LINE
%    WITH CROSS-SECTION W X H
%    THE STRIPS MAITAINED AT 1 VOLT AND -1 VOLT.
%
%    ONE STRIP IS LOCATED ON THE Y = H/2 PLANE WHILE THE OTHER
%    IS LOCATED ON THE Y = -H/2 PLANE.
%
%    ALL DIMENSIONS ARE IS S.I. UNITS
%
%    N IS THE NUMBER OF SUBSECTIONS INTO WHICH EACH STRIP IS DIVIDED
%    ****************************************************************

%    FIRST, SPECIFY THE PARAMETERS

NN = [3 7 11 18 39 59];

disp(' ')
disp('n        Zo')

for N = NN
    CL=3.0e8;  % SPEED OF LIGHT IN FREE SPACE
    ER=1.0;
    EO=8.8541878176E-12;
    H=2.0;
    W=5.0;

    NT=2*N;
    DELTA = W/(N);

% SECOND, CALCULATE THE MATCH POINTS
%    AND THE COEFFICIENT MATRIX [A]

    for K=1:N
        X(K) = DELTA*(K - .5);
        Y(K) = -H/2.0;
        X(K+N) = X(K);
        Y(K+N) = H/2.0;
    end

    FACTOR = DELTA/(2.0*pi*EO);

    for I=1:NT
        for J=1:NT
            if(I==J)        %eqn (5.98)
                A(I,J) = -(log(DELTA) - 1.5)*FACTOR;
            else
                R = sqrt( (X(I) - X(J))^2 + (Y(I) - Y(J))^2 );
                A(I,J) = -log(R)*FACTOR;
            end
        end
    end
```

FIGURE 5.9

MATLAB program for Example 5.8.   *(Continued)*

```
%    NOW DETERMINE THE MATRIX OF CONSTANT VECTOR [B]

     for K=1:N
         B(K)=1.0;
         %The line below was commented!
         B(K+N)= -1.0;        %rja -this line needed to be un-commented
     end

%    INVERT MATRIX A(I,J) AND CALCULATE MATRIX RO(N)
%    CONSISTING OF THE UNKNOWN ELEMENTS
%    ALSO CALCULATE THE TOTAL CHARGE Q,
%    THE CAPACITANCE C, AND THE CHARACTERISTIC IMPEDANCE Zo.

     NIV=NT;
     NMAX=100;
     A = inv(A);
     for I=1:NT
         RO(I)=0.0;
         %The line below had an error N instead of NT
         for M=1:NT
             RO(I)=RO(I) + A(I,M)*B(M);
         end
     end
     SUM=0.0;
     for I=1:N
         SUM = SUM + RO(I);
     end
     Q=SUM*DELTA;
     VD=2.0;
     C=abs(Q)/VD;
     ZO = sqrt(ER)/(CL*C);

 disp(num2str([N ZO ]))

 end
```

FIGURE 5.9 (Continued)
MATLAB program for Example 5.8..

TABLE 5.2

Characteristic Impedance of
a Strip Transmission Line

n	Z_o (in Ω)
3	94.382
7	96.755
11	97.366
18	97.755
39	98.163
59	98.3011

5.6 Applications II: Scattering Problems

The purpose of this section is to illustrate, with two examples, how the MoM can be applied to solve electromagnetic scattering problems. The first example is on scattering of a plane wave by a perfectly conducting cylinder [3], while the second is on scattering of a plane wave by an arbitrary array of parallel wires [34].

5.6.1 Scattering by Conducting Cylinder

Consider an infinitely long, perfectly conducting cylinder located at a far distance from a radiating source. Assuming a time-harmonic field with time factor $e^{j\omega t}$, Maxwell's equations can be written in phasor form as

$$\nabla \cdot \mathbf{E}_s = 0 \tag{5.101a}$$

$$\nabla \cdot \mathbf{H}_s = 0 \tag{5.101b}$$

$$\nabla \times \mathbf{E}_s = -j\omega\mu\mathbf{H}_s \tag{5.101c}$$

$$\nabla \times \mathbf{H}_s = \mathbf{J}_s + j\omega\epsilon\mathbf{E}_s \tag{5.101d}$$

where the subscript s denotes phasor or complex quantities. Henceforth, we will drop subscript s for simplicity and use the same symbols for the frequency-domain quantities and time-domain quantities. It is assumed that the reader can differentiate between the two quantities. Taking the curl of Equation 5.101c and applying Equation 5.101d, we obtain

$$\nabla \times \nabla \times \mathbf{E} = -j\omega\mu\nabla \times \mathbf{H} = -j\omega\mu(\mathbf{J} + j\omega\epsilon\mathbf{E}) \tag{5.102}$$

Introducing the vector identity

$$\nabla \times \nabla \times \mathbf{A} = \nabla(\nabla \cdot \mathbf{A}) - \nabla^2\mathbf{A}$$

into Equation 5.102 gives

$$\nabla(\nabla \cdot \mathbf{E}) - \nabla^2\mathbf{E} = -j\omega\mu(\mathbf{J} + j\omega\epsilon\mathbf{E})$$

In view of Equation 5.101a, $\nabla(\nabla \cdot \mathbf{E}) = 0$ so that

$$\nabla^2\mathbf{E} + k^2\mathbf{E} = j\omega\mu\mathbf{J} \tag{5.103}$$

where $k = \omega(\mu\epsilon)^{1/2} = 2\pi/\lambda$ is the wave number and λ is the wavelength. Equation 5.103 is the vector form of the Helmholtz wave equation. If we assume a TM wave ($H_z = 0$) with $\mathbf{E} = E_z(x, y)\mathbf{a}_z$, the vector equation (5.103) becomes a scalar equation, namely,

$$\nabla^2 E_z + k^2 E_z = j\omega\mu J_z \tag{5.104}$$

where $\mathbf{J} = J_z\mathbf{a}_z$ is the source current density. The integral solution to Equation 5.104 is

$$E_z(x, y) = E_z(\rho) = -\frac{k\eta_o}{4} \int_S J_z(\rho')H_0^{(2)}(k|\rho - \rho'|)dS' \tag{5.105}$$

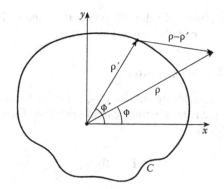

FIGURE 5.10
Cross section of the cylinder.

where $\rho = x\mathbf{a}_x + y\mathbf{a}_y$ is the field point, $\rho' = x'\mathbf{a}_x + y'\mathbf{a}_y$ is the source point, $\eta_o = (\mu_o/\epsilon_o)^{1/2} \simeq 377\ \Omega$ is the intrinsic impedance of free space, and $H_0^{(2)} =$ Hankel function of the second kind of zero order since an outward-traveling wave is assumed. The integration in Equation 5.105 is over the cross section of the cylinder shown in Figure 5.10.

If field E_z^i is incident on a perfectly conducting cylinder, it induces surface current J_z on the conducting cylinder, which in turn produces a scattered field E_z^s. The scattered field E_z^s due to J_z is expressed by Equation 5.105. On the boundary C, the tangential component of the total field must vanish. Thus,

$$E_z^i + E_z^s = 0 \text{ on } C \tag{5.106}$$

Substitution of Equation 5.105 into Equation 5.106 yields

$$E_z^i(\rho) = \frac{k\eta_o}{4} \int_C J_z(\rho')H_0^{(2)}(k|\rho - \rho'|)\,dl' \tag{5.107}$$

In IE (5.107), the induced surface current density J_z is the only unknown. We determine J_z using the moment method.

We divide the boundary C into N segments and apply the point matching technique. On a segment ΔC_n, Equation 5.107 becomes

$$E_z^i(\rho_n) = \frac{k\eta_o}{4} \sum_{m=1}^{N} J_z(\rho_m)H_0^{(2)}(k|\rho_n - \rho_m|)\,\Delta C_m \tag{5.108}$$

where the integration in Equation 5.107 has been replaced by summation. On applying Equation 5.108 to all segments, a system of simultaneous equations results. The system of equations can be cast in matrix form as

$$\begin{bmatrix} E_z^i(\rho_1) \\ E_z^i(\rho_2) \\ \vdots \\ E_z^i(\rho_N) \end{bmatrix} = \begin{bmatrix} A_{11} & A_{12} & \cdots & A_{1N} \\ A_{21} & A_{22} & \cdots & A_{2N} \\ \vdots & & \vdots & \\ A_{N1} & A_{N2} & \cdots & A_{NN} \end{bmatrix} \begin{bmatrix} J_z(\rho_1) \\ J_z(\rho_2) \\ \vdots \\ J_z(\rho_N) \end{bmatrix} \tag{5.109a}$$

or

$$[E] = [A][J] \tag{5.109b}$$

Hence,

$$[J] = [A]^{-1}[E] \tag{5.110}$$

To determine the exact values of elements of matrix $[A]$ may be difficult. Approximately [6],

$$A_{mn} \simeq \begin{cases} \dfrac{\eta_o k}{4} \Delta C_n H_0^{(2)} \{k(x_n - x_m)^2 + (y_n - y_m)^2]^{(1/2)}\}, & m \neq n \\[3mm] \dfrac{\eta_o k}{4} \left[1 - j\dfrac{2}{\pi} \log_{10}\left(\dfrac{\gamma k \Delta C_n}{4e}\right)\right], & m = n \end{cases} \tag{5.111}$$

where (x_n, y_n) is the midpoint of ΔC_n, $e = 2.718 \ldots$, and $\gamma = 1.781 \ldots$. Thus for a given cross section and specified incident field E_z^i the induced surface current density J_z can be found from Equation 5.110. To be specific, assume the propagation vector $\mathbf{k}$ is directed as shown in Figure 5.11 so that

$$E_z^i = E_o e^{j\mathbf{k}\cdot\mathbf{r}}$$

where $\mathbf{r} = x\mathbf{a}_x + y\mathbf{a}_y$, $\mathbf{k} = k(\cos\phi_i\mathbf{a}_x + \sin\phi_i\mathbf{a}_y)$, $k = 2\pi/\lambda$, and ϕ_i is the incidence angle. Taking $E_o = 1$ so that $|E_z^i| = 1$,

$$E_z^i = e^{jk(x\cos\phi_i + y\sin\phi_i)} \tag{5.112}$$

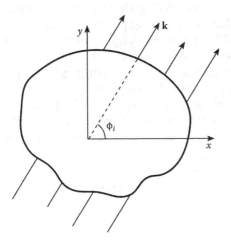

FIGURE 5.11
Typical propagation of vector *k*.

Given any C (dictated by the cross section of the cylinder), we can substitute Equations 5.111 and 5.112 into Equation 5.109 and determine $[J]$ from Equation 5.110. Once J_z, the induced current density, is known, we calculate the *scattering cross section* σ defined by

$$\sigma(\phi,\phi_i) = 2\pi\rho \left| \frac{E_z^s(\phi)}{E_z^s(\phi_i)} \right|^2$$

$$= \frac{k\eta_o^2}{4} \left| \int_C J_z(x',y')e^{jk(x'\cos\phi+y'\sin\phi)}dl' \right|^2 \tag{5.113}$$

where ϕ is the angle at the observation point, the point at which σ is evaluated. In matrix form,

$$\sigma(\phi_i,\phi) = \frac{k\eta^2}{4} \, |[V_n^s][Z_{nm}]^{-1}[V_m^i]|^2 \tag{5.114}$$

where

$$V_m^i = \Delta C_m e^{jk(x_m\cos\phi_i+y_m\cos\phi_i)}, \tag{5.115a}$$

$$V_n^s = \Delta C_n e^{jk(x_n\cos\phi+y_n\cos\phi)}, \tag{5.115b}$$

and

$$Z_{mn} = \Delta C_m \, A_{mn} \tag{5.115c}$$

5.6.2 Scattering by an Arbitrary Array of Parallel Wires

This problem is of more general nature than the one just described. As a matter of fact, any infinitely long, perfectly conducting, thin metal can be modeled as an array of parallel wires. It will be shown that the scattering pattern due to an arbitrary array of line sources approaches that of a solid conducting cylinder of the same cross-sectional geometry if a sufficiently large number of wires are present and they are arrayed on a closed curve. Hence, the problem of scattering by a conducting cylinder presented above can also be modeled with the techniques to be described here.

Consider an arbitrary array of N parallel, infinitely long wires placed parallel to the z-axis [34]. Three of such wires are illustrated in Figure 5.12. Let a harmonic TM wave be incident on the wires. Assuming a time factor $e^{j\omega t}$, the incident wave in phasor form is given by

$$E_z^i = E_i(x,y)e^{-jhz} \tag{5.116}$$

where

$$E_i(x,y) = E_o \, e^{-jk(x\sin\theta_i\cos\phi_i+y\sin\theta_i\sin\phi_i)} \tag{5.117a}$$

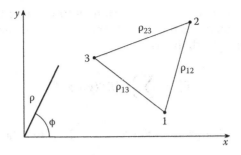

FIGURE 5.12
An array of three wires parallel to the z-axis.

$$h = k\cos\theta_i, \tag{5.117b}$$

$$k = \frac{2\pi}{\lambda} = \omega(\mu\epsilon)^{1/2}, \tag{5.117c}$$

and θ_i and ϕ_i define the axis of propagation as illustrated in Figure 5.13. The incident wave induces current on the surface of wire n. The induced current density has only z component.

It can be shown that the field due to a harmonic current I_n uniformly distributed on a circular cylinder of radius a_n has a z component given by

$$E_n = -I_n' H_0^{(2)}(g\rho_n)e^{-jhz}, \quad \rho_n > a_n \tag{5.118}$$

where

$$I_n' = \frac{\omega\mu g^2}{4k^2} I_n J_0(ga_n), \tag{5.119}$$

$$g^2 + h^2 = k^2, \tag{5.120}$$

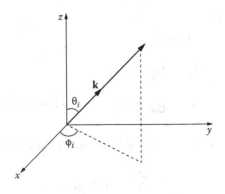

FIGURE 5.13
Propagation vector k.

J_0 is Bessel function of order zero, and H_0 is Hankel function of the second kind of order zero. By induction theorem, if I_n is regarded as the induced current, Equation 5.118 may be considered as the scattered field, that is,

$$E_z^s = -\sum_{n=1}^{N} I_n' H_0^{(2)}(g\rho_n)e^{-jhz} \tag{5.121}$$

where the summation is taken over all the N wires. On the surface of each wire (assumed perfectly conducting),

$$E_z^i + E_z^s = 0$$

or

$$E_z^i = -E_z^s, \quad \rho = \rho_n \tag{5.122}$$

Substitution of Equations 5.116 and 5.121 into Equation 5.122 leads to

$$\sum_{n=1}^{N} I_n' H_0^{(2)}(g\rho_{mn}) = E_i(x_m, y_m) \tag{5.123}$$

where

$$\rho_{mn} = \begin{cases} \sqrt{(x_m - x_n)^2 + (y_m - y_m)^2}, & m \neq n \\ a_m, & m = n \end{cases} \tag{5.124}$$

and a_m is the radius of the mth wire. In matrix form, Equation 5.123 can be written as

$$[A][I] = [B]$$

or

$$[I] = [A]^{-1}[B] \tag{5.125}$$

where

$$I_n = I_n', \tag{5.126a}$$

$$A_{mn} = H_0^{(2)}(g\rho_{mn}), \tag{5.126b}$$

$$B_m = E_o e^{-jk(x_m \sin\theta_i \cos\phi_i + y_m \sin\theta_i \sin\phi_i)} \tag{5.126c}$$

Once I_n' is calculated from Equation 5.125, the scattered field can be obtained as

$$E_z^s = -\sum_{n=1}^{N} I_n' H_0^{(2)}(g\rho_n)e^{-jhz} \tag{5.127}$$

Finally, we may calculate the "distant scattering pattern," defined as

$$E(\phi) = \sum_{n=1}^{N} I'_n e^{jg(x_n \cos\phi + y_n \sin\phi)} \tag{5.128}$$

The following example, taken from Richmond's work [34], will be used to illustrate the techniques discussed in the latter half of this section.

EXAMPLE 5.9

Consider the two arrays shown in Figure 5.14. For Figure 5.14a, take

Number of wires,	$N = 15$
Wire radius,	$ka = 0.05$
Wire spacing,	$ks = 1.0$

$\theta_o = 90°, \phi_o = 40°, 270° < \phi < 90°$

and for Figure 5.14b, take

Number of wires,	$N = 30$
Wire radius,	$ka = 0.05$
Cylinder radius,	$R = 1.12\lambda$

$\theta_o = 90°, \phi_o = 0, 0 < \phi < 180°$

For the two arrays, calculate and plot the scattering pattern as a function of ϕ.

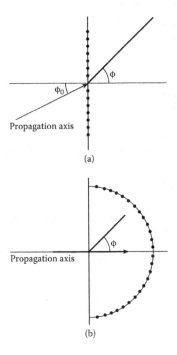

(a)

(b)

FIGURE 5.14
For Example 5.8: (a) A plane array of 15 parallel wires, (b) a semicircular array of 30 parallel wires.

Solution

The MATLAB code for calculating the scattering pattern $E(\phi)$ based on Equation 5.128 is shown in Figure 5.15. The same program can be used for the two arrays in Figure 5.14, except that the input data on N, ka, ks and the locations (x_n, y_n), $n = 1, 2, ..., N$ of the wires are different. The program essentially calculates I_n required in Equation 5.128 using Equations 5.125 and 5.126. The plots of $E(\phi)$ against ϕ are portrayed in Figures 5.16 and 5.17 for the arrays in Figure 5.14a and b, respectively.

```
%====================================================
% THIS PROGRAM CALCULATES THE SCATTERING PATTERN
% OF AN ARRAY OF PARALLEL WIRES
% ====================================================

clear; format compact;

thearray = 1; % thearray equals 1 for a plane array and 2 for a semicircu-
lar array

    %PHIO and THETAO define the axis of propagation
    THETAO = pi/2.0;    %elevation angle pi/2
    LAMBDA=1.0;         %wavelength
    EO = 1.0;           %E field
    R = 1.125*LAMBDA;   %Radius of semicircle array
    K = 2.0*pi/LAMBDA;  %Propagation constant
    AA = 0.05/K;        %wire radius
    S = 1.0/K;          %wire spacing
    H = K*cos(THETAO);  %H & G are part of the incident wave definition
    G = sqrt( K^2 - H^2 );

% DEFINE WIRE LOCATIONS
    if thearray ==1
        NN = 15;        %Number of wires
        PHIO = 40;
        X = zeros(1,NN);
        Y = linspace(-1,1,NN);
    elseif thearray ==2
        NN = 30;
        PHIO = 0.0;
        PHI2 = linspace(-pi/2,pi/2,NN);
        X = R*cos(PHI2);
        Y = R*sin(PHI2);
    end
% CALCULATE RHO
    [Mx,My]=meshgrid(X,Y);
    RHO = sqrt((Mx-Mx').^2+(My-My').^2)+diag(AA*ones(1,NN));

% CONSTRUCT MATRIX [A]
    A = besselh(0,2,G*RHO);

% CONSTRUCT MATRIX [B]
    ALPHA = X*sin(THETAO)*cos(PHIO)+ Y*sin(THETAO)*sin(PHIO);
    B = EO*exp( -i*K*ALPHA).';

% SOLVE FOR MATRIX[I] CONSISTING OF "MODIFIED CURRENT" OR CURRENT COEFFICIENTS
    A = inv(A);
    I = A*B;

% FINALLY, CALCULATE THE SCATTERING PATTERN E(PHI)
    PHI = linspace(0,pi,128);
    ALP = cos(PHI.')*X + sin(PHI.')*Y;
    E = abs(exp( i*G*ALP)*I);
```

FIGURE 5.15
Computer program for Example 5.9.

(Continued)

```
% Plot E(PHI)
if thearray ==2
    figure(1),plot(PHI*180/pi,E)
        title('Scattering Pattern')
        xlabel('\phi (Degrees)')
        ylabel('E(\phi)')
        grid on
elseif thearray ==1

%       ang = PHI*180/pi-90;
%       field = fftshift(E);

    figure(2),plot(PHI*180/pi,E)
        title('Example 5.8 compare to Figure 5.16')
        xlabel('\phi (Degrees)')
        ylabel('E(\phi)')
        grid on

end
```

FIGURE 5.15 (Continued)
Computer program for Example 5.9.

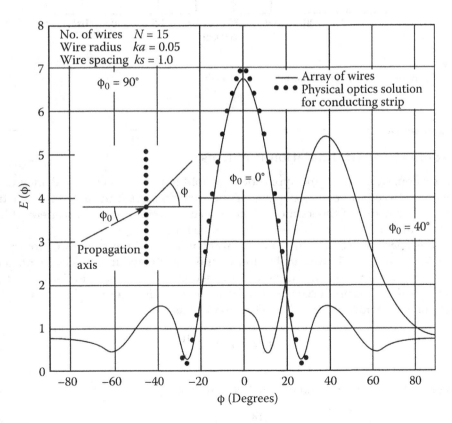

FIGURE 5.16
Scattering pattern for the plane array of Figure 5.14a.

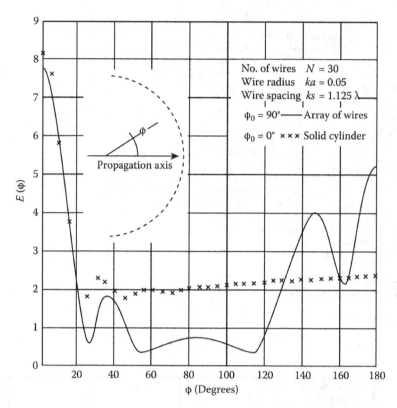

FIGURE 5.17
Scattering pattern for the semicircular array of Figure 5.14b.

5.7 Applications III: Radiation Problems

In this section, we consider the application of MoM to wires or cylindrical antennas. The distinction between scatterers considered in the previous section and antennas to be treated here is primarily that of the location of the source. An object acts as a scatterer if it is far from the source; it acts as an antenna if the source is on it [3].

Consider a perfectly conducting cylindrical antenna of radius a, extending from $z = -\ell/2$ to $z = \ell/2$ as shown in Figure 5.18. Let the antenna be situated in a lossless homogeneous dielectric medium ($\sigma = 0$). Assuming a z-directed current on the cylinder ($\mathbf{J} = J_z a_z$), only axial electric field E_z is produced due to axial symmetry. The electric field can be expressed in terms of the retarded potentials of Equation 1.38 as

$$E_z = -j\omega A_z - \frac{\partial V}{\partial z} \tag{5.129}$$

Applying the Lorentz condition of Equation 1.41, namely,

$$\frac{\partial A_z}{\partial z} = -j\omega\mu\epsilon V, \tag{5.130}$$

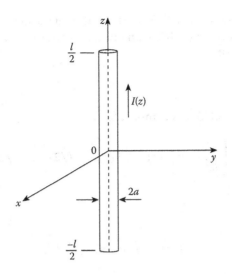

FIGURE 5.18
Cylindrical antenna of length *l* and radius *a*.

Equation 5.129 becomes

$$E_z = -j\omega\left(1 + \frac{1}{k^2}\frac{\partial^2}{\partial z^2}\right)A_z \tag{5.131}$$

where $k = \omega(\mu\epsilon)^{1/2} = 2\pi/\lambda$, ω is the angular frequency of the suppressed harmonic time variation $e^{j\omega t}$. From Equation 1.44

$$A_z = \mu\int_{-\ell/2}^{\ell/2} I(z')G(x,y,z;x',y',z')dz' \tag{5.132}$$

where $G(x, y, z; x', y', z')$ is the free space Greens' function, that is,

$$G(x,y,z;x',y',z') = \frac{e^{-jkR}}{4\pi R} \tag{5.133}$$

and R is the distance between observation point (x, y, z) and source point (x', y', z') or

$$R = [(x-x')^2 + (y-y')^2 + (z-z')^2]^{1/2} \tag{5.134}$$

Combining Equations 5.131 and 5.132 gives

$$E_z = -j\omega\mu\left(1 + \frac{1}{k^2}\frac{d^2}{dz^2}\right)\int_{-\ell/2}^{\ell/2} I(z')G(x,y,z;x',y',z')dz' \tag{5.135}$$

This integro-differential equation is not convenient for numerical analysis because it requires evaluation of the second derivative with respect to *z* of the integral. We will now consider two types of modification of Equation 5.135 leading to Hallen's (magnetic vector potential) and Pocklington's (electric field) IEs. Either of these IEs can be used to determine the current distribution on a cylindrical antenna or scatterer and subsequently calculate all other quantities of interest.

5.7.1 Hallen's IE

We can rewrite Equation 5.135 in a compact form as

$$\left(\frac{d^2}{dz^2} + k^2\right) F(z) = k^2 S(z), \quad -\ell/2 < z < \ell/2 \tag{5.136}$$

where

$$F(z) = \int_{-\ell/2}^{\ell/2} I(z') G(z, z') dz', \tag{5.137a}$$

$$S(z) = -\frac{E_z}{j\omega\mu} \tag{5.137b}$$

Equation 5.136 is a second-order linear ordinary differential equation. The general solution to the homogeneous equation

$$\left(\frac{d^2}{dz^2} + k^2\right) F(z) = 0,$$

which is consistent with the boundary condition that the current must be zero at the wire ends ($z = \pm\ell/2$), is

$$F_h(z) = c_1 \cos kz + c_2 \sin kz \tag{5.138}$$

where c_1 and c_2 are integration constants. The particular solution of Equation 5.136 can be obtained, for example, by the Lagrange method of varying constants [35] as

$$F_p(z) = k \int_{-\ell/2}^{\ell/2} S(z') \sin k |z - z'| dz' \tag{5.139}$$

Thus from Equations 5.137 through 5.139, the solution to Equation 5.136 is

$$\int_{-\ell/2}^{\ell/2} I(z') G(z, z') dz' = c_1 \cos kz + c_2 \sin kz$$

$$\tag{5.140}$$

$$-\frac{j}{\eta} \int_{-\ell/2}^{\ell/2} E_z(z') \sin k |z - z'| dz'$$

where $\eta = \sqrt{\mu/\epsilon}$ is the intrinsic impedance of the surrounding medium. Equation 5.140 is referred to as *Hallen's integral equation* [36] for a perfectly conducting cylindrical antenna or scatterer. The equation has been generalized by Mei [37] to perfectly conducting wires of arbitrary shape. Hallen's IE is computationally convenient since its kernel contains only ℓ/r terms. Its major advantage is the ease with which a converged solution may be obtained, while its major drawback lies in the additional work required in finding the integration constants c_1 and c_2 [35,38].

5.7.2 Pocklington's IE

We can also rewrite Equation 5.135 by introducing the operator in parentheses under the integral sign so that

$$\int_{-\ell/2}^{\ell/2} I(z')\left(\frac{\partial^2}{\partial z^2}+k^2\right)G(z,z')\,dz' = j\omega\epsilon\, E_z \tag{5.141}$$

This is known as *Pocklington's integral equation* [39]. Note that Pocklington's IE has E_z, which represents the field from the source on the right-hand side. Both Pocklington's and Hallen's IEs can be used to treat wire antennas. The third type of IE derivable from Equation 5.135 is the Schelkunoff's IE, found in Reference 35.

5.7.3 Expansion and Weighting Functions

Having derived suitable IEs, we can now find solutions for a variety of wire antennas or scatterers. This usually entails reducing the IEs to a set of simultaneous linear equations using the MoM. The unknown current $I(z)$ along the wire is approximated by a finite set $u_n(z)$ of basis (or expansion) functions with unknown amplitudes as discussed in the last chapter. That is, we let

$$I(z) = \sum_{n=1}^{N} I_n u_n(z), \tag{5.142}$$

where N is the number of basis functions needed to cover the wire and the expansion coefficients I_n are to be determined. The functions u_n are chosen to be linearly independent. The basis functions commonly used in solving antenna or scattering problems are of two types: entire domain functions and subdomain functions. The entire domain basis functions exist over the full domain $-\ell/2 < z < \ell/2$. Typical examples are [8,40]

1. Fourier:

$$u_n(z) = \cos(n-1)v/2, \tag{5.143a}$$

2. Chebychev:

$$u_n(z) = T_{2n-2}(v), \tag{5.143b}$$

3. Maclaurin:

$$u_n(z) = v^{2n-2}, \tag{5.143c}$$

4. Legendre:

$$u_n(z) = P_{2n-2}(v),$$ (5.143d)

5. Hermite:

$$u_n(z) = H_{2n-2}(v),$$ (5.143e)

where $v = 2z/\ell$ and $n = 1, 2, ..., N$. The subdomain basis functions exist only on one of the N nonoverlapping segments into which the domain is divided. Typical examples are [41,42]

1. Piecewise constant (pulse) function:

$$u_n(z) = \begin{cases} 1, & z_{n-1/2} < z < z_{n+1/2} \\ 0, & \text{otherwise}, \end{cases}$$ (5.144a)

2. Piecewise linear (triangular) function:

$$u_n(z) = \begin{cases} \dfrac{\Delta - |z - z_n|}{\Delta}, & z_{n-1} < z < z_{n+1} \\ 0, & \text{otherwise}, \end{cases}$$ (5.144b)

3. Piecewise sinusoidal function:

$$u_n(z) = \begin{cases} \dfrac{\sin k(z - |z - z_n|)}{\sin k \Delta}, & z_{n-1} < z < z_{n+1} \\ 0, & \text{otherwise}, \end{cases}$$ (5.144c)

where $\Delta = \ell/N$, assuming equal subintervals although this is unnecessary. Figure 5.19 illustrates these subdomain functions. The entire domain basis functions are of limited applications since they require a prior knowledge of the nature of the function to be represented. The subdomain functions are the most commonly used, particularly in developing general-purpose user-oriented computer codes for treating wire problems. For this reason, we will focus on using subdomain functions as basis functions.

Substitution of the approximate representation of current $I(z)$ in Equation 5.142 into Pocklington's IE of Equation 5.141 gives

$$\int_{-\ell/2}^{\ell/2} \sum_{n=1}^{N} I_n u_n(z') K(z_m, z') dz' \simeq E_z(z_m)$$ (5.145)

where

$$K(z_m, z') = \frac{1}{j\omega\epsilon} \left(\frac{\partial^2}{\partial z^2} + k^2 \right) G(z_m, z')$$

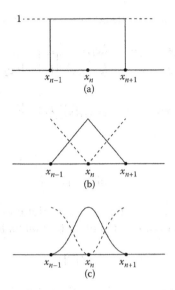

FIGURE 5.19
Typical subdomain weighting functions: (a) piecewise uniform function, (b) piecewise linear function, (c) piecewise sinusoidal function.

is the kernel, $z = z_m$ on segment m is the point on the wire at which the IE is being enforced. Equation 5.145 may be written as

$$\sum_{n=1}^{N} I_n \int_{\Delta z_n} K(z_m, z') u_n(z') dz' \simeq E_z(z_m)$$

or

$$\sum_{n=1}^{N} I_n g_m = E_z(z_m) \tag{5.146}$$

where

$$g_m = \int_{\Delta z_n'} K(z_m, z') u_n(z') dz' \tag{5.147}$$

In order to solve for the unknown current amplitudes I_n ($n = 1, 2, ..., N$), N equations need to be derived from Equation 5.146. We achieve this by multiplying Equation 5.146 by weighting (or testing) functions w_n ($n = 1, 2, ..., n$) and integrating over the wire length. In other words, we let Equation 5.146 be satisfied in an average sense over the entire domain. This leads to forming the inner product between each of the weighting functions and g_m so that Equation 5.146 is reduced to

$$\sum_{n=1}^{N} I_n \langle w_n, g_m \rangle = \langle w_n, E_z \rangle, \quad m = 1, 2, ..., N \tag{5.148}$$

Thus, we have a set of N simultaneous equations which can be written in matrix form as

$$\begin{bmatrix} \langle w_1, g_1 \rangle & \cdots & \langle w_1, g_N \rangle \\ \langle w_2, g_1 \rangle & \cdots & \langle w_2, g_N \rangle \\ \vdots & & \vdots \\ \langle w_N, g_1 \rangle & \cdots & \langle w_N, g_N \rangle \end{bmatrix} \begin{bmatrix} I_1 \\ I_2 \\ \vdots \\ I_N \end{bmatrix} = \begin{bmatrix} \langle w_1, E_{z1} \rangle \\ \langle w_2, E_{z2} \rangle \\ \vdots \\ \langle w_N, E_{zN} \rangle \end{bmatrix}$$

or

$$[Z][I] = [V] \tag{5.149}$$

where $z_{mn} = \langle w_n, g_m \rangle$ and $V_m = \langle w_m, E_z \rangle$. The desired solution for the current is then obtained by solving the simultaneous equations (5.149) or by matrix inversion, that is,

$$[I] = [Z]^{-1}[V] \tag{5.150}$$

Because of the similarity of Equation 5.149 to the network equations, the matrices $[Z]$, $[V]$, and $[I]$ are referred to as *generalized* impedance, voltage, and current matrices, respectively [6]. Once the current distribution $I(z')$ is determined from Equation 5.149 or Equation 5.150, parameters of practical interest such as input impedance and radiation patterns are readily obtained.

The weighting functions $\{w_n\}$ must be chosen so that each Equation 5.148 is linearly independent and computation of the necessary numerical integration is minimized. Evaluation of the integrals in Equation 5.149 is often the most time-consuming portion of scattering or radiation problems. Sometimes we select similar types of functions for both weighting and expansion. As discussed in the previous chapter, choosing $w_n = u_n$ leads to Galerkin's method, while choosing $w_n = \delta(z - z_n)$ results in point matching (or colocation) method. The point matching method is simpler than Galerkin's method and is sufficiently adequate for many EM problems. However, it tends to be a slower converging method. The general rules that should be followed in selecting the weighting functions are addressed in Reference 43. The following examples are taken from References 41,44–46.

EXAMPLE 5.10

Solve the Hallen's IE

$$\int_{-\ell/2}^{\ell/2} I(z')G(z,z')dz' = -\frac{j}{\eta_o}(A\cos kz + B\sin k|z|)$$

where $k = 2n/\lambda$ is the phase constant and $\eta_o = 377\ \Omega$ is the intrinsic impedance of free space. Consider a straight wire dipole with length $L = 0.5\ \lambda$ and radius $a = 0.005\lambda$.

Solution

The IE has the form

$$\int_{-\ell/2}^{\ell/2} I(z')K(z,z')dz' = D(z) \tag{5.151}$$

which is a Fredholm IE of the first kind. In Equation 5.151,

$$K(z,z') = G(z,z') = \frac{e^{-jkR}}{4\pi R},$$ (5.152a)

$$R = \sqrt{a^2 + (z - z')^2},$$ (5.152b)

and

$$D(z) = -\frac{j}{\eta_o}[A\cos(kz) + B\sin(k|z|)]$$ (5.152c)

If the terminal voltage of the wire antenna is V_T, the constant $B = V_T/2$. The absolute value in $\sin k|z|$ expresses the assumption of antenna symmetry, that is, $I(-z') = I(z')$. Thus,

$$\int_{-\ell/2}^{\ell/2} I(z') \frac{e^{-jkR}}{4\pi R} dz' = -\frac{j}{\eta_o}\left[A\cos kz + \frac{V_T}{2}\sin k|z|\right]$$ (5.153)

If we let

$$I(z) = \sum_{n=1}^{N} I_n u_n(z),$$ (5.154)

Equation 5.153 will contain N unknown variables I_n and the unknown constant A. To determine the $N + 1$ unknowns, we divide the wire into N segments. For the sake of simplicity, we choose segments of equal lengths $\Delta z = \ell/N$ and select $N + 1$ matching points such as

$$z = -\ell/2, -\ell/2 + \Delta z, \ldots, 0, \ldots, \ell/2 - \Delta z, \ell/2$$

At each match point $z = z_m$,

$$\int_{-\ell/2}^{\ell/2} \sum_{n=1}^{N} I_n u_n(z') K(z_m, z') dz' = D(z_m)$$ (5.155)

Taking the inner products (moments) by multiplying either side with a weighting function $w_m(z)$ and integrating both sides,

$$\int_{-\ell/2}^{\ell/2} \int_{-\ell/2}^{\ell/2} \sum_{n=1}^{N} I_n u_n(z') K(z_m, z') dz' w_m(z) dz$$

$$= \int_{-\ell/2}^{\ell/2} D(z_m) w_m(z) dz$$ (5.156)

By reversing the order of the summation and integration,

$$\sum_{n=1}^{N} I_n \int_{-\ell/2}^{\ell/2} u_n(z') \int_{-\ell/2}^{\ell/2} K(z_m, z') w_m(z) \, dz \, dz'$$

$$= \int_{-\ell/2}^{\ell/2} D(z_m) w_m(z) \, dz \tag{5.157}$$

The integration on either side of Equation 5.157 can be carried out numerically or analytically if possible. If we use the point matching method by selecting the weighting function as delta function, then

$$w_m(z) = \delta(z - z_m)$$

Since the integral of any function multiplied by $\delta(z - z_m)$ gives the value of the function at $z = z_m$, Equation 5.157 becomes

$$\sum_{n=1}^{N} I_n \int_{-\ell/2}^{\ell/2} u_n(z') K(z_m, z') \, dz' = D(z_m), \tag{5.158}$$

where $m = 1, 2, \ldots, N + 1$. Also, if we choose pulse function as the basis or expansion function,

$$u_n(z) = \begin{cases} 1, & z_n - \Delta z/2 < z < z_n + \Delta z/2 \\ 0, & \text{elsewhere,} \end{cases}$$

and Equation 5.158 yields

$$\sum_{n=1}^{N} I_n \int_{z_n - \Delta z/2}^{z_n + \Delta z/2} K(z_m, z') \, dz' = D(z_m) \tag{5.159}$$

Substitution of Equation 5.152 into Equation 5.159 gives

$$\sum_{n=1}^{N} I_n \int_{z_n - \Delta z/2}^{z_n + \Delta z/2} \frac{e^{jkR_m}}{4\pi R_m} \, dz' = -\frac{j}{\eta_o} \left[A \cos k z_m + \frac{V_T}{2} \sin k |z_m| \right] \tag{5.160}$$

where $m = 1, 2, \ldots, N + 1$ and $R_m = [a^2 + (z_m - z')^2]^{1/2}$. Thus, we have a set of $N + 1$ simultaneous equations, which can be cast in matrix form as

$$\begin{bmatrix} F_{11} & F_{12} & \cdots & F_{1,N} & \dfrac{j}{\eta} \cos(kz_1) \\[2ex] F_{21} & F_{22} & \cdots & F_{2,N} & \dfrac{j}{\eta} \cos(kz_2) \\[2ex] \vdots & & & & \vdots \\[2ex] F_{N+1,1} & F_{N+1,2} & \cdots & F_{N+1,N} & \dfrac{j}{\eta} \cos(kz_{N+1}) \end{bmatrix} \begin{bmatrix} I_1 \\ I_2 \\ \vdots \\ A \end{bmatrix}$$

$$= \begin{bmatrix} -\dfrac{j}{2\eta} V_T \sin k \, |z_1| \\[2mm] -\dfrac{j}{2\eta} V_T \sin k \, |z_2| \\[1mm] \vdots \\[1mm] -\dfrac{j}{2\eta} V_T \sin k \, |z_{N+1}| \end{bmatrix} \qquad (5.161a)$$

or

$$[F][X] = [Q] \qquad (5.161b)$$

where

$$F_{mn} = \int\limits_{z_n - \Delta z/2}^{z_n + \Delta z/2} \frac{e^{-jkR_m}}{4\pi R_m} dz' \qquad (5.162)$$

The $N + 1$ unknowns are determined by solving Equation 5.161 in the usual manner. To evaluate F_{mn} analytically rather than numerically, let the integrand in Equation 5.162 be separated into its real (RE) and imaginary (IM) parts,

$$\frac{e^{-jk R_m}}{R_m} = \text{RE} + j\text{IM}$$

$$= \frac{\cos kR_m}{R_m} - j \frac{\sin kR_m}{R_m} \qquad (5.163)$$

IM as a function of z' is a smooth curve so that

$$\int\limits_{z_n - \Delta z/2}^{z_n + \Delta z/2} \text{IM}(z') dz' = -\int\limits_{z_n - \Delta z/2}^{z_n + \Delta z/2} \frac{\sin k[a^2 + (z_m - z')^2]^{1/2}}{[a^2 + (z_m - z')^2]^{1/2}} dz'$$

$$\simeq -\frac{\Delta z \sin k[a^2 + (z_m - z_n)^2]^{1/2}}{[a^2 + (z_m - z_n)^2]^{1/2}} \qquad (5.164)$$

The approximation is accurate as long as $\Delta z < 0.05\lambda$. On the other hand, RE changes rapidly as $z' \to z_m$ due to R_m. Hence,

$$\int\limits_{z_n - \Delta z/2}^{z_n + \Delta z/2} \text{RE}(z') dz' = -\int\limits_{z_n - \Delta z/2}^{z_n + \Delta z/2} \frac{\cos k[a^2 + (z_m - z')^2]^{1/2}}{[a^2 + (z_m - z')^2]^{1/2}} dz'$$

$$\simeq \cos k[a^2 + (z_m - z_n)^2]^{1/2} \int\limits_{z_n - \Delta z/2}^{z_n + \Delta z/2} \frac{dz'}{[a^2 + (z_m - z')^2]^{1/2}}$$

$$= \cos k[a^2 + (z_m - z_n)^2]^{1/2}$$

$$\ln \left[\frac{z_m + \Delta z/2 - z_n + [a^2 + (z_m - z_n + \Delta z/2)^2]^{1/2}}{z_m - \Delta z/2 - z_n + [a^2 + (z_m - z_n - \Delta z/2)^2]^{1/2}} \right] \qquad (5.165)$$

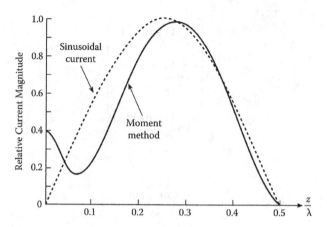

FIGURE 5.20
Current distribution of straight center-fed dipole.

Thus,

$$
\begin{aligned}
F_{mn} &\simeq \frac{1}{4\pi}\cos k[a^2 + (z_m-z_n)^2]^{1/2} \\
&\times \ln\left[\frac{z_m + \Delta z/2 - z_n + [a^2 + (z_m - z_n + \Delta z/2)^2]^{1/2}}{z_m - \Delta z/2 - z_n + [a^2 + (z_m - z_n - \Delta z/2)^2]^{1/2}}\right] \\
&- \frac{j\Delta z \sin k[a^2 + (z_m - z_n)^2]^{1/2}}{4\pi [a^2 + (z_m - z_n)^2]^{1/2}}
\end{aligned}
\tag{5.166}
$$

A typical example of the current distribution obtained for $\ell = \lambda$, $a = 0.01\lambda$ is shown in Figure 5.20, where the sinusoidal distribution commonly assumed for wire antennas is also shown for comparison. Notice the remarkable difference between the two near the dipole center.

EXAMPLE 5.11

Consider a perfectly conducting scatterer or antenna of a cylindrical nature shown in Figure 5.21. Determine the axial current $I(z)$ on the structure by solving the electric field integral equation (EFIE)

$$
\frac{j\eta}{4k\pi}\left(\frac{d^2}{dz^2} + k^2\right)\int_{-h}^{h} I(z')G(z,z')dz' = E_z^i(z)
\tag{5.167}
$$

where

$$
G(z,z') = \frac{1}{2\pi}\int_0^{2\pi} \frac{e^{-jkR}}{R}d\phi',
$$

$$
R = \left[(z-z')^2 + 4a^2\sin^2\frac{\phi'}{2}\right]^{1/2},
$$

$$
\eta = \sqrt{\frac{\mu}{\epsilon}}, \quad \text{and} \quad k = \frac{2\pi}{\lambda}
$$

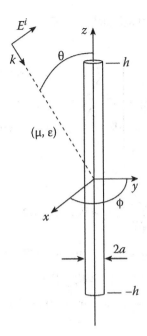

FIGURE 5.21
Cylindrical scatterer or antenna.

Solution

If the radius $a \ll \lambda$ (the wavelength) and $a \ll 2h$ (the length of the wire), the structure can be regarded as a "thin-wire" antenna or scatterer. As a scatterer, we may consider a plane wave excitation

$$E_z^i(z) = E_o \sin\theta e^{jkz\cos\theta} \tag{5.168a}$$

where θ is the angle of incidence. As an antenna, we may assume a delta-gap generator

$$E_z^i = V\delta(z - z_g) \tag{5.168b}$$

where V is the generator voltage and $z = z_g$ is the location of the generator.

In order to apply the MoM to the given IE (5.167), we expand the currents in terms of pulse basis function as

$$I(z) = \sum_{n=1}^{N} I_n u_n(z) \tag{5.169}$$

where

$$u_n(z) = \begin{cases} 1, & z_{n-1/2} < z < z_{n+1/2} \\ 0, & \text{elsewhere} \end{cases}$$

Substituting Equation 5.169 into Equation 5.167 and weighting the result with triangular functions

$$
w_m(z) = \begin{cases}
\dfrac{z - z_{m-1}}{\Delta}, & z_{m-1} < z < z_m \\[2mm]
-\dfrac{z - z_{m+1}}{\Delta}, & z_{m-1} < z < z_{m+1} \\[2mm]
0, & \text{elsewhere,}
\end{cases}
\tag{5.170}
$$

where $\Delta = 2\, h/N$, leads to

$$
\sum_{n=1}^{N} Z_{mn} I_n = V_m, \quad m = 1, 2, \ldots, N
\tag{5.171}
$$

Figure 5.22 illustrates $u_n(z)$ and $w_m(z)$. Equation 5.171 can be cast in matrix form as

$$
[Z][I] = [V]
\tag{5.172}
$$

where $[I]$ can be solved using any standard method. For the impedance matrix $[Z]$, the elements are given by

$$
Z_{mn} = \frac{j\eta}{4\pi k} \frac{2}{\Delta} \left[\frac{1}{2} G_{m-1,n} - \left(1 - \frac{k^2 \Delta^2}{2} \right) G_{m,n} + \frac{1}{2} G_{m+1,n} \right]
\tag{5.173}
$$

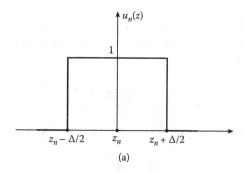

(a)

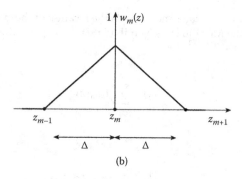

(b)

FIGURE 5.22

For Example 5.10: (a) Pulse basis function, (b) triangular weighting function.

where

$$G_{m,n} = \int\limits_{z_n-\Delta/2}^{z_n+\Delta/2} G(z_m, z') dz' \tag{5.174}$$

To obtain Equation 5.173, we have used the approximation

$$\int\limits_{z_{m-1}}^{z_{m+1}} w_m(z) f(z) dz = \Delta f(z_m)$$

For the plane wave excitation, the elements of the forcing vector $[V]$ are

$$V_m = \Delta E_0 e^{jkz_m \cos\theta} \tag{5.175a}$$

For delta-gap generator,

$$V_m = V\delta_{mg} \tag{5.175b}$$

where g is the index of the feed zone pulse.

Solving Equation 5.172 requires that we incorporate a method to perform numerically the integration in Equation 5.174. The kernel $G(z, z')$ exhibits a logarithmic singularity as $|z - z'| \to 0$, and therefore care must be exercised. To circumvent the difficulty, we let

$$G(z, z') = \frac{1}{2\pi} \int\limits_0^{2\pi} \frac{e^{-jkR}}{R} d\phi' = G_o(z, z') + G_1(z, z') \tag{5.176}$$

where

$$G_o(z, z') = \frac{1}{2\pi} \int\limits_0^{2\pi} \frac{d\phi'}{R} \tag{5.177}$$

and

$$G_1(z, z') = \frac{1}{2\pi} \int\limits_0^{2\pi} \frac{e^{-jkR} - 1}{R} d\phi' \tag{5.178}$$

We note that

$$G_o(z, z') \xrightarrow{\frac{(z-z')}{2a} \to 0} -\frac{1}{\pi a} \ln\frac{|z - z'|}{8a}$$

and hence we replace $G_o(z, z')$ by

$$\left[G_o(z, z') + \frac{1}{\pi a} \ln\frac{|z - z'|}{8a} \right] - \frac{1}{\pi a} \ln\frac{|z - z'|}{8a} \tag{5.179}$$

The term $G_1(z, z')$ is nonsingular, while the singularity of $G_o(z, z')$ can be avoided by using Equation 5.179. Thus, the double integral involved in evaluating Z_{mn} is easily done numerically. It is interesting to note that Z_{mn} would remain the same if we had chosen the triangular basis function and pulse weighting function [46].

5.8 Applications IV: EM Absorption in the Human Body

The interest in hyperthermia (or electromagnetic heating of deep-seated tumors) and in the assessment of possible health hazards due to EM radiation have prompted the development of analytical and numerical techniques for evaluating EM power deposition in the interior of the human body or a biological system [47]. The overall need is to provide a scientific basis for the establishment of an EM radiation safety standard. Since human experimentation is not possible, irradiation experiments must be performed on animals. Theoretical models are required to interpret and confirm the experiment, develop an extrapolation process, and thereby develop a radiation safety standard for humans [48].

The mathematical complexity of the problem has led researchers to investigate simple models of tissue structures such as plane slab, dielectric cylinder homogeneous and layered spheres, and prolate spheroid. A review of these earlier efforts is given in References 49,50. Although spherical models are still being used to study the power deposition characteristics of the head of humans and animals, realistic block model composed of cubical cells is being used to simulate the whole body.

The key issue in this bioelectromagnetic effort is how much EM energy is absorbed by a biological body and where is it deposited. This is usually quantified in terms of the specific absorption rate (SAR), which is the mass normalized rate of energy absorbed by the body. At a specific location, SAR may be defined by

$$\text{SAR} = \frac{\sigma}{\rho} |E|^2 \tag{5.180}$$

where σ = tissue conductivity, ρ = tissue mass density, E = rms value of the internal field strength. Thus, the localized SAR is directly related to the internal electric field and the major effort involves the determination of the electric field distribution within the biological body. The MoM has been extensively utilized to calculate localized SARs in block model representation of humans and animals.

As mentioned in Section 5.1, an application of MoM to EM problems usually involves four steps:

- Deriving the appropriate IE,
- Transforming the IE into a matrix equation (discretization),
- Evaluating the matrix elements, and
- Solving the resulting set of simultaneous equations.

We will apply these steps for calculating the electric field induced in an arbitrary human body or a biological system illuminated by an incident EM wave.

5.8.1 Derivation of IEs

In general, the induced electric field inside a biological body was found to be quite complicated even for the simple case of assuming the plane wave as the incident field. The complexity is due to the irregularity of the body geometry, and the fact that the body is finitely conducting. To handle the complexity, the so-called *tensor integral-equation* (TIE) was developed by Livesay and Chen [51]. Only the essential steps will be provided here; the interested reader is referred to References 51–53.

Consider a biological body of an arbitrary shape, with constitutive parameters ϵ, μ, σ illuminated by an incident (or impressed) plane EM wave as shown in Figure 5.23. The induced current in the body gives rise to a scattered field $\mathbf{E}^s$, which may be accounted for by replacing the body with an equivalent free-space current density $\mathbf{J}_{eq}$ given by

$$\mathbf{J}_{eq}(\mathbf{r}) = (\sigma(\mathbf{r}) + j\omega[\epsilon(\mathbf{r}) - \epsilon_o])\mathbf{E}(\mathbf{r}) = \tau(\mathbf{r})\mathbf{E}(\mathbf{r}) \tag{5.181}$$

where a time factor $e^{j\omega t}$ is assumed. The first term in Equation 5.181 is the conduction current density, while the second term is the polarization current density. With the equivalent current density $\mathbf{J}_{eq}$, we can obtain the scattered fields $\mathbf{E}^s$ and $\mathbf{H}^s$ by solving Maxwell's equations

$$\nabla \times \mathbf{E}^s = -\mathbf{J}_{eq} - j\omega\mathbf{H}^s \tag{5.182a}$$

$$\nabla \times \mathbf{H}^s = j\omega\mathbf{E}^s \tag{5.182b}$$

where $\mathbf{E}^s$, $\mathbf{H}^s$, and $\mathbf{J}_{eq}$ are all in phasor (complex) form. Elimination of $\mathbf{E}^s$ or $\mathbf{H}^s$ in Equation 5.182 leads to

$$\nabla \times \nabla \times \mathbf{E}^s - k_o^2 \mathbf{E}^s = -j\omega\mu_o \mathbf{J}_{eq} \tag{5.183a}$$

$$\nabla \times \nabla \times \mathbf{H}^s - k_o^2 \mathbf{H}^s = \nabla \times \mathbf{J}_{eq} \tag{5.183b}$$

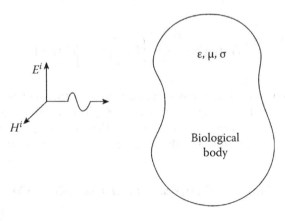

FIGURE 5.23
A biological body illuminated by a plane EM wave.

where $k_o^2 = \omega^2 \mu_o \epsilon_o$. The solutions to Equations 5.183a and 5.183b are

$$\mathbf{E}^s = -j\omega \left[1 + \frac{1}{k_o^2} \nabla \nabla \cdot \right] \mathbf{A} \qquad (5.184a)$$

$$\mathbf{H}^s = \frac{1}{\mu_o} \nabla \times \mathbf{A} \qquad (5.184b)$$

where

$$\mathbf{A} = \mu_o \int_v G_o(\mathbf{r}, \mathbf{r}') \mathbf{J}_{eq}(\mathbf{r}') dv' \qquad (5.185)$$

and

$$G_o(\mathbf{r}, \mathbf{r}') = \frac{e^{-jk_o(\mathbf{r} - \mathbf{r}')}}{4\pi |\mathbf{r} - \mathbf{r}'|} \qquad (5.186)$$

is the free-space scalar Green's function. By the operator $\nabla \nabla \cdot$, we mean that $\nabla \nabla \cdot \mathbf{A} = \nabla(\nabla \cdot \mathbf{A})$. It is evident from Equations 5.184 through 5.186 that $\mathbf{E}^s$ and $\mathbf{H}^s$ depend on $\mathbf{J}_{eq}$. Suppose $\mathbf{J}_{eq}$ is an infinitesimal, elementary source at $\mathbf{r}'$ pointed in the x direction so that

$$\mathbf{J}_{eq} = \delta(\mathbf{r} - \mathbf{r}')\mathbf{a}_x, \qquad (5.187)$$

the corresponding vector potential is obtained from Equation 5.185 as

$$\mathbf{A} = \mu_o G_o(\mathbf{r}, \mathbf{r}')\mathbf{a}_x \qquad (5.188)$$

If $\mathbf{G}_{ox}(\mathbf{r}, \mathbf{r}')$ is the electric field produced by the elementary source, then $\mathbf{G}_{ox}(\mathbf{r}, \mathbf{r}')$ must satisfy

$$\nabla \times \nabla \times \mathbf{G}_{ox}(\mathbf{r}, \mathbf{r}') - k_o^2 \mathbf{G}_{ox}(\mathbf{r}, \mathbf{r}') = -j\omega\mu_o \delta(\mathbf{r}, \mathbf{r}') \qquad (5.189)$$

with solution

$$\mathbf{G}_{ox}(\mathbf{r}, \mathbf{r}') = -j\omega\mu_o \left[1 + \frac{1}{k^2} \nabla \nabla \cdot \right] G_o(\mathbf{r}, \mathbf{r}') \qquad (5.190)$$

$\mathbf{G}_{ox}(\mathbf{r}, \mathbf{r}')$ is referred to as a free-space vector Green's function with a source pointed in the x direction. We could also have $\mathbf{G}_{oy}(\mathbf{r}, \mathbf{r}')$ and $\mathbf{G}_{oz}(\mathbf{r}, \mathbf{r}')$ corresponding to infinitesimal, elementary sources pointed in the y and z direction, respectively. We now introduce a dyadic function* which can store the three vector Green functions $\mathbf{G}_{ox}(\mathbf{r}, \mathbf{r}')$, $\mathbf{G}_{oy}(\mathbf{r}, \mathbf{r}')$, and $\mathbf{G}_{oz}(\mathbf{r}, \mathbf{r}')$, that is,

$$\mathbf{G}_o(\mathbf{r}, \mathbf{r}') = \mathbf{G}_{ox}(\mathbf{r}, \mathbf{r}')\mathbf{a}_x + \mathbf{G}_{oy}(\mathbf{r}, \mathbf{r}')\mathbf{a}_y + \mathbf{G}_{oz}(\mathbf{r}, \mathbf{r}')\mathbf{a}_z \qquad (5.191)$$

---

* A dyad is a group of two or a pair of quantities. A dyadic function, denoted by $\tilde{D}$, is formed by two functions, that is, $\tilde{D} = \mathbf{AB}$. See Tai [53] or Balanis [28] for an exposition on dyadic functions.

This is called free-space dyadic Green's function [53]. It is a solution to the dyadic differential equation

$$\nabla \times \nabla \times \mathbf{G}_o(\mathbf{r}, \mathbf{r}') - k_o^2 \mathbf{G}_o(\mathbf{r}, \mathbf{r}') = \tilde{I}\delta(\mathbf{r} - \mathbf{r}') \tag{5.192}$$

where $\tilde{I}$ denotes the unit dyad (or idem factor) defined by

$$\tilde{I} = \mathbf{a}_x\mathbf{a}_x + \mathbf{a}_y\mathbf{a}_y + \mathbf{a}_z\mathbf{a}_z \tag{5.193}$$

The physical meaning of $\mathbf{G}_o(\mathbf{r}, \mathbf{r}')$ is rather obvious. $\mathbf{G}_o(\mathbf{r}, \mathbf{r}')$ is the electric field at a field point $\mathbf{r}$ due to an infinitesimal source at $\mathbf{r}'$.

From Equations 5.184a and 5.192, the solution of E is

$$\mathbf{E}^s(\mathbf{r}) = -j\omega\mu_o \int \mathbf{G}_o(\mathbf{r}, \mathbf{r}') \cdot \mathbf{J}_{eq}(\mathbf{r}')dv' \tag{5.194}$$

Since $\mathbf{G}_o(\mathbf{r}, \mathbf{r}')$ has a singularity of the order $|\mathbf{r} - \mathbf{r}'|^3$, the integral in Equation 5.194 diverges if the field point $\mathbf{r}$ is inside the volume v of the body (or source region). This difficulty is overcome by excluding a small volume surrounding the field point first and then letting the small volume approach zero. The process entails defining the principal value (*PV*) and adding a correction term needed to yield the correct solution. Thus,

$$\mathbf{E}^s(\mathbf{r}) = PV \int_v \mathbf{J}_{eq}(\mathbf{r}) \cdot \mathbf{G}(\mathbf{r}, \mathbf{r}')dv' + [\mathbf{E}^s(\mathbf{r})]_{\text{correction}} \tag{5.195}$$

The correction term has been evaluated [51,52] to be $-\mathbf{J}_{eq}/j3\omega\epsilon_o$ so that

$$\mathbf{E}^s(\mathbf{r}) = PV \int_v \mathbf{J}_{eq}(\mathbf{r}) \cdot \mathbf{G}(\mathbf{r}, \mathbf{r}')dv' - \frac{\mathbf{J}_{eq}(\mathbf{r})}{j3\omega\epsilon_o} \tag{5.196}$$

The total electric field inside the body is the sum of the incident field $\mathbf{E}^i$ and scattered field $\mathbf{E}^s$, that is,

$$\mathbf{E}(\mathbf{r}) = \mathbf{E}^i(\mathbf{r}) + \mathbf{E}^s(\mathbf{r}) \tag{5.197}$$

Combining Equations 5.181, 5.196, and 5.197 gives the desired tensor IE for E(r):

$$\left[1 + \frac{\tau(\mathbf{r})}{3j\omega\epsilon_o}\right]\mathbf{E}(\mathbf{r}) - PV \int_v \tau(\mathbf{r}')\mathbf{E}(\mathbf{r}) \cdot \mathbf{G}(\mathbf{r}, \mathbf{r}')dv' = \mathbf{E}^i(\mathbf{r}) \tag{5.198}$$

In Equation 5.198, $\tau(\mathbf{r}) = \sigma(\mathbf{r}) + j\omega[\epsilon(\mathbf{r}) - \epsilon_o]$ and the incident electric field $\mathbf{E}^i$ are known quantities; the total electric field E inside the body is unknown and is to be determined by MoM.

5.8.2 Transformation to Matrix Equation (Discretization)

The inner product $\mathbf{E}(\mathbf{r}) \cdot \mathbf{G}(\mathbf{r}, \mathbf{r}')$ in Equation 5.198 may be represented as

$$
\mathbf{E}(\mathbf{r}) \cdot \mathbf{G}(\mathbf{r}, \mathbf{r}') = \begin{bmatrix} \mathbf{G}_{xx}(\mathbf{r}, \mathbf{r}') \, \mathbf{G}_{xy}(\mathbf{r}, \mathbf{r}') \, \mathbf{G}_{xz}(\mathbf{r}, \mathbf{r}') \\ \mathbf{G}_{yx}(\mathbf{r}, \mathbf{r}') \, \mathbf{G}_{yy}(\mathbf{r}, \mathbf{r}') \, \mathbf{G}_{yz}(\mathbf{r}, \mathbf{r}') \\ \mathbf{G}_{zx}(\mathbf{r}, \mathbf{r}') \, \mathbf{G}_{zy}(\mathbf{r}, \mathbf{r}') \, \mathbf{G}_{zz}(\mathbf{r}, \mathbf{r}') \end{bmatrix} \begin{bmatrix} E_x(\mathbf{r}') \\ E_y(\mathbf{r}') \\ E_z(\mathbf{r}') \end{bmatrix} \tag{5.199}
$$

showing that $\mathbf{G}(\mathbf{r}, \mathbf{r}')$ is a symmetric dyad. If we let

$$
x_1 = x, \quad x_2 = y, \quad x_3 = z,
$$

then $G_{x_p x_q}(\mathbf{r}, \mathbf{r}')$ can be written as

$$
G_{x_p x_q}(\mathbf{r}, \mathbf{r}') = -j\omega\mu_o \left[\delta_{pq} + \frac{1}{k_o^2} \frac{\partial^2}{\partial x_q \partial x_p} \right] G_o(\mathbf{r}, \mathbf{r}'), \quad p, q = 1, 2, 3 \tag{5.200}
$$

We now apply MoM to transform Equation 5.198 into a matrix equation. We partition the body into N subvolumes or cells, each denoted by v_m ($m = 1, 2, \ldots, N$), and assume that $\mathbf{E}(\mathbf{r})$ and $\tau(\mathbf{r})$ are constant within each cell. If $\mathbf{r}_m$ is the center of the mth cell, requiring that each scalar component of Equation 5.198 be satisfied at $\mathbf{r}_m$ this leads to

$$
\left[1 + \frac{\tau(\mathbf{r})}{3j\omega\epsilon_o} \right] E_{x_p}(\mathbf{r}_m) - \sum_{q=1}^{3} \left[\sum_{q=1}^{3} \tau(\mathbf{r}_n) PV \int_{vm} G_{x_p x_q}(\mathbf{r}_m, \mathbf{r}') dv' E_{x_q}(\mathbf{r}_n) \right] = E_{x_p}^i(\mathbf{r}_m) \tag{5.201}
$$

If we let $[G_{x_p x_q}]$ be an $N \times N$ matrix with elements

$$
G_{x_p x_q}^{mm} = \tau(\mathbf{r}_n) PV \int_{\nu_n} G_{x_p x_q}(\mathbf{r}_m, \mathbf{r}') dv' - \delta_{pq}\delta_{mm} \left[1 + \frac{\tau(\mathbf{r})}{3j\omega\epsilon_o} \right], \tag{5.202}
$$

where $m, n = 1, 2, \ldots, N$, $p, q = 1, 2, 3$, and let $[E_{x_p}]$ and $[E_{x_p}^i]$ be column matrices with elements

$$
E_{x_p} = \begin{bmatrix} E_{x_p}(\mathbf{r}_1) \\ \vdots \\ E_{x_p}(\mathbf{r}_N) \end{bmatrix}, \quad E_{x_p}^i = \begin{bmatrix} E_{x_p}^i(\mathbf{r}_1) \\ \vdots \\ E_{x_p}^i(\mathbf{r}_N) \end{bmatrix}, \tag{5.203}
$$

then from Equations 5.198 and 5.201, we obtain $3N$ simultaneous equations for E_x, E_y, and E_z at the centers of N cells by the point matching technique. These simultaneous equations can be written in matrix form as

$$
\begin{bmatrix} [G_{xx}] & [G_{xy}] & [G_{xz}] \\ \hline [G_{yx}] & [G_{yy}] & [G_{yz}] \\ \hline [G_{zx}] & [G_{zy}] & [G_{zz}] \end{bmatrix} \begin{bmatrix} [E_x] \\ \hline [E_y] \\ \hline [E_z] \end{bmatrix} = - \begin{bmatrix} [E_x^i] \\ \hline [E_y^i] \\ \hline [E_z^i] \end{bmatrix} \tag{5.204a}
$$

or simply

$$[G][E] = -[E^i] \tag{5.204b}$$

where $[G]$ is $3N \times 3N$ matrix and $[E]$ and $[E^i]$ are $3N$ column matrices.

5.8.3 Evaluation of Matrix Elements

Although the matrix $[E^i]$ in Equation 5.204 is known, while the matrix $[E]$ is to be determined, the elements of the matrix $[G]$, defined in Equation 5.202, are yet to be calculated. For the off-diagonal elements of $[G_{x_p x_q}]$, $\mathbf{r}_m$ is not in the nth cell ($\mathbf{r}_m$ is not in v_n) so that $G_{x_p x_q}(\mathbf{r}_m, \mathbf{r}')$ is continuous in v_n and the principal value operation can be dropped. Equation 5.202 becomes

$$G_{x_p x_q}^{mn} = \tau(\mathbf{r}_n) \int_{v_n} G_{x_p x_q}(\mathbf{r}_m, \mathbf{r}') dv', \quad m \neq n \tag{5.205}$$

As a first approximation,

$$G_{x_p x_q}^{mn} = \tau(\mathbf{r}_n) G_{x_p x_q}(\mathbf{r}_m, \mathbf{r}') \Delta v_n, \quad m \neq n \tag{5.206}$$

where Δv_n is the volume of cell v_n. Incorporating Equations 5.190 and 5.200 into Equation 5.206 yields

$$G_{x_p x_q}^{mn} = \frac{-j\omega\mu k_o \Delta v_n \tau(\mathbf{r}_n) \exp(-j\alpha_{mn})}{4\pi \alpha_{mn}^3}$$
$$\times \left[(\alpha_{mn}^2 - 1 - j\alpha_{mn})\delta_{pq} + \cos\theta_{x_p}^{mn} \cos\theta_{x_q}^{mn} (3 - \alpha_{mn}^2 + 3j\alpha_{mn}) \right], \quad m \neq n \tag{5.207}$$

where

$$\alpha_{mn} = k_o R_{mn}, \quad R_{mn} = |\mathbf{r}_m - \mathbf{r}_n|,$$
$$\cos\theta_{x_p}^{mn} = \frac{x_p^m - x_p^n}{R_{mn}}, \quad \cos\theta_{x_q}^{mn} = \frac{x_q^m - x_q^n}{R_{mn}},$$
$$\mathbf{r}_m = (x_1^m, x_2^m, x_3^m), \quad \mathbf{r}_n = (x_1^n, x_2^n, x_3^n)$$

The approximation in Equation 5.207 yields adequate results provided N is large. If greater accuracy is desired, the integral in Equation 5.205 must be evaluated numerically.

For the diagonal terms ($m = n$), Equation 5.202 becomes

$$G_{x_p x_q}^{nn} = \tau(\mathbf{r}_n) PV \int_{v_n} G_{x_p x_q}(\mathbf{r}_n, \mathbf{r}') dv' - \delta_{pq}\left[1 + \frac{\tau(\mathbf{r})}{3j\omega\epsilon_o} \right] \tag{5.208}$$

To evaluate this integral, we approximate cell v_n by an equivolumic sphere of radius a_n centered at $\mathbf{r}_n$, that is,

$$\Delta v = \frac{4}{3}\pi a_n^3$$

or

$$a_n = \left(\frac{3\Delta v}{4\pi}\right)^{1/3}$$ (5.209)

After a lengthy calculation, we obtain [51]

$$G_{x_p x_q}^{nn} = \delta_{pq}\left[\frac{-2j\omega\mu_o\tau(r_n)}{3k_o^2}(\exp[-jk_o a_n](1+jk_o a_n)-1]-\left(1+\frac{\tau(r_n)}{3j\omega\epsilon_o}\right)\right], \quad m=n$$ (5.210)

In case the shape of cell v_n differs considerably from that of a sphere, the approximation in Equation 5.210 may yield poor results. To have a greater accuracy, a small cube, cylinder, or sphere is created around $\mathbf{r}_n$ to evaluate the correction term, and the integration through the remainder of v_n is performed numerically.

5.8.4 Solution of the Matrix Equation

Once the elements of matrix [G] are evaluated, we are ready to solve Equation 5.204, namely,

$$[G][E] = -[E^i]$$ (5.204)

With the known incident electric field represented by [E^i], the total induced electric field represented by [E] can be obtained from Equation 5.204 by inverting [G] or by employing a Gauss–Jordan elimination method. If matrix inversion is used, the total induced electric field inside the biological body is obtained from

$$[E] = -[G]^{-1}[E^i]$$ (5.211)

Guru and Chen [54] have developed computer programs that yield accurate results on the induced electric field and the absorption power density in various biological bodies irradiated by various EM waves. The validity and accuracy of their numerical results were verified by experiments.

In the following examples, we illustrate the accuracy of the numerical procedure with one simple elementary shape and one advanced shape of biological bodies. The examples are taken from the works of Chen and others [52,55–57].

EXAMPLE 5.12

Determine the distribution of the energy absorption rate or EM heating induced by plane EM waves of 918 MHz in spherical models of animal brain having radius 3 cm. Assume the $\mathbf{E}^i$ field expressed as

$$\mathbf{E}^i = E_o e^{-jk_o z}\mathbf{a}_x = \mathbf{a}_x E_o(\cos k_o z - j\sin k_o z)\text{V/m}$$ (5.212)

where $k_o = 2\pi/\lambda = 2\pi f/c$, $E_o = \sqrt{2\eta_o P_i}$, P_i is the incident power in mW/cm^2 and $\eta_o = 377\ \Omega$ is the intrinsic impedance of free space. Take $P_i = 1$ mW/cm^2 ($E_o = 86.83$ V/m), $\epsilon_r = 35$, $\sigma = 0.7$ s/m.

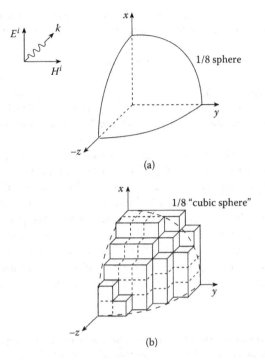

(a)

(b)

FIGURE 5.24
For Example 5.12: (a) One eighth of a sphere, (b) a "cubic sphere" constructed from 73 cubic cells.

Solution

In order to apply MoM, we first approximate the spherical model by a "cubic sphere." Figure 5.24 portrays an example in which one eighth of a sphere is approximated by 40 or 73 cubic cells. The center of each cell, for the case of 40 cells, is determined from Figure 5.25. E^i at the center of each cell can be calculated using Equation 5.212. With the computed E^i and the elements of the matrix $[G_{x_p x_q}]$ calculated using Equations 5.207 and

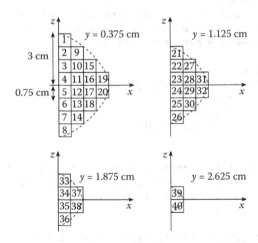

FIGURE 5.25
Geometry and dimensions of one half of the spherical model of the brain constructed from 40 cells. The cell numbering is used in the program of Figure 5.26.

5.210, the induced electric field **E** in each cell is computed from Equation 5.211. Once **E** is obtained, the absorption rate of the EM energy is determined using

$$P = \frac{\sigma}{2} |\mathbf{E}|^2 \tag{5.213}$$

The average heating is obtained by averaging P in the brain. The curve showing relative heating as a function of location is obtained by normalizing the distribution of P with respect to the maximum value of P at a certain location in the brain.

The computer program for the computation is shown in Figure 5.26. It is a modified version of the one developed by Jongakiem [57]. The numerical results are shown in Figure 5.27a, where relative heating along the x-, y-, and z-axis in the brain is plotted. The three curves identified by X, Y, and Z correspond with the distributions of the relative heating along x-, y-, and z-axis, respectively. Observe the strong standing wave patterns with peak heating located somewhere near the center of the brain. The average and maximum heating are found to be 0.3202 and 0.885 in mW/cm³. The exact solution obtained from Mie theory (see Section 2.8) is shown in Figure 5.27b. The average and maximum heating from exact solution are 0.295 and 0.814 mW/cm³, respectively. A comparison of Figure 5.27a and b confirms the accuracy of the numerical procedure.

```
% This program determines the power absorption of bilogical bodies.
% bodies are assumed to have the permeability of freespace
% eo = permittivity of free space
% er = relative permittivity of the body
% f = frequency in Hz, w in rad/s
% sig = conductivity
% xd, yd and zd have the x, y z coordiantes of the cells
% N is the number of cells
% Ei = incident electric field
% Rmn = distance between cell m and n.
% cxpmn and cxqmn = cos(theta_xp_mn) and cos(theta_xq_mn)
% We compute the 9 block matrices separately al-
though one might have a
% function to that and combine them into the G matrix.
% For the diagonal blocks we compute the diagonal elements.
% for the off-diagonal blocks they are zeros.

clear all;    close all
R=3e-2;    H = R/4;    an = H*(0.75/pi)^(1/3);
eo = 8.854e-12;    muo = pi*4e-7;
f = 918e6; er = 35; sig=0.7;
w = 2*pi*f;    ko = w*sqrt(muo*eo);
N=40; %

xd = [0.5*H*ones(1,8) 1.5*H*ones(1,6) 2.5*H*ones(1,4) 3.5*H 3.5*H ...
      0.5*H*ones(1,6) 1.5*H*ones(1,4) 2.5*H*ones(1,2) ...
      0.5*H*ones(1,4) 1.5*H*ones(1,2) ...
      0.5*H*ones(1,2)];

yd=  [0.5*H*ones(1,20) 1.5*H*ones(1,12) 2.5*H*ones(1,6) 3.5*H*ones(1,2)];
```

FIGURE 5.26
Computer program for Example 5.12.   *(Continued)*

```
zd=[(3.5*H:-H:-3.5*H) (2.5*H:-H:-2.5*H) (1.5*H:-H:-1.5*H) 0.5*H -
0.5*H ...
      (2.5*H:-H:-2.5*H) (1.5*H:-H:-1.5*H) 0.5*H -0.5*H ...
      (1.5*H:-H:-1.5*H) 0.5*H -0.5*H ...
       0.5*H -0.5*H];
r=[xd(:) yd(:) zd(:)];

tow=[(sig+j*w*eo*(er-1))*ones(N,1)];

A=-j*w*muo*ko*H^3/4/pi;
C=-2*j*w*muo/3/ko^2*(exp(-j*ko*an)*(1+j*ko*an)-1)-1/3/j/w/eo;
Ei = [86.83*exp(-j*ko*zd(:));0*exp(-j*ko*zd(:));0*exp(-j*ko*zd(:))];

%Gxx
for m=1:N
    for n=1:N
        if m==n
            Gxx(m,n) = C*tow(n)-1;
        else
        Rmn=norm(r(m,:)-r(n,:));
        almn = ko*Rmn;
        cxpmn =(r(m,1)-r(n,1))/Rmn;
        cxqmn =(r(m,1)-r(n,1))/Rmn;
            B = (almn^2-1-j*almn)+cxpmn*cxqmn*(3-almn^2+3*j*almn);
            Gxx(m,n)= A* tow(n)*exp(-j*almn)/almn^3*B;
        end
    end
end

%Gyy
for m=1:N
    for n=1:N
        if m==n
            Gyy(m,n) = C*tow(n)-1;
        else
        Rmn=norm(r(m,:)-r(n,:));
        almn = ko*Rmn;
        cxpmn =(r(m,2)-r(n,2))/Rmn;
        cxqmn =(r(m,2)-r(n,2))/Rmn;

            B = (almn^2-1-j*almn)+cxpmn*cxqmn*(3-almn^2+3*j*almn);
            Gyy(m,n)= A* tow(n)*exp(-j*almn)/almn^3*B;
        end
    end
end
```

FIGURE 5.26 (Continued)
Computer program for Example 5.12.   (*Continued*)

```
%Gzz
for m=1:N
    for n=1:N
        if m==n
            Gzz(m,n) = C*tow(n)-1;
        else
        Rmn=norm(r(m,:)-r(n,:));
        almn = ko*Rmn;
        cxpmn =(r(m,3)-r(n,3))/Rmn;
        cxqmn =(r(m,3)-r(n,3))/Rmn;

            B = (almn^2-1-j*almn)+cxpmn*cxqmn*(3-almn^2+3*j*almn);
            Gzz(m,n)= A* tow(n)*exp(-j*almn)/almn^3*B;
        end
    end
end


%Gxy
for m=1:N
    for n=1:N
        if m~=n
        Rmn=norm(r(m,:)-r(n,:));
        almn = ko*Rmn;
        cxpmn =(r(m,1)-r(n,1))/Rmn;
        cxqmn =(r(m,2)-r(n,2))/Rmn;
            B = cxpmn*cxqmn*(3-almn^2+3*j*almn);
            Gxy(m,n)= A* tow(n)*exp(-j*almn)/almn^3*B;
        end
    end
end


%Gxz
for m=1:N
    for n=1:N
        if m~=n
        Rmn=norm(r(m,:)-r(n,:));
        almn = ko*Rmn;
        cxpmn =(r(m,1)-r(n,1))/Rmn;
        cxqmn =(r(m,3)-r(n,3))/Rmn;

            B = cxpmn*cxqmn*(3-almn^2+3*j*almn);
            Gxz(m,n)= A* tow(n)*exp(-j*almn)/almn^3*B;
        end
    end
end
```

FIGURE 5.26 (Continued)
Computer program for Example 5.12.   *(Continued)*

```
%Gyx
for m=1:N
    for n=1:N
        if m~=n
        Rmn=norm(r(m,:)-r(n,:));
        almn = ko*Rmn;
        cxpmn =(r(m,2)-r(n,2))/Rmn;
        cxqmn =(r(m,1)-r(n,1))/Rmn;

            B = cxpmn*cxqmn*(3-almn^2+3*j*almn);
            Gyx(m,n)= A* tow(n)*exp(-j*almn)/almn^3*B;
        end
    end
end

%Gyz
for m=1:N
    for n=1:N
        if m~=n
        Rmn=norm(r(m,:)-r(n,:));
        almn = ko*Rmn;
        cxpmn =(r(m,2)-r(n,2))/Rmn;
        cxqmn =(r(m,3)-r(n,3))/Rmn;

            B = cxpmn*cxqmn*(3-almn^2+3*j*almn);
            Gyz(m,n)= A* tow(n)*exp(-j*almn)/almn^3*B;
        end
    end
end

%Gzx
for m=1:N
    for n=1:N
        if m~=n
        Rmn=norm(r(m,:)-r(n,:));
        almn = ko*Rmn;
        cxpmn =(r(m,3)-r(n,3))/Rmn;
        cxqmn =(r(m,1)-r(n,1))/Rmn;
            B = cxpmn*cxqmn*(3-almn^2+3*j*almn);
            Gzx(m,n)= A* tow(n)*exp(-j*almn)/almn^3*B;
        end
    end
end

%Gzy
for m=1:N
    for n=1:N
```

FIGURE 5.26 (Continued)
Computer program for Example 5.12.   *(Continued)*

```
         if m~=n
         Rmn=norm(r(m,:)-r(n,:));
         almn = ko*Rmn;
         cxpmn =(r(m,3)-r(n,3))/Rmn;
         cxqmn =(r(m,2)-r(n,2))/Rmn;

             B = cxpmn*cxqmn*(3-almn^2+3*j*almn);
             Gzy(m,n)= A* tow(n)*exp(-j*almn)/almn^3*B;
         end
     end
end
G = [Gxx Gxy Gxz;Gyx Gyy Gyz;Gzx Gzy Gzz];
E = -G\Ei;
Nx=[4 11 16 19];
Ny=[4 23 34 39];
Nz= 8:-1:1;;


Exaxis = (abs(E(Nx)).^2+abs(E(N+Nx)).^2+abs(E(2*N+Nx)).^2);
Exaxis = [Exaxis(4:-1:1);Exaxis];
Eyaxis = (abs(E(Ny)).^2+abs(E(N+Ny)).^2+abs(E(2*N+Ny)).^2);
Eyaxis = [Eyaxis(4:-1:1);Eyaxis];
Ezaxis = 2*(abs(E(Nz)).^2+abs(E(N+Nz)).^2+abs(E(2*N+Nz)).^2);


EMAX=max([Exaxis;Eyaxis;Ezaxis]);
 Exaxis=Exaxis/EMAX;
 Eyaxis=Eyaxis/EMAX;
 Ezaxis=Ezaxis/EMAX;
pd=(-2.625:.75:2.625);
plot(pd,Exaxis,'ok');hold on
plot(pd,Eyaxis,'ok','MarkerFaceColor','k');
plot(pd,Ezaxis,'sk');
pdi=(-2.625:5.25/199:2.625);
Exaxisi=interp1(pd,Exaxis,pdi,'spline');
Eyaxisi=interp1(pd,Eyaxis,pdi,'spline');
Ezaxisi=interp1(pd,Ezaxis,pdi,'spline');
plot(pdi,Exaxisi,'k','LineWidth',2);
plot(pdi,Eyaxisi,'-.k','LineWidth',2);
plot(pdi,Ezaxisi,':k','LineWidth',2);
xlabel('cm','FontSize',12)
axis([-3 3 0 1.1]);grid
text(-1.5,.28,'y')
text(-1.5,.45,'z')
text(-1.5,.92,'x')
```

FIGURE 5.26 (Continued)
Computer program for Example 5.12.

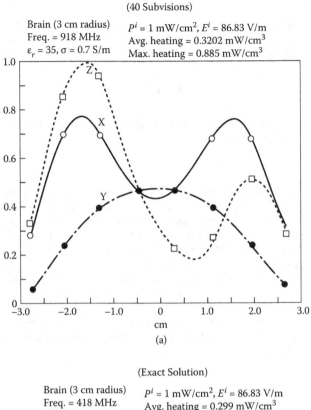

(40 Subvisions)

Brain (3 cm radius)
Freq. = 918 MHz
$\varepsilon_r = 35$, $\sigma = 0.7$ S/m

$P^i = 1$ mW/cm^2, $E^i = 86.83$ V/m
Avg. heating = 0.3202 mW/cm^3
Max. heating = 0.885 mW/cm^3

(a)

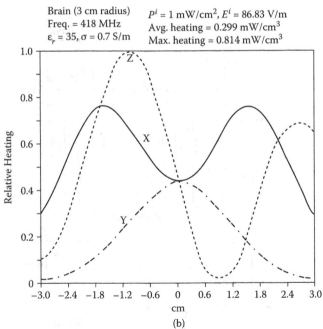

(Exact Solution)

Brain (3 cm radius)
Freq. = 418 MHz
$\varepsilon_r = 35$, $\sigma = 0.7$ S/m

$P^i = 1$ mW/cm^2, $E^i = 86.83$ V/m
Avg. heating = 0.299 mW/cm^3
Max. heating = 0.814 mW/cm^3

(b)

FIGURE 5.27
Distributions of heating along the *x*-, *y*-, and *z*-axis of a spherical model of an animal brain: (a) MoM solution, (b) exact solution. (From R. Rukspollmuang and K.M. Chen, *Radio Sci.*, vol. 14, no. 6S, Nov.–Dec. 1979, pp. 51–62.)

EXAMPLE 5.13

Having validated the accuracy of the tensor-integral-equation (TIE) method, determine the induced electric field and SAR of EM energy inside a model of typical human body irradiation (Figure 5.28), by EM wave at 80 MHz with vertical polarization, that is,

$$\mathbf{E} = \mathbf{a}_x \text{ V/m}$$

at normal incidence. Assume the body at 80 MHz is that of a high-water content tissue with $\epsilon = 80\epsilon_o$, $\mu = \mu_o$, $\sigma = 0.84$ s/m.

Solution

The body is partitioned into 108 cubic cells of various sizes ranging from 5 to 12 cm³. To ensure accurate results, the cell size is kept smaller than a quarter-wavelength (of the medium). With the coordinates of the center of each cell figured out from Figure 5.28, the program in Figure 5.26 can be used to find induced electric field components E_x, E_y and E_z at the centers of the cells due to an incident electric field 1 V/m (maximum value) at normal incidence. The SAR is calculated from $(\sigma/2)$ $(E_x^2 + E_y^2 + E_z^2)$. Figures 5.29 through 5.31 illustrate E_x, E_y, and E_z at the center of each cell. Observe that E_y and E_z are much smaller than E_x at this frequency due to the polarization of the incident wave.

As mentioned earlier, the model of the human body shown in Figure 5.28 is due to Chen and Guru [55]. An improved, more realistic model due to Gandhi et al. [58–60] is shown in Figure 5.32.

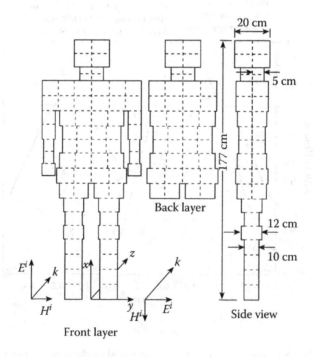

FIGURE 5.28
Geometry and dimensions of a model of typical human body of height 1.77 m. (Adapted from J.A. Kong (ed.), *Research Topics in Electromagnetic Theory*. New York: John Wiley, 1981, pp. 290–355.)

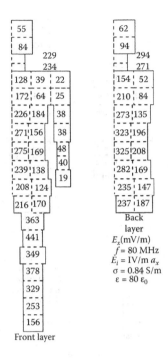

FIGURE 5.29

Induced E_x (in mV/m) at the center of each cell due to E_x^i of 1 V/m. (Adapted from J.A. Kong (ed.), *Research Topics in Electromagnetic Theory*. New York: John Wiley, 1981, pp. 290–355.)

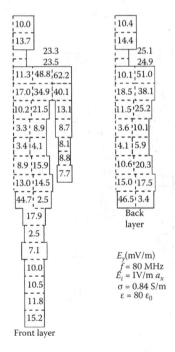

FIGURE 5.30

Induced E_y (in mV/m) at the center of each cell due to E_x^i of 1 V/m. (Adapted from J.A. Kong (ed.), *Research Topics in Electromagnetic Theory*. New York: John Wiley, 1981, pp. 290–355.)

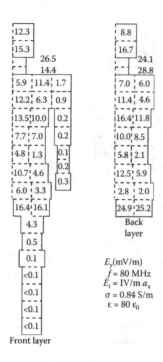

E_y(mV/m)
f = 80 MHz
E_i = IV/m a_x
σ = 0.84 S/m
ε = 80 ε$_0$

FIGURE 5.31

Induced E_z (in mV/m) at the center of each cell due to E_x^i of 1 V/m. (Adapted from J.A. Kong (ed.), *Research Topics in Electromagnetic Theory*. New York: John Wiley, 1981, pp. 290–355.)

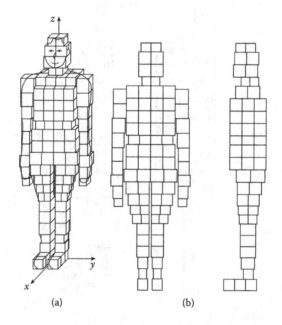

(a)　　　　　　　　　　　　(b)

FIGURE 5.32

A more realistic block model of the human body: (a) In three dimensions, (b) front and side views. (Adapted from O.P. Gandhi, *Bioelectromagnetics*, vol. 3, 1982, pp. 81–90.)

5.9 Concluding Remarks

The MoM is a powerful numerical method capable of applying weighted residual techniques to reduce an IE to a matrix equation. The solution of the matrix equation is usually carried out via inversion, elimination, or iterative techniques. Although MoM is commonly applied to open problems such as those involving radiation and scattering, it has been successfully applied to closed problems such as waveguides and cavities.

Needless to say, the issues on MoM covered in this chapter have been carefully selected. We have only attempted to cover the background and reference material upon which the reader can easily build. The interested reader is referred to the literature for more in-depth treatment of each subject. General concepts on MoM are covered in References 3,61. Clear and elementary discussions on IEs and Green's functions may be found in References 10–12,28–30,61,62. For further study on the theory of the MoM, one should see References 6,9,10,28,40. The error analysis of MoM solutions is provided in References 63,64.

The number of problems that can be treated by MoM is endless, and the examples given in this chapter just scratch the surface. The following problems represent typical EM-related application areas:

- Electrostatic problems [31,65–68]
- Wire antennas and scatterers [34,37,42,44,69,70]
- Scattering and radiation from bodies of revolution [71,72]
- Scattering and radiation from bodies of arbitrary shapes [38,73,74]
- Transmission lines [18–20,23,24,75–78]
- Aperture problems [79–81]
- Biomagnetic problems [47–52,82–84].

A number of user-oriented computer programs have evolved over the years to solve electromagnetic IEs by the MoM. These codes can handle radiation and scattering problems in both the frequency and time domains. Reviews of the codes may be found in References 38,85. The most popular of these codes is the Numerical Electromagnetic Code (NEC) developed at the Lawrence Livermore National Laboratory [7,86]. NEC is a frequency domain antenna modeling FORTRAN code applying the MoM to IEs for wire and surface structures. Its most notable features are probably that it is user friendly, includes documentation, and is available; for these reasons, it is being used in public and private institutions. A compact version of NEC is the mini-numerical electromagnetic code (MININEC) [87], which is intended to be used in personal computers.

It is important that we recognize the fact that MoM is limited in application to radiation and scattering from bodies that are electrically large. The size of the scatterer or radiator must be of the order λ^3. This is because the cost of storing, inverting, and computing matrix elements becomes prohibitively large. At high frequencies, asymptotic techniques such as the geometrical theory of diffraction (GTD) are usually employed to derive approximate but accurate solutions [46,88,89]. Some hybrid techniques have been developed to improve the accuracy of MoM [98–100].

PROBLEMS

5.1 Show that in Example 5.1,

$$A_{mn} = \langle w_m(x), L[u_n(x)] \rangle = \frac{-mn}{m+n+1}$$

$$B_m = \langle w_m(x), g(x) \rangle = \frac{-m}{4(m+4)}$$

5.2 Repeat Example 5.1 for

$$-\frac{d^2U}{dx^2} = 1 + 2x^2, \quad 0 < x < 1$$

subject to $U(0) = 0 = U(1)$.

5.3 Classify the following IEs and show that they have the stated solutions:

a. $\Phi(x) = \dfrac{5x}{6} + \dfrac{1}{2} \displaystyle\int_0^1 xt\Phi(t)dt$ [solution $\Phi(x) = x$],

b. $\Phi(x) = \cos x - \sin x + 2 \displaystyle\int_0^x \sin(x-t)\Phi(t)dt$ [solution $\Phi(x) = e^{-x}$],

c. $\Phi(x) = -\cosh x + \lambda \displaystyle\int_{-1}^1 \cosh(x+t)\Phi(t)dt$ $\left[\text{solution } \Phi(x) = \dfrac{\cosh x}{\dfrac{\lambda}{2}\sinh 2 + \lambda - 1} \right]$

5.4 Classify the following IEs:

a. $u(x) = \displaystyle\int_{-\infty}^{\infty} e^{ix\lambda} u(\lambda) d\lambda$

b. $u(x) = e^x - \lambda \displaystyle\int_0^1 G(x,y)u(y)dy$

c. $u(x) = \displaystyle\int_x^1 \frac{f(\lambda)d\lambda}{(\lambda^2 - x^2)^{1/2}}$

d. $u(x) = \sin x + 2 \displaystyle\int_0^x \cos(x-y)u(y)dy$

5.5 Solve the following Volterra IEs:

a. $\Phi(x) = 5 + 2 \int_0^x t\Phi(t)dt,$

b. $\Phi(x) = x + \int_0^x (t-x)\Phi(t)dt$

5.6 Find the IE corresponding to each of the following differential equations:
a. $y'' = -y, \quad y(0) = 0, \quad y'(1) = 1,$
b. $y'' + y = \cos x, \quad y(0) = 0, \quad y'(0) = 1$

5.7 Show that Green's function,

$$G = \frac{e^{-jkr}}{4\pi r}$$

where $r = \sqrt{x^2 + y^2 + z^2}$, satisfies Helmholtz equation

$$\nabla^2 G + k^2 G = -\delta(r)$$

5.8 Find Green's function for the scalar one-dimensional Helmholtz equation

$$\frac{d^2U}{dx^2} + k^2 U = \delta(x), \quad 0 < x < 1$$

subject to a homogenous Dirichlet boundary condition.

5.9 Obtain Green's function for the Helmholtz equation

$$\frac{d^2 G(x|x_o)}{dx^2} + k^2 G(x|x_o) = \delta(x - x_o), \quad -L < x < L$$

subject to Neumann and mixed boundary condition

$$G(-L|x_o) = 0 = \frac{dG}{dx}\bigg|_{x=L}$$

5.10 Show that

$$G(x, z; x', z') = \frac{j}{a} \sum_{n=1}^{\infty} \frac{\sin(n\pi x/a)\sin(n\pi x'/a)}{k_n} e^{jk_n(z-z')},$$

where $k_n^2 = k^2 - (n\pi/a)^2$ is Green's function for Helmholtz's equation.

5.11 Derive Green's function for

$$\nabla^2\Phi = f, \quad 0 < x, \quad y < 1$$

subject to zero boundary conditions.

5.12 Find Green's function satisfying

$$G_{xx} + G_{yy} + 2G_x = \delta(x - x')\delta(y - y'), \quad 0 < x < a, \quad 0 < y < b$$

and

$$G(0, y) = G(a, y) = G(x, 0) = G(x, b) = 0$$

5.13 a. Verify by Fourier expansion that Equations 5.79 and 5.80 in Example 5.5 are equivalent.

b. Show that another form of expressing Equation 5.79 is

$$G(x,y;x',y') = \begin{cases} -\dfrac{2}{\pi}\displaystyle\sum_{m=1}^{\infty}\sinh\dfrac{m\pi(b-y')}{a}\sin\dfrac{m\pi y}{a}\dfrac{m\pi x}{a}\dfrac{m\pi x'}{a}, & y < y' \\[4mm] -\dfrac{2}{\pi}\displaystyle\sum_{m=1}^{\infty}\sinh\dfrac{m\pi y'}{a}\sin\dfrac{m\pi(b-y)}{a}\dfrac{m\pi x}{a}\dfrac{m\pi x'}{a}, & y > y' \end{cases}$$

5.14 The two-dimensional delta function expressed in cylindrical coordinates reads

$$\delta(\rho - \rho') = \frac{1}{\rho}\delta(\rho - \rho')\delta(\phi - \phi')$$

Obtain Green's function for the potential problem

$$\nabla^2 G = \frac{1}{\rho}\delta(\rho - \rho')\delta(\phi - \phi')$$

with the region defined in Figure 5.33. Assume homogeneous Dirichlet boundary conditions.

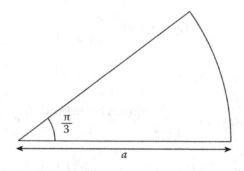

FIGURE 5.33
For Problem 5.14.

5.15 Consider the transmission line with cross section as shown in Figure 5.34. In a TEM wave approximation, the potential distribution satisfies Poisson's equation

$$\nabla^2 V = -\frac{\rho_s}{\epsilon}$$

subject to the following continuity and boundary conditions:

$$\frac{\partial}{\partial x} V(x, h_1 - 0) = \frac{\partial}{\partial x} V(x, h_1 + 0)$$

$$\frac{\partial}{\partial x} V(x, h_1 + h_2 - 0) = \frac{\partial}{\partial x} V(x, h_1 + h_2 + 0)$$

$$\epsilon_1 \frac{\partial}{\partial y} V(x, h_1 - 0) = \epsilon_2 \frac{\partial}{\partial y} V(x, h_1 + 0)$$

$$\epsilon_2 \frac{\partial}{\partial y} V(x, h_1 + h_2 - 0) = \epsilon_3 \frac{\partial}{\partial y} V(x, h_1 + h_2 + 0) - \rho_s(x, h_1 + h_2)$$

$$V(0, y) = V(a, y) = V(x, 0) = V(x, b) = 0$$

Using series expansion method, evaluate Green's function at $y = h_1 + h_2$, that is, $G(x, y; x', h_1 + h_2)$.

5.16 Show that the free-space Green's function for $L = \nabla^2 + k^2$ in two-dimensional space is $-\frac{j}{4} H_0^{(1)}(k\rho)$.

5.17 The spherical Green's function $h_0^{(2)}(|\mathbf{r} - \mathbf{r}'|)$ can be expanded in terms of spherical Bessel functions and Legendre polynomials. Show that

$$h_0^{(2)}(|\mathbf{r} - \mathbf{r}'|) = \begin{cases} \displaystyle\sum_{n=0}^{\infty} (2n+1) h_n^2(\mathbf{r}') j_n(r) P_n(\cos\alpha), & r < r' \\[2mm] \displaystyle\sum_{n=0}^{\infty} (2n+1) h_n^2(\mathbf{r}) j_n(r') P_n(\cos\alpha), & r < r' \end{cases}$$

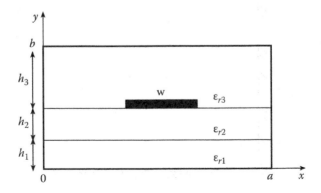

FIGURE 5.34
For Problem 5.15.

where $\cos\alpha = \cos\theta\cos\theta' + \sin\theta\sin\theta'\cos(\phi - \phi')$. From this, derive the plane wave expansion

$$e^{-j\mathbf{k}\cdot\mathbf{r}} = \sum_{n=0}^{\infty}(-j)^n(2n+1)j_n(kr)P_n(\cos\alpha)$$

5.18 Given the kernel

$$K(x,y) = \begin{cases} (1-x)y, & 0 \le y \le x \le 1 \\ (1-y)x, & 0 \le x \le y \le 1 \end{cases}$$

Show that

$$K(x,y) = 2\sum_{n=1}^{\infty}\frac{\sin n\pi x \sin n\pi y}{n^2\pi^2}$$

and that

$$\frac{\pi^2}{4} = \sum_{n=1}^{\infty}\frac{1}{n^2}$$

5.19 Derive the closed-form solution for Poisson's equation

$$\nabla^2 V = g$$

in the quarter-plane shown in Figure 5.35 with

$$V = f(x), \quad y = 0, \quad \frac{\partial V}{\partial x} = h(y), \quad x = 0$$

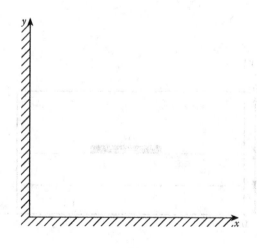

FIGURE 5.35
For Problem 5.19.

5.20 Consider the cross section of a microstrip transmission line shown in Figure 5.36. Let $G_{ij}\, \rho_j$ be the potential at the field point i on the center conductor due to the charge on subsection j. (It is assumed that the charge is concentrated in the filament along the center of the subsection.) G_{ij} is Green's function for this problem and is given by

$$G_{ij} = \frac{1}{4\pi\epsilon_r} \sum_{n=1}^{\infty} \left[k^{2(n-1)} \ln \frac{A_{ij}^2 + (4n-2)^2}{A_{ij}^2 + (4n-4)^2} + k^{2n-1} \ln \frac{A_{ij}^2 + (4n-2)^2}{A_{ij}^2 + (4n)^2} \right]$$

where

$$A_{ij} = \frac{\Delta}{H} |2(i-1) - 2(j-1) - 1|, \quad k = \frac{\epsilon_r - 1}{\epsilon_r + 1},$$

$\Delta = W/N$, and N is the number of equal subsections into which the center conductor is divided. By setting the potential equal to unity on the center conductor, one can find

$$C = \sum_{j=1}^{\infty} \rho_j \quad \text{(F/m)}$$

and

$$Z_0 = \frac{1}{c\sqrt{C_o C}}$$

where $c = 3 \times 10^8$ m/s and C_o is the capacitance per unit length for an air-filled transmission line (i.e., set $k = 1$ in G_{ij}). Find Z_0 for $N = 30$ and

a. $\epsilon_r = 6.0$, $\quad W = 4$ cm, $\quad H = 4$ cm
b. $\epsilon_r = 16.0$, $\quad W = 8$ cm, $\quad H = 4$ cm.

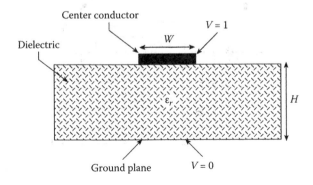

FIGURE 5.36
For Problem 5.20.

5.21 Consider the three-charge model shown in Figure 5.37. The radius of each charged sphere is a and the separation distances are equal; that is, $d_1 = d_2 = d$. The potential system results in

$$\begin{bmatrix} +V \\ 0 \\ -V \end{bmatrix} = \frac{1}{4\pi\varepsilon_o} \begin{bmatrix} \dfrac{1}{a} & \dfrac{1}{d} & \dfrac{1}{2d} \\[2mm] \dfrac{1}{d} & \dfrac{1}{a} & \dfrac{1}{d} \\[2mm] \dfrac{1}{2a} & \dfrac{1}{d} & \dfrac{1}{a} \end{bmatrix} \begin{bmatrix} Q_1 \\ Q_2 \\ Q_3 \end{bmatrix}$$

Let $V = 1$, $a/d = 1/10$. Use MATLAB to compute Q_1, Q_2, and Q_3 in terms of $4\pi\varepsilon_o d$.

5.22 Consider the sheet model for representing the p-n junction as shown in Figure 5.38. In matrix form,

$$\begin{bmatrix} V(-2d) \\ V(-d) \\ V(0) \\ V(+d) \\ V(+2d) \end{bmatrix} = \frac{-d}{2\varepsilon} \begin{bmatrix} 0 & 1 & 2 & 3 & 2 \\ 1/2 & 0 & 1 & 2 & 3/2 \\ 1 & 1 & 0 & 1 & 1 \\ 3/2 & 2 & 1 & 0 & 1/2 \\ 2 & 3 & 2 & 1 & 0 \end{bmatrix} \begin{bmatrix} \sigma_1 \\ \sigma_2 \\ \sigma_3 \\ \sigma_4 \\ \sigma_5 \end{bmatrix}$$

Let $V(-2d) = -2$, $V(-d) = -1.5$, $V(0) = 0$, $V(+d) = 1.5$, $V(+2d) = 2$. Using MATLAB, obtain the charges σ_1, σ_2, σ_3, σ_4, and σ_5.

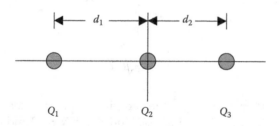

FIGURE 5.37
For Problem 5.21.

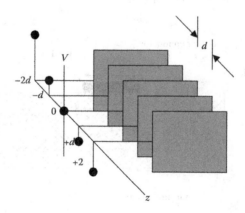

FIGURE 5.38
For Problem 5.22.

5.23 A rectangular section of microstrip transmission line of length L, width W, and height H above the ground plane is shown in Figure 5.39. The section is subdivided into N subsections. A typical subsection ΔS_j of sides Δx_j and Δy_j is assumed to bear a uniform surface charge density ρ_j. The potential V_i at ΔS_i due to a uniform charge density ρ_j on $\Delta S_j (j = 1, 2, ..., N)$ is

$$V_i = \sum_{j=1}^{N} G_{ij}\rho_j$$

where

$$G_{ij} = \sum_{n=1}^{\infty} \frac{k^{n-1}(-1)^{n+1}}{2\pi\epsilon_0(\epsilon_r + 1)}$$

$$
\begin{aligned}
&\Bigg[(x_j - x_i)\ln \frac{(y_j - y_i) + \sqrt{(x_j - x_i)^2 + (y_j - y_i)^2 + (2n-2)^2 H^2}}{(y_j + \Delta y_j - y_i) + \sqrt{(x_j - x_i)^2 + (y_j + \Delta y_j - y_i)^2 + (2n-2)^2 H^2}} \\
&+ (x_j + \Delta x_j - x_i) \\
&\quad \ln \frac{(y_j + \Delta y_j - y_i) + \sqrt{(x_j + \Delta x_j - x_i)^2 + (y_j + \Delta y_j - y_i)^2 + (2n-2)^2 H^2}}{(y_j - y_i) + \sqrt{(x_j + \Delta x_j - x_i)^2 + (y_j - y_i)^2 + (2n-2)^2 H^2}} \\
&+ (y_j - y_i)\ln \frac{(x_j - x_i) + \sqrt{(x_j - x_i)^2 + (y_j - y_i)^2 + (2n-2)^2 H^2}}{(x_j + \Delta x_j - x_i) + \sqrt{(x_j + \Delta x_j - x_i)^2 + (y_j - y_i)^2 + (2n-2)^2 H^2}} \\
&+ (y_j + \Delta y_j - y_i) \\
&\quad \ln \frac{(x_j + \Delta x_j - x_i) + \sqrt{(x_j + \Delta x_j - x_i)^2 + (y_j + \Delta y_j - y_i)^2 + (2n-2)^2 H^2}}{(x_j - x_i) + \sqrt{(x_j - x_i)^2 + (y_j + \Delta y_j - y_i)^2 + (2n-2)^2 H^2}} \\
&- (2n-2)H\tan^{-1}\frac{(x_j - x_i)(y_j - y_i)}{(2n-2)H\sqrt{(x_j - x_i)^2 + (y_j - y_i)^2 + (2n-2)^2 H^2}} - (2n-2) \\
&H\tan^{-1}\frac{(x_j + \Delta x_j - x_i)(y_j + \Delta y_j - y_i)}{(2n-2)H\sqrt{(x_j + \Delta x_j - x_i)^2 + (y_j + \Delta y_j - y_i)^2 + (2n-2)^2 H^2}} \\
&+ (2n-2)H\tan^{-1}\frac{(x_j - x_i)(y_j + \Delta y_j - y_i)}{(2n-2)H\sqrt{(x_j - x_i)^2 + (y_j + \Delta y_j - y_i)^2 + (2n-2)^2 H^2}} \\
&+ (2n-2)H\tan^{-1}\frac{(x_j + \Delta x_j - x_i)(y_j - y_i)}{(2n-2)H\sqrt{(x_j + \Delta x_j - x_i)^2 + (y_j - y_i)^2 + (2n-2)^2 H^2}} \Bigg]
\end{aligned}
$$

and $k = (\epsilon_r - 1)/(\epsilon_r + 1)$. If the ground plane is assumed to be at zero potential while the conducting strip at 1 V potential, we can find

$$C = \sum_{j=1}^{N} \rho_j$$

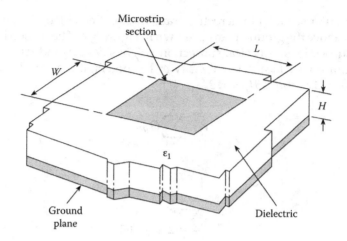

FIGURE 5.39
For Problem 5.23.

Find C for

a. $\epsilon_r = 9.6$, $W = L = H = 2$ cm,

b. $\epsilon_r = 9.6$, $W = H = 2$ cm, $L = 1$ cm.

5.24 For a conducting elliptic cylinder with cross section in Figure 5.40a, write a program to determine the scattering cross section $\sigma(\phi_i, \phi)$ due to a plane TM wave.

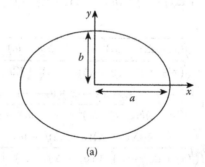

(a)

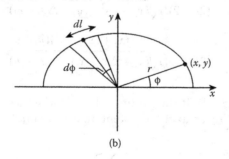

(b)

FIGURE 5.40
For Problem 5.24.

Consider $\phi = 0°, 10°, ..., 180°$ and cases $\phi_i = 0°, 30°$, and $90°$. Plot $\sigma(\phi_i, \phi)$ against ϕ for each ϕ_i. Take $\lambda = 1$ m, $2a = \lambda/2$, $2b = \lambda$, $N = 18$.

   Hint: Due to symmetry, consider only one half of the cross section as in Figure 5.40b. An ellipse is described by

$$\frac{x^2}{a^2} + \frac{y^2}{b^2} = 1$$

With $x = r \cos \phi$, $y = r \sin \phi$, it is readily shown that

$$r = \frac{a}{\sqrt{\cos^2 \phi + v^2 \sin^2 \phi}}, \quad v = a/b, \quad dl = r d\phi.$$

5.25 Use the program in Figure 5.15 (or develop your own program) to calculate the scattering pattern for each array of parallel wires shown in Figure 5.41.

5.26 Repeat Problem 5.24 using the techniques of Section 5.6.2. That is, consider the cylinder in Figure 5.41a as an array of parallel wires.

5.27 Consider the scattering problem of a dielectric cylinder with cross section shown in Figure 5.42. It is illuminated by a TM wave. To obtain the field $[E]$ inside the dielectric cylinder, MoM formulation leads to the matrix equation

$$[A][E] = [E^i]$$

where

$$A_{mn} = \begin{cases} \epsilon_m + j\dfrac{\pi}{2}(\epsilon_m - 1)ka_n H_1^{(2)}(ka_m), & m = n \\[2mm] j\dfrac{\pi}{2}(\epsilon_m - 1)ka_n J_1(ka_n) H_0^{(2)}(k\rho_{mn}), & m \neq n \end{cases}$$

$$E_m^i = e^{jk(x_m \cos \phi_i + y_m \sin \phi_i)}$$

$$\rho_{mn} = \sqrt{(x_m - x_n)^2 + (y_m - y_n)^2}, \quad m, n = 1, 2, ..., N$$

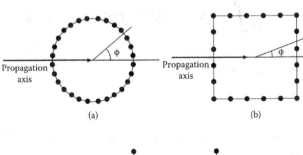

(a)   (b)

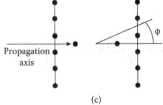

(c)

FIGURE 5.41
Arrays of parallel wires: (a) cylinder, (b) square, (c) I-beam, for Problem 5.25.

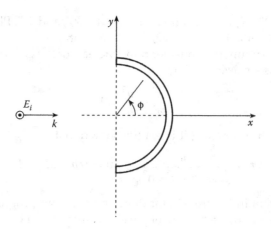

FIGURE 5.42
For Problem 5.27.

N is the number of cells the cylinder is divided into, ϵ_m is the average dielectric constant of cell m, a_m is the radius of the equivalent circular cell which has the same cross section as cell m. Solve the above matrix equation and obtain E_n, $n = 1$, 2, ..., N. Use E_n to obtain the echo width of the dielectric cylinder, that is,

$$W(\phi) = \frac{\pi^2 k}{|E^i|^2} \left| \sum_{n=1}^{N} (\epsilon_n - 1) E_n a_n J_1(ka_n) e^{jk(x_n \cos\phi + y_n \sin\phi)} \right|^2$$

for $\phi = 0°$, $5°$, $10°$, ..., $180°$. Plot $W(\phi)$ versus ϕ. For the dielectric cylinder, take $\mu = \mu_o$, $\epsilon = 4\epsilon_o$, inner radius is 0.25λ, outer radius is 0.4λ, and $\lambda = 1$ m.

5.28 The IE

$$-\frac{1}{2\pi} \int_{-w}^{w} I(z') \ln|z - z'| dz' = f(z), \quad -w < z < w$$

can be cast into matrix equation

$$[S][I] = [F]$$

using pulse basis function and delta expansion function (point matching).

a. Show that

$$S_{mn} = \frac{\Delta}{2\pi} \left[1 - \ln\Delta - \frac{1}{2}\ln|(m-n)^2 - \frac{1}{4}| - (m-n)\ln\frac{|m-n+1/2|}{|m-n-1/2|} \right]$$

$$F_n = f(z_n)$$

where $z_n = -w + \Delta(n - 1/2)$, $n = 1, 2, ..., N$, $\Delta = 2w/N$. Note that $[S]$ is a Toepliz matrix having only N distinct elements.

b. Determine the unknowns $\{I_m\}$ with $f(z) = 1$, $N = 10$, $2w = 1$.

c. Repeat part (b) with $f(z) = z$, $N = 10$, $2w = 1$.

5.29 For a dipole antenna $[Z][I] = [V]$ or

$$\begin{bmatrix} 0.9869 - j3324 & 0.4935 + j1753 \\ 0.9869 + j3506 & 2.576 - j1849 \end{bmatrix} \begin{bmatrix} I_1 \\ I_2 \end{bmatrix} = \begin{bmatrix} 0.02505 \\ 0.02505 \end{bmatrix}$$

Obtain the current vector.

5.30 Derive Equation 5.141 from Equation 5.135.

5.31 A two-term representation of the current distribution on a thin, center-fed half-wavelength dipole antenna is given by

$$I(z) = \sum_{n=1}^{2} B_n \sin\left(\frac{2\pi n}{\lambda}(\lambda/4 - |z|)\right)$$

Substituting this into Hallen's IE gives

$$\sum_{n=1}^{2} B_n \int_{-\lambda/4}^{\lambda/4} \sin\left[\frac{2\pi n}{\lambda}(\lambda/4 - |z'|)\right] G(z,z')dz' + \frac{jC_1}{\eta_o}\cos k_o z$$

$$= -\frac{j}{\eta_o} V_T \sin k_o |z|$$

where $\eta_o = 120\ \pi$, $k_o = 2\pi/\lambda = 2\pi f/c$, and $G(z, z')$ is given by Equation 5.152. Taking $V_T = 1$ V, $\lambda = 1$ m, $a/\lambda = 7.022 \times 10^{-3}$, and match points at $z = 0$, $\lambda/8$, $\lambda/4$, determine the constants B_1, B_2, and C_1. Plot the real and imaginary parts of $I(z)$ against z.

5.32 Using Hallen's IE, determine the current distribution $I(z)$ on a straight dipole of length ℓ. Plot $|I| = |I_r + j\ I_i|$ against z. Assume excitation by a unit voltage, $N = 51$, $\Omega = 2\ln\dfrac{\ell}{a} = 12.5$, and consider cases (a) $\ell = \lambda/2$, (b) $\ell = 1.5\lambda$.

5.33 a. Show that Pocklington's IE (5.141) can be written as

$$-E_z^i = \frac{\lambda\sqrt{\mu/\epsilon}}{8j\pi a^2} \int_{-\ell/2}^{\ell/2} \frac{I(z')e^{-jkR}}{R^5}[(1 + jkR)(2R^2 - 3a^2) + k^2 a^2 R^2]dz'$$

b. By changing variables, $z' - z = a\tan\theta$, show that

$$-E_z^i = \frac{\lambda\sqrt{\mu/\epsilon}}{8j\pi^2 a^2} \int_{\theta_1}^{\theta_2} I(\theta')e^{-jka/\cos\theta'}$$

$$\cdot[(jka + \cos\theta')(2 - 3\cos^2\theta') + k^2 a^2\cos\theta']d\theta'$$

where

$$\theta_1 = -\tan^{-1}\frac{\ell/2+z}{a}, \quad \theta_2 = \tan^{-1}\frac{\ell/2+z}{a}.$$

5.34 Using the program in Figure 5.26 (or your own self-developed program), calculate the electric field inside a thin conducting layer ($\mu = \mu_o$, $\epsilon = 70\epsilon_o$, $\sigma = 1$ mho/m) shown in Figure 5.43. Assume plane wave with electric field perpendicular to the plane of the layer, that is,

$$\mathbf{E}^i = e^{-jk_0z}\mathbf{a}_x \text{ V/m}$$

where $k_o = 2\pi f/c$. Consider only one half of the layer. Calculate $|E_x|/|E^i|$ and neglect E_y and E_z at the center of the cells since they are very small compared with E_x. Take $a = 0.5$ cm, $b = 4$ cm, $c = 6$ cm.

5.35 Consider an adult torso with a height 1.7 m and a shape shown in Figure 5.44. If the torso is illuminated by a vertically polarized EM wave of 80 MHz with an incident electric field of 1 V/m, calculate the absorbed power density given by

$$\frac{\sigma}{2}\left(E_x^2 + E_y^2 + E_z^2\right)$$

at the center of each cell. Take $\mu = \mu_o$, $\epsilon = 80\epsilon_o$, $\sigma = 0.84$ mhos/m.

5.36 Suppose the dielectric cylinder in Problem 5.23 is a biological body modeled by a cylinder of cross-section 75×50 cm², shown in Figure 5.45. A TM wave of frequency $f = 300$ MHz is normally incident on the body. Compute the fields inside the body using the MoM formulation of Problem 5.23. In this case, take ϵ_m as complex permittivity of cell m, that is,

$$\epsilon_m = \epsilon_{rm} - j\left(\sigma_m/\omega\epsilon_0\right), \quad m = 1, 2, \ldots, N = 150$$

To make the body inhomogeneous, take $\epsilon_{rm} = 8$ and $\sigma_m = 0.03$ for cells 65, 66, 75, 85, and 86; take $\epsilon_{rm} = 7$ and $\sigma_m = 0.04$ for cells 64, 67, 74, 77, 84, and 87; and take $\epsilon_{rm} = 5$ and $\sigma_m = 0.02$ for all the other cells. Compute E_n.

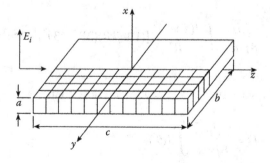

FIGURE 5.43
For Problem 5.34.

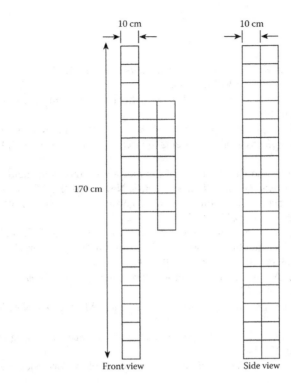

FIGURE 5.44
An adult torso: for Problem 5.35.

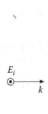

141	142	143	144	145	146	147	148	149	150
131									140
121									130
111									120
101									110
91									100
81			84	85	86	87			90
71			74	75	76	77			80
61			64	65	66	67			70
51									60
41									50
31									40
21									30
11									20
1	2	3	4	5	6	7	8	9	10

FIGURE 5.45
For Problem 5.36.

References

1. L.V. Kantorovich and V.I. Krylov, *Approximate Methods of Higher Analysis (translated from Russian by C.D. Benster)*. New York: John Wiley, 1964.
2. Y.U. Vorobev, *Method of Moments in Applied Mathematics (translated from Russian by Seckler)*. New York: Gordon & Breach, 1965.
3. R.F. Harrington, *Field Computation by Moment Methods*. Malabar, FL: Krieger, 1968.
4. B.J. Strait, *Applications of the Method of Moments to Electromagnetics*. St. Cloud, FL: SCEEE Press, 1980.
5. R.F. Harrington, "Origin and development of the method moments for field computation," in E.K. Miller et al. (ed.), *Computational Electromagnetics*. New York: IEEE Press, 1992, pp. 43–47.
6. J.H. Richmond, "Digital computer solutions of the rigorous equations for scattering problems," *Proc. IEEE*, vol. 53, Aug. 1965, pp. 796–804.
7. R.F. Harrington, "Matrix methods for field problems," *Proc. IEEE*, vol. 55, Feb. 1967, pp. 136–149.
8. M.M. Ney, "Method of moments as applied to electromagnetics problems," *IEEE Trans. Micro. Theo. Tech.*, vol. MTT-33, no. 10, Oct. 1985, pp. 972–980.
9. E. Arvas and L. Sevgi, "A tutorial on the method of moments," *IEEE Antennas Propag. Mag.*, vol. 54, no. 3, June 2012, pp. 260–275.
10. R.C. Booton, *Computational Methods for Electromagnetics and Microwaves*. New York: John Wiley & Sons, 1992, p. 117.
11. A. Ishimaru, *Electromagnetic Wave Propagation, Radiation, and Scattering*. Englewood Cliffs, NJ: Prentice-Hall, 1991, pp. 121–148.
12. P.M. Morse and H. Feshbach, *Methods of Theoretical Physics*. New York: McGraw-Hill, 1953, Part I, Chap. 7, pp. 791–895.
13. T. Myint-U, *Partial Differential Equations of Mathematical Physics*, 2nd Edition. New York: North-Holland, 1980, Chap. 10, pp. 282–305.
14. R.F. Harrington, *Time-Harmonic Electromagnetic Fields*. New York: McGraw-Hill, 1961, p. 232.
15. I.V. Bewley, *Two-Dimensional Fields in Electrical Engineering*. New York: Dover Publ., 1963, pp. 151–166.
16. K.C. Gupta et al., *Computer-aided Design of Microwave Circuits*. Dedham, MA: Artech House, 1981, pp. 237–261.
17. T. Itoh (ed.), *Numerical Techniques for Microwaves and Millimeterwave Passive Structures*. New York: John Wiley & Sons, 1989, pp. 221–250.
18. D.W. Kammler, "Calculation of characteristic admittance and coupling coefficients for strip transmission lines," *IEEE Trans. Micro. Tech.*, vol. MTT-16, no. 11, Nov. 1968, pp. 925–937.
19. Y.M. Hill et al., "A general method for obtaining impedance and coupling characteristics of practical microstrip and triplate transmission line configurations," *IBM J. Res. Dev.*, vol. 13, May 1969, pp. 314–322.
20. W.T. Weeks, "Calculation of coefficients of capacitance of multiconductor transmission lines in the presence of a dielectric interface," *IEEE Trans. Micro. Theo. Tech.*, vol. MTT-18, no. 1, Jan. 1970, pp. 35–43.
21. R. Chadha and K.C. Gupta, "Green's functions for triangular segments in planar microwave circuits," *IEEE Trans. Micro. Theo. Tech.*, vol. MTT-28, no. 10, Oct. 1980, pp. 1139–1143.
22. R. Terras and R. Swanson, "Image methods for constructing Green's functions and eigenfunctions for domains with plane boundaries," *J. Math. Phys.*, vol. 21, no. 8, Aug. 1980, pp. 2140–2153.
23. E. Yamashita and K. Atsuki, "Stripline with rectangular outer conductor and three dielectric layers," *IEEE Trans. Micro. Theo. Tech.*, vol. MTT-18, no. 5, May 1970, pp. 238–244.
24. R. Crampagne et al., "A simple method for determining the Green's function for a large class of MIC lines having multilayered dielectric structures," *IEEE Trans. Micro. Theo. Tech.*, vol. MTT-26, no. 2, Feb. 1978, pp. 82–87.

25. R. Chadha and K.C. Gupta, "Green's functions for circular sectors, annular rings, and annular sectors in planar microwave circuits," *IEEE Trans. Micro. Theo. Tech.*, vol. MTT-29, no. 1, Jan. 1981, pp. 68–71.

26. P.H. Pathak, "On the eigenfunction expansion of electromagnetic dydadic Green's functions," *IEEE Trans. Ant. Prog.*, vol. AP-31, no. 6, Nov. 1983, pp. 837–846.

27. R.F. Harrington and J.R. Mautz, "Green's functions for surfaces of revolution," *Radio Sci.*, vol. 7, no. 5, May 1972, pp. 603–611.

28. C.A. Balanis, *Advanced Engineering Electromagnetics*. New York: John Wiley, 1989, pp. 670–742, 851–916.

29. E. Butkov, *Mathematical Physics*. New York: Addison-Wesley, 1968, pp. 503–552.

30. J.D. Jackson, *Classical Electrodynamics*, 2nd Edition. New York: John Wiley, 1975, pp. 119–135.

31. L.L. Tsai and C.E. Smith, "Moment methods in electromagnetics undergraduates," *IEEE Trans. Educ.*, vol. E-21, no. 1, Feb. 1978, pp. 14–22.

32. P.P. Silvester and R.L. Ferrari, *Finite Elements for Electrical Engineers*. Cambridge, UK: Cambridge University Press, 1983, pp. 103–105.

33. H.A. Wheeler, "Transmission-line properties of parallel strips separated by a dielectric sheet," *IEEE Trans. Micro. Theo. Tech.*, vol. MTT-13, Mar. 1965, pp. 172–185.

34. J.H. Richmond, "Scattering by an arbitrary array of parallel wires," *IEEE Trans. Micro. Theo. Tech.*, vol. MTT-13, no. 4, July 1965, pp. 408–412.

35. B.D. Popovic et al., *Analysis and Synthesis of Wire Antennas*. Chichester, UK: Research Studies Press, 1982, pp. 3–21.

36. E. Hallen, "Theoretical investigations into the transmitting and receiving qualities of antennae," *Nova Acta Regiae Soc. Sci. Upsaliensis*, Ser. IV, no. 11, 1938, pp. 1–44.

37. K.K. Mei, "On the integral equations of thin wire antennas," *IEEE Trans. Ant. Prog.*, vol. AP-13, 1965, pp. 374–378.

38. J. Moore and P. Pizer (eds.), *Moment Methods in Electromagnetics: Techniques and Applications*. Letchworth, UK: Research Studies Press, 1984.

39. H.C. Pocklington, "Electrical oscillations in wire," *Cambridge Phil. Soc. Proc.*, vol. 9, 1897, pp. 324–332.

40. R. Mittra (ed.), *Computer Techniques for Electromagnetics*. Oxford: Pergamon Press, 1973, pp. 7–95.

41. C.A. Balanis, *Antenna Theory: Analysis and Design*. New York: Harper & Row, 1982, pp. 283–321.

42. C.M. Butler and D.R. Wilton, "Analysis of various numerical techniques applied to thin-wire scatterers," *IEEE Trans. Ant. Prog.*, vol. AP-23, no. 4, July 1975, pp. 524–540. Also in [46, pp. 46–52].

43. T.K. Sarkar, "A note on the choice of weighting functions in the method of moments," *IEEE Trans. Ant. Prog.*, vol. AP-33, no. 4, April 1985, pp. 436–441.

44. F.M. Landstorfer and R.F. Sacher, *Optimisation of Wire Antenna*. Letchworth, UK: Research Studies Press, 1985, pp. 18–33.

45. K.A. Michalski and C.M. Butler, "An efficient technique for solving the wire integral equation with non-uniform sampling," *Conf. Proc. IEEE Southeastcon.*, April 1983, pp. 507–510.

46. R.F. Harrington et al. (eds.), *Lectures on Computational Methods in Electromagnetics*. St. Cloud, FL: SCEEE Press, 1981.

47. R. Kastner and R. Mittra, "A new stacked two-dimensional spectral iterative technique (SIT) for analyzing microwave power deposition in biological media," *IEEE Trans. Micro. Theo. Tech.*, vol. MTT-31, no. 1, Nov. 1983, pp. 898–904.

48. P.W. Barber, "Electromagnetic power deposition in prolate spheroidal models of man and animals at resonance," *IEEE Trans. Biomed. Engr.*, vol. BME-24, no. 6, Nov. 1977, pp. 513–521.

49. J.M. Osepchuk (ed.), *Biological Effects of Electromagnetic Radiation*. New York: IEEE Press, 1983.

50. R.J. Spiegel, "A review of numerical models for predicting the energy deposition and resultant thermal response of humans exposed to electromagnetic fields," *IEEE Trans. Micro. Theo. Tech.*, vol. MTT-32, no. 8, Aug. 1984, pp. 730–746.

51. D.E. Livesay and K.M. Chen, "Electromagnetic fields induced inside arbitrary shaped biological bodies," *IEEE Trans. Micro. Theo. Tech.*, vol. MTT-22, no. 12, Dec. 1974, pp. 1273–1280.

52. J.A. Kong (ed.), *Research Topics in Electromagnetic Theory*. New York: John Wiley, 1981, pp. 290–355.

53. C.T. Tai, *Dyadic Green's Functions in Electromagnetic Theory*. Scranton, PA: Intex Educational Pub., 1971, pp. 46–54.

54. B.S. Guru and K.M. Chen, *A Computer Program for Calculating the Induced EM Field Inside an Irradiated Body*. East Lansing, MI: Dept. of Electrical Engineering and System Science, Michigan State Univ., 48824, 1976.

55. K.M. Chen and B.S. Guru, "Internal EM field and absorbed power density in human torsos induced by 1–500 MHz EM waves," *IEEE Micro. Theo. Tech.*, vol. MTT-25, no. 9, Sept. 1977, pp. 746–756.

56. R. Rukspollmuang and K.M. Chen, "Heating of spherical versus realistic models of human and infrahuman heads by electromagnetic waves," *Radio Sci.*, vol. 14, no. 6S, Nov.–Dec. 1979, pp. 51–62.

57. R. Jongakiem, "Electromagnetic absorption in biological bodies," *M.Sc. Thesis*, Dept. of Electrical and Computer Engineering, Florida Atlantic Univ., Boca Raton, Aug. 1988.

58. O.P. Gandhi, "Electromagnetic absorption in an inhomogeneous model of man for realistic exposure conditions," *Bioelectromagnetics*, vol. 3, 1982, pp. 81–90.

59. O.P. Gandhi et al., "Part-body and multibody effects on absorption of radio-frequency electromagnetic energy by animals and by models of man," *Radio Sci.*, vol. 14, no. 6S, Nov.–Dec. 1979, pp. 15–21.

60. M.J. Hagmann, O.P. Gandhi, and C.H. Durney, "Numerical calculation of electromagnetic energy deposition for a realistic model of man," *IEEE Trans. Micro. Theo. Tech.*, vol. MTT-27, no. 9, Sept. 1979, pp. 804–809.

61. W.C. Gibson, *The Method of Moments in Electromagnetics*. Boca Raton, FL: CRC Press, 2008.

62. G. Goertzel and N. Tralli, *Some Mathematical Methods of Physics*. New York: McGraw-Hill, 1960.

63. W.C. Chew et al., *Fast and Efficient Algorithms in Computational Electromagnetics*. Norwood, MA: Artech House, 2001, pp. 203–282.

64. K.F. Warnick, *Numerical Analysis for Electromagnetic Integral Equations*. Norwood, MA: Artech House, 2008.

65. R.F. Harrington et al., "Computation of Laplacian potentials by an equivalent source method," *Proc. IEEE*, vol. 116, no. 10, Oct. 1969, pp. 1715–1720.

66. J.R. Mautz and R.F. Harrington, "Computation of rotationally symmetric Laplacian," *Proc. IEEE*, vol. 117, no. 4, April 1970, pp. 850–852.

67. S.M. Rao et al., "A simple numerical solution procedure for static problems involving arbitrary-shaped surfaces," *IEEE Trans. Ant. Prog.*, vol. AP-27, no. 5, Sept. 1979, pp. 604–608.

68. K. Adamiak, "Application of integral equations for solving inverse problems in stationary electromagnetic fields," *Int. J. Num. Meth. Engr.*, vol. 21, 1985, pp. 1447–1485.

69. A.W. Glisson, "An integral equation for electromagnetic scattering from homogeneous dielectric bodies," *IEEE Trans. Ant. Prog.*, vol. AP-33, no. 2, Sept. 1984, pp. 172–175.

70. J. Perini and D.J. Buchanan, "Assessment of MOM techniques for shipboard applications," *IEEE Trans. Elect. Comp.*, vol. EMC-24, no. 1, Feb. 1982, pp. 32–39.

71. J.R. Mautz and R.F. Harrington, "Radiation and scattering from bodies of revolution," *Appl. Sci. Res.*, vol. 20, June 1969, pp. 405–435.

72. A.W. Glisson and C.M. Butler, "Analysis of a wire antenna in the presence of a body of revolution," *IEEE Trans. Ant. Prog.*, vol. AP-28, Sept. 1980, pp. 604–609.

73. J.H. Richmond, "A wire-grid model for scattering by conducting bodies," *IEEE Trans. Ant. Prog.*, vol. AP-14, Nov. 1966, pp. 782–786.

74. S.M. Rao et al., "Electromagnetic scattering by surfaces of arbitrary shape," *IEEE Trans. Ant. Prog.*, vol. AP-30, May 1966, pp. 409–418.

75. A. Farrar and A.T. Adams, "Matrix methods for microstrip three-dimensional problems," *IEEE Trans. Micro. Theo. Tech.*, vol. MTT-20, no. 8, Aug. 1972, pp. 497–505.

76. A. Farrar and A.T. Adams, "Computation of propagation constants for the fundamental and higher-order modes in microstrip," *IEEE Trans. Micro. Theo. Tech.*, vol. MTT-24, no. 7, July 1972, pp. 456–460.

77. A. Farrar and A.T. Adams, "Characteristic impedance of microstrip by the method of moments," *IEEE Trans. Micro. Theo. Tech.*, vol. MTT-18, no. 1, Jan 1970, pp. 68–69.

78. A. Farrar and A.T. Adams, "Computation of lumped microstrip capacitance by matrix methods—Rectangular sections and end effects," *IEEE Trans. Micro. Theo. Tech.*, vol. MTT-19, no. 5, May 1971, pp. 495–496.

79. R.F. Harrington and J.R. Mautz, "A generalized network formulation for aperture problems," *IEEE Trans. Ant. Prog.*, vol. AP-24, Nov. 1976, pp. 870–873.

80. R.H. Harrington and D.T. Auckland, "Electromagnetic transmission through narrow slots in thick conducting screens," *IEEE Trans. Ant. Prog.*, vol. AP-28, Sept. 1980, pp. 616–622.

81. C.M. Butler and K.R. Umashankar, "Electromagnetic excitation of a scatterer coupled to an aperture in a conducting screen," *Proc. IEEE*, vol. 127, Pt. H, June 1980, pp. 161–169.

82. "Special issue on electromagnetic wave interactions with biological systems," *IEEE Trans. Micro. Theo. Tech.*, vol. MTT-32, no. 8, Aug. 1984.

83. "Special issue on effects of electromagnetic radiation," *IEEE Engr. Med. Biol. Mag.*, March 1987.

84. "Helsinki symposium on biological effects of electromagnetic radiation," *Radio Sci.*, vol. 17, no. 5S, Sept.–Oct. 1982.

85. R.M. Bevensee, "Computer codes for EMP interaction and coupling," *IEEE Trans. Ant. Prog.*, vol. AP-26, no. 1, Jan. 1978, pp. 156–165.

86. G.J. Burke and A.J. Poggio, *Numerical Electromagnetic Code (NEC)—Method of Moments*. Livermore, CA: Lawrence Livermore National Lab., Jan. 1981.

87. J.W. Rockway et al., *The MININEC System: Microcomputer Analysis of Wire Antennas*. Norwood, MA: Artech House, 1988.

88. R. Mittra (ed.), *Numerical and Asymptotic Techniques in Electromagnetics*. New York: Springer-Verlag, 1975.

89. G.L. James, *Geometrical Theory of Diffraction for Electromagnetic Waves*, 3rd Edition. London: Peregrinus, 1986.

90. A.B. Manić et al., "Diakoptic approach combining finite-element method and method of moments in analysis of inhomogeneous anisotropic dielectric and magnetic scatterers," *Electromagnetics*, vol. 34, no. 3–4, 2014, pp. 222–238.

91. S.E. Sandströmand and I.K. Akeab, "Accurate hybrid techniques for the method of moments in 2D," *Int. J. Electron. Commun. (AEÜ)*, vol. 70, 2016, pp. 539–543.

92. J. Alvarez et al., "Fully coupled multi-hybrid finite element method–method of moments–physical optics method for scattering and radiation problems," *Electromagnetics*, vol. 30, no. 1–2, 2010, pp. 3–22.

93. J. Van Bladel, "Some remarks on Green's dyadic infinite space," *IRE Trans. Ant. Prog.*, vol. AP-9, Nov. 1961, pp. 563–566.

94. E. Max, "Integral equation for scattering by a dielectric," *IEEE Trans. Ant. Prog.*, vol. AP-32, no. 2, Feb. 1984, pp. 166–172.

95. A.T. Adams et al., "Near fields of wire antennas by matrix methods," *IEEE Trans. Ant. Prog.*, vol. AP-21, no. 5, Sept. 1973, pp. 602–610.

96. R.F. Harrington and J.R. Mautz, "Electromagnetic behavior of circular wire loops with arbitrary excitation and loading," *Proc. IEEE*, vol. 115, Jan. 1969, pp. 68–77.

97. S.A. Adekola and O.U. Okereke, "Analysis of a circular loop antenna using moment methods," *Int. J. Elect.*, vol. 66, no. 5, 1989, pp. 821–834.

98. E.K. Miller et al., "Computer evaluation of large low-frequency antennas," *IEEE Trans. Ant. Prog.*, vol. AP-21, no. 3, May 1973, pp. 386–389.

99. K.S.H. Lee et al., "Limitations of wire-grid modeling of a closed surface," *IEEE Trans. Elect. Comp.*, vol. EMC-18, no. 3, Aug. 1976, pp. 123–129.

100. E.H. Newman and D.M. Pozar, "Considerations for efficient wire/surface modeling," *IEEE Trans. Ant. Prog.*, vol. AP-28, no. 1, Jan. 1980, pp. 121–125.

6

Finite Element Method

Prayer without action is hypocrisy and action without prayer is arrogance.

—Unknown

6.1 Introduction

The finite element method (FEM) has its origin in the field of structural analysis. Although the earlier mathematical treatment of the method was provided by Courant [1] in 1943, the method was not applied to electromagnetic (EM) problems until 1968. Since then the method has been employed in diverse areas such as waveguide problems, electric machines, semiconductor devices, microstrips, and absorption of EM radiation by biological bodies.

Although the finite difference method (FDM) and the method of moments (MoM) are conceptually simpler and easier to program than the FEM, FEM is a more powerful and versatile numerical technique for handling problems involving complex geometries and inhomogeneous media. The systematic generality of the method makes it possible to construct general-purpose computer programs for solving a wide range of problems. Consequently, programs developed for a particular discipline have been applied successfully to solve problems in a different field with little or no modification. A brief history of the beginning of FEM is provided in Reference 2.

The finite element analysis of any problem involves basically four steps [3]:

- Discretizing the solution region into a finite number of nonoverlap *subregions* or *elements*,
- Deriving governing equations for a typical element,
- Assembling of all elements in the solution region, and
- Solving the system of equations obtained.

Discretization of the continuum involves dividing the solution region into subdomains, called *finite elements*. Figure 6.1 shows some typical elements for one-, two-, and three-dimensional problems. The problem of discretization will be fully treated in Sections 6.5 and 6.6. The other three steps will be described in detail in the subsequent sections.

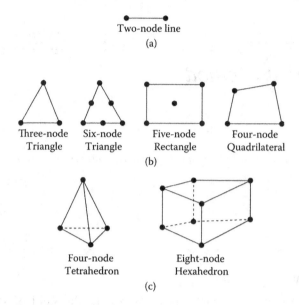

Two-node line

(a)

Three-node   Six-node   Five-node   Four-node
Triangle   Triangle   Rectangle   Quadrilateral

(b)

Four-node   Eight-node
Tetrahedron   Hexahedron

(c)

FIGURE 6.1
Typical finite elements: (a) one-dimensional, (b) two-dimensional, (c) three-dimensional.

6.2 Solution of Laplace's Equation

As an application of FEM to electrostatic problems, let us apply the four steps mentioned above to solve Laplace's equation, $\nabla^2 V = 0$. For the purpose of illustration, we will strictly follow the four steps mentioned above.

6.2.1 Finite Element Discretization

To find the potential distribution $V(x, y)$ for the two-dimensional solution region shown in Figure 6.2a, we divide the region into a number of finite elements as illustrated in Figure 6.2b. In Figure 6.2b, the solution region is subdivided into nine nonoverlapping *finite elements*; elements 6, 8, and 9 are four-node quadrilaterals, while other elements are three-node triangles. In practical situations, however, it is preferred, for ease of computation, to have elements of the same type throughout the region. That is, in Figure 6.2b, we could have split each quadrilateral into two triangles so that we would have 12 triangular elements altogether. The subdivision of the solution region into elements is usually done by hand, but in situations where a large number of elements is required, automatic schemes to be discussed in Sections 6.5 and 6.6 are used.

We seek an approximation for the potential V_e within an element e and then interrelate the potential distribution in various elements such that the potential is continuous across interelement boundaries. The approximate solution for the whole region is

$$V(x, y) \simeq \sum_{e=1}^{N} V_e(x, y), \tag{6.1}$$

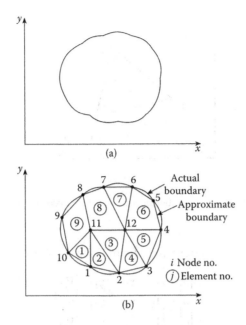

FIGURE 6.2
(a) The solution region; (b) its finite element discretization.

where N is the number of triangular elements into which the solution region is divided. The most common form of approximation for V_e within an element is polynomial approximation, namely,

$$V_e(x, y) = a + bx + cy \tag{6.2}$$

for a triangular element and

$$V_e(x, y) = a + bx + cy + dxy \tag{6.3}$$

for a quadrilateral element. The constants a, b, c, and d are to be determined. The potential V_e in general is nonzero within element e but zero outside e. In view of the fact that quadrilateral elements do not conform to curved boundary as easily as triangular elements, we prefer to use triangular elements throughout our analysis in this chapter. Notice that our assumption of linear variation of potential within the triangular element as in Equation 6.2 is the same as assuming that the electric field is uniform within the element, that is,

$$\mathbf{E}_e = -\nabla V_e = -(b\mathbf{a}_x + c\mathbf{a}_y) \tag{6.4}$$

6.2.2 Element Governing Equations

Consider a typical triangular element shown in Figure 6.3. The potential V_{e1}, V_{e2}, and V_{e3} at nodes 1, 2, and 3, respectively, are obtained using Equation 6.2, that is,

$$\begin{bmatrix} V_{e1} \\ V_{e2} \\ V_{e2} \end{bmatrix} = \begin{bmatrix} 1 & x_1 & y_1 \\ 1 & x_2 & y_2 \\ 1 & x_3 & y_3 \end{bmatrix} \begin{bmatrix} a \\ b \\ c \end{bmatrix} \tag{6.5}$$

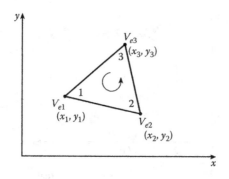

FIGURE 6.3
Typical triangular element; local node numbering 1-2-3 must proceed counterclockwise as indicated by the arrow.

The coefficients *a*, *b*, and *c* are determined from Equation 6.5 as

$$
\begin{bmatrix} a \\ b \\ c \end{bmatrix} = \begin{bmatrix} 1 & x_1 & y_1 \\ 1 & x_2 & y_2 \\ 1 & x_3 & y_3 \end{bmatrix}^{-1} \begin{bmatrix} V_{e1} \\ V_{e2} \\ V_{e3} \end{bmatrix} \tag{6.6}
$$

Substituting this into Equation 6.2 gives

$$
V_e = [1\,x\,y] \frac{1}{2A} \begin{bmatrix} (x_2y_3 - x_3y_2) & (x_3y_1 - x_1y_3) & (x_1y_2 - x_2y_1) \\ (y_2 - y_3) & (y_3 - y_1) & (y_1 - y_2) \\ (x_3 - x_2) & (x_1 - x_3) & (x_2 - x_1) \end{bmatrix} \begin{bmatrix} V_{e1} \\ V_{e2} \\ V_{e3} \end{bmatrix}
$$

or

$$
\boxed{V_e = \sum_{i=1}^{3} \alpha_i(x, y) V_{ei}} \tag{6.7}
$$

where

$$
\alpha_1 = \frac{1}{2A}[(x_2y_3 - x_3y_2) + (y_2 - y_3)x + (x_3 - x_2)y], \tag{6.8a}
$$

$$
\alpha_2 = \frac{1}{2A}[(x_3y_1 - x_1y_3) + (y_3 - y_1)x + (x_1 - x_3)y], \tag{6.8b}
$$

$$
\alpha_3 = \frac{1}{2A}[(x_1y_2 - x_2y_1) + (y_1 - y_2)x + (x_2 - x_1)y], \tag{6.8c}
$$

and *A* is the area of the element *e*, that is,

$$
2A = \begin{vmatrix} 1 & x_1 & y_1 \\ 1 & x_2 & y_2 \\ 1 & x_3 & y_3 \end{vmatrix}
$$

$$
= (x_1y_2 - x_2y_1) + (x_3y_1 - x_1y_3) + (x_2y_3 - x_3y_2)
$$

or

$$A = \frac{1}{2}[(x_2 - x_1)(y_3 - y_1) - (x_3 - x_1)(y_2 - y_1)] \tag{6.9}$$

The value of A is positive if the nodes are numbered counterclockwise (starting from any node) as shown by the arrow in Figure 6.3. Note that Equation 6.7 gives the potential at any point (x, y) within the element provided that the potentials at the vertices are known. This is unlike finite difference analysis, where the potential is known at the grid points only. Also note that α_i are linear interpolation functions. They are called the *element shape functions* and they have the following properties [4]:

$$\alpha_i(x_j, y_i) = \begin{cases} 1, & i = j \\ 0, & i \neq j \end{cases} \tag{6.10a}$$

$$\sum_{i=1}^{3} \alpha_i(x, y) = 1 \tag{6.10b}$$

The shape functions α_1, α_2, and α_3 are illustrated in Figure 6.4.
The functional corresponding to Laplace's equation, $\nabla^2 V = 0$, is given by

$$W_e = \frac{1}{2} \int \epsilon |\mathbf{E}_e|^2 \, dS = \frac{1}{2} \int \epsilon |\nabla V_e|^2 \, dS \tag{6.11}$$

(Physically, the functional W_e is the energy per unit length associated with the element e.) From Equation 6.7,

$$\nabla V_e = \sum_{i=1}^{3} V_{ei} \nabla \alpha_i \tag{6.12}$$

Substituting Equation 6.12 into Equation 6.11 gives

$$W_e = \frac{1}{2} \sum_{i=1}^{3} \sum_{j=1}^{3} \epsilon V_{ei} \left[\int \nabla \alpha_i \cdot \nabla \alpha_j \, dS \right] V_{ej} \tag{6.13}$$

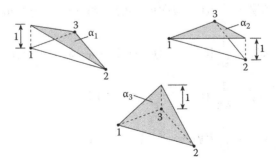

FIGURE 6.4
Shape functions α_1, α_2, and α_3 for a triangular element.

If we define the term in brackets as

$$C_{ij}^{(e)} = \int \nabla \alpha_i \cdot \nabla \alpha_j \, dS, \tag{6.14}$$

we may write Equation 6.13 in matrix form as

$$W_e = \frac{1}{2} \epsilon [V_e]^t [C^{(e)}][V_e] \tag{6.15}$$

where the superscript t denotes the transpose of the matrix,

$$[V_e] = \begin{bmatrix} V_{e1} \\ V_{e2} \\ V_{e3} \end{bmatrix} \tag{6.16a}$$

and

$$[C^{(e)}] = \begin{bmatrix} C_{11}^{(e)} & C_{12}^{(e)} & C_{13}^{(e)} \\ C_{21}^{(e)} & C_{22}^{(e)} & C_{23}^{(e)} \\ C_{31}^{(e)} & C_{32}^{(e)} & C_{33}^{(e)} \end{bmatrix} \tag{6.16b}$$

The matrix $[C^{(e)}]$ is usually called the *element coefficient matrix* (or "stiffness matrix" in structural analysis). The element $C_{ij}^{(e)}$ of the coefficient matrix may be regarded as the coupling between nodes i and j; its value is obtained from Equations 6.8 and 6.14. For example,

$$\begin{aligned} C_{12}^{(e)} &= \int \nabla \alpha_1 \cdot \nabla \alpha_2 \, dS \\ &= \frac{1}{4A^2} [(y_2 - y_3)(y_3 - y_1) + (x_3 - x_2)(x_1 - x_3)] \int dS \\ &= \frac{1}{4A} [(y_2 - y_3)(y_3 - y_1) + (x_3 - x_2)(x_1 - x_3)] \end{aligned} \tag{6.17a}$$

Similarly,

$$C_{13}^{(e)} = \frac{1}{4A} [(y_2 - y_3)(y_1 - y_2) + (x_3 - x_2)(x_2 - x_1)], \tag{6.17b}$$

$$C_{23}^{(e)} = \frac{1}{4A} [(y_3 - y_1)(y_1 - y_2) + (x_1 - x_3)(x_2 - x_1)], \tag{6.17c}$$

$$C_{11}^{(e)} = \frac{1}{4A} [(y_2 - y_3)^2 + (x_3 - x_2)^2], \tag{6.17d}$$

$$C_{22}^{(e)} = \frac{1}{4A} [(y_3 - y_1)^2 + (x_1 - x_3)^2], \tag{6.17e}$$

$$C_{33}^{(e)} = \frac{1}{4A} [(y_1 - y_2)^2 + (x_2 - x_1)^2], \tag{6.17f}$$

Also

$$C_{21}^{(e)} = C_{12}^{(e)}, \quad C_{31}^{(e)} = C_{13}^{(e)}, \quad C_{32}^{(e)} = C_{23}^{(e)} \tag{6.18}$$

6.2.3 Assembling of All Elements

Having considered a typical element, the next step is to assemble all such elements in the solution region. The energy associated with the assemblage of elements is

$$W = \sum_{e=1}^{N} W_e = \frac{1}{2} \epsilon [V]^t [C][V] \tag{6.19}$$

where

$$[V] = \begin{bmatrix} V_1 \\ V_2 \\ V_3 \\ \vdots \\ V_n \end{bmatrix}, \tag{6.20}$$

n is the number of nodes, N is the number of elements, and $[C]$ is called the overall or *global coefficient matrix*, which is the assemblage of individual element coefficient matrices. Notice that to obtain Equation 6.19, we have assumed that the whole solution region is homogeneous so that ϵ is constant. For an inhomogeneous solution region such as shown in Figure 6.5, for example, the region is discretized such that each finite element is homogeneous. In this case, Equation 6.11 still holds, but Equation 6.19 does not apply since $\epsilon(=\epsilon_r\epsilon_o)$ or simply ϵ_r varies from element to element. To apply Equation 6.19, we may replace ϵ by ϵ_o and multiply the integrand in Equation 6.14 by ϵ_r.

The process by which individual element coefficient matrices are assembled to obtain the global coefficient matrix is best illustrated with an example. Consider the finite element mesh consisting of three finite elements as shown in Figure 6.6. Observe the numberings of the mesh. The numbering of nodes 1, 2, 3, 4, and 5 is called *global* numbering. The numbering

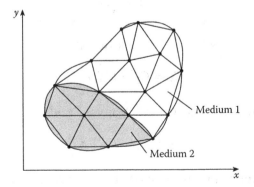

FIGURE 6.5
Discretization of an inhomogeneous solution region.

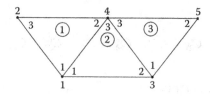

FIGURE 6.6
Assembly of three elements; *i- j-k* corresponds to local numbering (1-2-3) of the element in Figure 6.3.

i-j-k is called *local* numbering, and it corresponds with 1-2-3 of the element in Figure 6.3. For example, for element 3 in Figure 6.6, the global numbering 3-5-4 corresponds with local numbering 1-2-3 of the element in Figure 6.3. (Note that the local numbering must be in counterclockwise sequence starting from any node of the element.) For element 3, we could choose 4-3-5 instead of 3-5-4 to correspond with 1-2-3 of the element in Figure 6.3. Thus, the numbering in Figure 6.6 is not unique. But whichever numbering is used, the global coefficient matrix remains the same. Assuming the particular numbering in Figure 6.6, the global coefficient matrix is expected to have the form

$$[C] = \begin{bmatrix} C_{11} & C_{12} & C_{13} & C_{14} & C_{15} \\ C_{21} & C_{22} & C_{23} & C_{24} & C_{25} \\ C_{31} & C_{32} & C_{33} & C_{34} & C_{35} \\ C_{41} & C_{42} & C_{43} & C_{44} & C_{45} \\ C_{51} & C_{52} & C_{53} & C_{54} & C_{55} \end{bmatrix} \tag{6.21}$$

which is a 5×5 matrix since five nodes ($n = 5$) are involved. Again, C_{ij} is the coupling between nodes i and j. We obtain C_{ij} by using the fact that the potential distribution must be continuous across interelement boundaries. The contribution to the i, j position in $[C]$ comes from all elements containing nodes i and j. For example, in Figure 6.6, elements 1 and 2 have global node 1 in common; hence,

$$C_{11} = C_{11}^{(1)} + C_{11}^{(2)} \tag{6.22a}$$

Node 2 belongs to element 1 only; hence,

$$C_{22} = C_{33}^{(1)} \tag{6.22b}$$

Node 4 belongs to elements 1, 2, and 3; consequently

$$C_{44} = C_{22}^{(1)} + C_{33}^{(2)} + C_{33}^{(3)} \tag{6.22c}$$

Nodes 1 and 4 belong simultaneously to elements 1 and 2; hence,

$$C_{14} = C_{41} = C_{12}^{(1)} + C_{13}^{(2)} \tag{6.22d}$$

Since there is no coupling (or direct link) between nodes 2 and 3,

$$C_{23} = C_{32} = 0 \tag{6.22e}$$

Continuing in this manner, we obtain all the terms in the global coefficient matrix by inspection of Figure 6.6 as

$$
\begin{bmatrix}
C_{11}^{(1)}+C_{11}^{(2)} & C_{13}^{(1)} & C_{12}^{(2)} & C_{12}^{(1)}+C_{13}^{(2)} & 0 \\
C_{31}^{(1)} & C_{33}^{(1)} & 0 & C_{32}^{(2)} & 0 \\
C_{21}^{(2)} & 0 & C_{22}^{(2)}+C_{11}^{(3)} & C_{23}^{(2)}+C_{13}^{(3)} & C_{12}^{(3)} \\
C_{21}^{(1)}+C_{31}^{(2)} & C_{23}^{(1)} & C_{32}^{(2)}+C_{31}^{(3)} & C_{22}^{(1)}+C_{33}^{(2)}+C_{33}^{(3)} & C_{32}^{(3)} \\
0 & 0 & C_{21}^{(3)} & C_{23}^{(3)} & C_{22}^{(3)}
\end{bmatrix}
\tag{6.23}
$$

Note that element coefficient matrices overlap at nodes shared by elements and that there are 27 terms (nine for each of the three elements) in the global coefficient matrix $[C]$. Also note the following properties of the matrix $[C]$:

1. It is symmetric ($C_{ij} = C_{ji}$) just as the element coefficient matrix.
2. Since $C_{ij} = 0$ if no coupling exists between nodes i and j, it is expected that for a large number of elements $[C]$ becomes sparse. Matrix $[C]$ is also banded if the nodes are carefully numbered. It can be shown using Equation 6.17 that

$$
\sum_{i=1}^{3} C_{ij}^{(e)} = 0 = \sum_{j=1}^{3} C_{ij}^{(e)}
$$

3. It is singular. Although this is not so obvious, it can be shown using the element coefficient matrix of Equation 6.16b.

6.2.4 Solving the Resulting Equations

Using the concepts developed in Chapter 4, it can be shown that Laplace's equation is satisfied when the total energy in the solution region is minimum. Thus, we require that the partial derivatives of W with respect to each nodal value of the potential be zero, that is,

$$
\frac{\partial W}{\partial V_1} = \frac{\partial W}{\partial V_2} = \cdots = \frac{\partial W}{\partial V_n} = 0
$$

or

$$
\frac{\partial W}{\partial V_k} = 0 \quad k = 1, 2, \ldots, n
\tag{6.24}
$$

For example, to get $\partial W/\partial V_1 = 0$ for the finite element mesh of Figure 6.6, we substitute Equation 6.21 into Equation 6.19 and take the partial derivative of W with respect to V_1. We obtain

$$
0 = \frac{\partial W}{\partial V_1} = 2V_1 C_{11} + V_2 C_{12} + V_3 C_{13} + V_4 C_{14} + V_5 C_{15}
$$

$$
+ V_2 C_{21} + V_3 C_{31} + V_4 C_{41} + V_5 C_{51}
$$

or

$$0 = V_1 C_{11} + V_2 C_{12} + V_3 C_{13} + V_4 C_{14} + V_5 C_{15} \tag{6.25}$$

In general, $\partial W / \partial V_k = 0$ leads to

$$0 = \sum_{i=1}^{n} V_i C_{ik} \tag{6.26}$$

where n is the number of nodes in the mesh. By writing Equation 6.26 for all nodes $k = 1, 2, \ldots, n$, we obtain a set of simultaneous equations from which the solution of $[V]^t = [V_1, V_2, \ldots, V_n]$ can be found. This can be done in two ways similar to those used in solving finite difference equations obtained from Laplace's equation in Section 3.5.

1. *Iteration Method*: Suppose node 1 in Figure 6.6, for example, is a free node. A free node is when the potential is unknown, whereas a fixed node is when the potential is prescribed. From Equation 6.25,

$$V_1 = -\frac{1}{C_{11}} \sum_{i=2}^{5} V_i C_{1i} \tag{6.27}$$

Thus, in general, at node k in a mesh with n nodes

$$\boxed{V_k = -\frac{1}{C_{kk}} \sum_{i=1, i \neq k}^{n} V_i C_{ki}} \tag{6.28}$$

where node k is a free node. Since $C_{ki} = 0$ if node k is not directly connected to node i, only nodes that are directly linked to node k contribute to V_k in Equation 6.28. Equation 6.28 can be applied iteratively to all the free nodes. The iteration process begins by setting the potentials of fixed nodes (where the potentials are prescribed or known) to their prescribed values and the potentials at the free nodes (where the potentials are unknown) equal to zero or to the average potential [5]

$$V_{ave} = \frac{1}{2}(V_{min} + V_{max}) \tag{6.29}$$

where V_{min} and V_{max} are the minimum and maximum values of V at the fixed nodes. With these initial values, the potentials at the free nodes are calculated using Equation 6.28. At the end of the first iteration, when the new values have been calculated for all the free nodes, they become the old values for the second iteration. The procedure is repeated until the change between subsequent iterations is negligible enough.

2. *Band Matrix Method*: If all free nodes are numbered first and the fixed nodes last, Equation 6.19 can be written such that [4]

$$W = \frac{1}{2} \epsilon [V_f \ V_p] \begin{bmatrix} C_{ff} & C_{fp} \\ C_{pf} & C_{pp} \end{bmatrix} \begin{bmatrix} V_f \\ V_p \end{bmatrix} \tag{6.30}$$

where subscripts f and p, respectively, refer to nodes with free and fixed (or prescribed) potentials. Since V_p is constant (it consists of known, fixed values), we differentiate only with respect to V_f so that applying Equations 6.24 through 6.30 yields

$$[C_{ff} C_{fp}] \begin{bmatrix} V_f \\ V_p \end{bmatrix} = 0$$

or

$$\boxed{[C_{ff}][V_f] = -[C_{fp}][V_p]} \tag{6.31}$$

This equation can be written as

$$[A][V] = [B] \tag{6.32a}$$

or

$$[V] = [A]^{-1}[B] \tag{6.32b}$$

where $[V] = [V_f]$, $[A] = [C_{ff}]$, $[B] = -[C_{fp}][V_p]$. Since $[A]$ is, in general, nonsingular, the potential at the free nodes can be found using Equation 6.32. We can solve for $[V]$ in Equation 6.32a using Gaussian elimination technique. We can also solve for $[V]$ in Equation 6.32b using matrix inversion if the size of the matrix to be inverted is not large.

It is sometimes necessary to impose Neumann condition ($\partial V / \partial n = 0$) as a boundary condition or at the line of symmetry when we take advantage of the symmetry of the problem. Suppose, for concreteness, that a solution region is symmetric along the y-axis as in Figure 6.7. We impose condition ($\partial V / \partial x = 0$) along the y-axis by making

$$V_1 = V_2, \quad V_4 = V_5, \quad V_7 = V_8 \tag{6.33}$$

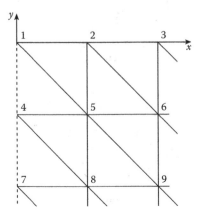

FIGURE 6.7
A solution region that is symmetric along the y-axis.

Notice that as from Equation 6.11 onward, the solution has been restricted to a two-dimensional problem involving Laplace's equation, $\nabla^2 V = 0$. The basic concepts developed in this section will be extended to finite element analysis of problems involving Poisson's equation ($\nabla^2 V = -\rho_v/\epsilon$, $\nabla^2 \mathbf{A} = -\mu \mathbf{J}$) or wave equation ($\nabla^2 \Phi - \gamma^2 \Phi = 0$) in the next sections.

The following two examples were solved in Reference 3 using the band matrix method; here they are solved using the iterative method.

EXAMPLE 6.1

Consider the two-element mesh shown in Figure 6.8a. Using the FEM, determine the potentials within the mesh.

Solution

The element coefficient matrices can be calculated using Equations 6.17 and 6.18. However, our calculations will be easier if we define

$$P_1 = (y_2 - y_3), \qquad P_2 = (y_3 - y_1), \qquad P_3 = (y_1 - y_2),$$
$$Q_1 = (x_3 - x_2), \qquad Q_2 = (x_1 - x_3), \qquad Q_3 = (x_2 - x_1) \tag{6.34}$$

With P_i and Q_i ($i = 1, 2, 3$ are the local node numbers), each term in the element coefficient matrix is found as

$$C_{ij}^{(e)} = \frac{1}{4A}(P_i P_j + Q_i Q_j) \tag{6.35}$$

where $A = 1/2(P_2 Q_3 - P_3 Q_2)$. It is evident that Equation 6.35 is more convenient to use than Equations 6.17 and 6.18. For element 1 consisting of global nodes 1-2-4 corresponding to the local numbering 1-2-3 as in Figure 6.8b,

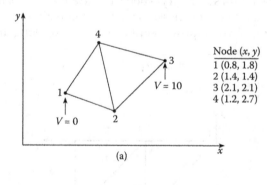

Node (x, y)
1 (0.8, 1.8)
2 (1.4, 1.4)
3 (2.1, 2.1)
4 (1.2, 2.7)

(a)

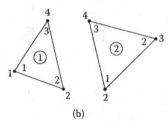

(b)

FIGURE 6.8

For Example 6.1: (a) Two-element mesh, (b) local and global numbering at the elements.

$$P_1 = -1.3, \quad P_2 = 0.9, \quad P_3 = 0.4,$$

$$Q_1 = -0.2, \quad Q_2 = -0.4, \quad Q_3 = 0.6,$$

$$A = \frac{1}{2}(0.54 + 0.16) = 0.35$$

Substituting all of these into Equation 6.35 gives

$$[C^{(1)}] = \begin{bmatrix} 1.2357 & -0.7786 & -0.4571 \\ -0.7786 & 0.6929 & 0.0857 \\ -0.4571 & 0.0857 & 0.3714 \end{bmatrix} \tag{6.36}$$

Similarly, for element 2 consisting of nodes global 2-3-4 corresponding to local numbering 1-2-3 as in Figure 6.8b,

$$P_1 = -0.6, \quad P_2 = 1.3, \quad P_3 = -0.7,$$

$$Q_1 = -0.9, \quad Q_2 = 0.2, \quad Q_3 = 0.7,$$

$$A = \frac{1}{2}(0.91 + 0.14) = 0.525$$

Hence,

$$[C^{(2)}] = \begin{bmatrix} 0.5571 & -0.4571 & -0.1 \\ -0.4571 & 0.8238 & -0.3667 \\ -0.1 & -0.3667 & 0.4667 \end{bmatrix} \tag{6.37}$$

The terms of the global coefficient matrix are obtained as follows:

$$C_{22} = C_{22}^{(1)} + C_{11}^{(2)} = 0.6929 + 0.5571 = 1.25$$
$$C_{24} = C_{23}^{(1)} + C_{13}^{(2)} = 0.0857 - 0.1 = -0.0143$$
$$C_{44} = C_{33}^{(1)} + C_{33}^{(2)} = 0.3714 + 0.4667 = 0.8381$$
$$C_{21} = C_{21}^{(1)} = -0.7786$$
$$C_{23} = C_{31}^{(2)} = -0.4571$$
$$C_{41} = C_{31}^{(1)} = -0.4571$$
$$C_{43} = C_{32}^{(2)} = -0.3667$$

Note that we follow local numbering for the element coefficient matrix and global numbering for the global coefficient matrix. Thus,

$$[C] = \begin{bmatrix} C_{11}^{(1)} & C_{12}^{(1)} & 0 & C_{13}^{(1)} \\ C_{21}^{(1)} & C_{22}^{(1)} + C_{11}^{(2)} & C_{12}^{(2)} & C_{23}^{(1)} + C_{12}^{(2)} \\ 0 & C_{21}^{(2)} & C_{22}^{(2)} & C_{23}^{(2)} \\ C_{31}^{(1)} & C_{32}^{(1)} + C_{31}^{(2)} & C_{32}^{(2)} & C_{33}^{(1)} + C_{33}^{(2)} \end{bmatrix}$$

$$= \begin{bmatrix} 1.2357 & -0.7786 & 0 & -0.4571 \\ -0.7786 & 1.25 & -0.4571 & -0.0143 \\ 0 & -0.4571 & 0.8238 & -0.3667 \\ -0.4571 & -0.0143 & -0.3667 & 0.8381 \end{bmatrix} \tag{6.38}$$

Note that $\sum_{i=1}^{4} C_{ij} = 0 = \sum_{j=1}^{4} C_{ij}$. This may be used to check if C is properly obtained. We now apply Equation 6.28 to the free nodes 2 and 4, that is,

$$V_2 = -\frac{1}{C_{22}}(V_1 C_{12} + V_3 C_{32} + V_4 C_{42})$$

$$V_4 = -\frac{1}{C_{44}}(V_1 C_{14} + V_3 C_{24} + V_3 C_{34})$$

or

$$V_2 = -\frac{1}{1.25}(-4.571 - 0.0143 V_4) \qquad (6.39a)$$

$$V_4 = -\frac{1}{0.8381}(-0.143 V_2 - 3.667) \qquad (6.39b)$$

By initially setting $V_2 = 0 = V_4$, we apply Equations 6.39a and 6.39b iteratively. The first iteration gives $V_2 = 3.6568$, $V_4 = 4.4378$ and at the second iteration $V_2 = 3.7075$, $V_4 = 4.4386$. Just after two iterations, we obtain the same results as those from the band matrix method [3]. Thus, the iterative technique is faster and is usually preferred for a large number of nodes. Once the values of the potentials at the nodes are known, the potential at any point within the mesh can be determined using Equation 6.7.

EXAMPLE 6.2

Write a MATLAB program to solve Laplace's equation using the FEM. Apply the program to the two-dimensional problem shown in Figure 6.9a.

Solution

The solution region is divided into 25 three-node triangular elements with total number of nodes being 21 as shown in Figure 6.9b. This is a necessary step in order to have input data defining the geometry of the problem. Based on the discussions in Section 6.2, a general program for solving problems involving Laplace's equation using three-node triangular elements is developed as shown in Figure 6.10. The development of the program basically involves four steps indicated in the program and explained as follows.

Step 1: This involves inputting the necessary data defining the problem. This is the only step that depends on the geometry of the problem at hand. We input the number of elements, the number of nodes, the number of fixed nodes, the prescribed values of the potentials at the free nodes, the x and y coordinates of all nodes, and a list identifying the nodes belonging to each element in the order of the local numbering 1-2-3. For the problem in Figure 6.9, the three sets of data for coordinates, element–node relationship, and prescribed potentials at fixed nodes are shown in Tables 6.1 through 6.3, respectively.

Step 2: This step entails finding the element coefficient matrix $[C^{(e)}]$ for each element and using the terms to form the global matrix $[C]$.

Step 3: At this stage, we first find the list of free nodes using the given list of prescribed nodes. We now apply Equation 6.28 iteratively to all the free nodes. The solution converges at 50 iterations or less since only six nodes are involved in this case. The solution obtained is exactly the same as those obtained using the band matrix method [3].

Step 4: This involves outputting the result of the computation. The output data for the problem in Figure 6.9 is presented in Table 6.4. The validity of the result in Table 6.4 is checked using the FDM. From the finite difference analysis, the potentials at the free nodes are obtained as.

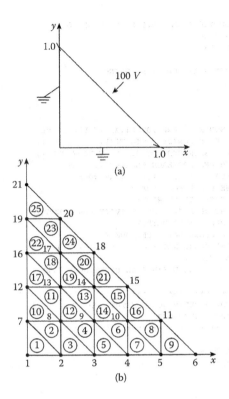

FIGURE 6.9
For Example 6.2: (a) Two-dimensional electrostatic problem, (b) solution region divided into 25 triangular elements.

$$V_8 = 15.41, \quad V_9 = 26.74, \quad V_{10} = 56.69,$$
$$V_{13} = 34.88, \quad V_{14} = 65.41, \quad V_{17} = 58.72V$$

   Although the result obtained using finite difference is considered more accurate in this problem, increased accuracy of finite element analysis can be obtained by dividing the solution region into a greater number of triangular elements, or using higher-order elements to be discussed in Section 6.8. As alluded to earlier, the FEM has two major advantages over the FDM. Field quantities are obtained only at discrete positions in the solution region using FDM; they can be obtained at any point in the solution region in FEM. Also, it is easier to handle complex geometries using FEM than using FDM.

6.3 Solution of Poisson's Equation

To solve the two-dimensional Poisson's equation,

$$\nabla^2 V = -\frac{\rho_v}{\epsilon} \tag{6.40}$$

using FEM, we take the same steps as in Section 6.2. Since the steps are essentially the same as in Section 6.2 except that we must include the source term, only the major differences will be highlighted here.

```
%       FINITE ELEMENT SOLUTION OF LAPLACE'S EQUATION FOR
%       TWO-DIMENSIONAL PROBLEMS
%       TRIANGULAR ELEMENTS ARE USED
%
%       THE UNKNOWN POTENTIALS ARE OBTAINED USING
%       ITERATION METHOD
%
%       ND = NO. OF NODES
%       NE = NO. OF ELEMENTS
%       NP = NO. OF FIXED NODES (WHERE POTENTIAL IS PRESCRIBED)
%       NDP(I) = NODE NO. OF PRESCRIBED POTENTIAL, I = 1,2,...NP
%       VAL(I)) = VALUE OF PRESCRIBED POTENTIAL AT NODE NDP(I)
%       NL(I,J) = LIST OF NODES FOR EACH ELEMENT I, WHERE
%       LF(I) = LIST OF FREE NODES I = 1,2,...,NF=ND-NP
%       J = 1, 2, 3  IS THE LOCAL NODE NUMBER
%       CE(I,J) = ELEMENT COEFFICIENT MATRIX
%       ER(I) = VALUE OF THE RELATIVE PERMITTIVITY FOR ELEMENT I
%       C(I,J)  = GLOBAL COEFFICIENT MATRIX
%       X(I), Y(I) = GLOBAL COORDINATES OF NODE I
%       XL(J), YL(J) = LOCAL COORDINATES OF NODE J = 1,2,3
%       V(I) = POTENTIAL AT NODE I
%       MATRICES P(I) AND Q(I) ARE DEFINED IN EQUATION(6.1.1)

%       ****************************************************
%       FIRST STEP - INPUT DATA DEFINING GEOMETRY AND
%                       BOUNDARY CONDITIONS
%       ****************************************************
clear;
NI = 50;   %! NO. OF ITERATIONS
NE=25;     %Number of Elements
ND=21;
NP=15;

X = [0 0.2 0.4 0.6 0.8 1.0 0 0.2 0.4 0.6 0.8 0 0.2 0.4 0.6 0 0.2 0.4 0 0.2 0];
Y = [0 0 0 0 0 0 0.2 0.2 0.2 0.2 0.2 0.4 0.4 0.4 0.4 0.6 0.6 0.6 0.8 0.8 1.0];

NDP = [1 2 3 4 5 6 7 11 12 15 16 18 19 20 21];
VAL = [0 0 0 0 0 50 0 100 0 100 0 100 0 100 50];

NL = [ 1 2 7 ;2 8 7;2 3 8; 3 9 8; 3 4 9; 4 10 9; 4 5 10; 5 11 10; 5 6 11;...
    7 8 12;8 13 12; 8 9 13;9 14 13;9 10 14; 10 15 14; 10 11 15;...
    12 13 16; 13 17 16; 13 14 17; 14 18 17; 14 15 18; 16 17 19; 17 20 19; 17 18 20; 19 20 21];

EO = 1.0E-9/(36.0*pi);
ER = ones(1,NE);
%       ****************************************************
%       SECOND STEP - EVALUATE COEFFICIENT MATRIX FOR EACH
%                       ELEMENT AND ASSEMBLE GLOBALLY
%       ****************************************************
        C = zeros(ND,ND);
        for I = 1: NE
%       FIND LOCAL COORDINATES XL, YL FOR ELEMENT I
            XL = X(NL(I,:));
            YL = Y(NL(I,:));

            P = [YL(2) - YL(3),YL(3) - YL(1),YL(1) - YL(2)];
            Q = [XL(3) - XL(2),XL(1) - XL(3),XL(2) - XL(1)];
            AREA = 0.5*abs( P(2)*Q(3) - Q(2)*P(3) );
```

FIGURE 6.10
Computer program for Example 6.2.

(*Continued*)

```
%       DETERMINE COEFFICIENT MATRIX FOR ELEMENT I
            CE = ER(I)*(P'*P+Q'*Q)/(4*AREA);
%       ASSEMBLE GLOBALLY - FIND C(I,J)
            J=1:3;
            L=1:3;
            IR = NL(I,J);
            IC - NL(I,L);
            C(IR,IC) = C(IR,IC) + CE(J,L);
        end

%       **************************************************
%       THIRD STEP - SOLVE THE RESULTING SYSTEM
%                      ITERATIVELY
%       **************************************************
%
%       DETERMINE LF - LIST OF FREE NODES
        LF = setdiff(1:ND, NDP);
        V = zeros(1,ND);
        V(NDP) = VAL;
        NF = length(LF);

    for N = 1:NI
        for ii = 1:NF
            rowC = C(LF(ii),:);
            rowC(LF(ii)) = 0;
            V(LF(ii)) = -1/(C(LF(ii),LF(ii)))*rowC*V.';
        end
        figure(1),stem(V),drawnow
    end

    disp([{'Node','X','Y','Potential'};num2cell([(1:ND)',X',Y',V'])]);
```

FIGURE 6.10 (Continued)
Computer program for Example 6.2.

6.3.1 Deriving Element-Governing Equations

After the solution region is divided into triangular elements, we approximate the potential distribution $V_e(x, y)$ and the source term ρ_{ve} (for two-dimensional problems) over each triangular element by linear combinations of the local interpolation polynomial α_i, that is,

$$V_e = \sum_{i=1}^{3} V_{ei}\alpha_i(x, y) \tag{6.41}$$

TABLE 6.1

Nodal Coordinates of the Finite Element Mesh in Figure 6.9

Node	x	y	Node	x	y
1	0.0	0.0	12	0.0	0.4
2	0.2	0.0	13	0.2	0.4
3	0.4	0.0	14	0.4	0.4
4	0.6	0.0	15	0.6	0.4
5	0.8	0.0	16	0.0	0.6
6	1.0	0.0	17	0.2	0.6
7	0.0	0.2	18	0.4	0.6
8	0.2	0.2	19	0.0	0.8
9	0.4	0.2	20	0.2	0.8
10	0.6	0.2	21	0.0	1.0
11	0.8	0.2			

TABLE 6.2

Element–Node Identification

Element	Local 1	Node 2	No. 3	Element	Local 1	Node 2	No. 3
1	1	2	7	14	9	10	14
2	2	8	7	15	10	15	14
3	2	3	8	16	10	11	15
4	3	9	8	17	12	13	16
5	3	4	9	18	13	17	16
6	4	10	9	19	13	14	17
7	4	5	10	20	14	18	17
8	5	11	10	21	14	15	18
9	5	6	11	22	16	17	19
10	7	8	12	23	17	20	19
11	8	13	12	24	17	18	20
12	8	9	13	25	19	20	21
13	9	14	13				

$$\rho_{ve} = \sum_{i=1}^{3} \rho_{ei}\alpha_i(x,y) \tag{6.42}$$

The coefficients V_{ei} and ρ_{ei}, respectively, represent the values of V and ρ_v at node i of element e as in Figure 6.3. The values of ρ_{ei} are known since $\rho_v(x, y)$ is prescribed, while the values of V_{ei} are to be determined.

From Table 4.1, an energy functional whose associated Euler equation is Equation 6.40 is

$$F(V_e) = \frac{1}{2} \int_S [\epsilon |\nabla V_e|^2 - 2\rho_{ve}V_e] dS \tag{6.43}$$

$F(V_e)$ represents the total energy per length within element e. The first term under the integral sign, $\frac{1}{2}\mathbf{D} \cdot \mathbf{E} = \frac{1}{2}\epsilon |\nabla V_e|^2$, is the energy density in the electrostatic system, while the

TABLE 6.3

Prescribed Potentials at Fixed Nodes

Node	Prescribed Potential	Node	Prescribed Potential
1	0.0	18	100.0
2	0.0	20	100.0
3	0.0	21	50.0
4	0.0	19	0.0
5	0.0	16	0.0
6	50.0	12	0.0
11	100.0	7	0.0
15	100.0		

TABLE 6.4

Output Data of the Program in Figure 6.10

Node	X	Y	Potential
1	0.00	0.00	0.000
2	0.20	0.00	0.000
3	0.40	0.00	0.000
4	0.60	0.00	0.000
5	0.80	0.00	0.000
6	1.00	0.00	50.000
7	0.00	0.20	0.000
8	0.20	0.20	18.182
9	0.40	0.20	36.364
10	0.60	0.20	59.091
11	0.80	0.20	100.000
12	0.00	0.40	0.000
13	0.20	0.40	36.364
14	0.40	0.40	68.182
15	0.60	0.40	100.000
16	0.00	0.60	0.000
17	0.20	0.60	59.091
18	0.40	0.60	100.000
19	0.00	0.80	0.000
20	0.20	0.80	100.000
21	0.00	1.00	50.00

Number of nodes = 21, number of elements = 25, number of fixed nodes = 15.

second term, $\rho_{se}V_e\,dS$, is the work done in moving the charge $\rho_{se}\,dS$ to its location at potential V_e. Substitution of Equations 6.41 and 6.42 into Equation 6.43 yields

$$F(V_e) = \frac{1}{2}\sum_{i=1}^{3}\sum_{j=1}^{3}\epsilon V_{ei}\left[\int \nabla\alpha_i \cdot \nabla\alpha_j\,dS\right]V_{ej}$$
$$-\sum_{i=1}^{3}\sum_{j=1}^{3}V_{ei}\left[\int \alpha_i\alpha_j\,dS\right]\rho_{ej}$$

This can be written in matrix form as

$$F(V_e) = \frac{1}{2}\epsilon[V_e]^t[C^{(e)}][V_e] - [V_e]^t[T^{(e)}][\rho_e] \tag{6.44}$$

where

$$C_{ij}^{(e)} = \int \nabla\alpha_i \cdot \nabla\alpha_j\,dS \tag{6.45}$$

which is already defined in Equation 6.17 and

$$T_{ij}^{(e)} = \int \alpha_i \alpha_j \, dS \tag{6.46}$$

It will be shown in Section 6.8 that

$$T_{ij}^{(e)} = \begin{cases} A/12, & i \neq j \\ A/6, & i = j \end{cases} \tag{6.47}$$

where A is the area of the triangular element.

Equation 6.44 can be applied to every element in the solution region. We obtain the discretized functional for the whole solution region (with N elements and n nodes) as the sum of the functionals for the individual elements, that is, from Equation 6.44,

$$F(V) = \sum_{e=1}^{N} F(V_e) = \frac{1}{2}\epsilon[V]^t[C][V] - [V]^t[T][\rho] \tag{6.48}$$

where t denotes transposition. In Equation 6.48, the column matrix $[V]$ consists of the values of V_{ei}, while the column matrix $[\rho]$ contains n values of the source function ρ_v at the nodes. The functional in Equation 6.48 is now minimized by differentiating with respect to V_{ei} and setting the result equal to zero.

6.3.2 Solving the Resulting Equations

The resulting equations can be solved by either the iteration method or the band matrix method as discussed in Section 6.2.4.

Iteration Method: Consider a solution region in Figure 6.6 having five nodes so that $n = 5$. From Equation 6.48,

$$F = \frac{1}{2}\epsilon[V_1 V_2 \cdots V_5] \begin{bmatrix} C_{11} & C_{12} & \cdots & C_{15} \\ C_{21} & C_{22} & \cdots & C_{25} \\ \vdots & & & \vdots \\ C_{51} & C_{52} & \cdots & C_{55} \end{bmatrix} \begin{bmatrix} V_1 \\ V_2 \\ \vdots \\ V_5 \end{bmatrix}$$

$$- [V_1 V_2 \cdots V_5] \begin{bmatrix} T_{11} & T_{12} & \cdots & T_{15} \\ T_{21} & T_{22} & \cdots & T_{25} \\ \vdots & & & \vdots \\ T_{51} & T_{52} & \cdots & T_{55} \end{bmatrix} \begin{bmatrix} \rho_1 \\ \rho_2 \\ \vdots \\ \rho_5 \end{bmatrix} \tag{6.49}$$

We minimize the energy by applying

$$\frac{\partial F}{\partial V_k} = 0, \quad k = 1, 2, \ldots, n \tag{6.50}$$

From Equation 6.49, we get $\partial F/\partial V_1 = 0$, for example, as

$$\frac{\partial F}{\partial V_1} = \epsilon[V_1 C_{11} + V_2 C_{21} + \cdots + V_5 C_{51}] - [T_{11}\rho_1 + T_{21}\rho_2 + \cdots + T_{51}\rho_5] = 0$$

or

$$V_1 = -\frac{1}{C_{11}} \sum_{i=2}^{5} V_i C_{i1} + \frac{1}{\epsilon C_{11}} \sum_{i=1}^{5} T_{i1}\rho_i \tag{6.51}$$

Thus, in general, for a mesh with n nodes

$$V_k = -\frac{1}{C_{kk}} \sum_{i=1, i \ne k}^{n} V_i C_{ki} + \frac{1}{\epsilon C_{kk}} \sum_{i=1}^{n} T_{ki}\rho_i \tag{6.52}$$

where node k is assumed to be a free node.

By fixing the potential at the prescribed nodes and setting the potential at the free nodes initially equal to zero, we apply Equation 6.52 iteratively to all free nodes until convergence is reached.

Band Matrix Method: If we choose to solve the problem using the band matrix method, we let the free nodes be numbered first and the prescribed nodes last. By doing so, Equation 6.48 can be written as

$$F(V) = \frac{1}{2}\epsilon[V_f V_p]\begin{bmatrix} C_{ff} & C_{fp} \\ C_{pf} & C_{pp} \end{bmatrix}\begin{bmatrix} V_f \\ V_p \end{bmatrix} - [V_f V_p]\begin{bmatrix} T_{ff} & T_{fp} \\ T_{pf} & T_{pp} \end{bmatrix}\begin{bmatrix} \rho_f \\ \rho_p \end{bmatrix} \tag{6.53}$$

Minimizing $F(V)$ with respect to V_f, that is,

$$\frac{\partial F}{\partial V_f} = 0$$

gives

$$0 = \epsilon(C_{ff}V_f + C_{pf}V_p) - (T_{ff}\rho_f + T_{fp}\rho_p)$$

or

$$[C_{ff}][V_f] = -[C_{fp}][V_p] + \frac{1}{\epsilon}[T_{ff}][\rho_f] + \frac{1}{\epsilon}[T_{fp}][\rho_p] \tag{6.54}$$

This can be written as

$$[A][V] = [B] \tag{6.55}$$

where $[A] = [C_{ff}]$, $[V] = [V_f]$, and $[B]$ is the right-hand side of Equation 6.54. Equation 6.55 can be solved to determine $[V]$ either by matrix inversion or Gaussian elimination technique discussed in Appendix C. There is little point in giving examples on applying FEM to Poisson's problems, especially when it is noted that the difference between Equations 6.28 and 6.52 or Equations 6.54 and 6.31 is slight. See Reference 6 for an example.

6.4 Solution of the Wave Equation

A typical wave equation is the inhomogeneous scalar Helmholtz's equation

$$\nabla^2 \Phi + k^2 \Phi = g \tag{6.56}$$

where Φ is the field quantity (for waveguide problem, $\Phi = H_z$ for TE mode or E_z for TM mode) to be determined, g is the source function, and $k = \omega\sqrt{\mu\epsilon}$ is the wave number of the medium. The following three distinct special cases of Equation 6.56 should be noted:

   i. $k = 0 = g$: Laplace's equation;
  ii. $k = 0$: Poisson's equation; and
 iii. k is an unknown, $g = 0$: homogeneous, scalar Helmholtz's equation.

We know from Chapter 4 that the variational solution to the operator equation

$$L\Phi = g \tag{6.57}$$

is obtained by extremizing the functional

$$I(\Phi) = \langle L, \Phi \rangle - 2 \langle \Phi, g \rangle \tag{6.58}$$

Hence, the solution of Equation 6.56 is equivalent to satisfying the boundary conditions and minimizing the functional

$$I(\Phi) = \frac{1}{2} \int \int [\,|\nabla\Phi|^2 - k^2\Phi^2 + 2\Phi g\,]dS \tag{6.59}$$

If other than the natural boundary conditions (i.e., Dirichlet or homogeneous Neumann conditions) must be satisfied, appropriate terms must be added to the functional as discussed in Chapter 4.

We now express potential Φ and source function g in terms of the shape functions α_i over a triangular element as

$$\Phi_e(x, y) = \sum_{i=1}^{3} \alpha_i \Phi_{ei} \tag{6.60}$$

$$g_e(x, y) = \sum_{i=1}^{3} \alpha_i g_{ei} \tag{6.61}$$

where Φ_{ei} and g_{ei} are, respectively, the values of Φ and g at nodal point i of element e.

Substituting Equations 6.60 and 6.61 into Equation 6.59 gives

$$
\begin{aligned}
I(\Phi_e) = & \frac{1}{2} \sum_{i=1}^{3} \sum_{j=1}^{3} \Phi_{ei} \Phi_{ej} \iint \nabla \alpha_i \cdot \nabla \alpha_j \, dS \\
& - \frac{k^2}{2} \sum_{i=1}^{3} \sum_{j=1}^{3} \Phi_{ei} \Phi_{ej} \iint \alpha_i \alpha_j \, dS \\
& + \sum_{i=1}^{3} \sum_{j=1}^{3} \Phi_{ei} g_{ej} \iint \alpha_i \alpha_j \, dS \\
= & \frac{1}{2} [\Phi_e]^t [C^{(e)}][\Phi_e] \\
& - \frac{k^2}{2} [\Phi_e]^t [T^{(e)}][\Phi_e] + [\Phi_e]^t [T^{(e)}][G_e]
\end{aligned}
\tag{6.62}
$$

where $[\Phi_e] = [\Phi_{e1}, \Phi_{e2}, \Phi_{e3}]^t$, $[G_e] = [g_{e1}, g_{e2}, g_{e3}]^t$, and $[C^{(e)}]$ and $[T^{(e)}]$ are defined in Equations 6.17 and 6.47, respectively.

Equation 6.62, derived for a single element, can be applied for all N elements in the solution region. Thus,

$$
I(\Phi) = \sum_{e=1}^{N} I(\Phi_e)
\tag{6.63}
$$

From Equations 6.62 and 6.63, $I(\Phi)$ can be expressed in matrix form as

$$
I(\Phi) = \frac{1}{2} [\Phi]^t [C][\Phi] - \frac{k^2}{2} [\Phi]^t [T][\Phi] + [\Phi]^t [T][G]
\tag{6.64}
$$

where

$$
[\Phi] = [\Phi_1, \Phi_2, \dots, \Phi_N]^t,
\tag{6.65a}
$$

$$
[G] = [g_1, g_2, \dots, g_N]^t,
\tag{6.65b}
$$

and $[C]$ and $[T]$ are global matrices consisting of local matrices $[C^{(e)}]$ and $[T^{(e)}]$, respectively.

Consider the special case in which the source function $g = 0$. Again, if free nodes are numbered first and the prescribed nodes last, we may write Equation 6.64 as

$$
\begin{aligned}
I = & \frac{1}{2} [\Phi_f \Phi_p] \begin{bmatrix} C_{ff} & C_{fp} \\ C_{pf} & C_{pp} \end{bmatrix} \begin{bmatrix} \Phi_f \\ \Phi_p \end{bmatrix} \\
& - \frac{k^2}{2} [\Phi_f \Phi_p] \begin{bmatrix} T_{ff} & T_{fp} \\ T_{pf} & T_{pp} \end{bmatrix} \begin{bmatrix} \Phi_f \\ \Phi_p \end{bmatrix}
\end{aligned}
\tag{6.66}
$$

Setting $\partial I / \partial \Phi_f$ equal to zero gives

$$
[C_{ff} C_{fp}] \begin{bmatrix} \Phi_f \\ \Phi_p \end{bmatrix} - k^2 [T_{ff} T_{fp}] \begin{bmatrix} \Phi_f \\ \Phi_p \end{bmatrix} = 0
\tag{6.67}
$$

For TM modes, $\Phi_p = 0$ and hence

$$[C_{ff} - k^2 T_{ff}]\Phi_f = 0 \tag{6.68}$$

Premultiplying by T_{ff}^{-1} gives

$$\boxed{[T_{ff}^{-1}C_{ff} - k^2 I]\Phi_f = 0} \tag{6.69}$$

Letting

$$A = T_{ff}^{-1}C_{ff}, \quad k^2 = \lambda, \quad X = \Phi_f \tag{6.70a}$$

we obtain the standard eigenproblem

$$(A - \lambda I)X = 0 \tag{6.70b}$$

where I is a unit matrix. Any standard procedure [7] may be used to obtain some or all of the eigenvalues $\lambda_1, \lambda_2, ..., \lambda_{nf}$ and eigenvectors $X_1, X_2, ..., X_{nf}$ where n_f is the number of free nodes. The eigenvalues are always real since C and T are symmetric.

Solution of the algebraic eigenvalue problems in Equation 6.70 furnishes eigenvalues and eigenvectors, which form good approximations to the eigenvalues and eigenfunctions of the Helmholtz problem, that is, the cuttoff wavelengths and field distribution patterns of the various modes possible in a given waveguide.

The solution of the problem presented in this section, as summarized in Equation 6.69, can be viewed as the finite element solution of homogeneous waveguides. The idea can be extended to handle inhomogeneous waveguide problems [8–11]. However, in applying FEM to inhomogeneous problems, a serious difficulty is the appearance of spurious, nonphysical solutions. Several techniques have been proposed to overcome the difficulty [12–18].

EXAMPLE 6.3

To apply the ideas presented in this section, we use the finite element analysis to determine the lowest (or dominant) cutoff wavenumber k_c of the TM_{11} mode in waveguides with square $(a \times a)$ and rectangular $(a \times b)$ cross sections for which the exact results are already known as

$$k_c = \sqrt{\left(\frac{m\pi}{a}\right)^2 + \left(\frac{n\pi}{b}\right)^2}$$

where $m = n = 1$.

It may be instructive to try with hand calculation the case of a square waveguide with two divisions in the x and y directions. In this case, there are nine nodes, eight triangular elements, and one free node ($n_f = 1$). Equation 6.68 becomes

$$C_{11} - k^2 T_{11} = 0$$

where C_{11} and T_{11} are obtained from Equations 6.34, 6.35, and 6.47 as

$$C_{11} = \frac{a^2}{2A}, \quad T_{11} = \frac{A}{12}, \quad A = \frac{a^2}{8}$$

Hence,

$$k^2 = \frac{a^2}{2A^2} = \frac{32}{a^2}$$

or

$$ka = 5.656$$

which is about 27% off the exact solution. To improve the accuracy, we must use more elements.

The computer program in Figure 6.11 applies the ideas in this section to find k_c. The main program calls function GRID (to be discussed in Section 6.5) to generate the necessary input data from a given geometry. If n_x and n_y are the number of divisions in the x and y directions, the total number of elements $n_e = 2n_x n_y$. By simply specifying the values of a, b, n_x, and n_y, the program determines k_c. The results for the square ($a = b$) and rectangular ($b = 2a$) waveguides are presented in Tables 6.5 and 6.6, respectively.

6.5 Automatic Mesh Generation I: Rectangular Domains

One of the major difficulties encountered in the finite element analysis of continuum problems is the tedious and time-consuming effort required in data preparation. Efficient finite element programs must have node and element generating schemes, referred to collectively as *mesh generators*. Automatic mesh generation minimizes the input data required to specify a problem. It not only reduces the time involved in data preparation, it eliminates human errors introduced when data preparation is performed manually. Combining the automatic mesh generation program with computer graphics is particularly valuable since the output can be monitored visually. Since some applications of the FEM to EM problems involve simple rectangular domains, we consider the generation of simple meshes [6] here; automatic mesh generator for arbitrary domains will be discussed in Section 6.6.

Consider a rectangular solution region of size $a \times b$ as in Figure 6.12. Our goal is to divide the region into rectangular elements, each of which is later divided into two triangular elements. Suppose n_x and n_y are the number of divisions in x and y directions, the total number of elements and nodes are, respectively, given by

$$n_e = 2n_x n_y$$
$$n_d = (n_x + 1)(n_y + 1) \tag{6.71}$$

Thus it is easy to figure out from Figure 6.12 a systematic way of numbering the elements and nodes. To obtain the global coordinates (x, y) for each node, we need an array containing Δx_i, $i = 1, 2, \ldots, n_x$ and Δy_j, $j = 1, 2, \ldots, n_y$, which are the distances between nodes in the x and y directions, respectively. If the order of node numbering is from left to right along horizontal rows and from bottom to top along the vertical rows, then the first node is the origin $(0,0)$. The next node is obtained as $x \rightarrow x + \Delta x_1$ while $y = 0$ remains unchanged. The following node has $x \rightarrow x + \Delta x_2$, $y = 0$, and so on until Δx_i are exhausted. We start the second horizontal row by starting with $x = 0$, $y \rightarrow y + \Delta y_1$ and increasing x until Δx_i are exhausted. We repeat the process until the last node $(n_x + 1)(n_y + 1)$ is reached, that is, when Δx_i and Δy_i are exhausted simultaneously.

```
clear;
% ***********************************************************
%         FINITE ELEMENT SOLUTION OF THE WAVE EQUATION
%         TRIANGULAR ELEMENTS ARE USED
%
%         ND = NO. OF NODES
%         NE = NO. OF ELEMENTS
%         NL(I,J) = LIST OF NODES FOR EACH ELEMENT I, WHERE
%         CE(I,J) = ELEMENT COEFFICIENT MATRIX
%         C(I,J)  = GLOBAL COEFFICIENT MATRIX
%         X(I), Y(I) = GLOBAL COORDINATES OF NODE I
%         XL(J), YL(J) = LOCAL COORDINATES OF NODE J = 1,2,3
%         MATRICES P(I) AND Q(I) ARE DEFINED IN
%         LF(I) = LIST OF FREE NODES
%         ALAM(I) = CONTAINS EIGENVALUES


%         ***********************************************************
%         FIRST STEP - INPUT DATA DEFINING GEOMETRY AND
%                      BOUNDARY CONDITIONS (USE FUNCTION GRID)
%         ***********************************************************

NX = 10;
select = 1; %select=1 for square, select=2 for rectangular

if select == 1 %For Square Waveguide
    NY = 1*NX;      AA = 1.0;      BB = 1.0;
else %For Rectangular Waveguide
    NY = 2.0*NX;    AA = 1.0;      BB = 2.0;
end

DELTAX = AA/NX;
DELTAY = BB/NY;
DX = ones(1,NX)*DELTAX;
DY = ones(1,NY)*DELTAY;

[NE,ND,NP,NL,X,Y,NDP]=gridmesh(NX,NY,DX,DY);

% Plot Mesh - Grab X and Y values of element nodes. Note:
inserting NaN
% keeps the plot function from connecting different elements
together!
mx = reshape([X([NL,NL(:,1)]),NaN*zeros(NE,1)]',1,NE*5);
my = reshape([Y([NL,NL(:,1)]),NaN*zeros(NE,1)]',1,NE*5);

figure(2),plot(X,Y,'o',mx,my,':'),xlabel('x'),     ylabel('y')

%         ***********************************************************
%         SECOND STEP - EVALUATE COEFFICIENT MATRIX FOR EACH
%                       ELEMENT AND ASSEMBLE GLOBALLY
%         ***********************************************************

    C = zeros(ND,ND);
    T = zeros(ND,ND);

for I = 1: NE
%     FIND LOCAL COORDINATES XL, YL FOR ELEMENT I
        XL = X(NL(I,:));
        YL = Y(NL(I,:));
        P = [YL(2) - YL(3),YL(3) - YL(1),YL(1) - YL(2)];
Q = [XL(3) - XL(2),XL(1) - XL(3),XL(2) - XL(1)];
AREA = 0.5*abs( P(2)*Q(3) - Q(2)*P(3) );
```

FIGURE 6.11
Computer program for Example 6.3.

(Continued)

```
%       DETERMINE COEFFICIENT MATRIX FOR ELEMENT I
        CE = (P'*P+Q'*Q)/(4*AREA);

%       ASSEMBLE GLOBALLY - FIND C(I,J) AND T(I,J)
        J=1:3;          L=1:3;
        IR = NL(I,J);
        IC = NL(I,L);
C(IR,IC) = C(IR,IC) + CE(J,L);
T(IR,IC) = T(IR,IC) + AREA/12*[2 1 1;1 2 1;1 1 2];
end

%       ****************************************************
%       THIRD STEP - SOLVE THE RESULTING SYSTEM
%       ****************************************************
%       DETERMINE LF(I) - LIST OF FREE NODES

    LF = setdiff(1:ND, NDP);
    NF = length(LF);

%       FROM GLOBAL C AND T, FIND C_ff AND T_ff (f for free nodes)

C_ff = C(LF,LF);
    T_ff = T(LF,LF);

    A = inv(T_ff)*C_ff;

    ALAM  = eig(A);
%       ****************************************************
%       FOURTH STEP - OUTPUT THE RESULTS
%       ****************************************************

Kcalc = sqrt(min(ALAM));

    Kc = sqrt( (pi/AA)^2+(pi/BB)^2 );
theerror = 100*(Kcalc-Kc)/Kc;

    s1 = sprintf('\n  Nx   Ne    Kcalc  %% error');
    s2 = sprintf('\n   %-3.5g  %-3.5g  %-6.5g   %-3.5g', NX,NE,
        Kcalc,theerror);

disp([s1,s2])
```

FIGURE 6.11 (Continued)
Computer program for Example 6.3.

The procedure presented here allows for generating uniform and nonuniform meshes. A mesh is uniform if all Δx_i are equal and all Δy_i are equal; it is nonuniform otherwise. A nonuniform mesh is preferred if it is known in advance that the parameter of interest varies rapidly in some parts of the solution domain. This allows a concentration of relatively small elements in the regions where the parameter changes rapidly, particularly since these

TABLE 6.5

Lowest Wavenumber for a Square Waveguide ($b = a$)

n_x	n_e	$k_c a$	% Error
2	8	5.656	27.3
3	18	5.030	13.2
5	50	4.657	4.82
7	98	4.553	2.47
10	200	4.497	1.22

Exact: $k_c a = 4.4429$, $n_y = n_x$.

TABLE 6.6

Lowest Wavenumber for a Rectangular Waveguide ($b = 2a$)

n_x	n_e	$k_c a$	% Error
2	16	4.092	16.5
4	64	3.659	4.17
6	144	3.578	1.87
8	256	3.549	1.04

Exact: $k_c a = 3.5124$, $n_y = 2n_x$.

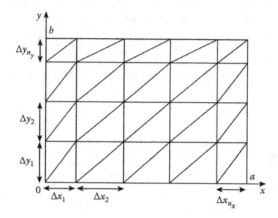

FIGURE 6.12
Discretization of a rectangular region into a nonuniform mesh.

regions are often of greatest interest in the solution. Without the preknowledge of the rapid change in the unknown parameter, a uniform mesh can be used. In that case, we set

$$\Delta x_1 = \Delta x_2 = \cdots = h_x$$
$$\Delta y_1 = \Delta y_2 = \cdots = h_y \tag{6.72}$$

where $h_x = a/n_x$ and $h_y = b/n_y$.

In some cases, we also need a list of prescribed nodes. If we assume that all boundary points have prescribed potentials, the number n_p of prescribed node is given by

$$n_p = 2(n_x + n_y) \tag{6.73}$$

A simple way to obtain the list of boundary points is to enumerate points on the bottom, right, top, and left sides of the rectangular region in that order.

The ideas presented here are implemented in the function GRID in Figure 6.13. The subroutine can be used for generating a uniform or nonuniform mesh out of a given rectangular region. If a uniform mesh is desired, the required input parameters are a, b, n_x, and n_y. If, on the other hand, a nonuniform mesh is required, we need to supply n_x, n_y, Δx_i, $i = 1,2, \ldots, n_x$, and Δy_j, $j = 1,2, \ldots, n_y$. The output parameters are n_e, n_d, n_p, connectivity list, the global coordinates (x, y) of each node, and the list of prescribed nodes. It is needless to say that the subroutine GRID is not useful for a nonrectangular solution region. See the program in Figure 6.11 as an example on how to use subroutine GRID. A more general program for discretizing a solution region of any shape will be presented in the next section.

```
function [NE,ND,NP,NL,X,Y,NDP]=gridmesh(NX,NY,DX,DY)
% function [NE,ND,NP,NL,X,Y,NDP]=gridmesh(NX,NY,DX,DY)
%  ********************************************************
%    THIS PROGRAM DIVIDES A RECTANGULAR DOMAIN INTO
%    TRIANGULAR ELEMENTS   (NY BY NY   NONUNIFORM
%    MESH IN GENERAL)
%    NX &  NY  ARE THE NUMBER OF SUBDIVISIONS ALONG X & Y AXES
%    NE = NO. OF ELEMENTS IN THE MESH
%    ND = NO. OF NODES IN THE MESH
%    NP = NO. OF BOUNDARY (PRESCRIBED) NODES
%    X(I) & Y(I) ARE GLOBAL COORDINATES  OF NODE I
%    DX(I) & DY(I) ARE DISTANCES BETWEEN NODES ALONG X & Y AXES
%    NL(I,J) IS THE LIST OF NODES FOR ELEMENT I,   J = 1, 2, 3 ARE
%         LOCAL NUMBERS
%    NDP(I) = LIST OF PRESCRIBED NODES I
%
%    REF:  J. N. REDDY, "AN INTRODUCTION TO THE FINITE ELEMENT
%                         METHOD."  NEW YORK: MCGRAW-HILL, 1984, P. 436

%  CALCULATE NE, ND, AND NP
     NE = 2*NX*NY;          %Number of elements
     NP = 2*(NX + NY); %Number of prescribed nodes (edges)
     NX1 = NX+1;            %Number of nodes in the X direction
     NY1 = NY +1;           %Number of nodes in the Y direction
     ND = NX1 * NY1;        %Number of nodes in mesh

%  Determine Node List which follows a regular pattern utilized below.
     NL1 = 1:(NX1*NY);
     NL1(NX1:NX1:end) = [];

     NL3 = (NX+2):(NX+NX1*NY+1);
     NL3(NX1:NX1:end) = [];

     NL = zeros(NE,3);
     NL(1:2:end,1) = NL1;
     NL(2:2:end,1) = NL1;
     NL(1:2:end,2) = NL3+1;
     NL(2:2:end,2) = NL1+1;
     NL(1:2:end,3) = NL3;
     NL(2:2:end,3) = NL3+1;

%    DETERMINE X AND Y (look saw tooth waves)
     x1 = [0 cumsum(DX)]'*ones(1,NY1);
     X = reshape(x1,1,numel(x1));

     y1 = ones(NX1,1)*[0 cumsum(DY)];
     Y = reshape(y1,1,numel(y1));

%    DETERMINE  NDP, [bottom, right, top, left]
NDP = [1:(NX1), 2*(NX1):(NX1):ND, (ND-1):-1:(ND-NX), ((ND-2*NX-1)):-
(NX1):(NX+2)];
```

FIGURE 6.13
Program GRID.

6.6 Automatic Mesh Generation II: Arbitrary Domains

As the solution regions become more complex than the ones considered in Section 6.5, the task of developing mesh generators becomes more tedious. A number of mesh generation algorithms (e.g., [19–31]) of varying degrees of automation have been proposed for arbitrary solution domains. Reviews of various mesh generation techniques can be found in Reference 32.

The basic steps involved in a mesh generation are as follows [33,34]:

- Subdivide solution region into a few quadrilateral blocks,
- Separately subdivide each block into elements,
- Connect individual blocks.

Each step is explained as follows.

6.6.1 Definition of Blocks

The solution region is subdivided into quadrilateral blocks. Subdomains with different constitutive parameters (σ, μ, ϵ) must be represented by separate blocks. As input data, we specify block topologies and the coordinates at eight points describing each block. Each block is represented by an eight-node quadratic isoparametric element. With natural coordinate system (ζ, η), the x and y coordinates are represented as

$$x(\zeta, \eta) = \sum_{i=1}^{8} \alpha_i(\zeta, \eta) x_i \tag{6.74}$$

$$y(\zeta, \eta) = \sum_{i=1}^{8} \alpha_i(\zeta, \eta) y_i \tag{6.75}$$

where $\alpha_i(\zeta, \eta)$ is a shape function associated with node i, and (x_i, y_i) are the coordinates of node i defining the boundary of the quadrilateral block as shown in Figure 6.14. The shape functions are expressed in terms of the quadratic or parabolic isoparametric elements shown in Figure 6.15. They are given by

$$\alpha_i = \frac{1}{4}(1+\zeta\zeta_i)(1+\eta\eta_i)(\zeta\zeta_i + \eta\eta_i + 1), \quad i = 1, 3, 5, 7 \tag{6.76}$$

for corner nodes,

$$\alpha_i = \frac{1}{2}\zeta_i^2(1+\zeta\zeta_i)(1-\eta^2)$$
$$+ \frac{1}{2}\eta_i^2(1+\eta\eta_i+1)(1-\zeta^2), \quad i = 2, 4, 6, 8 \tag{6.77}$$

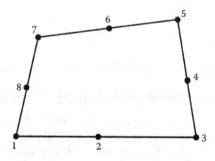

FIGURE 6.14
Typical quadrilateral block.

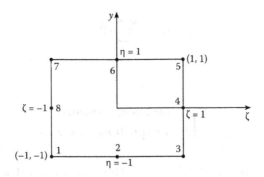

FIGURE 6.15
Eight-node serendipity element.

for midside nodes. Note the following properties of the shape functions:

1. They satisfy the conditions

$$\sum_{i=1}^{n} \alpha_i(\zeta, \eta) = 1 \tag{6.78a}$$

$$\alpha_i(\zeta_j, \eta_j) = \begin{cases} 1, & i = j \\ 0, & i \neq j \end{cases} \tag{6.78b}$$

2. They become quadratic along element edges ($\zeta = \pm 1, \eta = \pm 1$).

6.6.2 Subdivision of Each Block

For each block, we specify $N\ DIV\ X$ and $N\ DIV\ Y$, the number of element subdivisions to be made in the ζ and η directions, respectively. Also, we specify the weighting factors $(W_\zeta)_i$ and $(W_\eta)_i$ allowing for graded mesh within a block. In specifying $N\ DIV\ X$, $N\ DIV\ Y$, W_ζ, and W_η care must be taken to ensure that the subdivision along block interfaces (for adjacent blocks) are compatible. We initialize ζ and η to a value of -1 so that the natural coordinates are incremented according to

$$\zeta_i = \zeta_i + \frac{2(W_\zeta)_i}{W_\zeta^T \cdot F} \tag{6.79}$$

$$\eta_i = \eta_i + \frac{2(W_\eta)_i}{W_\eta^T \cdot F} \tag{6.80}$$

where

$$W_\zeta^T = \sum_{j=1}^{NDIVX} (W_\zeta)_j \tag{6.81a}$$

$$W_\eta^T = \sum_{j=1}^{NDIVX} (W_\eta)_j \qquad\qquad (6.81b)$$

and

$$F = \begin{cases} 1, & \text{for linear elements} \\ 2, & \text{for quadratic elements} \end{cases}$$

Three element types are permitted: (a) linear four-node quadrilateral elements, (b) linear three-node triangular elements, and (c) quadratic eight-node isoparametric elements.

6.6.3 Connection of Individual Blocks

After subdividing each block and numbering its nodal points separately, it is necessary to connect the blocks and have each node numbered uniquely. This is accomplished by comparing the coordinates of all nodal points and assigning the same number to all nodes having identical coordinates. That is, we compare the coordinates of node 1 with all other nodes, and then node 2 with other nodes, etc., until all repeated nodes are eliminated. The listing of the MATLAB code for automatic mesh generation is shown in Figure 6.16. The following example taken from Reference 34 illustrates the application of the code.

EXAMPLE 6.4

Utilize the distmesh2d function by Persson and Strang [34] in Figure 6.16 to discretize the geometry shown in Figure 6.17. This geometry is composed of the union of two rectangles whose bottom left and top right coordinates (x_1, y_1), (x_2, y_2) are $(0, 0)$, $(5, 10)$ and $(5, 5)$, $(8, 10)$, and a circle void centered at $(2.5, 7.5)$ with a radius of 1.5.

Solution

The code in Figure 6.16 describes the basic geometry of the structure with a signed distance function "fd" which is a MATLAB inline function composed of the union of two rectangles and a circular void. The parameter "box" defines the limit of the solution space of the mesh and the parameter "fix" contains the pre-determined nodes in the mesh. Distmesh2d uses an iterative mesh generation technique based on the physical analogy between a simplex mesh and a truss structure. Meshpoints are the nodes of the truss. The main program in Figure 6.16a needs the functions ddiff.m, dunion.m, dcirc.m, and drectangle.m all provided in Figure 6.16b. The completed mesh is shown in Figure 6.18.

6.7 Bandwidth Reduction

Since most of the matrices involved in FEM are symmetric, sparse, and banded, we can minimize the storage requirements and the solution time by storing only the elements involved in half bandwidth instead of storing the whole matrix. To take the fullest advantage of the benefits from using a banded matrix solution technique, we must make sure that the matrix bandwidth is as narrow as possible.

If we let d be the maximum difference between the lowest and the highest node numbers of any single element in the mesh, we define the semi-bandwidth B (which includes the diagonal term) of the coefficient matrix $[C]$ as

$$B = (d + 1)f \tag{6.82}$$

```
function ch6_FIG6_16

fd=inline('ddiff(dunion(drectangle(p,0,5,0,10),drectangle(p,5,8,5,10)),
dcircle(p,2.5,7.5,1.5) )','p');

box=[-.1 -.1; 8.1 10.1];
fix=[0 0; 0 5; 0 10; 5 0; 5 5; 5 10; 8 5; 8 10];
[p,t]=distmesh2d(fd,@huniform,0.7,box,fix);
disp('Mesh Complete')

function [p,t]=distmesh2d(fd,fh,h0,bbox,pfix,varargin)
%DISTMESH2D 2-D Mesh Generator using Distance Functions.
%    [P,T]=DISTMESH2D(FD,FH,H0,BBOX,PFIX,FPARAMS)
%
%      P:         Node positions (Nx2)
%      T:         Triangle indices (NTx3)
%      FD:        Distance function d(x,y)
%      FH:        Scaled edge length function h(x,y)
%      H0:        Initial edge length
%      BBOX:      Bounding box [xmin,ymin; xmax,ymax]
%      PFIX:      Fixed node positions (NFIXx2)
%      FPARAMS:   Additional parameters passed to FD and FH
%
%   Example: (Uniform Mesh on Unit Circle)
%      fd=inline('sqrt(sum(p.^2,2))-1','p');
%      [p,t]=distmesh2d(fd,@huniform,0.2,[-1,-1;1,1],[]);
%
%   Example: (Rectangle with circular hole, refined at circle boundary)
%      fd=inline('ddiff(drectangle(p,-1,1,-1,1),dcircle(p,0,0,0.5))','p');
%      fh=inline('min(4*sqrt(sum(p.^2,2))-1,2)','p');
%      [p,t]=distmesh2d(fd,fh,0.05,[-1,-1;1,1],[-1,-1;-1,1;1,-1;1,1]);
%
%   See also: MESHDEMO2D, DISTMESHND, DELAUNAYN, TRIMESH.

%   Copyright (C) 2004-2006 Per-Olof Persson.

dptol=.001; ttol=.1; Fscale=1.2; deltat=.2; geps=.001*h0; deps=sqrt(eps)*h0;

% 1. Create initial distribution in bounding box (equilateral triangles)
[x,y]=meshgrid(bbox(1,1):h0:bbox(2,1),bbox(1,2):h0*sqrt(3)/2:bbox(2,2));
x(2:2:end,:)=x(2:2:end,:)+h0/2;           % Shift even rows
p=[x(:),y(:)];                            % List of node coordinates

% 2. Remove points outside the region, apply the rejection method
p=p(feval(fd,p,varargin{:})<geps,:);      % Keep only d<0 points
r0=1./feval(fh,p,varargin{:}).^2;         % Probability to keep point
p=[pfix; p(rand(size(p,1),1)<r0./max(r0),:)]; % Rejection method
N=size(p,1);                              % Number of points N
pold=inf;                                 % For first iteration
while 1
  % 3. Retriangulation by the Delaunay algorithm
  if max(sqrt(sum((p-pold).^2,2))/h0)>ttol % Any large movement?
    pold=p;                               % Save current positions
    t=delaunayn(p);                       % List of triangles
```

FIGURE 6.16

MATLAB code for automatic mesh generation: (a) main program, (b) functions that the main function will need.

(Continued)

```
    pmid=(p(t(:,1),:)+p(t(:,2),:)+p(t(:,3),:))/3;      % Compute centroids
    t=t(feval(fd,pmid,varargin{:})<-geps,:);           % Keep interior triangles
    % 4. Describe each bar by a unique pair of nodes
    bars=[t(:,[1,2]);t(:,[1,3]);t(:,[2,3])];           % Interior bars duplicated
    bars=unique(sort(bars,2),'rows');                  % Bars as node pairs
    % 5. Graphical output of the current mesh
    trimesh(t,p(:,1),p(:,2),zeros(N,1))
    view(2),axis equal,axis off,drawnow
end

% 6. Move mesh points based on bar lengths L and forces F
barvec=p(bars(:,1),:)-p(bars(:,2),:);                  % List of bar vectors
L=sqrt(sum(barvec.^2,2));                               % L = Bar lengths
hbars=feval(fh,(p(bars(:,1),:)+p(bars(:,2),:))/2,varargin{:});
L0=hbars*Fscale*sqrt(sum(L.^2)/sum(hbars.^2));          % L0 = Desired lengths
F=max(L0-L,0);                                          % Bar forces (scalars)
Fvec=F./L*[1,1].*barvec;                                % Bar forces (x,y components)
Ftot=full(sparse(bars(:,[1,1,2,2]),ones(size(F))*[1,2,1,2],[Fvec,-Fvec],N,2));
Ftot(1:size(pfix,1),:)=0;                               % Force = 0 at fixed points
p=p+deltat*Ftot;                                        % Update node positions

% 7. Bring outside points back to the boundary
d=feval(fd,p,varargin{:}); ix=d>0;                     % Find points outside (d>0)
dgradx=(feval(fd,[p(ix,1)+deps,p(ix,2)],varargin{:})-d(ix))/deps; % Numerical
dgrady=(feval(fd,[p(ix,1),p(ix,2)+deps],varargin{:})-
d(ix))/deps; %     gradient
p(ix,:)=p(ix,:)-[d(ix).*dgradx,d(ix).*dgrady];         % Project back to boundary

% 8. Termination criterion: All interior nodes move less than dptol (scaled)
if max(sqrt(sum(deltat*Ftot(d<-geps,:).^2,2))/h0)<dptol, break; end
end

function h=huniform(p,varargin)
h=ones(size(p,1),1);

(a)

function d=drectangle(p,x1,x2,y1,y2)

d=-min(min(min(-y1+p(:,2),y2-p(:,2)),-x1+p(:,1)),x2-p(:,1));


function d=dcirc(p,xc,yc,r)

d=sqrt((p(:,1)-xc).^2+(p(:,2)-yc).^2)-r;


function d=dunion(d1,d2), d=min(d1,d2);


function d=ddiff(d1,d2), d=max(d1,-d2);

(b)
```

FIGURE 6.16 (Continued)
MATLAB code for automatic mesh generation: (a) main program, (b) functions that the main function will need.

where f is the number of degrees of freedom (or number of parameters) at each node. If, for example, we are interested in calculating the electric field intensity **E** for a three-dimensional problem, then we need E_x, E_y, and E_z at each node, and $f = 3$ in this case. Assuming that there is only one parameter per node,

$$B = d + 1 \tag{6.83}$$

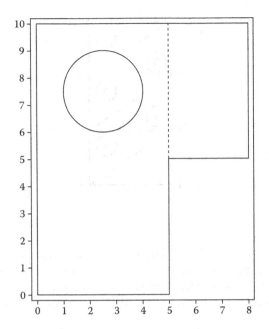

FIGURE 6.17
Solution region of Example 6.4.

The semi-bandwidth, which does not include the diagonal term, is obtained from Equation 6.82 or 6.83 by subtracting one from the right-hand side, that is, for $f = 1$,

$$B = d \tag{6.84}$$

Throughout our discussion in this section, we will stick to the definition of semi-bandwidth in Equation 6.84. The total bandwidth may be obtained from Equation 6.84 as $2B + 1$.

FIGURE 6.18
Generated mesh for Example 6.4.

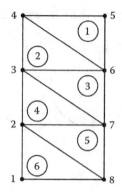

FIGURE 6.19
Original mesh with $B = 7$.

The bandwidth of the global coefficient matrix depends on the node numbering. Hence, to minimize the bandwidth, the node numbering should be selected to minimize d. Good node numbering is usually such that nodes with widely different numbers are widely separated. To minimize d, we must number nodes across the narrowest part of the region.

Consider, for the purpose of illustration, the mesh shown in Figure 6.19. If the mesh is numbered originally as in Figure 6.19, we obtain d_e for each element e as

$$d_1 = 2, \ d_2 = 3, \ d_3 = 4, \ d_4 = 5, \ d_5 = 6, \ d_6 = 7 \tag{6.85}$$

From this, we obtain

$$d = \text{maximum } d_e = 7$$

or

$$B = 7 \tag{6.86}$$

Alternatively, the semi-bandwidth may be determined from the coefficient matrix, which is obtained by mere inspection of Figure 6.19 as

$$
[C] =
\begin{array}{c}
\\
1 \\
2 \\
3 \\
4 \\
5 \\
6 \\
7 \\
8
\end{array}
\begin{array}{c}
\overset{\displaystyle \overset{B=7}{\longleftrightarrow}}{1 \quad 2 \quad 3 \quad 4 \quad 5 \quad 6 \quad 7 \quad 8} \\
\begin{bmatrix}
x & x & & & & & & x \\
x & x & x & & & & x & x \\
 & x & x & x & & x & & \\
 & & x & x & x & x & & \\
 & & & x & x & x & & \\
 & & x & x & x & x & x & \\
 & x & & & & x & x & x \\
x & x & & & & & & x
\end{bmatrix}
\end{array}
\tag{6.87}
$$

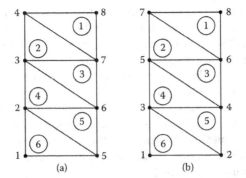

FIGURE 6.20
Renumbered nodes: (a) $B = 4$, (b) $B = 2$.

where x indicates a possible nonzero term and blanks are zeros (i.e., $C_{ij} = 0$, indicating no coupling between nodes i and j). If the mesh is renumbered as in Figure 6.20a,

$$d_1 = 4 = d_2 = d_3 = d_4 = d_5 = d_6 \tag{6.88}$$

and hence

$$d = \text{maximum } d_e = 4$$

or

$$B = 4 \tag{6.89}$$

Finally, we may renumber the mesh as in Figure 6.20b. In this case

$$d_1 = 2 = d_2 = d_3 = d_4 = d_5 = d_6 \tag{6.90}$$

and

$$d = \text{maximum } d_e = 2 \tag{6.91}$$

or

$$B = 2 \tag{6.92}$$

The value $B = 2$ may also be obtained from the coefficient matrix for the mesh in Figure 6.20b, namely,

$$
[C] = \begin{array}{c}
\\
\\
\\
\\
\\
\end{array}
\overset{\overset{\displaystyle \xleftarrow{\hspace{0.5cm} B=2 \hspace{0.5cm}}}{\text{P} \quad \text{Q}}}{
\begin{array}{c}
1 \; 2 \; 3 \; 4 \; 5 \; 6 \; 7 \; 8 \\
\begin{array}{c}
1 \\ 2 \\ 3 \\ 4 \\ 5 \\ 6 \\ 7 \\ 8
\end{array}
\left[
\begin{array}{cccccccc}
x & x & x & & & & & \\
x & x & x & x & & & & \\
x & x & x & x & x & & & \\
 & x & x & x & x & & & \\
 & & x & x & x & x & x & \\
 & & & x & x & x & x & x \\
 & & & & x & x & x & x \\
 & & & & & x & x & x
\end{array}
\right]
\begin{array}{c}
\\ \\ \\ \\ \\ \text{R} \\ \\ \text{S}
\end{array}
\end{array}
}
\tag{6.93}
$$

From Equation 6.93, one immediately notices that [C] is symmetric and that terms are clustered in a band about the diagonal. Hence, [C] is sparse and banded so that only the data within the area **PQRS** of the matrix need to be stored—a total of 33 terms out of 64. This illustrates the savings in storage by a careful nodal numbering.

For a simple mesh, hand-labeling coupled with a careful inspection of the mesh (as we have done so far) can lead to a minimum bandwidth. However, for a large mesh, a hand-labeling technique becomes a tedious, time-consuming task, which in most cases may not be successful. It is particularly desirable that an automatic relabeling scheme is implemented within a mesh generation program. A number of algorithms have been proposed for bandwidth reduction by automatic mesh renumbering [35–38]. A simple, efficient algorithm is found in Collins [35].

6.8 Higher-Order Elements

The finite elements we have used so far have been the linear type in that the shape function is of the order one. A higher-order element is one in which the shape function or interpolation polynomial is of the order two or more.

The accuracy of a finite element solution can be improved by using finer mesh or using higher-order elements or both. In general, fewer higher-order elements are needed to achieve the same degree of accuracy in the final results. The higher-order elements are particularly useful when the gradient of the field variable is expected to vary rapidly. They have been applied with great success in solving EM-related problems [4,39–44].

6.8.1 Pascal Triangle

Higher-order triangular elements can be systematically developed with the aid of the so-called Pascal triangle given in Figure 6.21. The family of finite elements generated in this manner with the distribution of nodes illustrated in Figure 6.22. Note that in higher-order elements, some secondary (side and/or interior) nodes are introduced in addition to the primary (corner) nodes so as to produce exactly the right number of nodes required to define the shape function of that order. The Pascal triangle contains terms of the basis functions of various degrees in variables x and y. An arbitrary

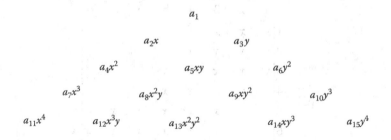

FIGURE 6.21
The Pascal triangle. The first row is (constant, $n = 0$), the second (linear, $n = 1$), the third (quadratic, $n = 2$), the fourth (cubic, $n = 3$), and the fifth (quartic, $n = 4$).

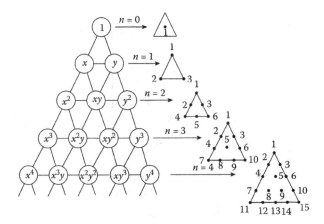

FIGURE 6.22
Pascal triangle and the associated polynomial basis function for degree $n = 1$–4.

function $\Phi_i(x, y)$ can be approximated in an element in terms of a complete nth-order polynomial as

$$\Phi(x, y) = \sum_{i=1}^{m} \alpha_i \Phi_i \tag{6.94}$$

where

$$m = \frac{1}{2}(n+1)(n+2) \tag{6.95}$$

is the number of terms in complete polynomials (also the number of nodes in the triangle). For example, for second order ($n = 2$) or quadratic (six-node) triangular elements,

$$\Phi_e(x, y) = a_1 + a_2 x + a_3 y + a_4 xy + a_5 x^2 + a_6 y^2 \tag{6.96}$$

This equation has six coefficients, and hence the element must have six nodes. It is also complete through the second-order terms. A systematic derivation of the interpolation function a for the higher-order elements involves the use of the local coordinates.

6.8.2 Local Coordinates

The triangular local coordinates (ξ_1, ξ_2, ξ_3) are related to Cartesian coordinates (x, y) as

$$x = \xi_1 x_1 + \xi_2 x_2 + \xi_3 x_3 \tag{6.97}$$

$$y = \xi_1 y_1 + \xi_2 y_2 + \xi_3 y_3 \tag{6.98}$$

The local coordinates are dimensionless with values ranging from 0 to 1. By definition, ξ_i at any point within the triangle is the ratio of the perpendicular distance from the point to the side opposite to vertex i to the length of the altitude drawn from vertex i. Thus,

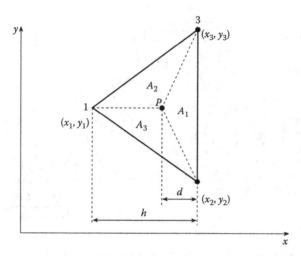

FIGURE 6.23
Definition of local coordinates.

from Figure 6.23 the value of ξ_1 at P, for example, is given by the ratio of the perpendicular distance d from the side opposite vertex 1 to the altitude h of that side, that is,

$$\xi_1 = \frac{d}{h} \tag{6.99}$$

Alternatively, from Figure 6.23, ξ_i at P can be defined as

$$\xi_i = \frac{A_i}{A} \tag{6.100}$$

so that

$$\xi_1 + \xi_2 + \xi_3 = 1 \tag{6.101}$$

since $A_1 + A_2 + A_3 = A$. In view of Equation 6.100, the local coordinates ξ_i are also called *area coordinates*. The variation of (ξ_1, ξ_2, ξ_3) inside an element is shown in Figure 6.24. Although the coordinates ξ_1, ξ_2, and ξ_3 are used to define a point P, only two are independent since they must satisfy Equation 6.101. The inverted form of Equations 6.97 and 6.98 is

$$\xi_i = \frac{1}{2A} [c_i + b_i x + a_i y] \tag{6.102}$$

where
$a_i = x_k - x_j,$
$b_i = y_j - y_k,$
$c_i = x_j y_k - x_k y_j$

$$A = \text{area of the triangle} = \frac{1}{2}(b_1 a_2 - b_2 a_1), \tag{6.103}$$

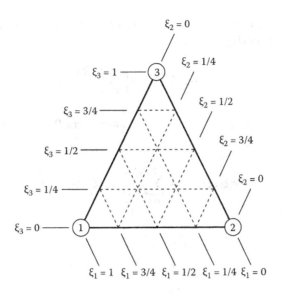

FIGURE 6.24
Variation of local coordinates.

and (i, j, k) is an even permutation of $(1, 2, 3)$. (Notice that a_i and b_i are the same as Q_i and P_i in Equation 6.34.) The differentiation and integration in local coordinates are carried out using [45]:

$$\frac{\partial f}{\partial \xi_1} = a_2 \frac{\partial f}{\partial x} - b_2 \frac{\partial f}{\partial y} \tag{6.104a}$$

$$\frac{\partial f}{\partial \xi_2} = -a_1 \frac{\partial f}{\partial x} + b_1 \frac{\partial f}{\partial y} \tag{6.104b}$$

$$\frac{\partial f}{\partial x} = \frac{1}{2A} \left(b_1 \frac{\partial f}{\partial \xi_1} + b_2 \frac{\partial f}{\partial \xi_2} \right) \tag{6.104c}$$

$$\frac{\partial f}{\partial y} = \frac{1}{2A} \left(a_1 \frac{\partial f}{\partial \xi_1} + a_2 \frac{\partial f}{\partial \xi_2} \right) \tag{6.104d}$$

$$\iint f \, dS = 2A \int_0^1 \left[\int_0^{1-\xi_2} f(\xi_1, \xi_2) d\xi_1 \right] d\xi_2 \tag{6.104e}$$

$$\iint \xi_1^i \xi_2^j \xi_3^k dS = \frac{i! j! k!}{(i+j+k+2)!} 2A \tag{6.104f}$$

$$dS = 2A \, d\xi_1 \, d\xi_2 \tag{6.104g}$$

6.8.3 Shape Functions

We may now express the shape function for higher-order elements in terms of local coordinates. Sometimes, it is convenient to label each point in the finite elements in

Figure 6.22 with three integers i, j, and k from which its local coordinates (ξ_1, ξ_2, ξ_3) can be found or vice versa. At each point P_{ijk}

$$(\xi_1, \xi_2, \xi_3) = \left(\frac{i}{n}, \frac{j}{n}, \frac{k}{n} \right) \tag{6.105}$$

Hence if a value of Φ, say Φ_{ijk}, is prescribed at each point P_{ijk}, Equation 6.94 can be written as

$$\Phi(\xi_1, \xi_2, \xi_3) = \sum_{i=1}^{m} \sum_{j=1}^{m-i} \alpha_{ijk} (\xi_1, \xi_2, \xi_3) \Phi_{ijk} \tag{6.106}$$

where

$$\alpha_\ell = \alpha_{ijk} = p_i (\xi_1) \, p_j (\xi_2) \, p_k (\xi_3), \quad \ell = 1, 2, \dots \tag{6.107}$$

$$p_r(\xi) = \begin{cases} \dfrac{1}{r!} \displaystyle\prod_{t=0}^{r-1} (n\xi - t), & r > 0 \\[2ex] 1, & r = 0 \end{cases} \tag{6.108}$$

and $r \in (i, j, k)$. $p_r(\xi)$ may also be written as

$$p_r(\xi) = \frac{(n\xi - r + 1)}{r} p_{r-1}(\xi). \quad r > 0 \tag{6.109}$$

where $p_0(\xi) = 1$.

The relationships between the subscripts $q \in \{1, 2, 3\}$ on $\xi_q, \ell \in \{1, 2, \dots, m\}$ on α_ℓ, and $r \in (i, j, k)$ on p_r and P_{ijk} in Equations 6.107 through 6.109 are illustrated in Figure 6.25 for n ranging from 1 to 4. Henceforth, point P_{ijk} will be written as P_n for conciseness.

Notice from Equation 6.108 or Equation 6.109 that

$$\begin{aligned} p_0(\xi) &= 1 \\ p_1(\xi) &= n\xi \\ p_2(\xi) &= \frac{1}{2}(n\xi - 1)n\xi \\ p_3(\xi) &= \frac{1}{6}(n\xi - 2)(n\xi - 1)n\xi \\ p_4(\xi) &= \frac{1}{24}(n\xi - 3)(n\xi - 2)(n\xi - 1)n\xi, \text{ etc.} \end{aligned} \tag{6.110}$$

Substituting Equation 6.110 into Equation 6.107 gives the shape functions α_ℓ for nodes $\ell = 1, 2, \dots, m$, as shown in Table 6.7 for $n = 1$–4. Observe that each α_ℓ takes the value of 1 at node ℓ and value of 0 at all other nodes in the triangle. This is easily verified using Equation 6.105 in conjunction with Figure 6.25.

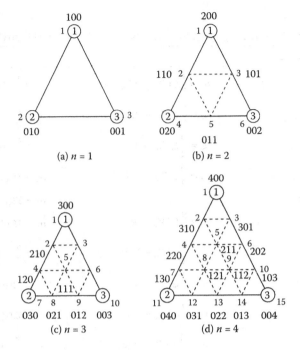

FIGURE 6.25
Distribution of nodes over triangles for $n = 1$–4. The triangles are in standard position.

6.8.4 Fundamental Matrices

The fundamental matrices $[T]$ and $[Q]$ for triangular elements can be derived using the shape functions in Table 6.7. (For simplicity, the brackets [ ] denoting a matrix quantity will be dropped in the remaining part of this section.) In Equation 6.46, the T matrix is defined as

$$T_{ij} = \iint \alpha_i \alpha_j \, dS \tag{6.46}$$

From Table 6.7, we substitute α_ℓ in Equation 6.46 and apply Equations 6.104e and 6.104f to obtain elements of T. For example, for $n = 1$,

$$T_{ij} = 2A \int\limits_{0}^{1} \int\limits_{0}^{1-\xi_2} \xi_i \xi_j \, d\xi_1 \, d\xi_2$$

when $i \neq j$,

$$T_{ij} = \frac{2A(1!)(1!)(0!)}{4!} = \frac{A}{12}, \tag{6.111a}$$

when $i = j$,

$$T_{ij} = \frac{2A(2!)}{4!} = \frac{A}{6} \tag{6.111b}$$

TABLE 6.7

Polynomial Basis Function $\alpha_\ell(\xi_1, \xi_2, \xi_3, \xi_4)$ for First-, Second-, Third-, and Fourth Order

$n = 1$	$n = 2$	$n = 3$	$n = 4$
$\alpha_1 = \xi_1$	$\alpha_1 = \xi_1(2\xi_1 - 1)$	$\alpha_1 = \dfrac{1}{2}\xi_1(3\xi_1 - 2)(3\xi_1 - 1)$	$\alpha_1 = \dfrac{1}{6}\xi_1(4\xi_1 - 3)(4\xi_1 - 2)(4\xi_1 - 1)$
$\alpha_2 = \xi_2$	$\alpha_2 = 4\xi_1\xi_2$	$\alpha_2 = \dfrac{9}{2}\xi_1(3\xi_1 - 1)\xi_2$	$\alpha_2 = \dfrac{8}{3}\xi_1(4\xi_1 - 2)(4\xi_1 - 1)\xi_2$
$\alpha_3 = \xi_3$	$\alpha_3 = 4\xi_1\xi_3$	$\alpha_3 = \dfrac{9}{2}\xi_1(3\xi_1 - 1)\xi_3$	$\alpha_3 = \dfrac{8}{3}\xi_1(4\xi_1 - 2)(4\xi_1 - 1)\xi_3$
	$\alpha_4 = \xi_2(2\xi_2 - 1)$	$\alpha_4 = \dfrac{9}{2}\xi_1(3\xi_2 - 1)\xi_2$	$\alpha_4 = 4\xi_1(4\xi_1 - 1)(4\xi_2 - 1)\xi_2$
	$\alpha_5 = 4\xi_2\xi_3$	$\alpha_5 = 27\xi_1\xi_2\xi_3$	$\alpha_5 = 32\xi_1(4\xi_1 - 1)\xi_2\xi_3$
	$\alpha_6 = \xi_3(2\xi_3 - 1)$	$\alpha_6 = \dfrac{9}{2}\xi_1(3\xi_3 - 1)\xi_3$	$\alpha_6 = 4\xi_1(4\xi_1 - 1)(4\xi_3 - 1)\xi_3$
		$\alpha_7 = \dfrac{1}{2}\xi_2(3\xi_2 - 2)(3\xi_2 - 1)$	$\alpha_7 = \dfrac{8}{3}\xi_1(4\xi_2 - 2)(4\xi_2 - 1)\xi_2$
		$\alpha_8 = \dfrac{9}{2}\xi_2(3\xi_2 - 1)\xi_3$	$\alpha_8 = 32\xi_1(4\xi_2 - 1)\xi_2\xi_3$
		$\alpha_9 = \dfrac{9}{2}\xi_2(3\xi_3 - 1)\xi_3$	$\alpha_9 = 32\xi_1\xi_2(4\xi_3 - 1)\xi_3$
		$\alpha_{10} = \dfrac{1}{2}\xi_3(3\xi_3 - 2)(3\xi_3 - 1)$	$\alpha_{10} = \dfrac{8}{3}\xi_1(4\xi_3 - 2)(4\xi_3 - 1)\xi_3$
			$\alpha_{11} = \dfrac{1}{6}\xi_2(4\xi_2 - 3)(4\xi_2 - 2)(4\xi_2 - 1)$
			$\alpha_{12} = \dfrac{8}{3}\xi_2(4\xi_2 - 2)(4\xi_2 - 1)\xi_3$
			$\alpha_{13} = 4\xi_2(4\xi_2 - 1)(4\xi_3 - 1)\xi_3$
			$\alpha_{14} = \dfrac{8}{3}\xi_2(4\xi_3 - 2)(4\xi_3 - 1)\xi_3$
			$\alpha_{15} = \dfrac{1}{6}\xi_3(4\xi_3 - 3)(4\xi_3 - 2)(4\xi_3 - 1)$

Hence,

$$T = \frac{A}{12}\begin{bmatrix} 2 & 1 & 1 \\ 1 & 2 & 1 \\ 1 & 1 & 2 \end{bmatrix} \tag{6.112}$$

By following the same procedure, higher-order T matrices can be obtained. The T matrices for orders up to $n = 4$ are tabulated in Table 6.8, where the factor A, the area of the element, has been suppressed. The actual matrix elements are obtained from Table 6.8 by multiplying the tabulated numbers by A and dividing by the indicated common denominator. The following properties of the T matrix are noteworthy:

a. T is symmetric with positive elements;

b. Elements of T all add up to the area of the triangle, that is, $\sum_i^m \sum_j^m T_{ij} = A$, since by definition $\sum_{\ell=1}^m \alpha_\ell = 1$ at any point within the element;

TABLE 6.8

Table of T Matrices for $n = 1$–4

$n = 1$ Common denominator: 12

$$\begin{pmatrix} 2 & 1 & 1 \\ 1 & 2 & 1 \\ 1 & 1 & 2 \end{pmatrix}$$

$n = 2$ Common denominator: 180

$$\begin{pmatrix} 6 & 0 & 0 & -1 & -4 & -1 \\ 0 & 32 & 16 & 0 & 16 & -4 \\ 0 & 16 & 32 & -4 & 16 & 0 \\ -1 & 0 & -4 & 6 & 0 & -1 \\ -4 & 16 & 16 & 0 & 32 & 0 \\ -1 & -4 & 0 & -1 & 0 & 6 \end{pmatrix}$$

$n = 3$ Common denominator: 6720

$$\begin{pmatrix} 76 & 18 & 18 & 0 & 36 & 0 & 11 & 27 & 27 & 11 \\ 18 & 540 & 270 & -189 & 162 & -135 & 0 & -135 & -54 & 27 \\ 18 & 270 & 540 & -135 & 162 & -189 & 27 & -54 & -135 & 0 \\ 0 & -189 & -135 & 540 & 162 & -54 & 18 & 270 & -135 & 27 \\ 36 & 162 & 162 & 162 & 1944 & 162 & 36 & 162 & 162 & 36 \\ 0 & -135 & -189 & -54 & 162 & 540 & 27 & -135 & 270 & 18 \\ 11 & 0 & 27 & 18 & 36 & 27 & 76 & 18 & 0 & 11 \\ 27 & -135 & -54 & 270 & 162 & -135 & 18 & 540 & -189 & 0 \\ 27 & -54 & -135 & -135 & 162 & 270 & 0 & -189 & 540 & 18 \\ 11 & 27 & 0 & 27 & 36 & 18 & 11 & 0 & 18 & 76 \end{pmatrix}$$

$n = 4$ Common denominator: 56,700

290	160	160	−80	160	80	0	−160	−160	0	−27	−112	−12	−112	−27
160	2560	1280	−1280	1280	−960	768	256	−256	512	0	512	64	256	−112
160	1280	2560	−960	1280	−1280	512	−256	256	768	−112	256	64	512	0
−80	−1280	−960	3168	384	48	−1280	384	−768	64	−80	−960	48	64	−12
160	1280	1280	384	10752	384	256	−1536	−1536	256	−160	−256	−768	−256	−160
−80	−960	−1280	48	384	3168	64	−768	384	−1280	−12	64	48	−960	−80
0	768	512	−1280	256	64	2560	1280	−256	256	160	1280	−960	512	−112
−160	256	−256	384	−1536	−768	1280	10752	−1536	−256	160	1280	384	256	−160
−160	−256	256	−768	−1536	384	−256	−1536	10,752	1280	−160	256	384	1280	160
0	512	768	64	256	−1280	256	−256	1280	2560	−112	512	−960	1280	160
−27	0	−112	−80	−160	−12	160	160	−160	−112	290	160	−80	0	−27
−112	512	256	−960	−256	64	1280	1280	256	512	160	2560	−1280	768	0
−12	64	64	48	−768	48	−960	384	384	−960	−80	−1280	3168	−1280	−80
−112	256	512	64	−256	−960	512	256	1280	1280	0	768	−1280	2560	160
−27	−112	0	−12	−160	−80	−112	−160	160	160	−27	0	−80	160	290

    c. Elements for which the two triple subscripts form similar permutations are equal, that is, $T_{ijk,prq} = T_{ikj,prq} = T_{kij,rpq} = T_{kji,rqp} = T_{jki,qrp} = T_{jik,qpr}$; this should be obvious from Equations 6.46 and 6.107.

    These properties are not only useful in checking the matrix, they have proved useful in saving computer time and storage. It is interesting to know that the properties are independent of coordinate system [44].

In Equation 6.14 or Equation 6.45, elements of [C] matrix are defined by

$$C_{ij} = \iint \left(\frac{\partial \alpha_i}{\partial x} \frac{\partial \alpha_j}{\partial x} + \frac{\partial \alpha_i}{\partial y} \frac{\partial \alpha_j}{\partial y} \right) dS \tag{6.113}$$

By applying Equations 6.104a through 6.104d to Equation 6.113, it can be shown that [4,41]

$$C_{ij} = \frac{1}{2A} \sum_{q=1}^{3} \cot \theta_q \iint \left(\frac{\partial \alpha_i}{\partial \xi_{q+1}} - \frac{\partial \alpha_i}{\partial \xi_{q-1}} \right) \left(\frac{\partial \alpha_j}{\partial \xi_{q+1}} - \frac{\partial \alpha_j}{\partial \xi_{q-1}} \right) dS$$

or

$$\boxed{C_{ij} = \sum_{q=1}^{3} Q_{ij}^{(q)} \cot \theta_q} \tag{6.114}$$

where θ_q is the included angle of vertex $q \in \{1, 2, 3\}$ of the triangle and

$$\boxed{Q_{ij}^{(q)} = \iint \left(\frac{\partial \alpha_i}{\partial \xi_{q+1}} - \frac{\partial \alpha_i}{\partial \xi_{q-1}} \right) \left(\frac{\partial \alpha_j}{\partial \xi_{q+1}} - \frac{\partial \alpha_j}{\partial \xi_{q-1}} \right) d\xi_1 \, d\xi_2} \tag{6.115}$$

    We notice that matrix C depends on the triangle shape, whereas the matrices $Q^{(q)}$ do not. The $Q^{(1)}$ matrices for $n = 1$–4 are tabulated in Table 6.9. The following properties of Q matrices should be noted:

    a. They are symmetric;
    b. The row and column sums of any Q matrix are zero, that is, $\sum_{i=1}^{m} Q_{ij}^{(q)} = 0 = \sum_{j=1}^{m} Q_{ij}^{(q)}$ so that the C matrix is singular.

    $Q^{(2)}$ and $Q^{(3)}$ are easily obtained from $Q^{(1)}$ by row and column permutations so that the matrix C for any triangular element is constructed easily if $Q^{(1)}$ is known. One approach [45] involves using a rotation matrix R similar to that in Silvester and Ferrari [4], which is essentially a unit matrix with elements rearranged to correspond to one rotation of the triangle about its centroid in a counterclockwise direction. For example, for $n = 1$, the rotation matrix is basically derived from Figure 6.26 as

$$R = \begin{bmatrix} 0 & 0 & 1 \\ 1 & 0 & 0 \\ 0 & 1 & 0 \end{bmatrix} \tag{6.116}$$

where $R_{ij} = 1$ if node i is replaced by node j after one counter clockwise rotation, or $R_{ij} = 0$ otherwise. Table 6.10 presents the R matrices for $n = 1$–4. Note that each row or column

TABLE 6.9

Table of Q Matrices for $n = 1\text{--}4$

$n = 1$ Common denominator: 2

$$\begin{pmatrix} 0 & 0 & 0 \\ 0 & 1 & -1 \\ 0 & -1 & 1 \end{pmatrix}$$

$n = 2$ Common denominator: 6

$$\begin{pmatrix} 0 & 0 & 0 & 0 & 0 & 0 \\ 0 & 8 & -8 & 0 & 0 & 0 \\ 0 & -8 & 8 & 0 & 0 & 0 \\ 0 & 0 & 0 & 3 & -4 & 1 \\ 0 & 0 & 0 & -4 & 8 & -4 \\ 0 & 0 & 0 & 1 & -4 & 3 \end{pmatrix}$$

$n = 3$ Common denominator: 80

$$\begin{pmatrix} 0 & 0 & 0 & 0 & 0 & 0 & 0 & 0 & 0 & 0 \\ 0 & 135 & -135 & -27 & 0 & 27 & 3 & 0 & 0 & -3 \\ 0 & -135 & 135 & 27 & 0 & -27 & -3 & 0 & 0 & 3 \\ 0 & -27 & 27 & 135 & -162 & 27 & 3 & 0 & 0 & -3 \\ 0 & 0 & 0 & -162 & 324 & -162 & 0 & 0 & 0 & 0 \\ 0 & 27 & -27 & 27 & -162 & 135 & -3 & 0 & 0 & 3 \\ 0 & 3 & -3 & 3 & 0 & -3 & 34 & -54 & 27 & -7 \\ 0 & 0 & 0 & 0 & 0 & 0 & -54 & 135 & -108 & 27 \\ 0 & 0 & 0 & 0 & 0 & 0 & 27 & -108 & 135 & -54 \\ 0 & -3 & 3 & -3 & 0 & 3 & -7 & 27 & -54 & 34 \end{pmatrix}$$

$n = 4$ Common denominator: 1890

0	0	0	0	0	0	0	0	0	0	0	0	0	0	0
0	3968	−3968	−1440	0	1440	640	0	0	−640	−80	0	0	0	80
0	−3968	3968	1440	0	−1440	−640	0	0	640	80	0	0	0	−80
0	−1440	1440	4632	−5376	744	−1248	768	768	−288	80	−128	96	−128	80
0	0	0	−5376	10,752	−5376	1536	−1536	−1536	1536	−160	256	−192	256	−160
0	1440	−1440	744	−5376	4632	−288	768	768	−1248	80	−128	96	−128	80
0	640	−640	−1248	1536	−288	3456	−4608	1536	−384	240	−256	192	−256	80
0	0	0	768	−1536	768	−4608	10,752	−7680	1536	−160	256	−192	256	−160
0	0	0	768	−1536	768	1536	−7680	10,752	−4608	−160	256	−192	256	−160
0	−640	640	−288	1536	−1248	−384	1536	−4608	3456	80	−256	192	−256	240
0	−80	80	80	−160	80	240	−160	−160	80	705	−1232	884	−464	107
0	0	0	−128	256	−128	−256	256	256	−256	−1232	3456	−3680	1920	−464
0	0	0	96	−192	96	192	−192	−192	192	884	−3680	5592	−3680	884
0	0	0	−128	256	−128	−256	256	256	−256	−464	1920	−3680	3456	−1232
0	80	−80	80	−160	80	80	−160	−160	240	107	−464	884	−1232	705

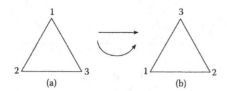

FIGURE 6.26
One counterclockwise rotation of the triangle in (a) gives the triangle in (b).

TABLE 6.10

R Matrix for $n = 1$–4, $n = 1$

$n = 1$

$$\begin{bmatrix} 0 & 0 & 1 \\ 1 & 0 & 0 \\ 0 & 1 & 0 \end{bmatrix}$$

$n = 2$

$$\begin{bmatrix} 0 & 0 & 0 & 0 & 0 & 1 \\ 0 & 0 & 1 & 0 & 0 & 0 \\ 0 & 0 & 0 & 0 & 1 & 0 \\ 1 & 0 & 0 & 0 & 0 & 0 \\ 0 & 1 & 0 & 0 & 0 & 0 \\ 0 & 0 & 0 & 1 & 0 & 0 \end{bmatrix}$$

$n = 3$

$$\begin{bmatrix} 0 & 0 & 0 & 0 & 0 & 0 & 0 & 0 & 0 & 1 \\ 0 & 0 & 0 & 0 & 0 & 1 & 0 & 0 & 0 & 0 \\ 0 & 0 & 0 & 0 & 0 & 0 & 0 & 0 & 1 & 0 \\ 0 & 0 & 1 & 0 & 0 & 0 & 0 & 0 & 0 & 0 \\ 0 & 0 & 0 & 0 & 1 & 0 & 0 & 0 & 0 & 0 \\ 0 & 0 & 0 & 0 & 0 & 0 & 0 & 1 & 0 & 0 \\ 1 & 0 & 0 & 0 & 0 & 0 & 0 & 0 & 0 & 0 \\ 0 & 1 & 0 & 0 & 0 & 0 & 0 & 0 & 0 & 0 \\ 0 & 0 & 0 & 1 & 0 & 0 & 0 & 0 & 0 & 0 \\ 0 & 0 & 0 & 0 & 0 & 0 & 1 & 0 & 0 & 0 \end{bmatrix}$$

$n = 4$

$$\begin{bmatrix} 0 & 0 & 0 & 0 & 0 & 0 & 0 & 0 & 0 & 0 & 0 & 0 & 0 & 0 & 1 \\ 0 & 0 & 0 & 0 & 0 & 0 & 0 & 0 & 0 & 1 & 0 & 0 & 0 & 0 & 0 \\ 0 & 0 & 0 & 0 & 0 & 0 & 0 & 0 & 0 & 0 & 0 & 0 & 0 & 1 & 0 \\ 0 & 0 & 0 & 0 & 0 & 1 & 0 & 0 & 0 & 0 & 0 & 0 & 0 & 0 & 0 \\ 0 & 0 & 0 & 0 & 0 & 0 & 0 & 0 & 1 & 0 & 0 & 0 & 0 & 0 & 0 \\ 0 & 0 & 0 & 0 & 0 & 0 & 0 & 0 & 0 & 0 & 0 & 0 & 1 & 0 & 0 \\ 0 & 0 & 1 & 0 & 0 & 0 & 0 & 0 & 0 & 0 & 0 & 0 & 0 & 0 & 0 \\ 0 & 0 & 0 & 0 & 1 & 0 & 0 & 0 & 0 & 0 & 0 & 0 & 0 & 0 & 0 \\ 0 & 0 & 0 & 0 & 0 & 0 & 0 & 1 & 0 & 0 & 0 & 0 & 0 & 0 & 0 \\ 0 & 0 & 0 & 0 & 0 & 0 & 0 & 0 & 0 & 0 & 0 & 1 & 0 & 0 & 0 \\ 1 & 0 & 0 & 0 & 0 & 0 & 0 & 0 & 0 & 0 & 0 & 0 & 0 & 0 & 0 \\ 0 & 1 & 0 & 0 & 0 & 0 & 0 & 0 & 0 & 0 & 0 & 0 & 0 & 0 & 0 \\ 0 & 0 & 0 & 1 & 0 & 0 & 0 & 0 & 0 & 0 & 0 & 0 & 0 & 0 & 0 \\ 0 & 0 & 0 & 0 & 0 & 0 & 1 & 0 & 0 & 0 & 0 & 0 & 0 & 0 & 0 \\ 0 & 0 & 0 & 0 & 0 & 0 & 0 & 0 & 0 & 0 & 1 & 0 & 0 & 0 & 0 \end{bmatrix}$$

of R has only one nonzero element since R is essentially a unit matrix with rearranged elements.

Once the R is known, we obtain

$$\boxed{Q^{(2)} = RQ^{(1)}R^t}$$ (6.117a)

$$\boxed{Q^{(3)} = RQ^{(2)}R^t}$$ (6.117b)

where R^t is the transpose of R.

EXAMPLE 6.5

For $n = 2$, calculate $Q^{(1)}$ and obtain $Q^{(2)}$ from $Q^{(1)}$ using Equation 6.117a.

Solution

By definition,

$$Q_{ij}^{(1)} = \iint \left(\frac{\partial \alpha_i}{\partial \xi_2} - \frac{\partial \alpha_i}{\partial \xi_3} \right) \left(\frac{\partial \alpha_j}{\partial \xi_2} - \frac{\partial \alpha_j}{\partial \xi_3} \right) d\xi_1 \, d\xi_2$$

For $n = 2$, $i, j = 1, 2, ..., 6$, and α_i are given in terms of the local coordinates in Table 6.7. Since $Q^{(1)}$ is symmetric, only some of the elements need to be calculated. Substituting for α_ℓ from Table 6.7 and applying Equations 6.104e and 6.104f, we obtain

$$Q_{1j} = 0, \quad j = 1 \text{ to } 6,$$
$$Q_{i1} = 0, \quad i = 1 \text{ to } 6,$$
$$Q_{22} = \frac{1}{2A} \iint (4\xi_1)^2 \, d\xi_1 \xi_2 = \frac{8}{6},$$
$$Q_{23} = \frac{1}{2A} \iint (4\xi_1)(-4\xi_1) \, d\xi_1 \xi_2 = -\frac{8}{6},$$
$$Q_{24} = \frac{1}{2A} \iint (4\xi_1)(4\xi_1 - 1) \, d\xi_1 \xi_2 = 0 = Q_{26},$$
$$Q_{25} = \frac{1}{2A} \iint (4\xi_1)(4\xi_3 - 4\xi_2) \, d\xi_1 \xi_2 = 0,$$
$$Q_{33} = \frac{1}{2A} \iint (-4\xi_1)^2 \, d\xi_1 \xi_2 = \frac{8}{6},$$
$$Q_{34} = \frac{1}{2A} \iint (-4\xi_1)(4\xi_2 - 1) \, d\xi_1 \xi_2 = 0 = Q_{36},$$
$$Q_{35} = \frac{1}{2A} \iint (-4\xi_1)(4\xi_3 - 4\xi_2) \, d\xi_1 \xi_2 = 0,$$
$$Q_{44} = \frac{1}{2A} \iint (4\xi_2 - 1)^2 \, d\xi_1 \xi_2 = \frac{3}{6},$$
$$Q_{45} = \frac{1}{2A} \iint (4\xi_2 - 1)(4\xi_3 - 4\xi_2) \, d\xi_1 \xi_2 = -\frac{4}{6},$$
$$Q_{46} = \frac{1}{2A} \iint (4\xi_2 - 1)(4\xi_3 - 1)(-1) \, d\xi_1 \xi_2 = \frac{1}{6},$$
$$Q_{55} = \frac{1}{2A} \iint (4\xi_3 - 4\xi_2)^2 \, d\xi_1 \xi_2 = \frac{8}{6},$$
$$Q_{56} = \frac{1}{2A} \iint (4\xi_3 - 4\xi_2)(-1)(4\xi_3 - 1) \, d\xi_1 \xi_2 = -\frac{4}{6},$$
$$Q_{66} = \frac{1}{2A} \iint (-1)(4\xi_3 - 1)^2 \, d\xi_1 \xi_2 = \frac{3}{6}$$

Hence,

$$Q^{(1)} = \frac{1}{6}\begin{bmatrix} 0 & 0 & 0 & 0 & 0 & 0 \\ 0 & 8 & -8 & 0 & 0 & 0 \\ 0 & -8 & 8 & 0 & 0 & 0 \\ 0 & 0 & 0 & 3 & -4 & 1 \\ 0 & 0 & 0 & -4 & 8 & -4 \\ 0 & 0 & 0 & 1 & -4 & 3 \end{bmatrix}$$

We now obtain $Q^{(2)}$ from

$$Q^{(2)} = RQ^{(1)}R^t$$

$$= \frac{1}{6}R\begin{bmatrix} 0 & 0 & 0 & 0 & 0 & 0 \\ 0 & 8 & -8 & 0 & 0 & 0 \\ 0 & -8 & 8 & 0 & 0 & 0 \\ 0 & 0 & 0 & 3 & -4 & 1 \\ 0 & 0 & 0 & -4 & 8 & -4 \\ 0 & 0 & 0 & 1 & -4 & 3 \end{bmatrix}\begin{bmatrix} 0 & 0 & 0 & 1 & 0 & 0 \\ 0 & 0 & 0 & 0 & 1 & 0 \\ 0 & 1 & 0 & 0 & 0 & 0 \\ 0 & 0 & 0 & 0 & 0 & 1 \\ 0 & 0 & 1 & 0 & 0 & 0 \\ 1 & 0 & 0 & 0 & 0 & 0 \end{bmatrix}$$

$$= \frac{1}{6}\begin{bmatrix} 0 & 0 & 0 & 0 & 0 & 1 \\ 0 & 0 & 1 & 0 & 0 & 0 \\ 0 & 0 & 0 & 0 & 1 & 0 \\ 1 & 0 & 0 & 0 & 0 & 0 \\ 0 & 1 & 0 & 0 & 0 & 0 \\ 0 & 0 & 0 & 1 & 0 & 0 \end{bmatrix}\begin{bmatrix} 0 & 0 & 0 & 0 & 0 & 0 \\ 0 & -8 & 0 & 0 & 8 & 0 \\ 0 & 8 & 0 & 0 & -8 & 0 \\ 1 & 0 & -4 & 0 & 0 & 3 \\ -4 & 0 & 8 & 0 & 0 & -4 \\ 3 & 0 & 4 & 0 & 0 & 1 \end{bmatrix}$$

$$Q^{(2)} = \frac{1}{6}\begin{bmatrix} 3 & 0 & -4 & 0 & 0 & 1 \\ 0 & 8 & 0 & 0 & -8 & 0 \\ -4 & 0 & 8 & 0 & 0 & -4 \\ 0 & 0 & 0 & 0 & 0 & 0 \\ 0 & -8 & 0 & 0 & 8 & 0 \\ 1 & 0 & -4 & 0 & 0 & 3 \end{bmatrix}$$

6.9 Three-Dimensional Elements

The finite element techniques developed in the previous sections for two-dimensional elements can be extended to three-dimensional elements. One would expect three-dimensional problems to require a large total number of elements to achieve an accurate result and demand a large storage capacity and computational time. For the sake of completeness, we will discuss the finite element analysis of Helmholtz's equation in three dimensions, namely,

$$\nabla^2 \Phi + k^2 \Phi = g \tag{6.118}$$

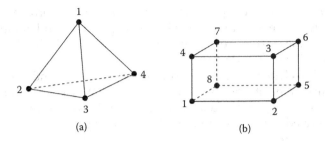

FIGURE 6.27
Three-dimensional elements: (a) Four-node or linear-order tetrahedral, (b) eight-node or linear-order hexahedral.

We first divide the solution region into tetrahedral or hexahedral (rectangular prism) elements as in Figure 6.27. Assuming a four-node tetrahedral element, the function Φ is represented within the element by

$$\Phi_e = a + bx + cy + dz \tag{6.119}$$

The same applies to the function g. Since Equation 6.119 must be satisfied at the four nodes of the tetrahedral elements,

$$\Phi_{ei} = a + bx_i + cy_i + dz_i, \quad i = 1, \ldots, 4 \tag{6.120}$$

Thus, we have four simultaneous equations (similar to Equation 6.5) from which the coefficients a, b, c, and d can be determined. The determinant of the system of equations is

$$\det = \begin{vmatrix} 1 & x_1 & y_1 & z_1 \\ 1 & x_2 & y_2 & z_2 \\ 1 & x_3 & y_3 & z_3 \\ 1 & x_4 & y_4 & z_4 \end{vmatrix} = 6v, \tag{6.121}$$

where v is the volume of the tetrahedron. By finding a, b, c, and d, we can write

$$\boxed{\Phi_e = \sum_{i=1}^{4} \alpha_i(x,y)\Phi_{ei}} \tag{6.122}$$

where

$$\alpha_1 = \frac{1}{6v} \begin{vmatrix} 1 & x & y & z \\ 1 & x_2 & y_2 & z_2 \\ 1 & x_3 & y_3 & z_3 \\ 1 & x_4 & y_4 & z_4 \end{vmatrix}, \tag{6.123a}$$

$$\alpha_2 = \frac{1}{6v} \begin{vmatrix} 1 & x_1 & y_1 & z_1 \\ 1 & x & y & z \\ 1 & x_3 & y_3 & z_3 \\ 1 & x_4 & y_4 & z_4 \end{vmatrix}, \tag{6.123b}$$

with α_3 and α_4 having similar expressions. For higher-order approximation, the matrices for *as* become large in size and we resort to local coordinates. Another motivation for using local coordinates is the existence of integration equations which simplify the evaluation of the fundamental matrices T and Q.

For the tetrahedral element, the local coordinates are ξ_1, ξ_2, ξ_3, and ξ_4, each perpendicular to a side. They are defined at a given point as the ratio of the distance from that point to the appropriate apex to the perpendicular distance from the side to the opposite apex. They can also be interpreted as volume ratios, that is, at a point P

$$\xi_i = \frac{v_i}{v} \tag{6.124}$$

where v_i is the volume bound by P and face i. It is evident that

$$\sum_{i=1}^{4} \xi_i = 1 \tag{6.125a}$$

or

$$\xi_4 = 1 - \xi_1 - \xi_2 - \xi_3 \tag{6.125b}$$

The following properties are useful in evaluating integration involving local coordinates [46]:

$$dv = 6v \, d\xi_1 \, d\xi_2 \, d\xi_3, \tag{6.126a}$$

$$\iiint f \, dv = 6v \int_0^1 \left[\int_0^{1-\xi_3} \left(\int_0^{1-\xi_2-\xi_3} f \, d\xi_1 \right) d\xi_2 \right] d\xi_3, \tag{6.126b}$$

$$\iiint \xi_1^i \xi_2^j \xi_3^k \xi_4^\ell \, dv = \frac{i! \, j! \, k! \, \ell!}{(i+j+k+\ell+3)!} 6v \tag{6.126c}$$

In terms of the local coordinates, an arbitrary function $\Phi(x, y, z)$ can be approximated within an element in terms of a complete nth-order polynomial as

$$\boxed{\Phi_e(x,y,z) = \sum_{i=1}^{m} \alpha_i(x,y,z) \Phi_{ei}} \tag{6.127}$$

where $m = 1/6 \, (n+1)(n+2)(n+3)$ is the number of nodes in the tetrahedron or the number of terms in the polynomial. The terms in a complete three-dimensional polynomial may be arrayed as shown in Figure 6.28.

Each point in the tetrahedral element is represented by four integers i, j, k, and ℓ which can be used to determine the local coordinates $(\xi_1, \xi_2, \xi_3, \xi_4)$. That is at $P_{ijk\ell}$,

$$(\xi_1, \xi_2, \xi_3, \xi_4) = \left(\frac{i}{n}, \frac{j}{n}, \frac{k}{n}, \frac{\ell}{n} \right) \tag{6.128}$$

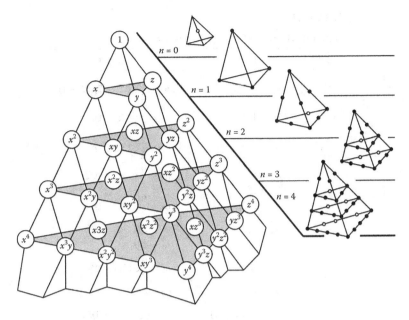

FIGURE 6.28
Pascal tetrahedron and associated array of terms.

Hence at each node,

$$\alpha_q = \alpha_{ijk\ell} = p_i\,(\xi_1)\,p_j\,(\xi_2)\,p_k(\xi_3)\,p_\ell\,(\xi_4), \tag{6.129}$$

where $q = 1, 2, \ldots, m$ and p_r is defined in Equation 6.108 or Equation 6.109. The relationship between the node numbers q and $ijk\ell$ is illustrated in Figure 6.29 for the second-order tetrahedron ($n = 2$). The shape functions obtained by substituting Equation 6.108 into Equation 6.129 are presented in Table 6.11 for $n = 1$–3.

The expressions derived from the variational principle for the two-dimensional problems in Sections 6.2 through 6.4 still hold except that the fundamental matrices $[T]$ and $[Q]$ now involve triple integration. For Helmholtz equation (6.56), for example, Equation 6.68 applies, namely,

$$[C_{ff} - k^2 T_{ff}]\Phi_f = 0 \tag{6.130}$$

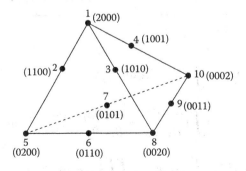

FIGURE 6.29
Numbering scheme for second-order tetrahedron.

TABLE 6.11

Shape Functions $\alpha_q(\xi_1, \xi_2, \xi_3, \xi_4)$ for $n = 1\text{--}3$

$n = 1$	$n = 2$	$n = 3$
$\alpha_1 = \xi_1$	$\alpha_1 = \xi_1(2\xi_1 - 1)$	$\alpha_1 = \dfrac{1}{2}\xi_1(3\xi_1 - 2)(3\xi_1 - 1)$
$\alpha_2 = \xi_2$	$\alpha_2 = 4\xi_1\xi_2$	$\alpha_2 = \dfrac{9}{2}\xi_1(3\xi_1 - 1)\xi_2$
$\alpha_3 = \xi_3$	$\alpha_3 = 4\xi_1\xi_3$	$\alpha_3 = \dfrac{9}{2}\xi_1(3\xi_1 - 1)\xi_3$
$\alpha_4 = \xi_4$	$\alpha_4 = 4\xi_1\xi_4$	$\alpha_4 = \dfrac{9}{2}\xi_1(3\xi_1 - 1)\xi_4$
	$\alpha_5 = \xi_2(2\xi_2 - 1)$	$\alpha_5 = \dfrac{9}{2}\xi_1(3\xi_3 - 1)\xi_2$
	$\alpha_6 = 4\xi_2\xi_3$	$\alpha_6 = 27\xi_1\xi_2\xi_3$
	$\alpha_7 = 4\xi_2\xi_4$	$\alpha_7 = 27\xi_1\xi_2\xi_4$
	$\alpha_8 = \xi_2(2\xi_3 - 1)$	$\alpha_8 = \dfrac{9}{2}\xi_1(3\xi_3 - 1)\xi_3$
	$\alpha_9 = 4\xi_3\xi_4$	$\alpha_9 = 27\xi_1\xi_3\xi_4$
	$\alpha_{10} = \xi_4(2\xi_4 - 1)$	$\alpha_{10} = \dfrac{9}{2}\xi_1(3\xi_4 - 1)\xi_4$
		$\alpha_{11} = \dfrac{1}{2}\xi_2(3\xi_2 - 1)(3\xi_2 - 2)$
		$\alpha_{12} = \dfrac{9}{2}\xi_2(3\xi_2 - 1)\xi_3$
		$\alpha_{13} = \dfrac{9}{2}\xi_2(3\xi_2 - 1)\xi_4$
		$\alpha_{14} = \dfrac{9}{2}\xi_2(3\xi_3 - 1)\xi_3$
		$\alpha_{15} = 27\xi_2\xi_3\xi_4$
		$\alpha_{16} = \dfrac{9}{2}\xi_2(3\xi_3 - 1)\xi_3$
		$\alpha_{17} = \dfrac{1}{2}\xi_3(3\xi_3 - 1)(3\xi_3 - 2)$
		$\alpha_{18} = \dfrac{9}{2}\xi_3(3\xi_3 - 1)\xi_4$
		$\alpha_{19} = \dfrac{9}{2}\xi_3(3\xi_4 - 1)\xi_4$
		$\alpha_{20} = \dfrac{1}{2}\xi_4(3\xi_4 - 1)(3\xi_4 - 2)$

except that

$$C_{ij}^{(e)} = \int_v \nabla\alpha_i \cdot \nabla\alpha_j \, dv$$

$$= \int_v \left(\frac{\partial\alpha_i}{\partial x}\frac{\partial\alpha_j}{\partial x} + \frac{\partial\alpha_i}{\partial y}\frac{\partial\alpha_j}{\partial y} + \frac{\partial\alpha_i}{\partial z}\frac{\partial\alpha_j}{\partial z} \right) dv, \tag{6.131}$$

$$T_{ij}^{(e)} = \int_v \alpha_i\alpha_j \, dv = v \iiint \alpha_i\alpha_j \, d\xi_1 \, d\xi_2 \, d\xi_3 \tag{6.132}$$

For further discussion on three-dimensional elements, one should consult Silvester and Ferrari [4]. Applications of three-dimensional elements to EM-related problems can be found in References 47–51.

6.10 FEMs for Exterior Problems

Thus far in this chapter, the FEM has been presented for solving interior problems. To apply the FEM to exterior or unbounded problems such as open-type transmission lines (e.g., microstrip), scattering, and radiation problems poses certain difficulties. To overcome these difficulties, several approaches [52–80] have been proposed, all of which have strengths and weaknesses. We will consider three common approaches: the infinite element method, the boundary element method (BEM), and absorbing boundary condition.

6.10.1 Infinite Element Method

Consider the solution region shown in Figure 6.30a. We divide the entire domain into a near-field (n.f.) region, which is bounded, and a far-field (f.f.) region, which is unbounded. The n.f. region is divided into finite triangular elements as usual, while the f.f. region is divided into *infinite elements*. Each infinite element shares two nodes with a finite element. Here, we are mainly concerned with the infinite elements.

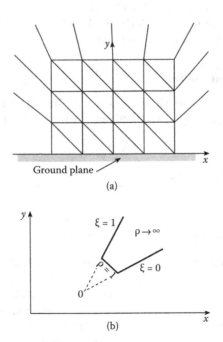

(a)

(b)

FIGURE 6.30
(a) Division of solution region into finite and infinite elements; (b) typical infinite element.

Consider the infinite element in Figure 6.30b with nodes 1 and 2 and radial sides intersecting at point (x_o, y_o). We relate triangular polar coordinates (ρ, ξ) to the global Cartesian coordinates (x, y) as [60]

$$x = x_o + \rho[(x_1 - x_o) + \xi(x_2 - x_1)]$$
$$y = y_o + \rho[(y_1 - y_o) + \xi(y_2 - y_1)]$$ (6.133)

where $1 \leq \rho < \infty, 0 \leq \xi \leq 1$. The potential distribution within the element is approximated by a linear variation as

$$V = \frac{1}{\rho}[V_1(1 - \xi) + V_2\xi]$$

or

$$V = \sum_{i=1}^{2} \alpha_i V_i$$ (6.134)

where V_1 and V_2 are potentials at nodes 1 and 2 of the infinite elements, α_1 and α_2 are the interpolation or shape functions, that is,

$$\alpha_1 = \frac{1 - \xi}{\rho}, \quad \alpha_2 = \frac{\xi}{\rho}$$ (6.135)

The infinite element is compatible with the ordinary first-order finite element and satisfies the boundary condition at infinity. With the shape functions in Equation 6.135, we can obtain the $[C^{(e)}]$ and $[T^{(e)}]$ matrices. We obtain solution for the exterior problem by using a standard finite element program with the $[C^{(e)}]$ and $[T^{(e)}]$ matrices of the infinite elements added to the $[C]$ and $[T]$ matrices of the n.f. region.

6.10.2 Boundary Element Method

A comparison between FEM and MoM is shown in Table 6.12. From the table, it is evident that the two methods have properties that complement each other. In view of this, hybrid

TABLE 6.12

Comparison between MoM and FEM [81]

MoM	FEM
Conceptually easy	Conceptually involved
Requires problem-dependent Green's functions	Avoids difficulties associated with singularity of Green's functions
Few equations; $O(n)$ for 2-D, $O(n^2)$ for 3-D	Many equations; $O(n^2)$ for 2-D, $O(n^3)$ for 3-D
Only boundary is discretized	Entire domain is discretized
Open boundary easy	Open boundary difficult
Fields by integration	Fields by differentiation
Good representation of far-field condition	Good representation of boundary conditions
Full matrices result	Sparse matrices result
Nonlinearity, inhomogeneity difficult	Nonlinearity, inhomogeneity easy

methods have been proposed. These methods allow the use of both MoM and FEM with the aim of exploiting the strong points in each method.

One of these hybrid methods is the so-called BEM. It is a finite element approach for handling exterior problems [66–78]. It basically involves obtaining the integral equation formulation of the boundary value problem [82] and solving this by a discretization procedure similar to that used in regular finite element analysis. Since the BEM is based on the boundary integral equivalent to the governing differential equation, only the surface of the problem domain needs to be modeled. Thus, the dimension of the problem is reduced by one as in MoM. For 2-D problems, the boundary elements are taken to be straight line segments, whereas for 3-D problems, they are taken as triangular elements. Thus the shape or interpolation functions corresponding to sub-sectional bases in the MoM are used in the finite element analysis.

6.10.3 Absorbing Boundary Condition

To apply the finite element approach to open region problems, such as for scattering or radiation, an artificial boundary is introduced in order to bound the region and limit the number of unknowns to a manageable size. The artificial boundary is set far from the guided region. One would expect that as the boundary approaches infinity, the approximate solution tends to the exact one. But the closer the boundary to the radiating or scattering object, the less computer memory is required. To avoid the error caused by this truncation, an *absorbing boundary condition* (ABC) is imposed on the artificial boundary S, as typically portrayed in Figure 6.31. The ABC minimizes the nonphysical reflections from the boundary. Several ABCs have been proposed [83–89]. The major challenge of these ABCs is to bring the truncation boundary as close as possible to the object without sacrificing accuracy and to absorb the outgoing waves with little or no reflection. A popular approach is the PML-based ABC discussed in Section 3.8.3 for FDTD. The finite element technique is used in enforcing the condition as a tool for mesh truncation [85].

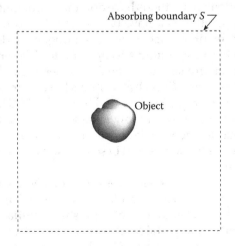

Absorbing boundary S

Object

FIGURE 6.31
A radiating (or scattering) object surrounded by an absorbing boundary.

Another popular ABC by derived Bayliss, Gunzburger, and Turkel (BGT) employs asymptotic analysis [89]. For example, for the solution of a 3-D problem, an expansion of the scalar Helmholtz equation is [88]

$$\Phi(r,\theta,\phi) = \frac{e^{-jkr}}{kr} \sum_{i=0}^{\infty} \frac{F_i(\theta,\phi)}{(kr)^i} \tag{6.136}$$

The sequence of BGT operators is obtained by the recursion relation

$$B_1 = \left(\frac{\partial}{\partial r} + jk + \frac{1}{r}\right)$$
$$B_m = \left(\frac{\partial}{\partial r} + jk + \frac{2m-1}{r}\right) B_{m-1}, \quad m = 2,3,\ldots \tag{6.137}$$

Since Φ satisfies the higher-order radiation condition

$$B_m \Phi = O\left(\frac{1}{r^{2m+1}}\right) \tag{6.138}$$

imposing the mth-order boundary condition

$$B_m \Phi = 0 \text{ on } S \tag{6.139}$$

will compel the solution Φ to match the first $2m$ terms of the expansion in Equation 6.136. Equation 6.139 along with other appropriate equations is solved for Φ using the FEM.

6.11 Finite-Element Time-Domain Method

Traditionally, frequency-domain methods have dominated computational electromagnetics, while time-domain computation is a novelty. The trend is to solve Maxwell's equations directly in the time domain. (It is much easier to get frequency-domain results from time-domain data than the other way around.) To date, finite-difference time-domain (FDTD) techniques have received the greatest attention due to their algorithmic simplicity. In recent years, finite-element time-domain (FETD) algorithms have increased in popularity because of their ability to approximate physical boundaries. The FDTD method is the method of choice when modeling geometries of low complexity, while FETD methods are most appropriate when complicated geometries need to be modeled. Since FETD formulations have not received as much attention as FDTD schemes, they are lacking in both maturity and variety of applications.

Numerous FETD methods have been proposed for EM computation [89–97]. These methods can be classified into two categories. One class of approaches directly solves Maxwell's equations and operates in a leapfrog fashion similar to the FDTD technique. The approaches are conditionally stable. Another class of FETD methods tackles the second-order vector wave equation obtained by eliminating one of the field variables from Maxwell's equations. We follow the approach in Reference 89.

Consider Maxwell's equations in space–time:

$$\nabla \times E = -\mu \frac{\partial H}{\partial t} \tag{6.140a}$$

$$\nabla \times H = J + \varepsilon \frac{\partial E}{\partial t} \tag{6.140b}$$

where E and H are electric field intensity and magnetic field intensity, respectively, and J is the current density. From Equation 6.140, we can derive an initial value problem in terms of the magnetic field H.

$$\nabla \times \frac{1}{\varepsilon} \nabla \times H + \frac{\mu\sigma}{\varepsilon} \frac{\partial H}{\partial t} + \mu \frac{\partial^2 H}{\partial t^2} = \frac{1}{\varepsilon} \nabla \times J \tag{6.141}$$

where σ is the conductivity, $\varepsilon = \varepsilon_r \varepsilon_o$ and $\mu = \mu_r \mu_o$ are, respectively, the permittivity and permeability of the medium. For 2-D region, the scalar wave equation for the longitudinal component of the magnetic field is

$$\frac{1}{\varepsilon_r \varepsilon_o} \nabla^2 H_z - \sigma \frac{\mu_r \mu_o}{\varepsilon_r \varepsilon_o} \frac{\partial H_z}{\partial t} - \mu_r \mu_o \frac{\partial^2 H_z}{\partial t^2} = -\frac{1}{\varepsilon_r \varepsilon_o} (\nabla \times J)_z \tag{6.142}$$

By defining the carrier frequency ω_c, the field component and the current density can be written as

$$\begin{aligned} H_z(t) &= V(t)e^{j\omega_c t} \\ J(t) &= j(t)e^{j\omega_c t} \end{aligned} \tag{6.143}$$

where $V(t)$ is the time-varying complex envelope of the field at the carrier frequency. The application of Galerkin's process (global weak formulation) produces a set of ordinary differential equations

$$[T]\frac{d^2 v}{dt^2} + [B]\frac{dv}{dt} + [G]v + [F] = 0 \tag{6.144}$$

where v is the coefficient vector of V. Matrices $[T]$, $[B]$, and $[G]$ are time-independent and are given by

$$T_{ij} = \int\int_S \frac{\mu_r}{c^2} W_i W_j \, dS \tag{6.145a}$$

$$B_{ij} = \int\int_S \frac{\mu_r}{c^2} (2j\omega_c + \alpha) W_i W_j \, dS \tag{6.145b}$$

$$G_{ij} = \int\int_S \frac{1}{\varepsilon_r} \nabla W_i \cdot \nabla W_j - \frac{\mu_r}{c^2}(\omega_c^2 - j\alpha\omega_c) W_i W_j \, dS \tag{6.145c}$$

$$F_i = \int \int_S \frac{1}{\varepsilon_r} W_i \cdot (\nabla \times j)_z \, dS \qquad (6.145\text{d})$$

where c is the speed of light in free space, $\alpha = \sigma/\varepsilon_r\varepsilon_o$ is a constant, W_j are 2-D FEM basis functions, and S is a 2-D area bounded by the boundary Γ.

In order to solve Equation 6.144, the time derivatives must be discretized.

$$\frac{d^2v}{dt^2} = \frac{1}{\Delta t^2}[v(n+1) - 2v(n) + v(n-1)] \qquad (6.146\text{a})$$

$$\frac{dv}{dt} = \frac{1}{2\Delta t}[v(n+1) - v(n-1)] \qquad (6.146\text{b})$$

$$v(n) = \beta v(n+1) + (1 - 2\beta)v(n) + \beta v(n-1) \qquad (6.146\text{c})$$

where $v(n) = v(n\Delta t)$ is the discrete-time version of $v(t)$ and β is a constant that determines the stability and the accuracy of the scheme. It is recommended that $\beta = 1/4$ which results in an unconditionally stable scheme. Thus, Equation 6.144 becomes

$$\left(\frac{[T]}{\Delta t^2} + \frac{B}{2\Delta t} + \frac{G}{4}\right)v(n+1) = \left(\frac{2[T]}{\Delta t^2} - \frac{[G]}{2}\right)v(n)$$

$$+ \left(-\frac{[T]}{\Delta t^2} + \frac{B}{2\Delta t} - \frac{[G]}{4}\right)v(n-1) - f(n) \qquad (6.147)$$

To solve these equations, we need to invert the matrix on the left-hand side. Since this matrix is time independent, it needs to be filled and solved only once.

6.12 Applications: Microstrip Lines

The reader will benefit from the numerous finite element software packages that are freely or commercially available. These include:

- COMSOL (www.comsol.com)
- Quickfield (www.quickfield.com/free_soft.htm)
- FEMM (femm.foster-millercom/index.htm)
- NASTRAN (https://www.autodesk.com/education/free-software/nastran)
- Abaqus (https://academy.3ds.com/en/software/abaqus-student-edition)
- ANSYS (www.ansys.com/Products/Electronics/ANSYS-HFSS)
- MaxFem (http://downloads.informer.com/maxfem/0.3/)

An extensive description of some of these codes and their capabilities can be found in References 98,99.

Although some of the codes were developed for one field of engineering or another, they can be applied to problems in a different field with little or no modification. Here, we illustrate solving microstrip problems using COMSOL multiphysics package.

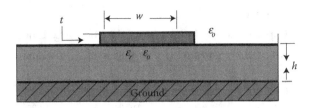

FIGURE 6.32
Cross-section of a single-strip microstrip line.

The microstrip transmission lines have received a lot of attention in microwave circuit design. In this section, we briefly illustrate how to use COMSOL to determine the capacitance, inductance, and characteristic impedance of an open microstrip line [97]. Figure 6.32 shows the cross section of the open single-strip microstrip line with the following parameters:

t = thickness of the conducting strip = 0.01 mm

w = width of the conductor (variable)

h = height of the dielectric material = 1 mm

ε_r = dielectric constant = 6

The simulation is done twice: one to get the capacitance per unit length C of the microstrip when the dielectric material is in place and the other to get C_o when the dielectric material is removed. The inductance per unit length L is given by

$$L = \frac{\mu_0 \varepsilon_0}{C_o} \tag{6.148}$$

and the characteristic impedance is

$$Z_o = \frac{1}{u\sqrt{C_o C}} \tag{6.149}$$

We now are using COMSOL to determine C_o and C by taking the following steps:

1. Construct the geometry of the line, as shown in Figure 6.32 with the microstrip surrounded by a 30w × 14h shield.
2. Take the difference between the conductor and air. Consider two cases—one for air and the other for dielectric material.
3. For the dielectric region, specify the relative permittivity.
4. For the boundary, select the outer conductor (shield) as ground and the single strip as port.
5. Generate the finite element mesh and solve the model.
6. As post-processing, select Point Evaluation and choose capacitance elements to find the capacitance per unit length of the line.

Table 6.13 shows the finite element results for the capacitance per unit length and inductance per unit length. These results are used in obtaining Table 6.14 for the characteristic impedance of the line.

TABLE 6.13

Capacitance and Inductance of the Microstrip Line

w/h	C_o (pF/m)	C (pF/m)	L (mh/m)
0.4	18.976	73.785	585.53
0.7	23.203	92.681	478.90
1.0	26.946	110.041	412.35
2.0	37.919	166.78	293.02
4.0	59.785	278.4	185.85
10.0	117.085	600.697	94.90

TABLE 6.14

Characteristic Impedance of the Microstrip Line

w/h	Present Work	MoM	Conformal Mapping
0.4	89.085	91.172	89.909
0.7	71.885	73.613	71.995
1.0	61.215	62.713	60.970
2.0	41.916	43.149	41.510
4.0	25.837	27.301	26.027
10.0	12.569	13.341	12.485

6.13 Concluding Remarks

An introduction to the basic concepts and applications of the FEM has been presented. It is by no means an exhaustive exposition of the subject. However, we have given the flavor of the way in which the ideas may be developed; the interested reader may build on this by consulting the references. Several introductory texts have been published on FEM. Although most of these texts are written for civil or mechanical engineers, the texts by Silvester and Ferrari [4], Chari and Silvester [39], Steele [100], Hoole [101], Itoh [102], and Jin [103] are for electrical engineers. More texts and monographs have been published on FEM than any other numerical technique.

The FEM has the following advantages [104]:

1. It is flexible and versatile in modeling complex or inhomogeneous solution regions
2. It can handle a wide variety of engineering problems such as in EM, solid mechanics, fluid mechanics, dynamics, and heat
3. It produces accurate and stable solutions
4. Boundary conditions are incorporated in the functional formulation
5. It can handle nonlinear problems

The limitations or disadvantages include:

1. Its sparsity patterns are highly unstructured and this makes it very difficult to efficiently parallelize a FEM code.

2. Due to inherent errors, the FEM produces only approximate solutions.

3. It produces spurious or extraneous nonphysical modes for some vector formulations (6.123b).

4. It is more directly suited for closed problems, but can be extended to open problems using ABCs.

To address the limitations and increase the accuracy of FEM, some hybrid methods have been proposed [98,105–110]. Due to its flexibility and versatility, the FEM has become a powerful tool throughout engineering disciplines. It has been applied with great success to numerous EM-related problems. Such applications include the following:

- Transmission line problems [103,111,112],
- Optical and microwave waveguide problems [8–17,113–118],
- Electric machines [39,119–121],
- Scattering problems [69,70,73,122],
- Human exposition to EM radiation [123–126], and
- Others [127–130].

For other issues on FEM not covered in this chapter, one is referred to introductory texts on FEM such as [2,4,34,39,46,90–92,99,131–139]. The issues of edge elements and absorbing boundary are covered in Reference 106. Estimating error in finite element solution is discussed in References 50,131,132. The reader may benefit from the numerous finite element codes that are commercially available. An extensive description of these systems and their capabilities can be found in References 133,140–142. Although the codes were developed for one field of engineering or another, they can be applied to problems in a different field with little or no modification.

PROBLEMS

6.1 For the triangular elements in Figure 6.33, determine the element coefficient matrices.

6.2 Obtain the global coefficient matrix for the three-element mesh shown in Figure 6.34.

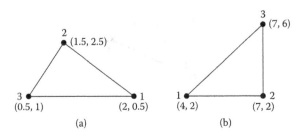

FIGURE 6.33
For Problem 6.1.

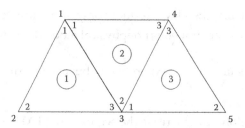

FIGURE 6.34
For Problem 6.2.

6.3   The triangular element shown in Figure 6.35 is part of a finite element mesh. If $V_1 = 8$ V, $V_2 = 12$, and $V_3 = 10$ V, find the potential at: (a) (1,2), (b) the center of the element.

6.4   Find the coefficient matrix for the two-element mesh of Figure 6.36. Given that $V_2 = 10$ and $V_4 = -10$, determine V_1 and V_3.

6.5   Determine the shape functions α_1, α_2, and α_3 for the element in Figure 6.37.

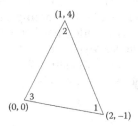

FIGURE 6.35
For Problems 6.3.

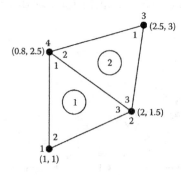

FIGURE 6.36
For Problem 6.4.

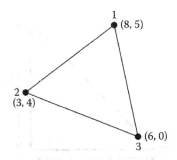

FIGURE 6.37
For Problem 6.5.

6.6 A triangular element has nodes at (0,1), (1,0), and (0,1). Construct the shape function α_1.

6.7 Show that the shape function α_1 evaluates to unity at node 1 and to zero at all other nodes for the first-order elements.

6.8 Consider the mesh shown in Figure 6.38. The shaded region is conducting and has no finite elements. Calculate the global elements $C_{3,10}$ and $C_{3,3}$.

6.9 With reference to the finite element in Figure 6.39, calculate the energy per unit length associated with the element.

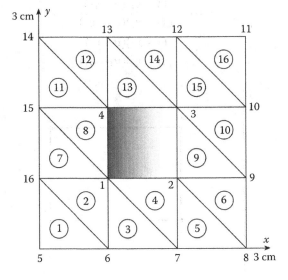

FIGURE 6.38
For Problem 6.8.

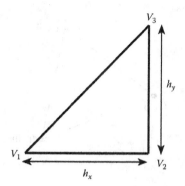

FIGURE 6.39
For Problem 6.9.

6.10 Consider the element whose sides are parallel to the x and y axis, as shown in Figure 6.40. Verify that the potential distribution within the elements can be expressed as

$$V(x, y) = \alpha_1 V_1 + \alpha_2 V_2 + \alpha_3 V_3 + \alpha_4 V_4$$

where V_i are the nodal potentials and α_i are local interpolating functions defined as

$$\alpha_1 = \frac{(x - x_2)(y - y_4)}{(x_1 - x_2)(y_1 - y_4)}$$

$$\alpha_2 = \frac{(x - x_1)(y - y_3)}{(x_2 - x_1)(y_2 - y_3)}$$

$$\alpha_3 = \frac{(x - x_4)(y - y_2)}{(x_3 - x_4)(y_3 - y_2)}$$

$$\alpha_4 = \frac{(x - x_3)(y - y_1)}{(x_4 - x_3)(y_4 - y_1)}$$

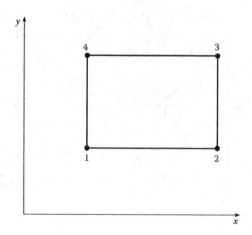

FIGURE 6.40
For Problem 6.10.

6.11 The cross section of an infinitely long rectangular trough is shown in Figure 6.41; develop a program using FEM to find the potential at the center of the cross section. Take $\epsilon_r = 4.5$.

6.12 Solve the problem in Example 3.3 using the FEM.

6.13 Once the potential distribution is obtained, the electric field intensity can be obtained from

$$E(x, y) = E_x a_x + E_y a_y = -\nabla V(x, y)$$

For each triangular element, show that

$$E_x = -\frac{1}{2A}[(y_j - y_k)V_i + (y_k - y_i)V_j + (y_i - y_k)V_k]$$

$$E_y = -\frac{1}{2A}[(x_k - x_j)V_i + (x_i - x_k)V_j + (x_j - x_i)V_k]$$

where A is the area of the element and $V_{i,j,k}$ represent the electric potentials of three nodes (i, j, k) of each element.

6.14 A potential field is defined over a triangular three-node element by

Node i	V_i (V)	x_i (cm)	y_i (cm)
1	40	4	6
2	−10	2	2
3	20	6	2

Calculate the potential and potential gradient at (4,4) cm.

6.15 Modify the program in Figure 6.10 to calculate the electric field intensity **E** at any point in the solution region.

6.16 The program in Figure 6.10 applies the iteration method to determine the potential at the free nodes. Modify the program and use the band matrix method to determine the potential. Test the program using the data in Example 6.2.

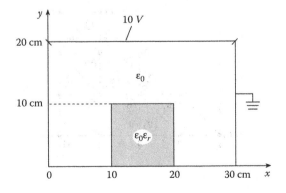

FIGURE 6.41
For Problem 6.11.

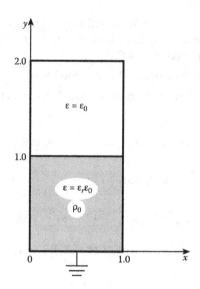

FIGURE 6.42
For Problem 6.17.

6.17 A grounded rectangular pipe with the cross section in Figure 6.42 is half-filled with hydrocarbons ($\epsilon = 2.5\,\epsilon_o$, $\rho_o = 10^{-5}\,\text{C/m}^3$). Use FEM to determine the potential along the liquid–air interface. Plot the potential versus x.

6.18 Solve the problem in Example 3.4 using the FEM.

6.19 The cross section of an isosceles right-triangular waveguide is discretized as in Figure 6.43. Determine the first 10 TM cutoff wavelengths of the guide.

6.20 Using FEM, determine the first 10 cutoff wavelengths of a rectangular waveguide of cross section 2 cm × 1 cm. Compare your results with the exact solution. Assume the guide is air-filled.

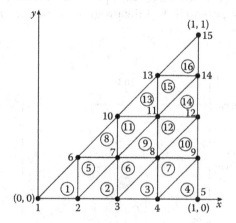

FIGURE 6.43
For Problem 6.19.

6.21 Use the mesh generation program in Figure 6.16 to subdivide the solution regions in Figure 6.44. Subdivide into as many triangular elements as you choose.

6.22 Determine the semi-bandwidth of the mesh shown in Figure 6.45. Renumber the mesh so as to minimize the bandwidth.

6.23 Find the semi-bandwidth B of the mesh in Figure 6.46. Renumber the mesh to minimize B and determine the new value of B.

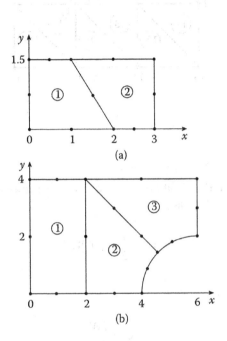

FIGURE 6.44
For Problem 6.21.

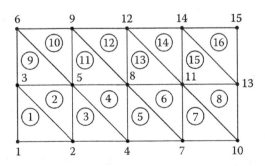

FIGURE 6.45
For Problem 6.22.

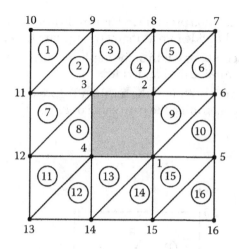

FIGURE 6.46
For Problem 6.23.

6.24 Rework Problem 3.22 using the FEM.

Hint: After calculating V at all free nodes with ϵ lumped with C_{ij}, use Equation 6.19 to calculate W, that is,

$$W = \frac{1}{2}[V]^t[C][V]$$

Then find the capacitance from

$$C = \frac{2W}{V_d^2}$$

where V_d is the potential difference between inner and outer conductors.

6.25 Using the area coordinates (ξ_1, ξ_2, ξ_3) for the triangular element in Figure 6.3, evaluate

a. $\int_S x \, dS$

b. $\int_S x \, dS$

c. $\int_S xy \, dS$

6.26 Evaluate the following integrals:

a. $\int_S \alpha_2^2 \, dS$

b. $\int_S \alpha_1 \alpha_5 \, dS$

c. $\int_S \alpha_1 \alpha_2 \alpha_3 \, dS$

6.27 Evaluate the shape functions $\alpha_1, \ldots, \alpha_6$ for the second-order elements in Figure 6.47.

6.28 Derive matrix T for $n = 2$.

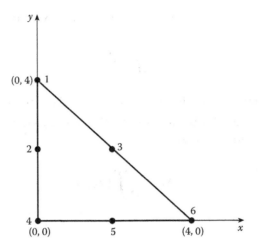

FIGURE 6.47
For Problem 6.27.

6.29 By hand calculation, obtain $Q^{(2)}$ and $Q^{(3)}$ for $n = 1$ and $n = 2$.

6.30 The $D^{(q)}$ matrix is an auxilliary matrix used along with the T matrix to derive other fundamental matrices. An element of D is defined in Reference 41 as the partial derivative of α_i with respect to ξ_q evaluated at node P_j, that is,

$$D_{ij}^{(q)} = \left. \frac{\partial \alpha_i}{\partial \xi_q} \right|_{P_i} , \quad i, j = 1, 2, \ldots, m$$

where $q \in \{1, 2, 3\}$. For $n = 1$ and 2, derive $D^{(1)}$. From $D^{(1)}$, derive $D^{(2)}$ and $D^{(3)}$.

6.31 a. The matrix $K^{(pq)}$ can be defined as

$$K_{ij}^{(pq)} = \iint \frac{\partial \alpha_i}{\partial \xi_p} \frac{\partial \alpha_j}{\partial \xi_q} dS$$

where $p, q = 1, 2, 3$. Using the $D^{(q)}$ matrix of the previous problem, show that

$$K^{(pq)} = D^{(p)} T\, D^{(q)t}$$

where t denotes transposition.

b. Show that the $Q^{(q)}$ matrix can be written as

$$Q^{(q)} = [D^{(q+1)} - D^{(q-1)}] T [D^{(q+1)} - D^{(q-1)}]^t$$

Use this formula to derive $Q^{(1)}$ for $n = 1$ and 2.

6.32 Verify the interpolation function for the 10-node tetrahedral element.

6.33 The (x, y, z) coordinates of nodes 1, 2, 3, and 4 of a three-dimensional simplex element are $(0, 0, 0)$, $(2, 4, 2)$, $(4, 0, 0)$, and $(2, 0, 6)$, respectively. Determine the shape functions of the element.

6.34 Using the volume coordinates for a tetrahedron, evaluate

$$\int z^2 \, dv$$

Assume that the origin is located at the centroid of the tetrahedron.

6.35 Obtain the T matrix for the first-order tetrahedral element.

6.36 Use Equation 6.104f to show that if

$$M_{ij} = \iint \xi_1^i \xi_2^j \, dS$$

then

$$M = \frac{A}{12} \begin{bmatrix} 2 & 1 & 1 \\ 1 & 2 & 1 \\ 1 & 1 & 2 \end{bmatrix}$$

6.37 For the two-dimensional problem, the BGI sequence of operators are defined by the recurrence relation

$$B_m = \left(\frac{\partial}{\partial \rho} + jk + \frac{4m-3}{2\rho} \right) B_{m-1}$$

where $B_o = 1$. Obtain B_1 and B_2.

6.38 Figure 6.48 shows the cross-section of an open double-strip microstrip line. Let

t = thickness of strip = 1 mm

w = width of the strip = 3 mm

h = height of dielectric material = 1 mm

d = distance between the two strips = 3 mm

ε_r = dielectric constant = 2

Use any finite element software to calculate the capacitance values of the line.

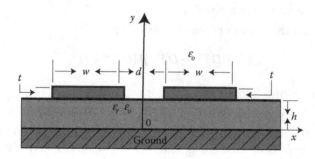

FIGURE 6.48
For Problem 6.38.

6.39 Use any finite element package to determine the characteristic impedance of the shielded microstrip line whose cross-section is shown in Figure 6.49. Let $a = 2.02$, $b = 7.0$, $h = 1.0 = w$, $t = 0.01$.

6.40 Consider the cross-section of a double-strip shielded microstrip line shown in Figure 6.50. Use any finite element package to compute the capacitance matrix of the line. Take $w = 3$ mm, $t = 1$ mm, $s = 2$ mm, $h = 1$ mm, $a = 11$ mm, $b = 2.7$ mm, for the dielectric $\varepsilon_r = 2$.

6.41 For the shielded broadside-coupled suspended microstrip line shown in Figure 6.51, use a finite element software to find the capacitance matrix. Consider the following parameters:

a = width of the shield = 40 mm

b = height of the shield = 20 mm

t = thickness of the strip = 0.01 mm

h = thickness of the dielectric material = 2 mm

w = width of the strip = 2 h, 4 h, and 8 h

ε_r = dielectric constant = 2.22

that is, consider three cases with $w = 2$ h, 4 h, and 8 h

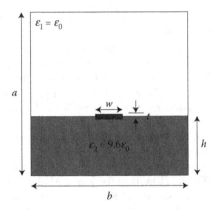

FIGURE 6.49
For Problem 6.39.

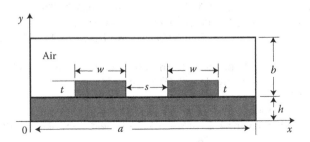

FIGURE 6.50
For Problem 6.40.

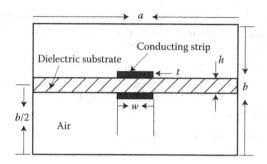

FIGURE 6.51
For Problem 6.41.

References

1. R. Courant, "Variational methods for the solution of problems of equilibrium and vibrations," *Bull. Am. Math. Soc.*, vol. 49, 1943, pp. 1–23.
2. K. K. Gupta and J. L. Meek, "A brief history of the beginning of the finite element method," *Int. J. Numer. Methods Eng.*, vol. 39, 1996, pp. 3761–3774.
3. M.N.O. Sadiku, "A simple introduction to finite element analysis of electromagnetic problems," *IEEE Trans. Educ.*, vol. 32, no. 2, May 1989, pp. 85–93.
4. P.P. Silvester and R.L. Ferrari, *Finite Elements for Electrical Engineers*, 3rd Edition. New York: Cambridge University Press, 1996.
5. O.W. Andersen, "Laplacian electrostatic field calculations by finite elements with automatic grid generation," *IEEE Trans. Power App. Syst.*, vol. PAS-92, no. 5, Sept./Oct. 1973, pp. 1485–1492.
6. M.N.O. Sadiku et al., "A further introduction to finite element analysis of electromagnetic problems," *IEEE Trans. Educ.*, vol. 34, no. 4, Nov. 1991, pp. 322–329.
7. B.S. Garbow, *Matrix Eigensystem Routine—EISPACK Guide Extension*. Berlin: Springer-Verlag, 1977.
8. S. Ahmed and P. Daly, "Finite-element methods for inhomogeneous waveguides," *Proc. IEEE*, vol. 116, no. 10, Oct. 1969, pp. 1661–1664.
9. Z.J. Csendes and P. Silvester, "Numerical solution of dielectric loaded waveguides: I—Finite-element analysis," *IEEE Trans. Micro. Theo. Tech.*, vol. MTT-18, no. 12, Dec. 1970, pp. 1124–1131.
10. Z.J. Csendes and P. Silvester, "Numerical solution of dielectric loaded waveguides: II—Modal approximation technique," *IEEE Trans. Micro. Theo. Tech.*, vol. MTT-19, no. 6, June 1971, pp. 504–509.
11. M. Hano, "Finite-element analysis of dielectric-loaded waveguides," *IEEE Trans. Micro. Theo. Tech.*, vol. MTT-32, no. 10, Oct. 1984, pp. 1275–1279.
12. A. Konrad, "Vector variational formulation of electromagnetic fields in anisotropic media," *IEEE Trans. Micro. Theo. Tech.*, vol. MTT-24, Sept. 1976, pp. 553–559.
13. M. Koshiba et al., "Improved finite-element formulation in terms of the magnetic field vector for dielectric waveguides," *IEEE Trans. Micro. Theo. Tech.*, vol. MTT-33, no. 3, March 1985, pp. 227–233.
14. M. Koshiba et al., "Finite-element formulation in terms of the electric-field vector for electromagnetic waveguide problems," *IEEE Trans. Micro. Theo. Tech.*, vol. MTT-33, no. 10, Oct. 1985, pp. 900–905.
15. K. Hayata et al., "Vectorial finite-element method without any spurious solutions for dielectric waveguiding problems using transverse magnetic-field component," *IEEE Trans. Micro. Theo. Tech.*, vol. MTT-34, no. 11, Nov. 1986.
16. K. Hayata et al., "Novel finite-element formulation without any spurious solutions for dielectric waveguides," *Elect. Lett.*, vol. 22, no. 6, March 1986, pp. 295, 296.

17. S. Dervain, "Finite element analysis of inhomogeneous waveguides," *M.S. thesis*, Department of Electrical and Computer Engineering, Florida Atlantic University, Boca Raton, April 1987.

18. J.R. Winkler and J.B. Davies, "Elimination of spurious modes in finite element analysis," *J. Comp. Phys.*, vol. 56, no. 1, Oct. 1984, pp. 1–14.

19. M. Kono, "A generalized automatic mesh generation scheme for finite element method," *Inter. J. Num. Method Engr.*, vol. 15, 1980, pp. 713–731.

20. J.C. Cavendish, "Automatic triangulation of arbitrary planar domains for the finite element method," *Inter. J. Num. Meth. Engr.*, vol. 8, 1974, pp. 676–696.

21. A.O. Moscardini et al., "AGTHOM—Automatic generation of triangular and higher order meshes," *Inter. J. Num. Meth. Engr.*, vol. 19, 1983, pp. 1331–1353.

22. C.O. Frederick et al., "Two-dimensional automatic mesh generation for structured analysis," *Inter. J. Num. Meth. Engr.*, vol. 2, no. 1, 1970, pp. 133–144.

23. E.A. Heighway, "A mesh generation for automatically subdividing irregular polygon into quadrilaterals," *IEEE Trans. Mag.*, vol. MAG-19, no. 6, Nov. 1983, pp. 2535–2538.

24. C. Kleinstreuer and J.T. Holdeman, "A triangular finite element mesh generator for fluid dynamic systems of arbitrary geometry," *Inter. J. Num. Meth. Engr.*, vol. 15, 1980, pp. 1325–1334.

25. A. Bykat, "Automatic generation of triangular grid. I—Subdivision of a general polygon into convex subregions. II—Triangulation of convex polygons," *Inter. J. Num. Meth. Engr.*, vol. 10, 1976, pp. 1329–1342.

26. N.V. Phai, "Automatic mesh generator with tetrahedron elements," *Inter. J. Num. Meth. Engr.*, vol. 18, 1982, pp. 273–289.

27. F.A. Akyuz, "Natural coordinates systems—An automatic input data generation scheme for a finite element method," *Nuclear Engr. Design*, vol. 11, 1970, pp. 195–207.

28. P. Girdinio et al., "New developments of grid optimization by the grid iteration method," in Z.J. Csendes (ed.), *Computational Electromagnetism*. New York: North-Holland, 1986, pp. 3–12.

29. M. Yokoyama, "Automated computer simulation of two-dimensional elastostatic problems by finite element method," *Inter. J. Num. Meth. Engr.*, vol. 21, 1985, pp. 2273–2287.

30. G.F. Carey, "A mesh-refinement scheme for finite element computations," *Comp. Meth. Appl. Mech. Engr.*, vol. 7, 1976, pp. 93–105.

31. K. Preiss, "Checking the topological consistency of a finite element mesh," *Inter. J. Meth. Engr.*, vol. 14, 1979, pp. 1805–1812.

32. W.C. Thacker, "A brief review of techniques for generating irregular computational grids," *Inter. J. Num. Meth. Engr.*, vol. 15, 1980, pp. 1335–1341.

33. E. Hinto and D.R. Owen, *An Introduction to Finite Element Computations*. Swansea, U.K.: Pineridge Press, 1979, pp. 247, 328–346.

34. P. O. Persson and G. Strang, "A simple mesh generator in MATLAB," *SIAM Rev.*, vol. 46, 2004, pp. 329–345.

35. R.J. Collins, "Bandwidth reduction by automatic renumbering," *Inter. J. Num. Meth. Engr.*, vol. 6, 1973, pp. 345–356.

36. E. Cuthill and J. McKee, "Reducing the bandwidth of sparse symmetric matrices," *ACM Nat. Conf.*, San Francisco, 1969, pp. 157–172.

37. G.A. Akhras and G. Dhatt, "An automatic node relabelling scheme for minimizing a matrix or network bandwidth," *Inter. J. Num. Meth. Engr.*, vol. 10, 1976, pp. 787–797.

38. F.A. Akyuz and S. Utku, "An automatic node-relabelling scheme for bandwidth minimization of stiffness matrices," *J. Amer. Inst. Aero. Astro.*, vol. 6, no. 4, 1968, pp. 728–730.

39. M.V.K. Chari and P.P. Silvester (eds.), *Finite Elements for Electrical and Magnetic Field Problems*. Chichester: John Wiley, 1980, pp. 125–143.

40. P. Silvester, "Construction of triangular finite element universal matrices," *Inter. J. Num. Meth. Engr.*, vol. 12, 1978, pp. 237–244.

41. P. Silvester, "High-order polynomial triangular finite elements for potential problems," *Inter. J. Engr. Sci.*, vol. 7, 1969, pp. 849–861.

42. G.O. Stone, "High-order finite elements for inhomogeneous acoustic guiding structures," *IEEE Trans. Micro. Theory Tech.*, vol. MTT-21, no. 8, Aug. 1973, pp. 538–542.

43. A. Konrad, "High-order triangular finite elements for electromagnetic waves in anistropic media," *IEEE Trans. Micro. Theory Tech.*, vol. MTT-25, no. 5, May 1977, pp. 353–360.

44. P. Daly, "Finite elements for field problems in cylindrical coordinates," *Inter. J. Num. Meth. Engr.*, vol. 6, 1973, pp. 169–178.

45. M. Sadiku and L. Agba, "News rules for generating finite elements fundamental matrices," *Proc. IEEE Southeastcon*, 1989, pp. 797–801.

46. C.A. Brebbia and J.J. Connor, *Fundamentals of Finite Element Technique*. London: Butterworth, 1973, pp. 114–118, 150–163, 191.

47. R.L. Ferrari and G.L. Maile, "Three-dimensional finite element method for solving electromagnetic problems," *Elect. Lett.*, vol. 14, no. 15, 1978, pp. 467, 468.

48. M. de Pourcq, "Field and power-density calculation by three-dimensional finite elements," *IEEE Proc.*, vol. 130, Pt. H, no. 6, Oct. 1983, pp. 377–384.

49. M.V.K. Chari et al., "Finite element computation of three-dimensional electrostatic and magnetostatic field problems," *IEEE Trans. Mag.*, vol. MAG-19, no. 16, Nov. 1983, pp. 2321–2324.

50. O.A. Mohammed et al., "Validity of finite element formulation and solution of three dimensional magnetostatic problems in electrical devices with applications to transformers and reactors," *IEEE Trans. Pow. App. Syst.*, vol. PAS-103, no. 7, July 1984, pp. 1846–1853.

51. J.S. Savage and A.F. Peterson, "Higher-order vector finite elements for tetrahedral cells," *IEEE Trans. Micro. Theo. Tech.*, vol. 44, no. 6, June 1996, pp. 874–879.

52. J.F. Lee and Z.J. Cendes, "Transfinite elements: A highly efficient procedure for modeling open field problems," *J. Appl. Phys.*, vol. 61, no. 8, April 1987, pp. 3913–3915.

53. B.H. McDonald and A. Wexler, "Finite-element solution of unbounded field problems," *IEEE Trans. Micro. Theo. Tech.*, vol. MTT-20, no. 12, Dec. 1972, pp. 841–847.

54. P.P. Silvester et al., "Exterior finite elements for 2-dimensional field problems with open boundaries," *Proc. IEEE*, vol. 124, no. 12, Dec. 1977, pp. 1267–1270.

55. S. Washisu et al., "Extension of finite-element method to unbounded field problems," *Elect. Lett.*, vol. 15, no. 24, Nov. 1979, pp. 772–774.

56. P. Silvester and M.S. Hsieh, "Finite-element solution of 2-dimensional exterior-field problems," *Proc. IEEE*, vol. 118, no. 12, Dec. 1971, pp. 1743–1747.

57. Z.J. Csendes, "A note on the finite-element solution of exterior-field problems," *IEEE Trans. Micro. Theo. Tech.*, vol. MTT-24, no. 7, July 1976, pp. 468–473.

58. T. Corzani et al., "Numerical analysis of surface wave propagation using finite and infinite elements," *Alta Frequenza*, vol. 51, no. 3, June 1982, pp. 127–133.

59. O.C. Zienkiewicz et al., "Mapped infinite elements for exterior wave problems," *Inter. J. Num. Meth. Engr.*, vol. 21, 1985.

60. F. Medina, "An axisymmetric infinite element," *Int. J. Num. Meth. Engr.*, vol. 17, 1981, pp. 1177–1185.

61. S. Pissanetzky, "A simple infinite element," *Int. J. Comp. Math. Elect. Engr.*, (COMPEL), vol. 3, no. 2, 1984, pp. 107–114.

62. J. K. Byun and J. M. Jin, "Finite-element analysis of scattering from a complex BOR using spherical infinite elements," *Electromagnetics*, vol. 25, 2005, pp. 267–304.

63. K. Hayata et al., "Self-consistent finite/infinite element scheme for unbounded guided wave problems," *IEEE Trans. Micro. Theo. Tech.*, vol. MTT-36, no. 3, Mar. 1988, pp. 614–616.

64. P. Petre and L. Zombory, "Infinite elements and base functions for rotationally symmetric electromagnetic waves," *IEEE Trans. Ant. Prog.*, vol. 36, no. 10, Oct. 1988, pp. 1490, 1491.

65. Z.J. Csendes and J.F. Lee, "The transfinite element method for modeling MMIC devices," *IEEE Trans. Micro. Theo. Tech.*, vol. 36, no. 12, Dec. 1988, pp. 1639–1649.

66. K.H. Lee et al., "A hybrid three-dimensional electromagnetic modeling scheme," *Geophys.*, vol. 46, no. 5, May 1981, pp. 796–805.

67. S.J. Salon and J.M. Schneider, "A hybrid finite element-boundary integral formulation of Poisson's equation," *IEEE Trans. Mag.*, vol. MAG-17, no. 6, Nov. 1981, pp. 2574–2576.

68. S.J. Salon and J. Peng, "Hybrid finite-element boundary-element solutions to axisymmetric scalar potential problems," in Z.J. Csendes (ed.), *Computational Electromagnetics*. New York: North-Holland/Elsevier, 1986, pp. 251–261.

69. J.M. Lin and V.V. Liepa, "Application of hybrid finite element method for electromagnetic scattering from coated cylinders," *IEEE Trans. Ant. Prop.*, vol. 36, no. 1, Jan. 1988, pp. 50–54.

70. J.M. Lin and V.V. Liepa, "A note on hybrid finite element method for solving scattering problems," *IEEE Trans. Ant. Prop.*, vol. 36, no. 10, Oct. 1988, pp. 1486–1490.

71. M.H. Lean and A. Wexler, "Accurate field computation with boundary element method," *IEEE Trans. Mag.*, vol. MAG-18, no. 2, Mar. 1982, pp. 331–335.

72. R.F. Harrington and T.K. Sarkar, "Boundary elements and the method of moments," in C.A. Brebbia et al. (eds.), *Boundary Elements*. Southampton: CML Publ., 1983, pp. 31–40.

73. M.A. Morgan et al., "Finite element-boundary integral formulation for electromagnetic scattering," *Wave Motion*, vol. 6, no. 1, 1984, pp. 91–103.

74. S. Kagami and I. Fukai, "Application of boundary-element method to electromagnetic field problems," *IEEE Trans. Micro. Theo. Tech.*, vol. 32, no. 4, Apr. 1984, pp. 455–461.

75. Y. Tanaka et al., "A boundary-element analysis of TEM cells in three dimensions," *IEEE Trans. Elect. Comp.*, vol. EMC-28, no. 4, Nov. 1986, pp. 179–184.

76. N. Kishi and T. Okoshi, "Proposal for a boundary-integral method without using Green's function," *IEEE Trans. Micro. Theo. Tech.*, vol. MTT-35, no. 10, Oct. 1987, pp. 887–892.

77. D.B. Ingham et al., "Boundary integral equation analysis of transmission-line singularities," *IEEE Trans. Micro. Theo. Tech.*, vol. MTT-29, no. 11, Nov. 1981, pp. 1240–1243.

78. S. Washiru et al., "An analysis of unbounded field problems by finite element method," *Electr. Comm. Japan*, vol. 64-B, no. 1, 1981, pp. 60–66.

79. T. Yamabuchi and Y. Kagawa, "Finite element approach to unbounded Poisson and Helmholtz problems using hybrid-type infinite element," *Electr. Comm. Japan, Pt. I*, vol. 68, no. 3, 1986, pp. 65–74.

80. K.L. Wu and J. Litva, "Boundary element method for modelling MIC devices," *Elect. Lett.*, vol. 26, no. 8, April 1990, pp. 518–520.

81. M.N.O. Sadiku and A.F. Peterson, "A comparison of numerical methods for computing electromagnetic fields," *Proc. IEEE Southeastcon*, April 1990, pp. 42–47.

82. P.K. Kythe, *An Introduction to Boundary Element Methods*. Boca Raton, FL: CRC Press, 1995, p. 2.

83. J.M. Jin et al., "Fictitious absorber for truncating finite element meshes in scattering," *IEEE Proc. H*, vol. 139, Oct. 1992, pp. 472–476.

84. R. Mittra and O. Ramahi, "Absorbing bounding conditions for direct solution of partial differential equations arising in electromagnetic scattering problems," in M.A. Morgan (ed.), *Finite Element and Finite Difference Methods in Electromagnetics*. New York: Elsevier, 1990, pp. 133–173.

85. U. Pekel and R. Mittra, "Absorbing boundary conditions for finite element mesh truncation," in T. Itoh et al. (eds.), *Finite Element Software for Microwave Engineering*. New York: John Wiley & Sons, 1996, pp. 267–312.

86. U. Pekel and R. Mittra, "A finite element method frequency domain application of the perfectly matched layer (PML) concept," *Micro. Opt. Tech. Lett.*, vol. 9, 1995, pp. 117–122.

87. A. Boag and R. Mittra, "A numerical absorbing boundary condition for finite difference and finite element analysis of open periodic structures," *IEEE Trans. Micro. Theo. Tech.*, vol. 43, no. 1, Jan. 1995, pp. 150–154.

88. P.P. Silvester and G. Pelosi (eds.), *Finite Elements for Wave Electromagnetics: Methods and Techniques*. New York: IEEE Press, 1994, pp. 351–490.

89. Y. Wang and T. Itoh, "Envelope-finite-element (EVFE) technique—A more efficient time-domain scheme," *IEEE Trans. Micro. Theo. Tech.*, vol. 49, no. 12, Dec. 2001, pp. 2241–2247.

90. J. F. Lee, R. Lee, and A. Cangellaris, "Time-domain finite-element methods," *IEEE Trans. Ant. Prop.*, vol. 45, no. 3, March 1997, pp. 430–442.

91. D. Jiao et al., "A fast time-domain finite element-boundary integral method for electromagnetic analysis," *IEEE Trans. Ant. Prop.*, vol. 49, no. 10, Oct. 2001.

92. D. Jiao and J. M. Jin, "A general approach for the stability analysis of the time-domain finite-element method for electromagnetic simulations," *IEEE Trans. Ant. Prop.*, vol. 50, no. 11, Nov. 2002, pp. 1624–1632.

93. S. M. Rao (ed.), *Time Domain Electromagnetics*. San Diego, CA: Academic Press, 1999, pp. 279–305.

94. S. Pernet, X. Ferrieres, and G. Cohen, "High spatial order finite element method to solve Maxwell's equations in time domain," *IEEE Trans. Ant. Prop.*, vol. 53, no. 9, Sept. 2005, pp. 2889–2899.

95. J.F. Lee, "WETD – a finite element time-domain approach for solving Maxwell's equations," *IEEE Micro. Guided Wave Lett.*, vol. 4, no. 1, Jan. 1994, pp. 11–13.

96. A.C. Cangellaris, C. C. Lin, and K. K. Mei, "Point-matched time domain finite element methods for electromagnetic radiation and scattering," *IEEE Trans. Ant. Prop.*, vol. AP-35, no. 10, Oct. 1987, pp. 1160–1173.

97. S. M. Musa and M. N. O. Sadiku, "Using finite element method to calculate capacitance, inductance, characteristic impedance of open microstrip lines," *Microw. Opt. Technol. Lett.*, vol. 50, no. 3, March 2008, pp. 611–614.

98. J. Alvarez et al., "Fully coupled multi-hybrid finite element method–method of moments–physical optics method for scattering and radiation problems," *Electromagnetics*, vol. 30, no. 1–2, 2010, pp. 3–22.

99. J.N. Reddy, *An Introduction to the Finite Element Method*, 2nd Edition. New York: McGraw-Hill, 1993, pp. 293–403.

100. C.W. Steele, *Numerical Computation of Electric and Magnetic Fields*. New York: Van Nostrand Reinhold, 1987.

101. S.R. Hoole, *Computer-Aided Analysis and Design of Electromagnetic Devices*. New York: Elsevier, 1989.

102. T. Itoh (ed.), *Numerical Technique for Microwave and Millimeterwave Passive Structure*. New York: John Wiley, 1989.

103. J. M. Jin, *The Finite Element Method in Electromagnetics*, 3rd Edition. New York, NY: Wiley, 2014.

104. M. V. K. Chari and S. J. Salon, *Numerical Methods in Electromagnetism*. San Diego, CA: Academic Press, 2000, p. 283.

105. G. Mur and I. E. Lager, "On the causes of spurious solutions in electromagnetics," *Electromagnetics*, vol. 22, no. 4, 2002, pp. 357–367.

106. S. J. Sajon, "The hybrid finite element-boundary element method in electromagnetic," *IEEE Trans. Magn.*, vol. 21, no. 5, Sept. 1985, pp. 1829–1834.

107. A. B. Manić et al., "Diakoptic approach combining finite-element method and method of moments in analysis of inhomogeneous anisotropic dielectric and magnetic scatterers," *Electromagnetics*, vol. 34, no. 3–4, 2014, pp. 222–238.

108. B. K. Li et al., "Hybrid numerical techniques for the modelling of radiofrequency coils in MRI," *NMR Biomed*, vol. 22, 2009, pp. 937–951.

109. J. Chen, "A hybrid spectral-element/finite-element time-domain method for multiscale electromagnetic simulations," *Doctoral Dissertation*, Department of Electrical and Computer Engineering, Duke University, 2010.

110. X. Q. Sheng and W. Song, *Essentials of Computational Electromagnetics*. Singapore: John Wiley & Sons, 2012, Chap. 5, pp. 243–276.

111. P. Daly, "Upper and lower bounds to the characteristic impedance of transmission lines using the finite method," *Inter. J. Comp. Math. Elect. Engr., (COMPEL)*, vol. 3, no. 2, 1984, pp. 65–78.

112. A. Khebir et al., "An absorbing boundary condition for quasi-TEM analysis of microwave transmission lines via the finite element method," *J. Elect. Waves Appl.*, vol. 4, no. 2, 1990, pp. 145–157.

113. N. Mabaya et al., "Finite element analysis of optical waveguides," *IEEE Trans. Micro. Theo. Tech.*, vol. MTT-29, no. 6, June 1981, pp. 600–605.

114. M. Ikeuchi et al., "Analysis of open-type dielectric waveguides by the finite-element iterative method," *IEEE Trans. Micro. Theo. Tech.*, vol. MTT-29, no. 3, Mar. 1981, pp. 234–239.

115. C. Yeh et al., "Single model optical waveguides," *Appl. Optics*, vol. 18, no. 10, May 1979, pp. 1490–1504.

116. J. Katz, "Novel solution of 2-D waveguides using the finite element method," *Appl. Optics*, vol. 21, no. 15, Aug. 1982, pp. 2747–2750.

117. B.A. Rahman and J.B. Davies, "Finite-element analysis of optical and microwave waveguide problems," *IEEE Trans. Micro. Theo. Tech.*, vol. MTT-32, no. 1, Jan. 1984, pp. 20–28.

118. X.Q. Sheng and S. Xu, "An efficient high-order mixed-edge rectangular-element method for lossy anisotropic dielectric waveguide," *IEEE Micro. Theo. Tech.*, vol. 45, no. 7, July 1997, pp. 1009–1013.

119. C.B. Rajanathan et al., "Finite-element analysis of the Xi-core leviator," *IEEE Proc.*, vol. 131, no. 1, Pt. A, Jan. 1984, pp. 62–66.

120. T.L. Ma and J.D. Lavers, "A finite-element package for the analysis of electromagnetic forces and power in an electric smelting furnace," *IEEE Trans. Indus. Appl.*, vol. IA-22, no. 4, July/Aug. 1986, pp. 578–585.

121. C.O. Obiozor and M.N.O. Sadiku, "Finite element analysis of a solid rotor induction motor under stator winding effects," *Proc. IEEE Southeastcon*, 1991, pp. 449–453.

122. J.L. Mason and W.J. Anderson, "Finite element solution for electromagnetic scattering from two-dimensional bodies," *Inter. J. Num. Meth. Engr.*, vol. 21, 1985, pp. 909–928.

123. A. Chiba et al., "Application of finite element method to analysis of induced current densities inside human model exposed to 60 Hz electric field," *IEEE Trans. Power App. Sys.*, vol. PAS-103, no. 7, July 1984, pp. 1895–1902.

124. Y. Yamashita and T. Takahashi, "Use of the finite element method to determine epicardial from body surface potentials under a realistic torso model," *IEEE Trans. Biomed. Engr.*, vol. BME-31, no. 9, Sept. 1984, pp. 611–621.

125. M.A. Morgan, "Finite element calculation of microwave absorption by the cranial structure," *IEEE Trans. Biomed. Engr.*, vol. BME-28, no. 10, Oct. 1981, pp. 687–695.

126. D.R. Lynch et al., "Finite element solution of Maxwell's equation for hyperthermia treatment planning," *J. Comp. Phys.*, vol. 58, 1985, pp. 246–269.

127. J.R. Brauer et al., "Dynamic electric fields computed by finite elements," *IEEE Trans. Ind. Appl.*, vol. 25, no. 6, Nov./Dec. 1989, pp. 1088–1092.

128. C.H. Chen and C.D. Lien, "A finite element solution of the wave propagation problem for an inhomogeneous dielectric slab," *IEEE Trans. Ant. Prop.*, vol. AP-27, no. 6, Nov. 1979, pp. 877–880.

129. T.L.W. Ma and J.D. Lavers, "A finite-element package for the analysis of electromagnetic forces and power in an electric smelting furnace," *IEEE Trans. Ind. Appl.*, vol. IA-22, no. 4, July/Aug., 1986, pp. 578–585.

130. M.A. Kolbehdari and M.N.O. Sadiku, "Finite element analysis of an array of rods or rectangular bars between ground," *J. Franklin Inst.*, vol. 335B, no. 1, 1998, pp. 97–107.

131. J. Cushman, "Difference schemes or element schemes," *Int. J. Num. Meth. Engr.*, vol. 14, 1979, pp. 1643–1651.

132. A.J. Baker and M.O. Soliman, "Utility of a finite element solution algorithm for initial-value problems," *J. Comp. Phys.*, vol. 32, 1979, pp. 289–324.

133. C.A. Brebbia (ed.), *Applied Numerical Modelling*. New York: John Wiley, 1978, pp. 571–586.

134. C.A. Brebbia (ed.), *Finite Element Systems: A Handbook*. Berlin: Springer-Verlag, 1985.

135. O.C. Zienkiewicz, *The Finite Element Method*. New York: McGraw-Hill, 1977.

136. A.J. Davies, *The Finite Element Method: A First Approach*. Oxford: Clarendon, 1980.

137. C. Martin and G.F. Carey, *Introduction to Finite Element Analysis: Theory and Application*. New York: McGraw-Hill, 1973.

138. T.J. Chung, *Finite Element Analysis in Fluid Dynamics*. New York: McGraw-Hill, 1978.

139. D.H. Norris and G. deVries, *An Introduction to Finite Element Analysis*. New York: Academic Press, 1978.

140. T. Itoh et al. (eds.), *Finite Element Software for Microwave Engineering*. New York: John Wiley & Sons, 1996.

141. S. Nakamura, *Computational Methods in Engineering and Science*. New York: John Wiley, 1977, pp. 446, 447.

142. The IMSL Libraries. *Problem-Solving Software Systems for Numerical FORTRAN Programming*, Houston, TX: IMSL, 1984.

7

Transmission-Line-Matrix Method

Excuses are the most important tools of non-achievers.

—Unknown

7.1 Introduction

The link between field and circuit theories has been exploited in developing numerical techniques to solve certain types of partial differential equations (PDEs) arising in field problems with the aid of equivalent electrical networks [1]. There are three ranges in the frequency spectrum for which numerical techniques for field problems in general have been developed. In terms of the wavelength λ and the approximate dimension ℓ of the apparatus, these ranges are [2]:

$$\lambda \gg \ell$$
$$\lambda \approx \ell$$
$$\lambda \ll \ell$$

In the first range, the special analysis techniques are known as *circuit theory*; in the second, as *microwave theory*; and in the third, as *geometric optics* (frequency independent). Hence, the fundamental laws of circuit theory can be obtained from Maxwell's equations by applying an approximation valid when $\lambda \gg \ell$. However, it should be noted that circuit theory was not developed by approximating Maxwell's equations, but rather was developed independently from experimentally obtained laws. The connection between circuit theory and Maxwell equations (summarizing field theory) is important; it adds to the comprehension of the fundamentals of electromagnetics. According to Silvester and Ferrari, circuits are mathematical abstractions of physically real fields; nevertheless, electrical engineers at times feel they understand circuit theory more clearly than fields [3].

The idea of replacing a complicated electrical system by a simple equivalent circuit goes back to Kirchhoff and Helmholtz. As a result of Park's [4], Kron's [5,6] and Schwinger's [7,8] works, the power and flexibility of equivalent circuits become more obvious to engineers. The recent applications of this idea to scattering problems, originally due to Johns and Beurle [9], have made the method more popular and attractive.

Transmission-line modeling (TLM), otherwise known as the *transmission-line-matrix method*, is a numerical technique time-domain for solving field problems using circuit equivalent. It is based on the equivalence between Maxwell's equations and the equations for voltages and currents on a mesh of continuous two-wire transmission lines. The main feature of this method is the simplicity of formulation and programming for a

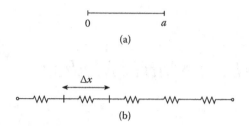

FIGURE 7.1
(a) 1-D conducting system, (b) discretized equivalent.

wide range of applications [10,11]. As compared with the lumped network model, the transmission-line model is more general and performs better at high frequencies where the transmission and reflection properties of geometrical discontinuities cannot be regarded as lumped [7].

Like other numerical techniques, the TLM method is a discretization process. Unlike other methods such as finite difference and finite element methods, which are mathematical discretization approaches, the TLM is a physical discretization approach. In the TLM, the discretization of a field involves replacing a continuous system by a network or array of lumped elements. For example, consider the one-dimensional (1-D) system (a conducting wire) with no energy storage as in Figure 7.1a. The wire can be replaced by a number of lumped resistors providing a discretized equivalent in Figure 7.1b. The discretization of the two-dimensional (2-D), distributed field is shown in Figure 7.2. More general systems containing energy-reservoir elements as well as dissipative elements will be considered later.

The TLM method involves dividing the solution region into a rectangular mesh of transmission lines. Junctions are formed where the lines cross forming impedance discontinuities. A comparison between the transmission-line equations and Maxwell's

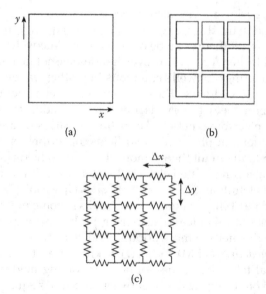

FIGURE 7.2
(a) 2-D conductive sheet, (b) partially discretized equivalent, (c) fully discretized equivalent.

equations allows equivalences to be drawn between voltages and currents on the lines and electromagnetic fields in the solution region. Thus, the TLM method involves two basic steps [12]:

- Replacing the field problem by the equivalent network and deriving analogy between the field and network quantities and
- Solving the equivalent network by iterative methods.

Before we apply the method, it seems fit to briefly review the basic concepts of transmission lines and then show how the TLM method can be applied to a wide range of EM-related problems.

7.2 Transmission-Line Equations

Consider an elemental portion of length $\Delta\ell$ of a two-conductor transmission line. We intend to find an equivalent circuit for this line and derive the line equations. An equivalent circuit of a portion of the line is shown in Figure 7.3, where the line parameters R, L, G, and C are resistance per unit length, inductance per unit length, conductance per unit length, and capacitance per unit length of the line, respectively. The model in Figure 7.3 may represent any two-conductor line. The model is called the T-type equivalent circuit; other types of equivalent circuits are possible, but we end up with the same set of equations. In the model of Figure 7.3, we assume without loss of generality that wave propagates in the $+z$ direction, from the generator to the load.

By applying Kirchhoff's voltage law to the left loop of the circuit in Figure 7.3, we obtain

$$V(z,t) = R\frac{\Delta\ell}{2}I(z,t) + L\frac{\Delta\ell}{2}\frac{\partial I}{\partial t}(z,t) + V(z+\Delta\ell/2,t)$$

or

$$-\frac{V(z+\Delta\ell/2,t) - V(z,t)}{\Delta\ell/2} = RI(z,t) + L\frac{\partial I}{\partial t}(z,t) \tag{7.1}$$

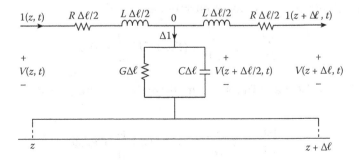

FIGURE 7.3
T-type equivalent circuit model of a differential length of a two-conductor transmission line.

Taking the limit of Equation 7.1 as $\Delta\ell \to 0$ leads to

$$-\frac{\partial V(z,t)}{\partial z} = RI(z,t) + L\frac{\partial I}{\partial t}(z,t) \tag{7.2}$$

Similarly, applying Kirchhoff's current law to the main node of the circuit in Figure 7.3 gives

$$I(z,t) = I(z+\Delta\ell,t) + \Delta I$$

$$= I(z+\Delta\ell,t) + G\Delta\ell V(z+\Delta\ell/2,t) + C\Delta\ell\frac{\partial V}{\partial t}(z+\Delta\ell/2,t)$$

or

$$-\frac{I(z+\Delta\ell,t) - I(z,t)}{\Delta\ell} = GV(z+\Delta\ell/2,t) + C\frac{\partial V}{\partial t}(z+\Delta\ell/2,t) \tag{7.3}$$

As $\Delta\ell \to 0$, Equation 7.3 becomes

$$-\frac{\partial I}{\partial z}(z,t) = GV(z,t) + C\frac{\partial V}{\partial t}(z,t) \tag{7.4}$$

Differentiating Equation 7.2 with respect to z and Equation 7.4 with respect to t, the two equations become

$$-\frac{\partial^2 V}{\partial z^2} = R\frac{\partial I}{\partial z} + L\frac{\partial^2 I}{\partial z\partial t} \tag{7.2a}$$

and

$$-\frac{\partial^2 I}{\partial t\partial z} = G\frac{\partial V}{\partial t} + C\frac{\partial^2 V}{\partial t^2} \tag{7.4a}$$

Substituting Equations 7.4 and 7.4a into Equation 7.2a gives

$$\frac{\partial^2 V}{\partial z^2} = LC\frac{\partial^2 V}{\partial t^2} + (RC+GL)\frac{\partial V}{\partial t} + RGV \tag{7.5}$$

Similarly, we obtain the equation for current I as

$$\frac{\partial^2 I}{\partial z^2} = LC\frac{\partial^2 I}{\partial t^2} + (RC+GL)\frac{\partial I}{\partial t} + RGI \tag{7.6}$$

Equations 7.5 and 7.6 have the same mathematical form, which in general may be written as

$$\frac{\partial^2 \Phi}{\partial z^2} = LC\frac{\partial^2 \Phi}{\partial t^2} + (RC+GL)\frac{\partial \Phi}{\partial t} + RG\Phi \tag{7.7}$$

where $\Phi(z, t)$ has replaced either $V(z, t)$ or $I(z, t)$.

Ignoring certain transmission-line parameters in Equation 7.7 leads to the following special cases [13]:

a. $L = C = 0$ yields

$$\frac{\partial^2 \Phi}{\partial z^2} = k_1 \Phi \tag{7.8}$$

where $k_1 = RG$. Equation 7.8 is the 1-D elliptic PDE called Poisson's equation.

b. $R = C = 0$ or $G = L = 0$ yields

$$\frac{\partial^2 \Phi}{\partial z^2} = k_2 \frac{\partial \Phi}{\partial t} \tag{7.9}$$

where $k_2 = GL$ or RC. Equation 7.9 is the 1-D parabolic PDE called the diffusion equation.

c. $R = G = 0$ (lossless line) yields

$$\frac{\partial^2 \Phi}{\partial z^2} = k_3 \frac{\partial^2 \Phi}{\partial t^2} \tag{7.10}$$

where $k_3 = LC$. This is the 1-D hyperbolic PDE called the Helmholtz equation, or simply the wave equation. Thus, under certain conditions, the 1-D transmission line can be used to model problems involving an elliptic, parabolic, or hyperbolic partial differential equation (PDE). The transmission line of Figure 7.3 reduces to those in Figure 7.4 for these three special cases.

Apart from the equivalent models, other transmission-line parameters are of interest. A detailed explanation of these parameters can be found in standard field theory texts, for example, Reference 14. We briefly present these important parameters. For the lossless line in Figure 7.4c, the characteristic resistance

$$R_o = \sqrt{\frac{L}{C}}, \tag{7.11a}$$

the wave velocity

$$u = \frac{1}{\sqrt{LC}}, \tag{7.11b}$$

and the reflection coefficient at the load

$$\Gamma = \frac{R_L - R_o}{R_L + R_o}, \tag{7.11c}$$

where R_L is the load resistance.

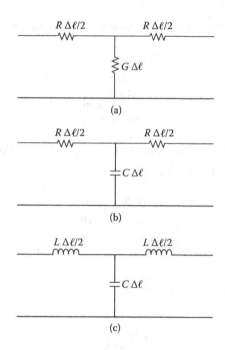

FIGURE 7.4

Transmission-line equivalent models for: (a) elliptic PDE, Poisson's equation, (b) parabolic PDE, diffusion equation, (c) hyperbolic PDE, wave equation.

The generality of the TLM method has been demonstrated in this section. In the following sections, the method is applied specifically to diffusion [15,16] and wave propagation problems [10–13,17,18].

7.3 Solution of Diffusion Equation

We now apply the TLM method to the diffusion problem arising from current density distribution within a wire [15]. If the wire has a circular cross section with radius a and is infinitely long, then the problem becomes 1-D. We will assume sinusoidal source or harmonic fields (with time factor $e^{j\omega t}$).

The analytical solution of the problem has been treated in Example 2.3. For the TLM solution, consider the equivalent network of the cylindrical problem in Figure 7.5,

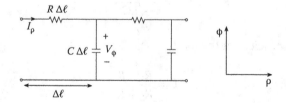

FIGURE 7.5

RC equivalent network.

where $\Delta\ell$ is the distance between nodes or the mesh size. Applying Kirchhoff's laws to the network in Figure 7.5 gives

$$\frac{\partial I_\rho}{\partial \rho} = -j\omega C V_\phi \tag{7.12a}$$

$$\frac{\partial V_\phi}{\partial \rho} = -R I_\rho \tag{7.12b}$$

where R and C are the resistance and capacitance per unit length. Within the conductor, Maxwell's curl equations ($\sigma \gg \omega\varepsilon$) are

$$\nabla \times \mathbf{E} = -j\omega\mu\mathbf{H} \tag{7.13a}$$

$$\nabla \times \mathbf{H} = \sigma\mathbf{E} \tag{7.13b}$$

where $\mathbf{E}$ and $\mathbf{H}$ are assumed to be in phasor forms. In cylindrical coordinates, Equation 7.13 becomes

$$-\frac{\partial E_z}{\partial \rho} = -j\omega\mu H_\phi$$

$$\frac{1}{\rho}\frac{\partial}{\partial \rho}(\rho H_\phi) = \sigma E_z$$

These equations can be written as

$$\frac{\partial E_z}{\partial \rho} = j\omega\left(\frac{\mu}{\rho}\right)(\rho H_\phi) \tag{7.14a}$$

$$\frac{\partial}{\partial \rho}(\rho H_\phi) = (\sigma\rho)E_z \tag{7.14b}$$

Comparing Equation 7.12 with Equation 7.14 leads to the following analogy between the network and field quantities:

$$I_\rho \equiv -E_z \tag{7.15a}$$

$$V_\phi \equiv \rho H_\phi \tag{7.15b}$$

$$C \equiv \mu/\rho \tag{7.15c}$$

$$R \equiv \sigma\rho \tag{7.15d}$$

Therefore, solving the impedance network is equivalent to solving Maxwell's equations. We can solve the overall impedance network in Figure 7.6 by an iterative method. Since the network in Figure 7.6 is in the form of a ladder, we apply the *ladder* method. By applying

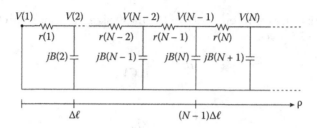

FIGURE 7.6
The overall equivalent network.

Kirchhoff's current law, the Nth nodal voltage ($N > 2$) is related to ($N - 1$)th and ($N - 2$)th voltages according to

$$V(N) = \frac{r(N-1)}{r(N-2)}[V(N-1) - V(N-2)]$$
$$+ jB(N-1)r\ (N-1)V(N-1) + V(N-1) \qquad (7.16)$$

where the resistance r and susceptance B are given by

$$r(N) = R(\Delta)\ell = \sigma(N-0.5)(\Delta\ell)^2, \qquad (7.17a)$$

$$B(N) = \omega C \Delta\ell = \frac{\omega\mu\Delta\ell}{(N-1)\Delta\ell} = \frac{\omega\mu}{N-1} \qquad (7.17b)$$

We note that $V(1) = 0$ because the magnetic field at the center of the conductor ($\rho = 0$) is zero. Also $V(2) = I(1) \cdot r(1)$, where $I(1)$ can be arbitrarily chosen, say $I(1) = 1$. Once $V(1)$ and $V(2)$ are known, we can use Equation 7.16 to scan all nodes in Figure 7.6 once from left to right to determine all nodal voltages ($\equiv \rho H_\phi$) and currents ($\equiv E_z = J_z/\sigma$).

EXAMPLE 7.1

Develop a computer program to determine the relative (or normalized) current density $J_z(\rho)/J_z(a)$ in a round copper wire operated at 1 GHz. Plot the relative current density against the radical position ρ/a for cases $a/\delta = 1$, 2, and 4. Take $\Delta\ell/\delta = 0.1$, $\mu = \mu_0$, $\sigma = 5.8 \times 10^7$ S/m.

Solution

The computer program is presented in Figure 7.7. It calculates the voltage at each node using Equations 7.16 and 7.17. The current on each $r(N)$ is found from Figure 7.7 as

$$I(N-1) = \frac{V(N) - V(N-1)}{r(N-1)}$$

Since $J = \sigma E$, we obtain $J_z(\rho)/J_z(a)$ as the ratio of $I(N)$ and $I(N\ MAX)$, where $I\ (N\ MAX)$ is the current at $\rho = a$.

To verify the accuracy of the TLM solution, we also calculate the exact $J_z(\rho)/J_z(a)$ using Equation 2.120. (For further details, see Example 2.3.) Table 7.1 shows a comparison between TLM results and exact results for the case $a/\delta = 4.0$. It is noticed that the percentage error is maximum (about 8%) at the center of the wire and diminishes to zero as we approach the surface of the wire. Figure 7.8 portrays the plot of the relative current density versus the radial position for cases $a/\delta = 1$, 2, and 4.

```
%================================================================
% USING THE TLM METHOD, THIS PROGRAM CALCULATES THE RELATIVE
% CURRENT DENSITY JR IN A ROUND COPPER WIRE
% THE EXACT SOLUTION JRE IS ALSO INCLUDED
% ================================================================

clear

% SPECIFY INPUT DATA

F = 1.0E+9;
SIGMA = 5.8E+7;
PIE = 4.0*atan(1.0);
MIU = 4.0*PIE*1.0E-7;
OMEGA = 2.0*PIE*F;

DELTA = sqrt( 2.0/(SIGMA*MIU*OMEGA) );
H = 0.1*DELTA;
A = 4.0*DELTA;
NMAX = A/H;

% INITIALIZE AND CALCULATE RELATIVE CURRENT DENSITY JR USING TLM

V(1) = 0;
V(2) = 0.1*(SIGMA*( (1) - 0.5 )*(H^2));

for N = 3:NMAX+1
    V(N) = (SIGMA*(N-1-0.5)*H^2)/(SIGMA*(N-2-0.5)*H^2)*(V(N-1) - V(N-
2))...
            + i*((OMEGA*MIU/(N-2)))*(SIGMA*(N-1-0.5)*H^2)*V(N-
1) + V(N-1);
end

N = 2:NMAX+1;
I = ( V(N) - V(N-1) )./(SIGMA*(N-1-0.5)*H^2);
JR = abs(I/I(NMAX));

% CALCULATE THE CURRENT DENSITY JRE FROM THE EXACT SOLUTION

K = 0:20;
NRO = 1:NMAX;
X = ( (NRO)/(NMAX) )*A*sqrt(2.0)/DELTA;
[K2,X2] = meshgrid(K,X);

FB = ( (X2/2).^(2*K2) )./(factorial(K2).^2);
BEO = FB*exp(i*K*pi/2).';

JRE = abs(BEO);
JRE = JRE/JRE(NMAX);   % RELATIVE CURRENT DENSITY

RHO = (1:NMAX)/(NMAX);

%Output Result

disp(sprintf('\n%s    %s     %s','RADIAL POSITION','TLM J','EXACT J'));

for N=4:4:NMAX
    s2 = sprintf('%10g %16g %12g',RHO(N), JR(N), JRE(N));
    disp(s2)
end

    figure(1),plot(RHO,JR,RHO,JRE)
    ylabel('Relative current Density');
    xlabel('Radial position')
```

FIGURE 7.7
Computer program for Example 7.1.

TABLE 7.1

Comparison of Relative Current Density
Obtained from TLM and Exact Solutions
$(a/\delta = 4.0)$

Radial Position (ρ/a)	TLM Result	Exact Result
0.1	0.11581	0.10768
0.2	0.11765	0.11023
0.3	0.12644	0.12077
0.4	0.14953	0.14612
0.5	0.19301	0.19138
0.6	0.26150	0.26082
0.7	0.36147	0.36115
0.8	0.50423	0.50403
0.9	0.70796	0.70786
1.0	1.0	1.0

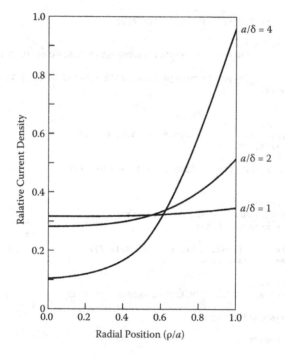

FIGURE 7.8
Relative current density versus radial position.

7.4 Solution of Wave Equations

In order to show how Maxwell's equations may be represented by the transmission-line equations, the differential length of the lossless transmission line between two nodes of the mesh is represented by lumped inductors and capacitors as shown in Figure 7.9 for 2-D wave propagation problems [17,18]. At the nodes, pairs of transmission lines form impedance

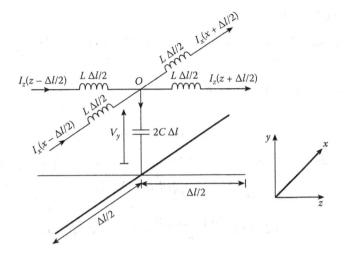

FIGURE 7.9
Equivalent network of a 2-D TLM shunt node.

discontinuity. The complete network of transmission-line matrix is made up of a large number of such building blocks as depicted in Figure 7.10. Notice that in Figure 7.10 single lines are used to represent a transmission-line pair. Also, a uniform internodal distance of $\Delta\ell$ is assumed throughout the matrix (i.e., $\Delta\ell = \Delta x = \Delta z$). We shall first derive equivalences between network and field quantities.

7.4.1 Equivalence between Network and Field Parameters

We refer to Figure 7.9 and apply Kirchhoff's current law at node O to obtain

$$I_x(x-\Delta\ell/2)-I_x(x+\Delta\ell/2)+I_z(z-\Delta\ell/2)-I_z(z+\Delta\ell/2)=2C\Delta\ell\frac{\partial V_y}{\partial t}$$

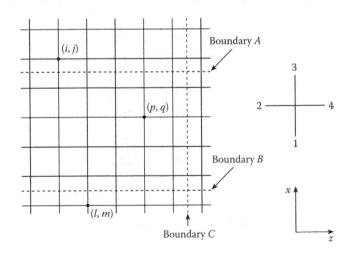

FIGURE 7.10
TLM and boundaries.

Dividing through by $\Delta\ell$ gives

$$\frac{I_x(x-\Delta\ell/2)-I_x(x+\Delta\ell/2)}{\Delta\ell}+\frac{I_z(z-\Delta\ell/2)-I_z(z+\Delta\ell/2)}{\Delta\ell}=2C\frac{\partial V_y}{\partial t}$$

Taking the limit as $\Delta\ell \to 0$ results in

$$-\frac{\partial I_z}{\partial z}-\frac{\partial I_x}{\partial x}=2C\frac{\partial V_y}{\partial t} \qquad (7.18a)$$

Applying Kirchhoff's voltage law around the loop in the $x - y$ plane gives

$$V_y(x-\Delta\ell/2)-L\Delta\ell/2\frac{\partial I_x(x-\Delta\ell/2)}{\partial t}$$
$$-L\Delta\ell/2\frac{\partial I_x(x+\Delta\ell/2)}{\partial t}-V_y(x+\Delta\ell/2)=0$$

Upon rearranging and dividing by $\Delta\ell$, we have

$$\frac{V_y(x-\Delta\ell/2)-V_y(x+\Delta\ell/2)}{\Delta\ell}=\frac{L}{2}\frac{\partial I_x(x-\Delta\ell/2)}{\partial t}+\frac{L}{2}\frac{\partial I_x(x+\Delta\ell/2)}{\partial t}$$

Again, taking the limit as $\Delta\ell \to 0$ gives

$$\frac{\partial V_y}{\partial x}=-L\frac{\partial I_x}{\partial t} \qquad (7.18b)$$

Taking similar steps on the loop in the $y - z$ plane yields

$$\frac{\partial V_y}{\partial z}=-L\frac{\partial I_z}{\partial t} \qquad (7.18c)$$

These equations will now be combined to form a wave equation. Differentiating Equation 7.18a with respect to t, Equation 7.18b with respect to x, and Equation 7.18c with respect to z, we have

$$-\frac{\partial^2 I_z}{\partial z\partial t}-\frac{\partial^2 I_x}{\partial x\partial t}=2C\frac{\partial^2 V_y}{\partial t^2} \qquad (7.19a)$$

$$\frac{\partial^2 V_y}{\partial x^2}=-L\frac{\partial^2 I_x}{\partial t\partial x} \qquad (7.19b)$$

$$\frac{\partial^2 V_y}{\partial z^2}=-L\frac{\partial^2 I_z}{\partial t\partial z} \qquad (7.19c)$$

Substituting Equations 7.19b and 7.19c into Equation 7.19a leads to

$$\frac{\partial^2 V_y}{\partial x^2}+\frac{\partial^2 V_y}{\partial z^2}=2LC\frac{\partial^2 V_y}{\partial t^2} \qquad (7.20)$$

Equation 7.20 is the Helmholtz wave equation in 2-D space.

In order to show the field theory equivalence of Equations 7.19 and 7.20, consider Maxwell's equations

$$\nabla \times \mathbf{E} = -\mu \frac{\partial \mathbf{H}}{\partial t} \tag{7.21a}$$

and

$$\nabla \times \mathbf{H} = \epsilon \frac{\partial \mathbf{E}}{\partial t} \tag{7.21b}$$

Expansion of Equation 7.21 in the rectangular coordinate system yields

$$\frac{\partial E_z}{\partial y} - \frac{\partial E_y}{\partial z} = -\mu \frac{\partial H_x}{\partial t}, \tag{7.22a}$$

$$\frac{\partial E_x}{\partial z} - \frac{\partial E_z}{\partial x} = -\mu \frac{\partial H_y}{\partial t}, \tag{7.22b}$$

$$\frac{\partial E_y}{\partial x} - \frac{\partial E_x}{\partial y} = -\mu \frac{\partial H_z}{\partial t}, \tag{7.22c}$$

$$\frac{\partial H_z}{\partial y} - \frac{\partial H_y}{\partial z} = \epsilon \frac{\partial E_x}{\partial t}, \tag{7.22d}$$

$$\frac{\partial H_x}{\partial z} - \frac{\partial H_z}{\partial x} = \epsilon \frac{\partial E_y}{\partial t}, \tag{7.22e}$$

$$\frac{\partial H_y}{\partial x} - \frac{\partial H_x}{\partial y} = \epsilon \frac{\partial E_z}{\partial t} \tag{7.22f}$$

Consider the situation for which $E_x = E_z = H_y = 0$, $\partial/\partial y = 0$. It is noticed at once that this mode is a transverse electric (TE) mode with respect to the z-axis but a transverse magnetic (TM) mode with respect to the y-axis. Thus by the principle of duality, the network in Figure 7.9 can be used for E_y, H_x, H_z fields as well as for E_x, E_z, H_y fields. A network capable of reproducing TE waves is also capable of reproducing TM waves. For TE waves, Equation 7.22 reduces to

$$\frac{\partial H_x}{\partial z} - \frac{\partial H_z}{\partial x} = \epsilon \frac{\partial E_y}{\partial t}, \tag{7.23a}$$

$$\frac{\partial E_y}{\partial x} = -\mu \frac{\partial H_z}{\partial t}, \tag{7.23b}$$

$$\frac{\partial E_y}{\partial z} = \mu \frac{\partial H_x}{\partial t} \tag{7.23c}$$

Taking similar steps on Equations 7.23a through 7.23c as were taken for Equations 7.18a through 7.18c results in another Helmholtz equation

$$\frac{\partial^2 E_y}{\partial x^2} + \frac{\partial^2 H_y}{\partial z^2} = \mu \epsilon \frac{\partial^2 E_y}{\partial t^2} \tag{7.24}$$

Comparing Equations 7.23 and 7.24 with Equations 7.18 and 7.20 yields the following equivalence between the parameters

$$\boxed{\begin{aligned} E_y &\equiv V_y \\ H_x &\equiv -I_z \\ H_z &\equiv I_x \\ \mu &\equiv L \\ \epsilon &\equiv 2C \end{aligned}} \tag{7.25}$$

Thus in the equivalent circuit:

- The voltage at shunt node is E_y,
- The current in the z direction is $-H_x$,
- The current in the x direction is H_z,
- The inductance per unit length represents the permeability of the medium,
- Twice the capacitance per unit length represents the permittivity of the medium.

7.4.2 Dispersion Relation of Propagation Velocity

For the basic transmission line in the TLM which has $\mu_r = \varepsilon_r = 1$, the inductance and capacitance per unit length are related by Reference 8

$$\frac{1}{\sqrt{(LC)}} = \frac{1}{\sqrt{(\mu_0 \epsilon_0)}} = c \tag{7.26}$$

where $c = 3 \times 10^8$ m/s is the speed of light in vacuum. Notice from Equation 7.26 that for the equivalence made in Equation 7.25, if voltage and current waves on each transmission line component propagate at the speed of light c, the complete network of intersecting transmission lines represents a medium of relative permittivity twice that of free space. This implies that as long as the equivalent circuit in Figure 7.9 is valid, the propagation velocity in the TLM mesh is $1/\sqrt{2}$ of the velocity of light. The manner in which wave propagates on the mesh is now derived.

If the ratio of the length of the transmission-line element to the free-space wavelength of the wave is $\theta/2\pi = \Delta \ell / \lambda$ (θ is called the *electrical length* of the line), the voltage and current at node i are related to those at node $i + 1$ by the transfer-matrix equation (see Problem 7.2) given as [19]

$$\begin{bmatrix} V_i \\ I_i \end{bmatrix} = \begin{bmatrix} (\cos\theta/2) & (j\sin\theta/2) \\ (j\sin\theta/2) & (\cos\theta/2) \end{bmatrix} \begin{bmatrix} 1 & 0 \\ (2j\tan\theta/2) & 1 \end{bmatrix} \begin{bmatrix} (\cos\theta/2) & (j\sin\theta/2) \\ (j\sin\theta/2) & (\cos\theta/2) \end{bmatrix} \begin{bmatrix} V_{i+1} \\ I_{i+1} \end{bmatrix} \tag{7.27}$$

If the waves on the periodic structure have a propagation constant $\gamma_n = \alpha_n + j\beta_n$, then

$$\begin{bmatrix} V_i \\ I_i \end{bmatrix} = \begin{bmatrix} e^{\gamma_n \Delta \ell} & 0 \\ 0 & e^{\gamma_n \Delta \ell} \end{bmatrix} \begin{bmatrix} V_{i+1} \\ I_{i+1} \end{bmatrix} \tag{7.28}$$

Solution of Equations 7.27 and 7.28 gives

$$\cosh(\gamma_n \Delta \ell) = \cos(\theta) - \tan(\theta/2)\sin(\theta) \tag{7.29}$$

This equation describes the range of frequencies over which propagation can take place (passbands), that is,

$$|\cos(\theta) - \tan(\theta/2)\sin(\theta)| < 1, \tag{7.30a}$$

and the range of frequencies over which propagation cannot occur (stop-bands), that is,

$$|\cos(\theta) - \tan(\theta/2)\sin(\theta)| > 1, \tag{7.30b}$$

For the lowest frequency propagation region,

$$\gamma_n = j\beta_n \tag{7.31a}$$

and

$$\theta = \frac{2\pi \Delta \ell}{\lambda} = \frac{\omega}{c} \Delta \ell \tag{7.31b}$$

Introducing Equation 7.31 in Equation 7.29, we obtain

$$\cos(\beta_n \Delta \ell) = \cos\left(\frac{\omega \Delta \ell}{c}\right) - \tan\left(\frac{\omega \Delta \ell}{2c}\right)\sin\left(\frac{\omega \Delta \ell}{c}\right) \tag{7.32}$$

Applying trigonometric identities

$$\sin(2A) = 2\sin(A)\cos(A)$$

and

$$\cos(2A) = 1 - 2\sin^2(A)$$

to Equation 7.32 results in

$$\sin\left(\frac{\beta_n \Delta \ell}{2}\right) = \sqrt{2}\sin\left(\frac{\omega \Delta \ell}{2c}\right) \tag{7.33}$$

which is a transcendental equation. If we let r be the ratio of the velocity u_n of the waves on the network to the free-space wave velocity c, then

$$r = u_n/c = \frac{\omega}{\beta_n c} = \frac{2\pi}{\lambda \beta_n} \tag{7.34a}$$

or

$$\beta_n = \frac{2\pi}{\lambda r} \tag{7.34b}$$

Substituting Equations 7.34 into Equation 7.33, the transcendental equation becomes

$$\sin\left(\frac{\pi}{r} \cdot \frac{\Delta\ell}{\lambda}\right) = \sqrt{2}\sin\left(\frac{\pi\Delta\ell}{\lambda}\right) \tag{7.35}$$

By selecting different values of $\Delta\ell/\lambda$, the corresponding values of $r = u_n/c$ can be obtained numerically as in Figure 7.11 for 2-D problems. From Figure 7.11, we conclude that the TLM can represent Maxwell's equations only over the range of frequencies from zero to the first network cutoff frequency, which occurs at $\omega\Delta\ell/c = \pi/2$ or $\Delta\ell/\lambda = 1/4$. Over this range, the velocity of the waves behaves according to the characteristic of Figure 7.11. For frequencies much smaller than the network cutoff frequency, the propagation velocity approximates to $1/\sqrt{2}$ of the free-space velocity.

Following the same procedure, the dispersion relation for three-dimensional (3-D) problems can be derived as

$$\sin\left(\frac{\pi}{r} \cdot \frac{\Delta\ell}{\lambda}\right) = 2\sin\left(\pi\frac{\Delta\ell}{\lambda}\right) \tag{7.36}$$

Thus for low frequencies ($\Delta\ell/\lambda < 0.1$), the network propagation velocity in 3-D space may be considered constant and equal to $c/2$.

7.4.3 Scattering Matrix

If $_kV_n^i$ and $_kV_n^r$ are the voltage impulses incident upon and reflected from terminal n of a node at time $t = k\Delta\ell/c$, we derive the relationship between the two quantities as follows. Let us assume that a voltage impulse function of unit magnitude is launched into terminal 1 of a node, as shown in Figure 7.12a, and that the characteristic resistance of the line is normalized. A unit-magnitude delta function of voltage and current will then travel toward the junction with unit energy ($S_i = 1$). Since line 1 has three other lines joined to it, its

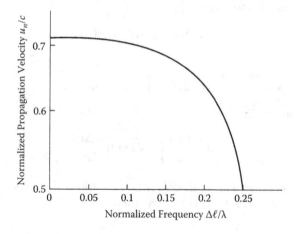

FIGURE 7.11
Dispersion of the velocity of waves in a 2-D TLM network.

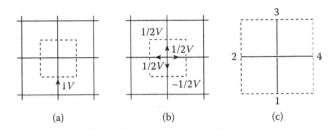

FIGURE 7.12
Impulse response of a node in a matrix.

effective terminal resistance is 1/3. With the knowledge of its reflection coefficient, both the reflected and transmitted voltage impulses can be calculated. The reflection coefficient is

$$\Gamma = \frac{R_L - R_o}{R_L + R_o} = \frac{1/3 - 1}{1/3 + 1} = -\frac{1}{2} \tag{7.37}$$

so that the reflected and transmitted energies are

$$S_r = S_i \Gamma^2 = \frac{1}{4} \tag{7.38a}$$

$$S_t = S_i(1 - \Gamma^2) = \frac{3}{4} \tag{7.38b}$$

where subscripts *i*, *r*, and *t* indicate incident, reflected, and transmitted quantities, respectively. Thus, a voltage impulse of −1/2 is reflected back in terminal 1, while a voltage impulse of $1/2 = [\frac{3}{4} \div 3]^{1/2}$ will be launched into each of the other three terminals as shown in Figure 7.12b.

The more general case of four impulses being incident on four branches of a node can be obtained by applying the superposition principle to the previous case of a single pulse. Hence, if at time $t = k\Delta\ell/c$, voltage impulses $_kV_1^i$, $_kV_2^i$, $_kV_3^i$, and $_kV_4^i$ are incident on lines 1–4, respectively, at any node as in Figure 7.12c, the combined voltage reflected along line 1 at time $t = (k + 1)\Delta\ell/c$ will be [9,10]

$$_{k+1}V_1^r = \frac{1}{2}\left\langle {}_kV_2^i + {}_kV_3^i + {}_kV_4^i - {}_kV_1^i \right\rangle \tag{7.39}$$

In general, the total voltage impulse reflected along line *n* at time $t = (k + 1)\Delta\ell/c$ will be

$$_{k+1}V_n^r = \frac{1}{2}\left[\sum_{m=1}^{4} {}_kV_m^i\right] - {}_kV_n^i, \quad n = 1, 2, 3, 4 \tag{7.40}$$

This idea is conveniently described by a *scattering matrix* equation relating the reflected voltages at time $(k + 1)\Delta\ell/c$ to the incident voltages at the previous time step $k\Delta\ell/c$:

$$\begin{bmatrix} V_1 \\ V_2 \\ V_3 \\ V_4 \end{bmatrix}_{k+1}^r = \frac{1}{2}\begin{bmatrix} -1 & 1 & 1 & 1 \\ 1 & -1 & 1 & 1 \\ 1 & 1 & -1 & 1 \\ 1 & 1 & 1 & -1 \end{bmatrix}\begin{bmatrix} V_1 \\ V_2 \\ V_3 \\ V_4 \end{bmatrix}_k^i \tag{7.41a}$$

or

$$_{k+1}V^r = S_k\,{}_kV^i \tag{7.41b}$$

where

$$S_k = \frac{1}{2}\begin{bmatrix} -1 & 1 & 1 & 1 \\ 1 & -1 & 1 & 1 \\ 1 & 1 & -1 & 1 \\ 1 & 1 & 1 & -1 \end{bmatrix}$$

Also an impulse emerging from a node at position (z, x) in the mesh (reflected impulse) becomes automatically an incident impulse at the neighboring node. Hence,

$$\begin{aligned} _{k+1}V_1^i(z, x + \Delta\ell) &= {}_{k+1}V_3^r(z, x) \\ _{k+1}V_2^i(z + \Delta\ell, x) &= {}_{k+1}V_4^r(z, x) \\ _{k+1}V_3^i(z, x - \Delta\ell) &= {}_{k+1}V_1^r(z, x) \\ _{k+1}V_4^i(z - \Delta\ell, x) &= {}_{k+1}V_2^r(z, x) \end{aligned} \tag{7.42}$$

Thus by applying Equations 7.41 and 7.42, the magnitudes, positions, and directions of all impulses at time $(k + 1)\Delta\ell/c$ can be obtained at each node in the network provided that their corresponding values at time $k\Delta\ell/c$ are known. The impulse response may, therefore, be found by initially fixing the magnitude, position, and direction of travel of impulse voltages at time $t = 0$, and then calculating the state of the network at successive time intervals. The scattering process forms the basic algorithm of the TLM method [10,17].

The propagation of pulses in the TLM model is illustrated in Figure 7.13, where the first two iterations following an initial excitation pulse in a 2-D shunt-connected TLM are shown. We have assumed free-space propagation for the sake of simplicity.

7.4.4 Boundary Representation

Boundaries are usually placed halfway between two nodes in order to ensure synchronism. In practice, this is achieved by making the mesh size $\Delta\ell$ an integer fraction of the structure's dimensions.

Any resistive load at boundary C (see Figure 7.10) may be simulated by introducing a reflection coefficient Γ

$$_{k+1}V_4^i(p, q) = {}_kV_2^r(p + 1, q) = \Gamma[{}_kV_4^r(p, q)] \tag{7.43}$$

where

$$\Gamma = \frac{R_s - 1}{R_s + 1} \tag{7.44}$$

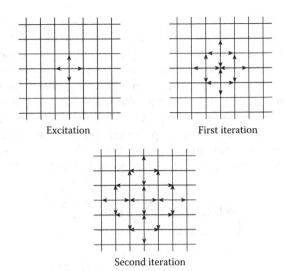

FIGURE 7.13
Scattering in a 2-D TLM network excited by a Dirac impulse.

and R_s is the surface resistance of the boundary normalized by the line characteristic impedance. If, for example, a perfectly conducting wall ($R_s = 0$) is to be simulated along boundary C, Equation 7.44 gives $\Gamma = -1$, which represents a short circuit, and

$$_{k+1}V_4^i(p,q) = -_kV_4^r(p,q) \tag{7.45}$$

is used in the simulation. For waves striking the boundary at arbitrary angles of incidence, a method for modeling free-space boundaries is discussed in Reference 20.

7.4.5 Computation of Fields and Frequency Response

We continue with the TE mode of Equation 7.23 as our example and calculate E_y, H_x, and H_z. E_y at any point corresponds to the node voltage at the point, H_z corresponds to the net current entering the node in the x direction (see Equation 7.25), while H_x is the net current in the negative z direction. For any point ($z = m$, $x = n$) on the grid of Figure 7.10, we have for each kth transient time

$$_kE_y(m,n) = \frac{1}{2}[_kV_1^i(m,n) + _kV_2^i(m,n) + _kV_3^i(m,n) + _kV_4^i(m,n)] \tag{7.46}$$

$$-_kH_x(m,n) = _kV_2^i(m,n) - _kV_4^i(m,n), \tag{7.47}$$

and

$$_kH_z(m,n) = _kV_3^i(m,n) - _kV_1^i(m,n) \tag{7.48}$$

Thus, a series of discrete delta function of magnitudes E_y, H_x, and H_z corresponding to time intervals of $\Delta\ell/c$ are obtained by the iteration of Equations 7.41 and 7.42. (Notice that reflections at the boundaries A and B in Figure 7.10 will cancel out, thus $H_z = 0$.) Any point in the mesh can serve as an output or observation point. Equations 7.46 through 7.48 provide the output-impulse functions for the point representing the response of the system to an impulsive excitation. These output functions may be used to obtain the output waveform. For example, the output waveform corresponding to a pulse input may be obtained by convolving the output-impulse function with the shape of the input pulse.

Sometimes we are interested in the response to a sinusoidal excitation. This is obtained by taking the Fourier transform of the impulse response. Since the response is a series of delta functions, the Fourier transform integral becomes a summation, and the real and imaginary parts of the output spectrum are given by References 9,10

$$\text{Re}[F(\Delta\ell/\lambda)] = \sum_{k=1}^{N} {}_k I \cos\left(\frac{2\pi k \Delta\ell}{\lambda}\right) \tag{7.49a}$$

$$\text{Im}[F(\Delta\ell/\lambda)] = \sum_{k=1}^{N} {}_k I \sin\left(\frac{2\pi k \Delta\ell}{\lambda}\right) \tag{7.49b}$$

where $F(\Delta\ell/\lambda)$ is the frequency response, ${}_k I$ is the value of the output-impulse response at time $t = k\Delta\ell/c$, and N is the total number of time intervals for which the calculation is made. Henceforth, N will be referred to as the number of iterations.

7.4.6 Output Response and Accuracy of Results

The output-impulse function, in terms of voltage or current, may be taken from any point in the TLM mesh. It consists of a train of impulses of varying magnitude in the time domain separated by a time interval $\Delta\ell/c$. Thus, the frequency response obtained by taking the Fourier transform of the output response consists of series of delta functions in the frequency domain corresponding to the discrete modal frequencies for which a solution exists. For practical reasons, the output response has to be truncated, and this results in a spreading of the solution delta function $\sin x/x$ type of curves.

To investigate the accuracy of the result, let the output response be truncated after N iterations. Let $V_{\text{out}}(t)$ be the output-impulse function taken within $0 < t < N\,\Delta\ell/c$. We may regard $V_{\text{out}}(t)$ as an impulse function $V_\infty(t)$ taken within $0 < t < \infty$, multiplied by a unit pulse function $V_p(t)$ of width $N\Delta\ell/c$, that is,

$$V_{\text{out}}(t) = V_\infty(t) \times V_p(t) \tag{7.50}$$

where

$$V_p = \begin{cases} 1, & 0 \le t \le N\Delta\ell/c \\ 0, & \text{elsewhere} \end{cases} \tag{7.51}$$

Let $S_{\text{out}}(f)$, $S_\infty(f)$, and $S_p(f)$ be the Fourier transform of $V_{\text{out}}(t)$, $V_\infty(t)$, and $V_p(t)$, respectively. The Fourier transform of Equation 7.50 is the convolution of $S_\infty(f)$ and $S_p(f)$. Hence,

$$S_{\text{out}}(f) = \int_{-\infty}^{\infty} S_\infty(\alpha) S_p(f - \alpha)\, d\alpha \tag{7.52}$$

where

$$S_p(f) = \frac{N\Delta\ell}{c} \frac{\sin\dfrac{\pi N\Delta\ell f}{c}}{\dfrac{\pi N\Delta\ell f}{c}} e^{-(\pi N\Delta\ell f)/c} \tag{7.53}$$

which is of the form $\sin x/x$. Equations 7.52 and 7.53 show that $S_p(f)$ is placed in each of the positions of the exact response $S_\infty(f)$. Since the greater the number of iterations N the sharper the maximum peak of the curve, the accuracy of the result depends on N. Thus, the solution of a wave equation by TLM method involves the following four steps [21]:

1. *Space discretization*: The solution region is divided into a number of blocks to fit the geometrical and frequency requirements. Each block is replaced by a network of transmission lines interconnected to form a "node." Transmission lines from adjacent nodes are connected to form a mesh describing the entire solution region.

2. *Excitation*: This involves imposing the initial conditions and source terms.

3. *Scattering*: With the use of the scattering matrix, pulses propagate along transmission lines toward each node. At each new time step, a multiple of propagation time δt, scattered pulses from each node become incident on adjacent nodes. The scattering and connection processes may be repeated to simulate propagation for any desired length of time.

4. *Output*: At any time step, voltages and currents on transmission lines are available. These represent the electric and magnetic fields corresponding to the particular problem and excitation. The quantities available at each time step are the solution in the time domain—there is no need for an iterative solution procedure. If desired, frequency-domain information may be obtained by using Fourier transform techniques.

The following examples are taken from Johns's work [9,18].

EXAMPLE 7.2

The MATLAB program in Figure 7.14 is for the numerical calculations of 1-D TEM wave problems. It should be mentioned that the computer program in this example and the following ones are modified versions of those in Agba [22]. The calculations were carried out on a 25×11 rectangular matrix. TEM field-continuation boundaries were fixed along $x = 2$ and $x = 10$, producing boundaries, in effect, along the lines $x = 1.5$ and $x = 10.5$. The initial impulse excitation was on all points along the line $z = 4$, and the field along this line was set to zero at all subsequent time intervals. In this way, interference from boundaries to the left of the excitation line was avoided. Calculations in the z direction were terminated at $z = 24$, so that no reflections were received from points at $z = 25$ in the matrix, and the boundary C in Figure 7.10, situated at $z = 24.5$, was therefore matched to free space. The output-impulse response for E_y and H_x was taken at the point $z = 14$, $x = 6$, which is 10.5 mesh points away from the boundary C, for 100, 150, and 200 iterations.

Since the velocity of waves on the matrix is less than that in free space by a factor u_n/c (see Figure 7.11), the effective intrinsic impedance presented by the network matrix is less by the same factor. The magnitude of the wave impedance on the matrix, normalized to the intrinsic impedance of free space, is given by $Z = |E_y|/|H_x|$ and is tabulated in Table 7.2, together with Arg(Z), for the various numbers of iterations made. A comparison is made with the exact impedance values [14].

```
clear
% *********************************************************************
% THIS PROGRAM APPLIES THE TLM METHOD TO SOLVE
% ONE-DIMENSIONAL WAVE PROBLEMS. THE SPECIFIC EXAMPLE
% SOLVED HERE IS DESCRIBED AS FOLLOWS:-
%
% THE TEM WAVES ON A 25 x 11 MATRIX
% THE BOUNDARIES ARE AT X = 2 AND X = 10
% INITIAL IMPULSE EXCITATION IS ALONG Z=4 AT t = 0
% AND SUBSEQUENTLY THIS LINE IS SET TO ZERO. THE GRID
% IS TERMINATED AT Z=25. OUTPUT IS TAKEN AT Z = 14,
% X = 6 FOR Ey AND Hx FOR 100,150,200 ITERATIONS
% VI(IT,I,J,K) -- ARRAY FOR INCIDENT VOLTAGE
% VR(IT,I,J,K) -- ARRAY FOR REFLECTED VOLTAGE
% IT = 1        -- FOR PREVIOUS PULSE VALUE
% IT = 2        -- FOR CURRENT PULSE VALUE
% I,J           -- CORRESPOND TO NODE LOCATION (Z,X)
% K = 1..4      -- FOR TERMINALS
% NX            -- INDEX OF NODES IN X-DIRECTION
% NZ            -- INDEX OF NODES IN Z-DIRECTION
% NX/NZ B,E     -- INDEX OF BEGINNING, END NODE
% NX/NX O       -- INDEX OF OUTPUT NODE
% GAMMA         -- RELFLECTION COEFFICIENT AT THE BOUNDARY C
% DELTA         -- MESH SIZE DIVIDED BY LAMBA
% ITRATE        -- NO. OF ITERATIONS
% *********************************************************************

VI = zeros(2,25,11,4); VR = zeros(2,25,11,4);
EFI = zeros(1,20); EFR = zeros(1,20);
HFI = zeros(1,20); HFR = zeros(1,20);

CEF = zeros(1,20);    CHF = zeros(1,20);
OUT = zeros(20,9);

NXB=2;NXE=10;NZB=4;NZE=24;NT=4;ITRATE=200;
NXO=6;NZO=14;PIE=3.1415927;GAMMA=0;DELTA=.002;

% STEP #1 ************************************************************
% INSERT INITIAL PUSLE EXCITATION ALONG LINE Z = 4

VI(1,NZB+1,NXB:NXE,2) = 1.0;

% STEP #2 ************************************************************
% Using EQUATIONS (7.40) TO (7.42), CALCULATE THE
% REFLECTED VOLTAGE AND SUBMIT IT DIRECTLY
% TO THE NEIGHBORING NODE.

K = 1:20; II = 0;
DEL = DELTA:DELTA:DELTA*20;
OUT(:,1) = DEL;
for ITIME = 1: ITRATE
    IT = 2;
    for I = (NZB+1):NZE
        for J = NXB:NXE

            VR(IT,I,J,1:NT) = 0.5*sum(VI(IT-1,I,J,1:NT)) - VI(IT-1,I,J,1:NT);
            VI(IT,I,J-1,3) = VR(IT,I,J,1);
            VI(IT,I-1,J,4) = VR(IT,I,J,2);
            VI(IT,I,J+1,1) = VR(IT,I,J,3);
            VI(IT,I+1,J,2) = VR(IT,I,J,4);
```

FIGURE 7.14
Computer program for Example 7.2.   (*Continued*)

```
    % STEP #3 **********************************************************
    % Using EQUATIONS (7.43) AND (7.44), INSERT BOUNDARY CONDITIONS

        if (J==NXE), VI(IT,I,NXE,3) = VR(IT,I,NXE,1); end
        if (J==NXB), VI(IT,I,NXB,1) = VR(IT,I,NXB,3); end
        if (I==NZE), VI(IT,NZE,J,4) = GAMMA*VR(IT,NZE,J,4); end
      end
   end

% STEP #4 **********************************************************
% Using EQUATIONS (7.46) - (7.49), CALCULATE IMPULSE RESPONSE
% OF Ey and Hx AT Z=NZO,X=NXO

   EI = sum(VI(IT,NZO,NXO,1:NT)) * 0.5;
   HI = VI(IT,NZO,NXO,2) - VI(IT,NZO,NXO,4);

% SUM THE FREQUENCY RESPONSE FOR DIFFERENT VALUES OF MESH-SIZE DIVIDED
% BY WAVELENGTH

   CEF = CEF + EI*exp(j*2*pi*ITIME*DEL); %Complex E Field
   CHF = CHF + HI*exp(j*2*pi*ITIME*DEL); %Complex H Field

% SAVE THE CURRENT PULSE MAGNITUDE FOR NEXT ITERATION
   VI(IT-1,:,:,:) = VI(IT,:,:,:);
   VR(IT-1,:,:,:) = VR(IT,:,:,:);

   IT = ITIME;
   if((IT==100)||(IT==150)||(IT==200))
      II = II + 2;
   % STEP #5 **********************************************************
   % CALCULATE MAGNITUDE & ARGUMENT OF IMPEDANCE
      OUT(K,II) = abs(CEF)./abs(CHF); % MAGNITUDE
      ZARG = CEF./CHF;
      OUT(K,II+1) = -atan(imag(ZARG)./real(ZARG));
   end
  end
   % STEP #6 **********************************************************
   % CALCULATE EXACT VALUE OF IMPEDANCE [REF. 13]

   R2 = 1./(pi*DEL./(asin(sqrt(2).*sin(pi*DEL)))));
   R3 = tan(21.0*R2*pi.*DEL);
   CNUM = R2+R3*i;
   CDEM = 1 + j*real(CNUM).*imag(CNUM);
   OUT(K,8) = abs(CNUM)./(abs(CDEM).*real(CNUM));
   ZARG = CNUM./CDEM;
   OUT(K,9) = atan(imag(ZARG)./real(ZARG));
   END

   % disp('Compare to Table 7.2')
   % disp(OUT(1:9,:))

   hdr2 = {'','|Z|','Arg(Z)','|Z|','Arg(Z)','|Z|','Arg(Z)','|Z|','Arg(Z)'};
   hdr3 = {'iteration','100','','150','','200','','Exact',''};
   disp([hdr3;hdr2;num2cell(OUT(1:9,:))])
```

FIGURE 7.14 (Continued)
Computer program for Example 7.2.

EXAMPLE 7.3

The second example was on a rectangular waveguide with a simple load. The MATLAB program used for the numerical analysis was basically similar to that of 1-D simulation. A 25×11 matrix was used for the numerical analysis of the waveguide. Short-circuit boundaries were placed at $x = 2$ and $x = 10$, the width between the waveguide walls

TABLE 7.2

Normalized Impedance of a TEM Wave with Free-Space Discontinuity

	TLM Results						Exact Results	
$\Delta\ell/\lambda$	$|Z|$	Arg(Z)	$|Z|$	Arg(Z)	$|Z|$	Arg(Z)	$|Z|$	Arg(Z)
Number of Iterations	100		150		200			
0.002	0.9789	−0.1368	0.9730	−0.1396	0.9781	−0.1253	0.9747	−0.1282
0.004	0.9028	−0.2432	0.8980	−0.2322	0.9072	−0.2400	0.9077	−0.2356
0.006	0.8114	−0.3068	0.8229	−0.2979	0.8170	−0.3046	0.8185	−0.3081
0.008	0.7238	−0.3307	0.7328	−0.3457	0.7287	−0.3404	0.7256	−0.3390
0.010	0.6455	−0.3201	0.6367	−0.3350	0.6396	−0.3281	0.6414	−0.3263
0.012	0.5783	−0.2730	0.5694	−0.2619	0.5742	−0.2680	0.5731	−0.2707
0.014	0.5272	−0.1850	0.5313	−0.1712	0.5266	−0.1797	0.5255	−0.1765
0.016	0.4993	−0.0609	0.5043	−0.0657	0.5009	−0.0538	0.5018	−0.0545
0.018	0.5002	−0.0790	0.4987	−0.0748	0.5057	−0.0785	0.5057	0.0768

thus being 9 mesh points. The system was excited at all points along the line $z = 2$, and the impulse function of the output was taken from the point $(x = 6, z = 12)$. The C boundary at $z = 24$ represented an abrupt change to the intrinsic impedance of free space. The minor changes in the program of Figure 7.14 are shown in Figure 7.15. The cutoff frequency for the waveguide occurs [19] at $\Delta\ell/\lambda_n = 1/18$, λ_n is the network-matrix wavelength, which corresponds to $\Delta\ell/\lambda = \sqrt{2}/18$ since

$$\frac{\lambda_n}{\lambda} = \frac{u_n}{c} = \frac{\sqrt{\mu_0\epsilon_0}}{\sqrt{\mu_n\epsilon_n}} = \frac{\sqrt{LC}}{\sqrt{2LC}} = \frac{1}{\sqrt{2}}$$

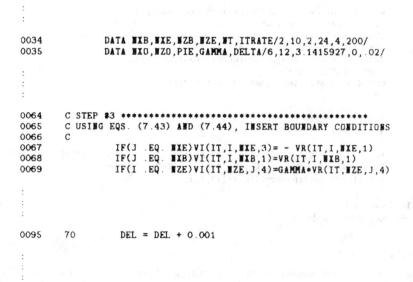

```
       :
       :
       :

0034        DATA NXB,NXE,NZB,NZE,NT,ITRATE/2,10,2,24,4,200/
0035        DATA NXO,NZO,PIE,GAMMA,DELTA/6,12,3.1415927,0,.02/

       :
       :
       :


0064   C STEP #3 ************************************************
0065   C USING EQS. (7.43) AND (7.44), INSERT BOUNDARY CONDITIONS
0066   C
0067          IF(J .EQ. NXE)VI(IT,I,NXE,3)= - VR(IT,I,NXE,1)
0068          IF(J .EQ. NXB)VI(IT,I,NXB,1)=VR(IT,I,NXB,1)
0069          IF(I .EQ. NZE)VI(IT,NZE,J,4)=GAMMA*VR(IT,NZE,J,4)

       :
       :
       :


0095   70     DEL = DEL + 0.001

       :
       :
       :
```

FIGURE 7.15

Modification in the program in Figure 7.14 for simulating waveguide problem in Example 7.3.

TABLE 7.3

Normalized Impedance of a Rectangular Waveguide with Simple Load

	TLM Results		Exact Results	
$\Delta\ell/\lambda$	$\|Z\|$	Arg(Z)	$\|Z\|$	Arg(Z)
0.020	1.9391	0.8936	1.9325	0.9131
0.021	2.0594	0.6175	2.0964	0.6415
0.022	1.9697	0.3553	2.0250	0.3603
0.023	1.7556	0.1530	1.7800	0.1438
0.024	1.5173	0.0189	1.5132	0.0163
0.025	1.3036	−0.0518	1.2989	−0.0388
0.026	1.1370	−0.0648	1.1471	−0.0457
0.027	1.0297	−0.0350	1.0482	−0.0249
0.028	0.9776	0.0088	0.9900	0.0075
0.029	0.9620	0.0416	0.9622	0.0396
0.030	0.9623	0.0554	0.9556	0.0632

A comparison between the results for the normalized guide impedance using this method is made with exact results in Table 7.3.

7.5 Inhomogeneous and Lossy Media in TLM

In our discussion on the transmission-line-matrix (TLM) method in the last section, it was assumed that the medium in which wave propagates was homogeneous and lossless. In this section, we consider media that are inhomogeneous or lossy or both. This necessitates that we modify the equivalent network of Figure 7.9 and the corresponding TLM of Figure 7.10. Also, we need to draw the corresponding equivalence between the network and Maxwell's equations and derive the scattering matrix. We will finally consider how lossy boundaries are represented.

7.5.1 General 2-D Shunt Node

To account for the inhomogeneity of a medium (where ϵ is not constant), we introduce additional capacitance at nodes to represent an increase in permittivity [17,23–25]. We achieve this by introducing an additional length of line or stub to the node as shown in Figure 7.16a. The stub of length $\Delta\ell/2$ is open circuited at the end and is of variable characteristic admittance Y_o relative to the unity characteristic admittance assumed for the main transmission line. At low frequencies, the effect of the stub is to add to each node an additional lumped shunt capacitance $CY_o\Delta\ell/2$, where C is the shunt capacitance per unit length of the main lines that are of unity characteristic admittance. Thus at each node, the total shunt capacitance becomes

$$C' = 2C\Delta\ell + CY_o\Delta\ell/2$$

or

$$C' = 2C\Delta\ell(1 + Y_o/4) \tag{7.54}$$

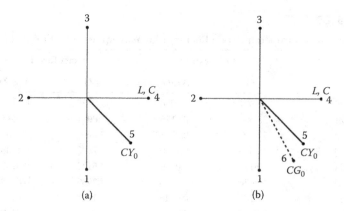

FIGURE 7.16
A 2-D node with: (a) permittivity stub, (b) permittivity and loss stub.

To account for the loss in the medium, we introduce a power-absorbing line at each node, lumped into a single resistor, and this is simulated by an infinite or matched line of characteristic admittance G_o normalized to the characteristic impedance of the main lines as illustrated in Figure 7.16b.

Due to these additional lines, the equivalent network now becomes that shown in Figure 7.17 (cf. Figure 7.17 with Figure 7.9). Applying Kirchhoff's current law to shunt node O in the $x - z$ plane in Figure 7.17 and taking limits as $\Delta \ell \to 0$ results in

$$-\frac{\partial I_z}{\partial z} - \frac{\partial I_x}{\partial x} = \frac{G_o V_y}{Z_o \Delta \ell} + 2C(1 + Y_o/4)\frac{\partial V_y}{\partial t} \tag{7.55}$$

Expanding Maxwell's equations $\nabla \times \mathbf{E} = -\mu \dfrac{\partial \mathbf{H}}{\partial t}$ and $\nabla \times \mathbf{H} = \sigma \mathbf{E} + \epsilon \dfrac{\partial \mathbf{E}}{\partial t}$ for $\partial/\partial y \equiv 0$ leads to

$$\frac{\partial H_x}{\partial z} - \frac{\partial H_z}{\partial x} = \sigma E_y + \epsilon_o \epsilon_r \frac{\partial E_y}{\partial t} \tag{7.56}$$

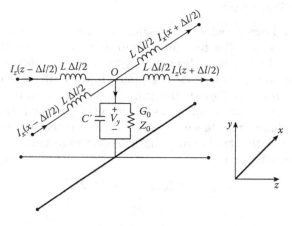

FIGURE 7.17
General 2-D shunt node.

This may be considered as denoting TE$_{m0}$ modes with field components H_z, H_x, and E_y. From Equations 7.55 and 7.56, the following equivalence between the TLM equations and Maxwell's equations can be drawn:

$$
\begin{vmatrix}
E_y \equiv V_y \\
H_x \equiv -I_z \\
H_z \equiv I_x \\
\epsilon_o \equiv 2C \\
\epsilon_r \equiv \dfrac{4+Y_o}{4} \\
\sigma \equiv \dfrac{G_o}{Z_o \Delta \ell}
\end{vmatrix}
\tag{7.57}
$$

where $Z_o = \sqrt{L/C}$. From Equation 7.57, the normalized characteristic admittance G_o of the loss stub is related to the conductivity of the medium by

$$
G_o = \sigma \Delta \ell Z_o \tag{7.58}
$$

Thus losses on the matrix can be varied by altering the value of G_o. Also from Equation 7.57, the variable characteristic admittance Y_o of the permittivity stub is related to the relative permittivity of the medium as

$$
Y_o = 4(\epsilon_r - 1) \tag{7.59}
$$

7.5.2 Scattering Matrix

We now derive the impulse response of the network comprising the interconnection of many generalized nodes such as that in Figure 7.17. As in the previous section, if $kV_n(z, x)$ is unit voltage impulse reflected from the node at (z, x) into the nth coordinate direction $(n = 1, 2, \ldots, 5)$ at time $k\Delta \ell/c$, then at node (z, x),

$$
{}_{k+1}\begin{bmatrix} V_1(z,x) \\ V_2(z,x) \\ V_3(z,x) \\ V_4(z,x) \\ V_5(z,x) \end{bmatrix}^r = [S] \; {}_{k}\begin{bmatrix} V_3(z,x-\Delta \ell) \\ V_4(z-\Delta \ell,x) \\ V_1(z,x+\Delta \ell) \\ V_2(z+\Delta \ell,x) \\ V_5(z,x+\Delta \ell) \end{bmatrix}^i
\tag{7.60}
$$

where $[S]$ is the scattering matrix given by

$$
[S] = \frac{2}{Y}\begin{bmatrix}
1 & 1 & 1 & 1 & Y_o \\
1 & 1 & 1 & 1 & Y_o \\
1 & 1 & 1 & 1 & Y_o \\
1 & 1 & 1 & 1 & Y_o \\
1 & 1 & 1 & 1 & Y_o
\end{bmatrix} - [I]
\tag{7.61}
$$

$[I]$ is a unit matrix and $Y = 4 + Y_o + G_o$. The coordinate directions 1, 2, 3, and 4 correspond to $-x$, $-z$, $+x$, and $+z$, respectively (as in the last section), and 5 refers to the permittivity stub. Notice that the voltage V_6 (see Figure 7.16) scattered into the loss stub is dropped across G_o and not returned to the matrix. We apply Equation 7.60 just as Equation 7.41.

As in the last section, the output-impulse function at a particular node in the mesh can be obtained by recording the amplitude and the time of the stream of pulses as they pass through the node. By taking the Fourier transform of the output-impulse function using Equation 7.49, the required information can be extracted.

The dispersion relation can be derived in the same manner as in the last section. If $\gamma_n = \alpha_n + j\beta_n$ is the network propagation constant and $\gamma = \alpha + j\beta$ is the propagation constant of the medium, the two propagation constants are related as

$$\frac{\beta}{\beta_n} = \frac{\theta/2}{\sin^{-1}\left[\sqrt{2(1+Y_o/4)}\sin\theta/2\right]} \tag{7.62a}$$

$$\frac{\alpha}{\alpha_n} = \frac{\sqrt{1-2(1+Y_o/4)\sin^2\theta/2}}{\sqrt{2(1+Y_o/4)}\cos\theta/2} \tag{7.62b}$$

where $\theta = 2\pi\Delta\ell/\lambda$ and

$$\alpha = \frac{G_o}{8\Delta\ell(1+Y_o/4)} \tag{7.63}$$

In arriving at Equation 7.62, we have assumed that $\alpha_n\Delta\ell \ll 1$. For low frequencies, the attenuation constant α_n and phase constant β_n of the network are fairly constant so that Equation 7.62 reduces to

$$\gamma_n = \sqrt{2(1+Y_o/4)}\gamma \tag{7.64}$$

From this, the network velocity $u_n(=\omega/\beta_n = \beta c/\beta_n)$ of waves on the matrix is readily obtained as

$$u_n^2 = \frac{c^2}{2(1+Y_o/4)} \tag{7.65}$$

where c is the free-space velocity of waves.

7.5.3 Representation of Lossy Boundaries

The above analysis has incorporated conductivity σ of the medium in the TLM formulation. To account for a lossy boundary [25–27], we define the reflection coefficient

$$\Gamma = \frac{Z_s - Z_o}{Z_s + Z_o} \tag{7.66}$$

where $Z_o = \sqrt{\mu_o/\epsilon_o}$ is the characteristic impedance of the main lines and Z_s is the surface impedance of the lossy boundary given by

$$Z_s = \sqrt{\frac{\mu\omega}{2\sigma_c}}(1+j) \tag{7.67}$$

where μ and σ_c are the permeability and conductivity of the boundary. It is evident from Equations 7.66 and 7.67 that the reflection coefficient Γ is in general complex. However, complex Γ implies that the shape of the pulse functions is altered on reflection at the conducting boundary, and this cannot be accounted for in the TLM method [22]. Therefore, assuming that Z_s is small compared with Z_o and that the imaginary part of Γ is negligible,

$$\Gamma \simeq -1 + \sqrt{\frac{2\epsilon_o \omega}{\sigma_c}} \tag{7.68}$$

where $\mu = \mu_o$ is assumed. We notice that Γ is slightly less than -1. Also, we notice that Γ depends on the frequency ω and hence calculations involving lossy boundaries are only accurate for the specific frequency; calculations must be repeated for a different value of $\Delta\ell/\lambda$. The following example is taken from Akhtarzad and Johns [24].

EXAMPLE 7.4

Consider the lossy homogeneous filled waveguide shown in Figure 7.18. The guide is 6 cm wide and 13 cm long. It is filled with a dielectric of relative permittivity $\epsilon_r = 4.9$ and conductivity $\sigma = 0.05$ mhos/m and terminated in an open-circuit discontinuity. Calculate the normalized wave impedance.

Solution

The computer program for this problem is in Figure 7.19. It is an extension of the program in Figure 7.14 with the incorporation of new concepts developed in this section. Enough comments are added to make it self-explanatory. The program is suitable for a 2-D TE$_{m0}$ mode.

The waveguide geometry shown in Figure 7.18 is simulated on a matrix of 12×26 nodes. The matrix is excited at all points along line $z = 1$ with impulses corresponding to E_y. The impulse function of the output at point $(z, x) = (6, 6)$ is taken after 700 iterations. Table 7.4 presents both the TLM and theoretical values of the normalized wave impedance and shows a good agreement between the two.

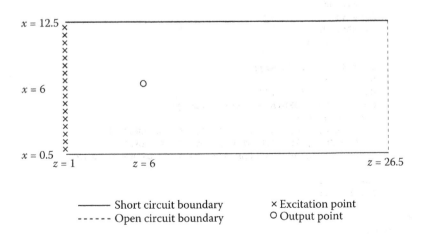

FIGURE 7.18
A lossy homogeneously filled waveguide.

```
clear
% ===========================================================
% THIS PROGRAM SOLVES A TYPICAL TWO-DIMENSIONAL
% WAVE PROPOGATION PROBLEM AT STATED BELOW:
% A WAVE GUIDE 6cm X 13cm IS FILLED WITH A
% DIELECTRIC OF RELATIVE PERMITTIVITY EQUAL TO 4.9
% AND CONDUCTIVITY 0.05 MHO/M. THIS GEOMETRY IS
% BEING SIMULATED ON A MATRIX OF 12 X 26 NODES
% THE MATRIX WAS EXCITED AT ALL POINTS ALONG THE
% LINE Z = 1 WITH IMPULSES CORRESPONDING TO Ey.
% THE IMPLUSE FUNCTION OF THE OUTPUT WAS TAKEN FROM
% THE POINT (Z = 6, X=6) AFTER 750 ITERATIONS.
% THE BOUNDARIES WERE SHORT CIRCUITED AT X = 0.5
% AND X = 12.5 AND TERMINATED AT Z = 26.5 IN AN
% OPEN CIRCUIT DISCONTINUITY.
% VI -- ARRAY FOR INCIDENT VOLTAGE
% VR -- ARRAY FOR REFLECTED VOLTAGE
% NX -- INDEX OF NODES IN X-DIRECTION
% NZ -- INDEX OF NODES IN Z-DIRECTION
% RO -- REFLECTION COEFFICIENT
% DELTA -- MESH SIZE (SPACING)


VI=zeros(2,27,13,5);
VR=zeros(2,27,13,5);
EFI=zeros(1,50);
EFR=zeros(1,50);
HFI=zeros(1,50);
HFR=zeros(1,50);
NXB=1;
NXE=12;
NZB=1;
NZE=26;
NT=5;
ITRATE=750;
NXI=0;
NZI=1;
NXO=6;
NZO=6;
RoS=-1;
RoC=1;
Eo=8.854E-12;
Er=4.9;
SIGMA=0.05;
DELTA=0.005;

Yo = 4.0 * (Er/2 -1);
Ur = 1.0;
Uo = 4.0 * pi * 1.0E-07;
Go = SIGMA * DELTA * sqrt( Uo/Eo );
Y = 4.0 + Yo + Go;
% SINCE REFLECTION COEFFICIENT, RO, DEPENDS ON
% FREQUENCY, THE ITERATIONS MUST BE REPEATED FOR
% EACH VALUE OF THE MESH-SIZE DIVIDED BY WAVELENGTH.

DELTA = 0.0;
for L = 1:10
```

FIGURE 7.19
Computer program for Example 7.4.

(Continued)

```
DELTA = DELTA + 0.003;

% INITIALIZE ALL NODES (BOTH FREE AND FIXED)
VI = zeros(2,NZE,NXE,NT);
VR = zeros(2,NZE,NXE,NT);

% START THE ITERATION
for ITime = 0: ITRATE

    IT = 1;
    for I = NZB: NZE
        for J = NXB: NXE
            SUM = 0.0;
            for K = 1:4
                SUM = SUM + VI(IT-1+1,I,J,K);
            end
            SUM = (SUM + Yo*VI(IT-1+1,I,J,5)) * 2/Y;

            % INSERT THE INITIAL CONDITION AT I = 1

            if ((ITime==0)&&(I==NZI)), SUM = 1.0; end

            for K = 1: NT
                VR(IT+1,I,J,K) = SUM - VI(IT-1+1,I,J,K);
            end

            if (J~=NXB), VI(IT+1,I,J-1,3) = VR(IT+1,I,J,1);end
            if (I~=NZB), VI(IT+1,I-1,J,4) = VR(IT+1,I,J,2);end
            if (J~=NXE), VI(IT+1,I,J+1,1) = VR(IT+1,I,J,3);end
            if (I~=NZE), VI(IT+1,I+1,J,2) = VR(IT+1,I,J,4);end
            VI (IT+1,I,J,5) = VR(IT+1,I,J,5);

            % INSERT THE BOUNDARY CONDITIONS
            % FOR THE SHORT CIRCUIT AT X = 0.5
            if (J==NXB), VI(IT+1,I,1,1) = RoS*VR(IT+1,I,1,1);end

            % FOR THE SHORT CIRCUIT AT X = 12.5
            if (J==NXE), VI(IT+1,I,J,3) = RoS*VR(IT+1,I,J,3);end

            % FOR THE OPEN CIRCUIT DISCONTINUITY AT Z = 26.5
            if (I==NZE), VI(IT+1,I,J,4) = RoC*VR(IT+1,I,J,4);end

        end
    end
            % IN ORDER TO CONSERVE SPACE, THE ARRAYS
            % HAVE TO BE UPDATED

            VI(IT-1+1,1:NZE,1:NXE,1:NT) = VI(IT+1,1:NZE,1:NXE,1:NT);
            VR(IT-1+1,1:NZE,1:NXE,1:NT) = VR(IT+1,1:NZE,1:NXE,1:NT);

            % CALCULATE IMPULSE RESPONCE AT Z=NZO, X=NXO
            EI = (sum(VI(IT+1,NZO,NXO,1:4)) + Yo * VI(IT+1,NZO,NXO,5)) * 2.0/Y;
            HI = -(VI(IT+1,NZO,NXO,2)-VI(IT+1,NZO,NXO,4));
```

FIGURE 7.19 (Continued)
Computer program for Example 7.4.   (*Continued*)

```
% SUM THE FREQUENCY RESPONSE (imaginary and
% real parts) FOR DIFFERENT VALUES OF
% MESH-SIZE DIVIDED BY WAVELENGTH

T = (ITime);
EFI(L) = EFI(L)+EI*sin(2.0 * pi * T * DELTA);
EFR(L) = EFR(L)+EI*cos(2.0 * pi * T * DELTA);
HFI(L) = HFI(L)+HI*sin(2.0 * pi * T * DELTA);
HFR(L) = HFR(L)+HI*cos(2.0 * pi * T * DELTA);
OUT(L,1) = DELTA;
if (ITime==ITRATE)
    CEF = EFR(L)+EFI(L)*i;
    CHF = HFR(L)+HFI(L)*i;
    OUT(L,2) = abs(CEF)/abs(CHF);

    % CALCULATE ARGUMENT Z

    ZARG = CEF/CHF;
    XZ = real(ZARG);
    YZ = imag(ZARG);
    XYZ = YZ / XZ;
    OUT(L,3) = -atan(XYZ);
end
        end
    end

format short g
disp('        DELTA        |Z|         Arg(Z)')
disp(OUT)
```

FIGURE 7.19 (Continued)
Computer program for Example 7.4.

TABLE 7.4

Impedance of a Homogeneously Filled Waveguide with Losses

	TLM Results		Exact Results	
$\Delta\ell/\lambda$	|Z|	Arg(Z)	|Z|	Arg(Z)
0.003	0.0725	1.5591	0.0729	1.5575
0.006	0.1511	1.5446	0.1518	1.5420
0.009	0.2446	1.5243	0.2453	1.5205
0.012	0.3706	1.4890	0.3712	1.4840
0.015	0.5803	1.4032	0.5792	1.3977
0.018	1.0000	1.0056	0.9979	1.0065
0.021	1.1735	0.5156	1.1676	0.5121
0.024	0.5032	−0.1901	0.5093	−0.2141
0.027	0.6766	0.6917	0.6609	0.6853
0.030	0.8921	−0.3869	0.8921	−0.4185

7.6 3-D TLM Mesh

The TLM mesh considered in Sections 7.4 and 7.5 is 2-D. The choice of shunt-connected nodes to represent the 2-D wave propagation was quite arbitrary; the TLM mesh could have equally been made up of series-connected nodes. To represent a 3-D space, however, we must apply a hybrid TLM mesh consisting of three shunt and three series nodes to simultaneously describe all the six field components. First of all, we need to understand what a series-connected node is.

7.6.1 Series Nodes

Figure 7.20 portrays a lossless series-connected node that is equipped with a short-circuited stub called the permeability stub. The corresponding network representation is illustrated in Figure 7.21. The input impedance of the short-circuited stub is

$$Z_{in} = jZ_o\sqrt{\frac{L}{C}} \tan\left(\frac{\omega\Delta\ell}{2c}\right) \simeq j\omega LZ_o\Delta\ell/2 \tag{7.69}$$

where Equation 7.26 has been applied. This represents an impedance with inductance

$$L' = L\frac{\Delta\ell}{2}Z_o \tag{7.70}$$

Hence, the total inductance on the side in which the stub is inserted is $L\Delta\ell(1 + Z_o)/2$ as in Figure 7.21. We now apply Kirchhoff's voltage law around the series node of Figure 7.21 and obtain

$$V_z + L\frac{\Delta\ell}{2}(1+Z_o)\frac{\partial I_x}{\partial t} + V_y + \frac{\partial V_y}{\partial z}\Delta\ell - \left(V_z + \frac{\partial V_z}{\partial y}\Delta\ell\right) + 3L\frac{\Delta\ell}{2}\frac{\partial I_x}{\partial t} - V_y = 0$$

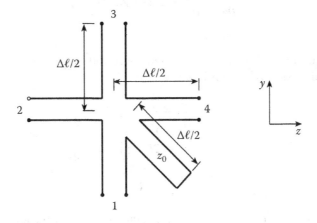

FIGURE 7.20
A lossless series connected node with permeability stub.

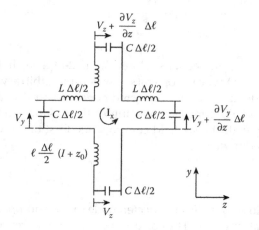

FIGURE 7.21
Network representation of a series node.

Dividing through by $\Delta\ell$ and rearranging terms leads to

$$\frac{\partial V_z}{\partial y} - \frac{\partial V_y}{\partial z} = 2L(1 + Z_o/4)\frac{\partial I_x}{\partial t} \tag{7.71}$$

Note that the series node is oriented in the $y - z$ plane. Equations for series nodes in the $x - y$ and $x - z$ planes can be obtained in a similar manner as

$$\frac{\partial V_y}{\partial x} - \frac{\partial V_x}{\partial y} = 2L(1 + Z_o/4)\frac{\partial I_z}{\partial t} \tag{7.72}$$

and

$$\frac{\partial V_x}{\partial z} - \frac{\partial V_z}{\partial x} = 2L(1 + Z_o/4)\frac{\partial I_y}{\partial t}, \tag{7.73}$$

respectively.

Comparing Equations 7.71 through 7.73 with Maxwell's equations in Equation 7.22, the following equivalences can be identified:

$$\begin{vmatrix} E_x \equiv V_x \\ E_z \equiv V_z \\ \mu_o \equiv 2L \\ \mu_r \equiv \dfrac{4 + Z_o}{4} \end{vmatrix} \tag{7.74}$$

A series-connected 2-D TLM mesh is shown in Figure 7.22a, while the equivalent 1-D mesh is in Figure 7.22b. A voltage impulse incident on a series node is scattered in accordance with Equation 7.60, where the scattering matrix is now

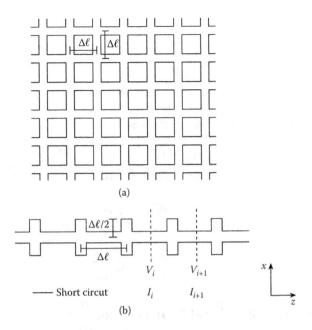

FIGURE 7.22
(a) A 2-D series-connected TLM mesh. (b) A 1-D series-connected TLM mesh.

$$[S] = \frac{2}{Z} \begin{bmatrix} -1 & 1 & 1 & -1 & -1 \\ 1 & -1 & -1 & 1 & 1 \\ 1 & -1 & -1 & 1 & 1 \\ -1 & 1 & 1 & -1 & -1 \\ -Z_o & Z_o & Z_o & -Z_o & -Z_o \end{bmatrix} + [I]$$

(7.75)

where $Z = 4 + Z_o$, and $[I]$ is the unit matrix. The velocity characteristic for the 2-D series matrix is the same as for the shunt node [24]. For low frequencies ($\Delta\ell/\lambda < 0.1$) the velocity of the waves on the matrix is approximately $1/\sqrt{2}$ of the free-space velocity. This is due to the fact that the stubs have twice the inductance per unit length, while the capacitance per unit length remains unchanged. This is the dual of the 2-D shunt case in which the capacitance was doubled and the inductance was unchanged.

7.6.2 3-D Node

A 3-D TLM node [27] consists of three shunt nodes in conjunction with three series nodes. The voltages at the three shunt nodes represent the three components of the **E** field, while the currents of the series nodes represent the three components of the **H** field. In the $x - z$ plane, for example, the voltage–current equations for the shunt node are

$$\frac{\partial I_x}{\partial z} - \frac{\partial I_z}{\partial x} = 2C \frac{\partial V_y}{\partial t}$$

(7.76a)

$$\frac{\partial V_y}{\partial x} = -L\frac{\partial I_x}{\partial t} \tag{7.76b}$$

$$\frac{\partial V_y}{\partial z} = -L\frac{\partial I_z}{\partial t} \tag{7.76c}$$

and for the series node in the $x - z$ plane, the equations are

$$\frac{\partial V_x}{\partial z} - \frac{\partial V_z}{\partial x} = 2L\frac{\partial I_y}{\partial t} \tag{7.77a}$$

$$\frac{\partial I_y}{\partial x} = -C\frac{\partial V_z}{\partial t} \tag{7.77b}$$

$$\frac{\partial I_y}{\partial z} = -C\frac{\partial V_x}{\partial t} \tag{7.77c}$$

Maxwell's equations $\nabla \times \mathbf{E} = \dfrac{\partial \mathbf{B}}{\partial t}$ and $\nabla \times \mathbf{H} = \epsilon\dfrac{\partial \mathbf{E}}{\partial t}$ for $\dfrac{\partial}{\partial y} \equiv 0$ give

$$\frac{\partial H_x}{\partial z} - \frac{\partial H_z}{\partial x} = \epsilon\frac{\partial E_y}{\partial t} \tag{7.78a}$$

$$\frac{\partial E_y}{\partial x} = \mu\frac{\partial H_x}{\partial t} \tag{7.78b}$$

$$\frac{\partial E_y}{\partial z} = -\mu\frac{\partial H_z}{\partial t} \tag{7.78c}$$

and

$$\frac{\partial E_x}{\partial z} - \frac{\partial E_z}{\partial x} = -\mu\frac{\partial H_y}{\partial t} \tag{7.79a}$$

$$\frac{\partial H_y}{\partial x} = -\epsilon\frac{\partial E_x}{\partial t} \tag{7.79b}$$

$$\frac{\partial H_y}{\partial z} = -\epsilon\frac{\partial E_z}{\partial t} \tag{7.79c}$$

A similar analysis for shunt and series nodes in the $x - y$ and $y - z$ planes will yield the voltage–current equations and the corresponding Maxwell's equations. The three sets of 2-D shunt and series nodes oriented in the $x - y$, $y - z$, and $z - x$ planes form a 3-D model. The 2-D nodes must be connected in such a way as to correctly describe Maxwell's equations at each 3-D node. Each of the shunt and series nodes has a spacing of $\Delta\ell/2$ so that like nodes are spaced $\Delta\ell$ apart.

Figure 7.23 illustrates a 3-D node representing a cubical volume of space $\Delta\ell/2$ long in each direction. A close examination shows that if the voltage between lines represents the

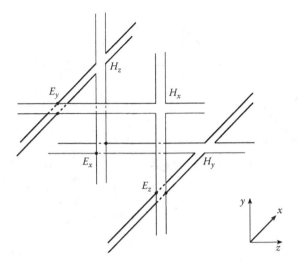

FIGURE 7.23
A 3-D node consisting of three shunt nodes and three series nodes.

E field and the current in the lines represents the **H** field, then the following Maxwell's equations are satisfied:

$$\frac{\partial H_x}{\partial z} - \frac{\partial H_z}{\partial x} = \epsilon \frac{\partial E_y}{\partial t} \tag{7.80a}$$

$$\frac{\partial E_z}{\partial y} - \frac{\partial E_y}{\partial z} = -\mu \frac{\partial H_x}{\partial t} \tag{7.80b}$$

$$\frac{\partial E_y}{\partial x} - \frac{\partial E_x}{\partial y} = -\mu \frac{\partial H_z}{\partial t} \tag{7.80c}$$

$$\frac{\partial E_x}{\partial z} - \frac{\partial E_z}{\partial x} = -\mu \frac{\partial H_y}{\partial t} \tag{7.80d}$$

$$\frac{\partial H_z}{\partial y} - \frac{\partial H_y}{\partial z} = \epsilon \frac{\partial E_x}{\partial t} \tag{7.80e}$$

$$\frac{\partial H_y}{\partial x} - \frac{\partial H_x}{\partial y} = \epsilon \frac{\partial E_z}{\partial t} \tag{7.80f}$$

In the upper half of the node in Figure 7.23, we have a shunt node in the $x - z$ plane (representing Equation 7.80a) connected to a series node in the $y - z$ plane (representing Equation 7.80b) and a series node in the $x - y$ plane (representing Equation 7.80c). In the lower half of the node, a series node in the $x - z$ plane (representing Equation 7.80d) is connected to a shunt node in the $y - z$ plane (representing Equation 7.80e) and a shunt node in the $x - y$ plane (representing Equation 7.80f). Thus Maxwell's equations are completely satisfied at the 3-D node. A 3-D TLM mesh is obtained by stacking similar nodes in x, y, and z directions (see Figure 7.25, for example).

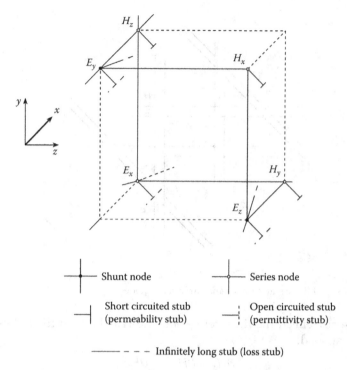

FIGURE 7.24
A general 3-D node.

The wave characteristics of the 3-D mesh are similar to those of the 2-D mesh with the difference that low-frequency velocity is now $c/2$ instead of $c/\sqrt{2}$.

Figure 7.24 illustrates a schematic diagram of a 3-D node using single lines to represent pairs of transmission lines. It is more general than the representation in Figure 7.23 in that it includes the permittivity, permeability, and loss stubs. Note that the dotted lines making up the corners of the cube are guidelines and do not represent transmission lines or stubs. It can be shown that for the general node the following equivalences apply [28]:

$$
\begin{array}{l}
E_x \equiv \text{the common voltage at shunt node } E_x \\
E_y \equiv \text{the common voltage at shunt node } E_y \\
E_z \equiv \text{the common voltage at shunt node } E_z \\
H_x \equiv \text{the common current at series node } H_x \\
H_y \equiv \text{the common current at series node } H_y \\
H_z \equiv \text{the common current at series node } H_z \\
\epsilon_0 \equiv C \text{ (the capacitance per unit length of lines)} \\
\epsilon_r \equiv 2(1 + Y_o/4) \\
\mu_0 \equiv L \text{ (the inductance per unit length of lines)} \\
\mu_r \equiv 2(1 + Z_o/4) \\
\sigma \equiv \dfrac{G_o}{\Delta\ell \dfrac{L}{C}}
\end{array}
\qquad (7.81)
$$

where Y_o, Z_o, and G_o remain as defined in Sections 7.4 and 7.5. Interconnection of many of such 3-D nodes forms a TLM mesh representing any inhomogeneous media. The TLM method for 3-D problems is therefore concerned with applying Equation 7.60 in conjunction with Equations 7.61 and 7.75 and obtaining the impulse response. Any of the field components may be excited initially by specifying initial impulses at the appropriate nodes. Also, the response at any node may be monitored by recording the pulses that pass through the node.

7.6.3 Boundary Conditions

Boundary conditions are simulated by short-circuiting shunt nodes (electric wall) or open-circuiting series nodes (magnetic wall) situated on a boundary. The tangential components of **E** must vanish in the plane of an electric wall, while the tangential components of **H** must be zero in the plane of a magnetic wall. For example, to set E_x and E_y equal to zero in a particular plane, all shunt nodes E_x and E_y lying in that plane are shorted. Similarly, to set H_y and H_z equal to zero in some plane, the series nodes H_y and H_z in that plane are simply open-circuited.

The continuity of the tangential components of **E** and **H** fields across a dielectric/dielectric boundary is automatically satisfied in the TLM mesh. For example, for a dielectric/dielectric boundary in the $x - z$ plane such as shown in Figure 7.25, the following equations valid for a transmission-line element joining the nodes on either side of the boundaries are applicable:

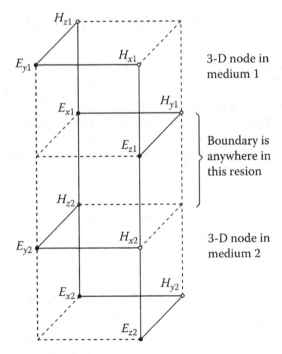

FIGURE 7.25
A dielectric/dielectric boundary in TLM mesh.

$$E_{z1} = E_{z2} + \frac{\partial E_{z2}}{\partial y} \Delta \ell$$

$$E_{x1} = E_{x2} + \frac{\partial E_{x2}}{\partial y} \Delta \ell$$

$$H_{x1} = H_{x2} + \frac{\partial H_{x2}}{\partial y} \Delta \ell \tag{7.82}$$

$$H_{z1} = H_{z2} + \frac{\partial H_{z2}}{\partial y} \Delta \ell$$

Finally, wall losses are included by introducing imperfect reflection coefficients as discussed in Section 7.5. The 3-D TLM mesh will be applied in solving the 3-D problems of resonant cavities in the following examples, taken from Akhtarzad and Johns [27].

EXAMPLE 7.5

Determine the resonant frequency of an $a \times b \times d$ empty rectangular cavity using the TLM method. Take $a = 12\Delta\ell$, $b = 8\Delta\ell$, and $d = 6\Delta\ell$.

Solution

The exact solution [13,14] for TE_{mnp} or TM_{mnp} mode is

$$f_r = \frac{c}{2} \sqrt{(m/a)^2 + (n/b)^2 + (p/d)^2}$$

from which we readily obtain

$$k_c = \frac{w_r}{c} = \frac{2\pi f_r}{c} = \pi \sqrt{(m/a)^2 + (n/b)^2 + (p/d)^2}$$

The TLM program, the modified version of the program in Reference 22, is shown in Figure 7.26. The program initializes all field components by setting them equal to zero at all nodes in the $12\Delta\ell \times 6\Delta\ell \times 6\Delta\ell$ TLM mesh and exciting one field component. With subroutine COMPUTE, it applies Equation 7.60 in conjunction with Equations 7.61 and 7.75 to calculate the reflected $\mathbf{E}$ and $\mathbf{H}$ field components at all nodes. It applies the boundary conditions and calculates the impulse response at a particular node in the mesh.

The results of the computation along with the exact analytical values for the first few modes in the cavity are shown in Table 7.5.

EXAMPLE 7.6

Modify the TLM program in Figure 7.26 to calculate the resonant wavenumber $k_c a$ of the inhomogeneous cavities in Figure 7.27. Take $\epsilon_r = 16$, $a = \Delta\ell$, $b = 3a/10$, $d = 4a/10$, $s = 7a/12$.

Solution

The main program in Figure 7.26 can be used to solve this example. Only the subroutine COMPUTE requires slight modification to take care of the inhomogeneity of the cavity. The modifications in the subprogram for the cavities in Figure 7.27a and b are shown in Figure 7.28a and b, respectively. For each modification, the few lines in Figure 7.28 are inserted in between lines 15 and 17 in subroutine COMPUTE of Figure 7.26. The results are shown in Table 7.6 for TE_{101} mode.

```
0001      C***********************************************************
0002      C  THIS PROGRAM ANLYZES THREE-DIMENSIONAL WAVE PROBLEMS
0003      C  USING THE TLM METHOD
0004      C  THE SPECIFIC EXAMPLE SOLVED HERE IS THE DETERMINATION
0005      C  OF THE TM DOMINANT NODE OF RECTANGULAR LOSSLESS CAVITY
0006      C  OF DIMENSION 12 X 8 X 6
0007      C        E    ...  E - field
0008      C        H    ...  H - field
0009      C        X    ...  X - component
0010      C        Y    ...  Y - component
0011      C        Z    ...  Z - component
0012      C        I    ...  incident impulse
0013      C        R    ...  reflected impulse
0014      C        NX   ...  number of nodes in X-direction
0015      C        NY   ...  number of nodes in Y-direction
0016      C        NZ   ...  number of nodes in Z-direction
0017      C  SHUNT and SERIES  ...  scattering matrix
0018      C    DELTA = DELTA/LAMDA  (FREE SPACE)
0019
0020              IMPLICIT  REAL*8(A,B,D-H,O-Z),COMPLEX*8(C),
0021           1      INTEGER*2(I-N)
0022              DIMENSION EXR(2,7,9,13,5), EXI(2,7,9,13,5),
0023           1            EYR(2,7,9,13,5), EYI(2,7,9,13,5),
0024           2            EZR(2,7,9,13,5), EZI(2,7,9,13,5),
0025           3            HXR(2,7,9,13,5), HXI(2,7,9,13,5),
0026           4            HYR(2,7,9,13,5), HYI(2,7,9,13,5),
0027           5            HZR(2,7,9,13,5), HZI(2,7,9,13,5),
0028           6            EX(2,1000),EY(2,1000),EZ(2,1000),
0029           7            HX(2,1000),HY(2,1000),HZ(2,1000),
0030           8            OUT(12,1000),SHUNT(5,5),SERIES(5,5)
0031
0032              COMMON / PART1 /NT,NXB,NXE,NYB,NYE,NZB,NZE,IT
0033              SQMAG(CM) = ( CABS(CM) ) ** 2
0034              DATA ITRATE,NXB,NXE,NYB,NYE,NZB,NZE,NT,DELTA
0035           1       /1000, 1,  12, 1,  8,  1,  6,  4, 1.0/
0036              DATA NXI,NYI,NZI,NXO,NYO,NZO,RoH,NUMPT,DEL
0037           1       /11, 7,  0,  2,  2, 2,-1.0,1000,0.0001/
0038              DATA PI,       Eo,       Er, Ur,SIGMA,DELTA
0039           1       /3.1415927,8.854E-12,1.0,1.0,0.0,0.01 /
0040              DATA SERIES /-1.0,1.0,1.0,-1.0,-1.0,
0041           1               1.0,-1.0,-1.0,1.0,1.0,
0042           2               1.0,-1.0,-1.0,1.0,1.0,
0043           3              -1.0,1.0,1.0,-1.0,-1.0,
0044           4              -1.0,1.0,1.0,-1.0,-1.0/
0045              DATA SHUNT / 1.0,1.0,1.0,1.0,1.0,
0046           1               1.0,1.0,1.0,1.0,1.0,
0047           2               1.0,1.0,1.0,1.0,1.0,
0048           3               1.0,1.0,1.0,1.0,1.0,
0049           4               1.0,1.0,1.0,1.0,1.0/
0050
0051              Uo =  4.0 * PI * 1.0E-07
0052              Yo = 4.0 * (Er - 1.0)
0053              Zo = 4.0 * (Ur - 1.0)
0054              Go = SIGMA * DELTA * SQRT(Uo/Eo)
0055              Y = 4.0 + Yo + Go
0056              Z = 4.0 + Zo
0057              DO 5 I = 1,5
0058              SHUNT(I,5)  = SHUNT(I,5) * Yo
0059              SERIES(5,I) = SERIES(5,I) * Zo
0060        5     CONTINUE
0061      C
0062      C  INITIALIZE ALL THE NODES
0063      C
0064              DO 10 IT = 1,2
0065              DO 10 K = NZB, NZE
0066              DO 10 J = NYB, NYE
0067              DO 10 I = NXB, NXE
0068              DO 10 L = 1, NT
0069                  EXI(IT,K,J,I,L) = 0.0
0070                  EYI(IT,K,J,I,L) = 0.0
```

FIGURE 7.26
Computer program for Example 7.5.

(Continued)

```
0071                        EZI(IT,K,J,I,L) = 0.0
0072                        HXI(IT,K,J,I,L) = 0.0
0073                        HYI(IT,K,J,I,L) = 0.0
0074                        HZI(IT,K,J,I,L) = 0.0
0075                        EXR(IT,K,J,I,L) = 0.0
0076                        EYR(IT,K,J,I,L) = 0.0
0077                        EZR(IT,K,J,I,L) = 0.0
0078                        HXR(IT,K,J,I,L) = 0.0
0079                        HYR(IT,K,J,I,L) = 0.0
0080     10                 HZR(IT,K,J,I,L) = 0.0
0081     C
0082     C INSERT THE INITIAL EXCITATION
0083     C
0084                DO 15 K = NZB, NZE
0085                DO 15 L = 1, 4
0086                  EZR(2,K,NYI,NXI,L) = 1.0
0087     15         CONTINUE
0088                IT = 2
0089     C
0090     C START ITERATION
0091     C
0092                DO 60 ITime = 1, ITRATE
0093                  IF(ITIME .EQ. 1) GOTO 25
0094     C
0095     C COMPUTE THE REFLECTED E-FIELD AT ALL NODES
0096     C
0097                ISIGN = -1
0098                CALL COMPUTE(EXR,EYR,EZR,SHUNT
0099     1                     ,Y,EXI,EYI,EZI,ISIGN)
0100     C
0101     C RE-INITIALIZE MATRIX AND FORCE TANGENTIAL
0102     C E-FIELD TO ZERO AT BOUNDARY
0103     C
0104                DO 20 K = NZB, NZE
0105                DO 20 J = NYB, NYE
0106                DO 20 I = NXB, NXE
0107                DO 20 L = 1, NT
0108                  HXI(IT-1,K,J,I,L) = 0.0
0109                  HYI(IT-1,K,J,I,L) = 0.0
0110                  HZI(IT-1,K,J,I,L) = 0.0
0111                IF(I .EQ. NXB) THEN
0112                  EYR(IT,K,J,I,L) = 0.0
0113                  EZR(IT,K,J,I,L) = 0.0
0114                  HXR(IT,K,J,I,L) = 0.0
0115                END IF
0116                IF(J .EQ. NYB)THEN
0117                  EXR(IT,K,J,I,L) = 0.0
0118                  EZR(IT,K,J,I,L) = 0.0
0119                  HYR(IT,K,J,I,L) = 0.0
0120                END IF
0121                IF(K .EQ. NZB)THEN
0122                  EXR(IT,K,J,I,L) = 0.0
0123                  EYR(IT,K,J,I,L) = 0.0
0124                  HZR(IT,K,J,I,L) = 0.0
0125                END IF
0126     20       CONTINUE
0127     25       CONTINUE
0128                DO 30 K = NZB, NZE
0129                DO 30 J = NYB, NYE
0130                DO 30 I = NXB, NXE
0131                IF(J .NE. NYB)HZI(IT-1,K,J-1,I,4)=EXR(IT,K,J,I,1)
0132                IF(K .NE. NZB)HYI(IT-1,K-1,J,I,4)=EXR(IT,K,J,I,2)
0133                              HZI(IT-1,K,J,I,2)  =EXR(IT,K,J,I,3)
0134                              HYI(IT-1,K,J,I,2)  =EXR(IT,K,J,I,4)
0135                IF(I .NE. NXB)HZI(IT-1,K,J,I-1,3)=EYR(IT,K,J,I,1)
0136                IF(K .NE. NZB)HXI(IT-1,K-1,J,I,4)=EYR(IT,K,J,I,2)
0137                              HZI(IT-1,K,J,I,1)  =EYR(IT,K,J,I,3)
0138                              HXI(IT-1,K,J,I,2)  =EYR(IT,K,J,I,4)
0139                IF(I .NE. NXB)HYI(IT-1,K,J,I-1,3)=EZR(IT,K,J,I,1)
0140                IF(J .NE. NYB)HXI(IT-1,K,J-1,I,3)=EZR(IT,K,J,I,2)
```

FIGURE 7.26 (Continued)
Computer program for Example 7.5.

(Continued)

```
0141                             HYI(IT-1,K,J,I,1)  =EZR(IT,K,J,I,3)
0142                             HXI(IT-1,K,J,I,1)  =EZR(IT,K,J,I,4)
0143      30      CONTINUE
0144      C
0145      C INSERT THE BOUNDARY CONDITIONS AT ALL BOUNDARIES
0146      C
0147              DO 35 K = NZB, NZE
0148              DO 35 J = NYB, NYE
0149              DO 35 I = NXB, NXE
0150              IF(I .EQ. NXB)THEN
0151                HYI(IT-1,K,J,I,1) = RoH * HYR(IT,K,J,I,1)
0152                HZI(IT-1,K,J,I,1) = RoH * HZR(IT,K,J,I,1)
0153              END IF
0154              IF(I .EQ. NXE)THEN
0155                HZI(IT-1,K,J,I,3) = RoH * HZR(IT,K,J,I,3)
0156                HYI(IT-1,K,J,I,3) = RoH * HYR(IT,K,J,I,3)
0157              END IF
0158              IF(J .EQ. NYB)THEN
0159                HXI(IT-1,K,J,I,1) = RoH * HXR(IT,K,J,I,1)
0160                HZI(IT-1,K,J,I,2) = RoH * HZR(IT,K,J,I,2)
0161              END IF
0162              IF(J .EQ. NYE)THEN
0163                HXI(IT-1,K,J,I,3) = RoH * HXR(IT,K,J,I,3)
0164                HZI(IT-1,K,J,I,4) = RoH * HZR(IT,K,J,I,4)
0165              END IF
0166              IF(K .EQ. NZB)THEN
0167                HXI(IT-1,K,J,I,2) = RoH * HXR(IT,K,J,I,2)
0168                HYI(IT-1,K,J,I,2) = RoH * HYR(IT,K,J,I,2)
0169              END IF
0170              IF(K .EQ. NZE)THEN
0171                HXI(IT-1,K,J,I,4) = RoH * HXR(IT,K,J,I,4)
0172                HYI(IT-1,K,J,I,4) = RoH * HYR(IT,K,J,I,4)
0173              END IF
0174      35      CONTINUE
0175      C
0176      C  COMPUTE THE H-FIELDS AT ALL THE NODES
0177      C
0178              ISIGN = 1
0179              CALL COMPUTE(HXR,HYR,HZR,SERIES,
0180      1                     Z,HXI,HYI,HZI,ISIGN)
0181      C
0182      C  RE-INITIALIZE ALL THE NODES
0183      C
0184              DO 40 K = NZB, NZE
0185              DO 40 J = NYB, NYE
0186              DO 40 I = NXB, NXE
0187              DO 40 L = 1, NT
0188                EXI(IT-1,K,J,I,L) = 0.0
0189                EYI(IT-1,K,J,I,L) = 0.0
0190                EZI(IT-1,K,J,I,L) = 0.0
0191      40      CONTINUE
0192              DO 45 K = NZB, NZE
0193              DO 45 J = NYB, NYE
0194              DO 45 I = NXB, NXE
0195                     EZI(IT-1,K,J,I,4)  =HXR(IT,K,J,I,1)
0196                     EYI(IT-1,K,J,I,4)  =HXR(IT,K,J,I,2)
0197              IF(J .NE. NYE)EZI(IT-1,K,J+1,I,2)=HXR(IT,K,J,I,3)
0198              IF(K .NE. NZE)EYI(IT-1,K+1,J,I,2)=HXR(IT,K,J,I,4)
0199                     EZI(IT-1,K,J,I,3)  =HYR(IT,K,J,I,1)
0200                     EXI(IT-1,K,J,I,4)  =HYR(IT,K,J,I,2)
0201              IF(I .NE. NXE)EZI(IT-1,K,J,I+1,1)=HYR(IT,K,J,I,3)
0202              IF(K .NE. NZE)EXI(IT-1,K+1,J,I,2)=HYR(IT,K,J,I,4)
0203                     EYI(IT-1,K,J,I,3)  =HZR(IT,K,J,I,1)
0204                     EXI(IT-1,K,J,I,3)  =HZR(IT,K,J,I,2)
0205              IF(I .NE. NXE)EYI(IT-1,K,J,I+1,1)=HZR(IT,K,J,I,3)
0206              IF(J .NE. NYE)EXI(IT-1,K,J+1,I,1)=HZR(IT,K,J,I,4)
0207      45      CONTINUE
0208      C
0209      C  CALCULATE THE IMPULSE RESPONSE AT NXO,NYO,NZO
0210      C
```

FIGURE 7.26 (Continued)
Computer program for Example 7.5.

(Continued)

```
0211                EXT = 0.0
0212                EYT = 0.0
0213                EZT = 0.0
0214                HXT = 0.0
0215                HYT = 0.0
0216                HZT = 0.0
0217                DO 50 L = 1, NT
0218                  EXT = EXT + EXI(IT-1,NZO,NYO,NXO,L) * (2.0/Y)
0219                  EYT = EYT + EYI(IT-1,NZO,NYO,NXO,L) * (2.0/Y)
0220                  EZT = EZT + EZI(IT-1,NZO,NYO,NXO,L) * (2.0/Y)
0221                  HXT = HXT + HXI(IT-1,NZO,NYO,NXO,L) * (2.0/Y)
0222                  HYT = HYT + HYI(IT-1,NZO,NYO,NXO,L) * (2.0/Y)
0223        50        HZT = HZT + HZI(IT-1,NZO,NYO,NXO,L) * (2.0/Y)
0224     C
0225     C  SUM THE FREQUENCY RESPONSE(imaginary and real
0226     C  parts) FOR DIFFERENT VALUES OF MESH-SIZE DIVIDED
0227     C  BY WAVELENGTH BUT FIRST CONVOLVE IMPULSE RESPONSE
0228     C  WITH HANNING PROFILE
0229     C
0230                DINCR = DELTA
0231                DO 55 L = 1, NUMPT
0232                  T = DFLOAT(ITIME)
0233                  AMT = DFLOAT(ITRATE)
0234                  EXTH = EXT * (1.0 + DCOS(PI * T/AMT)) * 0.5
0235                  EYTH = EYT * (1.0 + DCOS(PI * T/AMT)) * 0.5
0236                  EZTH = EZT * (1.0 + DCOS(PI * T/AMT)) * 0.5
0237                  HXTH = HXT * (1.0 + DCOS(PI * T/AMT)) * 0.5
0238                  HYTH = HYT * (1.0 + DCOS(PI * T/AMT)) * 0.5
0239                  HZTH = HZT * (1.0 + DCOS(PI * T/AMT)) * 0.5
0240             EX(1,L) = EX(1,L) + EXTH * DCOS(2.0*PI*T*DINCR)
0241             EX(2,L) = EX(2,L) + EXTH * DSIN(2.0*PI*T*DINCR)
0242             EY(1,L) = EY(1,L) + EYTH * DCOS(2.0*PI*T*DINCR)
0243             EY(2,L) = EY(2,L) + EYTH * DSIN(2.0*PI*T*DINCR)
0244             EZ(1,L) = EZ(1,L) + EZTH * DCOS(2.0*PI*T*DINCR)
0245             EZ(2,L) = EZ(2,L) + EZTH * DSIN(2.0*PI*T*DINCR)
0246             HX(1,L) = HX(1,L) + HXTH * DCOS(2.0*PI*T*DINCR)
0247             HX(2,L) = HX(2,L) + HXTH * DSIN(2.0*PI*T*DINCR)
0248             HY(1,L) = HY(1,L) + HYTH * DCOS(2.0*PI*T*DINCR)
0249             HY(2,L) = HY(2,L) + HYTH * DSIN(2.0*PI*T*DINCR)
0250             HZ(1,L) = HZ(1,L) + HZTH * DCOS(2.0*PI*T*DINCR)
0251             HZ(2,L) = HZ(2,L) + HZTH * DSIN(2.0*PI*T*DINCR)
0252                  OUT(1,L) = DINCR
0253        55        DINCR = DINCR + DEL
0254        60      CONTINUE
0255                DO L = 1, NUMPT
0256                  CEX = CMPLX(EX(1,L),EX(2,L))
0257                  CEY = CMPLX(EY(1,L),EY(2,L))
0258                  CEZ = CMPLX(EZ(1,L),EZ(2,L))
0259                  CHX = CMPLX(HX(1,L),HX(2,L))
0260                  CHY = CMPLX(HY(1,L),HY(2,L))
0261                  CHZ = CMPLX(HZ(1,L),HZ(2,L))
0262                  OUT(2,L) = CABS(CEX)
0263                  OUT(3,L) = CABS(CEY)
0264                  OUT(4,L) = CABS(CEZ)
0265                  OUT(5,L) = CABS(CHX)
0266                  OUT(6,L) = CABS(CHY)
0267                  OUT(7,L) = CABS(CHZ)
0268             OUT(8,L)=SQRT(SQMAG(CEX)+SQMAG(CEY)+SQMAG(CEZ))
0269             OUT(9,L)=SQRT(SQMAG(CHX)+SQMAG(CHY)+SQMAG(CHZ))
0270                END DO
0271     C
0272     C  PICK THE MODES
0273     C
0274                DO 65 L = 2, 9
0275                DO 65 K = 1, NUMPT
0276                IF(K .NE. 1 .AND. K .NE. NUMPT) THEN
0277                  IF(OUT(L,K) .GT. OUT(L,K-1) .AND.
0278        1            OUT(L,K) .GT. OUT(L,K+1)) THEN
0279                    WRITE(6,80) L,K,(OUT(J,K),J=1,9)
0280                END IF
```

FIGURE 7.26 (Continued)
Computer program for Example 7.5.

(Continued)

```
0281                      END IF
0282      65            CONTINUE
0283      C
0284      C  WRITE OUT DATA FOR PLOTTING
0285      C
0286                    DO 70 L = 2, 9
0287                    DO 70 K = 1, NUMPT
0288                    WRITE(6,75) K, OUT(1,K), OUT(L,K)
0289      70            CONTINUE
0290      75            FORMAT(I5,4F15.8)
0291      80            FORMAT(2I5,10F8.4)
0292                    STOP
0293                    END

0001      C********************************************************
0002      C  THIS SUBROUTINE COMPUTES THE REFLECTED PULSES
0003      C
0004                    SUBROUTINE COMPUTE(AXR,AYR,AZR,SCATTER,
0005           1                          W,AXI,AYI,AZI,ISIGN)
0006                    IMPLICIT REAL*8(A-H,O-Z), INTEGER*2(I-N)
0007                    DIMENSION AXR(2,NZE,NYE,NXE,NT),
0008           1        AYR(2,NZE,NYE,NXE,NT),SCATTER(5,5),
0009           2        AZR(2,NZE,NYE,NXE,NT),AXI(2,NZE,NYE,NXE,NT),
0010           3        AYI(2,NZE,NYE,NXE,NT),AZI(2,NZE,NYE,NXE,NT)
0011
0012                    COMMON / PART1 /NT,NXB,NXE,NYB,NYE,NZB,NZE,IT
0013
0014                    DO 20 K = NZB, NZE
0015                    DO 20 J = NYB, NYE
0016                    DO 20 I = NXB, NXE
0017      C  INSERT FEW LINES HERE FOR INHOMOGENEOUS CAVITY
0018                    DO 15 L = 1, NT
0019                    AXRS = 0.0
0020                    AYRS = 0.0
0021                    AZRS = 0.0
0022                    DO 10 M = 1, NT
0023              AXRS = AXRS+2./W*SCATTER(L,M)*AXI(IT-1,K,J,I,M)
0024              AYRS = AYRS+2./W*SCATTER(L,M)*AYI(IT-1,K,J,I,M)
0025              AZRS = AZRS+2./W*SCATTER(L,M)*AZI(IT-1,K,J,I,M)
0026      10            CONTINUE
0027              AXR(IT,K,J,I,L)=AXRS+ISIGN*AXI(IT-1,K,J,I,L)
0028              AYR(IT,K,J,I,L)=AYRS+ISIGN*AYI(IT-1,K,J,I,L)
0029              AZR(IT,K,J,I,L)=AZRS+ISIGN*AZI(IT-1,K,J,I,L)
0030      15            CONTINUE
0031      20            CONTINUE
0032                    RETURN
0033                    END
```

FIGURE 7.26 (Continued)
Computer program for Example 7.5.

TABLE 7.5

Resonant Wavenumber ($k_r a$) of an Empty Rectangular Cavity

Modes	Exact Results	TLM Results	Error %
TM_{110}	5.6636	5.6400	0.42
TE_{101}	7.0249	6.9819	0.61
TM_{210}, TE_{011}	7.8540	7.8112	0.54

Where $k_r a = 4\pi a/c$ and λ is the free-space wavelength.

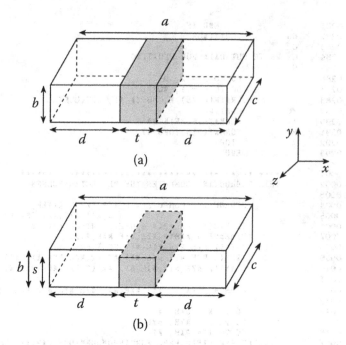

FIGURE 7.27
Rectangular cavity loaded with dielectric slab.

```
0001    C******************************************************
0002    C  THIS SUBROUTINE COMPUTES THE REFLECTED PULSES
0003    C
0004           SUBROUTINE COMPUTE(AXR,AYR,AZR,SCATTER,
0005    1                         WW,AXI,AYI,AZI,ISIGN)

  :
  :
  :

0017    C  INSERT FEW LINES HERE FOR INHOMOGENEOUS CAVITY
0018       IF ((I.GE.9).AND.(I.LE.13)
0019    1                    .AND.(ISIGN.EQ.-1)) THEN
0020         NT = 5
0021         W = WW
0022                  ELSE
0023         NT = 4
0024         W = 4.0
0025       END IF

  :
  :
  :
```

FIGURE 7.28
(a) Modification in subroutine COMPUTE for the inhomogeneous cavity of Figure 7.27a.   *(Continued)*

```
0001    C**************************************************
0002    C   THIS SUBROUTINE COMPUTES THE REFLECTED PULSES
0003    C
0004            SUBROUTINE COMPUTE(AXR,AYR,AZR,SCATTER,
0005        1                      WW,AXI,AYI,AZI,ISIGN)

        :
        :
        :

0017    C   INSERT FEW LINES HERE FOR INHOMOGENEOUS CAVITY
0018        IF (J.GT.4) GO TO 5
0019        IF ((I.GE.9).AND.(I.LE.13)
0020        1              .AND.(ISIGN.EQ.-1)) THEN
0021          NT = 5
0022          W = WW
0023              GO TO 6
0024        END IF
0025    5   CONTINUE
0026          NT = 4
0027          W = 4.0
0028    6   CONTINUE
0029            DO 15 L = 1, NT

        :
        :
```

FIGURE 7.28 (Continued)
(b) Modification in subroutine COMPUTE for the inhomogeneous cavity of Figure 7.27b.

TABLE 7.6

Resonant Wavenumber ($k_r a$) for TE$_{101}$ Mode of Inhomogeneous Rectangular Cavities

Modes	Exact Results	TLM Results	Error %
Figure 7.27a	2.589	2.5761	0.26
Figure 7.27b	(none)	3.5387	

Where $k_r a = 4\pi a/c$, and λ is the free-space wavelength.

7.7 Error Sources and Correction

As in all approximate solutions such as the TLM technique, it is important that the error in the final result be minimal. In the TLM method, four principal sources of error can be identified [10,28,29]:

- Truncation error
- Coarseness error
- Velocity error
- Misalignment error

Each of these sources of error and ways of minimizing it will be discussed.

7.7.1 Truncation Error

The truncation error is due to the need to truncate the impulse response in time. As a result of the finite duration of the impulse response, its Fourier transform is not a line spectrum but rather a superposition of sin x/x functions, which may interfere with each other and cause a slight shift in their maxima. The maximum truncation error is given by

$$e_T = \frac{\Delta S}{\Delta \ell / \lambda c} = \frac{3\lambda_c}{SN^2\pi^2\Delta\ell} \tag{7.83}$$

where λ_c is the cutoff wavelength to be calculated, ΔS is the absolute error in $\Delta\ell/\lambda_c$, S is the frequency separation (expressed in terms of $\Delta\ell/\lambda_c$, λ_c being the free-space wavelength) between two neighboring peaks as shown in Figure 7.29, and N is the number of iterations. Equation 7.83 indicates that e_T decreases with increasing N and increasing S. It is therefore desirable to make N large and suppress all unwanted modes close to the desired mode by carefully selecting the input and output points in the TLM mesh. An alternative way of reducing the truncation error is to use a Hanning window in the Fourier transform. For further details on this, one should consult [10,30].

7.7.2 Coarseness Error

This occurs when the TLM mesh is too coarse to resolve highly nonuniform fields as can be found at corners and edges. An obvious solution is to use a finer mesh ($\Delta\ell \to 0$), but this would lead to large memory requirements and there are limits to this refinement. A better approach is to use variable mesh size so that a higher resolution can be obtained in the nonuniform field region [31]. This approach requires more complicated programming.

7.7.3 Velocity Error

This stems from the assumption that propagation velocity in the TLM mesh is the same in all directions and equal to $u_n = u/\sqrt{2}$, where u is the propagation velocity in the medium filling the structure. The assumption is only valid if the wavelength λ_n in the TLM mesh is large compared with the mesh size $\Delta\ell(\Delta\ell/\lambda_n < 0.1)$. Thus, the cutoff frequency f_{cn} in the TLM mesh is related to the cutoff frequency f_c of the real structure according to $f_c = f_{cn}\sqrt{2}$. If $\Delta\ell$ is comparable with λ_n, the velocity of propagation depends on the direction and the

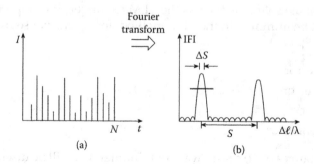

FIGURE 7.29
Source of truncation error: (a) Truncated output-impulse, (b) resulting truncation error in the frequency domain.

assumption of constant velocity results in a velocity error in f_c. Fortunately, a measure to reduce the coarseness error takes care of the velocity error as well.

7.7.4 Misalignment Error

This error occurs in dielectric interfaces in 3-D inhomogeneous structures such as microstrip or fin line. It is due to the manner in which boundaries are simulated in a 3-D TLM mesh; dielectric interfaces appear halfway between nodes, while electric and magnetic boundaries appear across such nodes. If the resulting error is not acceptable, one must make two computations, one with recessed and one with protruding dielectric, and take the average of the results.

7.8 Absorbing Boundary Conditions

Just like FDTD and FEM, the TLM method requires absorbing boundary conditions (ABCs) at the limit of the solution region. Several ABCs have been proposed for TLM simulations [32–37]. It has been recognized that the perfectly matched-layer (PML) technique, discussed for FDTD in Section 3.9, has excellent absorbing performances that are significantly superior to other techniques. So, only PML will be discussed here.

Consider the PML region and the governing Maxwell's equations. Each field component is split into two. For example, $E_x = E_{xy} + E_{xz}$. In 3-D, Maxwell's equations become 12 [38]:

$$\mu_o \frac{H_{xy}}{\partial t} + \sigma_y^* H_{xy} = -\frac{\partial(E_{zx} + E_{zy})}{\partial y} \tag{7.84a}$$

$$\mu_o \frac{H_{xz}}{\partial t} + \sigma_z^* H_{xz} = -\frac{\partial(E_{yx} + E_{yz})}{\partial z} \tag{7.84b}$$

$$\mu_o \frac{H_{yz}}{\partial t} + \sigma_z^* H_{yz} = -\frac{\partial(E_{xy} + E_{xz})}{\partial z} \tag{7.84c}$$

$$\mu_o \frac{H_{yx}}{\partial t} + \sigma_x^* H_{yx} = \frac{\partial(E_{zx} + E_{zy})}{\partial x} \tag{7.84d}$$

$$\mu_o \frac{H_{zx}}{\partial t} + \sigma_x^* H_{zx} = -\frac{\partial(E_{yx} + E_{yz})}{\partial x} \tag{7.84e}$$

$$\mu_o \frac{H_{zy}}{\partial t} + \sigma_y^* H_{zy} = \frac{\partial(E_{xy} + E_{xz})}{\partial y} \tag{7.84f}$$

$$\epsilon_o \frac{E_{xy}}{\partial t} + \sigma_y E_{xy} = \frac{\partial(H_{zx} + H_{zy})}{\partial y} \tag{7.84g}$$

$$\epsilon_o \frac{E_{xz}}{\partial t} + \sigma_z E_{xz} = -\frac{\partial(H_{yx} + H_{yz})}{\partial z} \tag{7.84h}$$

$$\epsilon_o \frac{E_{yz}}{\partial t} + \sigma_z E_{yz} = \frac{\partial(H_{xy} + H_{xz})}{\partial z} \tag{7.84i}$$

$$\epsilon_o \frac{E_{yx}}{\partial t} + \sigma_x E_{yx} = -\frac{\partial(H_{zx} + H_{zy})}{\partial x} \tag{7.84j}$$

$$\epsilon_o \frac{E_{zx}}{\partial t} + \sigma_x E_{zx} = \frac{\partial(H_{yx} + H_{yz})}{\partial x} \tag{7.84k}$$

$$\epsilon_o \frac{E_{zy}}{\partial t} + \sigma_y E_{zy} = -\frac{\partial(H_{xy} + H_{xz})}{\partial y} \tag{7.84l}$$

in which (σ_i, σ_i^*) where $i \in \{x, y, z\}$ are, respectively, the electric and magnetic conductivities of the PML region and they satisfy

$$\frac{\sigma_i}{\epsilon_o} = \frac{\sigma_i^*}{\mu_o} \tag{7.85}$$

Using the usual Yees's notation, the field samples are expressed as

$$
\begin{aligned}
E_x^n(i,j,k) &= E_x[(i+1/2)\delta, j\delta, k\delta, (n+1/2)\delta t] \\
E_y^n(i,j,k) &= E_y[i\delta, (j+1/2)\delta, k\delta, (n+1/2)\delta t] \\
E_z^n(i,j,k) &= E_z[i\delta, j\delta, (k+1/2)\delta, (n+1/2)\delta t] \\
H_x^n(i,j,k) &= H_x[(i\delta, (j+1/2)\delta, (k+1/2)\delta, n\delta t] \\
H_y^n(i,j,k) &= H_y[(i+1/2)\delta, j\delta, (k+1/2)\delta, n\delta t] \\
H_z^n(i,j,k) &= H_z[(i+1/2)\delta, (j+1/2)\delta, k\delta, n\delta t]
\end{aligned}
\tag{7.86}
$$

where $\delta = \Delta x = \Delta y = \Delta z = \Delta \ell$. Without loss of generality, we set $\delta t = \delta/2c$. Since we want to interface the FDTD algorithm with the TLM, we express the fields in terms of voltages. For a cubic cell,

$$V_{ers}^n(i,j,k) = \delta E_{rs}^n(i,j,k) \quad \text{with } r \in \{x,y\}, \quad s \in \{x,z\} \tag{7.87a}$$

$$V_{ms}^n(i,j,k) = \sqrt{\frac{\mu_o}{\epsilon_o}} \delta H_s^n(i,j,k) \quad \text{with } s \in \{y,z\} \tag{7.87b}$$

where the subscripts e and m denote electric and magnetic, respectively. By applying the central-difference scheme to Equation 7.84, we obtain, after some algebraic manipulations,

$$V_{exy}^n(i,j,k) = \left(\frac{4 - G_{ey}}{4 + G_{ey}}\right) V_{exy}^{n-1}(i,j,k)$$
$$+ \left(\frac{2}{4 + G_{ey}}\right)\left(V_{mz}^n(i,j,k) - V_{mz}^n(i,j-1,k)\right) \tag{7.88a}$$

$$V_{exz}^n(i,j,k) = \left(\frac{4-G_{ez}}{4+G_{ez}}\right)V_{exz}^{n-1}(i,j,k)$$

$$-\left(\frac{2}{4+G_{ez}}\right)\left(V_{my}^n(i,j,k)-V_{my}^n(i,j,k-1)\right) \quad (7.88b)$$

$$V_{ex}^n(i,j,k) = V_{exy}^n(i,j,k) + V_{exz}^n(i,j,k) \quad (7.88c)$$

where $G_{es} = \delta\sigma_s(i,j,k)\sqrt{\mu_0\epsilon_0}$ with $s \in \{y, z\}$. Applying this TLM FDTD-PML algorithm has been found to yield excellent performance with reflection level below -55 dB [37].

7.9 Concluding Remarks

This chapter has described the TLM method which is a modeling process rather than a numerical method for solving differential equations. The flexibility, versatility, and generality of the time-domain method have been demonstrated. Our discussion in this chapter has been introductory, and one is advised to consult [10,39–41] for a more in-depth treatment. A generalized treatment of TLM in the curvilinear coordinate system is presented in Reference 42, while a theoretical basis of TLM is derived in Reference 43. Further developments in TLM can be found in References 44–50.

Although the application of the TLM method in this chapter has been limited to diffusion and wave propagation problems, the method has a wide range of applications. The technique has been applied to other problems including:

- Cutoff frequencies in fin lines [29,51]
- Transient analysis of striplines [52,53]
- Linear and nonlinear lumped networks [54–59]
- Microstrip lines and resonators [17,60,61]
- Diffusion problems [62–64]
- Electromagnetic compatibility problems [21,65–68]
- Antenna problems [43,54,69,70]
- Induced currents in biological bodies exposed to EM fields [71]
- Cylindrical and spherical waves [31,54,72]
- Thin wires [73–75]
- Metamaterials [76]
- Bio-heat transfer [77,78], and
- Others [79–92].

A major advantage of the TLM method, as compared with other numerical techniques, is the ease with which even the most complicated structures can be analyzed. The great flexibility and versatility of the method reside in the fact that the TLM mesh incorporates the properties of EM fields and their interaction with the boundaries and material media. Hence, the EM problem need not be formulated for every new structure. Thus a general-purpose program

TABLE 7.7

A Comparison of TLM and FDTD Methods

FDTD	TLM
A mathematical model based on Maxwell's equations	A physical model based on Huygen's principle
E and H are shifted with respect to space and time	E and H are calculated at the same time and position
Requires less memory and one-half the CPU time	Needs more memory and more CPU time
Provides solution at each time step	Requires some iterative procedure

such as in Reference 79 can be developed such that only the parameters of the structure need be entered for computation. Another advantage of using the TLM method is that certain stability properties can be deduced by inspection of the circuit. There are no problems with convergence, stability, or spurious solutions. The method is limited only by the amount of memory storage required, which depends on the complexity of the TLM mesh. Also, being an explicit numerical solution, the TLM method is suitable for nonlinear or inhomogeneous problems since any variation of material properties may be updated at each time step.

Perhaps the best way to conclude this chapter is to compare the TLM method with the finite difference method, especially FDTD [80–86]. While TLM is a physical model based on Huygens' principle using interconnected transmission lines, the FDTD is an approximate mathematical model directly based on Maxwell's equations. In the 2-D TLM, the magnetic and electric field components are located at the same position with respect to space and time, whereas in the corresponding 2-D FDTD cell, the magnetic field components are shifted by half an interval in space and time with respect to the electric field components. Due to this displacement between electric and magnetic field components in Yee's FDTD, Chen et al. [83] derived a new FDTD and demonstrated that the new FDTD formulation is exactly equivalent to the symmetric condensed node model used in the TLM method. This implies that the TLM algorithm can be formulated in FDTD form and vice versa. However, both algorithms retain their unique advantages. For example, the FDTD model has a simpler algorithm where constitutive parameters are directly introduced, while the TLM has certain advantages in the modeling of boundaries and the partitioning of the solution region. Furthermore, the FDTD requires less than one-half of the CPU time spent by the equivalent TLM program under identical conditions. While the TLM scheme requires 22 real memory stores per node, the FDTD method requires only seven real memory stores per 3-D node in an isotropic dielectric medium [81]. Although both are time-domain schemes, the quantities available at each time step are the solution in TLM model and there is no need for an iterative procedure. The dispersion relations for TLM and FDTD are identical for 2-D but are different for 3-D problems. The comparison is summarized in Table 7.7. According to Johns, the two methods complement each other rather than compete with each other [80]. More information on TLM can be found in References 11,93.

PROBLEMS

7.1 A conductor has a uniform resistance R per unit length and leakage conductance G per unit length. Show that the potential V at a point distant x from one end satisfies the differential equation

$$\frac{d^2V}{dx^2} - RGV = 0$$

7.2 For the two-port network in Figure 7.30a, the relation between the input and output variables can be written in matrix form as

$$\begin{bmatrix} V_1 \\ I_1 \end{bmatrix} = \begin{bmatrix} A & B \\ C & D \end{bmatrix} \begin{bmatrix} V_2 \\ -I_2 \end{bmatrix}$$

For the lossy line in Figure 7.30b, show that the ABCD matrix (also called the cascaded matrix) is

$$\begin{bmatrix} \cosh \gamma \ell & Z_0 \sinh \gamma \ell \\ \dfrac{1}{Z_0} \sinh \gamma \ell & \cosh \gamma \ell \end{bmatrix}$$

7.3 The circuit in Figure 7.31 is used to model diffusion processes and presents a Δz section of a lossy transmission line. Show that

$$\frac{\partial^2 V}{\partial z^2} = -Ri + RC \frac{\partial V}{\partial t} - L \frac{\partial i}{\partial t} + LC \frac{\partial^2 V}{\partial t^2}$$

where $i = I_m/\Delta z$, the current density.

7.4 Consider an EM wave propagation in a lossless medium in TEM mode ($E_y = 0 = E_z = H_z = H_x$) along the z direction. Using 1-D TLM mesh, derive the equivalencies between network and field quantities.

7.5 Modify the program in Figure 7.14 to calculate the cutoff frequency (expressed in terms of $\Delta \ell/\lambda$) in a square section waveguide of size $10\Delta \ell$. Perform the calculation for the TM_{11} mode by using open-circuit symmetry boundaries to

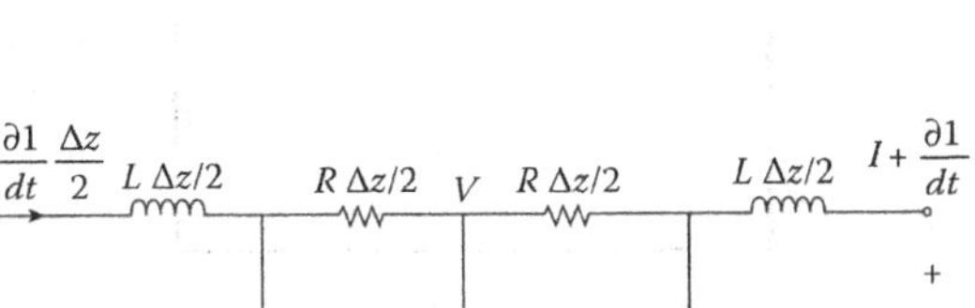

FIGURE 7.30
For Problem 7.2.

FIGURE 7.31
For Problem 7.3.

suppress even-order modes and by taking the excitation and output points as in Figure 7.32 to suppress the TM_{13}, TM_{33}, and TM_{15} modes. Use $N = 500$.

7.6 Repeat Problem 7.5 of higher-order modes but take excitation and output points as in Figure 7.33.

7.7 For the waveguide with a free space discontinuity considered in Example 7.2, plot the variation of the magnitude of the normalized impedance of the guide with $\Delta\ell/\lambda$. The plot should be for frequencies above and below the cutoff frequency, that is, including both evanescent and propagating modes.

7.8 Rework Example 7.4, but take the output point at $(x = 6, z = 13)$.

7.9 Verify Equation 7.62.

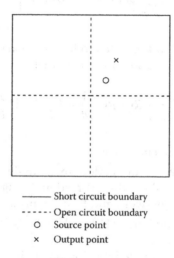

— Short circuit boundary
----- Open circuit boundary
O   Source point
×   Output point

FIGURE 7.32
Square cross-section waveguide of Problem 7.5.

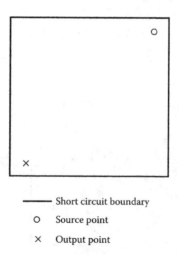

— Short circuit boundary
O   Source point
×   Output point

FIGURE 7.33
Square cross-section waveguide of Problem 7.6.

7.10 For transverse waves on a stub-loaded TLM, the dispersion relation is given by

$$\sin^2\left(\frac{\beta_n \Delta \ell}{2}\right) = 2(1 + Y_o/4)\sin^2\left(\frac{\omega\Delta\ell}{2c}\right)$$

Plot the velocity characteristic similar to that in Figure 7.11 for $Y_o = 0, 1, 2, 10, 20, 100$.

7.11 Verify Equation 7.68.

7.12 The transmission equation for one cell in a stub-loaded 3-D TLM network is

$$\begin{bmatrix} V_1 \\ I_1 \end{bmatrix} = T \cdot \begin{bmatrix} 1 & j(2+z_o)\tan\theta/2 \\ 0 & 1 \end{bmatrix} \cdot T \cdot T \cdot \begin{bmatrix} 1 & 0 \\ g_o + j(2+y_o)\tan\theta/2 & 1 \end{bmatrix} \cdot T \cdot \begin{bmatrix} V_{i+1} \\ I_{i+1} \end{bmatrix}$$

where

$$T = \begin{bmatrix} \cos\theta/4 & j\sin\theta/4 \\ j\sin\theta/4 & \cos\theta/4 \end{bmatrix}$$

$\theta = 2\pi\,\Delta\ell/\lambda$, $y_o = 4(\epsilon_r - 1)$, $z_o = 4(\mu_r - 1)$, and $g_o = \sigma\Delta\ell\sqrt{L/C}$. Assuming small losses $\alpha_n\Delta\ell \ll 1$, show that the transmission equation can be reduced to

$$\begin{bmatrix} V_i \\ I_i \end{bmatrix} = \begin{bmatrix} e^{\gamma_n\Delta\ell} & 0 \\ 0 & e^{\gamma_n\Delta\ell} \end{bmatrix}\begin{bmatrix} V_{i+1} \\ I_{i+1} \end{bmatrix}$$

where $\gamma_n = \alpha_n + j\beta_n$ is the propagation constant and

$$\cos(\beta_n\Delta\ell) = 1 - 8(1 + y_o/4)(1 + z_o/4)\sin^2\theta/2$$

$$\alpha_n\Delta\ell\sin(\beta_n\Delta\ell) = \frac{g_o}{2}(4 + z_o)\sin\theta/2\cos\theta/2$$

7.13 In the $y - z$ plane of a symmetric condensed node of the TLM mesh, the normalized characteristic impedance of the inductive stub is

$$Z_x = \frac{2\mu_r}{u_o\Delta t} \cdot \frac{\Delta y\Delta z}{\Delta x} - 4$$

Assuming that $\Delta x = \Delta y = \Delta z = 0.1$ m, determine the stubs required to model a medium with $\epsilon_r = 4$, $\mu_r = 1$, $u_o = c$, and the value of Δt for stability.

7.14 Consider the 61×8 rectangular matrix with boundaries at $x = 0.5$ and $x = 8.5$ as in Figure 7.34. By making one of the boundaries, say $x = 8.5$, an open circuit, a waveguide of twice the width can be simulated. For the TE_{m0} family of modes, excite the system at all points on line $z = 1$ with impulses corresponding to E_y and take the impulse function of the output at point $x = 7$, $z = 6$. Calculate the normalized wave impedance $Z = E_y/H_x$ for frequencies above cutoff, that is, $\Delta\ell/\lambda = 0.023, 0.025, 0.027, \ldots, 0.041$. Take $\sigma = 0$, $\epsilon_r = 2$, $\mu_r = 1$.

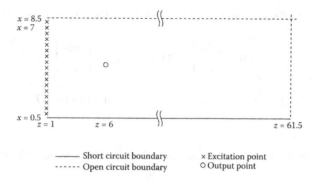

FIGURE 7.34

The 61 × 8 TLM mesh of Problem 7.14.

7.15 Repeat Problem 7.14 for a lossy waveguide with $\sigma = 278$ mhos/m, $\epsilon_r = 1$, $\mu_r = 1$.

7.16 Using the TLM method, determine the cutoff frequency (expressed in terms of $\Delta\ell/\lambda$) of the lowest order TE and TM modes for the square waveguide with cross section shown in Figure 7.35. Take $\epsilon_r = 2.45$.

7.17 For the dielectric ridge waveguide of Figure 7.36, use the TLM method to calculate the cutoff wavenumber k_c of the dominant mode. Express the result in terms of $k_c a(=\omega a/c)$ and try $\epsilon_r = 2$ and $\epsilon_r = 8$. Take $a = 10\Delta\ell$.

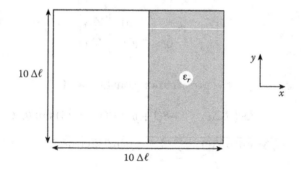

FIGURE 7.35

For Problem 7.16.

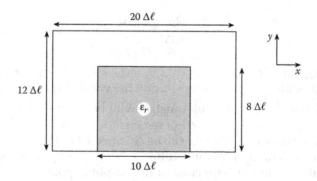

FIGURE 7.36

For Problem 7.17.

7.18 Rework Example 7.6 for the inhomogeneous cavity of Figure 7.37. Take $\epsilon_r = 16$, $a = 12\Delta\ell$, $b = 3a/10$, $d = 4a/10$, $s = 7a/12$, $u = 3d/8$.

7.19 Consider a single microstrip line shown in Figure 7.38. Dispersion analysis of the line by the TLM method involves resonating a section of the transmission line by placing shorting planes along the axis of propagation (the z-axis in this case). Write a TLM computer program and specify the input data as

$$E_x = 0 = E_z \text{ along } y = 0, y = b,$$

$$E_x = 0 = E_z \text{ along } x = 2a,$$

$$E_x = 0 = E_z \text{ for } y = H, -W \leq x \leq W,$$

$$H_y = 0 = H_z \text{ along } x = 0$$

Plot the dispersion curves depicting the phase constant β as a function of frequency f for cases when the line is air-filled and dielectric-filled. The distance $L(=\pi/\beta)$ between the shorting planes is the variable. Assume the dielectric substrate and the walls of the enclosure are lossless. Take $\epsilon_r = 4.0$, $a = 2$ mm, $H = 1.0$ mm, $W = 1.0$ mm, $b = 2$ mm, $\Delta\ell = a/8$.

7.20 For the cubical cavity of Figure 7.39, use the TLM technique to calculate the time taken for the total power in the lossy dielectric cavity to decay to $1/e$ of its

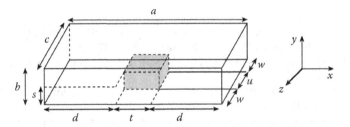

FIGURE 7.37
Inhomogeneous cavity of Problem 7.18.

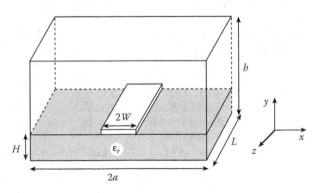

FIGURE 7.38
Microstrip line of Problem 7.19.

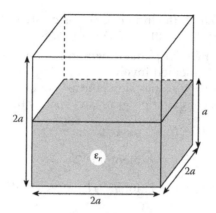

FIGURE 7.39
Lossy cavity of Problem 7.20.

original value. Consider cases when the cavity is completely filled with dielectric material and half-filled. Take $\epsilon_r = 2.45$, $\sigma = 0.884$ mhos/m, $\mu_r = 1$, $\Delta\ell = 0.3$ cm, $2a = 7\Delta\ell$.

References

1. G. Kron, "Numerical solution of ordinary and partial differential equations by means of equivalent circuits," *J. Appl. Phys.*, vol. 16, Mar. 1945, pp. 172–186.
2. C.H. Durney and C.C. Johnson, *Introduction to Modern Electromagnetics*. New York: McGraw-Hill, 1969, pp. 286–287.
3. P.P. Silvester and F.L. Ferrari, *Finite Elements for Electrical Engineers*. New York: Cambridge University Press, 1983, p. 24.
4. R.H. Park, "Definition of an ideal synchronous machine and formulators for armature flux linkages," *Gen. Elect. Rev.*, vol. 31, 1928, pp. 332–334.
5. G. Kron, "Equivalent circuit of the field equations of Maxwell," *Proc. IRE*, May 1944, pp. 289–299.
6. G. Kron, *Equivalent Circuits of Electrical Machinery*. New York: John Wiley, 1951.
7. N. Marcovitz and J. Schwinger, "On the reproduction of the electric and magnetic fields produced by currents and discontinuity in wave guides, I," *J. Appl. Phys.*, vol. 22, no. 6, June 1951, pp. 806–819.
8. J. Schwinger and D.S. Saxon, *Discontinuities in Waveguides*. New York: Gordon and Breach, 1968.
9. P.B. Johns and R.L. Beurle, "Numerical solution of 2-dimensional scattering problems using a transmission line matrix," *Proc. IEEE*, vol. MTT-118, no. 9, Sept. 1971, pp. 1203–1208.
10. W.J.R. Hoefer, "The transmission-line matrix method—Theory and applications," *IEEE Trans. Microwave Theory Tech.*, vol. 33, no. 10, Oct. 1985, pp. 882–893.
11. C. Christopoulos, *The Transmission-Line Modeling Method (TLM)*. New York: IEEE Press, 1995.
12. M.N.O. Sadiku and L.C. Agba, "A simple introduction to the transmission-line modeling," *IEEE Trans. Cir. Sys.*, vol. CAS-37, no. 8, Aug. 1990, pp. 991–999.
13. B.J. Ley, *Computer Aided Analysis and Design for Electrical Engineers*. New York: Holt, Rinehart and Winston, 1970, pp. 815–817.
14. M.N.O. Sadiku, *Elements of Electromagnetics*, 4th Edition. New York: Oxford Univ. Press, 2007.
15. C.C. Wong, "Solution of the network analog of one-dimensional field equations using the ladder method," *IEEE Trans. Educ.*, vol. E-28, no. 3, Aug. 1985, pp. 176–179.

16. C.C. Wong and W.S. Wong, "Multigrid TLM for diffusion problems," *Int. J. Num. Model.*, vol. 2, no. 2, 1989, pp. 103–111.

17. G.E. Marike and G. Yek, "Dynamic three-dimensional T.L.M. analysis of microstrip lines on anisotropic substrate," *IEEE Trans. Micro. Theo. Tech.*, vol. MTT-33, no. 9, Sept. 1985, pp. 789–799.

18. P.B. Johns, "Applications of the transmission-line matrix method to homogeneous waveguides of arbitrary cross-section," *Proc. IEEE*, vol. 119, no. 8, Aug. 1972, pp. 1086–1091.

19. R.A. Waldron, *Theory of Guided Electromagnetic Waves*. London: Van Nostrand Reinhold Co., 1969, pp. 157–172.

20. N.R.S. Simons and E. Bridges, "Method for modeling free space boundaries in TLM situations," *Elect. Lett.*, vol. 26, no. 7, March 1990, pp. 453–455.

21. C. Christopoulos and J.L. Herring, "The application of the transmission-line modeling (TLM) to electromagnetic compatibility problems," *IEEE Trans. Elect. Magn. Comp.*, vol. 35, no. 2, May 1993, pp. 185–191.

22. L.C. Agba, "Transmission-line-matrix modeling of inhomogeneous rectangular waveguides and cavities," *M.Sc. thesis*, Department of Electrical and Computer Engineering, Florida Atlantic University, Boca Raton, Aug. 1987.

23. P.B. Johns, "The solution of inhomogeneous waveguide problems using a transmission line matrix," *IEEE Trans. Micro. Theo. Tech.*, vol. MTT-22, no. 3, Mar. 1974, pp. 209–215.

24. S. Akhtarzad and P.B. Johns, "Generalized elements for t.l.m. method of numerical analysis," *Proc. IEEE*, vol. 122, no. 12, Dec. 1975, pp. 1349–1352.

25. S. Akhtarzad and P.B. Johns, "Numerical solution of lossy waveguides: T.L.M. computer programs," *Elec. Lett.*, vol. 10, no. 15, July 25, 1974, pp. 309–311.

26. S. Akhtarzad and P.B. Johns, "Transmission line matrix solution of waveguides with wall losses," *Elec. Lett.*, vol. 9, no. 15, July 1973, pp. 335–336.

27. S. Akhtarzad and P.B. Johns, "Solution of Maxwell's equations in three space dimensional and time by the T.L.M. method of numerical analysis," *Proc. IEEE*, vol. 122, no. 12, Dec. 1975, pp. 1344–1348.

28. S. Akhtarzad and P.B. Johns, "Three-dimensional transmission-line matrix computer analysis of microstrip resonators," *IEEE Trans. Micro. Theo. Tech.*, vol. MTT-23, no. 12, Dec. 1975, pp. 990–997.

29. Y.C. Shih and W.J.R. Hoefer, "The accuracy of TLM analysis of finned rectangular waveguides," *IEEE Trans. Micro. Theo. Tech.*, vol. MTT-28, no. 7, July 1980, pp. 743–746.

30. N. Yoshida et al., "Transient analysis of two-dimensional Maxwell's equations by Bergeron's method," *Trans. IECE (Japan)*, vol. J62B, June 1979, pp. 511–518.

31. D.A. Al-Mukhtar and T.E. Sitch, "Transmission-line matrix method with irregularly graded space," *IEEE Proc.*, vol. 128, Pt. H, no. 6, Dec. 1981, pp. 299–305.

32. J.A. Morente, J.A. Porti, and M. Khalladi, "Absorbing boundary conditions for the TLM method," *IEEE Trans. Micro. Theo. Tech.*, vol. 40, no. 11, Nov. 1992, pp. 2095–2099.

33. S.C. Pomeroy, G. Zhang, and C. Wykes, "Variable coefficient absorbing boundary condition for TLM," *Elect. Lett.*, vol. 29, no. 13, June 1993, pp. 1198–1200.

34. C. Eswarappa and W.J.R. Hoefer, "One-way equation absorbing boundary conditions for 3-D TLM analysis of planar and quasi-planar structures," *IEEE Trans. Micro. Theo. Tech.*, vol. 42, no. 9, Sept. 1994, pp. 1669–1677.

35. N. Kukutsu and R. Konno, "Super absorption boundary conditions for guided waves in the 3-D TLM simulation," *IEEE Micro. Guided Wave Lett.*, vol. 5, Sept. 1995, pp. 299–301.

36. N. Pena and M.M. Ney, "A new TLM node for Berenger's perfectly matched layer (PML)," *IEEE Micro. Guided Wave Lett.*, vol. 6, Nov. 1996, pp. 410–412.

37. N. Pena and M.M. Ney, "Absorbing-boundary conditions using perfectly matched-layer (PML) technique for three-dimensional TLM simulations," *IEEE Trans. Micro. Theo. Tech.*, vol. 45, no. 10, Oct. 1997, pp. 1749–1755.

38. A. Taflove, *Computational Electrodynamics*. Boston: Artech House, 1995, pp. 189–190.

39. P.B. Johns, "Numerical modeling by the TLM method," in A. Wexler (ed.), *Large Engineering Systems*. Oxford: Pergamon, 1977.

40. T. Itoh (ed.), *Numerical Techniques for Microwave and Millimeterwave Passive Structure*. New York: John Wiley, 1989, pp. 496–591.

41. P.B. Johns, "Simulation of electromagnetic wave interactions by transmission-line modeling (TLM)," *Wave Motion*, vol. 10, no. 6, 1988, pp. 597–610.

42. A.K. Bhattacharyya and R. Garg, "Generalized transmission line model for microstrip patches," *IEEE Proc.*, vol. 132, Pt. H, no. 2, April 1985, pp. 93–98.

43. M. Krumpholz and P. Russer, "A field theoretical derivation of TLM," *IEEE Trans. Micro. Theo. Tech.*, vol. 42, no. 9, Sept. 1994, pp. 1660–1668.

44. P. Naylor and C. Christopoulos, "A comparison between the time-domain and the frequency-domain diakopic methods of solving field problems by transmission-line modeling," *Int. J. Num. Model.*, vol. 2, no. 1, 1989, pp. 17–30.

45. P. Saguet, "The 3-D transmission-line matrix method theory and comparison of the processes," *Int. J. Num. Model.*, vol. 2, 1989, pp. 191–201.

46. W.J.R. Hoefer, "The discrete time domain Green's function or Johns' matrix–a new powerful concept in transmission line modeling (TLM)," *Int. J. Num. Model.*, vol. 2, no. 4, 1989, pp. 215–225.

47. L.R.A.X. de Menezes and W.J.R. Hoefer, "Modeling of general constitutive relationship in SCN TLM," *IEEE Trans. Micro. Theo. Tech.*, vol. 44, no. 6, June 1996, pp. 854–861.

48. V. Trenkic, C. Christopoulos, and T.M. Benson, "Simple and elegant formulation of scattering in TLM nodes," *Elec. Lett.*, vol. 29, no. 18, Sept. 1993, pp. 1651–1652.

49. J.L. Herring and C. Christopoulos, "Solving electromagnetic field problems using a multiple grid transmission-line modeling method," *IEEE Trans. Ant. Prop.*, vol. 42, no. 12, Dec. 1994, pp. 1655–1658.

50. J.A. Porti, J.A. Morente, and M.C. Carrion, "Simple derivation of scattering matrix for TLM nodes," *Elect. Lett.*, vol. 34, no. 18, Sept. 1998, pp. 1763–1764.

51. Y.C. Shih and W.J.R. Hoefer, "Dominant and second-order mode cutoff frequencies in fin lines calculated with a two-dimensional TLM program," *IEEE Trans. Micro. Theo. Tech.*, vol. MTT-28, no. 12, Dec. 1980, pp. 1443–1448.

52. N. Yoshida, I. Fukai, and J. Fukuoka, "Transient analysis of three-dimensional electromagnetic fields by nodal equations," *Trans. Inst. Electron. Comm. Engr. Japan*, vol. J63B, Sept. 1980, pp. 876–883.

53. N. Yoshida and I. Fukai, "Transient analysis of a stripline having corner in three-dimensional space," *IEEE Trans. Micro. Theo. Tech.*, vol. MTT-32, no. 5, May 1984, pp. 491–498.

54. W.R. Zimmerman, "Network analog of Maxwell's field equations in one and two dimensions," *IEEE Trans. Educ.*, vol. E-25, no. 1, Feb. 1982, pp. 4–9.

55. P.B. Johns and M. O'Brien, "Use of the transmission-line modeling (T.L.M.) method to solve non-linear lumped networks," *Radio Electron. Engr.*, vol. 50, no. 1/2, Jan./Feb. 1980, pp. 59–70.

56. J.W. Bandler et al., "Transmission-line modeling and sensitivity evaluation for lumped network simulation and design in the time domain," *J. Franklin Inst.*, vol. 304, no. 1, 1971, pp. 15–23.

57. C.R. Brewitt-Taylor and P.B. Johns, "On the construction and numerical solution of transmission line and lumped network models of Maxwell's equations," *Int. J. Num. Meth. Engr.*, vol. 15, 1980, pp. 13–30.

58. P. Saguet and W.J.R. Hoefer, "The modeling of multiaxial discontinuities in quasi-planar structures with the modified TLM method," *Int. J. Num. Model.*, vol. 1, 1988, pp. 7–17.

59. E.M. El-Sayed and M.N. Morsy, "Analysis of microwave ovens loaded with lossy process materials using the transmission-line matrix method," *Int. J. Num. Meth. Engr.*, vol. 20, 1984, pp. 2213–2220.

60. N.G. Alexopoulos, "Integrated-circuit structures on anisotropic substrates," *IEEE Trans. Micro. Theo. Tech.*, vol. MTT-33, no. 10, Oct. 1985, pp. 847–881.

61. S. Akhatarzad and P.B. Johns, "TLMRES-the TLM computer program for the analysis of microstrip resonators," *IEEE Trans. Micro. Theo. Tech.*, vol. 35, no. 1, Jan. 1987, pp. 60–61.

62. P.B. Johns, "A simple explicit and unconditionally stable numerical routine for the solution of the diffusion equation," *Int. J. Num. Methods Engr.*, vol. 11, 1977, pp. 1307–1328.

63. P.B. Johns and G. Butler, "The consistency and accuracy of the TLM method for diffusion and its relationship to existing methods," *Int. J. Num. Methods Engr.*, vol. 19, 1983, pp. 1549–1554.

64. P.W. Webb, "Simulation of thermal diffusion in transistors using transmission line matrix modeling," *Electron. Comm. Engr. Jour.*, vol. 4, no. 6, Dec. 1992, pp. 362–366.

65. P. Naylor, C. Christopoulos, and P.B. Johns, "Coupling between electromagnetic field and wires using transmission-line modeling," *IEEE Proc., Pt. A*, vol. 134, no. 8, 1987, pp. 679–686.

66. P. Naylor and C. Christopoulos, "A new wire node for modeling thin wires in electromagnetic field problems solved by transmission line modeling," *IEEE Trans. Micro. Theo. Tech.*, vol. 38, no. 3, March 1990, pp. 328–330.

67. J.A. Porti et al., "Comparison of the thin-wire models for TLM method," *Elect. Lett.*, vol. 28, no. 20, Sept. 1992, pp. 1910–1911.

68. P. Russer and J.A. Russer, "Some remarks on the transmission line matrix (TLM) method and its application to transient EM fields and to EMC problems," in I. Ahmed and Z. D. Chen (eds.), *Computational Elecromagnetics – Retrospective and Outlook: In Honor of Wolfgang J. R. Hoefer.* Singapore: Springer, 2015, chapter 2, pp. 29–56.

69. I. Palocz and N. Marcovitz, "A network-oriented approach in the teaching of electromagnetics," *IEEE Trans. Educ.*, vol. E-28, no. 3, Aug. 1985, pp. 150–154.

70. H. Pues and A. Van de Capelle, "Accurate transmission-line model for the rectangular microstrip antenna," *IEEE Proc.*, vol. 131, Pt. H, no. 6, Dec. 1984, pp. 334–340.

71. J.F. Deford and O.P. Gandhi, "An impedance method to calculate currents induced in biological bodies exposed to quasi-static electromagnetic fields," *IEEE Trans. Elect. Comp.*, vol. EMC-27, no. 3, Aug. 1985, pp. 168–173.

72. H.L. Thal, "Exact circuit analysis of spherical waves," *IEEE Trans. Ant. Prog.*, vol. AP-26, no. 2, Mar. 1978, pp. 282–287.

73. A.P. Duffy et al., "New methods for accurate modeling of wires using TLM," *Elect. Lett.*, vol. 29, no. 2, Jan. 1993, pp. 224–226.

74. P. Sewell, Y.K. Choong, and C. Christopoulos, "An accurate thin-wire model for 3-D TLM simulations," *IEEE Transactions on Electromagnetic Compatibility*, vol. 45, no. 2, May 2003, pp. 207–217.

75. M. Zedler, C. Caloz, and P. Russer, "A 3-D isotropic left-handed metamaterial based on the rotated transmission-line matrix (TLM) scheme," *IEEE Transactions on Microwave Theory and Techniques*, vol. 55, no. 12, December 2007, pp. 2930–2941.

76. G.V. Eleftheriades, "Transmission line metamaterials and their relation to the transmission line matrix method," *IEEE MTT-S International Microwave Symposium (IMS)*, 2016, pp. 1–3.

77. H.F.M. Milan and K.G. Gebremedhin, "Triangular node for transmission-line modeling (TLM) applied to bio-heat transfer," *Journal of Thermal Biology*, vol. 62, 2016, pp. 116–122.

78. H.F.M. Milan and K.G. Gebremedhin, "Tetrahedral node for transmission-line modeling (TLM) applied to bioheat transfer," *Computers in Biology and Medicine*, vol. 79, 2016, pp. 243–249.

79. C.V. Jones and D.L. Prior, "Unification of fields and circuit theories of electrical machines," *Proc. IEEE*, vol. 119, no. 7, July 1972, pp. 871–876.

80. P. Hammond and G.J. Rogers, "Use of equivalent circuits in electrical-machine studies," *Proc. IEEE*, vol. 121, no. 6, June 1974, pp. 500–507.

81. E.M. Freeman, "Equivalent circuits from electromagnetic theory: Low-frequency induction devices," *Proc. IEEE*, vol. 121, no. 10, Oct. 1974, pp. 1117–1121.

82. W.J. Karplus and W.W. Soroka, *Analog Methods: Computation and Simulation.* New York: McGraw-Hill, 1959.

83. G.L. Ragan (ed.), *Microwave Transmission Circuits.* New York: McGraw-Hill, 1948, pp. 544–547.

84. R.H. MacNeal, *Electric Circuit Analogies for Elastic Structures*, vol. 2. New York: John Wiley & Sons, 1962.

85. S. Akhtarzad, "Analysis of lossy microwave structures and microstrip resonators by the TLM method," *Ph.D. thesis*, University of Nottingham, England, July 1975.

86. P.B. Johns, "On the relationship between TLM and finite-difference methods for Maxwell's equations," *IEEE Trans. Micro. Theo. Tech.*, vol. MTT-35, no. 1, Jan. 1987, pp. 60, 61.

87. D.H. Choi and W.J.R. Hoefer, "The finite-difference time-domain method and its application to eingenvalue problems," *IEEE Trans. Micro. Theo. Tech.*, vol. 34, no. 12, Dec. 1986, pp. 1464–1472.

88. D.H. Choi, "A comparison of the dispersion characteristics associated with the TLM and FD-TD methods," *Int. J. Num. Model.*, vol. 2, 1989, pp. 203–214.

89. Z. Chen, M. Ney, and W.J.R. Hoefer, "A new finite-difference time-domain formulation and its equivalence with the TLM symmetrical condensed node," *IEEE Trans. Micro. Theo. Tech.*, vol. 39, no. 12, Dec. 1992, pp. 2160–2169.

90. M. Krumpholz and P. Russer, "Two-dimensional FDTD and TLM," *Int. J. Num. Model.*, vol. 7, no. 2, 1994, pp. 141–143.

91. M. Krumpholz and P. Russer, "On the dispersion of TLM and FDTD," *IEEE Trans. Micro. Theo. Tech.*, vol. 42, no. 7, July 1994, pp. 1275–1279.

92. M. Krumpholz, C. Huber, and P. Russer, "A field theoretical comparison of FDTD and TLM," *IEEE Trans. Micro. Theo. Tech.*, vol. 43, no. 8, Aug. 1995, pp. 1935–1950.

93. P. Saguet, *Numerical Analysis in Electromagnetics: The TLM Method*. London, UK: ISTE and John Wiley & Sons, 2012.

8

Monte Carlo Methods

Written in Chinese the word CRISIS is composed of two characters: danger and opportunity.

—J. F. Kennedy

8.1 Introduction

Unlike the deterministic numerical methods covered in the foregoing chapters, Monte Carlo methods (MCMs) are nondeterministic (probabilistic or stochastic) numerical methods employed in solving mathematical and physical problems. The Monte Carlo method (MCM), also known as the *method of statistical trials*, is the marriage of two major branches of theoretical physics: the probabilistic theory of random process dealing with Brownian motion or random-walk experiments and potential theory, which studies the equilibrium states of a homogeneous medium [1]. It is a method of approximately solving problems using sequences of random numbers. It is a means of treating mathematical problems by finding a probabilistic analog and then obtaining approximate answers to this analog by some experimental sampling procedure. The solution of a problem by this method is closer in spirit to physical experiments than to classical numerical techniques.

It is generally accepted that the development of Monte Carlo techniques as we presently use them dates from about 1944, although there are a number of undeveloped instances on much earlier occasions. Credit for the development of MCM goes to a group of scientists, particularly von Neumann and Ulam, at Los Alamos during the early work on nuclear weapons. The groundwork of the Los Alamos group stimulated a vast outpouring of literature on the subject and encouraged the use of MCM for a variety of problems [2–4]. The name "Monte Carlo" comes from the city in Monaco, famous for its gambling casinos.

MCMs are applied in two ways: simulation and sampling. Simulation refers to methods of providing mathematical imitation of real random phenomena. A typical example is the simulation of a neutron's motion into a reactor wall, its zigzag path being imitated by a random walk. Sampling refers to methods of deducing properties of a large set of elements by studying only a small, random subset. For example, the average value of $f(x)$ over $a < x < b$ can be estimated from its average over a finite number of points selected randomly in the interval. This amounts to a MCM of numerical integration. MCMs have been applied successfully for solving differential and integral equations, for finding eigenvalues, for inverting matrices, and particularly for evaluating multiple integrals.

The simulation of any process or system in which there are inherently random components requires a method of generating or obtaining numbers that are random. Examples of such simulation occur in random collisions of neutrons, in statistics, in queueing models, in games of strategy, and in other competitive enterprises. Monte Carlo calculations require having available sequences of numbers which appear to be drawn at random from particular probability distributions.

8.2 Generation of Random Numbers and Variables

Various techniques for generating random numbers are discussed fully in [5–12]. The almost universally used method of generating random numbers is to select a function $g(x)$ that map integers into random numbers. Select x_0 somehow, and generate the next random number as $x_{k+1} = g(x_k)$. The most common function $g(x)$ takes the form

$$g(x) = (ax + c) \bmod m \tag{8.1}$$

where

$$x_0 = \text{starting value or a seed } (x_0 > 0),$$
$$a = \text{multiplier } (a \geq 0),$$
$$c = \text{increment } (c \geq 0),$$
$$m = \text{the modulus}$$

The modulus m is usually 2^t for t-digit binary integers. For a 31-bit computer machine, for example, m may be 2^{31-1}. Here x_0, a, and c are integers in the same range as $m > a$, $m > c$, $m > x_0$. The desired sequence of random numbers $\{x_n\}$ is obtained from

$$\boxed{x_{n+1} = (ax_n + c) \bmod m} \tag{8.2}$$

This is called a *linear congruential sequence*. For example, if $x_0 = a = c = 7$ and $m = 10$, the sequence is

$$7, 6, 9, 0, 7, 6, 9, 0, \ldots \tag{8.3}$$

It is evident that congruential sequences always get into a loop; that is, there is ultimately a cycle of numbers that is repeated endlessly. The sequence in Equation 8.3 has a period of length 4. A useful sequence will of course have a relatively long period. The terms *multiplicative congruential method* and *mixed congruential method* are used by many authors to denote linear congruential methods with $c = 0$ and $c \neq 0$, respectively. Rules for selecting x_0, a, c, and m can be found in References 6,10.

Here, we are interested in generating random numbers from the uniform distribution in the interval (0,1). These numbers will be designated by the letter U and are obtained from Equation 8.2 as

$$U = \frac{x_{n+1}}{m} \tag{8.4}$$

Thus, U can assume values from only the set $\{0, 1/m, 2/m, \ldots, (m-1)/m\}$. (For random numbers in the interval (0,1), a quick test of the randomness is that the mean is 0.5. Other tests can be found in References 3,6.) In MATLAB, a command *rand* generates a random number in the interval (0,1). For generating random numbers X uniformly distributed in the interval (a, b), we use

$$\boxed{X = a + (b - a)U}$$ (8.5)

Random numbers produced by a computer code (using Equations 8.2 and 8.4) are not truly random; in fact, given the seed of the sequence, all numbers U of the sequence are completely predictable. Some authors emphasize this point by calling such computer-generated sequences *pseudorandom numbers*. However, with a good choice of a, c, and m, the sequences of U appear to be sufficiently random in that they pass a series of statistical tests of randomness. They have the advantage over truly random numbers of being generated in a fast way and of being reproducible, when desired, especially for program debugging.

It is usually necessary in a Monte Carlo procedure to generate random variable X from a given probability distribution $F(x)$. This can be accomplished using several techniques [6,13–15] including the *direct method* and *rejection method*.

The direct method, otherwise known as inversion or transform method, entails inverting the cumulative probability function $F(x) = \mathrm{Prob}(X \leq x)$ associated with the random variable X. The fact that $0 \leq F(x) \leq 1$ intuitively suggests that by generating random number U uniformly distributed over (0,1), we can produce a random sample X from the distribution of $F(x)$ by inversion. Thus to generate random X with probability distribution $F(x)$, we set $U = F(x)$ and obtain

$$X = F^{-1}(U)$$ (8.6)

where X has the distribution function $F(x)$. For example, if X is a random variable that is exponentially distributed with mean μ, then

$$F(x) = 1 - e^{-x/\mu}, \quad 0 < x < \infty$$ (8.7)

Solving for X in $U = F(X)$ gives

$$X = -\mu \ln(1 - U)$$ (8.8)

Since $(1 - U)$ is itself a random number in the interval (0,1), we simply write

$$X = -\mu \ln U$$ (8.9)

Sometimes the inverse $F^{-1}(x)$ required in Equation 8.6 does not exist or is difficult to obtain. This situation can be handled using the rejection method. Let $f(x) = dF(x)/dx$ be the probability density function of the random variable X. Let $f(x) = 0$ for $a > x > b$, and $f(x)$ is bounded by M (i.e., $f(x) \leq M$) as shown in Figure 8.1. We generate two random numbers (U_1, U_2) in the interval (0,1). Then

$$X_1 = a + (b - a)U_1 \quad \text{and} \quad f_1 = U_2 M$$ (8.10)

are two random numbers with uniform distributions in (a, b) and $(0, M)$, respectively. If

$$f_1 \leq f(X_1)$$ (8.11)

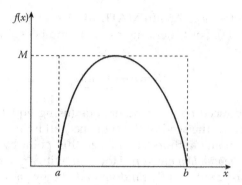

FIGURE 8.1
Rejection method of generating a random variable with probability density function $f(x)$.

then X_1 is accepted as a choice of X, otherwise X_1 is rejected and a new pair (U_1, U_2) is tried again. Thus in the rejection technique all points falling above $f(x)$ are rejected, while those points falling on or below $f(x)$ are utilized to generate X_1 through $X_1 = a + (b - a)U_1$.

EXAMPLE 8.1

Develop a subroutine for generating random number U uniformly distributed between 0 and 1. Using this subroutine, generate random variable Θ with probability distribution given by

$$T(\theta) = \frac{1}{2}(1 - \cos\theta), \quad 0 < \theta < \pi$$

Solution

The subroutine for generating θ is shown in Figure 8.2. MATLAB uses *rand* to produce random numbers uniformly distributed between 0 and 1.

To generate the random variable Θ, set

$$U = T(\Theta) = \frac{1}{2}(1 - \cos\Theta),$$

then

$$\Theta = T^{-1}(U) = \cos^{-1}(1 - 2U)$$

Using this, a sequence of random numbers Θ with the given distribution is generated in the main program of Figure 8.2.

```
for k = 1:100
r = rand;
theta(k) = acos( 1 - 2*r);
end;
```

FIGURE 8.2
Random number generator; for Example 8.1.

8.3 Evaluation of Error

Monte Carlo procedures give solutions which are averages over a number of tests. For this reason, the solutions contain fluctuations about a mean value, and it is impossible to ascribe a 100% confidence in the results. To evaluate the statistical uncertainty in Monte Carlo calculations, we must resort to various statistical techniques associated with random variables. We briefly introduce the concepts of expected value and variance, and utilize the central limit theorem to arrive at an error estimate [13,16].

Suppose that X is a random variable. The *expected value* or *mean value* $\bar{x}$ of X is defined as

$$\bar{x} = \int_{-\infty}^{\infty} x f(x) dx \tag{8.12}$$

where $f(x)$ is the probability density distribution of X. If we draw random and independent samples, $x_1, x_2, ..., x_N$ from $f(x)$, our estimate of x would take the form of the mean of N samples, namely,

$$\hat{x} = \frac{1}{N} \sum_{n=1}^{N} x_n \tag{8.13}$$

While $\bar{x}$ is the true mean value of X, $\hat{x}$ is the unbiased estimator of $\bar{x}$, an unbiased estimator being one with the correct expectation value. Although the expected value of $\hat{x}$ is equal to $\bar{x}$, $\hat{x} \neq \bar{x}$. Therefore, we need a measure of the spread in the values of $\hat{x}$ about $\bar{x}$.

To estimate the spread of values of X, and eventually of $\hat{x}$ about $\bar{x}$, we introduce the *variance* of X defined as the expected value of the square of the deviation of X from $\bar{x}$, that is,

$$\text{Var}(x) = \sigma^2 = \overline{(x-\bar{x})^2} = \int_{-\infty}^{\infty} (x-\bar{x})^2 f(x) dx \tag{8.14}$$

But $(x-\bar{x})^2 = x^2 - 2x\bar{x} + \bar{x}^2$. Hence,

$$\sigma^2(x) = \int_{-\infty}^{\infty} x^2 f(x) dx - 2\bar{x} \int_{-\infty}^{\infty} x f(x) dx + \bar{x}^2 \int_{-\infty}^{\infty} f(x) dx \tag{8.15}$$

or

$$\sigma^2(x) = \overline{x^2} - \bar{x}^2 \tag{8.16}$$

The square root of the variance is called the *standard deviation*, that is,

$$\sigma(x) = \left(\overline{x^2} - \bar{x}^2 \right)^{1/2} \tag{8.17}$$

The standard deviation provides a measure of the spread of x about the mean value $\bar{x}$; it yields the order of magnitude of the error. The relationship between the variance of $\hat{x}$ and the variance of x is

$$\sigma(\hat{x}) = \frac{\sigma(x)}{\sqrt{N}} \tag{8.18}$$

This shows that if we use $\hat{x}$ constructed from N values of x_n according to Equation 8.13 to estimate $\bar{x}$, then the spread in our results of $\hat{x}$ about $\bar{x}$ is proportional to $\sigma(x)$ and falls off as the number of N of samples increases.

In order to estimate the spread in $\hat{x}$, we define the *sample variance*

$$S^2 = \frac{1}{N-1}\sum_{n=1}^{N}(x_n - \hat{x})^2 \tag{8.19}$$

Again, it can be shown that the expected value of S^2 is equal to $\sigma^2(x)$. Therefore, the sample variance is an unbiased estimator of $\sigma^2(x)$. Multiplying out the square term in Equation 8.19, it is readily shown that the *sample standard deviation* is

$$S = \left(\frac{1}{N-1}\right)^{1/2}\left[\frac{1}{N}\sum_{n=1}^{N}x_n^2 - \hat{x}^2\right]^{1/2} \tag{8.20}$$

For large N, the factor $N/(N-1)$ is set equal to 1.

As a way of arriving at the *central limit theorem*, a fundamental result in probability theory, consider the binomial function

$$B(M) = \frac{N!}{M!(N-M)!}p^M q^{N-M} \tag{8.21}$$

which is the probability of M successes in N independent trials. In Equation 8.21, p is the probability of success in a trial, and $q = 1 - p$. If M and $N - M$ are large, we may use *Stirling's formula*

$$n! \sim n^n e^{-n}\sqrt{2\pi n} \tag{8.22}$$

so that Equation 8.21 is approximated [17] as the normal distribution:

$$B(M) \simeq f(\hat{x}) = \frac{1}{\sigma(\hat{x})\sqrt{2\pi}}\exp\left[-\frac{(\hat{x}-\bar{x})^2}{2\sigma^2(\hat{x})}\right] \tag{8.23}$$

where $\bar{x} = Np$ and $\sigma(\hat{x}) = \sqrt{Npq}$. Thus as $N \to \infty$, the central limit theorem states that the probability density function which describes the distribution of $\hat{x}$ that results from N Monte Carlo calculations is the normal distribution $f(\hat{x})$ in Equation 8.23. In other words, the sum of a large number of random variables tends to be normally distributed. Inserting Equation 8.18 into Equation 8.23 gives

$$f(\hat{x}) = \sqrt{\frac{N}{2\pi}}\frac{1}{\sigma(x)}\exp\left[-\frac{N(\hat{x}-\bar{x})^2}{2\sigma^2(x)}\right] \tag{8.24}$$

The normal (or Gaussian) distribution is very useful in various problems in engineering, physics, and statistics. The remarkable versatility of the Gaussian model stems from the central limit theorem. For this reason, the Gaussian model often applies to situations in which the quantity of interest results from the summation of many irregular and fluctuating components. In Example 8.2, we present an algorithm based on central limit theorem for generating Gaussian random variables.

Since the number of samples N is finite, absolute certainty in Monte Carlo calculations is unattainable. We try to estimate some limit or interval around $\bar{x}$ such that we can predict with some confidence that $\hat{x}$ falls within that limit. Suppose we want the probability that $\hat{x}$ lies between $\hat{x} - \delta$ and $\bar{x} + \delta$. By definition,

$$\text{Prob}\{\bar{x} - \delta < \hat{x} < \bar{x} + \delta\} = \int_{\bar{x}-\delta}^{\bar{x}+\delta} f(\hat{x}) d\hat{x} \tag{8.25}$$

By letting

$$\lambda = \frac{(\hat{x} - \bar{x})}{\sqrt{2/N}\sigma(x)},$$

$$\text{Prob}\{\bar{x} - \delta < \hat{x} < \bar{x} + \delta\} = \frac{2}{\sqrt{\pi}} \int_{0}^{(\sqrt{N/2})(\delta/\sigma)} e^{-\lambda^2} d\lambda$$

$$= \text{erf}\left(\sqrt{N/2}\frac{\delta}{\sigma(x)}\right) \tag{8.26a}$$

or

$$\text{Prob}\left\{\bar{x} - z_{\alpha/2}\frac{\sigma}{\sqrt{N}} \leq \hat{x} \leq \bar{x} + z_{\alpha/2}\frac{\sigma}{\sqrt{N}}\right\} = 1 - \alpha \tag{8.26b}$$

where erf(x) is the error function and $z_{\alpha/2}$ is the upper $\alpha/2 \times 100$ percentile of the standard normal deviation. Equation 8.26 may be interpreted as follows: if the Monte Carlo procedure of taking random and independent observations and constructing the associated random interval $\bar{x} \pm \delta$ is repeated for large N, approximately erf $(\sqrt{N/2}\delta/\sigma(x)) \times 100$ percent of these random intervals will contain $\hat{x}$. The random interval $\hat{x} \pm \delta$ is called a *confidence interval* and erf $(\sqrt{N/2}\delta/\sigma(x))$ is the *confidence level*. Most Monte Carlo calculations use error $\delta = \sigma(x)/\sqrt{N}$, which implies that $\hat{x}$ is within one standard deviation of $\bar{x}$, the true mean. From Equation 8.26, it means that the probability that the sample mean $\hat{x}$ lies within the interval $\hat{x} \pm \sigma(x)/\sqrt{N}$ is 0.6826 or 68.3%. If higher confidence levels are desired, two or three standard deviations may be used. For example,

$$\text{Prob}\left(\bar{x} - M\frac{\sigma(x)}{\sqrt{N}} < \hat{x} < \bar{x} + M\frac{\sigma(x)}{\sqrt{N}}\right) = \begin{cases} 0.6826, & M=1 \\ 0.954, & M=2 \\ 0.997, & M=3 \end{cases} \tag{8.27}$$

where M is the number of standard deviations.

In Equations 8.26 and 8.27, it is assumed that the population standard deviation σ is known. Since this is rarely the case, σ must be estimated by the sample standard deviation S calculated from Equation 8.20 so that the normal distribution is replaced by the Student's t-distribution. It is well known that the t-distribution approaches the normal distribution as N becomes large, say $N > 30$. Equation 8.26 is equivalent to

$$\text{Prob}\left(\bar{x} - \frac{St_{\alpha/2;N-1}}{\sqrt{N}} \le \hat{x} \le \bar{x} + \frac{St_{\alpha/2;N-1}}{\sqrt{N}}\right) = 1 - \alpha \tag{8.28}$$

where $t_{\alpha/2;N-1}$ is the upper $100 \times \alpha/2$ percentage point of the Student's t-distribution with $(N-1)$ degrees of freedom; and its values are listed in any standard statistics text. Thus, the upper and lower limits of a confidence interval are given by

$$\text{upper limit} = \bar{x} + \frac{St_{\alpha/2;N-1}}{\sqrt{N}} \tag{8.29}$$

$$\text{lower limit} = \bar{x} - \frac{St_{\alpha/2;N-1}}{\sqrt{N}} \tag{8.30}$$

For further discussion on error estimates in Monte Carlo computations, consult References 18,19.

EXAMPLE 8.2

A random variable X with Gaussian (or normal) distribution is generated using the central limit theorem. According to the central limit theorem, the sum of a large number of independent random variables about a mean value approaches a Gaussian distribution regardless of the distribution of the individual variables. In other words, for any random numbers, Y_i, $i = 1, 2, ..., N$ with mean $\bar{Y}$ and variance $\text{Var}(Y)$,

$$Z = \frac{\sum_{i=1}^{N} Y_i - N\bar{Y}}{\sqrt{N}\,\text{Var}(Y)} \tag{8.31}$$

converges asymptotically with N to a normal distribution with zero mean and a standard deviation of unity. If Y_i are uniformly distributed variables (i.e., $Y_i = U_i$), then $\bar{Y} = 1/2$, $\text{Var}(Y) = 1/\sqrt{12}$, and

$$Z = \frac{\sum_{i=1}^{N} U_i - N/2}{\sqrt{N/12}} \tag{8.32}$$

and the variable

$$X = \sigma Z + \mu \tag{8.33}$$

approximates the normal variable with mean μ and variance σ^2. A value of N as low as 3 provides a close approximation to the familiar bell-shaped Gaussian distribution. To ease computation, it is a common practice to set $N = 12$ since this choice eliminates the square root term in Equation 8.32. However, this value of N truncates the distribution at $\pm 6\sigma$ limits and is unable to generate values beyond 3σ. For simulations in which the tail of the distribution is important, other schemes for generating Gaussian distribution must be used [20–22].

Thus, to generate a Gaussian variable X with mean μ and standard deviation σ, we follow these steps:

1. Generate 12 uniformly distributed random numbers $U_1, U_2, ..., U_{12}$.

2. Obtain $Z = \sum\limits_{i=1}^{12} U_i - 6$.

3. Set $X = \sigma Z + \mu$.

In MATLAB, normal variable X is generated using command *randn*.

8.4 Numerical Integration

For one-dimensional integration, several quadrature formulas, such as presented in Section 3.11, exist. The numbers of such formulas are relatively few for multidimensional integration. It is for such multidimensional integrals that a Monte Carlo technique becomes valuable for at least two reasons. The quadrature formulas become very complex for multiple integrals while the MCM remains almost unchanged. The convergence of Monte Carlo integration is independent of dimensionality, which is not true for quadrature formulas. The statistical method of integration has been found to be an efficient way to evaluate two- or three-dimensional integrals in antenna problems, particularly those involving very large structures [23]. Two types of Monte Carlo integration procedures, the crude MCM and the MCM with antithetic variates, will be discussed. For other types, such as hit-or-miss and control variates, see References 24,25. The application of MCM to improper integrals will be covered briefly.

8.4.1 Crude Monte Carlo Integration

Suppose we wish to evaluate the integral

$$I = \int_R f \tag{8.34}$$

where R is an n-dimensional space. Let $\mathbf{X} = (X^1, X^2, ..., X^n)$ be a random variable that is uniformly distributed in R. Then $f(\mathbf{X})$ is a random variable whose mean value is given by [26,27]

$$\overline{f(\mathbf{X})} = \frac{1}{|R|} \int_R f = \frac{1}{|R|} \tag{8.35}$$

and the variance by

$$\mathrm{Var}(f(\mathbf{X})) = \frac{1}{|R|} \int_R f^2 - \left(\frac{1}{|R|} \int_R f \right)^2 \tag{8.36}$$

where

$$|R| = \int_R d\mathbf{X} \tag{8.37}$$

If we take N independent samples of $\mathbf{X}$, that is, $\mathbf{X}_1$, $\mathbf{X}_2$, ..., $\mathbf{X}_N$, all having the same distribution as $\mathbf{X}$ and form the average

$$\frac{f(\mathbf{X}_1) + f(\mathbf{X}_2) + \cdots + f(\mathbf{X}_N)}{N} = \frac{1}{N} \sum_{i=1}^{N} f(\mathbf{X}_i) \tag{8.38}$$

we might expect this average to be close to the mean of $f(\mathbf{X})$. Thus, from Equations 8.35 and 8.38,

$$\boxed{I = \frac{|R|}{N} \sum_{i=1}^{N} f(\mathbf{X}_i)} \tag{8.39}$$

This Monte Carlo formula applies to any integration over a finite region R. For the purpose of illustration, we now apply Equation 8.39 to one- and two-dimensional integrals.

For a one-dimensional integral, suppose

$$I = \int_a^b f(x)dx \tag{8.40}$$

Applying Equation 8.39 yields

$$\boxed{I = \frac{b-a}{N} \sum_{i=1}^{N} f(X_i)} \tag{8.41}$$

where X_i is a random number in the interval (a, b), that is,

$$X_i = a + (b-a)U, \quad 0 < U < 1 \tag{8.42}$$

For a two-dimensional integral

$$I = \int_a^b \int_c^d f(X^1, X^2)dX^1 dX^2, \tag{8.43}$$

the corresponding Monte Carlo formula is

$$\boxed{I = \frac{(b-a)(d-c)}{N} \sum_{i=1}^{N} f(X_i^1, X_i^2)} \tag{8.44}$$

where

$$\begin{aligned}
X_i^1 &= a + (b-a)U^1, \ 0 < U^1 < 1 \\
X_i^2 &= c + (d-c)U^2, \ 0 < U^2 < 1
\end{aligned} \tag{8.45}$$

The convergence behavior of the unbiased estimator I in Equation 8.39 is slow since the variance of the estimator is of the order $1/N$. Accuracy and convergence is increased by reducing the variance of the estimator using an improved method, the method of antithetic variates.

8.4.2 Monte Carlo Integration with Antithetic Variates

The term *antithetic variates* [28] is used to describe any set of estimators which mutually compensate each other's variations. For convenience, we assume that the integral is over the interval (0,1). Suppose we want an estimator for the single integral

$$I = \int_0^1 g(U)dU \tag{8.46}$$

We expect the quantity $1/2[g(U) + g(1 - U)]$ to have smaller variance than $g(U)$. If $g(U)$ is too small, then $g(1 - U)$ will have a good chance of being too large and conversely. Therefore, we define the estimator

$$I = \frac{1}{N}\sum_{i=1}^{N}\frac{1}{2}[g(U_i) + g(1 - U_i)] \tag{8.47}$$

where U_i are random numbers between 0 and 1. The variance of the estimator is of the order $1/N^4$, a tremendous improvement over Equation 8.39. For two-dimensional integral,

$$I = \int_0^1 \int_0^1 g(U^1, U^2)dU^1 dU^2, \tag{8.48}$$

and the corresponding estimator is

$$I = \frac{1}{N}\sum_{i=1}^{N}\frac{1}{4}[g(U_i^1, U_i^2) + g(U_i^1, 1 - U_i^2)$$
$$+ g(1 - U_i^1, U_i^2) + g(1 - U_i^1, 1 - U_i^2)] \tag{8.49}$$

Following similar lines, the idea can be extended to higher-order integrals. For intervals other than (0, 1), transformations such as in Equations 8.41 through 8.45 should be applied. For example,

$$\int_a^b f(x)dx = (b - a)\int_0^1 g(U)dU$$

$$\simeq \frac{b - a}{N}\sum_{i=1}^{N}\frac{1}{2}[g(U_i) + g(1 - U_i)] \tag{8.50}$$

where $g(U) = f(X)$ and $X = a + (b - a)U$. It is observed from Equations 8.47 and 8.49 that as the number of dimensions increases, the minimum number of antithetic variates per dimension required to obtain an increase in efficiency over crude Monte Carlo also increases. Thus, the crude MCM becomes preferable in many dimensions. Monte Carlo integration technique has been applied in the moment method solution of integral equations [29,30].

8.4.3 Improper Integrals

The integral

$$I = \int_0^\infty g(x)dx \tag{8.51}$$

may be evaluated using Monte Carlo simulations. For a random variable X having probability density function $f(x)$, where $f(x)$ integrates to 1 on interval $(0, \infty)$,

$$\int_0^\infty \frac{g(x)}{f(x)}dx = \int_0^\infty g(x)dx \tag{8.52}$$

Hence, to compute I in Equation 8.51, we generate N independent random variables distributed according to a probability density function $f(x)$ integrating to 1 on the interval $(0, \infty)$. The sample mean

$$\overline{g(x)} = \frac{1}{N}\sum_{i=1}^{N} \frac{g(x_i)}{f(x_i)} \tag{8.53}$$

gives an estimate for I.

EXAMPLE 8.3

Evaluate the integral

$$I = \int_0^1 \int_0^{2\pi} e^{j\alpha\rho\cos\phi}\rho\, d\rho\, d\phi$$

using the MCM.

Solution

This integral represents radiation from a circular aperture-antenna with a constant amplitude and phase distribution. It is selected because it forms at least part of every radiation integral. The solution is available in the closed form, which can be used to assess the accuracy of the Monte Carlo results. In closed form,

$$I(\alpha) = \frac{2\pi J_1(\alpha)}{\alpha}$$

where $J_1(\alpha)$ is Bessel's function of the first order.

A simple program for evaluating the integral employing Equations 8.44 and 8.45, where $a = 0$, $b = 1$, $c = 0$, and $d = 2\pi$, is shown in Figure 8.3. For different values of N, both the crude and antithetic variate MCMs are used in evaluating the radiation integral, and the results are compared with the exact value in Table 8.1 for $\alpha = 5$. In applying Equation 8.49, the following correspondences are used:

$$U_1 \equiv X^1, \ U^2 \equiv X^2, \ 1-U^1 \equiv b - X^1 = (b-a)(1-U^1),$$
$$1-U^2 \equiv d - X^2 = (d-c)(1-U^2)$$

0000000000000000000000000000000000

```
% Integration using crude Monte Carlo
% and antithetic methods
%
% Only few lines need be changed to use this
% program for any multi-dimensional integration
%
% the function fun.m is to be on a separate file

a =0; b= 1.0; c = 0;  % limits of integration
d=2*pi;
alpha = 5;
nrun = 10000;
sum1 = 0; sum2 = 0;
for i=1:nrun
    u1 = rand;
    u2 = rand;
    x1 = a + (b-a)*u1;
    x2 = c + (d-c)*u2;
    x3 = (b-a)*(1-u1);
    x4 = (d-c)*(1-u2);
    sum1 = sum1 + fun(x1,x2);
    sum2 = sum2 + fun(x1,x2) + fun(x1,x4) + fun(x3,x2) + fun(x3,x4);
end
area1 = (b-a)*(d-c)*sum1/nrun
area2 = (b-a)*(d-c)*sum2/(4*nrun)
(a)


function y=fun(rho,phi)
alpha = 5;
y=rho*exp(j*alpha*rho*cos(phi));

(b)
```

FIGURE 8.3
MATLAB program for Monte Carlo evaluation of a two-dimensional integral: (a) main program, (b) function fun.m kept in a separate file.

TABLE 8.1

Results of Example 8.3 on Monte Carlo Integration of Radiation Integral

N	Crude MCM	Antithetic Variates MCM
500	$-0.2892 - j0.0742$	$-0.2887 - j0.0585$
1000	$-0.5737 + j0.0808$	$-0.4982 - j0.0080$
2000	$-0.4922 - j0.0040$	$-0.4682 - j0.0082$
4000	$-0.3999 - j0.0345$	$-0.4216 - j0.0323$
6000	$-0.3608 - j0.0270$	$-0.3787 - j0.0440$
8000	$-0.4327 - j0.0378$	$-0.4139 - j0.0241$
10,000	$-0.4229 - j0.0237$	$-0.4121 - j0.0240$
	Exact: $-0.4116 + j0$.	

8.5 Solution of Potential Problems

The connection between potential theory and Brownian motion (or random walk) was first shown in 1944 by Kakutani [31]. Since then the resulting so-called probabilistic potential theory has been applied to problems in many disciplines such as heat conduction [32–37], electrostatics [38–45], and electrical power engineering [46,47]. An underlying concept of the probabilistic or Monte Carlo solution of differential equations is the random walk. Different types of random walk lead to different MCMs. The most popular types are the *fixed-random walk* and *floating random walk*. Other types that are less popular include the *Exodus method, shrinking boundary method, inscribed figure method,* and the *surface density method.*

8.5.1 Fixed Random Walk

Suppose, for concreteness, that the MCM with fixed random walk is to be applied to solve Laplace's equation

$$\nabla^2 V = 0 \text{ in region } R \tag{8.54a}$$

subject to Dirichlet boundary condition

$$V = V_p \text{ on boundary } B \tag{8.54b}$$

We begin by dividing R into mesh and replacing ∇^2 by its finite difference equivalent. The finite difference representation of Equation 8.54a in two-dimensional R is given by Equation 3.31, namely,

$$\boxed{\begin{aligned} V(x,\ y) = p_{x+}V(x+\Delta,\ y) + p_{x-}V(x-\Delta,\ y) \\ + p_{y+}V(x,\ y+\Delta) + p_{y-}V(x,\ y-\Delta) \end{aligned}} \tag{8.55a}$$

where

$$\boxed{p_{x+} = p_{x-} = p_{y+} = p_{y-} = \frac{1}{4}} \tag{8.55b}$$

In Equation 8.55, a square grid of mesh size Δ, such as in Figure 8.4, is assumed. The equation may be given a probabilistic interpretation. If a random walking particle is instantaneously at the point (x, y), it has probabilities $p_{x+}, p_{x-}, p_{y+},$ and p_{y-} of moving from (x, y) to $(x + \Delta, y), (x - \Delta, y), (x, y + \Delta),$ and $(x, y - \Delta)$, respectively. A means of determining which way the particle should move is to generate a random number U, $0 < U < 1$ and instruct the particle to walk as follows:

$$\boxed{\begin{aligned} (x,y) &\rightarrow (x+\Delta,y) & \text{if } 0 < U < 0.25 \\ (x,y) &\rightarrow (x-\Delta,y) & \text{if } 0.25 < U < 0.5 \\ (x,y) &\rightarrow (x,y+\Delta) & \text{if } 0.5 < U < 0.75 \\ (x,y) &\rightarrow (x,y-\Delta) & \text{if } 0.75 < U < 1 \end{aligned}} \tag{8.56}$$

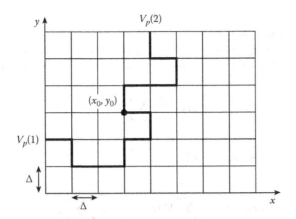

FIGURE 8.4
Configuration for fixed random walks.

If a rectangular grid rather than a square grid is employed, then $p_{x+} = p_{x-}$ and $p_{y+} = p_{y-}$, but $p_x \neq p_y$. Also for a three-dimensional problem in which cubical cells are used, $p_{x+} = p_{x-} = p_{y+} = p_{y-} = p_{z+} = p_{z-} = 1/6$. In both cases, the interval $0 < U < 1$ is subdivided according to the probabilities.

To calculate the potential at (x_0, y_0), a random-walking particle is instructed to start at that point. The particle proceeds to wander from node to node in the grid until it reaches the boundary. When it does, the walk is terminated and the prescribed potential V_p at that boundary point is recorded. Let the value of V_p at the end of the first walk be denoted by $V_p(1)$, as illustrated in Figure 8.4. Then a second particle is released from (x_0, y_0) and allowed to wander until it reaches a boundary point, where the walk is terminated and the corresponding value of V_p is recorded as $V_p(2)$. This procedure is repeated for the third, fourth, ..., and Nth particle released from (x, y), and the corresponding prescribed potential $V_p(3)$, $V_p(4)$, ..., $V_p(N)$ are noted. According to Kakutani [31], the expected value of $V_p(1)$, $V_P(2)$, ..., $V_p(N)$ is the solution of the Dirichlet problem at (x, y), that is,

$$V(x_0, y_0) = \frac{1}{N} \sum_{i=1}^{N} V_p(i) \qquad (8.57)$$

where N, the total number of walks, is large. The rate of convergence varies as $\sqrt{N}$ so that many random walks are required to ensure accurate results.

If it is desired to solve Poisson's equation

$$\nabla^2 V = -g(x, y) \text{ in } R \qquad (8.58a)$$

subject to

$$V = V_p \text{ on } B, \qquad (8.58b)$$

then the finite difference representation is in Equation 3.30, namely,

$$V(x, y) = p_{x+} V(x + \Delta, y) + p_{x-} V(x - \Delta, y)$$
$$+ p_{y+} V(x, y + \Delta) + p_{y-} V(x, y - \Delta) + \frac{\Delta^2 g}{4} \qquad (8.59)$$

where the probabilities remain as stated in Equation 8.55b. The probabilistic interpretation of Equation 8.59 is similar to that for Equation 8.55. However, the term $\Delta^2 g/4$ in Equation 8.59 must be recorded at each step of the random walk. If m_i steps are required for the ith random walk originating at (x_0, y_0) to reach the boundary, then one records

$$V_p(i) + \frac{\Delta^2}{4} \sum_{j=1}^{m_i-1} g(x_j, y_j) \tag{8.60}$$

Thus, the Monte Carlo result for $V(x_0, y_0)$ is

$$V(x_0, y_0) = \frac{1}{N} \sum_{i=1}^{N} V_p(i) + \frac{\Delta^2}{4N} \sum_{i=1}^{N} \left[\sum_{j=1}^{m_i-1} g(x_j, y_j) \right] \tag{8.61}$$

An interesting analogy to the MCM just described is the walking drunk problem [15,34]. We regard the random-walking particle as "drunk," the squares of the mesh as the "blocks in a city," the nodes as "crossroads," the boundary B as the "city limits," and the terminus on B as the "policeman." Though the drunk is trying to walk home, he is so intoxicated that he wanders randomly throughout the city. The job of the policeman is to seize the drunk in his first appearance at the city limits and ask him to pay a fine V_p. What is the expected fine the drunk will receive? The answer to this problem is in Equation 8.57.

On the dielectric boundary, the boundary condition $D_{1n} = D_{2n}$ is imposed. Consider the interface along $y = $ constant plane as shown in Figure 8.5. According to Equation 3.53, the finite difference equivalent of the boundary condition at the interface is

$$V_o = p_{x+}V_1 + p_{x-}V_2 + p_{y+}V_3 + p_{y-}V_4 \tag{8.62a}$$

where

$$p_{x+} = p_{x-} = \frac{1}{4}, \quad p_{y+} = \frac{\epsilon_1}{2(\epsilon_1 + \epsilon_2)}, \quad p_{y-} = \frac{\epsilon_2}{2(\epsilon_1 + \epsilon_2)} \tag{8.62b}$$

An interface along $x = $ constant plane can be treated in a similar manner.

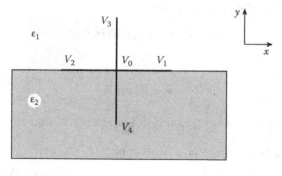

FIGURE 8.5
Interface between media of dielectric permittivities ϵ_1 and ϵ_2.

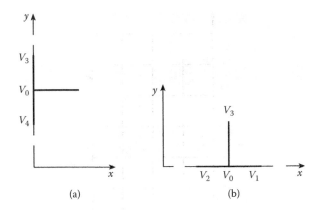

FIGURE 8.6
Satisfying symmetry conditions: (a) $\partial V/\partial x = 0$, (b) $\partial V/\partial y = 0$.

On a line of symmetry, the condition $\partial V/\partial n = 0$ must be imposed. If the line of symmetry is along the y-axis as in Figure 8.6a, according to Equation 3.55.

$$V_o = p_{x+}V_1 + p_{y+}V_3 + p_{y-}V_4 \tag{8.63a}$$

where

$$p_{x+} = \frac{1}{2}, \quad p_{y+} = p_{y-} = \frac{1}{4} \tag{8.63b}$$

The line of symmetry along the x-axis, shown in Figure 8.6b, is treated similarly following Equation 3.56.

For an axisymmetric solution region such as shown in Figure 8.7, $V = V(\rho, z)$. The finite difference equivalent of Equation 8.54a for $\rho \neq 0$ is obtained in Section 3.10 as

$$\begin{aligned} V(\rho, z) = {} & p_{\rho+}V(\rho + \Delta, z) + p_{\rho-}V(\rho - \Delta, z) \\ & + p_{z+}V(\rho, z + \Delta) + p_{z-}V(\rho, z - \Delta) \end{aligned} \tag{8.64}$$

where $\Delta\rho = \Delta z = \Delta$ and the random walk probabilities are given by

$$\begin{aligned} p_{z+} &= p_{z-} = \frac{1}{4} \\ p_{\rho+} &= \frac{1}{4} + \frac{\Delta}{8\rho} \\ p_{\rho-} &= \frac{1}{4} - \frac{\Delta}{8\rho} \end{aligned} \tag{8.65}$$

For $\rho = 0$, the finite difference equivalent of Equation 8.54a is Equation 3.120, namely

$$V(0, z) = p_{\rho+}V(\Delta, z) + p_{z+}V(0, z + \Delta) + p_{z-}V(0, z - \Delta) \tag{8.66}$$

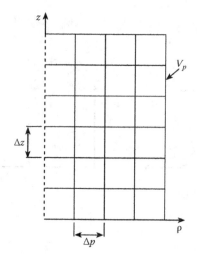

FIGURE 8.7
Typical axisymmetric solution region.

so that

$$p_{\rho+} = \frac{4}{6}, \quad p_{\rho-} = 0, \quad p_{z+} = \frac{1}{6} = p_{z-} \tag{8.67}$$

The random-walking particle is instructed to begin to walk at (ρ_o, z_o). It wanders through the mesh according to the probabilities in Equations 8.65 and 8.67 until it reaches the boundary where it is absorbed and the prescribed potential $V_p(1)$ is recorded. By sending out N particles from (ρ_o, z_o) and recording the potential at the end of each walk, we obtain the potential at (ρ_o, z_o) as [48]

$$V(\rho_o, z_o) = \frac{1}{N} \sum_{i=1}^{N} V_p(i) \tag{8.68}$$

This MCM is called *fixed random walk* type since the step size Δ is fixed and the steps of the walks are constrained to lie parallel to the coordinate axes. Unlike in the finite difference method (FDM), where the potentials at all mesh points are determined simultaneously, MCM is able to solve for the potential at one point at a time. One disadvantage of this MCM is that it is slow if potential at many points is required and is therefore recommended for solving problems for which only a few potentials are required. It shares a common difficulty with FDM in connection with irregularly shaped bodies having Neumann boundary conditions. This drawback is fully removed by employing MCM with floating random walk.

8.5.2 Floating Random Walk

The mathematical basis of the floating random walk method is the mean value theorem of potential theory. If S is a sphere of radius r, centered at (x, y, z), which lies wholly within region R, then

$$V(x, y, z) = \frac{1}{4\pi a^2} \int_S V(r') dS' \tag{8.69}$$

That is, the potential at the center of any sphere within R is equal to the average value of the potential taken over its surface. When the potential varies in two dimensions, $V(x, y)$ is given by

$$V(x,y) = \frac{1}{2\pi\rho} \oint_L V(\rho')dl' \tag{8.70}$$

where the integration is around a circle of radius ρ centered at (x, y). It can be shown that Equations 8.69 and 8.70 follow from Laplace's equation. Also, Equations 8.69 and 8.70 can be written as

$$V(x,y,z) = \int_0^1 \int_0^1 V(a, \theta, \phi)dF\, dT \tag{8.71}$$

$$V(x,y) = \int_0^1 V(a, \phi)dF \tag{8.72}$$

where

$$F = \frac{\phi}{2\pi}, \quad T = \frac{1}{2}(1 - \cos\theta) \tag{8.73}$$

and θ and ϕ are regular spherical coordinate variables. The functions F and T may be interpreted as the probability distributions corresponding to ϕ and θ. While $dF/d\phi = $ constant, $dT/d\theta = 1/2 \sin\theta$; that is, all angles ϕ are equally probable, but the same is not true for θ.

The floating random walk MCM depends on the application of Equations (8.69) and (8.70) in a statistical sense. For a two-dimensional problem, suppose that a random-walking particle is at some point (x_j, y_j) after j steps in the ith walk. The next $(j + 1)$th step is taken as follows. First, a circle is constructed with center at (x_j, y_j) and radius ρ_j, which is equal to the shortest distance between (x_j, y_j) and the boundary. The ϕ coordinate is generated as a random variable uniformly distributed over $(0, 2\pi)$, that is, $\phi = 2\pi U$, where $0 < U < 1$. Thus, the location of the random-walking particle after the $(j + 1)$th step is illustrated in Figure 8.8 and given as

$$\boxed{x_{j+1} = x_j + \rho_j \cos\phi_j} \tag{8.74a}$$
$$\boxed{y_{j+1} = y_j + \rho_j \sin\phi_j} \tag{8.74b}$$

The next random walk is executed by constructing a circle centered at (x_{j+1}, y_{j+1}) and of radius ρ_{j+1}, which is the shortest distance between (x_{j+1}, y_{j+1}) and the boundary. This procedure is repeated several times, and the walk is terminated when the walk approaches some prescribed small distance τ of the boundary. The potential $V_p(i)$ at the end of this ith walk is recorded as in fixed random walk MCM and the potential at (x, y) is eventually determined after N walks using Equation 8.57.

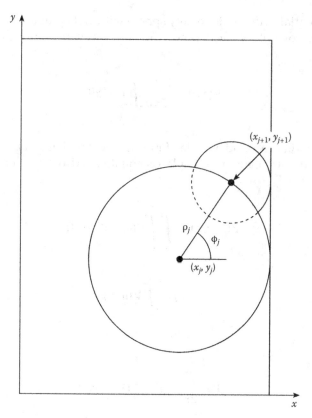

FIGURE 8.8
Configuration for floating random walks.

The floating random walk MCM can be applied to a three-dimensional Laplace problem by proceeding along lines similar to those outlined above. A random-walking particle at (x_j, y_j, z_j) will step to a new location on the surface of a sphere whose radius r_j is equal to the shortest distance between point (x_j, y_j, z_j) and the boundary. The ϕ coordinate is selected as a random number U between 0 and 1, multiplied by 2π. The coordinate θ is determined by selecting another random number U between 0 and 1 and solving for $\theta = \cos^{-1}(1 - 2U)$ as in Example 8.1. Thus, the location of the particle after its $(j + 1)$th step in the ith walk is

$$x_{j+1} = x_j + r_j \cos \phi_j \sin \theta_j \tag{8.75a}$$
$$y_{j+1} = y_j + r_j \sin \phi_j \sin \theta_j \tag{8.75b}$$
$$z_{j+1} = z_j + r_j \cos \theta_j \tag{8.75c}$$

Finally, we apply Equation 8.57.

Solving Poisson's equation (8.58) for a two-dimensional problem requires only a slight modification. For a three-dimensional problem, $V(a, \theta, \phi)$ in Equation 8.71 is replaced by $[V(a, \theta, \phi) + r^2 g/6]$. This requires that the term $gr_j^2/6$ at every jth step of the ith random walk be recorded.

An approach for handling a discretely inhomogeneous medium is presented in References 38,42,43,49.

It is evident that in the floating random walk MCM, neither the step sizes nor the directions of the walk are fixed in advance. The quantities may be regarded as "floating" and hence the designation *floating random walk*. A floating random walk bypasses many intermediate steps of a fixed random walk in favor of a long jump. Fewer steps are needed to reach the boundary, and so computation is much more rapid than in fixed random walk.

8.5.3 Exodus Method

The *Exodus method*, first suggested in Reference 50 and developed for electromagnetics in References 51,52, does not employ random numbers and is generally faster and more accurate than the fixed random walk. It basically consists of dispatching numerous walkers (say 10^6) simultaneously in directions controlled by the random walk probabilities of going from one node to its neighbors. As these walkers arrive at new nodes, they are dispatched according to the probabilities until a set number (say 99.999%) have reached the boundaries. The advantage of the Exodus method is its independence of the random number generator.

To implement the Exodus method, we first divide the solution region R into mesh, such as in Figure 8.4. Suppose p_k is the probability that a random walk starting from point (x, y) ends at node k on the boundary with prescribed potential $V_p(k)$. For M boundary nodes (excluding the corner points since a random walk never terminates at those points), the potential at the starting point (x_0, y_0) of the random walks is

$$\boxed{V(x_0, y_0) = \sum_{k=1}^{M} p_k V_p(k)}$$

(8.76)

If m is the number of different boundary potentials ($m = 4$ in Figure 8.4), Equation 8.76 can be simplified to

$$V(x_0, y_0) = \sum_{k=1}^{m} p_k V_p(k)$$

(8.77)

where p_k in this case is the probability that a random walk terminates on boundary k. Since $V_p(k)$ is specified, our problem is reduced to finding p_k. We find p_k using the Exodus method in a manner similar to the iterative process applied in Section 3.5.

Let $P(i, j)$ be the number of particles at point (i, j) in R. We begin by setting $P(i, j) = 0$ at all points (both fixed and free) except at point (x_0, y_0), where $P(i, j)$ assumes a large number N (say, $N = 10^6$ or more). By a scanning process, we dispatch the particles at each free node to its neighboring nodes according to the probabilities $p_{x+}, p_{x-}, p_{y+},$ and p_{y-} as illustrated in Figure 8.9. Note that in Figure 8.9b, new $P(i, j) = 0$ at that node, while old $P(i, j)$ is shared among the neighboring nodes. When all the free nodes in R are scanned as illustrated in Figure 8.9, we record the number of particles that have reached the boundary (i.e., the fixed nodes). We keep scanning the mesh until a set number of particles (say 99.99% of N) have

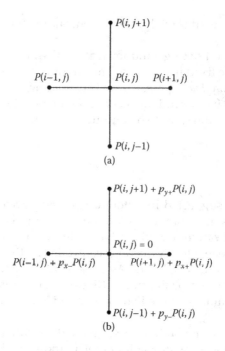

FIGURE 8.9
(a) Before the particles at (i, j) are dispatched, (b) after the particles at (i, j) are dispatched.

reached the boundary, where the particles are absorbed. If N_k is the number of particles that reached side k, we calculate

$$p_k = \frac{N_k}{N} \tag{8.78}$$

Hence, Equation 8.77 can be written as

$$V(x, y) = \frac{\sum_{k=1}^{m} N_k V_p(k)}{N} \tag{8.79}$$

Thus the problem is reduced to just finding N_k using the Exodus method, given N and $V_p(k)$. We notice that if $N \rightarrow \infty$, $\Delta \rightarrow 0$, and all the particles were allowed to reach the boundary points, the values of p_k and consequently $V(x, y)$ would be exact. It is easier to approach this exact solution using the Exodus method than any other MCMs or any other numerical techniques such as difference and finite element methods.

We now apply the Exodus method to Poisson's equation (see Equation 8.58a). To compute the solution of the problem defined in Equation 8.58, for example, at a specific point (x_o, y_o), we need the *transition probability* p_k and the *transient probability* q_ℓ. The transition probability p_k is already defined as the probability that a random walk starting at the point of interest (x_o, y_o) in R ends at a boundary point (x_k, y_k), where potential $V_p(k)$ is prescribed, that is,

$$p_k = Prob(x_o, y_o \rightarrow x_k, y_k) \tag{8.80}$$

The transient probability ql is the probability that a random walk starting at point (x_o, y_o) passes through point (x_ℓ, y_ℓ) on the way to the boundary, that is,

$$g_\ell = \text{Prob}\left(x_o, y_o \xrightarrow{x_\ell, y_\ell} \text{boundary } B\right) \tag{8.81}$$

If there are M_b boundary (or fixed) nodes (excluding the corner points since a random walk never terminates at those points) and M_f free nodes in the mesh, the potential at the starting point (x_o, y_o) of the random walks is

$$V(x_o, y_o) = \sum_{k=1}^{M_b} p_k V_p(k) + \sum_{\ell=1}^{M_f} q_l G_\ell, \tag{8.82}$$

where

$$G_\ell = \Delta^2 g(x_\ell, y_\ell)/4$$

If m_b is the number of different boundary potentials, the first term in Equation 8.82 can be simplified so that

$$V(x_o, y_o) = \sum_{k=1}^{m} p_k V_p(k) + \sum_{\ell=1}^{M_f} q_\ell G_\ell, \tag{8.83}$$

where p_k in this case is the probability that a random walk terminates on boundary k. Since $V_p(k)$ is specified and the source term G_ℓ is known, our problem is reduced to finding the probabilities p_k and q_ℓ. We notice from Equation 8.83 that the value of $V(x_o, y_o)$ would be "exact" if the transition probabilities p_k and the transient probabilities q_ℓ were known exactly. These probabilities can be obtained in one of two ways: either analytically or numerically. The analytical approach involves using an expansion technique described in Reference 53. But this approach is limited to homogeneous rectangular solution regions. For inhomogeneous or non-rectangular regions, we must resort to some numerical simulation. The Exodus method offers a numerical means of finding p_k and q_ℓ. The fixed random walk can also be used to compute the transient and transition probabilities.

To apply the Exodus method, let $P(i, j)$ be the number of particles at point (i, j) in R, while $Q(i, j)$ is the number of particles passing through the same point. We begin the application of the Exodus method by setting $P(i, j) = 0 = Q(i, j)$ at all nodes (both fixed and free) except at free node (x_o, y_o) where both $P(i, j)$ and $Q(i, j)$ are set equal to a large number N_p (say $N_p = 10^6$ or more). In other words, we inject a large number of particles at (x_o, y_o) to start with. By scanning the mesh iteratively as is usually done in finite difference analysis, we dispatch the particles at each free node to its neighboring nodes according to the random walk probabilities $p_{x+}, p_{x-}, p_{y+},$ and p_{y-} as illustrated in Figure 8.9. Note that in Figure 8.9b, new $P(i, j) = 0$ at that node, while old $P(i, j)$ is shared among the neighboring nodes. As shown in Figure 8.10, the value of $Q(i, j)$ does not change at that node, while Q at the neighboring nodes is increased by the old $P(i, j)$ that is shared by those nodes. While $P(i, j)$ keeps records of the number of particles at point (i, j) during each iteration, $Q(i, j)$ tallies the number of particles passing through that point.

At the end of each iteration (i.e., scanning of the free nodes in R as illustrated in Figures 8.9 and 8.10), we record the number of particles that have reached the boundary (i.e., the

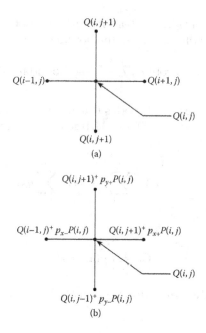

FIGURE 8.10
Number of particles passing through node (i, j) and its neighboring nodes: (a) before the particles at the node are dispatched, (b) after the particles at the node are dispatched.

fixed nodes) where the particles are absorbed. We keep scanning the mesh in a manner similar to the iterative process applied in finite difference solution until a set number of particles (say 99.99% of N_p) have reached the boundary. If N_k is the number of particles that reached boundary k, we calculate

$$p_k = \frac{N_k}{N_p} \tag{8.84}$$

Also, at each free node, we calculate

$$q_\ell = \frac{Q_\ell}{N_p}, \tag{8.85}$$

where $Q_\ell = Q(i, j)$ is now the total number of particles that have passed through that node on their way to the boundary. Hence Equation 8.83 can be written as

$$V(x_o, y_o) = \frac{\sum\limits_{k=1}^{m} N_k V_p(k)}{N_p} + \frac{\sum\limits_{\ell=1}^{M_f} Q_\ell G_\ell}{N_p} \tag{8.86}$$

Thus the problem is reduced to just finding N_k and Q_ℓ using the Exodus method, given N_p, $V_p(k)$, and G_ℓ. If $N_p \to \infty$, $\Delta \to 0$, and all the particles were allowed to reach the boundary points, the values of p_k and q_ℓ and consequently $V(x_o, y_o)$ would be exact. It is interesting to note that the accuracy of the Exodus method does not really depend on the number of particles N_p. The accuracy depends on the step size Δ and the number of iteration or the tolerance, the number of particles (say 0.001% of N_p), which are yet to reach the boundary before the iteration is terminated. However, a large value of N_p reduces the truncation error in the computation.

EXAMPLE 8.4

Give a probabilistic interpretation using the finite difference form of the energy equation

$$u\frac{\partial T}{\partial x}+v\frac{\partial T}{\partial y}=\alpha\left(\frac{\partial^2 T}{\partial x^2}+\frac{\partial^2 T}{\partial y^2}\right)$$

Assume a square grid of size Δ.

Solution

Applying a backward difference to the left-hand side and a central difference to the right-hand side, we obtain

$$u\frac{T(x,y)-T(x-\Delta,y)}{\Delta}+v\frac{T(x,y)-T(x,y-\Delta)}{\Delta}$$
$$=\alpha\frac{T(x+\Delta,y)-2T(x,y)+T(x-\Delta,y)}{\Delta^2}$$
$$+\alpha\frac{T(x,y+\Delta)-2T(x,y)+T(x,y-\Delta)}{\Delta^2} \tag{8.87}$$

Rearranging terms leads to

$$T(x,\,y)=p_{x+}T(x+\Delta,\,y)+\,p_{x-}T(x-\Delta,\,y)$$
$$+p_{y+}T(x,\,y+\Delta)+\,p_{y-}T(x,\,y-\Delta) \tag{8.88}$$

where

$$p_{x+}=p_{y+}=\frac{1}{\dfrac{u\Delta}{\alpha}+\dfrac{v\Delta}{\alpha}+4} \tag{8.89a}$$

$$p_{x-}=\frac{\left(1+\dfrac{\Delta u}{\alpha}\right)}{\dfrac{u\Delta}{\alpha}+\dfrac{v\Delta}{\alpha}+4} \tag{8.89b}$$

$$p_{y-}=\frac{\left(1+\dfrac{\Delta v}{\alpha}\right)}{\dfrac{u\Delta}{\alpha}+\dfrac{v\Delta}{\alpha}+4} \tag{8.89c}$$

Equation 8.88 is given probabilistic interpretation as follows: a walker at point (x, y) has probabilities p_{x+}, p_{x-}, p_{y+}, and p_{y-} of moving to point $(x + \Delta, y)$, $(x − \Delta, y)$, $(x, y + \Delta)$, and $(x, y − \Delta)$, respectively. With this interpretation, Equation 8.88 can be used to solve the differential equation with fixed random MCM.

EXAMPLE 8.5

Consider a conducting trough of infinite length with square cross section shown in Figure 8.11. The trough wall at $y = 1$ is connected to 100 V, while the other walls are grounded as shown. We intend to find the potential within the trough using the fixed random walk MCM.

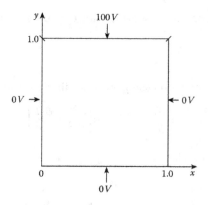

FIGURE 8.11
For Example 8.5.

Solution

The problem is solving Laplace's equation subject to

$$V(0, y) = V(1, y) = V(x, 0) = 0, \quad V(x, 1) = 100 \tag{8.90}$$

The exact solution obtained by the method of separation of variables is given in Equation 2.31, namely,

$$V(x, y) = \frac{400}{\pi} \sum_{n=0}^{\infty} \frac{\sin k\pi x \sinh k\pi y}{k \sinh k\pi}, \quad k = 2n + 1 \tag{8.91}$$

To apply the fixed random MCM, the flowchart in Figure 8.12 was developed. Based on the flowchart, the program of Figure 8.13 was developed. A built-in command *rand* was

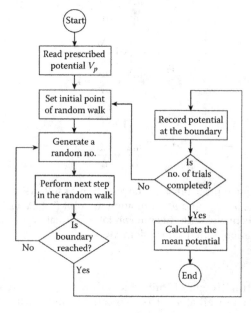

FIGURE 8.12
Flowchart for random walk of Example 8.5.

```
%  Using fixed random walk MCM to solve a potential
%  problem involving Laplace's equation; for Example 8.5

nrun = 2000;   % no. of runs, N
delta = 0.05;  % step size
A=1.0; B=1.0;
v1=0; v2=0; v3=100; v4=0;
xo=0.5;
yo=0.5;
io=xo/delta;      % location at which U is to be determined
jo=yo/delta;
imax=A/delta;
jmax=B/delta;
sum=0;
ms = 0;          % no. of steps before reaching the boundary
m1 = 0;          % no. of walks terminating at v1
m2 = 0;          % no. of walks terminating at v2
m3 = 0;          % no. of walks terminating at v3
m4 = 0;          % no. of walks terminating at v4
for k=1:nrun
    i=io;
    j=jo;
    while i<=imax & j<=jmax
        ms = ms + 1;
    r=rand; %random number between 0 and 1
    if (r >= 0.0 & r <= 0.25)
       i=i+1;
    end
    if (r >= 0.25 & r <= 0.5)
       i=i-1;
     end
     if (r  >= 0.5 & r <= 0.75)
        j=j+1;
     end
     if (r >= 0.75 & r <= 1.0)
        j=j-1;
     end
      % check if (i,j) is on the boundary
      if(i == 0)
         sum=sum+ v4;
         m4 = m4 +1;
         break;
      end
      if(i == imax)
         sum=sum+ v2;
          m2 = m2 +1;
          break;
      end
      if(j == 0)
              sum=sum+ v1;
              m1 = m1 + 1;
            break;
          end
      if(j == jmax)
              sum=sum+ v3;
              m3 = m3 + 1;
            break;
          end
        end % while
    end
    v=sum/nrun
    m=ms/nrun
```

FIGURE 8.13
MATLAB code for Example 8.5.

TABLE 8.2

Results of Example 8.5

x	y	N	$\bar{m}$	Monte Carlo Solution	Exact Solution
0.25	0.75	250	66.20	42.80	43.20
		500	69.65	41.80	
		750	73.19	41.60	
		1000	73.95	41.10	
		1250	73.67	42.48	
		1500	73.39	42.48	
		1750	74.08	42.67	
		2000	74.54	43.35	
0.5	0.5	250	118.62	21.60	25.00
		500	120.00	23.60	
		750	120.27	25.89	
		1000	120.92	25.80	
		1250	120.92	25.92	
		1500	120.78	25.27	
		1750	121.50	25.26	
		2000	121.74	25.10	
0.75	0.25	250	64.82	7.60	6.797
		500	68.52	6.60	
		750	68.56	6.93	
		1000	70.17	7.50	
		1250	72.12	8.00	
		1500	71.78	7.60	
		1750	72.40	7.43	
		2000	72.40	7.30	

used to generate random numbers U uniformly distributed between 0 and 1. The step size Δ was selected as 0.05. The results of the potential computation are listed in Table 8.2 for three different locations. The average number of random steps $\bar{m}$ taken to reach the boundary is also shown. It is observed from Table 8.2 that it takes a large number of random steps for a small step size and that the error in MCM results can be less than 1%.

Rather than using Equation 8.57, an alternative approach of determining $V(x_0, y_0)$ is to calculate the probability of a random walk terminating at a grid point located on the boundary. The information is easily extracted from the program used for obtaining the results in Table 8.2. To illustrate the validity of this approach, the potential at (0.25, 0.75) was calculated. For $N = 1000$ random walks, the number of walks terminating at $x = 0$, $x = 1$, $y = 0$ and $y = 1$ is 461, 62, 66, and 411, respectively. Hence, according to Equation 8.79

$$V(x_0, y_0) = \frac{461}{1000}(0) + \frac{62}{1000}(0) + \frac{66}{1000}(0) + \frac{411}{1000}(100) = 41.1 \tag{8.92}$$

The statistical error in the simulation can be found. In this case, the potential on the boundary takes values 0 or $V_o = 100$ so that $V(x_0, y_0)$ has a binomial distribution with mean $V(x_0, y_0)$ and variance

$$\sigma^2 = \frac{V(x,y)[V_o - V(x_0, y_0)]}{N} \tag{8.93}$$

At point $(0.5, 0.5)$, for example, $N = 1000$ gives $\sigma = 1.384$ so that at 68% confidence interval, the error is $\delta = \sigma / \sqrt{N} = 0.04375$.

EXAMPLE 8.6

Use the floating random walk MCM to determine the potential at points $(1.5, 0.5)$, $(1.0, 1.5)$, and $(1.5, 2.0)$ in the two-dimensional potential system in Figure 8.14.

Solution

To apply the floating random walk, we use the flowchart in Figure 8.12 except that we apply Equation 8.74 instead of Equation 8.57 at every step in the random walk. A program based on the modified flowchart was developed. The shortest distance ρ from (x, y) to the boundary was found by dividing the solution region in Figure 8.14 into three rectangles and checking

if $\{(x, y): 1 < x < 2, 0 < y < 1\}$, $\quad \rho = \text{minimum}\{x - 1, 2 - x, y\}$
if $\{(x, y): 0 < x < 1, 1 < y < 2.5\}$, $\quad \rho = \text{minimum}\{x, y - 1, 2.5 - y\}$
if $\{(x, y): 1 < x < 2, 1 < y < 2.5\}$,

$$\rho = \text{minimum}\left\{2 - x, 2.5 - y, \sqrt{(x-1)^2 + (y-1)^2}\right\}$$

A prescribed tolerance $\tau = 0.05$ was selected so that if the distance between a new point in the random walk and the boundary is less than τ, it is assumed that the boundary is reached and the potential at the closest boundary point is recorded.

Table 8.3 presents the Monte Carlo result with the average number of random steps $\bar{m}$. It should be observed that it takes fewer walks to reach the boundary in floating random walk than in fixed random walk. Since no analytic solution exists, we compare Monte Carlo results with those obtained using finite difference with $\Delta = 0.05$ and 500 iterations.

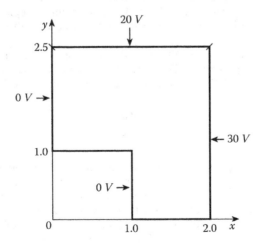

FIGURE 8.14
For Example 8.6.

TABLE 8.3

Results of Example 8.6

x	y	N	$\bar{m}$	Monte Carlo Solution ($V \pm \delta$)	Finite Difference Solution (V)
1.5	0.5	250	6.738	11.52 ± 0.8973	11.44
		500	6.668	11.80 ± 0.9378	
		750	6.535	11.83 ± 0.4092	
		1000	6.476	11.82 ± 0.6205	
		1250	6.483	11.85 ± 0.6683	
		1500	6.465	11.72 ± 0.7973	
		1750	6.468	11.70 ± 0.6894	
		2000	6.460	11.55 ± 0.5956	
1.0	1.5	250	8.902	10.74 ± 0.8365	10.44
		500	8.984	10.82 ± 0.3709	
		750	8.937	10.75 ± 0.5032	
		1000	8.928	10.90 ± 0.7231	
		1250	8.836	10.84 ± 0.7255	
		1500	8.791	10.93 ± 0.5983	
		1750	8.788	10.87 ± 0.4803	
		2000	8.811	10.84 ± 0.3646	
1.5	2.0	250	7.242	21.66 ± 0.7509	21.07
		500	7.293	21.57 ± 0.5162	
		750	7.278	21.53 ± 0.3505	
		1000	7.316	21.53 ± 0.2601	
		1250	7.322	21.53 ± 0.3298	
		1500	7.348	21.51 ± 0.3083	
		1750	7.372	21.55 ± 0.2592	
		2000	7.371	21.45 ± 0.2521	

As evidenced in Table 8.3, the Monte Carlo results agree well with the finite difference results even with 1000 walks. Also, by dividing the solution region into 32 elements, the finite element results at points (1.5, 0.5), (1.0, 1.5), and (1.5, 2.0) are 11.265, 9.788, and 21.05 V, respectively.

Unlike the program in Figure 8.13, where the error estimates are not provided for the sake of simplicity, the program in Figure 8.15 incorporates evaluation of error estimates in the Monte Carlo calculations. Using Equation 8.29, the error is calculated as

$$\delta = \frac{St_{\alpha/2;n-1}}{\sqrt{n}}$$

In the program in Figure 8.15, the number of trials n (the same of N in Section 8.3) is taken as 5 so that $t_{\alpha/2;n-1} = 2.776$. The sample variance S is calculated using Equation 8.19. The values of δ are also listed in Table 8.3. Notice that unlike in Table 8.2, where $\bar{m}$ and V are the mean values after N walks, $\bar{m}$ and V in Table 8.3 are the mean values of n trials, each of which involves N walks, that is, the "mean of the mean" values. Hence, the results in Table 8.3 should be regarded as more accurate than those in Table 8.2.

```
% Using floating random walk MCM
% to solve Laplace's equation; for Example 8.6

nrun = 500;
ntrials = 5;   % no. of trials
tol=0.005;     % tolerance
xo = 1.5;      % location at which potential
yo = 0.5;       % is to be determined
for n=1:ntrials
sum = 0.0;
m=0;  % no. of steps before reaching the boundary
for k=1:nrun
   x=xo;
   y=yo;
 while x >=0 & x <=2  & y >=0 & y<=2.5
      u=rand;  % generate a random no. and move to the next point
      phi=2.0*pi*u;
      % find the shortest distance r;
      rc = sqrt( (x-1)^2 + (y-1)^2 );
      if (x >1 & x < 2 & y >0 & y < 1)
         rx = min(x-1,2-x);
         r = min(rx,y);
      end
      if (x >0 & x < 1 & y >1 & y < 2.5)
         rx = min(x,y-1);
         r = min(rx,2.5-y);
      end
      if (x >1 & x < 2 & y >1 & y < 2.5)
         rx = min(2-x,2.5-y);
         r = min(rx,rc);
      end
      x=x+r*cos(phi);
      y=y+r*sin(phi);
      m=m+1;
      % check if (x,y) is on the boundary
      if ( x < (1+tol) & y < (1+tol) ) % corner
         sum=sum +0;
         break;
      end
        if ( x >= (2-tol) )
         sum=sum +30;
         break;
      end
      if ( y >= (2.5-tol) )
         sum=sum +20;
         break;
      end
      if ( x < tol & y >1 & y < 2.5)
         sum=sum +0;
         break;
      end
    if ( x > tol & x < (1-tol) & y <= (1+tol)   )
           sum=sum +0;
           break;
       end
```

FIGURE 8.15

MATLAB code for Example 8.6.   *(Continued)*

```
         if ( x < (1+tol) & y >= tol & y <= 1 )
            sum=sum +0;
            break;
            end
         if ( x >= (1+tol) & x <= (2.5-tol)  & y < tol )
            sum=sum +0;
            break;
            end
      end %while
   end % nrun
   vv(n)=sum/nrun;
   steps(n)=m/nrun;
   end % ntrials
   % find the mean value of V and mean no. of steps
   sum1 = 0.0;
   sum2 = 0.0;
   for n=1:ntrials
      sum1 = sum1 + vv(n);
      sum2 = sum2 + steps(n);
   end
   vmean=sum1/ntrials
   stepm=sum2/ntrials
   % calculate error
   sum3=0.0;
   for n=1:ntrials
      sum3 = sum3 + ( vv(n) - vmean )^2;
   end
   std=sqrt( sum3/(ntrials-1) );
   error = std*2.776/sqrt(ntrials)
```

FIGURE 8.15 (Continued)
MATLAB code for Example 8.6.

EXAMPLE 8.7

Apply the Exodus method to solve the potential problem shown in Figure 8.16. The potentials at $x = 0$, $x = a$, and $y = 0$ sides are zero while the potential at $y = b$ sides is V_o. Typically, let

$$V_o = 100, \quad \epsilon_1 = \epsilon_o, \quad \epsilon_2 = 2.25\epsilon_o, \quad a = 3.0, \quad b = 2.0, \quad c = 1.0$$

Solution

The analytic solution to this problem using series expansion technique discussed in Section 2.7 is

$$V = \begin{cases} \displaystyle\sum_{k=1}^{\infty} \sin\beta x [a_n \sinh\beta y + b_n \cosh\beta y], & 0 \leq y \leq c \\[2em] \displaystyle\sum_{k=1}^{\infty} c_n \sin\beta x \sinh\beta y, & c \leq y \leq b \end{cases} \tag{24}$$

where

$$\beta = \frac{n\pi}{a}, \quad n = 2k - 1$$

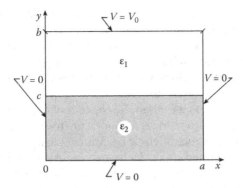

FIGURE 8.16
Potential system for Example 8.7.

TABLE 8.4

Results of Example 8.7

x	y	Exodus Method V	Fixed Random Walk ($V \pm \delta$)	Finite Difference V	Exact Solution V
0.5	1.0	13.41	13.40 ± 1.113	13.16	13.41
1.0	1.0	21.13	20.85 ± 1.612	20.74	21.13
1.5	1.0	23.43	23.58 ± 1.2129	22.99	23.43
1.5	0.5	10.52	10.13 ± 0.8789	10.21	10.52
1.5	1.5	59.36	58.89 ± 2.1382	59.06	59.34

$$a_n = 4V_o\left[\epsilon_1 \tanh\beta c - \epsilon_2 \coth\beta c\right]/d_n,$$
$$b_n = 4V_o(\epsilon_2 - \epsilon_1)/d_n,$$
$$c_n = 4V_o\left[\epsilon_1 \tanh\beta c - \epsilon_2 \coth\beta c + (\epsilon_2 - \epsilon_1)\coth\beta c\right]/d_n, \tag{25}$$
$$d_n = n\pi\sinh\beta b\left[\epsilon_1 \tanh\beta c - \epsilon_2 \coth\beta c + (\epsilon_2 - \epsilon_1)\coth\beta b\right]$$

The potentials were calculated at five typical points using the Exodus method, the fixed random walk MCM, and the analytic solution. The number of particles, N, was taken as 10^7 for the Exodus method and the step size $\Delta = 0.05$ was used. For the fixed random walk method, $\Delta = 0.05$ and 2000 walks were used. It was noted that 2000 walks were sufficient for the random walk solutions to converge. The results are displayed in Table 8.4. In the table, δ is the error estimate, which is obtained by repeating each computation five times and using statistical formulas provided in Reference 13. It should be noted from the table that the results of the Exodus method agree to four significant places with the exact solution. Thus, the Exodus method is more accurate than the random walk technique. It should also be noted that the Exodus method does not require the use of a random number routine and also the need of calculating the error estimate. The Exodus method, therefore, takes less computation time than the random walk method.

8.6 Markov Chain Regional MCM

A major limitation inherent with the standard MCMs discussed above is that they only permit single-point calculations. In view of this limitation, several techniques have been proposed for using Monte Carlo for whole-field computation. The popular ones are the *shrinking boundary method* [36] and *inscribed figure method* [37].

The shrinking boundary method is similar to the regular fixed random walk except that once the potential at an interior point is calculated, that point is treated as a boundary or absorbing point. This way, the random walking particles will have more points to terminate their walks and the walking time is reduced.

The inscribed figure method is based on the concept of subregion calculation. It involves dividing the solution region into standard shapes or inscribing standard shapes into the region. (By standard shapes is meant circles, squares, triangles, rectangles, etc. for which Green's function can be obtained analytically or numerically.) Then, an MCM is used in computing potential along the dividing lines between the shapes and the regions that have nonstandard shapes. Standard analytical methods are used to compute the potential in the subregions.

Both the shrinking boundary method and the inscribed figure method do not make MCMs efficient for whole-field calculation. They still require point-by-point calculations and a number of large tests as standard Monte Carlo techniques. Therefore, they offer no significant advantage over the standard MCMs. Using Markov chains for whole-field computations has been found to be more efficient than the shrinking boundary method and the inscribed figure method. Markov chains are named after Andrey Markov, a Russian mathematician, who invented them. The technique basically calculates the transition probabilities using absorbing Markov chains [54,55].

A Markov chain is a sequence of random variables $X^{(0)}$, $X^{(1)}$, ..., where the probability distribution for $X^{(n)}$ is determined entirely by the probability distribution of $X^{(n-1)}$. A Markov process is a type of random process that is characterized by the memoryless property [56–59]. It is a process evolving in time that remembers only the most recent past and whose conditional distributions are time invariant. Markov chains are mathematical models of this kind of process. The Markov chains of interest to us are *discrete-state* and *discrete-time*. In our case, the Markov chain is the random walk and the states are the grid nodes. The transition probability P_{ij} is the probability that a random-walking particle at node i moves to node j. It is expressed by the Markov property

$$P_{ij} = P(x_{n+1} = j \,|\, x_o, x_1, ..., x_n)$$
$$= P(x_{n+1} = j \,|\, x_n) \quad j \in \mathbf{X}, n = 0, 1, 2, ... \tag{8.94}$$

The Markov chain is characterized by its transition probability matrix **P**, defined by

$$\mathbf{P} = \begin{bmatrix} P_{00} & P_{01} & P_{02} & \cdots \\ P_{10} & P_{11} & P_{12} & \cdots \\ P_{20} & P_{21} & P_{22} & \cdots \\ \vdots & \vdots & \vdots & \cdots \end{bmatrix} \tag{8.95}$$

P is a stochastic matrix, meaning that the sum of the elements in each row is unity, that is,

$$\sum_{j \in \mathbf{X}} P_{ij} = 1 \quad i \in \mathbf{X} \tag{8.96}$$

We may also use the state transition diagram as a way of representing the evolution of a Markov chain. An example for a three-state Markov chain is shown in Figure 8.17.

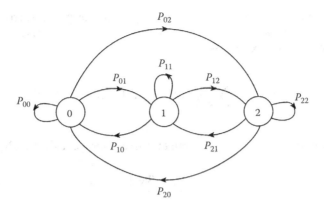

FIGURE 8.17
State transition diagram for a three-state Markov chain.

If we assume that there are n_f free (or nonabsorbing) nodes and n_p fixed (prescribed or absorbing) nodes, the size of the transition matrix $\mathbf{P}$ is n, where

$$n = n_f + n_p \tag{8.97}$$

If the absorbing nodes are numbered first and the nonabsorbing states are numbered last, the $n \times n$ transition matrix becomes

$$\mathbf{P} = \begin{bmatrix} \mathbf{I} & \mathbf{0} \\ \mathbf{R} & \mathbf{Q} \end{bmatrix} \tag{8.98}$$

where the $n_f \times n_p$ matrix $\mathbf{R}$ represents the probabilities of moving from nonabsorbing nodes to absorbing ones; the $n_f \times n_f$ matrix $\mathbf{Q}$ represents the probabilities of moving from one nonabsorbing node to another; $\mathbf{I}$ is the identity matrix representing transitions between the absorbing nodes ($P_{ii} = 1$ and $P_{ij} = 0$); and $\mathbf{0}$ is the null matrix showing that there are no transitions from absorbing to nonabsorbing nodes. For the solution of Laplace's equation, we obtain the elements of $\mathbf{Q}$ from Equation 8.55b as

$$Q_{ij} = \begin{cases} \dfrac{1}{4}, & \text{if } i \text{ is directly connected to } j, \\ 0, & \text{if } i = j \text{ or } i \text{ is not directly connected to } j \end{cases} \tag{8.99}$$

The same applies to R_{ij} except that j is an absorbing node.

For any absorbing Markov chain, $\mathbf{I} - \mathbf{Q}$ has an inverse. This is usually referred to as the fundamental matrix

$$\mathbf{N} = (\mathbf{I} - \mathbf{Q})^{-1} \tag{8.100}$$

where N_{ij} is the average number of times the random-walking particle starting from node i passes through node j before being absorbed. The absorption probability matrix $\mathbf{B}$ is

$$\mathbf{B} = \mathbf{NR} \tag{8.101}$$

where B_{ij} is the probability that a random-walking particle originating from a non-absorbing node i will end up at the absorbing node j. **B** is an $n_f \times n_p$ matrix and is stochastic like the transition probability matrix, that is,

$$\sum_{j=1}^{n_p} B_{ij} = 1, \quad i = 1, 2, \ldots, n_f \tag{8.102}$$

If $\mathbf{V}_f$ and $\mathbf{V}_p$ contain potentials at the free and fixed nodes, respectively, then

$$\boxed{\mathbf{V}_f = \mathbf{B}\mathbf{V}_p} \tag{8.103}$$

In terms of the prescribed potentials V_1, V_2, ..., V_{np}, Equation 8.103 becomes

$$V_i = \sum_{j=1}^{n_p} B_{ij} V_j, \quad i = 1, 2, \ldots, n_f \tag{8.104}$$

where V_i is the potential at any free node i. Unlike Equation 8.57, Equation 8.103 or Equation 8.104 provides the solution at all the free nodes at once.

An alternative way to obtain the solution in Equation 8.103 is to exploit a property of the transition probability matrix **P**. When **P** is multiplied by itself repeatedly for a large number of times, we obtain

$$\lim_{n \to \infty} \mathbf{P}^n = \begin{bmatrix} \mathbf{I} & 0 \\ \mathbf{B} & 0 \end{bmatrix} \tag{8.105}$$

Thus,

$$\begin{bmatrix} \mathbf{V}_p \\ \mathbf{V}_f \end{bmatrix} = \mathbf{P}^n \begin{bmatrix} \mathbf{V}_p \\ \mathbf{V}_f \end{bmatrix} = \begin{bmatrix} \mathbf{I} & 0 \\ \mathbf{B} & 0 \end{bmatrix} \begin{bmatrix} \mathbf{V}_p \\ \mathbf{V}_f \end{bmatrix} \tag{8.106}$$

Either Equation 8.103 or Equation 8.106 can be used to find $\mathbf{V}_f$ but it is evident that Equation 8.103 will be more efficient and accurate. From Equation 8.103 or Equation 8.104, it should be noticed that if **N** is calculated accurately, the solution is "exact."

There are several other procedures for whole-field computation [36,37,60–63]. One technique involves using Green's function in the floating random walk [41].

The random walk MCMs and the MCM method applied to elliptic PDEs in this chapter can be applied to parabolic PDEs as well [64–70].

The following two examples will corroborate Markov chain MCM. The first example requires no computer programming and can be done by hand, while the second one needs computer programming.

EXAMPLE 8.8

Rework Example 8.5 using Markov chain. The problem is shown in Figure 8.18. We wish to determine the potential at points $(a/3, a/3)$, $(a/3, 2a/3)$, $(2a/3, a/3)$, and $(2a/3, 2a/3)$. Although we may assume that $a = 1$, that is not necessary.

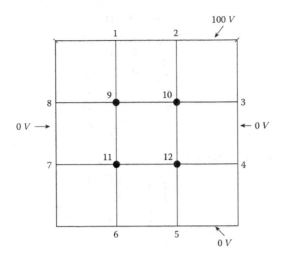

FIGURE 8.18
For Example 8.8.

Solution

In this case, there are four free nodes ($n_f = 4$) and eight fixed nodes ($n_p = 8$) as shown in Figure 8.18. The transition probability matrix is obtained by inspection as

$$
\mathbf{P} = \begin{array}{c} \\ 1 \\ 2 \\ 3 \\ 4 \\ 5 \\ 6 \\ 7 \\ 8 \\ 9 \\ 10 \\ 11 \\ 12 \end{array}
\begin{array}{c} \begin{array}{cccccccccccc} 1 & 2 & 3 & 4 & 5 & 6 & 7 & 8 & 9 & 10 & 11 & 12 \end{array} \\
\left[\begin{array}{cccccccccccc}
1 & & & & & & & & & & & \\
& 1 & & & & & & & & & & \\
& & 1 & & & & & & & & & \\
& & & 1 & & & & & & & & \\
& & & & 1 & & & & & & & \\
& & & & & 1 & & & & & & \\
& & & & & & 1 & & & & & \\
\frac{1}{4} & & & & & & & \frac{1}{4} & 0 & \frac{1}{4} & \frac{1}{4} & 0 \\
0 & \frac{1}{4} & \frac{1}{4} & & & & & & \frac{1}{4} & 0 & 0 & \frac{1}{4} \\
0 & 0 & 0 & & & \frac{1}{4} & \frac{1}{4} & & \frac{1}{4} & & & \frac{1}{4} \\
0 & & & \frac{1}{4} & \frac{1}{4} & & & 0 & 0 & 0 & \frac{1}{4} & \frac{1}{4} & 0
\end{array} \right]
\end{array}
$$

Other entries in **P** shown vacant are zeros.
From **P**, we obtain

$$
\mathbf{R} = \begin{array}{c} \\ 9 \\ 10 \\ 11 \\ 12 \end{array}
\begin{array}{c} \begin{array}{cccccccc} 1 & 2 & 3 & 4 & 5 & 6 & 7 & 8 \end{array} \\
\left[\begin{array}{cccccccc}
\frac{1}{4} & 0 & 0 & 0 & 0 & 0 & 0 & \frac{1}{4} \\
0 & \frac{1}{4} & \frac{1}{4} & 0 & 0 & 0 & 0 & 0 \\
0 & 0 & 0 & 0 & 0 & \frac{1}{4} & \frac{1}{4} & 0 \\
0 & 0 & 0 & \frac{1}{4} & \frac{1}{4} & 0 & 0 & 0
\end{array} \right]
\end{array}
$$

$$
\begin{array}{c}
\begin{array}{cccc} \quad 9 & 10 & 11 & 12 \end{array} \\
\mathbf{Q} = \begin{array}{c} 9 \\ 10 \\ 11 \\ 12 \end{array}\left[\begin{array}{cccc}
0 & \dfrac{1}{4} & \dfrac{1}{4} & 0 \\[8pt]
\dfrac{1}{4} & 0 & 0 & \dfrac{1}{4} \\[8pt]
\dfrac{1}{4} & 0 & 0 & \dfrac{1}{4} \\[8pt]
0 & \dfrac{1}{4} & \dfrac{1}{4} & 0
\end{array}\right]
\end{array}
$$

The fundamental matrix **N** is obtained as

$$
\mathbf{N} = (\mathbf{I} - \mathbf{Q})^{-1} = \begin{bmatrix}
1 & -\dfrac{1}{4} & -\dfrac{1}{4} & 0 \\[8pt]
-\dfrac{1}{4} & 1 & 0 & -\dfrac{1}{4} \\[8pt]
-\dfrac{1}{4} & 0 & 1 & -\dfrac{1}{4} \\[8pt]
0 & -\dfrac{1}{4} & -\dfrac{1}{4} & 1
\end{bmatrix}^{-1}
$$

or

$$
\mathbf{N} = \frac{1}{6}\begin{bmatrix}
7 & 2 & 2 & 1 \\
2 & 7 & 1 & 2 \\
2 & 1 & 7 & 2 \\
1 & 2 & 2 & 7
\end{bmatrix}
$$

The absorption probability matrix **B** is obtained as

$$
\begin{array}{c}
\begin{array}{cccccccc} \;1 & \;2 & \;3 & \;4 & \;5 & \;6 & \;7 & \;8 \end{array} \\
\mathbf{B} = \mathbf{NR} = \begin{array}{c} 9 \\ 10 \\ 11 \\ 12 \end{array}\left[\begin{array}{cccccccc}
\dfrac{7}{24} & \dfrac{1}{12} & \dfrac{1}{12} & \dfrac{1}{24} & \dfrac{1}{24} & \dfrac{1}{12} & \dfrac{1}{12} & \dfrac{7}{24} \\[8pt]
\dfrac{1}{12} & \dfrac{7}{24} & \dfrac{7}{24} & \dfrac{1}{12} & \dfrac{1}{12} & \dfrac{1}{24} & \dfrac{1}{24} & \dfrac{1}{12} \\[8pt]
\dfrac{1}{12} & \dfrac{1}{24} & \dfrac{1}{24} & \dfrac{1}{12} & \dfrac{1}{12} & \dfrac{7}{24} & \dfrac{7}{24} & \dfrac{1}{12} \\[8pt]
\dfrac{1}{24} & \dfrac{1}{12} & \dfrac{1}{12} & \dfrac{7}{24} & \dfrac{7}{24} & \dfrac{1}{12} & \dfrac{1}{12} & \dfrac{1}{24}
\end{array}\right]
\end{array}
$$

Notice that Equation 8.102 is satisfied. We now use Equation 8.104 to obtain the potentials at the free nodes. For example,

$$
V_9 = \frac{7}{24}V_1 + \frac{1}{12}V_2 + \frac{1}{12}V_3 + \frac{1}{24}V_4 + \frac{1}{24}V_5 + \frac{1}{12}V_6 + \frac{1}{12}V_7 + \frac{7}{24}V_8
$$

Since $V_1 = V_2 = 100$ while $V_3 = V_4 = \cdots = V_8 = 0$,

$$
V_9 = \left(\frac{7}{24} + \frac{1}{12}\right)100 = 37.5
$$

By symmetry, $V_{10} = V_9 = 37.5$. Similarly,

$$
V_{11} = V_{12} = \left(\frac{1}{24} + \frac{1}{12}\right)100 = 12.5
$$

Table 8.5 compares these results with the finite difference solution (with 10 iterations) and the exact solution using Equation 2.31b or Equation 8.91. It is evident that the Markov chain solution compares well.

TABLE 8.5

Results of Example 8.8

Node	Finite Difference Solution	Markov Chain Solution	Exact Solution
9	37.499	37.5	38.074
10	37.499	37.5	38.074
11	12.499	12.5	11.926
12	12.499	12.5	11.926

EXAMPLE 8.9

Consider the potential problem shown in Figure 8.19. Let

$$V_o = 100, \quad \epsilon_1 = \epsilon_o, \quad \epsilon_2 = 3\epsilon_o$$

$$a = b = 0.5, \quad h = w = 1.0$$

Solution

The Markov chain solution was implemented using MATLAB. The approach involved writing code that generated the transition probability matrices using the random walk probabilities, computing the appropriate inverse, and manipulating the solution matrix. The use of MATLAB significantly reduced the programming complexity by the way the software internally handles matrices. The Q-matrix was selected as a timing index since the absorbing Markov chain algorithm involves inverting it. In this example, the Q-matrix is 361×361 and the running time was 90 and 34 seconds on 486DX2 and Pentium, respectively. $\Delta = 0.05$ was assumed. At the corner point $(x, y) = (a, b)$, the random walk probabilities are

$$p_{x+} = p_{y+} = \frac{\epsilon_1}{3\epsilon_1 + \epsilon_2}, \quad p_{x-} = p_{y-} = \frac{\epsilon_1 + \epsilon_2}{2(3\epsilon_1 + \epsilon_2)}$$

The plot of the potential distribution is portrayed in Figure 8.20. Since the problem has no exact solution, the results at five typical points are compared with those from the

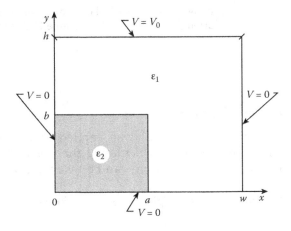

FIGURE 8.19
Potential system for Example 8.9.

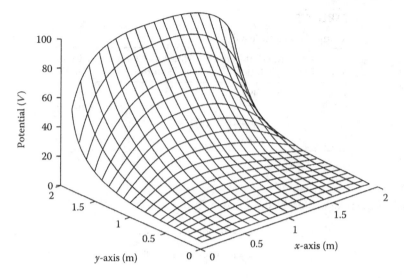

FIGURE 8.20
Potential distribution obtained by Markov chains; for Example 8.9.

TABLE 8.6

Results of Example 8.9

Node		Markov	Exodus	Finite
x	y	Chain	Method	Difference
0.25	0.5	10.2688	10.269	10.166
0.5	0.5	16.6667	16.667	16.576
0.75	0.5	15.9311	15.931	15.887
0.5	0.75	51.0987	51.931	50.928
0.5	0.25	6.2163	6.2163	6.1772

Exodus method and finite difference in Table 8.6. It should be observed that the Markov chain approach provides a solution that is close to that obtained by the Exodus method.

8.7 MCMC for Poisson's Equation

The Monte Carlo Markov chain (MCMC) was applied in the previous section to solve Laplace's equation. This means of calculating the potentials at all grid points simultaneously (or whole-field computation) can be extended to problems involving Poisson's equation.

Suppose the MCMC method is to be applied in solving Poisson's equation

$$\nabla^2 V = -g(x,y) = -\frac{\rho_s}{\varepsilon} \quad \text{in region R} \tag{8.107}$$

subject to Dirichlet boundary condition

$$V = V_p \text{ on boundary B} \tag{8.108}$$

We begin by dividing the solution region R into a mesh and derive the finite difference equivalent of Equation 8.107. For $V = V(x, y)$, the problem is reduced to a two-dimensional one and Equation 8.107 becomes

$$\frac{\partial^2 V}{\partial x^2} + \frac{\partial^2 V}{\partial y^2} = -g(x, y)$$

Going through the derivation we had in the previous section, we arrive at the solution as

$$\mathbf{V}_f = \mathbf{B}\mathbf{V}_p + \mathbf{N}\mathbf{G}_f \tag{8.109}$$

where $\mathbf{V}_f$ and $\mathbf{V}_p$ contain potentials at the free and fixed nodes, respectively, $\mathbf{G}_f$ is the evaluation of the term $(\Delta^2/4)g(x, y)$, and Δ is the step size. Unlike the classical random walk solution in Equation 8.61, the Markov chain solution in Equation 8.109 provides the solution at all the free nodes at once. Two simple examples will be used to illustrate the solution to Poisson's equation in Equation 8.109. The examples are done with hand calculation so that no computer programming is needed.

EXAMPLE 8.10

Consider an infinitely long conducting trough with square cross-section with the sides grounded, as shown in Figure 8.21. Let $\rho_s = x(y-1)nC/m^2$ and $\varepsilon = \varepsilon_o$ Then,

$$g(x, y) = \frac{\rho_s}{\varepsilon} = \frac{x(y-1) \times 10^{-9}}{10^{-9}\big/36\pi} = 36\pi x(y-1)$$

In this case [65],

$$Q = 0, \quad N = (I - Q)^{-1} = I$$

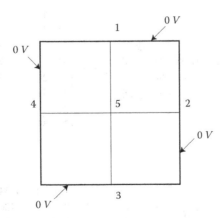

FIGURE 8.21
For Example 8.10.

Since $V_p = 0$, and there is only one free node (node 5), Equation 8.109 becomes

$$\mathbf{V}_f = \mathbf{N}\mathbf{G}_f$$

$$N = 1, \quad G_f = G_5 = \frac{\Delta^2}{4}g(x,y)$$

But

$$\Delta = \frac{1}{2}, x = \frac{1}{2}, y = \frac{1}{2},$$

$$G_5 = \frac{(1/2)^2}{4}36\pi(1/2)(-1/2) = -\frac{9\pi}{16} = -1.7671$$

$$V_5 = G_f = -1.7671$$

We can compare this with the finite difference solution. From Equation 3.30,

$$V_5 = \frac{1}{4}(0) + \frac{\Delta^2 g(x,y)}{4} = \frac{(1/2)^2}{4}36\pi\left(\frac{1}{2}\right)\left(-\frac{1}{2}\right) = -\frac{9\pi}{16} = -1.7671 \qquad (22)$$

The exact solution, based on series expansion in Chapter 2 is −2.086. The error is due to the fact that the step size Δ is large in this example.

EXAMPLE 8.11

This is the same problem as in Example 8.10 except that we now select $\Delta = 1/3$. We have four free nodes as shown in Figure 8.22. The fundamental matrix is obtained as [65]

$$N = \frac{1}{6}\begin{bmatrix} 7 & 2 & 2 & 1 \\ 2 & 7 & 1 & 2 \\ 2 & 1 & 7 & 2 \\ 1 & 2 & 2 & 7 \end{bmatrix} \qquad (8.11.1)$$

$$G_f = \frac{(1/3)^2}{4}\begin{bmatrix} g(x_9,y_9) \\ g(x_{10},y_{10}) \\ g(x_{11},y_{11}) \\ g(x_{12},y_{12}) \end{bmatrix} = \frac{1}{36}(36\pi)\begin{bmatrix} \frac{1}{3}\left(-\frac{1}{3}\right) \\ \frac{2}{3}\left(-\frac{1}{3}\right) \\ \frac{1}{3}\left(-\frac{2}{3}\right) \\ \frac{2}{3}\left(-\frac{2}{3}\right) \end{bmatrix} = -\frac{\pi}{9}\begin{bmatrix} 1 \\ 2 \\ 2 \\ 4 \end{bmatrix} \qquad (8.11.2)$$

Since $V_p = 0$,

$$V_f = NG_f = \frac{1}{6}\left(-\frac{\pi}{9}\right)\begin{bmatrix} 7 & 2 & 2 & 1 \\ 2 & 7 & 1 & 2 \\ 2 & 1 & 7 & 2 \\ 1 & 2 & 2 & 7 \end{bmatrix}\begin{bmatrix} 1 \\ 2 \\ 2 \\ 4 \end{bmatrix} = \begin{bmatrix} -1.1054 \\ -1.5126 \\ -1.5126 \\ -2.1526 \end{bmatrix} \qquad (8.11.3)$$

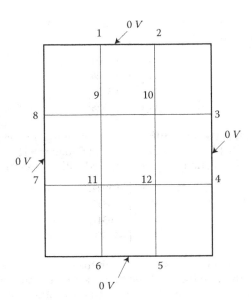

FIGURE 8.22
For Example 8.11.

We may compare this solution with the finite difference solution. Applying Equation 8.59 to node 9 in Figure 8.22, we obtain

$$V_9 = 0 + 0 + \frac{V_{10}}{4} + \frac{V_{11}}{4} + \frac{(1/3)^2}{4} 36\pi \left(\frac{1}{3}\right)\left(-\frac{1}{3}\right) \tag{8.11.4}$$

or

$$V_9 = 0.25V_{10} + 0.25V_{11} - \frac{\pi}{9} \tag{8.11.5}$$

Similarly, at node 10,

$$V_{10} = 0.25V_9 + 0.25V_{12} - \frac{2\pi}{9} \tag{8.11.6}$$

At node 11,

$$V_{11} = 0.25V_9 + 0.25V_{12} - \frac{2\pi}{9} \tag{8.11.7}$$

Ar node 12,

$$V_{12} = 0.25V_{10} + 0.25V_{11} - \frac{4\pi}{9} \tag{8.11.8}$$

Putting Equations 8.11.5 through 8.11.8 in matrix form yields

$$\underbrace{\begin{bmatrix} -1 & 0.25 & 0.25 & 0 \\ 0.25 & -1 & 0 & 0.25 \\ 0.25 & 0 & -1 & 0.25 \\ 0 & 0.25 & 0.25 & -1 \end{bmatrix}}_{A} \begin{bmatrix} V_9 \\ V_{10} \\ V_{11} \\ V_{12} \end{bmatrix} = \frac{\pi}{9} \begin{bmatrix} 1 \\ 2 \\ 2 \\ 4 \end{bmatrix} \tag{8.11.9}$$

Matrix A can easily be obtained by inspection using the band matrix method. By inverting the matrix A, we obtain

$$\begin{bmatrix} V_9 \\ V_{10} \\ V_{11} \\ V_{12} \end{bmatrix} = \begin{bmatrix} -1.1054 \\ -1.5126 \\ -1.5126 \\ -2.1526 \end{bmatrix} \tag{8.11.10}$$

It is not surprising that the finite difference solution is exactly the same as the Markov chain. It is easy to see that the inverse of matrix A in Equation 8.11.9 produces matrix N in Equaiton 8.11.1. The two solutions may be compared with the exact solution in Chapter 2:

$$\begin{bmatrix} V_9 \\ V_{10} \\ V_{11} \\ V_{12} \end{bmatrix} = \begin{bmatrix} -1.1924 \\ -1.6347 \\ -1.6347 \\ -2.373 \end{bmatrix} \tag{33}$$

The Markov chain solution agrees exactly with the finite difference solution. The two solutions differ slightly from the exact solution due to the large step size. By reducing the step size and using a computer, the Markov chain solution can be made accurate.

8.8 Time-Dependent Problems

MCM is well known for solving static problems such as Laplace's or Poisson's equation [1–4]. In this section, we extend the applicability of the conventional MCM to solve time-dependent (heat) problems [65–67]. We present results in 1-D and 2-D that agree with the exact solutions.

We may derive the diffusion equation from Maxwell's equations in Example 2.3. The result is

$$\nabla^2 J = \mu\sigma \frac{\partial J}{\partial t} \tag{8.110}$$

which is the diffusion equation. We will solve this in 1-D and 2-D both in Cartesian and polar cylindrical coordinates.

A. *One-Dimensional Diffusion Equation*

There are five concepts, consider the one-dimensional diffusion's equation:

$$U_{xx} = U_t, \quad 0 < x < 1, t > 0 \tag{8.111a}$$

Boundary conditions:

$$U(0,t) = 0 = U(1,t), \quad t > 0 \tag{8.111b}$$

Initial condition:

$$U(x,\ 0) = 100, \quad 0 < x < 1 \tag{8.111c}$$

In Equation (8.111a), U_{xx} indicates second partial derivative with respect to x, while U_t indicates partial derivative with respect to t. The problem models temperature distribution in a rod or eddy current in a conducting medium. In order to solve this problem using the MCM, we first need to obtain the finite difference equivalent of the partial differential equation in Equation 8.111a. Using the central-space and backward-time scheme, we obtain

$$\frac{U(i+1,n)-2U(i,n)+U(i-1,n)}{(\Delta x)^2} = \frac{U(i,n)-U(i,n-1)}{\Delta t} \tag{8.112}$$

where $x = i\Delta x$ and $t = n\Delta t$. If we let

$$\alpha = \frac{(\Delta x)^2}{\Delta t}$$

Equation 8.112 becomes

$$U(i,n) = p_{x+}U(i+1,n) + p_{x-}U(i-1,n) + p_{t-}U(i,n-1) \tag{8.113}$$

where

$$p_{x+} = p_{x-} = \frac{1}{2+\alpha}, \quad p_{t-} = \frac{\alpha}{2+\alpha}$$

Notice that $p_{x+} + p_{x-} + p_{t-} = 1$. Equation 8.113 can be given a probabilistic interpretation. If a random-walking particle is instantaneously at the point (x, y), it has probabilities $p_+, p_-,$ and p_{t-} of moving from (x, t) to $(x + \Delta x, t), (x–\Delta x, t),$ and $(x, t–\Delta t)$ respectively. The particle can only move toward the past, never toward the future. A means of determining which way the particle should move is to generate a random number $r, 0 < r < 1$, and instruct the particle to walk as follows:

$$
\begin{aligned}
(x,t) &\rightarrow (x+\Delta x,t) \quad \text{if } (0 < r < 0.25) \\
(x,t) &\rightarrow (x-\Delta x,t) \quad \text{if } (0.25 < r < 0.5) \\
(x,t) &\rightarrow (x,t-\Delta t) \quad \text{if } (0.5 < r < 1)
\end{aligned}
\tag{8.114}
$$

where it is assumed that $\alpha = 2$. Most modern software such as MATLAB have a random number generator to obtain r.

To calculate U at point (x_o, t_o), we follow the following random walk algorithm:

1. Begin a random walk at $(x, t) = (x_o, t_o)$.
2. Generate a random number $0 < r < 1$, and move to the next point using Equation 8.114.
3. (a) If the next point is not on the boundary, repeat step 2.

   (b) If the random walk hits the boundary, terminate the random walk. Record U_b at the boundary, go to step 1, and begin another random walk.
4. After N random walks, determine

$$U(x_o,t_o) = \frac{1}{N}\sum_{k=1}^{N} U_b(k) \tag{8.115}$$

where N, the number of random walks, is assumed large. A typical random walk is illustrated in Figure 8.23.

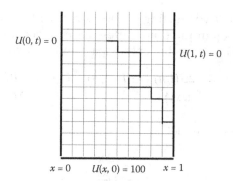

FIGURE 8.23
A typical random walk.

As a numerical example, consider the solution of the problem in Equation (8.111a). We select $a = 2$, $\Delta x = 0.1$, so that $\Delta t = 0.005$ and $p_{x+} = p_{x-} = 1/4$, $p_{t-} = 1/2$.

We calculate U at $x = 0.4$, $t = 0.01, 0.02, 0.03, \ldots$ As shown in Table 8.7, we compare the results with the finite different solution and exact solution:

$$U(x,t) = \frac{400}{\pi} \sum_{k=0}^{\infty} \frac{1}{n} \sin(n\pi x)\exp(-n^2\pi^2 t),$$

$$n = 2k + 1$$

(8.116)

B. Two-Dimensional Diffusion Equation

Suppose we are interested in the solution of the two-dimensional heat equation in cylindrical coordinates:

$$U_{\rho\rho} + \frac{U_\rho}{\rho} + U_{zz} = U_t, \quad 0 < \rho < 1, \ 0 < z < 1, \ t > 0 \tag{8.117}$$

Boundary conditions:

$$U(\rho, 0, t) = 0 = U(\rho, 1, t), \quad 0 < \rho < 1, t > 0 \tag{8.118a}$$

$$U(1, z, t) = 0, \quad 0 < z < 1, t > 0 \tag{8.118b}$$

Initial condition:

$$U(p, z, 0) = T_o, \quad 0 < \rho < 1, 0 < z < 1 \tag{8.118c}$$

TABLE 8.7

Comparing Monte Carlo (MCM) Solution with FD and Exact Solution ($x_o = 0.4$)

t	Exact	MCM	FD
0.01	99.53	98.44	100
0.02	95.18	93.96	96.87
0.03	88.32	87.62	89.84
0.04	80.88	81.54	82.03
0.10	45.13	46.36	45.18

This models the temperature distribution in a solid cylinder of unit height and unit radius.

Using the central-space and backward-time scheme, we obtain the finite difference equivalent as

$$\frac{U(i+1,j,n)-2U(i,j,n)+U(i-1,j,n)}{(\Delta\rho)^2}+\frac{U(i+1,j,n)-U(i-1,j,n)}{\rho 2\Delta\rho}$$
$$+\frac{U(i,j+1,n)-2U(i,j,n)+U(i,j-1,n)}{(\Delta z)^2}$$
$$=\frac{U(i,j,n)-U(i,j,n-1)}{\Delta t} \tag{8.119}$$

Let $\Delta x = \Delta z = h$ and $\rho = ih, z = jh, t = n\Delta t$

$$\alpha = \frac{h^2}{\Delta t} \tag{8.120}$$

Equation 8.119 becomes

$$U(i,j,n) = p_{\rho+}U(i+1,j,n)+p_{\rho-}U(i-1,j,n)$$
$$+p_{z+}U(i,j+1,n)+p_{z-}U(i,j-1,n)+p_{t-}U(i,j,n-1) \tag{8.121}$$

where

$$p_{\rho+} = \frac{1+1/2i}{4+\alpha}, \quad p_{\rho-} = \frac{1-1/2i}{4+\alpha}, \tag{8.122a}$$

$$p_{z+} = \frac{1}{4+\alpha}, \quad p_{z-} = \frac{1}{4+\alpha}, \quad p_{t-} = \frac{\alpha}{4+\alpha} \tag{8.122b}$$

Note that $p_{\rho+} + p_{\rho-} + p_{z+} + p_{z-} + p_{t-} = 1$ so that a probabilistic interpretation can be given to Equation 8.121. A random walking particle at point (ρ, z, t) moves to $(\rho + h, z, t)$, $(\rho - h, z, t)$, $(\rho, z + h, t)$, $(\rho, z - h, t)$, $(\rho, z, t - \Delta t)$ with probabilities $p_{\rho+}, p_{\rho-}, p_{z+}, p_{z-},$ and p_{t-}, respectively. By generating a random number $0 < r < 1$, we instruct the particle to move as follows:

$$(\rho, z, t) \rightarrow (\rho + h, z, t) \quad \text{if } (0 < r < prho)$$

$$(\rho, z, t) \rightarrow (\rho - h, z, t) \quad \text{if } (prho < r < 0.4)$$

$$(\rho, z, t) \rightarrow (\rho, z + h, t) \quad \text{if } (0.4 < r < 0.6)$$

$$(\rho, z, t) \rightarrow (\rho, z - h, t) \quad \text{if } (0.6 < r < 0.8)$$

$$(\rho, z, t) \rightarrow (\rho, z, t - \Delta t) \quad \text{if } (0.8 < r < 1) \tag{8.123}$$

assuming that $\alpha = 1$ and $prho = 0.2 * (1 + 1/(2 * i))$.

Equations 8.119 through 8.123 apply only for $\rho \neq 0$. For $\rho = 0$, we apply L'Hopital's rule in Equation 8.117 and obtain

$$2U_{\rho\rho} + U_{zz} = U_t \tag{8.124}$$

We now apply central-space and backward-time scheme to Equation 8.124 and noting that $U(h, z, t) = U(-h, z, t)$, we obtain

$$U(0, j, n) = p_{\rho+}U(1, j, n) + p_{z+}U(0, j+1, n) + p_{z-}U(0, j-1, n)$$
$$+ p_{t-}U(0, j, n-1) \tag{8.125}$$

where

$$p_{\rho+} = \frac{4}{6+\alpha}, \quad p_{z+} = p_{z-} = \frac{1}{6+\alpha}, \quad p_{t-} = \frac{\alpha}{6+\alpha} \tag{8.126}$$

A random walking particle that finds itself at $\rho = 0$ determines the next location by generating a random number r, $0 < r < 1$, and walking as follows:

$$(0, z, t) \rightarrow (h, z, t) \quad \text{if } (0 < r < 4 * pp)$$

$$(0, z, t) \rightarrow (0, z+h, t) \quad \text{if } (4* pp < r < 5 * pp)$$

$$(0, z, t) \rightarrow (0, z-h, t) \quad \text{if } (5* pp < r < 6 * pp) \tag{8.127}$$

$$(0, z, t) \rightarrow (0, z, t - \Delta t) \quad \text{if } (6* pp < r < 1)$$

where $pp = 1/(6 + \alpha)$ and it is assumed that $\alpha = 1$.

Therefore, we take the following steps to calculate U at point (ρ_o, z_o, t_o):

1. Begin a random walk at $(\rho, z, t) = (\rho_o, z_o, t_o)$.
2. Generate a random number $0 < r < 1$, and move the next point according to Equation 8.123 if $\rho \neq 0$ or Equation 8.127 if $\rho = 0$.
3. (a) If the next point is not on the boundary, repeat step 2.

   (b) If the random walk hits the boundary, terminate the random walk. Record U_b at the boundary, go to step 1, and begin another random walk.
4. After N random walks, determine

$$U(\rho_o, z_o, t_o) = \frac{1}{N} \sum_{k=1}^{N} U_b(k) \tag{8.128}$$

A typical random walk is shown in Figure 8.24. The only difference between 1-D and 2-D is that there are three kinds of displacement in 1-D while there are five displacements (four spatial ones and one temporal one) in 2-D.

As a numerical example, consider the solution of the problem in Equations 8.117 and 8.118. We select $\alpha = 1$, $T_o = 10$, $h = 0.1$, so that $\Delta t = 0.01$, and we calculate U at $\rho = 0.5$, $z = 0.5$,

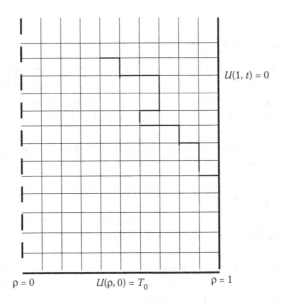

FIGURE 8.24
A typical random walk.

$t = 0.05, 0.1, 0.15, 0.2, 0.25, 0.3$. As shown in Table 8.8, we compare the results from the MCM with the finite difference (FD) solution and exact solution [67–69]:

$$U(\rho, z, t) = \frac{8T_o}{\pi} \sum_{m=1}^{\infty} \sum_{n=1,3,5}^{\infty} \frac{J_o(k_m\rho)}{nk_m J_1(k_m)} \sin(n\pi z)\exp(-\lambda_{mn}^2 t), \qquad (8.129)$$

where $\lambda_{mn}^2 = k_m^2 + (n\pi)^2$ and k_m is the mth root of Bessel's function $J_o(k_m)$.

Due to the randomness of the Monte Carlo solution, each MCM result in Tables 8.7 and 8.8 was obtained by running the simulation five times and taking the average.

In this section, the conventional MCM has been shown to be effectively applicable to time-dependent problems such as the heat equation in Cartesian and cylindrical coordinates. For 1-D and 2-D cases, we notice that the Monte Carlo solutions agree well with the finite difference solution and the exact analytical solutions and it is easier to understand and program than the finite difference method. The MCM does not require the need for solving large matrices and is trivially easy to program. The idea can be extended to wave equations.

TABLE 8.8

Comparing Monte Carlo Solution with FD
and Exact Solution

t	Exact	MCM	FD
0.05	6.2475	6.614	6.3848
0.10	2.8564	3.182	2.9123
0.15	1.3059	1.582	1.2975
0.20	0.5971	0.7760	0.5913
0.25	0.2730	0.4140	0.270
0.30	0.1248	0.156	0.1233

8.9 Concluding Remarks

The Monte Carlo technique is essentially a means of estimating expected values and hence is a form of numerical quadrature. It recasts deterministic problems in probabilistic terms. Although the technique can be applied to simple processes and estimating multidimensional integrals, the power of the technique rests in the fact that [77]

- It is often more efficient than other quadrature formulas for estimating multidimensional integrals,
- It is adaptable in the sense that variance reduction techniques can be tailored to the specific problem, and
- It can be applied to highly complex problems for which the definite integral formulation is not obvious and standard analytic techniques are ineffective.

MCMs are widely used in engineering, statistical physics, medicine, finance, economics, and other disciplines. For rigorous mathematical justification for the methods employed in Monte Carlo simulations, one is urged to read [31,71]. As is typical with current MCMs, other numerical methods of solutions appear to be preferable when they may be used. Monte Carlo techniques often yield numerical answers of limited accuracy and are therefore employed as a last resort. However, there are problems for which the solution is not feasible using other methods. Problems that are probabilistic and continuous in nature (e.g., neutron absorption, charge transport in semiconductors, and scattering of waves by random media) are ideally suited to these methods and represent the most logical and efficient use of the stochastic methods. Since the recent appearance of vector machines, the importance of the MCMs is growing.

It should be emphasized that in any Monte Carlo simulation, it is important to indicate the degree of confidence of the estimates or insert error bars in graphs illustrating Monte Carlo estimates. Without such information, Monte Carlo results are of questionable significance.

Applications of MCMs to other branches of science and engineering are summarized in References 14,15,25,72. EM-related problems, besides those covered in this chapter, to which Monte Carlo procedures have been applied include the following:

- Diffusion problems [61,63,73],
- Transmission lines [39,74–76],
- Random periodic arrays [77],
- Waveguide structures [78–83],
- Scattering of waves by random media [84–91],
- Noise in magnetic recording [92,93],
- Induced currents in biological bodies [94,95].

We conclude this chapter by referring to two new MCMs. One new MCM, known as the equilateral triangular mesh fixed random walk, has been proposed to handle Neumann problems [96,97]. Another new MCM, known as Neuro-Monte Carlo solution, is an attempt at whole-field computation [98]. It combines an artificial neural network and a MCM as a training data source. For further exposition on Monte Carlo techniques, one should consult [25,60,99,100,101].

PROBLEMS

8.1 Write a program to generate 1000 pseudorandom numbers U uniformly distributed between 0 and 1. Calculate their mean and compare the calculated mean with the expected mean (0.5) as a test of randomness.

8.2 Generate 10,000 random numbers uniformly distributed between 0 and 1. Find the percentage of numbers between 0 and 0.1, between 0.1 and 0.2, etc., and compare your results with the expected distribution of 10% in each interval.

8.3 a. Using the linear congruential scheme, generate 10 pseudorandom numbers with $a = 1573$, $c = 19$, $m = 10^3$, and seed value $X_0 = 89$.

   b. Repeat the generation with $c = 0$.

8.4 For $a = 13$, $m = 2^6 = 64$, and $X_0 = 1, 2, 3$, and 4, find the period of the random number generator using the multiplicative congruential method.

8.5 Develop a program that uses the inverse transformation method to generate a random number from a distribution with the probability density function

$$f(x) = \begin{cases} 0.25, & 0 \le x \le 1 \\ 0.75, & 1 \le x \le 1 \end{cases}$$

8.6 It is not easy to apply the inverse transform method to generate normal distribution. However, by making use of the approximation

$$e^{-x^2/2} \simeq \frac{2e^{-kx}}{(1+e^{-kx})^2}, \quad x > 0$$

where $k = \sqrt{8/\pi}$, the inverse transform method can be applied. Develop a program to generate normal deviates using inverse transform method.

8.7 Using the rejection method, generate a random variable from $f(x) = 5x^2, 0 \le x \le 1$.

8.8 Use the rejection method to generate Gaussian (or normal) deviates in the truncated region $-a \le X \le a$.

8.9 Use sample mean Monte Carlo integration to evaluate the following:

   a. $\displaystyle\int_0^1 4\sqrt{1-x^2}\,dx,$

   b. $\displaystyle\int_0^1 \sin x \, dx,$

   c. $\displaystyle\int_0^1 e^x \, dx,$

   d. $\displaystyle\int_0^1 \frac{1}{\sqrt{x}}\,dx$

8.10 Evaluate the following four-dimensional integrals:

   a. $\displaystyle\int_0^1\int_0^1\int_0^1\int_0^1 \exp(x^1 x^2 x^3 x^4 - 1)\,dx^1 dx^2 dx^3 dx^4,$

b. $\displaystyle\int_0^1\int_0^1\int_0^1\int_0^1 \sin(x^1+x^2+x^3+x^4)\,dx^1dx^2dx^3dx^4$

8.11 The radiation from a rectangular aperture with constant amplitude and phase distribution may be represented by the integral

$$I(\alpha,\beta)=\int_{-1/2}^{1/2}\int_{-1/2}^{1/2} e^{j(\alpha x+\beta y)}\,dx\,dy$$

Evaluate this integral using a Monte Carlo procedure and compare your result for $\alpha=\beta=\pi$ with the exact solution

$$I(\alpha,\beta)=\frac{\sin(\alpha/2)\sin(\beta/2)}{\alpha\beta/4}$$

8.12 Consider the differential equation

$$\frac{\partial^2 W}{\partial x^2}+\frac{\partial^2 W}{\partial y^2}+\frac{k}{y}\frac{\partial W}{\partial y}=0$$

where $k=$ constant. By finding its finite difference form, give a probabilistic interpretation to the equation.

8.13 Given the one-dimensional differential equation

$$y''=0,\quad 0\le x\le 1$$

subject to $y(0)=0$, $y(1)=10$, use an MCM to find $y(0.25)$ assuming $\Delta x=0.25$ and the following 20 random numbers:

0.1306, 0.0422, 0.6597, 0.7905, 0.7695, 0.5106, 0.2961, 0.1428, 0.3666, 0.6543, 0.9975, 0.4866, 0.8239, 0.8722, 0.1330, 0.2296, 0.3582, 0.5872, 0.1134, 0.1403.

8.14 Consider N equal resistors connected in series as in Figure 8.25. By making $V(0)=0$ and $V(N)=10$ V, find $V(k)$ using the fixed random walk for the following cases: (a) $N=5$, $k=2$, (b) $N=10$, $k=7$, (c) $N=20$, $k=11$.

8.15 Consider a parallel-plate geometry with dielectric interface half-way in between as shown in Figure 8.26. Fringing effects may be neglected since the lengths of

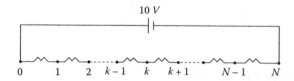

FIGURE 8.25
For Problem 8.14.

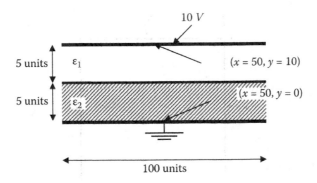

FIGURE 8.26
For Problem 8.15.

the plates are chosen large. The upper electrode is fixed at 10 V, while the lower electrode is fixed at 0 V. Take $\varepsilon_1 = \varepsilon_o$ and $\varepsilon_2 = 3.9\varepsilon_o$. Calculate the potential at nine different points ($y = 1, 2, ..., 9$) using fixed or floating random walk.

8.16 Use a Monte Carlo procedure to determine the potential at points (2, 2), (3, 3), and (4, 4) in the problem shown in Figure 8.27a. By virtue of double symmetry, it is sufficient to consider a quarter of the solution region as shown in Figure 8.27b.

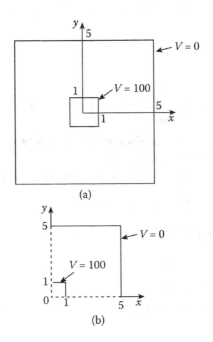

FIGURE 8.27
For Problem 8.16.

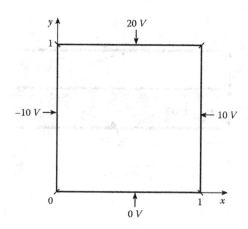

FIGURE 8.28
For Problem 8.17.

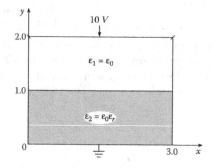

FIGURE 8.29
For Problem 8.18.

8.17 In the solution region of Figure 8.28, $\rho_v = x(y - 1)nC/m^3$. Find the potential at the center of the region using an MCM.

8.18 Consider the potential system shown in Figure 8.29. Determine the potential at the center of the solution region. Take $\epsilon_r = 2.25$.

8.19 Apply an MCM to solve Laplace's equation in the three-dimensional region

$$|x| \leq 1, \quad |y| \leq 0.5, \quad |z| \leq 0.5$$

subject to the boundary condition

$$V(x, y, z) = x + y + z + 0.5$$

Find the solution at $(0.5, 0.1, 0.1)$.

8.20 Consider the interface separating two homogeneous media in Figure 8.30. By applying Gauss's law

$$\oint_S \epsilon \frac{\partial V}{\partial n} \, dS = 0$$

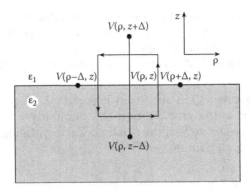

FIGURE 8.30
For Problem 8.20.

show that

$$V(\rho, z) = p_{\rho+}V(\rho + \Delta, z) + p_{\rho-}V(\rho - \Delta, z)$$
$$+ p_{z+}V(\rho, z + \Delta) + p_{z-}V(\rho, z - \Delta)$$

where

$$p_{z+} = \frac{\epsilon_1}{2(\epsilon_1 + \epsilon_2)}, \qquad p_{z-} = \frac{\epsilon_2}{2(\epsilon_1 + \epsilon_2)}$$
$$p_{\rho+} = p_{\rho-} = \frac{1}{4}$$

8.21 Consider the finite cylindrical conductor held at $V = 100$ enclosed in a larger grounded cylinder. The axial symmetric problem is portrayed in Figure 8.31 for your convenience. Using a Monte Carlo technique, write a program to determine the potential at points $(\rho, z) = (2,10)$, $(5,10)$, $(8,10)$, $(5,2)$, and $(5,18)$.

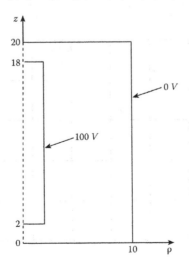

FIGURE 8.31
For Problem 8.21.

8.22 Figure 8.32 shows a prototype of an electrostatic particle focusing system employed in a recoil-mass time-of-flight spectrometer. It is essentially a finite cylindrical conductor that abruptly expands radius by a factor of 2. Write a program based on an MCM to calculate the potential at points $(\rho, z) = (5,18)$, $(5,10)$, $(5,2)$, $(10,2)$, and $(15,2)$.

8.23 Consider the square region shown in Figure 8.33. The transition probability $p(Q, S_i)$ is defined as the probability that a randomly walking particle leaving point Q will arrive at side S_i of the square boundary. Using the Exodus method, write a program to determine

   a. $p(Q_1, S_i)$, $i = 1, 2, 3, 4$,

   b. $p(Q_2, S_i)$, $i = 1, 2, 3, 4$.

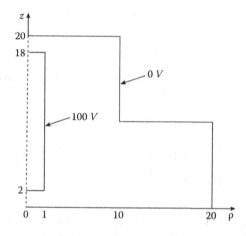

FIGURE 8.32
For Problem 8.22.

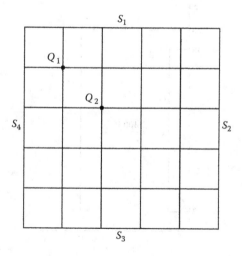

FIGURE 8.33
For Problem 8.23.

8.24 Given the one-dimensional differential equation

$$\frac{d^2\Phi}{dx^2} = 0, \quad 0 \le x \le 1$$

subject to $\Phi(0) = 0$, $\Phi(1) = 10$, use the Exodus method to find $\Phi(0.25)$ by injecting 256 particles at $x = 0.25$. You can solve this problem by hand calculation.

8.25 Use the Exodus method to find the potential at node 4 in Figure 8.34. Inject 256 particles at node 4 and scan nodes in the order 1, 2, 3, 4. You can solve this problem by hand calculation.

8.26 Using the Exodus method, write a program to calculate $V(0.25, 0.75)$ in Example 8.5.

8.27 Write a program to calculate $V(1.0, 1.5)$ in Example 8.6 using the Exodus method.

8.28 Consider the cross section of an infinitely long trough whose sides are maintained as shown in Figure 8.35. Write a MATLAB code using the Exodus method to calculate the potential at (0.5, 0.5), (0.8, 0.8), (1.0, 0.5), and (0.8, 0.2). Compare your results with exact results in Equations 2.44, 2.53–2.56.

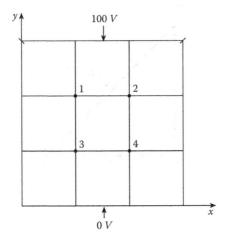

FIGURE 8.34
For Problem 8.25

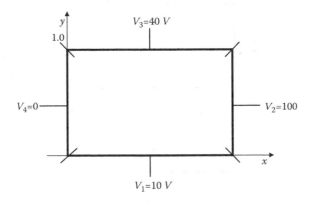

FIGURE 8.35
For Problem 8.28.

8.29 Write a program that will apply the Exodus method to determine the potential at point (0.2, 0.4) in the system shown in Figure 8.36.

8.30 Use Markov chain MCM to determine the potential at node 5 in Figure 8.37.

8.31 Rework Problem 8.19 using Markov chain.

8.32 Rework Problem 8.23 using Markov chain.

8.33 Consider the two-dimensional heat equation

$$U_{xx} + U_{yy} = U_t, \quad 0 < x < 1, 0 < y < 1, t > 0$$

with boundary conditions

$$U(0, y, t) = 0 = U(1, y, t), \quad 0 < y < 1, t > 0$$
$$U(x, 0, t) = 0 = U(x, 1, t), \quad 0 < x < 1, t > 0$$

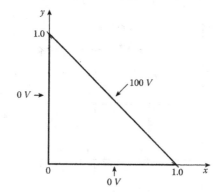

FIGURE 8.36
For Problem 8.29.

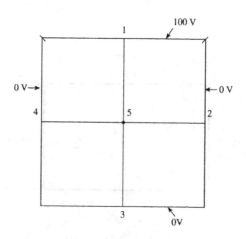

FIGURE 8.37
For Problem 8.30.

and initial condition

$$U(x, y, 0) = 10xy, \quad 0 < x < 1, 0 < y < 1$$

Select $\Delta - \Delta x - \Delta y - 0.1$, $\Delta t - 0.01$ and calculate U at $x = 0.5 = y$, $t = 0.05$, 0.1, 0.15, 0.2, 0.25, 0.3.

8.34 Consider the one-dimension heat equation in cylindrical coordinates

$$\nabla^2 U = \frac{\partial U}{\partial t} \rightarrow U_{\rho\rho} + \frac{1}{\rho} U_\rho = U_t, \quad 0 < \rho < 1, t > 0$$

with boundary conditions

$$U(1, t) = 0, t < 0$$

and initial condition

$$U(\rho, 0) = T_0 \quad \text{(constant)}$$

Use fixed random walk MCM to obtain the solution $U(\rho, t)$.

Select $T_0 = 10$, $\Delta\rho = 0.1$, and $\Delta t = 0.005$. Calculate U at $\rho = 0.5$, $t = 0.1, 0.2, 0.3, ..., 1.0$.

8.35 Rework Problem 8.35 using the Exodus method.

8.36 Use the Exodus method to solve the one-dimensional heat equation

$$U_{xx} = U_t, \quad 0 < x < 1, \ t > 0$$

subject to

$$U(0, t) = 0 = U(1, t), t > 0$$

$$U(x, 0) = 100, \quad 0 < x < 1$$

Let $\Delta x = 0.1$, $\Delta t = 0.005$, $x = x_0 = 0.4$, $t = 0.01, 0.02, 0.03, ...$

8.37 Using the Exodus method, find the solution of the 2D heat equation

$$U_{xx} + U_{yy} = U_t, \quad 0 < x < 1, 0 < y < 1, t > 0$$

subject to

$$U(0, y, t) = 0 = U(1, y, t), \quad 0 < y < 1, t > 0$$

$$U(x, 0, t) = 0 = U(x, 1, t), \quad 0 < x < 1, t > 0$$

$$U(x, y, 0) = 10xy, \quad 0 < x < 1, 0 < y < 1$$

Let $\Delta x = \Delta y = \Delta = 0.1$, $\Delta t = 0.01$, $x = y = 0.5$, $t = 0.1, 0.15, 0.2, 0.25, 0.3, ...$

8.38 Apply the Exodus method to determine the solution of the two-dimensional heat equation:

$$U_{\rho\rho} + \frac{U_\rho}{\rho} + U_{zz} = U_t, \quad 0 < \rho < 1, 0 < z < 1, t > 0$$

Boundary conditions:

$$U(\rho,0,t) = 0 = U(\rho,1,t), \quad 0 < \rho < 1, t > 0$$
$$U(1,z,t) = 0, \quad 0 < z < 1, t > 0$$

Initial condition:

$$U(\rho,z,0) = T_o, \quad 0 < \rho < 1, 0 < z < 1$$

Let $T_o = 10$, $\Delta\rho = \Delta z = h = 0.05$, $\Delta t = 0.0025$, $\rho = z = 0.5$, $t = 0.05, 0.1, 0.15, 0.2, 0.25, 0.3$.

References

1. R. Hersch and R.J. Griego, "Brownian motion and potential theory," *Sci. Am.*, Mar. 1969, pp. 67–74.
2. T.F. Irvine and J.P. Hartnett (eds.), *Advances in Heat Transfer*. New York: Academic Press, 1968.
3. H.A. Meyer (ed.), *Symposium on Monte Carlo Methods*. New York: John Wiley, 1956.
4. D.D. McCracken, "The Monte Carlo method," *Sci. Am.*, vol. 192, May 1955, pp. 90–96.
5. T.E. Hull and A.R. Dobell, "Random number generators," *SIAM Rev.*, vol. 4, no. 3, July 1982, pp. 230–254.
6. D.E. Knuth, *The Art of Computer Programming*, vol. 2. Reading, MA: Addison-Wesley, 1969, pp. 9, 10, 78, 155.
7. J. Banks and J. Carson, *Discrete Event System Simulation*. Englewood Cliffs, NJ: Prentice-Hall, 1984, pp. 257–288.
8. P.A.W. Lewis et al., "A pseudo-random number generator for the system/360," *IBM System J.*, vol. 8, no. 2, 1969, pp. 136–146.
9. S.S. Kuo, *Computer Applications of Numerical Methods*. Reading, MA: Addison-Wesley, 1972, pp. 327–345.
10. A.M. Law and W.D. Kelton, *Simulation Modeling and Analysis*. New York: McGraw-Hill, 1982, pp. 219–228.
11. D.E. Raeside, "An introduction to Monte Carlo methods," *Am. J. Phys.*, vol. 42, Jan. 1974, pp. 20–26.
12. C. Jacoboni and L. Reggiani, "The Monte Carlo method for the solution of charge transport in semiconductors with applications to covalent materials," *Rev. Mod. Phys.*, vol. 55, no. 3, July 1983, pp. 645–705.
13. H. Kobayashi, *Modeling and Analysis: An Introduction to System Performance Evaluation Methodology*. Reading, MA: Addison-Wesley, 1978, pp. 221–247.
14. I.M. Sobol, *The Monte Carlo Method*. Chicago: University of Chicago Press, 1974, pp. 24–30.
15. Y.A. Shreider, *Method of Statistical Testing (Monte Carlo Method)*. Amsterdam: Elsevier, 1964, pp. 39–83. Another translation of the same Russian text: Y.A. Shreider, The Monte Carlo Method (The Method of Statistical Trials). Oxford: Pergamon, 1966.

16. E.E. Lewis and W.F. Miller, *Computational Methods of Neutron Transport*. New York: John Wiley, 1984, pp. 296–360.

17. I.S. Sokolinkoff and R.M. Redheffer, *Mathematics of Physics and Modern Engineering*. New York: McGraw-Hill, 1958, pp. 644–649.

18. M.H. Merel and F.J. Mullin, "Analytic Monte Carlo error analysis," *J. Spacecraft*, vol. 5, no. 11, Nov. 1968, pp. 1304–1308.

19. A.J. Chorin, "Hermite expansions in Monte-Carlo computation," *J. Comp. Phys.*, vol. 8, 1971, pp. 472–482.

20. R.Y. Rubinstein and D.P. Kroese, *Simulation and the Monte Carlo Method*, 3rd edition. Hoboken, NJ: John Wiley & Sons, 2016.

21. W.J. Graybeal and U.W. Pooch, *Simulation: Principles and Methods*. Cambridge, MA: Winthrop Pub., 1980, pp. 77–97.

22. D. Landau and K. Binder, *A Guide to Monte Carlo Simulations in Statistical Physics*, 3rd Edition. Cambridge, UK: Cambridge University Press, 2009.

23. C.W. Alexion et al., "Evaluation of radiation fields using statistical methods of integration," *IEEE Trans. Ant. Prog.*, vol. AP-26, no. 2, Mar. 1979, pp. 288–293.

24. R.C. Millikan, "The magic of the Monte Carlo method," *BYTE*, vol. 8, Feb. 1988, pp. 371–373.

25. M.H. Kalos and P.A. Whitlock, *Monte Carlo Methods, 2nd Edition*. Verlag, Germany: Wiley-Blackwell, 2008.

26. S. Haber, "A modified Monte-Carlo quadrature II," *Math. Comp.*, vol. 21, July 1967, pp. 388–397.

27. S. Haber, "Numerical evaluation of multiple integrals," *SIAM Rev.*, vol. 12, no. 4, Oct. 1970, pp. 481–527.

28. J.H. Halton and D.C. Handscom, "A method for increasing the efficiency of Monte Carlo integration," *J. ACM*, vol. 4, 1957, pp. 329–340.

29. M. Mishra and N. Gupta, "Monte Carlo integration technique for method of moments solution of EFIE in scattering problems," *J. Electromagn. Analysis Appl.*, vol. 1, 2009, pp. 254–258.

30. M. Mishra et al., "Application of quasi Monte Carlo integration technique in efficient capacitance computation," *Prog. Electromagn. Res.*, vol. 90, 2009, pp. 309–322.

31. S. Kakutani, "Two-dimensional Brownian motion harmonic functions," *Proc. Imp. Acad. (Tokyo)*, vol. 20, 1944, pp. 706–714.

32. A. Haji-Sheikh and E.M. Sparrow, "The floating random walk and its application to Monte Carlo solutions of heat equations," *J. SIAM Appl. Math.*, vol. 14, no 2, Mar. 1966, pp. 370–389.

33. A. Haji-Sheikh and E.M. Sparrow, "The solution of heat conduction problems by probability methods," *J. Heat Transfer, Trans. ASME, Series C*, vol. 89, no. 2, May 1967, pp. 121–131.

34. G.E. Zinsmeiter, "Monte Carlo methods as an aid in teaching heat conduction," *Bull. Mech. Engr. Educ.*, vol. 7, 1968, pp. 77–86.

35. R. Chandler et al., "The solution of steady state convection problems by the fixed random walk method," *J. Heat Transfer, Trans. ASME, Series C*, vol. 90, Aug. 1968, pp. 361–363.

36. G.E. Zinsmeiter and S.S. Pan, "A method for improving the efficiency of Monte Carlo calculation of heat conduction problems," *Trans. ASME, C, J. Heat Transf.*, vol. 96, 1974, pp. 246–248.

37. G.E. Zinsmeiter and S.S. Pan, "A modification of the Monte Carlo method," *Int. J. Num. Meth. Engr.*, vol. 10, 1976, pp. 1057–1064.

38. G.M. Royer, "A Monte Carlo procedure for potential theory of problems," *IEEE Trans. Micro. Theo. Tech.*, vol. MTT-19, no. 10, Oct. 1971, pp. 813–818.

39. R.M. Bevensee, "Probabilistic potential theory applied to electrical engineering problems," *Proc. IEEE*, vol. 61, no. 4, April 1973, pp. 423–437.

40. R.L. Gibbs and J.D. Beason, "Solutions to boundary value problems of the potential type by random walk method," *Am. J. Phys.*, vol. 43, no. 9, Sept. 1975, pp. 782–785.

41. J.H. Pickles, "Monte Carlo field calculations," *Proc. IEEE*, vol. 124, no. 12, Dec. 1977, pp. 1271–1276.

42. F. Sanchez-Quesada et al., "Monte-Carlo method for discrete inhomogeneous problems," *Proc. IEEE*, vol. 125, no. 12, Dec. 1978, pp. 1400–1402.

43. R. Schlott, "A Monte Carlo method for the Dirichlet problem of dielectric wedges," *IEEE Trans. Micro. Theo. Tech.*, vol. 36, no. 4, April 1988, pp. 724–730.

44. M.N.O. Sadiku, "Monte Carlo methods in an introductory electromagnetic class," *IEEE Trans. Educ.*, vol. 33, no. 1, Feb. 1990, pp. 73–80.

45. M.D.R. Beasley et al., "Comparative study of three methods for computing electric fields," *Proc. IEEE*, vol. 126, no. 1, Jan. 1979, pp. 126–134.

46. J.R. Currie et al., "Monte Carlo determination of the frequency of lightning strokes and shielding failures on transmission lines," *IEEE Trans. Power Appl. Syst.*, vol. PAS-90, 1971, pp. 2305–2310.

47. R.S. Velazquez et al., "Probabilistic calculations of lightning protection for tall buildings," *IEEE Trans. Ind. Appl.*, vol. IA-18, no. 3, May/June 1982, pp. 252–259.

48. M.N.O. Sadiku, "Monte Carlo Solution of Axisymmetric Potential Problems," *IEEE Trans. Ind. Appl.*, vol. 29, no. 6, Nov./Dec. 1993, pp. 1042–1046.

49. J.N. Jere and Y.L.L. Coz, "An improved floating-random-walk algorithm for solving multi-dielectric Dirichlet problem," *IEEE Trans. Micro. Theo. Tech.*, vol. 41, no. 2, Feb. 1993, pp. 252–329.

50. A.F. Emery and W.W. Carson, "A modification to the Monte Carlo method—the Exodus method," *Trans. ASME, C, J. Heat Transf.*, vol. 90, 1968, pp. 328–332.

51. M.N.O. Sadiku and D. Hunt, "Solution of Dirichlet problems by the Exodus method," *IEEE Trans. Microw. Theory Tech.*, vol. 40, no. 1, Jan. 1992, pp. 89–95.

52. M.N.O. Sadiku, S.O. Ajose, and Z. Fu, "Applying the Exodus method to solve Poisson's equation," *IEEE Trans. Microw. Theory Tech.*, vol. 42, no. 4, April 1994, pp. 661–666.

53. W.H. McCrea and F.J.W. Whipple, "Random paths in two and three dimensions," *Proc. Roy. Soc. Edinb.*, vol. 60, 1940, pp. 281–298.

54. Fusco, V.F. and Linden, P.A., "A Markov chain approach for static field analysis," *Microw. Opt. Technol. Lett.*, vol. 1, no. 6, Aug. 1988, pp. 216–220.

55. M.N.O. Sadiku and R. Garcia, "Whole field computation using Monte Carlo Method," *Int. J. Num. Model.*, vol. 10, 1997, pp. 303–312.

56. M.E. Woodward, *Communication and Computer Networks*. Los Alamitos, CA: IEEE Computer Society Press, 1994, pp. 53–57.

57. J.G. Kemeny and J.L. Snell, *Finite Markov Chains*. New York: Springer–Verlag, 1976, pp. 43–68.

58. M. Iosifescu, *Finite Markov Processes and Their Applications*. New York: John Wiley & Sons, 1980, pp. 45, 99–106.

59. G.J. Anders, *Probability Concepts in Electric Power Systems*. New York: John Wiley & Sons, 1990, pp. 160–170.

60. T.J. Hoffman and N.E. Banks, "Monte Carlo surface density solution to the Dirichlet heat transfer problem," *Nucl. Sci. Engr.*, vol. 59, 1976, pp. 205–214.

61. T.J. Hoffman and N.E. Banks, "Monte Carlo solution to the Dirichlet problem with the double-layer potential density," *Trans. Am. Nucl. Sci.*, vol. 18, 1974, pp. 136, 137. See also vol. 19, 1974, p. 164; vol. 24, 1976, p. 181.

62. T.E. Booth, "Exact Monte Carlo solution of elliptic partial differential equations," *J. Comp. Phys.*, vol. 39, 1981, pp. 396–404.

63. T.E. Booth, "Regional Monte Carlo solution of elliptic partial differential equations," *J. Comp. Phys.*, vol. 47, 1982, pp. 281–290.

64. M.N.O. Sadiku, K. Gu, and C.N. Obiozor, "Regional Monte Carlo potential calculation using Markov chains," *Int. J. Eng. Educ.*, vol. 18, no. 6, 2002, pp. 745–752.

65. D. Netter, J. Levenque, P. Masson, and A. Rezzoug, "Monte Carlo method for transient eddy-current calculations," *IEEE Trans. Magn.*, vol. 40, no. 5, Sept. 2004, pp. 3450–3456.

66. M.N.O. Sadiku, C.M. Akujuobi, and S.M. Musa, "Monte Carlo analysis of time-dependent problems," *Proc. of IEEE Southeastcon*, 2006, pp. 7–10.

67. M.N.O. Sadiku, C.M. Akujuobi, S.M. Musa, and S.R. Nelatury, "Analysis of time-dependent cylindrical problems using Monte Carlo," *Micro. Opt. Tech. Lett.*, vol. 49, no. 10, Oct. 2007, pp. 2571–2573.

68. L.C. Andrews, *Elementary Partial Differential Equations with Boundary Value Problems*. Orlando, FL: Academic Press, 1986, pp. 459, 466–467.

69. L.C. Andrews, *Answer Booklet to Accompany Elementary Partial Differential Equations with Boundary Value Problems*. Orlando, FL: Academic Press, 1986, p. 33.

70. E.S. Troubetzkoy and N.E. Banks, "Solution of the heat diffusion equation by Monte Carlo," *Trans. Am. Nucl. Soc.*, vol. 19, 1974, pp. 163, 164.

71. A.W. Knapp, "Connection between Brownian motion and potential theory," *J. Math. Analy. Appl.*, vol. 12, 1965, pp. 328–349.

72. J.H. Halton, "A retrospective and prospective survey of the Monte Carlo method," *SIAM Rev.*, vol. 12, no. 1, Jan. 1970, pp. 1–61.

73. G.W. King, "Monte Carlo method for solving diffusion problems," *Ind. Engr. Chem.*, vol. 43, no. 11, Nov. 1951, pp. 2475–2478.

74. X.Q. Sheng et al., "Monte Carlo simulations of microstrip lines with random substrate impurity," *Int. J. RF Microw. Comput. Aided Eng.*, vol. 11, no. 4, 2001, pp. 177–187.

75. C.S. Indulkar, "Monte Carlo analysis of a transposed overhead line," *Int. J. Electr. Eng. Educ.*, vol. 37, no. 4, October 2000, pp. 368–373.

76. O. Ozgun and M. Kuzuoglu, "Monte Carlo simulations of Helmholtz scattering from randomly positioned array of scatterers by utilizing coordinate transformations in finite element method," *Wave Motion*, vol. 56, 2015, pp. 165–182.

77. Y.T. Lo, "Random periodic arrays," *Rad. Sci.*, vol. 3, no. 5, May 1968, pp. 425–436.

78. T. Troudet and R.J. Hawkins, "Monte Carlo simulation of the propagation of single-mode dielectric waveguide structures," *Appl. Opt.*, vol. 27, no. 24, Feb. 1988, pp. 765–773.

79. T.R. Rowbotham and P.B. Johns, "Waveguide analysis by random walks," *Elect. Lett.*, vol. 8, no. 10, May 1972, pp. 251–253.

80. P.B. Johns and T.R. Rowbothan, "Solution of resistive meshes by deterministic and Monte Carlo transmission-line modelling," *IEEE Proc.*, vol. 128, Part A, no. 6, Sept. 1981, pp. 453–462.

81. R.G. Olsen, "The application of Monte Carlo techniques to the study of impairments in the waveguide transmission system," *B. S. T. J.*, vol. 50, no. 4, April 1971, pp. 1293–1310.

82. H.E. Rowe and D.T. Young, "Transmission distortion in multimode random waveguides," *IEEE Trans. Micro. Theo. Tech.*, vol. MMT-20, no. 6, June 1972, pp. 349–365.

83. C. Huang et al., "Stationary phase Monte Carlo path integral analysis of electromagnetic wave propagation in graded-index waveguides," *IEEE Trans. Micro. Theo. Tech.*, vol. 42, no. 9, Sept. 1994, pp. 1709–1714.

84. H.T. Chou and J.T. Johnson, "A novel acceleration algorithm for the computation of scattering from a rough surfaces with the forward-backward method," *Radio Sci.*, vol. 33, no. 5, Sept./Oct. 1998, pp. 1277–1287.

85. M. Nieto-Vesperinas and J.M. Soto-Crespo, "Monte Carlo simulations for scattering of electromagnetic waves from perfectly conductive random rough surfaces," *Opt. Lett.*, vol. 12, no. 12, Dec. 1987, pp. 979–981.

86. G.P. Bein, "Monte Carlo computer technique for one-dimensional random media," *IEEE Trans. Ant. Prop.*, vol. AP-21, no. 1, Jan. 1973, pp. 83–88.

87. N. Garcia and E. Stoll, "Monte Carlo calculation for electromagnetic-wave scattering from random rough surfaces," *Phy. Rev. Lett.*, vol. 52, no. 20, May 1984, pp. 1798–801.

88. J. Nakayama, "Anomalous scattering from a slightly random surface," *Rad. Sci.*, vol. 17, no. 3, May–June 1982, pp. 558–564.

89. A. Ishimaru, *Wave Propagation and Scattering in Random Media*, vol. 2. New York: Academic Press, 1978.

90. A.K. Fung and M.F. Chen, "Numerical simulation of scattering from simple and composite random surfaces," *J. Opt. Soc. Am.*, vol. 2, no. 12, Dec. 1985, pp. 2274–2284.

91. L. Wang et al., "Electromagnetic scattering model for rice canopy based on Monte Carlo simulation," *Prog. Electromagn. Res.*, vol. 52, 2005, pp. 153–171.

92. R.A. Arratia and H.N. Bertram, "Monte Carlo simulation of particulate noise in magnetic recording," *IEEE Trans. Magn.*, vol. MAG-20, no. 2, Mar. 1984, pp. 412–420.

93. P.K. Davis, "Monte Carlo analysis of recording codes," *IEEE Trans. Magn.*, vol. MAG-20, no. 5, Sept. 1984, p. 887.

94. J.H. Pickles, "Monte-Carlo calculation of electrically induced human-body currents," *IEEE Proc.*, vol. 134, Pt. A, no. 9, Nov. 1987, pp. 705–711.

95. D. Voyer et al., "Probabilistic methods applied to 2D electromagnetic numerical dosimetry," *COMPEL - Int. J. Comput. Math. Electr. Electr. Eng.*, vol. 27, no. 3, 2008, pp. 651–667.
96. K. Gu and M.N.O. Sadiku, "A triangular mesh random walk for Dirichlet problems," *J. Franklin Inst.*, vol. 332B, no. 5, 1995, pp. 569–578.
97. M.N.O. Sadiku and K. Gu, "A new Monte Carlo method for Neumann problems," *Proceedings of IEEE Southeastcon*, 1996, pp. 88–91.
98. R.C. Garcia and M.N.O. Sadiku, "Neuro-Monte Carlo solution of electrostatic problems," *J. Franklin Inst.*, vol. 335B, no. 1, 1998, pp. 53–69.
99. M.N. Barber and B.W. Ninham, *Random and Restricted Walks.* New York: Gordon and Breach, 1970.
100. K.K. Sabelfeld, *Monte Carlo Methods in Boundary Value Problems.* New York: Springer-Verlag, 1991.
101. M.N.O. Sadiku, *Monte Carlo Methods for Electromagnetics.* Boca Raton, FL: CRC Press, 2009.

9

Method of Lines

The difficulties of life are intended to make us better, not bitter.

— Unknown

9.1 Introduction

The method of lines (MOL) is a well-established numerical technique (or rather a semianalytical method) for the analysis of transmission lines, waveguide structures, and scattering problems. The method was originally developed by mathematicians and used for boundary value problems in physics and mathematics (e.g., [1–5]). A review of these earlier uses (1930–1965) of MOL is found in Liskovets [6]. The method was introduced into the EM community around 1980 and further developed by Pregla et al. [7–15] and other researchers. Although the formulation of this modern application is different from the earlier approach, the basic principles are the same.

The MOL is regarded as a special finite difference method (FDM) but more effective with respect to accuracy and computational time than the regular FDM. It basically involves discretizing a given differential equation in one or two dimensions while using analytical solution in the remaining dimension. MOL has the merits of both the finite difference method and analytical method; it does not yield spurious modes nor does it have the problem of "relative convergence."

Besides, the MOL has the following properties that justify its use:

a. Computational efficiency: the semianalytical character of the formulation leads to a simple and compact algorithm, which yields accurate results with less computational effort than other techniques.

b. Numerical stability: by separating discretization of space and time, it is easy to establish stability and convergence for a wide range of problems.

c. Reduced programming effort: by making use of the state-of-the-art, well-documented, and reliable ordinary differential equations (ODE) solvers, programming effort can be substantially reduced.

d. Reduced computational time: since only a small amount of discretization lines is necessary in the computation, there is no need to solve a large system of equations; hence computing time is small.

To apply MOL usually involves the following five basic steps [18]:

1. Partitioning the solution region into layers
2. Discretization of the differential equation in one coordinate direction

3. Transformation to obtain decoupled ordinary differential equations

4. Inverse transformation and introduction of the boundary conditions

5. Solution of the equations

We begin to apply these steps to the problem of solving Laplace's equation. Since MOL involves many matrix manipulations, it is expedient that all computer codes in chapters are written in MATLAB.

9.2 Solution of Laplace's Equation

Although the MOL is commonly used in the EM community for solving hyperbolic (wave equation), it can be used to solve parabolic and elliptic equations [1,15–18]. In this section, we consider the application of MOL to solve Laplace's equation (elliptic problem) involving two-dimensional rectangular and cylindrical regions.

9.2.1 Rectangular Coordinates

Laplace's equation in Cartesian system is

$$\frac{\partial^2 V}{\partial x^2} + \frac{\partial^2 V}{\partial y^2} = 0 \tag{9.1}$$

Consider a two-dimensional solution shown in Figure 9.1. The first step is discretization of the x-variable. The region is divided into strips by N dividing straight lines (hence the name *method of lines*) parallel to the y-axis. Since we are discretizing along x, we replace the second derivative with respect to x with its finite difference equivalent. We apply the three-point central difference scheme,

$$\frac{\partial^2 V_i}{\partial x^2} = \frac{V_{i+1} - 2V_i + V_{i-1}}{h^2} \tag{9.2}$$

where h is the spacing between discretized lines, that is,

$$h = \Delta x = \frac{a}{N+1} \tag{9.3}$$

Replacing the derivative with respect to x by its finite difference equivalent, Equation 9.1 becomes

$$\frac{\partial^2 V_i}{\partial x^2} + \frac{1}{h^2}[V_{i+1}(y) - 2V_i(y) + V_{i-1}(y)] = 0 \tag{9.4}$$

Thus the potential V in Equation 9.1 can be replaced by a vector of size N, namely,

$$[V] = [V_1, V_2, \ldots, V_N]^t \tag{9.5a}$$

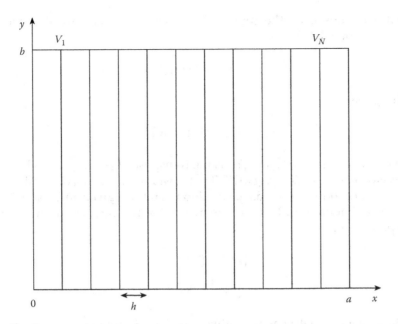

FIGURE 9.1
Illustration of discretization in the *x*-direction.

where *t* denotes the transpose,

$$V_i(y) = V(x_i, y), i = 1, 2, ..., N \tag{9.5b}$$

and $x_i = i\Delta x$. Substituting Equations 9.4 and 9.5 into Equation 9.1 yields

$$\frac{\partial^2[V(y)]}{\partial y^2} - \frac{1}{h^2}[P][V(y)] = [0] \tag{9.6}$$

where [0] is a zero column vector and [P] is an $N \times N$ tridiagonal matrix representing the discretized form of the second derivative with respect to *x*.

$$[P] = \begin{bmatrix} p_\ell & -1 & 0 & \cdots & 0 \\ -1 & 2 & -1 & \cdots & 0 \\ & \ddots & \ddots & \ddots & \\ 0 & \cdots & -1 & 2 & -1 \\ 0 & \cdots & 0 & -1 & p_r \end{bmatrix} \tag{9.7}$$

All the elements of matrix [P] are zeros except the tridiagonal terms; the elements of the first and the last row of [P] depend on the boundary conditions at $x = 0$ and $x = a$. $p_\ell = 2$ for Dirichlet boundary condition and $p_\ell = 1$ for Neumann boundary condition. The same is true of p_r.

The next step is to analytically solve the resulting equations along the *y* coordinate. To solve Equation 9.6 analytically, we need to obtain a system of uncoupled ordinary differential equations from the coupled Equation 9.6. To achieve this, we define the transformed potential $[\bar{V}_i]$ by letting

$$[V] = [T]\,[\bar{V}_i] \tag{9.8}$$

and requiring that

$$[T]^t\,[P][T] = [\lambda^2] \tag{9.9}$$

where $[\lambda^2]$ is a diagonal matrix and $[T]^t$ is the transpose of $[T]$. $[\lambda^2]$ and $[T]$ are eigenvalue and eigenvector matrices belonging to $[P]$. The transformation matrix $[T]$ and the eigenvalue matrix $[\lambda^2]$ depend on the boundary conditions and are given in Table 9.1 for various combinations of boundaries. It should be noted that the eigenvector matrix $[T]$ has the following properties:

$$[T]^{-1} = [T]^t$$
$$[T][T]^t = [T]^t\,[T] = [I] \tag{9.10}$$

where $[I]$ is an identity matrix. Substituting Equation 9.8 into Equation 9.6 gives

$$\frac{\partial^2 [T][\bar{V}]}{\partial y^2} - \frac{1}{h^2}\,[P][T][\bar{V}] = [0]$$

Multiplying through by $[T]^{-1} = [T]^t$ yields

$$\left(\frac{\partial^2}{\partial y^2} - \frac{1}{h^2}\,[\lambda^2] \right)[\bar{V}] = [0] \tag{9.11}$$

This is an ordinary differential equation with solution

$$\bar{V}_i = A_i \cosh \alpha_i y + B_i \sinh \alpha_i y \tag{9.12}$$

where $\alpha_i = \lambda_i / h$.

TABLE 9.1

Elements of Transformation Matrix $[T]$ and Eigenvalues

Left Boundary	Right Boundary	T_{ij}	λ_i
Dirichlet	Dirichlet	$\sqrt{\frac{2}{N+1}}\,\sin\frac{ij\pi}{N+1},[T_{DD}]$	$2\sin\frac{i\pi}{2(N+1)}$
Dirichlet	Neumann	$\sqrt{\frac{2}{N+0.5}}\,\sin\frac{i(j-0.5)\pi}{N+0.5},[T_{DN}]$	$2\sin\frac{(i-0.5)\pi}{2N+1}$
Neumann	Dirichlet	$\sqrt{\frac{2}{N+0.5}}\,\cos\frac{(i-0.5)(j-0.5)\pi}{N+0.5},[T_{ND}]$	$2\sin\frac{(i-0.5)\pi}{2N+1}$
Neumann	Neumann	$\sqrt{\frac{2}{N}}\,\cos\frac{(i-0.5)(j-1)\pi}{N},j>1,[T_{NN}]$ $\frac{1}{\sqrt{N}},j=1$	$2\sin\frac{(i-1)\pi}{2N}$

Note: Where $i, j = 1,2, \dots, N$ and subscripts D and N are for Dirichlet and Neumann conditions, respectively.

Thus, Laplace's equation is solved numerically using a finite difference scheme in the x-direction and analytically in the y-direction. However, we have only demonstrated three out of the five basic steps for applying MOL. There remain two more steps to complete the solution: imposing the boundary conditions and solving the resulting equations. Imposing the boundary conditions is problem dependent and will be illustrated in Example 9.1. The resulting equations can be solved using the existing packages for solving ODE or developing our own codes in FORTRAN, MATLAB, C, or any other programming language. We will take the latter approach in Example 9.1.

EXAMPLE 9.1

For the rectangular region in Figure 9.1, let

$$V(0, y) = V(a, y) = V(x, 0) = 0, \quad V(x, b) = 100$$

and $a = b = 1$. Find the potential at $(0.25, 0.75)$, $(0.5, 0.5)$, $(0.75, 0.25)$.

Solution

In this case, we have Dirichlet boundaries at $x = 0$ and 1, which are already indirectly taken care of in the solution in Equation 9.12. Hence, from Table 9.1,

$$\lambda_i = 2\sin\frac{i\pi}{2(N+1)} \tag{9.13}$$

and

$$T_{ij} = \sqrt{\frac{2}{N+1}}\sin\frac{i j\pi}{N+1} \tag{9.14}$$

Let $N = 15$ so that $h = \Delta x = 1/16$ and $x = 0.25, 0.5, 0.75$ will correspond to $i = 4, 8, 12$, respectively.

By combining Equations 9.8 and 9.12, we obtain the required solution. To get constants A_i and B_i, we apply boundary conditions at $y = 0$ and $y = b$ to V and perform inverse transformation. Imposing $V(x, y = 0) = 0$ to the combination of Equations 9.8 and 9.12, we obtain

$$\begin{bmatrix} V_1 \\ V_2 \\ \vdots \\ V_N \end{bmatrix} = [0] = \begin{bmatrix} T_{11} & T_{12} & \cdots & T_{1N} \\ T_{21} & T_{22} & \cdots & T_{2N} \\ \vdots & & \cdots & \vdots \\ T_{N1} & T_{N2} & \cdots & T_{NN} \end{bmatrix} \begin{bmatrix} A_1 \\ A_2 \\ \vdots \\ A_N \end{bmatrix}$$

which implies that

$$[A] = 0 \quad \text{or} \quad A_i = 0 \tag{9.15}$$

Imposing $V(x, y = b) = 100$ yields

$$\begin{bmatrix} 100 \\ 100 \\ \vdots \\ 100 \end{bmatrix} = [T] \begin{bmatrix} B_1 \sinh \alpha_1 b \\ B_2 \sinh \alpha_2 b \\ \vdots \\ B_N \sinh \alpha_N b \end{bmatrix}$$

If we let

$$[C] = \begin{bmatrix} B_1 \sinh \alpha_1 b \\ B_2 \sinh \alpha_2 b \\ \vdots \\ B_N \sinh \alpha_N b \end{bmatrix} = [T]^{-1} \begin{bmatrix} 100 \\ 100 \\ \vdots \\ 100 \end{bmatrix}$$

then

$$B_i = C_i / \sinh \alpha_i b \tag{9.16}$$

With A_i and B_i found in Equations 9.15 and 9.16, the potential $V(x, y)$ is determined as

$$V_i(y) = \sum_{j=1}^{N} T_{ij} B_j \sinh(\alpha_j y) \tag{9.17}$$

By applying Equations 9.13 through 9.17, the MATLAB code in Figure 9.2 was developed to obtain

```
AA = 1;
BB = 1;
N = 15;
% DETERMINE VECTOR ALPHA
H = AA/(N+1);
LAM = 2*sin ((1:N)*pi*0.5/(N+1) );
ALPHA = LAM/H;
% CALCULATE THE TRANSFORMATION MATRIX AND COEFFICIENT B
K = sqrt (2/(N+1));
T = zeros(N,N);
for I=1:N
  for J=1:N
    T(I,J) = K*sin (I*J*pi/(N+1));
  end
end
V = 100*ones(N,1);
C = inv(T)*V;
A = ALPHA';
B = C./sinh(BB*A);
% CALCULATE V AT THE GIVEN POINTS
V1 = 0;  V2 = 0;  V3 = 0;
for K=1:N
V1 = V1 + T(4,K)*B(K)*sinh(ALPHA(K)*0.75);
V2 = V2 + T(8,K)*B(K)*sinh(ALPHA(K)*0.5);
V3 = V3 + T(12,K)*B(K)*sinh(ALPHA(K)*0.25);
end
diary
V1, V2, V3
diary off
```

FIGURE 9.2
MATLAB code for Example 9.1.

$$V(0.25, 0.75) = 43.1, \quad V(0.5, 0.5) = 24.96, \quad V(0.75, 0.25) = 6.798$$

The result compares well with the exact solution:

$$V(0.25, 0.75) = 43.2, \quad V(0.5, 0.5) = 25.0, \quad V(0.75, 0.25) = 6.797$$

Notice that it is not necessary to invert the transformation matrix $[T]$ in view of Equation 9.10.

EXAMPLE 9.2

For Dirichlet–Neumann conditions, derive the transformation matrix $[T_{DN}]$ and the corresponding eigenvalues $[\lambda^2]$.

Solution

Let λ_k^2 be the elements of eigenvalue matrix $[\lambda^2]$ and $[t_k]$ be the column vectors of the transformation matrix $[T_{DN}]$ corresponding to matrix $[P]$. Then, by definition,

$$([P] - \lambda_k^2[I]) [t_k] = [0] \tag{9.18}$$

Substituting $[P]$ for Dirichlet–Neumann (DN) condition in Equation 9.7 into Equation 9.18 gives a second-order difference equation

$$-t_{i-1}^{(k)} + (2 - \lambda_k^2)t_i^{(k)} - t_{i-1}^{(k)} = 0 \tag{9.19}$$

except the first and last equations in Equation 9.18. If we let

$$t_i^{(k)} = A_k e^{ji\phi_k} + B_k e^{-ji\phi_k} \tag{9.20}$$

and substitute this into Equation 9.19, we obtain

$$0 = (A_k e^{ji\phi_k} + B_k e^{-ji\phi_k})(-2\cos\phi_k + 2 - \lambda_k^2)$$

from which we obtain the characteristic equation

$$\lambda_k^2 = 2(1 - \cos\phi_k) = 4\sin^2\frac{\phi_k}{2} \tag{9.21}$$

or

$$\lambda_k = 2\sin\frac{\phi_k}{2} \tag{9.22}$$

This is valid for all types of boundary combinations but ϕ_k will depend on the boundary conditions. To determine ϕ_k, A_k, and B_k, we use the first and the last equations in Equation 9.18. For DN conditions,

$$t_0^{(k)} = 0 \tag{9.23a}$$

$$-t_N^{(k)} + t_{N+1}^{(k)} = 0 \tag{9.23b}$$

Substituting this into Equation 9.20, we obtain

$$\begin{bmatrix} 1 & 1 \\ e^{jN\phi_k}(e^{j\phi_k} - 1) & e^{-jN\phi_k}(e^{-j\phi_k} - 1) \end{bmatrix} \begin{bmatrix} A_k \\ B_k \end{bmatrix} = [0] \tag{9.24}$$

For nontrivial solutions,

$$\phi_k = \frac{k - 0.5}{N + 0.5} \pi, \quad k = 1, 2, \dots, N \tag{9.25}$$

Also from Equations 9.23a and 9.20, $A_k = -B_k$ so that

$$t_i^{(k)} = A_k \sin(i\phi_k) \tag{9.26}$$

Thus, for Dirichlet–Neumann conditions, we obtain

$$\lambda_k = 2\sin\left(0.5\pi \frac{k - 0.5}{N + 0.5}\right) \tag{9.27a}$$

$$T_{ij} = \sqrt{\frac{2}{N + 0.5}} \sin\left(0.5\pi \frac{i(k - 0.5)}{N + 0.5}\right) \tag{9.27b}$$

9.2.2 Cylindrical Coordinates

Although MOL is not applicable to problems with complex geometry, the method can be used to analyze homogeneous and inhomogeneous cylindrical problems. The principal steps in applying MOL in cylindrical coordinates are the same as in Cartesian coordinates.

Here, we illustrate with the use of MOL to solve Laplace's equation in cylindrical coordinates [18]. We apply discretization procedure in the angular direction. The resulting coupled ordinary differential equations are decoupled by matrix transformation and solved analytically.

Assume that we are interested in finding the potential distribution in a cylindrical transmission line with a uniform but arbitrary cross section. We assume that the inner conductor is grounded while the outer conductor is maintained at constant potential V_o, as shown in Figure 9.3. In cylindrical coordinates (ρ, ϕ), Laplace's equation can be expressed as

$$\rho^2 \frac{\partial^2 V}{\partial \rho^2} + \rho \frac{\partial V}{\partial \rho} + \frac{\partial^2 V}{\partial \phi^2} = 0 \tag{9.28}$$

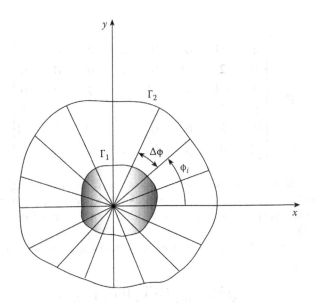

FIGURE 9.3
Discretization along ϕ-direction.

subject to

$$V(\rho) = 0, \quad \rho \in \Gamma_1 \tag{9.29a}$$

$$V(\rho) = V_{o'}, \quad \rho \in \Gamma_2 \tag{9.29b}$$

We discretize in the ϕ-direction by using N radial lines, as shown in Figure 9.3, such that

$$V_i(\rho) = V(\rho, \phi_i), \quad i = 1, 2, \ldots, N \tag{9.30}$$

where

$$\phi_i = ih = \frac{2\pi i}{N}, \, h = \Delta\phi = \frac{2\pi}{N} \tag{9.31}$$

and h is the angular spacing between the lines. We have subdivided the solution region into N subregions with boundaries at Γ_1 and Γ_2. In each subregion, $V(\rho, \phi)$ is approximated by $V_i = V(\rho, \phi_i)$, with ϕ_i being constant.

Applying the three-point central finite difference scheme yields

$$\frac{\partial^2 [V]}{\partial \phi^2} = -\frac{[P]}{h^2} [V] \tag{9.32}$$

where

$$[V] = [V_1, V_2, \ldots, V_N]^t \tag{9.33}$$

and

$$[P] = \begin{bmatrix} 2 & -1 & 0 & 0 & \cdots & 0 & 0 & -1 \\ -1 & 2 & -1 & 0 & \cdots & 0 & 0 & 0 \\ 0 & -1 & 2 & -1 & \cdots & 0 & 0 & 0 \\ \vdots & \vdots & \vdots & \vdots & \cdots & \vdots & \vdots & \vdots \\ 0 & 0 & 0 & 0 & \cdots & -1 & 2 & -1 \\ -1 & 0 & 0 & 0 & \cdots & 0 & -1 & 2 \end{bmatrix} \tag{9.34}$$

Notice that $[P]$ contains an element -1 in the lower left and upper right corners due to its angular periodicity. Also, notice that $[P]$ is a quasi-three-band symmetric matrix which is independent of the arbitrariness of the cross section as a result of the discretization over a finite interval $[0, 2\pi]$.

Introducing Equation 9.32 into Equation 9.28 leads to the following set of coupled differential equations

$$\rho^2 \frac{\partial^2 [V]}{\partial \rho^2} + \rho \frac{\partial [V]}{\partial \rho} - \frac{[P]}{h^2} [V] = 0 \tag{9.35}$$

To decouple Equation 9.35, we must diagonalize $[P]$ by an orthogonal matrix $[T]$ such that

$$[\lambda^2] = [T]^t [P][T] \tag{9.36}$$

with

$$[T]^t = [T] = [T]^{-1} \tag{9.37}$$

where $[\lambda^2]$ is a diagonal matrix of the eigenvalues λ_n^2 of $[P]$. The diagonalization is achieved using [19]

$$T_{ij} = \frac{\cos \alpha_{ij} + \sin \alpha_{ij}}{\sqrt{N}}, \quad \lambda_n^2 = 2(1 - \cos \alpha_n) \tag{9.38}$$

where

$$\alpha_{ij} = h \cdot i \cdot j, \quad \alpha_n = h \cdot n, \quad i, j, n = 1, 2, \ldots, N \tag{9.39}$$

If we introduce the transformed potential U that satisfies

$$[U] = [T][V] \tag{9.40}$$

Equation 9.35 becomes

$$\rho^2 \frac{\partial^2 [U]}{\partial \rho^2} + \rho \frac{\partial [U]}{\partial \rho} - [\mu^2][U] = 0 \tag{9.41}$$

where

$$[U] = [U_1, U_2, \ldots, U_N]^t \tag{9.42}$$

is a vector containing the transformed potential function and

$$\mu_n = \frac{\lambda_n}{h} = \frac{2}{h}\sin(\alpha_n/2) \tag{9.43}$$

Equation 9.41 is the Euler-type and has the analytical solution (see Section 2.4.1)

$$U_n = \begin{cases} A_n + B_n \ln \rho, & \mu_n = 0 \\ A_n \rho^{\mu_n} + B_n \rho^{-\mu_n}, & \mu_n \neq 0 \end{cases} \tag{9.44}$$

This is applied to each subregion. By taking the inverse transform using Equation 9.40, we obtain the potential $V_i(\rho)$ as

$$V_i(\rho) = \sum_{j=1}^{N} T_{ij} U_j \tag{9.45}$$

where T_{ij} are the elements of matrix $[T]$.

We now impose the boundary conditions in Equation 9.29, which can be rewritten as

$$V(\rho = r_i) = 0, \quad r_i \in \Gamma_1 \tag{9.46a}$$

$$V(\rho = R_i) = V_o, \quad R_i \in \Gamma_2 \tag{9.46b}$$

Applying these to Equations 9.44 and 9.45,

$$T_{ij}\left[A_j + B_j \ln r_i\right]\Big|_{\mu_j=0} + \sum_{j=1}^{N} T_{ij}\left[A_j r_i^{\mu_j} + B_j r_i^{-\mu_j}\right]\Big|_{\mu_j \neq 0} = 0, \quad i = 1, 2, \ldots, N \tag{9.47a}$$

$$T_{ij}\left[A_j + B_j \ln r_i\right]\Big|_{\mu_j=0} + \sum_{j=1}^{N} T_{ij}\left[A_j R_i^{\mu_j} + B_j R_i^{-\mu_j}\right]\Big|_{\mu_j \neq 0} = V_o, \quad i = 1, 2, \ldots, N \tag{9.47b}$$

Equation 9.47 is solved to determine the unknown coefficients A_i and B_i. The potential distribution is finally obtained from Equations 9.44 and 9.45.

EXAMPLE 9.3

Consider a coaxial cable with inner radius a and outer radius b. Let $b = 2a = 2$ cm and $V_o = 100$ V. This simple example is selected to be able to compare MOL solution with the exact solution.

Solution

From Equation 9.43, it is evident that $\mu_n = 0$ only when $n = N$. Hence, we may write U as

$$U_n = \begin{cases} A_n \rho^{\mu_n} + B_n \rho^{-\mu_n}, & n = 1, 2, \ldots, N-1 \\ A_n + B_n \ln \rho, & n = N \end{cases} \tag{9.48}$$

Equation 9.47 can be written as

$$\sum_{j=1}^{N-1} T_{ij}\left[A_j a_i^{\mu_j} + B_j a_i^{-\mu_j} \right] + T_{iN}[A_N + B_N \ln a] = 0, \quad i = 1, 2, \ldots, N \tag{9.49a}$$

for $\rho = a$, and

$$\sum_{j=1}^{N-1} T_{ij}\left[A_j b_i^{\mu_j} + B_j b_i^{-\mu_j} \right] + T_{iN}[A_N + B_N \ln b] = V_o, \quad i = 1, 2, \ldots, N \tag{9.49b}$$

for $\rho = b$. These $2N$ equations will enable us to find the $2N$ unknown coefficients A_i and B_i. They can be cast into a matrix form as

$$\begin{bmatrix} T_{11}a^{\mu_1} & \cdots & T_{1N} & T_{11}a^{-\mu_1} & \cdots & \ln a \\ \vdots & & & & & \vdots \\ T_{N1}a^{\mu_1} & \cdots & T_{NN} & T_{N1}a^{-\mu_1} & \cdots & \ln a \\ T_{11}b^{\mu_1} & \cdots & T_{1N} & T_{11}b^{-\mu_1} & \cdots & \ln b \\ \vdots & & & & & \vdots \\ T_{N1}b^{\mu_1} & \cdots & T_{NN} & T_{N1}b^{-\mu_1} & \cdots & \ln b \end{bmatrix} \begin{bmatrix} A_1 \\ A_2 \\ \vdots \\ A_N \\ B_1 \\ B_2 \\ \vdots \\ B_N \end{bmatrix} = \begin{bmatrix} 0 \\ 0 \\ \vdots \\ 0 \\ 100 \\ 100 \\ \vdots \\ 100 \end{bmatrix} \tag{9.50}$$

This can be written as

$$[D][C] = [F] \tag{9.51}$$

from which we obtain

$$[C] = [D]^{-1}[F] \tag{9.52}$$

where C_j corresponds to A_j when $j = 1, 2, \ldots, N$ and C_j corresponds to B_j when $j = N + 1, \ldots, 2N$.

Once A_j and B_j are known, we substitute them into Equation 9.48 to find U_j. We finally apply Equation 9.45 to find V. The exact analytical solution of the problem is

$$V(\rho) = V_o \frac{\ln \dfrac{\rho}{a}}{\ln \dfrac{b}{a}} \tag{9.53}$$

For $a < \rho < b$, we obtain V for both exact and MOL solutions using the MATLAB codes in Figure 9.4. The results of the two solutions are shown in Figure 9.5. The two solutions agree perfectly.

9.3 Solution of Wave Equation

The MOL is particularly suitable for modeling a wide range of transmission lines and planar waveguide structures with multiple layers [8,11,19–29]. This involves discretizing

EXAMPLE 9.3 SOLVED USING METHOD OF LINES

```
% OUR OBJECTIVE IS TO DETERMINE THE POTENTIAL
% DISTRIBUTION IN A COAXIAL CABLE OF INNER RADIUS a
% AND OUTER RADIUS b ASSUMING A POTENTIAL DIFFERENCE OF Vo

a= 0.01; b = 0.02; Vo = 100;
N = 15;
h = 2*pi/N;
K = 1/sqrt (N);
% COMPUTE THE TRANSFORMATION MATRIX T AND MIU
T = zeros(N,N);
miu = zeros(N,1);
for I=1:N
  miu(I) = 2*sin (I*h*0.5 )/h;
  for J=1:N
     alpha = I*J*h;
     T(I,J) = K*( cos(alpha) + sin(alpha) );
  end
end
% CALCULATE THE MATRIX D IN EQ. (9.50)
D = zeros(2*N,2*N);
for i=1:2*N
   for j=1:2*N
%Do the upper part of the matrix
     if (i <= N & j < N)
        D(i,j) = T(i,j)*a^miu(j);
     end
     if (i <= N & j == N)
        D(i,j) = T(i,j);
     end
     if (i <= N & j > N & j < 2*N)
        D(i,j) = T(i,j-N)*a^(-miu(j-N));
     end
     if (i <= N & j == 2*N)
        D(i,j) = T(i,j-N)*log(a);
     end
%Now do the lower part of the matrix
     if (i > N & i <= 2*N & j < N)
        D(i,j) = T(i-N,j)*b^miu(j);
     end
     if (i > N & i <= 2*N & j == N)
        D(i,j) = T(i-N,j);
     end
     if (i > N & i <= 2*N & j > N & j < 2*N)
        D(i,j) = T(i-N,j-N)*b^(-miu(j-N));
     end
     if (i > N & i <= 2*N & j == 2*N)
        D(i,j) = T(i-N,j-N)*log(b);
     end
   end
end
```

FIGURE 9.4
MATLAB code for Example 9.3. *(Continued)*

the Helmholtz's wave equation in one direction while the other direction is treated analytically. Here, we consider the general problem of two-layer structures covered on the top and bottom with perfectly conducting planes. The conducting strips are assumed to be thin. We will illustrate with two-layer planar and cylindrical microstrip structures.

```
% DETERMINE THE BOUNDARY POTENTIAL MATRIX
% AND THE COEFFICIENT MATRIX
F = zeros(2*N,1);
for i=1:2*N
   if i > N
       F(i) = Vo;
   end
end
C = inv(D)*F;
% WITH THE COEFFICIENTS DETERMINED,
% NOW FIND TRANSFORMED POTENTIAL U
% AND FINALLY DETERMINE THE POTENTIAL V USING EQ. (9.45)
% WE MAY SELECT ANY VALUE OF phi, say, phi = 0^o, i.e. i=1
rho = 0.01:0.001:0.02;
M = 10; % no. of divisions along rho
Vmol = zeros(M+1,1);
for k=1:M+1
   Vmol(k) = 0.0;
   for j=1:N
    if (j <N)
    U(j) = C(j)*(rho(k))^miu(j) + C(j+N)*(rho(k))^(-miu(j));
     end
    if (j == N)
       U(j) = C(j) + C(j+N)*log(rho(k));
    end
       Vmol(k) = Vmol(k) + T(10,j)*U(j);
%        Vmol(k) = Vmol(k) + T(1,j)*U(j);
     end
end
% ALSO, CALCULATE THE EXACT VALUE OF V
Vex = Vo* (log(rho/a))/log(b/a);
diary
Vmol, Vex'
diary off
hold off
plot (rho,Vmol);
title( 'Fig. 9.5  Comparison of exact and method of lines solutions.')
xlabel('rho'), ylabel('V')
hold on
plot (rho,Vex);
hold off
```

FIGURE 9.4 (Continued)
MATLAB code for Example 9.3.

9.3.1 Planar Microstrip Structures

Typical planar structures are shown in Figure 9.6. The two independent field components E_z and H_z in each separate layer must satisfy the Helmholtz's equation. Assuming the factor $e^{j(\omega t - \beta z)}$ and that wave propagates along z,

$$\frac{\partial^2 \psi}{\partial x} + \frac{\partial^2 \psi}{\partial y} + (k^2 - \beta^2)\psi = 0 \tag{9.54}$$

where ψ represents either E_z or H_z and

$$k^2 = \epsilon_r k_o^2, \quad k_o = \omega\sqrt{\mu_o \epsilon_o} = 2\pi / \lambda_o \tag{9.55}$$

Applying the MOL, we discretize the fields along the x direction by laying a family of straight lines parallel to the y axis and evaluating on the e-lines for E_z and h-lines for H_z,

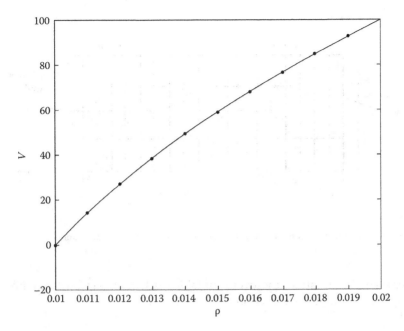

FIGURE 9.5
Comparison of exact and MOL solutions.

FIGURE 9.6
Typical planar structures.

as shown in Figure 9.7. The lines are evenly spaced although this is not necessary. If h is the spacing between adjacent lines, it is expedient to shift the e-lines and the h-lines by $h/2$ in order to guarantee a simple fitting of the literal boundary conditions. The potential in Equation 9.54 can now be replaced by a set $[\psi_1, \psi_2, ..., \psi_N]$ at lines

$$x_i = x_0 + ih, \quad i = 1, 2, ..., N \tag{9.56}$$

and $\partial\psi_i/\partial x$ can be replaced by their finite difference equivalents. Thus, Equation 9.54 becomes

$$\frac{\partial^2 \psi_i}{\partial y^2} + \frac{1}{h^2}[\psi_{i+1}(y) - 2\psi_i(y) + \psi_{i+1}(y)] + k_c^2\psi_i(y) = 0. \quad i = 1, 2, ..., N \tag{9.57}$$

where

$$k_c^2 = k^2 - \beta^2 \tag{9.58}$$

This is a system of N coupled ordinary differential equations. We cannot solve them in their present form because the equations are coupled due to the tridiagonal nature of $[P]$.

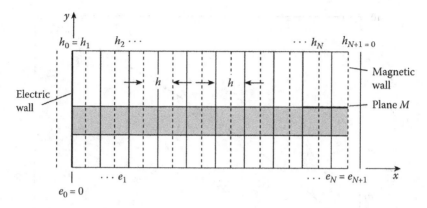

FIGURE 9.7
Cross-section of planar microstrip structure with discretization lines; —— for E_z and - - - - for H_z.

We can decouple the equations by several suitable mathematical transformations and then analytically solve along the y direction.

If we let

$$[\psi] = [\psi_1, \psi_2, ..., \psi_N]^t \tag{9.59}$$

where t denotes the transpose and

$$[P] = \begin{bmatrix} p_\ell & -1 & & & & \\ -1 & 2 & -1 & & & \\ \ddots & \ddots & \ddots & & & \\ & & -1 & 2 & -1 \\ & & & -1 & p_r \end{bmatrix} \tag{9.60}$$

which is the same as Equation 9.7, where p_ℓ and p_r are defined. Introducing the column vector $[\psi]$ and the matrix $[P]$ into Equation 9.57 leads to

$$h^2 \frac{\partial^2 [\psi]}{\partial y^2} - \left([P] - h^2 k_c^2 [I]\right)[\psi] = [0] \tag{9.61}$$

where $[I]$ is the identity matrix and $[0]$ is a zero column vector. Since $[P]$ is a real symmetric matrix, we can find an orthogonal matrix $[T]$ such that

$$[T]^t [P][T] = [\lambda^2] \tag{9.62}$$

where the elements λ_i^2 of the diagonal matrix $[\lambda^2]$ are the eigenvalues of $[P]$. With the orthogonal matrix $[T]$, we now introduce a transformed vector $[U]$ such that

$$[T]^t [\psi] = [U] \tag{9.63}$$

We can rewrite Equation 9.61 in terms of $[U]$ and obtain

$$h^2 \frac{\partial^2 U_i}{\partial y^2} - \left(\lambda_i^2 - h^2 k_c^2\right)U_i = 0, \quad i = 1, 2, \ldots, N \tag{9.64}$$

Since Equation 9.64 is uncoupled, it can be solved analytically for each homogeneous region. The solution is similar in form to the telegraph equation. It may be expressed as a relation between U_i and its normal derivative in a homogeneous dielectric layer from $y = y_1$ to $y = y_2$, that is,

$$\begin{bmatrix} U_i(y_1) \\ h\frac{\partial U_i(y_1)}{dy} \end{bmatrix} = \begin{bmatrix} \cosh\alpha_i(y_1 - y_2) & \frac{1}{k_i}\sinh\alpha_i(y_1 - y_2) \\ k_i\sinh\alpha_i(y_1 - y_2) & \cosh\alpha_i(y_1 - y_2) \end{bmatrix} \begin{bmatrix} U_i(y_2) \\ h\frac{\partial U_i(y_2)}{dy} \end{bmatrix} \tag{9.65}$$

where

$$k_i = \left(\lambda_i^2 - h^2 k_c^2\right)^{1/2}$$
$$\alpha_i = \frac{k_i}{h}, \quad i = 1, 2, \ldots, N \tag{9.66}$$

Equation 9.65 can be applied repeatedly to find the transformed potential $[U]$ from one homogeneous layer $y_1 < y < y_2$ to another. Keep in mind that each iteration will require that we recalculate the transformation matrix $[T]$ and its eigenvalues λ_i, which are given in Table 9.1. The field components E_z and H_z are derivable from the scalar potentials $\psi^{(e)}$ and $\psi^{(h)}$ as

$$E_z = \frac{k_c}{j\omega\epsilon}\psi^{(e)} \tag{9.67a}$$

$$H_z = \frac{k_c}{j\omega\mu}\psi^{(h)} \tag{9.67b}$$

To be concrete, consider the shielded microstrip line shown in Figure 9.8. Because of the symmetry, only half of the solution region needs to be considered. At the interface $y = d$, the continuity conditions with Equation 9.67 require that

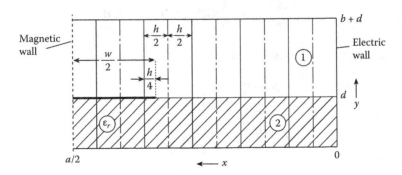

FIGURE 9.8
Half-cross-section of a shielded microstrip line.

$$\frac{\beta}{\omega\epsilon_o}\frac{\partial}{\partial x}\left(\psi_I^{(e)} - \frac{1}{\epsilon_r}\psi_{II}^{(e)}\right) = \frac{\partial\psi_{II}^{(h)}}{\partial y} - \frac{\partial\psi_I^{(h)}}{\partial y} \tag{9.68}$$

$$\left(k_o^2 - \beta^2\right)\psi_I^{(e)} = \frac{1}{\epsilon_r}\left(\epsilon_r k_o^2 - \beta^2\right)\psi_{II}^{(e)} \tag{9.69}$$

$$\frac{\partial\psi_I^{(h)}}{\partial y} - \frac{\partial\psi_{II}^{(h)}}{\partial y} = \frac{\beta}{\omega\mu}\frac{\partial}{\partial x}\left(\psi_I^{(h)} - \psi_{II}^{(h)}\right) - J_z \tag{9.70}$$

$$\left(k_o^2 - \beta^2\right)\psi_I^{(h)} = \left(\epsilon_r k_o^2 - \beta^2\right)\psi_{II}^{(h)} - j\omega\mu J_x \tag{9.71}$$

where the subscripts I and II refer to dielectric regions 1 and 2 and J_x and J_z are the current densities at the interface $y = d$.

We replace the partial derivative operator $\partial/\partial x$ with the difference operator $[D]$, where

$$[D] = \begin{bmatrix} 1 & -1 & 0 & \cdots & 0 \\ 0 & 1 & -1 & \cdots & 0 \\ \vdots & \ddots & \ddots & \ddots & \vdots \\ 0 & 0 & \cdots & 1 & -1 \end{bmatrix} \tag{9.72}$$

so that

$$\begin{aligned} \frac{\partial\psi^{(e)}}{\partial x} &\rightarrow \frac{1}{h}[D][\psi^{(e)}] \\ \frac{\partial\psi^{(h)}}{\partial x} &\rightarrow -\frac{1}{h}[D]^t[\psi^{(h)}] \end{aligned} \tag{9.73}$$

We replace the normal derivatives of $\partial\psi/\partial n$ at the interface $y = d$ with the following matrix operators.

$$\begin{aligned} \frac{\partial\psi_k^{(e)}}{\partial n} &\rightarrow \frac{1}{h}\left[G_k^{(e)}\right]\left[\psi_k^{(e)}\right], \quad k = I, II \\ \frac{\partial\psi_k^{(h)}}{\partial n} &\rightarrow \frac{1}{h}\left[G_k^{(h)}\right]\left[\psi_k^{(h)}\right], \quad k = I, II \end{aligned} \tag{9.74}$$

We can transform this into the diagonal form

$$\begin{aligned} h\frac{\partial\left[U_k^{(e)}\right]}{\partial n} &= \left[\gamma_k^{(e)}\right]\left[U_k^{(e)}\right], \quad k = I, II \\ h\frac{\partial\left[U_k^{(h)}\right]}{\partial n} &= \left[\gamma_k^{(h)}\right]\left[U_k^{(h)}\right], \quad k = I, II \end{aligned} \tag{9.75}$$

With the aid of Equation 9.65 and the boundary conditions at $y = 0$ and $y = b + d$, the diagonal matrices $[y_k]$ are determined analytically as

$$\left[\gamma_{\mathrm{I}}^{(e)}\right] = \mathrm{diag}\left[\chi_i \coth\left(\chi_i b/h\right)\right]$$

$$\left[\gamma_{\mathrm{I}}^{(h)}\right] = \mathrm{diag}\left[\chi_i \tanh\left(\chi_i b/h\right)\right]$$

$$\left[\gamma_{\mathrm{II}}^{(e)}\right] = \mathrm{diag}\left[\eta_i \coth\left(\eta_i d/h\right)\right] \tag{9.76}$$

$$\left[\gamma_{\mathrm{II}}^{(h)}\right] = \mathrm{diag}\left[\eta_i \tanh\left(\eta_i d/h\right)\right]$$

where

$$\chi_i = \left[4\sin^2\left(\frac{i-0.5}{2N+1}\pi\right) - h^2\left(k_o^2 - \beta^2\right)\right]^{1/2} \tag{9.77}$$

and

$$\eta_i = \left[4\sin^2\left(\frac{i-0.5}{2N+1}\pi\right) - h^2\left(\epsilon_r k_o^2 - \beta^2\right)\right]^{1/2} \tag{9.78}$$

We can discretize Equations 9.68 through 9.71 and eliminate $\psi_{\mathrm{II}}^{(e)}$ and $\psi_{\mathrm{II}}^{(h)}$ using $[T^{(e)}]$ and $[T^{(h)}]$ matrices. Equations 9.68 and 9.70 become

$$\frac{\beta}{\omega\epsilon_o}(1-\tau)[\delta]\left[U_{\mathrm{I}}^{(e)}\right] = \left(\left[\gamma_{\mathrm{I}}^{(h)}\right] + \tau\left[\gamma_{\mathrm{II}}^{(h)}\right]\right)\left[U_{\mathrm{I}}^{(h)}\right] \tag{9.79}$$

$$\left(\left[\gamma_{\mathrm{I}}^{(e)}\right] + \epsilon_r\tau\left[\gamma_{\mathrm{II}}^{(e)}\right]\right)\left[U_{\mathrm{I}}^{(e)}\right] = \frac{\beta}{\omega\mu}(1-\tau)[\delta]^t\left[U_{\mathrm{I}}^{(h)}\right] - [T^{(e)}]^t[J_z] \tag{9.80}$$

where

$$\tau = \frac{1-\epsilon_{\mathrm{eff}}}{\epsilon_r - \epsilon_{\mathrm{eff}}} \tag{9.81}$$

$$\epsilon_{\mathrm{eff}} = \frac{\beta^2}{k_o^2} \tag{9.82}$$

$$[\delta] = [T^{(h)}]^t[D][T^{(e)}] \tag{9.83}$$

and $[T^{(e)}] = [T_{ND}]$ and $[T^{(h)}] = [T_{DN}]$ as given in Table 9.1. Notice that $[\delta]$ is a diagonal matrix and is analytically determined as

$$\delta_i = \mathrm{diag}\left[2\sin\left(\frac{i-0.5}{2N+1}\pi\right)\right] \tag{9.84}$$

Since J_x is negligibly small compared with J_z, we solve Equations 9.79 and 9.80 to obtain

$$\left[U_{\mathrm{I}}^{(e)}\right] = [\rho][T^{(e)}]^t[J_z] \tag{9.85}$$

where

$$[\rho] = \left[\left[\gamma_I^{(e)}\right] + \epsilon_r \tau \left[\gamma_{II}^{(e)}\right] - \epsilon_{\text{eff}}(1-\tau)^2 [\delta]^t \left(\left[\gamma_I^{(h)}\right] + \tau \left[\gamma_{II}^{(h)}\right]\right)^{-1} [\delta]\right]^{-1} \quad (9.86)$$

which is a diagonal matrix. Using Equation 9.63, we now take the inverse transform of Equation 9.85 to obtain

$$\left[\psi_I^{(e)}\right] = [T^{(e)}][\rho][T^{(e)}]^t [J_z] \quad (9.87)$$

We finally impose the boundary condition on the strip, namely,

$$\left[\psi_I^{(e)}\right] = [0] \quad \text{on the strip} \quad (9.88)$$

which leads to a reduced matrix equation

$$[J_z] = \begin{cases} [J_z]_{\text{red}} & \text{on the strip} \\ 0 & \text{elsewhere} \end{cases} \quad (9.89)$$

and the corresponding characteristic equation

$$\left([T^{(e)}][\rho][T^{(e)}]^t\right)_{\text{red}} [J_z]_{\text{red}} = [0] \quad (9.90)$$

It is known from mathematics that a homogeneous linear matrix equation shows nontrivial solutions only when the determinant of the matrix is equal to zero. Thus, the propagation constant is determined by solving the determinant equation

$$\boxed{\det\left([T^{(e)}][\rho(\beta,\omega)][T^{(e)}]^t\right)_{\text{red}} = [0]} \quad (9.91)$$

The effective dielectric constant ϵ_{eff} is obtained from Equation 9.82. Notice that only the number of points on the strip determines the size of the matrix and that Equation 9.91 applies to a microstrip with more than one strip. We solve Equation 9.91 using a root-finding algorithm [28] in FORTRAN, Maple, or MATLAB. Although a microstrip example is considered here, the formulation is generally valid for any two-layer structures.

Once we solve Equation 9.91 to determine the effective dielectric constant, the current distribution on the strip, the potential functions ψ_e and ψ_h, the electric field E_z, and magnetic field H_z can be computed. Finally, the characteristic impedance is obtained from

$$Z_o = \frac{2P}{I^2} \quad (9.92)$$

where P is the average power transport along the line

$$P = \frac{1}{2} \int (\mathbf{E} \times \mathbf{H}^*) \cdot dx\, dy\, \mathbf{a}_z \quad (9.93)$$

and I is the total current flowing on the strip

$$I = \int J_z dx\, dy \tag{9.94}$$

Since the above analysis applies to multiple strips, the characteristic impedance to the mth strip is

$$Z_{om} = \frac{2P_m}{I_m^2} \tag{9.95}$$

EXAMPLE 9.4

Consider the shielded microstrip line shown in Figure 9.8. Using the MOL, find the effective dielectric constant of the line when $\epsilon_r = 9$, $w/d = 2$, $a/d = 7$, $b/d = 3$ and $d = 1$ mm.

Solution

The number of lines along the x-axis is selected as $N = 18$ and the number of lines crossing the strip is $M = 6$. These numbers are for only one potential, say $[\psi_e]$. Since only one half of the structure is considered due to symmetry, only three points on the strip are necessary. Hence, the size of the matrix associated with Equation 9.91 is 3×3.

Figure 9.9 shows the three MATLAB codes for solving Equation 9.91. The main program varies the values of d from 0.01 to 0.15, assuming that $\lambda_o = 1$, the wavelength in free space, since β or ϵ_{eff} are frequency-dependent. (Alternatively, we could keep d fixed and vary frequency, from, say, 1–50 GHz.) The program plots ϵ_{eff} with d/λ_o as shown in Figure 9.10.

The second M-file fun.m does the actual computation of the matrices involved using Equations 9.76 through 9.91. It eventually finds the determinant of matrix $[F]$, where

$$[F] = \left([T_e][\rho(\beta,\omega)][T_e]^t\right)_{\text{red}} \tag{9.96}$$

The third M-file root.m is a root-finding algorithm based on the secant method [28] and is used to determine the value of ϵ_{eff} that will satisfy

$$\det [F] = 0 \tag{9.97}$$

9.3.2 Cylindrical Microstrip Structures

The MOL can be used to analyze homogeneous and inhomogeneous cylindrical transmission structures [19,29–36] and circular and elliptic waveguides [37]. The principal steps involved in applying MOL in cylindrical coordinates are the same as in Cartesian coordinates. Here, we illustrate with the use of MOL to analyze the dispersion characteristics of the cylindrical microstrip transmission line using full-wave analysis.

We introduce the scalar potentials $\Phi^{(e)}$ and $\Phi^{(h)}$ to represent the electric and magnetic field components. In cylindrical coordinates (ρ, ϕ), the two scalar functions can be expressed as

$$\Psi^{(e,h)} = \Phi^{(e,h)}(\rho,\ \phi)e^{-j\beta z} \tag{9.98}$$

where β is the phase constant and the time-harmonic dependence has been suppressed. Substituting Equation 9.98 into the Helmholtz equation for the scalar potential functions yields

$$\rho^2 \frac{\partial^2 \Phi}{\partial \rho^2} + \rho \frac{\partial \Phi}{\partial \rho} + \frac{\partial^2 \Phi}{\partial \phi^2} + \rho^2(k^2 - \beta^2)\Phi = 0 \tag{9.99}$$

```
%  EXAMPLE 9.4 SOLVED USING METHOD OF LINES
%  THIS M-FILES REQUIRES TWO OTHER M-FILES
%  "FUN.M" AND "ROOT.M" TO WORK
%  FUN.M DETERMINES THE DETERMINANT OF MATRIX [F]
%  ROOT.M FINDS THE ROOT(S) OF THE det(F)
global d

Nmax = 50; tol = 1E-5;
ceff1 = 5; eeff2 = 10;% initial/guessed values
for i=1:15
    d = 0.01*i;
    x(i) = d;
    [eeff] = root('fun', eeff1, eeff2, tol, Nmax)
    y(i) = eeff;
%diary a:test.out
end
diary
  plot(x,y)
diary off
```

(a)

```
function determinant = fun(eeff)
%  THIS FUNCTION IS NEEDED FOR EXAMPLE 9.4
%  IT DETERMINES THE DETERMINANT OF MATRIX  [F]
%  FOR A GIVEN EFFECTIVE DIELECTRIC CONSTANT EEFF
global d

N = 9;
%d = 0.001;
a = 7*d;  b = 3*d; w = 2*d;
er = 9; h = a/N; M = w/h;
lambdao = 1; %assumed
ko = 2*pi/lambdao;
% First calculate the transformation matrices
% Te (= T_ND) and Th (=T_DN)
cons=sqrt(2/(N+0.5));
for i=1:N
    for j=1:N
        alj = (j - 0.5)*pi/(N + 0.5);
        if (j==1)
            te(i,j) = 1/sqrt(N);
        else
            te(i,j) = cons*cos( (i-0.5)*alj);
        end
    end
end
% Calculate matrices: chi, eta, delta, and gamma
beta = ko*sqrt(eeff);
tau = (1 - eeff)/(er - eeff);
chi = zeros(N,1);
eta = zeros(N,1);
%gamma1 = zeros(N,N)
for i=1:N
x = ((i -0.5)*pi)/(2*N + 1)
chi(i) = sqrt( 4*sin(x)*sin(x) -h^2*(ko^2 - beta^2) );
eta(i) = sqrt( 4*sin(x)*sin(x) -h^2*(er*ko^2 - beta^2) );
 end
```

FIGURE 9.9

For Example 9.4: (a) Main MATLAB code, (b) fun M-file for calculating F and its determinant, (c) root M-file for finding the roots of fun $(x) = 0$. *(Continued)*

```
for i=1:N
    for j=1:N
        if(i==j)
            del(i,i) = 2*sin (((i - 0.5)*pi)/(2*N + 1));
            gammae1(i,i) = chi(i)*coth(b*chi(i)/h);
            gammah1(i,i) = chi(i)*tanh(b*chi(i)/h);
            gammae2(i,i) = eta(i)*coth(d*eta(i)/h);
            gammah2(i,i) = eta(i)*tanh(d*eta(i)/h);
        else
            del(i,j) = 0;
            gammae1(i,j) = 0; gammae2(i,j) = 0;
            gammah1(i,j) = 0; gammah2(i,j) = 0;
        end
    end
end

% calculate rho matrix
rho = inv(gammae1 + er*tau*gammae2 - eeff*(1 - tau)^2*del'*(inv(gammah1 +
tau*gammah2))*del);
F = [te(1:3,1:3)*rho(1:3,1:3)*te(1:3,1:3)'];
determinant = det(F);
% Next, solve for the root det(X) = 0 using 'root.m'
```

(b)

```
function [x2] = root(F, x0, x1, tol, Nmax)
% THIS FUNCTION FINDS THE ROOTS OF fun(x) USING
% THE SECANT METHOD
% fun(x) - EXTERNAL FUNCTION THAT COMPUTES THE VALUES OF f(x)
% x0, x1 - LIMITS OF THE INITIAL RANGE
% x2     - ROOT RETURNED TO THE CALLING ROUTINE
% tol    - TOLERANCE VALUE USED IN DETERMINING CONVERGENCE
% Nmax   - MAXIMUM NO. OF ITERATIONS
%

%F0 = eval( [F, '(x0)']);
F0 = fun(x0);
dx = x1 - x0;
for l=1:Nmax
    F1 = fun(x1);
    %F1 = eval( [F, '(x1)']);
    dx = F1*dx/(F1 - F0);
    x2 = x1 - dx;
% CHECK IF TOLERANCE HAS BEEN MET
    if (abs(dx) <= tol)
    fprintf('root at x = %5.g\n', x2);
    fprintf('root found after this no. of iterations %5.g\n',l);
    break
    end
    dx = x2 - x1;
    x0 = x1;
    x1 = x2;
    F0 = F1;
end
```

(c)

FIGURE 9.9 (Continued)
For Example 9.4: (a) Main MATLAB code, (b) fun M-file for calculating F and its determinant, (c) root M-file for finding the roots of fun $(x) = 0$.

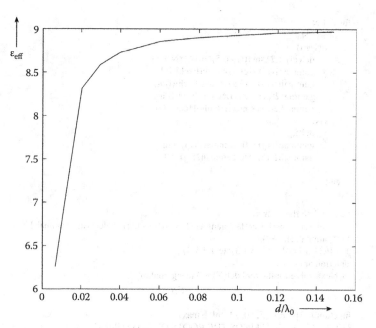

FIGURE 9.10
For Example 9.4: Effective dielectric constant of the microstrip line.

where $k^2 = \omega^2 \mu \epsilon$. Discretizing in the ϕ-direction by using N radial lines, as shown in Figure 9.11, such that

$$\phi_i = \phi_o + (i-1)h = \frac{2\pi i}{N}, \quad i = 1, 2, \ldots, N \tag{9.100}$$

where $h = \Delta\phi = 2\pi/N$ is the angular spacing between the lines. The discretization lines for the electric potential function $\Phi^{(e)}$ are shifted from the magnetic potential function $\Phi^{(h)}$ by $h/2$. Applying the central finite difference scheme yields

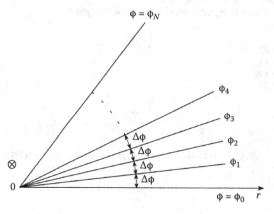

FIGURE 9.11
Discretization in the ϕ-direction.

$$\frac{\partial^2[\Phi]}{\partial\phi^2} = \frac{[P]}{h^2}[\Phi] \tag{9.101}$$

where

$$[\Phi] = [\Phi_1, \Phi_2, ..., \Phi_N]^t \tag{9.102}$$

and $[P]$ is given in Equation 9.34. Introducing Equation 9.101 into Equation 9.99 leads to N coupled differential equations:

$$\rho^2\frac{\partial^2[\Phi]}{\partial\rho^2} + \rho\frac{\partial[\Phi]}{\partial\rho} + \rho^2 k_c^2[\Phi] - \frac{[P]}{h^2}[\Phi] = 0 \tag{9.103}$$

where $k_c^2 = k^2 - \beta^2$ and $[P]$ is the same as in Equation 9.34 if ϕ goes from 0 to 2π, otherwise $[P]$ is as in Equation 9.7. Here we will assume $[P]$ in Equation 9.7. To decouple Equation 9.103, we must diagonalize $[P]$ by an orthogonal matrix $[T]$ given in Equation 9.38 and introduce the transformed potential U that satisfies

$$[U] = [T][\Phi] \tag{9.104}$$

Thus, Equation 9.103 becomes

$$\rho^2\frac{\partial^2[U]}{\partial\rho^2} + \rho\frac{\partial[U]}{\partial\rho} + [k_c^2\rho^2 - \mu_i^2][U] = 0 \tag{9.105}$$

where

$$[U] = [U_1, U_2, ..., U_N]^t \tag{9.106}$$

is a vector containing the transformed potential function and

$$\mu_i = \frac{\lambda_i}{h} \tag{9.107}$$

We notice that Equation 9.105 is essentially a Bessel equation and can be solved for every homogeneous region to produce Bessel function of order μ_n. The solution is

$$U_i(\rho) = A_i J_{\mu i}(k_c\rho) + B_i Y_{\mu i}(k_c\rho), \quad i = 1, 2, ..., N \tag{9.108}$$

where J and Y are Bessel functions of the first and second kind, respectively.

To be concrete, consider the cross section of a cylindrical microstrip line shown in Figure 9.12. Due to the symmetry of the structure, we need only consider half the cross section as in Figure 9.13. We have regions I and II and we apply Equation 9.108 to each region. On the boundaries $\rho = d$ and $\rho = b$ (electric walls), we have the boundary conditions

$$U_{Ii}^{(e)}(\rho = d) = 0$$
$$U_{IIi}^{(e)}(\rho = b) = 0 \tag{9.109}$$

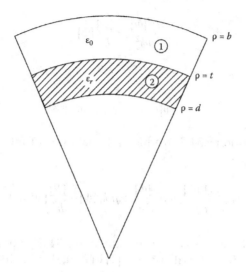

FIGURE 9.12
The cross section of a shielded cylindrical microstrip line.

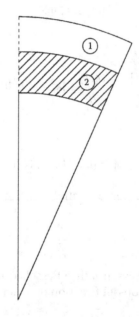

FIGURE 9.13
Half the cross section of the microstrip in Figure 9.12 (— electric wall; - - - magnetic wall).

Enforcing Equation 9.109 on Equation 9.108, we obtain

$$0 = A_i J_{\mu_i}(k_c d) + B_i Y_{\mu_i}(k_c d),$$
$$0 = C_i J_{\mu_i}\left(k_c' b\right) + D_i Y_{\mu_i}\left(k_c' b\right) \tag{9.110}$$

where $k_c = \sqrt{k_o^2 - \beta^2}$ and $k_c' = \sqrt{\epsilon_r k_o^2 - \beta^2}$, $k_o = 2\pi/\lambda_o$, and λ_o is the wavelength in free space. From Equation 9.110,

$$\frac{B_i}{A_i} = -\frac{J_{\mu_i}(k_c d)}{Y_{\mu_i}(k_c d)}$$

$$\frac{D_i}{C_i} = -\frac{J_{\mu_i}\left(k_c' b\right)}{Y_{\mu_i}\left(k_c' b\right)}$$

(9.111)

For $\Phi^{(h)}$, the boundary conditions are

$$\frac{\partial U_{Ii}^{(h)}}{\partial \rho}\bigg|_{\rho=a} = 0$$

$$\frac{\partial U_{IIIi}^{(h)}}{\partial \rho}\bigg|_{\rho=b} = 0$$

(9.112)

Enforcing this on Equation 9.108 yields

$$0 = E_i J_{\mu_i}'(k_c d) + F_i Y_{\mu_i}'(k_c d),$$

$$0 = G_i J_{\mu_i}'\left(k_c' \rho\right) + H_i Y_{\mu_i}'\left(k_c' b\right),$$

(9.113)

which leads to

$$\frac{F_i}{E_i} = -\frac{J_{\mu_i}'(k_c d)}{Y_{\mu_i}'(k_c d)}$$

$$\frac{H_i}{G_i} = -\frac{J_{\mu_i}'\left(k_c' b\right)}{Y_{\mu_i}'\left(k_c' b\right)}$$

(9.114)

At the interface $\rho = t$, both $\Phi^{(e)}$ and $\Phi^{(h)}$ are related by the continuity conditions of the tangential components of the electric and magnetic fields. Since

$$\Phi^{(e)} = \frac{j\omega\epsilon_o\epsilon_r}{k_o^2\epsilon_r - \beta^2} E_z$$

$$\Phi^{(h)} = \frac{j\omega\mu_r}{k_o^2\epsilon_r - \beta^2} H_z$$

(9.115)

the continuity conditions are

$$\frac{1}{t}\frac{\beta}{\omega\epsilon_o}\frac{\partial}{\partial\phi}\left(\Phi_I^{(e)} - \frac{1}{\epsilon_r}\Phi_{II}^{(e)}\right) = \frac{\partial\Phi_{II}^{(h)}}{\partial\rho} - \frac{\partial\psi_I^{(h)}}{\partial\rho}$$

(9.116)

$$\left(k_o^2 - \beta^2\right)\phi_I^{(e)} = \frac{1}{\epsilon_r}\left(\epsilon_r k_o^2 - \beta^2\right)\phi_{II}^{(e)}$$

(9.117)

$$\frac{\partial\Phi_I^{(h)}}{\partial\rho} - \frac{\partial\Phi_{II}^{(h)}}{\partial\rho} = \frac{\beta}{\omega\mu_o t}\frac{\partial}{\partial\phi}\left(\Phi_I^{(h)} - \Phi_{II}^{(h)}\right) - J_z$$

(9.118)

$$\left(k_o^2 - \beta^2\right)\phi_I^{(h)} = \left(\epsilon_r k_o^2 - \beta^2\right)\phi_{II}^{(h)} - j\omega\mu J_\phi$$

(9.119)

As we did in Section 9.3.1, we replace the derivative operator $\partial/\partial\phi$ with the difference operator $[D]$ and transform the resulting equations into the diagonal matrices. We obtain the elements of the diagonal matrices as [30]

$$\gamma_{1i}^{(e)} = S_o h \left[\frac{J'_{\mu_i}(S_o) + (B_i/A_i)Y'_{\mu_i}(S_o)}{J_{\mu_i}(S_o) + (B_i/A_i)Y_{\mu_i}(S_o)} \right] \tag{9.120a}$$

$$\gamma_{IIi}^{(e)} = -S'_o h \left[\frac{J'_{\mu_i}(S'_o) + (D_i/C_i)Y'_{\mu_i}(S'_o)}{J_{\mu_i}(S'_o) + (D_i/C_i)Y_{\mu_i}(S'_o)} \right] \tag{9.120b}$$

$$\gamma_{1i}^{(h)} = S_o h \left[\frac{J'_{\mu_i}(S_o) + (F_i/E_i)Y'_{\mu_i}(S_o)}{J_{\mu_i}(S_o) + (F_i/E_i)Y_{\mu_i}(S_o)} \right] \tag{9.120c}$$

$$\gamma_{IIi}^{(h)} = -S'_o h \left[\frac{J'_{\mu_i}(S'_o) + (H_i/G_i)Y'_{\mu_i}(S'_o)}{J_{\mu_i}(S'_o) + (H_i/G_i)Y_{\mu_i}(S'_o)} \right] \tag{9.120d}$$

where $h = \Delta\phi$ and

$$S_o = t\sqrt{k_o^2 - \beta^2}, \quad S'_o = t\sqrt{k_o^2\epsilon - \beta^2} \tag{9.121}$$

By ignoring J_ϕ and reducing J_z to what we have in Equation 9.89, we finally obtain the characteristic equation

$$\left([T^{(e)}][\rho][T^{(e)}]^t \right)_{\text{red}} [J_z]_{\text{red}} = [0] \tag{9.122}$$

where

$$[\rho] = \left[\left[\gamma_I^{(e)}\right] + \epsilon_r\tau\left[\gamma_{II}^{(e)}\right] - \epsilon_{\text{eff}}(1-\tau)^2[\delta]^t\left(\left[\gamma_I^{(h)}\right] + \tau\left[\gamma_{II}^{(h)}\right] \right)^{-1}[\delta] \right]^{-1} \tag{9.123}$$

$$\tau = \frac{1-\epsilon_{\text{eff}}}{\epsilon_r - \epsilon_{\text{eff}}}, \quad \epsilon_{\text{eff}} = \frac{\beta^2}{k_o^2}, \quad [\delta] = [T^{(h)}]^t[D][T^{(e)}] \tag{9.124}$$

With a root-finding algorithm, Equation 9.122 can be solved to obtain β or ϵ_{eff}. Notice that Equation 9.122 is of the same form as Equation 9.90 and only the number of points on the strip determines the size of the matrix. However, the expressions for $[\gamma_I^{(e)}], [\gamma_{II}^{(e)}], [\gamma_I^{(h)}]$, and $[\gamma_{II}^{(h)}]$ are given in Equation 9.120.

9.4 Time-Domain Solution

The frequency-domain version of the MOL covered in Section 9.3 can be extended to the time domain [38–43]. In fact, MOL can also be used to solve parabolic equations [1,44,45].

However, in this section, we will use MOL to solve hyperbolic Maxwell's equations in the time domain. Essentially, the MOL proceeds by leaving the derivatives along one selected axis untouched (usually in time), while all other partial derivatives (usually in space) are discretized using well-known techniques such as finite difference and finite element. The partial differential equation is reduced to a system of ordinary differential equations that can be solved numerically using standard methods.

Consider an empty rectangular waveguide which is infinite in the z-direction [38] and with cross-section $0 < x < a$, $0 < y < b$. We assume that the waveguide is excited by a uniform electric field E_z. The problem becomes a two-dimensional one. It corresponds to calculating the cutoff frequencies of various modes in the frequency domain. Such information can be obtained from the time-domain data.

Due to the excitation, only E_z, H_x, and H_y exist and $\partial/\partial z = 0$. Maxwell's equations become

$$-\mu \frac{\partial H_x}{\partial t} = \frac{\partial E_z}{\partial y}$$

$$\mu \frac{\partial H_y}{\partial t} = \frac{\partial E_z}{\partial x} \qquad (9.125)$$

$$\epsilon \frac{\partial E_z}{\partial t} = \frac{\partial H_y}{\partial x} - \frac{\partial H_x}{\partial y}$$

which can be manipulated to yield the wave equation

$$\frac{\partial^2 E_z}{\partial x^2} + \frac{\partial^2 E_z}{\partial y^2} - \mu\epsilon \frac{\partial^2 E_z}{\partial t^2} = 0 \qquad (9.126)$$

Discretizing in the x-direction only leads to

$$-\mu \frac{\partial [H_x]}{\partial t} = \frac{\partial [E_z]}{\partial y} \qquad (9.127a)$$

$$\mu \frac{\partial [H_y]}{\partial t} = \frac{\left[D_x^{(e)}\right][E_z]}{\Delta x} \qquad (9.127b)$$

$$\epsilon \frac{\partial [E_z]}{\partial t} = \frac{\left[D_x^{(h)}\right][H_y]}{\Delta x} - \frac{\partial [H_x]}{\partial y} \qquad (9.127c)$$

$$\frac{\left[D_{xx}^{(e)}\right][E_z]}{(\Delta x)^2} + \frac{\partial^2 [E_z]}{\partial y^2} - \mu\epsilon \frac{\partial^2 [E_z]}{\partial t^2} = 0 \qquad (9.127d)$$

where $[E_z]$, $[H_x]$, and $[H_y]$ are column vectors representing the fields along each line and are functions of y and t. As given in Section 9.3.1, matrices $[D_x^{(e)}], [D_x^{(h)}]$, and $[D_{xx}^{(e)}]$ represent difference operators in which the boundary conditions at the side walls are incorporated.

Due to the fact that $[D_{xx}^{(e)}]$ is a real symmetric matrix, there exists a real orthogonal matrix $[T_x^{(e)}]$ that transforms $[D_{xx}^{(e)}]$ into a diagonal matrix $[\lambda^2]$. We can transform $[E_z]$ into a transform

$$\left[\overline{E}_z\right] = \left[T_x^{(e)}\right][E_z] \qquad (9.128)$$

(and similarly $[H_x]$ and $[H_y]$) so that Equation 9.127d becomes

$$\frac{\left[\lambda^2\right][E_z]}{(\Delta x)^2} + \frac{\partial^2\left[\bar{E}_z\right]}{\partial y^2} - \mu\epsilon\frac{\partial^2\left[\bar{E}_z\right]}{\partial t^2} = 0 \tag{9.129}$$

This is a set of uncoupled partial differential equations. The solution for the ith line is

$$\bar{E}_{zi}(y,t) = \sum_n (A_{ni}\cos\omega_{ni}t + B_{ni}\sin\omega_{ni}t)\sin\alpha_n y \tag{9.130}$$

where

$$\omega_{ni} = \frac{u}{\sqrt{(n\pi/b)^2 - \lambda_i^2/(\Delta x)^2}} \tag{9.131}$$

$$\alpha_n = n\pi/b$$

and $u = 1/\sqrt{\mu\epsilon}$ is the wave velocity. Given the initial conditions for $\bar{E}_z$ and its time derivative, we can find A_{ni} and B_{ni}. The solution at any point at any time can be extracted from Equations 9.130, 9.127a, 9.127b, and 9.127c and the subsequent inverse transforms such as

$$[E_z(y,t)] = \left[T_x^{(e)}\right][\bar{E}_z] \tag{9.132}$$

This completes the solution process.

9.5 Concluding Remarks

MOL is a differential-difference approach of solving elliptic, parabolic, and hyperbolic PDEs. It involves a judicious combination of analysis and computation. Given a partial differential equation, all but one of the independent variables are discretized to obtain a system of ordinary differential equations.

MOL requires that the structures be at least piecewise uniform in one dimension. Also, the eigenmatrices and eigenvalues depend on the boundaries of the solution region. These requirements have limited the applications of the method. Although not applicable to problems with complex geometries, the MOL has been efficient for the analysis of compatible planar structures. Several approaches have been taken to make MOL more efficient [46]. Applications of the method include but are not limited to the following EM-related problems:

- Waveguides including optical types [47–65],
- Planar and cylindrical microstrip transmission lines [19–27,66,67],
- Scattering from discontinuities in planar structures [39,40,68],
- Antennas [32],
- Electro-optic modulator structures [17,69,70], and
- Other areas [71–75]

Originally, MOL was developed for problems with closed-solution domain. Recently, absorbing boundary conditions appropriate for MOL have been introduced [51,76–78]. With these conditions, it is now possible to simulate and model unbounded electromagnetic structures. The equivalence between the MOL and variational method is given in Reference 79.

PROBLEMS

9.1 In Equation 9.7, show that $p_\ell = 2$ for Dirichlet condition and $p_\ell = 1$ for Neumann condition.

9.2 If the first-order finite difference scheme can be written as

$$h \frac{\partial [V]}{\partial x} \simeq -[D_x]^t [V]$$

where the equidistance difference matrix $[D_x]$ is an $(N-1) \times N$ matrix given by

$$[D_x] = \begin{bmatrix} 1 & -1 & & \\ & \ddots & \ddots & \\ & & 1 & -1 \end{bmatrix}$$

show that the central finite difference scheme for second-order partial differential operator yields

$$h^2 \frac{\partial^2 [V]}{\partial x^2} \simeq [D_{xx}][V]$$

where $[D_{xx}] = -[D_x]^t [D_x] = -[D_x][D_x]^t$. Assume Neumann conditions at both side walls and obtain D_{xx}.

9.3 Obtain the transformation matrix $[T]$ and its corresponding eigenvalue matrix $[\lambda^2]$ for Neumann–Dirichlet boundary conditions. Assume that $t_0^{(k)} - t_1^{(k)} = 0$ and $t_{N+1}^{(k)} = 0$ on the boundaries.

9.4 Using MOL, solve Laplace's equation

$$\nabla^2 \Phi = 0$$

in a rectangular domain $0 \le x \le 1$, $-1 \le y \le 1$ with the following Dirichlet boundary conditions:

$$\Phi(0, y) = \Phi(1, y) = 0$$

$$\Phi(x, 1) = \Phi(x, -1) = \sin \pi x$$

Obtain Φ at $(0, 0.5)$, $(0.5, 0.25)$, $(0.5, 0.5)$, $(0.5, 0.75)$. Compare your solution with the exact solution

$$\Phi(x, y) = \frac{\cosh(\pi y)\sin(\pi x)}{\cosh(\pi b)}$$

9.5   Obtain the solution of Problem 2.4(a) using MOL.

9.6   Consider the coaxial cable of elliptical cylindrical cross section shown in Figure 9.14. Take $A = 2$ cm, $B = 4$ cm, $a = 1$ cm, and $b = 2$ cm. For the inner ellipse, for example,

$$r = \frac{a}{\sqrt{\sin^2 \phi + \nu^2 \cos^2 \phi}}, \quad \nu = \frac{a}{b}$$

By modifying the MOL codes used in Example 9.3, plot the potential for $\phi = 0$, $a < \rho < b$.

9.7   Solve Problem 2.10 using MOL and compare your result with the exact solution

$$V(\rho, z) = \frac{4V_o}{\pi} \sum_{n=\text{odd}} \frac{I_0(n\pi\rho/L)}{nI_0(n\pi a/L)} \sin\left(\frac{n\pi z}{L}\right)$$

Take $L = 2a = 1$ m and $V_o = 10$ V.

9.8   Rework Example 9.4 for a pair of coupled microstrips shown in Figure 9.15. Let $\epsilon_r = 10.2$, $w = 1.5$, $s/d = 1.5$, $a/d = 20$, $h/d = 19$, and $d = 1$ cm. Plot the effective dielectric constant versus d/λ_o.

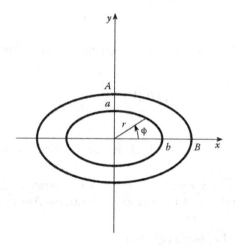

FIGURE 9.14
For Problem 9.6.

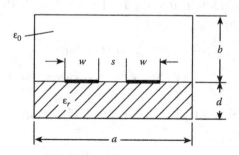

FIGURE 9.15
For Problem 9.8.

9.9 Given the difference operator

$$[P] = \begin{bmatrix} 2 & -s^{*2} & \cdots & & -s^2 \\ -s^2 & 2 & -s^{*2} & \cdots & \\ & \ddots & \ddots & \ddots & \\ & & & & -s^{*2} \\ -s^{*2} & \cdots & & -s^2 & 2 \end{bmatrix}$$

which is Hermitian, that is, $[P] = [P^*]$. Show that $[P]$ has the following eigenvalues

$$\lambda_k^2 = 4\sin^2 \frac{\phi_k \beta h}{2}, \ \phi_k = \frac{2\pi k}{N}, k = 1, 2, \ldots, N$$

and the eigenvector matrices

$$T_{ik}^{(e)} = \frac{1}{\sqrt{N}} e^{ji\phi_k}, \quad T_{ik}^{(h)} = \frac{1}{\sqrt{N}} e^{j(i+0.5)\phi_k}$$

where $s = e^{j\beta h/2}$, s^* is the complex conjugate of s, β is the propagation constant, and h is the step size.

9.10 Show that for

$$[P] = \begin{bmatrix} 2 & -1/s & \cdots & & -s \\ -s & 2 & -1/s & \cdots & \\ & \ddots & \ddots & \ddots & \\ & & & & -1/s \\ -1/s & \cdots & & -s & 2 \end{bmatrix}$$

the eigenvalue matrices remain the same as in the previous problem.

References

1. A. Zafarullah, "Application of the method of lines to parabolic partial differential equation with error estimates," *J. ACM*, vol. 17, no. 2, Apr. 1970, pp. 294–302.
2. M.B. Carver and H.W. Hinds, "The method of lines and the advection equation," *Simulation*, vol. 31, no. 2, 1978, pp. 59–69.
3. W.E. Schiesser, *The Numerical Method of Lines*. San Diego: Academic Press, 1991.
4. G.H. Meyer, "The method of lines for Poisson's equation with nonlinear or free boundary conditions," *Numer. Math.*, vol. 29, 1978, pp. 329–344.
5. V.N. Sushch, "Difference Poisson equation on a curvilinear mesh," *Diff. Equations*, vol. 12, no. 5, 1996, pp. 693–697.
6. O.A. Liskovets, "The method of lines (review)," *Diff. Equations*, vol. 1, no. 12, 1965, pp. 1308–1323.
7. R. Pregla and W. Pascher, "The method of lines," in T. Itoh (ed.), *Numerical Techniques for Microwave and Millimeter-Wave Passive Structures*. New York: John Wiley, 1989, pp. 381–446.

8. R. Pregla, "Analysis of planar microwave structures on magnetized ferrite substrate," *Arch. Elek. Ubertragung.*, vol. 40, 1986, pp. 270–273.

9. R. Pregla, "About the nature of the method of lines," *Arch. Elek. Ubertragung.*, vol. 41, no. 6, 1987, pp. 368–370.

10. R. Pregla, "Higher order approximation for the difference operators in the method of lines," *IEEE Micro. Guided Wave Lett.*, vol. 5, no. 2, Feb. 1995, pp. 53–55.

11. U. Schulz and R. Pregla, "A new technique for the analysis of the dispersion characteristics of planar waveguides and its application to microstrips with tuning septums," *Radio Sci.*, vol. 16, Nov./Dec. 1981, pp. 1173–1178.

12. R. Pregla and W. Pascher, "Diagonalization of difference operators and system matrices in the method of lines," *IEEE Micro. Guided Wave Lett.*, vol. 2, no. 2, Feb. 1992, pp. 52–54.

13. M.N.O. Sadiku and C.N. Obiozor, "A simple introduction to the method of lines," *IJEEE*, vol. 37, no. 3, July 2000, pp. 282–296.

14. A. Dreher and T. Rother, "New aspects of the method of lines," *IEEE Micro. Guided Wave Lett.*, vol. 5, no. 11, Nov. 1995, pp. 408–410.

15. D.J. Jones et al., "On the numerical solution of elliptic partial differential equations by the method of lines," *J. Comp. Phys.*, vol. 9, 1972, pp. 496–527.

16. B.H. Edwards, "The numerical solution of elliptic differential equation by the method of lines," *Revista Colombiana de Matematicas*, vol. 19, 1985, pp. 297–312.

17. A.G. Keen et al., "Quasi-static analysis of electrooptic modulators by the method of lines," *J. Lightwave Tech.*, vol. 8, no. 1, Jan. 1990, pp. 42–50.

18. J.G. Ma and Z. Chen, "Application of the method of lines to the Laplace equation," *Micro. Opt. Tech. Lett.*, vol. 14, no. 6, 1997, pp. 330–333.

19. S. Xiao et al., "Analysis of cylindrical transmission lines with the method of lines," *IEEE Trans. Micro. Theo. Tech.*, vol. 44, no. 7, July 1996, pp. 993–999.

20. H. Diestel, "Analysis of planar multiconductor transmission-line system with the method of lines," *AEU*, vol. 41, 1987, pp. 169–175.

21. Z. Chen and B. Gao, "Full-wave analysis of multiconductor coupled lines in MICs by the method of lines," *IEEE Proc., Pt. H*, vol. 136, no. 5, Oct. 1989, pp. 399–404.

22. W. Pascher and R. Pregla, "Full wave analysis of complex planar microwave structures," *Radio Sci.*, vol. 22, no. 6, Nov. 1987, pp. 999–1002.

23. M.A. Thorburn et al., "Application of the method of lines to planar transmission lines in waveguides with composite cross-section geometries," *IEEE Proc., Pt. H*, vol. 139, no. 6, Dec. 1992, pp. 542–544.

24. R. Pregla, "Analysis of a microstrip bend discontinuity by the method of lines," *Frequenza*, vol. 145, 1991, pp. 213–216.

25. Y.J. He and S.F. Li, "Analysis of arbitrary cross-section microstrip using the method of lines," *IEEE Trans. Micro. Theo. Tech.*, vol. 42, no. 1, Jan. 1994, pp. 162–164.

26. M. Thorburn et al., "Computation of frequency-dependent propagation characteristics of microstriplike propagation structures with discontinuous layers," *IEEE Trans. Micro. Theo. Tech.*, vol. 38, no. 2, Feb. 1990, pp. 148–153.

27. P. Meyer and P.W. van der Walt, "Closed-form expression for implementing the method of lines for two-layer boxed planar structures," *Elect. Lett.*, vol. 30, no. 18, Sept. 1994, pp. 1497–1498.

28. E.R. Champion, *Numerical Methods for Engineering Applications.* New York: Marcel Dekker, 1993, pp. 196–215.

29. L. Urshev and A. Stoeva, "Application of equivalent transmission line concept to the method of lines," *Micro. Opt. Tech. Lett.*, vol. 3, no. 10, Oct. 1990, pp. 341–343.

30. Y. Xu, "Application of method of lines to solve problems in the cylindrical coordinates," *Micro. Opt. Tech. Lett.*, vol. 1, no. 5, May 1988, pp. 173–175.

31. K. Gu and Y. Wang, "Analysis of dispersion characteristics of cylindrical microstrip line with method of lines," *Elect. Lett.*, vol. 26, no. 11, May 1990, pp. 748–749.

32. R. Pregla, "New approach for the analysis of cylindrical antennas by the method of lines," *Elect. Lett.*, vol. 30, no. 8, April 1994, pp. 614–615.

33. R. Pregla, "General formulas for the method of lines in cylindrical coordinates," *IEEE Trans. Micro. Theo. Tech.*, vol. 43, no. 7, July 1995, pp. 1617–1620.

34. R. Pregla, "The method of lines for the analysis of dielectric waveguide bends," *J. Lightwave Tech.*, vol. 14, no. 4, 1996, pp. 634–639.

35. D. Kremer and R. Pregla, "The method of lines for the hybrid analysis of multilayered cylindrical resonator structures," *IEEE Trans. Micro. Theo. Tech.*, vol. 45, no. 12, Dec. 1997, pp. 2152–2155.

36. M. Thorburn et al., "Application of the method of lines to cylindrical inhomogeneous propagation structures," *Elect. Lett.*, vol. 26, no. 3, Feb. 1990, pp. 170–171.

37. K. Wu and R. Vahldieck, "The method of lines applied to planar transmission lines in circular and elliptical waveguides," *IEEE Trans. Micro. Theo. Tech.*, vol. 37, no. 12, Dec. 1989, pp. 1958–1963.

38. S. Nam et al., "Time-domain method of lines," *Elect. Lett.*, vol. 24, no. 2, Jan. 1988, pp. 128–129.

39. S. Nam et al., "Time-domain method of lines applied to planar waveguide structures," *IEEE Trans. Micro. Theo. Tech.*, vol. 37, 1989, pp. 897–901.

40. S. Nam et al., "Characterization of uniform microstrip line and its discontinuities using the time-domain method of lines," *IEEE Trans. Micro. Theo. Tech.*, vol. 37, 1989, pp. 2051–2057.

41. H. Zhao et al., "Numerical solution of the power density distribution generated in a multimode cavity by using the method of lines technique to solve directly for the electric field," *IEEE Trans. Micro. Theo. Tech.*, vol. 44, no. 12, Dec. 1996, pp. 2185–2194.

42. W. Fu and A. Metaxas, "Numerical prediction of three-dimensional power density distribution in a multi-mode cavity," *J. Micro. Power Electro. Energy*, vol. 29, no. 2, 1994, pp. 67–75.

43. W.B. Fu and A.C. Metaxas, "Numerical solution of Maxwell's equations in three dimensions using the method of lines with applications to microwave heating in a multimode cavity," *Int. J. Appl. Engr. Mech.*, vol. 6, 1995, pp. 165–186.

44. J. Lawson and M. Berzins, "Towards an automatic algorithm for the numerical solution of parabolic partial differential equations using the method of lines," in J. R. Cash and I. Gladwell (eds.), *Computational Ordinary Differential Equations*. Oxford: Clarendon Press, 1992, pp. 309–322.

45. B. Zubik-Kowal, "The method of lines for parabolic differential-functional equations," *IMA J. Num. Analy.*, vol. 17, 1997, pp. 103–123.

46. M. Furqan and L. Vietzorreck, "New modelling aspects in the method of lines," *URSI International Symposium on Electromagnetic Theory*, 2010, pp. 696–698.

47. S.J. Chung and T.R. Chrang, "Full-wave analysis of discontinuities in conductor-backed coplanar waveguides using the method of lines," *IEEE Trans. Micro. Theo. Tech.*, vol. 41, no. 9, 1993, pp. 1601–1605.

48. A. Papachristoforos, "Method of lines for analysis of planar conductors with finite thickness," *IEEE Proc. Micro. Ant. Prog.*, vol. 141, no. 3, June 1994, pp. 223–228.

49. R. Pregla and W. Yang, "Method of lines for analysis of multilayered dielectric waveguides with Bragg grating," *Elect. Lett.*, vol. 29, no. 22, Oct. 1993, pp. 1962–1963.

50. R.R. Kumar et al., "Modes of a shielded conductor-backed coplanar waveguide," *Elect. Lett.*, vol. 30, no. 2, Jan. 1994, pp. 146–147.

51. A. Dreher and R. Pregla, "Full-wave analysis of radiating planar resonators with the method of lines," *IEEE Trans. Micro. Theo. Tech.*, vol. 41, no. 8, Aug. 1993, pp. 1363–1368.

52. U. Rogge and R. Pregla, "Method of lines for the analysis of strip-load optical waveguide," *J. Opt. Soc. Amer. B*, vol. 8, no. 2, pp. 463–489.

53. F.J. Schmuckle and R. Pregla, "The method of lines for the analysis of lossy planar waveguides," *IEEE Trans. Micro. Theo. Tech.*, vol. 38, no. 10, Oct. 1990, pp. 1473–1479.

54. F.J. Schmuckle and R. Pregla, "The method of lines for the analysis of planar waveguides with finite metallization thickness," *IEEE Trans. Micro. Theo. Tech.*, vol. 39, no. 1, Jan. 1991, pp. 107–111.

55. R.S. Burton and T.E. Schlesinger, "Least squares technique for improving three-dimensional dielectric waveguide analysis by the method of lines," *Elect. Lett.*, vol. 30, no. 13, June 1994, pp. 1071–1072.

56. R. Pregla and E. Ahlers, "Method of lines for analysis of arbitrarily curved waveguide bends," *Elect. Lett.*, vol. 30, no. 18, Sept. 1994, pp. 1478–1479.

57. R. Pregla and E. Ahlers, "Method of lines for analysis of discontinuities in optical waveguides," *Elect. Lett.*, vol. 29, no. 21, Oct. 1993, pp. 1845–1846.

58. J.J. Gerdes, "Bidirectional eigenmode propagation analysis of optical waveguides based on method of lines," *Elect. Lett.*, vol. 30, no. 7, March 1994, pp. 550–551.

59. S.J. Al-Bader and H.A. Jamid, "Method of lines applied to nonlinear guided waves," *Elect. Lett.*, vol. 31, no. 17, Aug. 1995, pp. 1455–1457.

60. W.D. Yang and R. Pregla, "Method of lines for analysis of waveguide structures with multidiscontinuities," *Elect. Lett.*, vol. 31, no. 11, May 1995, pp. 892–893.

61. J. Gerdes et al., "Three-dimensional vectorial eigenmode algorithm for nonparaxal propagation in reflecting optical waveguide structures," *Elect. Lett.*, vol. 31, no. 1, Jan. 1995, pp. 65–66.

62. W. Pascher and R. Pregla, "Vectorial analysis of bends in optical strip waveguides by the method of lines," *Radio Sci.*, vol. 28, no. 6, Nov./Dec. 1993, pp. 1229–1233.

63. S.B. Worm, "Full-wave analysis of discontinuities in planar waveguides by the method of lines using a source approach," *IEEE Trans. Micro. Theo. Tech.*, vol. 38, no. 10, Oct. 1990, pp. 1510–1514.

64. S.J. Al-Bader and H.A. Jamid, "Mode scattering by a nonlinear step-discontinuity in dielectric optical waveguide," *IEEE Trans. Micro. Theo. Tech.*, vol. 44, no. 2, Feb. 1996, pp. 218–224.

65. S.J. Chung and L.K. Wu, "Analysis of the effects of a resistively coated upper dielectric layer on the propagation characteristics of hybrid modes in a waveguide-shielded microstrip using the method of lines," *IEEE Trans. Micro. Theo. Tech.*, vol. 41, no. 8, Aug. 1993, pp. 1393–1399.

66. M.J. Webster et al., "Accurate determination of frequency dependent three element equivalent circuit for symmetric step microstrip discontinuity," *IEEE Proc.*, vol. 137, Pt. H, no. 1, Feb. 1990, pp. 51–54.

67. Y. Chen and B. Beker, "Study of microstrip step discontinuities on bianisotropic substrates using the method of lines and transverse resonance technique," *IEEE Trans. Micro. Theo. Tech.*, vol. 42, no. 10, Oct. 1994, pp. 1945–1950.

68. P. Meyer, "Solving microstrip discontinuities with a combined mode-matching and method-of-lines procedure," *Micro. Opt. Tech. Lett.*, vol. 8, no. 1, Jan. 1995, pp. 4–8.

69. J. Gerdes et al., "Full wave analysis of traveling-wave electrodes with finite thickness for electro-optic modulators by the method of lines," *J. Lightwave Tech.*, vol. 9, no. 4, Aug. 1991, pp. 461–467.

70. S.J. Chung, "A 3-dimensional analysis of electrooptic modulators by the method of lines," *IEEE Trans. Mag.*, vol. 29, no. 2, March 1993, pp. 1976–1980.

71. A. Kormatz and R. Pregla, "Analysis of electromagnetic boundary-value problems in inhomogeneous media with the method of lines," *IEEE Trans. Micro. Theo. Tech.*, vol. 44, no. 12, Dec. 1996, pp. 2296–2299.

72. B.M. Sherrill and N.G. Alexopoulos, "The method of lines applied to a finline/strip configuration on an anisotropic subtrate," *IEEE Trans. Micro. Theo. Tech.*, vol. 35, no. 6, June 1987, pp. 568–574.

73. L. Vietzorreck and R. Pregla, "Hybrid analysis of three-dimensional MMIC elements by the method of lines," *IEEE Trans. Micro. Theo. Tech.*, vol. 44, no. 12, Dec. 1996, pp. 2580–2586.

74. K. Suwan and A. Anderson, "Method of lines applied to Hyperbolic fluid transient equations," *Int. Jour. Num. Methods Engr.*, vol. 33, 1992, pp. 1501–1511.

75. A.G. Bratsos, "The solution of the Boussinesq equation using the method of lines," *Comp. Methods Appl. Mech. Engr.*, vol. 157, 1998, pp. 33–44.

76. A. Dreher and R. Pregla, "Analysis of planar waveguides with the method of lines and absorbing boundary conditions," *IEEE Micro. Guided Wave Lett.*, vol. 1, no. 6, June 1991, pp. 138–140.

77. R. Pregla, and L. Vietzorreck, "Combination of the source method with absorbing boundary conditions in the method of lines," *IEEE Micro. Guided Wave Lett.*, vol. 5, no. 7, July 1995, pp. 227–229.

78. K. Wu and X. Jiang, "The use of absorbing boundary conditions in the method of lines," *IEEE Micro. Guided Wave Lett.*, vol. 6, no. 5, May 1996, pp. 212–214.

79. W. Hong and W.X. Zhang, "On the equivalence between the method of lines and the variational method," *AEU*, vol. 45, no. 1, 1991, pp. 198–201.

Selected Bibliography

This bibliography includes only books (and few important papers) published on or after 1990. These are the best resources for learning more about computational electromagnetics.

Ahmed, I. and Z. D. Chen (eds.), *Computational Electromagnetics—Retrospective and Outlook: in Honor of Wolfgang J. R. Hoefer.* Singapore: Springer, 2015.

Baba, Y. and V. A. Rakov, *Electromagnetic Computation Methods for Lightning Surge Protection Studies.* Hoboken, NJ: Wiley-IEEE Press, 2016.

Bastos, J. P. A. and N. Sadowski, *Electromagnetic Modeling by Finite Element Methods.* New York: Marcel Dekker, 2003.

Bastos, J. P. A. and N. Sadowski, *Magnetic Materials and 3D Finite Element Modeling.* Boca Raton, FL: CRC Press, 2014.

Binns, K. J., P. J. Lawrence, and C. W. Trowbridge, *The Analytical and Numerical Solution of Electric and Magnetic Fields.* Chichester, UK: John Wiley & Sons, 1992.

Bondeson, A., T. Rylander, and P. Ingelstrom, *Computational Electromagnetics.* New York: Springer, 2005.

Booton, R. C., *Computational Methods for Electromagnetics and Microwaves.* New York: John Wiley & Sons, 1992.

Bossavit, A., *Computational Electromagnetism: Variational Formulations, Complimentarity, Edge Elements.* San Diego, CA: Academic Press, 1998.

Bourlier, C., N. Pinel, and G. Kubicke, *Method of Moments for 2D Scattering Problems: Basic Concepts and Applications.* Hoboken, NJ: Wiley-ISTE Press, 2013.

Cai, W., *Computational Methods for Electromagnetic Phenomena: Electrostatics in Solvation, Scattering, and Electron Transport.* Cambridge, UK: Cambridge University Press, 2012.

Cardoso, J. R., *Electromagnetics through the Finite Element Method: A Simplified Approach Using Maxwell's Equations.* Boca Raton, FL: CRC Press, 2017.

Chari, M. V. K. and S. J. Salon, *Numerical Methods in Electromagnetism.* San Diego, CA: Academic Press, 2000.

Chew, W. C. et al., *Fast and Efficient Algorithms in Computational Electromagnetics.* Norwood, MA: Artech House, 2001.

Christopoulos, C., *The Transmission-Line Modeling Method: TLM.* Hoboken, NJ: Wiley-IEEE Press, 1995.

Ciarlet, P. G. (ed.), *Essential Numerical Methods in Electromagnetics.* Elsevier Science, 2010.

Crutzen, Y. R., G. Molinari, and G. Rubinacci (eds.), *Industrial Application of Electromagnetic Computer Codes.* Dordrecht, Netherlands: Kluwer Academic Publishers, 1990.

Davidson, D. B., *Computational Electromagnetics for RF and Microwave Engineering.* Cambridge, UK: Cambridge University Press, 2005.

Davidson, D. B., "A personal selection of books on electromagnetics and computational electromagnetics," *IEEE Antennas and Propagation Magazine,* vol. 53, no. 6, December 2011, pp. 156–160.

Davies, A. J., *The Finite Element Method: An Introduction with Partial Differential Equations,* 2nd edition. Oxford, UK: Oxford University Press, 2011.

Elsherbeni, A. Z. and V. Demir, *The Finite-Difference Time-Domain Method for Electromagnetics with MATLAB Simulations.* Raleigh, NC: SciTech Publishing, 2009.

Eom, H. J., *Electromagnetic Wave Theory for Boundary-Value Problems: An Advanced Course on Analytical Methods.* Berlin: Springer-Verlag, 2004.

Garg, R. and R. Mittra, *Analytical and Computational Methods in Electromagnetics.* Norwood, MA: Artech House, 2008.

Gedney, S. D., *Introduction to the Finite-Difference Time-Domain (FDTD) Method for Electromagnetic*. San Rafael, CA: Morgan & Claypool Publishers, 2011.

Gibson, W. C., *The Method of Moments in Electromagnetics*. Boca Raton, FL: CRC Press, 2008.

Graglia, R. D. and A. F. Peterson, *Higher-Order Techniques in Computational Electromagnetics*. Herts, UK: Institution of Engineering and Technology, 2015.

Hafner, C., *The Generalized Multipole Technique for Computational Electromagnetics*. Boston, MA: Artech House, 1990.

Hahn, S. Y., *Advanced Computational and Design Techniques in Applied Electromagnetic Systems*. Amsterdam, Netherlands: Elsevier Science, 1995.

Hammond, P. and J. K. Sykulski, *Engineering Electromagnetism: Physical Processes and Computation*. New York: Oxford University Press, 1994.

Hao, Y. and R. Mittra, *FDTD Modeling of Metamaterials: Theory and Applications*. Norwood, MA: Artech House, 2009.

Harrington, R. F., *Field Computation by Moment Methods*. Hoboken, NJ: Wiley-IEEE Press, 1993.

Inan, U. S. and R. A. Marshall, *Numerical Electromagnetics: The FDTD Method*. Cambridge, UK: Cambridge University Press, 2011.

Itoh, T., G. Pelosi, and P. P. Silvester (eds.), *Finite Element Software for Microwave Engineering*. New York, NY: John Wiley and Sons, 1996.

Jarem, J. M. and P. P. Banerjee, *Computational Methods for Electromagnetic and Optical Systems*, 2nd edition. Boca Raton, FL: CRC Press, 2011.

Jin, J.-M., *The Finite Element Method in Electromagnetics*, 3rd edition. New York, NY: Wiley, 2014.

Jin, J.-M., *Theory and Computation of Electromagnetic Fields*, 2nd edition. Hoboken, NJ: John Wiley & Sons, 2015.

Jung, B. H. et al., *Time and Frequency Domain Solutions of EM Problems Using Integral Equations and a Hybrid Methodology*. Hoboken, NJ: John Wiley & Sons, 2010.

Kaliakin, V. N., *Introduction to Approximate Solution Techniques, Numerical Modeling, and Finite Element Methods*. New York: Marcel Dekker, 2002.

Kalluri, D. K., *Electromagnetic Waves, Materials, and Computation with MATLAB*. Boca Raton, FL: CRC Press, 2012.

Kunz, K. S. and R. J. Luebbers, *The Finite Difference Time Domain Method for Electromagnetics*. Boca Raton, FL: CRC Press, 1993.

Makarov, S. N., G. M. Noetscher, and A. Nazarian, *Low-Frequency Electromagnetic Modeling for Electrical and Biological Systems Using MATLAB*, 2nd edition. Hoboken, NJ: John Wiley & Sons, 2015.

Meunier, G., *The Finite Element Method for Electromagnetic Modeling*. London, UK: ISTE and John Wiley & Sons, 2008.

Morita, N., N. Kumagai, and J. R. Mautz, *Integral Equation Methods for Electromagnetics*. Boston, MA: Artech House, 1990.

Monk, P., *Finite Element Methods for Maxwell's Equations*. Oxford, UK: Oxford University Press, 2003.

Nuruzzaman, M., *Modern Approach to Solving Electromagnetics in MATLAB*. North Charleston, SC: Booksurge Publishing, 2009.

Peterson, A. F., S. L. Ray, and R. Mittra, *Computational Methods for Electromagnetics*. New York: IEEE Press, 1998.

Poljak, D., *Advanced Modeling in Computational Electromagnetic Compatibility*. Hoboken, NJ: John Wiley & Sons, 2007.

Pregla, R., *Analysis of Electromagnetic Fields and Waves: The Method of Lines*. West Sussex, UK: Wiley, 2008.

Rao, S. M. (ed.), *Time Domain Electromagnetics*. San Diego, CA: Academic Press, 1999.

Sabbagh, H. A. et al., *Computational Electromagnetics and Model-Based Inversion: A Modern Paradigm for Eddy-Current Nondestructive Evaluation*. New York: Springer, 2013.

Sabelfeld, K. K., *Monte Carlo Methods in Boundary Value Problems*. Berlin: Springer-Verlag, 1991.

Sadiku, M. N. O., *Monte Carlo Methods for Electromagnetics*. Boca Raton, FL: CRC Press, 2009.

Sadiku, M. N. O. and S. R. Nelatury, *Analytical Techniques in Electromagnetics*. Boca Raton, FL: CRC Press, 2016.

Saguet, P., *Numerical Analysis in Electromagnetics: The TLM Method*. Hoboken, NJ: Wiley-ISTE Press, 2012.

Schiesser, W. E., *The Numerical Method of Lines: Integration of Partial Differential Equations*. San Diego, CA: Academic Press, 1991.

Schiesser, W. E. and G. W. Griffiths, *A Compendium of Partial Differential Equation Models: Method of Lines Analysis with MATLAB*. Cambridge, UK: Cambridge University Press, 2009.

Schilders, W. H. A. and E. J. W. terMaten (eds.), *Numerical Methods in Electromagnetics*, Vol. XIII. Amsterdam, Netherlands: Elsevier, 2005.

Sevgi, L., *Complex Electromagnetic Problems and Numerical Simulation Approaches*. Piscataway, NJ: IEEE Press; Hoboken, NJ: Wiley-Interscience, 2003.

Sevgi, L., *Electromagnetic Modeling and Simulation*. Hoboken, NJ: Wiley-IEEE Press, 2014.

Shang, J. J. S., *Computational Electromagnetic-Aerodynamics*. Hoboken, NJ: Wiley-IEEE Press, 2016.

Shang, J. S., "Computational Electromagnetics," in A. B. Tucker (ed.), *The Computer Science and Engineering Handbook*. Boca Raton, FL: CRC Press, 1997, Chapter 39, pp. 912–934.

Sheng, X. Q. and W. Song, *Essentials of Computational Electromagnetics*. Singapore: John Wiley & Sons, 2012.

Silvester, P. P. and R. L. Ferrari, *Finite Elements for Electrical Engineers*, 3rd edition. Cambridge, UK: Cambridge University Press, 1996.

Solin, P. et al., *Integral Methods in Low-Frequency Electromagnetics*. New York: Wiley-Interscience, 2009.

Sullivan, D. M., *Electromagnetic Simulation Using the FDTD Method*, 2nd edition. Hoboken, NJ: Wiley-IEEE Press, 2013.

Sykulski, J. K., *Computational Magnetics*. London, UK: Chapman & Hall, 1994.

Taflove, A. and S. C. Hagness, *Computational Electrodynamics: The Finite-Difference Time-Domain Method*, 3rd edition. Boston, MA: Artech House, 2005.

Taflove, A., A. Oskooi, and S. G. Johnson (eds.), *Advances in FDTD Computational Electrodynamics: Photonics and Nanotechnology*. Boston, MA: Artech House, 2013.

Tarricone, L. and A. Esposito, *Grid Computing for Electromagnetics*. Boston, MA: Artech House, 2004.

Tesche, F. M., M. V. Ianoz, and T. Karisson, *EMC Analysis Methods and Computational Models*. New York: John Wiley & Sons, 1997.

Tretyakov, S., *Analytical Modeling in Applied Electromagnetics*. Norwood, MA: Artech House, 2003.

Umashankar, P. and A. Taflove, *Computational Electromagnetics*. Norwood, MA: Artech House, 1993.

Uzunoglu, N. K., K. S. Nikita, and D. I. Kaklamani, *Applied Computational Electromagnetics: State of the Art and Future Trends*. Berlin: Springer-Verlag, 2000.

Volakis, J. L. and K. Sertel, *Integral Equation Methods for Electromagnetics*. Raleigh, NC: SciTech Publishing, 2011.

Volakis, J. L., A. Chatterjee, and L. Kempel, *Finite Element Method for Electromagnetics: Antennas, Microwave Circuits and Scattering Applications*. Hoboken, NJ: Wiley-IEEE Press, June 1998.

Wang, J. J. H., *Generalized Moment Methods in Electromagnetics*. New York, NY: John Wiley & Sons, 1991.

Warnick, K. F., *Numerical Analysis for Electromagnetic Integral Equations*. Norwood, MA: Artech House, 2008.

Warnick, K. F., *Numerical Methods for Engineering: An Introduction Using MATLAB and Computational Electromagnetics Examples*. Raleigh, NC: Scitech Publishing, 2011.

White, J. F., *High Frequency Techniques: An Introduction to RF and Microwave Design and Computer Simulation*. Hoboken, NJ: Wiley-IEEE Press, 2003.

Wiak, S., A. Krawcyk, and I. Dolezel, *Intelligent Computer Techniques in Applied Electromagnetics*. Berlin: Springer, 2008.

Wiak, S., A. Krawczyk, and I. Dolezel (eds.), *Advanced Computer Techniques in Applied Electromagnetics*. Amsterdam, Netherlands: IOP Press, 2008.

Wiak, S. and E. Napieralska-Juszczak (eds.), *Computer Fields of Electromagnetic Devices*. Amsterdam, Netherlands: IOP Press, 2010.

Wolf, J. P. and C. Song, *Finite-Element Modelling of Unbounded Media*. Hoboken, NJ: John Wiley & Sons, 1996.

Wouwer, A. V., P. Saucez, and W. E. Schiesser (eds.), *Adaptive Method of Lines*. Boca Raton, FL: Chapman & Hall/CRC, 2001.

Yamashita, E., *Analysis Methods for Electromagnetic Wave Problems*. Norwood, MA: Artech House, 1990.

Yamashita, E., *Analysis Methods for Electromagnetic Wave Problems*, Vol. 2. Norwood, MA: Artech House, 1996.

Yu, W., *Electromagnetic Simulation Techniques based on the FDTD Method*. Hoboken, NJ: John Wiley & Sons, 2009.

Yu, W. et al., *Advanced Computational Electromagnetic Methods and Applications*. Boston, MA: Artech House, 2015.

Zhu, Y. and A. C. Cangellaris, *Multigrid Finite Element Methods for Electromagnetic Field Modeling*, New York, NY: IEEE Press, 2006.

Appendix A: Vector Relations

A.1 Vector Identities

If **A** and **B** are vector fields while U and V are scalar fields, then

$$\nabla V(U + V) = \nabla U + \nabla V$$

$$\nabla(UV) = U\nabla V + V\nabla U$$

$$\nabla\left(\frac{U}{V}\right) = \frac{\nabla(\nabla U) - U(\nabla V)}{V^2}$$

$$\nabla V^n = nV^{n-1}\nabla V \quad (n = \text{integer})$$

$$\nabla(\mathbf{A} \cdot \mathbf{B}) = (\mathbf{A} \cdot \nabla)\mathbf{B} + (\mathbf{B} \cdot \nabla)\mathbf{A} + \mathbf{A} \times (\nabla \times \mathbf{B}) + \mathbf{B} \times (\nabla \times \mathbf{A})$$

$$\nabla \cdot (\mathbf{A} + \mathbf{B}) = \nabla \cdot \mathbf{A} + \nabla \cdot \mathbf{B}$$

$$\nabla \cdot (\mathbf{A} \times \mathbf{B}) = \mathbf{B} \cdot (\nabla \times \mathbf{A}) - \mathbf{A} \cdot (\nabla \times \mathbf{B})$$

$$\nabla \cdot (V\mathbf{A}) = V\nabla \cdot \mathbf{A} + \mathbf{A} \cdot \nabla V$$

$$\nabla \cdot (\nabla V) = \nabla^2 V$$

$$\nabla \cdot (\nabla \times \mathbf{A}) = 0$$

$$\nabla \times (\mathbf{A} + \mathbf{B}) = \nabla \times \mathbf{A} + \nabla \times \mathbf{B}$$

$$\nabla \times (\mathbf{A} \times \mathbf{B}) = \mathbf{A}(\nabla \cdot \mathbf{B}) - \mathbf{B}(\nabla \cdot \mathbf{A}) + (\mathbf{B} \cdot \nabla)\mathbf{A} - (\mathbf{A} \cdot \nabla)\mathbf{B}$$

$$\nabla \times (V\mathbf{A}) = \nabla V \times \mathbf{A} + V(\nabla \times \mathbf{A})$$

$$\nabla \times (\nabla V) = 0$$

$$\nabla \times (\nabla \times \mathbf{A}) = \nabla(\nabla \cdot \mathbf{A}) - \nabla^2 \mathbf{A}$$

A.2 Vector Theorems

If v is the volume bounded by the closed surface S, and $\mathbf{a}_n$ is a unit normal to S, then

$$\oint_S \mathbf{A} \cdot d\mathbf{S} = \int_v \nabla \cdot \mathbf{A} \, dv \quad \text{(Divergence theorem)}$$

$$\oint_S V \, d\mathbf{S} = \int_v \nabla V \, dv \quad \text{(Gradient theorem)}$$

$$\oint_S \mathbf{A} \times d\mathbf{S} = -\int_v \nabla \times \mathbf{A} \, dv$$

$$\oint_S \left[(\mathbf{A} \cdot d\mathbf{S})\mathbf{A} - \frac{1}{2} A^2 \, d\mathbf{S} \right] = \int_v [(\nabla \times \mathbf{A}) \times \mathbf{A} + \mathbf{A} \nabla \cdot \mathbf{A}] \, dv$$

$$\oint_S U \nabla V \cdot d\mathbf{S} = \int_v [U \nabla^2 V + \nabla U \cdot \nabla V] \, dv \quad \text{(Green first identity)}$$

$$\oint_S [U \nabla V - V \nabla U] \cdot d\mathbf{S} = \int_v [U \nabla^2 V - V \nabla^2 U] \, dv \quad \text{(Green second identity)}$$

where $d\mathbf{S} = d S \, \mathbf{a}_n$.

If S is the area bounded by the closed path L and the positive directions of elements $d\mathbf{S}$ and $d\mathbf{l}$ are related by the right-hand rule, then

$$\oint_L \mathbf{A} \cdot d\mathbf{l} = \int_S \nabla \times \mathbf{A} \cdot d\mathbf{S} \quad \text{(Stokes' theorem)}$$

$$\oint_L V \, d\mathbf{l} = -\int_S \nabla V \times d\mathbf{S}$$

A.3 Orthogonal Coordinates

Rectangular coordinates (x, y, z)

$$\nabla V = \frac{\partial V}{\partial x} \mathbf{a}_x + \frac{\partial V}{\partial y} \mathbf{a}_y + \frac{\partial V}{\partial z} \mathbf{a}_z$$

$$\nabla \cdot \mathbf{A} = \frac{\partial A_x}{\partial x} + \frac{\partial A_y}{\partial y} + \frac{\partial A_z}{\partial z}$$

$$\nabla \times \mathbf{A} = \left[\frac{\partial A_z}{\partial y} - \frac{\partial A_y}{\partial z} \right] \mathbf{a}_x + \left[\frac{\partial A_x}{\partial z} - \frac{\partial A_z}{\partial x} \right] \mathbf{a}_y + \left[\frac{\partial A_y}{\partial x} - \frac{\partial A_x}{\partial y} \right] \mathbf{a}_z$$

$$\nabla^2 V = \frac{\partial^2 V}{\partial x^2} + \frac{\partial^2 V}{\partial y^2} + \frac{\partial^2 V}{\partial z^2}$$

$$\nabla^2 \mathbf{A} = \nabla^2 A_x \mathbf{a}_x + \nabla^2 A_y \mathbf{a}_y + \nabla^2 A_z \mathbf{a}_z$$

Cylindrical coordinates (ρ, ϕ, z)

$$\nabla V = \frac{\partial V}{\partial \rho}\mathbf{a}_\rho + \frac{1}{\rho}\frac{\partial V}{\partial \phi}\mathbf{a}_\phi + \frac{\partial V}{\partial z}\mathbf{a}_z$$

$$\nabla \cdot \mathbf{A} = \frac{1}{\rho}\frac{\partial}{\partial \rho}(\rho A_\rho) + \frac{1}{\rho}\frac{\partial}{\partial \phi}A_\phi + \frac{\partial}{\partial z}A_z$$

$$\nabla \times \mathbf{A} = \left[\frac{1}{\rho}\frac{\partial}{\partial \phi}A_z - \frac{\partial}{\partial z}A_\phi\right]\mathbf{a}_\rho + \left[\frac{\partial}{\partial z}A_\rho - \frac{\partial}{\partial \rho}A_z\right]\mathbf{a}_\phi$$

$$+ \frac{1}{\rho}\left[\frac{\partial}{\partial \rho}(\rho A_\phi) - \frac{\partial}{\partial \phi}A_\rho\right]\mathbf{a}_z$$

$$\nabla^2 V = \frac{1}{\rho}\frac{\partial}{\partial \rho}\left(\rho \frac{\partial V}{\partial \rho}\right) + \frac{1}{\rho^2}\frac{\partial^2 V}{\partial \phi^2} + \frac{\partial^2 V}{\partial z^2}$$

$$\nabla^2 \mathbf{A} = \left[\nabla^2 A_\rho - \frac{2}{\rho^2}\frac{\partial}{\partial \phi}A_\phi - \frac{A_\rho}{\rho^2}\right]\mathbf{a}_\rho$$

$$+ \left[\nabla^2 A_\phi + \frac{2}{\rho^2}\frac{\partial}{\partial \phi}A_\rho - \frac{A_\phi}{\rho^2}\right]\mathbf{a}_\phi + \nabla^2 A_z \mathbf{a}_z$$

Spherical coordinates (r, θ, ϕ)

$$\nabla V = \frac{\partial V}{\partial r}\mathbf{a}_r + \frac{1}{r}\frac{\partial V}{\partial \theta}\mathbf{a}_\theta + \frac{1}{r\sin\theta}\frac{\partial V}{\partial \phi}\mathbf{a}_\phi$$

$$\nabla \cdot \mathbf{A} = \frac{1}{r^2}\frac{\partial}{\partial r}(r^2 A_r) + \frac{1}{r\sin\theta}\frac{\partial}{\partial \theta}(\sin\theta A_\theta) + \frac{1}{r\sin\theta}\frac{\partial A_\phi}{\partial \phi}$$

$$\nabla \times \mathbf{A} = \frac{1}{r\sin\theta}\left[\frac{\partial}{\partial \theta}(\sin\theta A_\phi) - \frac{\partial}{\partial \phi}A_\theta\right]\mathbf{a}_r + \frac{1}{r}\left[\frac{1}{\sin\theta}\frac{\partial}{\partial \phi}A_r - \frac{\partial}{\partial r}(r A_\phi)\right]\mathbf{a}_\theta$$

$$+ \frac{1}{r}\left[\frac{\partial}{\partial r}(r A_\theta) - \frac{\partial}{\partial \theta}A_r\right]\mathbf{a}_\phi$$

$$\nabla^2 V = \frac{1}{r^2}\frac{\partial}{\partial r}\left(r^2 \frac{\partial V}{\partial r}\right) + \frac{1}{r^2\sin\theta}\frac{\partial}{\partial \theta}\left(\sin\theta \frac{\partial V}{\partial \theta}\right) + \frac{1}{r^2\sin^2\theta}\frac{\partial^2 V}{\partial \phi^2}$$

$$\nabla^2 \mathbf{A} = \left[\nabla^2 A_r - \frac{2}{r^2\sin\theta}\frac{\partial}{\partial \theta}(\sin\theta A_\theta) - \frac{2}{r^2\sin\theta}\frac{\partial}{\partial \phi}A_\phi - \frac{2}{r^2}A_r\right]\mathbf{a}_r$$

$$+ \left[\nabla^2 A_\theta + \frac{2}{r^2}\frac{\partial}{\partial \theta}A_r - \frac{2\cos\theta}{r^2\sin^2\theta}\frac{\partial}{\partial \phi}A_\phi - \frac{1}{r^2\sin^2\theta}A_\theta\right]\mathbf{a}_\theta$$

$$+ \left[\nabla^2 A_\phi + \frac{2\cos\theta}{r^2\sin^2\theta}\frac{\partial}{\partial \phi}A_\theta + \frac{2}{r^2\sin\theta}\frac{\partial}{\partial \phi}A_r - \frac{1}{r^2\sin^2\theta}A_\phi\right]\mathbf{a}_\phi$$

Appendix B: Programming in MATLAB

MATLAB has become a powerful tool of technical professionals worldwide. The term MATLAB is an abbreviation for MATrix LABoratory implying that MATLAB is a computational tool that enables one to perform engineering and scientific calculations and employs matrices and vectors/arrays to carry out numerical analysis, signal processing, and scientific visualization tasks. One can use MATLAB to graph functions, solve equations, perform numerical analysis, and do much more. Since MATLAB uses matrices as its fundamental building blocks, one can write mathematical expressions involving matrices just as easily as one would on paper. MATLAB is available for Macintosh, Unix, and Windows operating systems. A student version of MATLAB is available for PCs. A copy of MATLAB can be obtained from

The Mathworks, Inc.
3 Apple Hill Drive
Natick, MA 01760-2098
Phone: (508) 647-7000
Website: http://www.mathworks.com

A brief introduction to MATLAB is presented in this Appendix. What is presented is sufficient for solving problems in this book. Other information on MATLAB required in this book is provided on a chapter-to-chapter basis as needed. The best way to learn MATLAB is to work with it after you have learned the basics. It will be assumed that you are practicing each command while sitting at a computer running MATLAB.

B.1 MATLAB Fundamentals

The Command window is the primary area where you interact with MATLAB. A little later, we will learn how to use the text editor to create M-files, which allow executing sequences of commands. For now, we focus on how to work in the Command window. We will first learn how to use MATLAB as a calculator. We do so by using the algebraic operators in Table B.1.

To begin to use MATLAB, we use these operators. Type commands at the MATLAB prompt ">>" in the Command window (correct any mistakes by backspacing) and press the <Enter> key. For example,

```
>> a=2; b=4; c=-6;
>> dat = b^2 - 4*a*c
dat = 64
>> e = sqrt(dat)/10
e = 0.8000
```

The first command assigns the values 2, 4, and -6 to the variables a, b, and c, respectively. MATLAB does not respond because this line ends with a semicolon. The second command sets dat to $b^2 - 4ac$ and MATLAB returns the answer as 64. Finally, the third line sets e equal

TABLE B.1

Basic Operations

Operation	MATLAB Formula
Addition	a + b
Division (right)	a/b (means $a \div b$)
Division (left)	a\b (means $b \div a$)
Multiplication	a*b
Power	a^b
Subtraction	a − b

to the square root of *dat* divided by 10. MATLAB prints the answer as 0.8. As function *sqrt* is used here, other mathematical functions listed in Table B.2 can be used. Table B.2 provides just a small sample of MATLAB functions. Others can be obtained from the online help. To get help, type

```
>> help
[a long list of topics come up]
```

and for a specific topic, type the command name. For example, to get help on *log to base 2*, type

```
>> help log2
[a help message on the log function follows]
```

TABLE B.2

Typical Elementary Math Functions

Function	Remark
abs(x)	Absolute value or complex magnitude of x
acos, acosh(x)	Inverse cosine and inverse hyperbolic cosine of x in radians
acot, acoth (x)	Inverse cotangent and inverse hyperbolic cotangent of x in radians
angle(x)	Phase angle (in radian) of a complex number x
asin, asinh(x)	Inverse sine and inverse hyperbolic sine of x in radians
atan, atanh(x)	Inverse tangent and inverse hyperbolic tangent of x in radians
bessel	Bessel functions; besselj(n, x) and bessely(n, x) are of order n
conj(x)	Complex conjugate of x
cos, cosh(x)	Cosine and hyperbolic cosine of x in radians
cot, coth(x)	Cotangent and hyperbolic cotangent of x in radians
exp(x)	Exponential of x
fix	Round toward zero
gamma	Gamma function
imag(x)	Imaginary part of a complex number x
log(x)	Natural logarithm of x
log2(x)	Logarithm of x to base 2
log10(x)	Common logarithms (base 10) of x
real(x)	Real part of a complex number x
sin, sinh(x)	Sine and hyperbolic sine of x in radians
sqrt(x)	Square root of x
tan, tanh(x)	Tangent and hyperbolic tangent of x in radians

Note that MATLAB is case sensitive so that sin(a) is not the same as sin(A). MATLAB distinguishes between lower and upper case variables.

Try the following examples:

```
>> 3^(log10(25.6))
>> y=2* sin(pi/3)
>>exp(y+4-1)
```

In addition to operating on mathematical functions, MATLAB easily allows one to work with vectors and matrices. A vector (or array) is a special matrix with one row or one column. For example,

```
>> a = [ 1 -3 6 10 -8 11 14];
```

is a row vector. Defining a matrix is similar to defining a vector. For example, a 3 × 3 matrix can be entered as

```
>> A = [ 1 2 3; 4 5 6; 7 8 9]
    or as
>> A = [ 1 2 3
          4 5 6
          7 8 9]
```

In addition to the arithmetic operations that can be performed on a matrix, the operations in Table B.3 can be implemented.

Using the operations in Table B.3, we can manipulate matrices as follows.

```
>> B = A'
B =
    1   4   7
    2   5   8
    3   6   9

>> C = A + B
C =
    2    6   10
    6   10   14
   10   14   18
```

TABLE B.3

Matrix Operations

Operation	Remark
A′	Finds the transpose of matrix A
det(A)	Evaluates the determinant of matrix A
inv(A)	Calculates the inverse of matrix A
eig(A)	Determines the eigenvalues of matrix A
diag(A)	Finds the diagonal elements of matrix A
expm(A)	Exponential of matrix A
rank(A)	Determines the rank of matrix A

```
>> D = A^3 - B*C
D =
   372   432   492
   948  1131  1314
  1524  1830  2136

>> e = [1 2; 3 4]
e =
   1   2
   3   4

>> f = det(e)
f =
   -2

>> g = inv(e)
g =
  -2.0000  1.0000
   1.5000   -0.5000

>> H = eig(g)
H =
  -2.6861
   0.1861
```

Note that not all matrices can be inverted. A matrix can be inverted if and only if its determinant is nonzero. Special matrices, variables, and constants are listed in Table B.4. For example, type

```
>> eye(3)
ans =
   1  0  0
   0  1  0
   0  0  1
```

to get a 3×3 identity matrix.

TABLE B.4

Special Matrices, Variables, and Constants

Matrix/Variable/Constant	Remark
eye	Identity matrix
ones	An array of ones
zeros	An array of zeros
i or j	Imaginary unit or sqrt(−1)
pi	3.142
NaN	Not a number
inf	Infinity
eps	A very small number, 2.2e−16
rand	Generates random values between 0 and 1
randn	Generates random values from a normal distribution
num2str	Converts number to string

Note that MATLAB cannot have negative or zero index such as x(−5) or A(0); index must be a positive integer.

B.2 Using MATLAB to Plot

To plot using MATLAB is easy. For a two-dimensional plot, use the **plot** command with two arguments as

```
>> plot(xdata,ydata)
```

where *xdata* and *ydata* are vectors of the same length containing the data to be plotted.

For example, suppose we want to plot y = 10*sin(2*pi*x) from 0 to 5*pi, we will proceed with the following commands:

```
>>   x = 0:pi/100:5*pi;    %   x is a vector, 0 <= x <= 5*pi,
                               increments of pi/100
>>   y = 10*sin(2*pi*x);   %   create a vector y
>>   plot(x,y);            %   create the plot
```

With this, MATLAB responds with the plot in Figure B.1 MATLAB will let you graph multiple plots together and distinguish with different colors. This is obtained with the command **plot**(xdata, ydata, 'color'), where the color is indicated by using a character string from the options listed in Table B.5.

For example,

```
>> plot(x1, y1, 'r', x2,y2, 'b', x3,y3, '--');
```

will graph data (x1,y1) in red, data (x2,y2) in blue, and data (x3,y3) in dashed line all on the same plot.

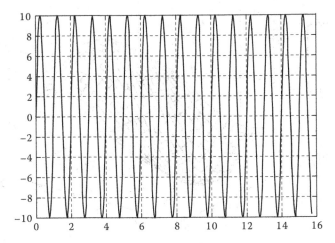

FIGURE B.1

MATLAB plot of y = 10*sin(2*pi*x).

TABLE B.5

Various Color and Line Types

y	yellow	.	point
m	magenta	o	circle
c	cyan	x	x-mark
r	red	+	plus
g	green	−	solid
b	blue	*	star
w	white	:	dotted
k	black	-.	dashdot
		−	dashed

MATLAB also allows for logarithm scaling. Rather than the **plot** command, we use

```
loglog       log(y) versus log(x)
semilogx     y versus log(x)
semilogy     log(y) versus x
```

Three-dimensional plots are drawn using the functions *mesh* and *meshdom* (mesh domain). For example, to draw the graph of $z = x*\exp(-x^2 - y^2)$ over the domain $-1 < x, y < 1$, we type the following commands:

```
>> xx = -1:.1:1;
>> yy = xx;
>> [x,y] = meshgrid(xx,yy);
>> z = x.*exp(-x.^2 -y.^2);
>> mesh(z);
```

(The dot symbol used in x. and y. allows element-by-element multiplication.) The result is shown in Figure B.2.

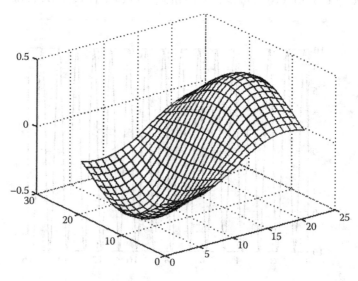

FIGURE B.2
A three-dimensional plot.

TABLE B.6

Other Plotting Commands

Command	Comments
bar(x,y)	A bar graph
contour(z)	A contour plot
errorbar(x,y,l,u)	A plot with error bars
hist(x)	A histogram of the data
plot3(x,y,z)	A three-dimensional version of plot()
polar(r, angle)	A polar coordinate plot
stairs(x,y)	A stairstep plot
stem(x)	Plots the data sequence as stems
subplot(m,n,p)	Multiple (m-by-n) plots per window
surf(x,y,x,c)	A plot of 3-D colored surface.

Other plotting commands in MATLAB are listed in Table B.6. The **help** command can be used to find out how each of these is used.

B.3 Programming with MATLAB

So far, MATLAB has been used as a calculator, you can also use MATLAB to create your own program. The command line editing in MATLAB can be inconvenient if you have several lines to execute. To avoid this problem, you create a program which is a sequence of statements to be executed. If you are in Command window, click **File/New/M-files** to open a new file in the MATLAB Editor/Debugger or text editor. Type the program and save the program in a file with an extension.m, that is, filename.m; it is for this reason it is called an M-file. Once the program is saved as an M-file, exit the Debugger window. You are now back in the Command window. Type the file without the extension.m to get results. For example, the plot that was made above can be improved by adding title and labels and typing as an M-files called example1.m

```
x = 0:pi/100:5*pi;          % x is a vector, 0 <= x
                            <= 5*pi, increments of pi/100
y = 10*sin(2*pi*x);         % create a vector y
plot(x,y);                  % create the plot
xlabel('x (in radians)');   % label the x-axis
ylabel('10*sin(2*pi*x)');   % label the y-axis
title('A sine functions');  % title the plot
grid                        % add grid
```

Once it is saved as *example1.m* and you exit text editor, type

```
>> example1
```

in the Command window and hit <Enter> to obtain the result shown in Figure B.3.

To allow flow control in a program, certain relational and logical operators are necessary. They are shown in Table B.7. Perhaps the most commonly used flow control statements are

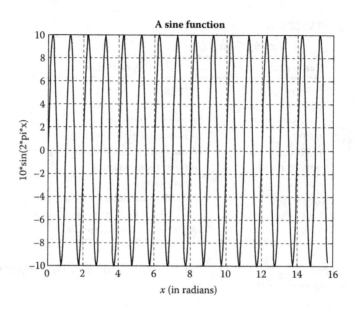

FIGURE B.3

MATLAB plot of y = 10*sin(2*pi*x) with title and labels.

for, if, and *while.* The *for* statement is used to create a loop or a repetitive procedure and has the general form

```
for x = array
  [commands]
end
```

The *if* statement is used when certain conditions need to be met before an expression is executed. It has the general form

```
if expression
  [commands if expression is True]
else
  [commands if expression is False]
end
```

TABLE B.7

Relational and Logical Operators

Operator	Remark
<	less than
<=	less than or equal
>	greater than
>=	greater than or equal
==	equal
~=	not equal
&	and
		or
~	not

A *while* loop consists of statements that are repeated indefinitely as long as some condition is met. The *while* loop has the form

```
while expression
  [statements]
end
```

A *while* loop may be terminated with the *break* statement, which passes control to the statement that follows the corresponding *end*. Consider the following example:

```
x = 2;
while x<1
  y=cos(x)
  if x==0, break, end
end
z=1;
```

For example, suppose we have an array y(x) and we want to determine the minimum value of y and its corresponding index x. This can be done by creating an M-file as shown below.

```
% example2.m
% This program finds the minimum y value and its corresponding x index
x = [1 2 3 4 5 6 7 8 9 10]; %the nth term in y
y = [3 9 15 8 1 0 -2 4 12 5];
min1 = y(1);
for k=1:10
  min2=y(k);
  if (min2 < min1)
    min1 = min2;
    xo = x(k);
  else
    min1 = min1;
  end
end
diary
min1, xo
diary off
```

Note the use of *for* and *if* statements. When this program is saved as example2.m, we execute it in the Command window and obtain the minimum value of y as –2 and the corresponding value of x as 7, as expected.

```
>> example2
min1 =
-2
xo =
7
```

If we are not interested in the corresponding index, we could do the same thing using the command

```
>> min(y)
```

To display output, use the *disp* or *input* command. While *disp* displays text or array, *input* prompts for user input.

```
>> disp('This will print without quotes.')
>> x = input('starting value:')
```

The command *disp* is often combined with *num2str* (convert a number to a string)

```
>> x = 2*cos(x);
.>> str = ['disp: x = ', num3str(x)];
>> disp(str);
```

Sometimes you want to label a plot with Greek letters, you can simply use \beta for β. Other selected Greek and mathematical symbols are shown in Table B.8. For example, you may label the title of a plot as follows:

```
>> title('Gain versus angle \theta');
```

The following tips are helpful in working effectively with MATLAB:

- Comment your M-file by adding lines beginning with a % character.
- To suppress output, end each command with a semi-colon (;), you may remove the semi-colon when debugging the file.
- Press up and down arrow keys to retrieve previously executed commands.
- If your expression does not fit on one line, use an ellipse (...) at the end of the line and continue on the next line. For example, MATLAB considers

```
y = sin (x + log10(2x + 3)) + cos(x +···
log10(2x + 3));
```

as one line of expression
- Keep in mind that variable and function names are case sensitive.

TABLE B.8

Selected Greek and Mathematical Symbols

Character	Symbol	Character	Symbol
\alpha	α	\Gamma	Γ
\beta	β	\Delta	Δ
\gamma	γ	\Lamdba	Λ
\delta	δ	\Pi	Π
\epsilon	ε	\Sigma	Σ
\eta	η	\Omega	Ω
\theta	θ	\int	$\int$
\lambda	λ	\sim	$\sim$
\nu	ν	\infty	∞
\pi	π	\pm	$\pm$
\phi	ϕ	\eq	$\leq$
\rho	ρ	\geq	$\geq$
\sigma	σ	\neq	$\neq$
\tau	τ	\div	$\div$
\omega	ω		

B.4 Functions

All of the M-files we have seen thus far are known as *script* files. A script file is a collection of MATLAB statements; it has no input argument and returns no results. In contrast, a MATLAB *function* is a special M-file that runs in its own independent workspace. It accepts some input data, performs some calculation, and returns results to the caller. It has the following general form:

```
function [outarg1, outarg2, ... ]
  = filename(inarg1, inarg2, ... )
[statements]
end
```

The *function* statement begins the function, and it is terminated by an *end* statement. The function name in the calling program must match the function statement name. Consider the function M-file, called *root.m*:

```
function z = root(x)
% root(x) returns the square root of x
z = sqrt(x)
end
```

Typing root(16) gives 4. As an example of a function that has multiple input and output arguments, consider function *polar1.m*.

```
function [rho, theta] = polar1(x,y)
% Given the rectangular coordinates (x,y), calculate the
% polar coordinates (rho, theta)
rho = sqrt(x^2 + y^2);
theta = atan2(y,x);
end
```

This function must be saved as polar1.m You can now use this function in a script or in another function. For example,

```
>> x = 3; y = -4;
>> [rho1,theta1] = polar1(x,y);
>> z = rho1*cos(theta1);
```

B.5 Solving Equations

Consider the general system of n simultaneous equations as

$$a_{11}x_1 + a_{12}x_2 + \cdots + a_{1n}x_n = b_1$$
$$a_{21}x_1 + a_{22}x_2 + \cdots + a_{2n}x_n = b_2$$
$$\cdots \qquad \cdots \qquad \cdots$$
$$a_{n1}x_1 + a_{n2}x_2 + \cdots + a_{nn}x_n = b_n$$

or in matrix form

$$AX = B$$

where

$$A = \begin{bmatrix} a_{11} & a_{12} & \cdots & a_{1n} \\ a_{21} & a_{22} & \cdots & a_{2n} \\ \cdots & \cdots & \cdots & \cdots \\ a_{n1} & a_{n2} & a_{n3} & a_{nn} \end{bmatrix}, \quad X = \begin{bmatrix} x_1 \\ x_2 \\ \cdots \\ x_n \end{bmatrix}, \quad B = \begin{bmatrix} b_1 \\ b_2 \\ \cdots \\ b_n \end{bmatrix}$$

A is a square matrix and is known as the coefficient matrix, while X and B are vectors. X is the solution vector we are seeking to obtain. There are two ways to solve for X in MATLAB. First, we can use the backslash operator (\) so that

$$X = A\backslash B$$

Second, we can solve for X as

$$X = A^{-1}B$$

which in MATLAB is the same as

$$X = inv(A) * B$$

We can also solve equations using the command **solve**. For example, given the quadratic equation $x^2 + 2x - 3 = 0$, we obtain the solution using the following MATLAB command:

```
>> [x]=solve('x^2 + 2*x - 3 =0')
x =
[ -3]
[ 1]
```

indicating that the solutions are $x = -3$ and $x = 1$. Of course, we can use the command **solve** for a case involving two or more variables. We will see this in the following example.

EXAMPLE B.1

Use MATLAB to solve the following simultaneous equations:

$$25x_1 - 5x_2 - 20x_3 = 50$$
$$-5x_1 + 10x_2 - 4x_3 = 0$$
$$-5x_1 - 4x_2 + 9x_3 = 0$$

Solution

We can use MATLAB to solve this in two ways.

Method 1:

The given set of simultaneous equations could be written as

$$\begin{bmatrix} 25 & -5 & -20 \\ -5 & 10 & -4 \\ -5 & -4 & 9 \end{bmatrix} \begin{bmatrix} x_1 \\ x_2 \\ x_3 \end{bmatrix} = \begin{bmatrix} 50 \\ 0 \\ 0 \end{bmatrix} \quad \text{or} \quad AX = B$$

We obtain matrix A and vector B and enter them in MATLAB as follows.

```
>> A = [25 -5 -20; -5 10 -4; -5 -4 9]
A =
25  -5  -20
-5  10  -4
-5  -4  9

>> B = [50 0 0]'
B =
50
0
0

>> X = inv(A)*B
X =
29.6000
26.0000
28.0000

>> X = A\B
X =
29.6000
26.0000
28.0000
```

Thus, $x_1 = 29.6$, $x_2 = 26$, and $x_3 = 28$.

Method 2:

Since the equations are not many in this case, we can use the command **solve** to obtain the solution of the simultaneous equations as follows:

```
[x1,x2,x3]=solve('25*x1 - 5*x2 - 20*x3=50',
        '-5*x1 + 10*x2 - 4*x3 =0', '-5*x1 - ...
        4*x2 + 9*x3=0')
x1 =
148/5
x2 =
26
x3 =
28
```

which is the same as before.

B.6 Programming Hints

A good program should be well documented, of reasonable size, and capable of performing some computation with reasonable accuracy within a reasonable amount of time. The

TABLE B.9

Other Useful MATLAB Commands

Command	Explanation
clear	Clears items from workspace
diary	Saves screen display output in text format
ezplot	Easy plot command for symbolic expressions
mean	Mean value of a vector
min(max)	Minimum (maximum) of a vector
grid	Adds a grid mark to the graphic window
poly	Converts a collection of roots into a polynomial
roots	Finds the roots of a polynomial
sort	Sort the elements of a vector
sound	Play vector as sound
std	Standard deviation of a data collection
sum	Sum of elements of a vector
exit, quit	Terminate a MATLAB session

following are some helpful hints that may make writing and running MATLAB programs easier:

- Use the minimum number of commands possible and avoid execution of extra commands. This is particularly true of loops.
- Use matrix operations directly as much as possible and avoid *for, do,* and/or *while* loops if possible.
- Make effective use of functions for executing a series of commands over several times in a program.
- When unsure about a command, take advantage of the help capabilities of the software.
- It takes much less time running a program using files on the hard disk than on an external drive.
- Start each file with comments to help you remember what it is all about later.
- When writing a long program, save frequently. If possible, avoid a long program; break it down into smaller functions.

B.7 Other Useful MATLAB Commands

Some common useful MATLAB commands which may be used in this book are provided in Table B.9.

Appendix C: Solution of Simultaneous Equations

Application of some numerical methods to EM problems often results in a set of simultaneous equations

$$
\begin{bmatrix}
a_{11} & a_{12} & \cdots & a_{1n} \\
a_{21} & a_{22} & \cdots & a_{2n} \\
\vdots & & & \vdots \\
a_{n1} & a_{n2} & \cdots & a_{nn}
\end{bmatrix}
\begin{bmatrix}
x_1 \\
x_2 \\
\vdots \\
x_n
\end{bmatrix}
=
\begin{bmatrix}
b_1 \\
b_2 \\
\vdots \\
b_n
\end{bmatrix}
\tag{C.1a}
$$

or

$$
[A][X] = [B] \tag{C.1b}
$$

where $[A]$ is the coefficient matrix, $[X]$ is the column matrix of the unknowns to be determined, and $[B]$ is the column matrix of constants. Familiarity with the various techniques for solving Equation C.1 is therefore vital. In this appendix, we provide a brief coverage of direct and iterative procedures for solving Equation C.1; direct methods are more versatile for linear problems, while iterative methods are suitable for non-linear problems. We also consider various techniques for solving eigenvalue systems $[A][X] = \lambda[X]$.

C.1 Elimination Methods

Elimination methods constitute the simplest direct approach to the solution of a set of simultaneous equations. They usually involve successive elimination of the unknowns by combining equations. Such methods include Gauss's method, Gauss–Jordan method, Cholesky's or Crout's method, and the square-root method. Only Gauss's and Cholesky's methods will be discussed. The reader should consult References 1–4 for the treatment of other methods.

C.1.1 Gauss's Method

This simple method involves eliminating one unknown at a time and proceeding with the remaining equations. This leads to a set of simultaneous equations in triangular form from which each unknown is determined by back substitution. To describe this method, consider Equation C.1, that is,

$$a_{11}x_1 + a_{12}x_2 + \cdots + a_{1n}x_n = b_1 \tag{C.2a}$$

$$a_{21}x_1 + a_{22}x_2 + \cdots + a_{2n}x_n = b_2 \tag{C.2b}$$

$$\vdots$$

$$a_{n1}x_1 + a_{n2}x_2 + \cdots + a_{nn}x_n = b_n \tag{C.2c}$$

We divide Equation C.2a by a_{11} to give

$$x_1 + a'_{12}x_2 + \cdots + a'_{1n}x_n = b'_1 \tag{C.3}$$

where the primes denote that the coefficients are new. We multiply Equation C.3 by $-a_{i1}$ for $i = 2, 3, \ldots, n$ and add Equation C.3 to the ith equation in Equation C.2 to eliminate x_1 from other equations so that Equation C.2 becomes

$$x_1 + a'_{12}x_2 + \cdots + a'_{1n}x_n = b'_1 \tag{C.4a}$$

$$a'_{22}x_2 + \cdots + a'_{2n}x_n = b'_2 \tag{C.4b}$$

$$\vdots$$

$$a'_{n2}x_2 + \cdots + a'_{nn}x_n = b'_n \tag{C.4c}$$

Equation C.2a used to eliminate x_1 from other equations is called the *pivot equation* and a_{11} is called the *pivot coefficient*. We now use Equation C.4b as the pivot equation and we take similar steps to eliminate x_2 from all equations following the pivot equation. Continuing this reduction procedure eventually leads to an equivalent triangular set of equations:

$$\begin{aligned}
x_1 + u_{12}x_2 + u_{13}x_3 + \cdots + u_{1n}x_n &= c_1 \\
x_2 + u_{23}x_2 + \cdots + u_{2n}x_n &= c_2 \\
x_3 + \cdots + u_{3n}x_n &= c_3 \\
&\vdots \\
x_n &= c_n
\end{aligned} \tag{C.5}$$

This completes the first phase known as *forward elimination* in the Gauss algorithm, and the system in Equation C.5 is said to be in *upper triangular* form. The second phase known as *back substitution* involves solving for the unknowns in Equation C.5 by starting at the bottom. That is,

$$\begin{aligned}
x_n &= c_n \\
x_{n-1} &= c_{n-1} - u_{n-1,n}x_n \\
&\vdots \\
x_1 &= c_1 - u_{12}x_2 - \cdots - u_{1n}x_n
\end{aligned} \tag{C.6}$$

In summary, this algorithm can be stated as follows:
Forward elimination:

$$a'_{kj} = a_{kj}/a_{kk}, \quad b'_k = b_k/a_{kk}, \quad j = k, k+1, \ldots, n$$
$$a'_{ij} = a_{ij} - a_{ik}a'_{kj}, \quad i = k+1, \ldots, n \qquad \text{(C.7a)}$$
$$b'_i = b_i - a_{ik}b'_k, \quad i = k+1, \ldots, n$$

Backward substitution:

$$x_n = b_n, \quad \text{for the last row}$$
$$x_i = b_i - \sum_{j=i+1}^{n} a_{ij}x_j, \quad i = n-1, \ldots, 1 \qquad \text{(C.7b)}$$

Based on the idea outlined above, a general MATLAB code for solving a set of simultaneous equations by Gaussian elimination is shown in Figure C.1.

```
% A set of simultaneous equations [A][X]=[B] is solved by a simple Gaussian
% elimination scheme (without pivoting) and by the MATLAB backward slash "\"
% operation.  The MATLAB backward slash "\" operation performs Gaussian
% elimination with pivoting as explained in "Numerical Computing with
% MATLAB" by Cleave Moler.  This excellent reference is available for
% download at the mathworks website.

clear;

%Input matrix
A = [10 -7 0;-3 2 6;5 -1 5];
B = [7;4;6];

%MATLAB's backward slash operation
x = A\B;

%Begin simple Gaussian Elimination
N = size(A,1);
%Forward Elimination
for kk = 1:(N-1);
    for ii = (kk+1):N
        q = A(ii,kk)/A(kk,kk);
        %Calculate Elements of the new matrix
        jj = (kk+1):N;
        A(ii,jj) = A(ii,jj) - q*A(kk,jj);
        A(ii,kk) = q;
        B(ii) = B(ii) - q*B(kk);
    end
end
%Backward Substituion
X(N) = B(N)/A(N,N);
for ii = (N-1):-1:1
    sum = A(ii,:)*X(:);
    X(ii) = ( B(ii)-sum )/A(ii,ii);
end

thedifference = x-X(:)
```

FIGURE C.1
Gauss elimination method of solving $[A][X] = [B]$.

C.1.2 Cholesky's Method

This method, also known as Crout's method or the method of matrix decomposition, involves determining a lower triangular matrix that will reduce the original system in Equation C.1 to a unit upper triangular matrix. If the original system

$$[A][X] = [B] \tag{C.1a}$$

or

$$\begin{bmatrix} a_{11} & a_{12} & \cdots & a_{1n} \\ a_{21} & a_{22} & \cdots & a_{2n} \\ \vdots & & & \vdots \\ a_{n1} & a_{n2} & \cdots & a_{nn} \end{bmatrix} \begin{bmatrix} x_1 \\ x_2 \\ \vdots \\ x_n \end{bmatrix} = \begin{bmatrix} b_1 \\ b_2 \\ \vdots \\ b_n \end{bmatrix} \tag{C.1b}$$

can be redefined in the upper unit triangular matrix $[T]$ such that

$$[T][X] = [C] \tag{C.8a}$$

or

$$\begin{bmatrix} 1 & T_{12} & \cdots & T_{1n} \\ 0 & 1 & \cdots & T_{2n} \\ \vdots & & & \vdots \\ 0 & 0 & \cdots & 1 \end{bmatrix} \begin{bmatrix} x_1 \\ x_2 \\ \vdots \\ x_n \end{bmatrix} = \begin{bmatrix} c_1 \\ c_2 \\ \vdots \\ c_n \end{bmatrix} \tag{C.8b}$$

the unknown x_i can be obtained by back substitution. Let $[A]$ be a product of an upper unit triangular matrix $[T]$ and a lower triangular matrix $[L]$, that is,

$$[L][T] = [A] \tag{C.9}$$

Since

$$[L][TX-C] = 0 = [AX-B], \tag{C.10}$$

it follows that

$$[L][C] = [B] \tag{C.11}$$

For computational reasons, it is convenient to work with the augmented form of the matrices. The augmented matrix is obtained by adding the column vector of constants to the square coefficient matrix. Equations C.8 and C.11 may be combined to give

$$\begin{bmatrix} a_{11} & a_{12} & \cdots & a_{1n} & \vdots & b_1 \\ a_{21} & a_{22} & \cdots & a_{2n} & \vdots & b_2 \\ \vdots & & & \vdots & & \\ a_{n1} & a_{n2} & \cdots & a_{nn} & \vdots & b_n \end{bmatrix}$$

$$= \begin{bmatrix} L_{11} & 0 & \cdots & 0 \\ L_{21} & L_{22} & \cdots & 0 \\ \vdots & & & \\ L_{n1} & L_{n2} & \cdots & L_{nn} \end{bmatrix} \begin{bmatrix} 1 & T_{12} & \cdots & T_{1n} & \vdots & c_1 \\ 0 & 1 & \cdots & T_{2n} & \vdots & c_2 \\ \vdots & & & & & \\ 0 & 0 & \cdots & 1 & \vdots & c_n \end{bmatrix} \tag{C.12a}$$

or

$$[A \vdots B] = [L][T \vdots C] \tag{C.12b}$$

The elements of $[L]$, $[T]$, and $[C]$ can be defined in terms of $[A]$ and $[B]$ as follows [1,2,5]:

$$L_{ij} = a_{ij} - \sum_{k=1}^{j-1} L_{ik} T_{kj}, \quad i \geq j, i = 1, 2, \ldots, n$$

$$L_{ij} = a_{i1}, \quad j = 1$$

$$T_{ij} = \frac{1}{L_{ii}} \left(a_{ij} - \sum_{k=1}^{i-1} L_{ik} T_{kj} \right), \quad i < j, j = 2, 3, \ldots, n \tag{C.13}$$

$$T_{ij} = a_{ij}/a_{11}, \quad i = 1$$

$$c_i = \frac{1}{L_{ii}} \left(b_i - \sum_{k=1}^{i-1} L_{ik} c_k \right), \quad i = 2, 3, \ldots, n$$

$$c_1 = b_1/L_{11}$$

The unknown x_i are obtained by back substitution as follows:

$$x_n = c_n$$

$$x_i = c_i - \sum_{j=i+1}^{n} T_{ij} x_j, \quad i = 1, 2, \ldots, n-1 \tag{C.14}$$

Cholesky's method can easily be applied in calculating the determinant of $[A]$. Since

$$\det[A] = \det[L] \det[T] \tag{C.15}$$

and $\det[T] = 1$ due to the fact that $T_{ii} = 1$, it follows that

$$\det[A] = \det[L] = L_{11} L_{22} \ldots L_{nn}$$

or

$$\det[A] = \prod_{i=1}^{n} L_{ii} \tag{C.16}$$

Figure C.2 shows a subroutine based on Cholesky's method of solving a set of simultaneous equations.

C.2 Iterative Methods

The direct or elimination method for solving a system of simultaneous equations can be used for $n = 25$–60. This number can be greater if the system is well conditioned or the

```
%  This program applies the Cholesky elimination method to solve the system
%  [A][X]=[B] where [AA] = augmented matrix [A,B] and compares the result to
%  the MATLAB backslash operation.

clear;

%Define A and B
A = [10 -7 0;-3 2 6;5 -1 5];
B = [7;4;6];

%MATLAB backslash operation
x = A\B;

%Begin Cholesky elimination
AA = [A,B];
M = size(AA,2);
N = size(AA,1);

%Calculate the first row of the upper triangular matrix [T]
for jj = 2:M
    AA(1,jj) = AA(1,jj)/AA(1,1);
end
%Calculate the other elements of [T] and lower triangular [L] matrices
for ii = 2:N
    for qq = ii:N
        sum = 0;
        for kk = 1:(ii-1)
            sum = sum + AA(qq,kk)*AA(kk,ii);
        end
        AA(qq,ii) = AA(qq,ii)-sum; %calculates [L]
    end
    for jj = ii+1:M
        sum = 0;
        for kk = 1:ii-1
            sum = sum + AA(ii,kk)*AA(kk,jj);
        end
        AA(ii,jj) = (AA(ii,jj)-sum )/AA(ii,ii); %calculates [T]
    end
end
%Solve for [X] by back substitution
X(N) = AA(N,M);
for nn  = 1:N-1
    sum = 0;
    ii = N-nn;
    for jj = ii+1:N
        sum = sum + AA(ii,jj)*X(jj);
    end
    X(ii) = AA(ii,M)-sum;
end

thedifference = x-X(:)
```

FIGURE C.2
Cholesky's elimination method of solving $[A][X] = [B]$.

matrix is sparse. For very large systems, say $n = 100$ or even 1000, elimination methods become time consuming and prove inadequate due to roundoff error. For these types of problems, indirect or iterative methods provide an alternative.

C.2.1 Jacobi's Method

This is the simplest iterative method. If the system in Equation C.1 is rearranged so that the ith equation is explicit in x_i, we obtain

$$x_1 = \frac{1}{a_{11}}[b_1 - a_{12}x_2 - a_{13}x_3 - \cdots - a_{1n}x_n] \tag{C.16a}$$

$$x_2 = \frac{1}{a_{22}}[b_2 - a_{21}x_1 - a_{23}x_3 - \cdots - a_{2n}x_n] \tag{C.16b}$$

$$\vdots$$

$$x_n = \frac{1}{a_{nn}}[b_n - a_{n1}x_1 - a_{n2}x_2 - \cdots - a_{n,n-1}x_{n-1}] \tag{C.16c}$$

assuming that the diagonal elements are all nonzero. We start the solution process by using guesses for the x's, say $x_1 = x_2 = \cdots = x_n = 0$. The first equation can be solved for x_1, the second for x_2, and so on. If we denote the estimates after the kth iteration as $x_1^k, x_2^k, \ldots, x_n^k$, the estimates after $(k+1)$th iteration can be obtained from Equation C.16 as

$$x_i^{k+1} = \frac{1}{a_{ii}}\left[b_i - \sum_{j=1, j \neq 1}^{n} a_{ij}x_j^k\right], \quad i = 1, 2, \ldots, n \tag{C.17}$$

The iteration process is continued until values of x_i at two successive iterations are within an allowable prescribed deviation.

Convergence is measured in terms of the change in x_i from the kth iteration to the next. If we compute

$$d_i = \left|\frac{x_i^{k+1} - x_i^k}{x_i^{k+1}}\right| \times 100\% \tag{C.18}$$

for each x_i, convergence can be checked using the criterion

$$d_i < \epsilon_s \tag{C.19}$$

where ϵ_s is a specified small quantity. A better test would be to compute

$$d = \frac{\sum_{i=1}^{n}|x_i^{k+1} - x_i^k|}{\sum_{i=1}^{n}|x_i^{k+1}|} \times 100\% \tag{C.20}$$

and require that $d < \epsilon_s$.

C.2.2 Gauss–Seidel Method

This is the most commonly used iterative method. In Jacobi's method, the entire set of x_i from the kth iteration is used in calculating the new set during the $(k+1)$th iteration, whereas the most recently calculated value of each variable is used at each step in the Gauss–Seidel method. This makes the Gauss–Seidel method converge more rapidly than

```
%  This program employs Gauss-Seidel iterative method to solve a set of
%  simultaneous equations [A][X] = [B] and compares the result to the MATLAB
%  backslash operator "\".  This method will converge if the matrix A is
%  diagonally dominant (i.e. the absolute value of the diagonal term is
%  greater than the sum of absolute values of the other terms in each row).

clear;

%Define A and B
A = [10 -7 0;-3 6 1;2 -1 5];
B = [7;4;6];

%MATLAB backslash operation
x = A\B;

%Begin Gauss-Seidel iterative method
N = size(A,1);
X = zeros(N,1);
K = 0; % iteration count
TOL = 1e-3; % tolerance for zero
converged = 0; %while loop condition
Xnew = zeros(N,1);

while converged == 0
    for ii = 1:N
        Xnew(ii) = 1/A(ii,ii)*( B(ii) - A(ii,1:(ii-1))*Xnew(1:(ii-1))...
             - A(ii,(ii+1):N)*X((ii+1):N));
    end
    %Convergent condition
    d = sum(abs(Xnew-X))/sum(abs(Xnew));
    %Replace old value with newly computed value
    X = Xnew;

    K = K + 1;
    disp([num2str(K),'              ',num2str(X(:)'),'            ',num2str(d)])
    if d<=TOL %convergence test
        converged = 1;
    end
end

thedifference = x-X(:)
```

FIGURE C.3
Gauss–Seidel iterative method of solving $[A][X] = [B]$.

(about twice as) Jacobi's method and is always used in preference to it. Instead of Equation C.17, we use

$$x_i^{k+1} = \frac{1}{a_{ii}}\left[b_i - \sum_{j=1}^{i-1} a_{ij} x_j^{k+1} - \sum_{j=i+1}^{n} a_{ij} x_j^{k}\right], \quad i = 1, 2, \ldots, n \qquad (C.21)$$

A computer program based on this method is displayed in Figure C.3.

C.2.3 Relaxation Method

This is a slight modification of the Gauss–Seidel method and is designed to enhance convergence. If x_i^k is added to the right-hand side of Equation C.21 and $(a_{ii} x_i^k)/a_{ii}$ is subtracted from it, we obtain

$$x_i^{k+1} = x_i^k + \frac{1}{a_{ii}}\left[b_i - \sum_{j=1}^{i-1} a_{ij} x_j^{k+1} - \sum_{j=i}^{n} a_{ij} x_j^{k}\right], \quad i = 1, 2, \ldots, n \qquad (C.22)$$

The second term on the right-hand side can be regarded as a correction term. The correction term tends to zero as convergence is approached. If this term is multiplied by ω, Equation C.22 becomes

$$x_i^{k+1} = x_i^k + \frac{\omega}{a_{ii}} \left[b_i - \sum_{j=1}^{i-1} a_{ij} x_j^{k+1} - \sum_{j=i}^{n} a_{ij} x_j^k \right], \quad i = 1, 2, \ldots, n \tag{C.23}$$

The *relaxation factor* ω is selected such that $1 < \omega < 2$. The choice of a proper value of ω is problem dependent and is often determined by trial and error. The added weight of ω is intended to improve the estimate by pushing it closer to the exact value.

C.2.4 Gradient Methods

The iterative methods considered above may be broadly classified as *stationary* while the ones to be presented now are *gradient* (or nonstationary) methods. The two common gradient methods are the *steepest method* and *conjugate gradients method* [6–8]. A major advantage gradient methods have over stationary methods is that convergence is faster; hence, gradient methods are particularly useful when the number of simultaneous equations is very large.

A set of n simultaneous equations may be solved by finding the position of the minimum of an error function defined over an n-dimensional space. In each step of a gradient method, a trial set of values for the variables is used to generate a new set corresponding to a lower value of the error function. If $\bar{X}$ is the trial vector, the vector residual is

$$R = B - A\bar{X} \tag{C.24}$$

where A is real, symmetric, and positive definite. If we define the error function as

$$e = R^t A^{-1} R, \tag{C.25}$$

then

$$e = \bar{X}^t A\bar{X} - 2B^t \bar{X} + B^t A^{-1} B \tag{C.26}$$

showing that e is quadratic in $\bar{X}$.

Starting from an arbitrary point X_o, we locate a sequence of points

$$X_{k+1} = X_k + \alpha_k D_k \tag{C.27}$$

which are successively closer to X, where X minimizes e in Equation C.26. The parameter α_k is proportional to the distance between X_i and X_{i+1} along the direction vector D_k. Substituting Equation C.27 into Equation C.26 and setting $\partial e/\partial \alpha_k$ equal to zero gives

$$\alpha_k = \frac{D_k^t R_k}{D_k^t A D_k} \tag{C.28}$$

Both the methods of steepest descent and conjugate gradients use Equation C.28 but differ in the choice of D_k.

In the method of descent, D_k is taken as the direction of maximum gradient of e at X_k. This direction is proportional to X_k so that the iterative algorithm has the form

  i. Select an arbitrary X_0
 ii. Compute $R_0 = B - AX_0$
iii. Determine successively

$$U_k = AR_k$$
$$\alpha_k = \frac{R_k^t R_k}{R_k^t U_k}$$
$$X_{k+1} = X_k + \alpha_k R_k \qquad\qquad \text{(C.29)}$$
$$R_{k+1} = R_k - \alpha_k U_k$$

 iv. Repeat step (iii) until residual vector $(R^T R)$ becomes sufficiently small.

In the method of conjugate gradients, D_k are selected as n vectors P_k which are mutually conjugate. The vectors P_k are conjugate or orthogonal to A, that is,

$$P_k^t A P_k = 0, \qquad i \neq j$$
$$\neq 0, \qquad i = j \qquad\qquad \text{(C.30)}$$

Thus, the conjugate gradients algorithm is as follows:

  i. Select an arbitrary X_0
 ii. Set $P_0 = R_0 = B - AX_0$
iii. Determine successively

$$U_k = AR_k$$
$$\alpha_k = \frac{P_k^t R_k}{P_k^t U_k}$$
$$X_{k+1} = X_k + \alpha_k R_k$$
$$R_{k+1} = R_k - \alpha_k U_k \qquad\qquad \text{(C.31)}$$
$$\beta_k = -\frac{R_{k+1}^t U_k}{P_k^t U_k}$$
$$P_{k+1} = R_k + \beta_k P_k$$

 iv. Repeat step (iii) until $k = n - 1$ or the residual vector $(R^T R)$ becomes sufficiently small.

This algorithm is guaranteed to yield the true solution in no more than n iterations—a condition known as *quadratic convergence*. Because of this, the conjugate gradients method has the advantage of an iterative scheme in that the roundoff error is limited to only the final step of the solution and also the advantage of a direct method in that it converges to the exact solution in a finite number of steps.

The MATLAB program in Figure C.4 applies the conjugate gradients method to solve a given set of simultaneous equations. Typical areas where the conjugate gradient methods have been applied in EM can be found in References 9–12.

```
% This program applies the conjugate gradients method to solve a set of
% simultaneous equations [A][X]=[B] and compares the result to the MATLAB
%  backslash operator "\".

clear;

%Define A and B
A = [10 -7 0;-3 6 1;2 -1 5];
B = [7;4;6];

%MATLAB backslash operation
x = A\B;

%Begin Conjugate Gradients Method
N = size(A,1);
kk = 0;% iteration count
giveup = 20;
TOL = 1e-3; %tolerance for zero
X = zeros(N,1);
P = B;
R = B;
converged = 0; %while loop condition

%Iteration begins here
while ~converged
    U = A*P;
    alpha = (P.'*R)/(P.'*U);
    X = X+alpha*P;
    R = R-alpha*U;

    %calculate residuals and direction vectors
    if R.'*R<TOL
        converged = 1;
        disp(['Result: X = ',num2str(X')]);
    elseif (kk>giveup)
        disp('max number of iterations reached.  exiting.')
        converged = 1;
    end
    beta = -(R.'*U)/(P.'*U);
    P = R+beta*P;

    kk = kk + 1;
end

thedifference = x-X
```

FIGURE C.4
The program applies the conjugate gradients method to solve $[A][X] = [B]$.

C.3 Matrix Inversion

If $[A]$ is a square matrix, there is another matrix $[A]^{-1}$, called the *inverse* of $[A]$, such that

$$[A][A]^{-1} = [A]^{-1}[A] = [I] \tag{C.32}$$

where I is the *identity* or *unit matrix*. Matrix inversion can be used to solve a set of simultaneous equations in Equation (C.1) as

$$[X] = [A]^{-1}[B] \tag{C.33}$$

The solution of a system of simultaneous equations by matrix inversion and multiplication is most valuable when several systems are to be solved, all of which have the same coefficient matrix but different column matrices of constants. This situation requires calculating the inverse matrix only once and using it as a premultiplier of each of the column matrices of constants [2,13].

The inversion of matrices is closely related to the solution of sets of simultaneous equations. The inverse of $[A]$ can be determined from Equation C.32. If we let $[C] = [A]^{-1}$, then

$$[A][C] = [I] \tag{C.34a}$$

or

$$
\begin{bmatrix}
a_{11} & a_{12} & \cdots & a_{1n} \\
a_{21} & a_{22} & \cdots & a_{2n} \\
\vdots & & & \\
a_{n1} & a_{n2} & \cdots & a_{nn}
\end{bmatrix}
\begin{bmatrix}
c_{11} & c_{12} & \cdots & c_{1n} \\
c_{21} & c_{22} & \cdots & c_{2n} \\
\vdots & & & \\
c_{n1} & c_{n2} & \cdots & c_{nn}
\end{bmatrix}
=
\begin{bmatrix}
1 & 0 & \cdots & 0 \\
0 & 1 & \cdots & 0 \\
\vdots & & & \\
0 & 0 & \cdots & 1
\end{bmatrix}
\tag{C.34b}
$$

This may be regarded as n sets of n simultaneous equations with identical coefficient matrix. The ith set of n simultaneous equations, for example, is

$$
\begin{bmatrix}
a_{11} & a_{12} & \cdots & a_{1n} \\
a_{21} & a_{22} & \cdots & a_{1n} \\
\vdots & & & \\
a_{ni} & a_{ni} & \cdots & a_{ni} \\
\vdots & & & \\
a_{n1} & a_{n2} & \cdots & a_{nn}
\end{bmatrix}
\begin{bmatrix}
c_{1i} \\
c_{2i} \\
\vdots \\
c_{ii} \\
\vdots \\
c_{ni}
\end{bmatrix}
=
\begin{bmatrix}
0 \\
0 \\
\vdots \\
1 \\
\vdots \\
0
\end{bmatrix}
\tag{C.35}
$$

Thus, the inversion of $[A]$ may be accomplished by solving n sets of equations such as Equation C.35. A common approach for matrix inversion is applying the elimination method, with or without pivotal compensation. This implies that any elimination technique (Gauss, Gauss–Jordan, or Cholesky's method) can be modified to calculate an inverse matrix. Here, we apply the Gauss–Jordan elimination method.

To apply the Gauss–Jordan method, we first augment the coefficient matrix by the identity matrix to obtain

$$
[A \vdots I] =
\begin{bmatrix}
a_{11} & a_{12} & \cdots & a_{1n} & \vdots & 1 & 0 & \cdots & 0 \\
a_{21} & a_{22} & \cdots & a_{2n} & \vdots & 0 & 1 & \cdots & 0 \\
\vdots & & & & \vdots & \vdots & & & \\
a_{n1} & a_{n2} & \cdots & a_{nn} & \vdots & 0 & 0 & \cdots & 1
\end{bmatrix}
\tag{C.36}
$$

The goal is to transform this augmented matrix to another augmented matrix of the form

$$[I \vdots C] = \begin{bmatrix} 1 & 0 & \dots & 0 & \vdots & c_{11} & c_{12} & \dots & c_{1n} \\ 0 & 1 & \dots & 0 & \vdots & c_{21} & c_{22} & \dots & c_{2n} \\ \vdots & & & & \vdots & \vdots & & & \\ 0 & 0 & \dots & 1 & \vdots & c_{n1} & c_{n2} & \dots & c_{nn} \end{bmatrix} \tag{C.37}$$

where $[C]$ is the inverse of $[A]$. The transformation is achieved using the Gauss–Jordan method, which involves applying the following equations in the order listed [2]:

$$\begin{aligned}
a'_{kj} &= a_{kj}/a_{kk}, \quad j = 1, 2, \dots, n, \quad j \neq k \\
a'_{kk} &= 1/a_{kk}, \\
a'_{ij} &= a_{ij} - a_{ik}a'_{kj}, \quad i = 1, 2, \dots, n, \quad i \neq k \\
&\qquad j = 1, 2, \dots, n, \quad j \neq k \\
a'_{ik} &= -a_{ik}a'_{kk}, \quad i = 1, 2, \dots, n, \quad i \neq k
\end{aligned} \tag{C.38}$$

We apply Equation C.38 for $k = 1, 2, \dots, n$. A computer program applying Equation C.38 is presented in Figure C.5. An iterative method of correcting the elements of the inverse matrix is available in Reference 14.

```
%This program performs matrix inversion using Gauss-Jordan elimination
%method and the result is compared to the MATLAB function 'inv'

clear;

%Define [A]
A = [10 -7 0;-3 6 1;2 -1 5];
%MATLAB operation
matinv = inv(A);

%Begin Gauss-Jordan Elimination method
N = size(A,1);
%Calculate elements of reduced matrix
for kk = 1:N
    ii = [1:(kk-1),(kk+1):N]; %non-pivot row index
    jj = [1:(kk-1),(kk+1):N]; %non-pivot column index

    %Calculate new elements of pivot row
    A(kk,jj) = A(kk,jj)/A(kk,kk);

    %Calculate element replacing pivot element
    A(kk,kk) = 1/A(kk,kk);

    %Calculate new elements not in povot row or pivot column
    A(ii,jj) = A(ii,jj) - A(ii,kk)*A(kk,jj);

    %Calculate replacement elements for pivot colum-
except pivot element
    A(ii,kk) = - A(ii,kk)*A(kk,kk);
end

anydiff = matinv-A
```

FIGURE C.5
Matrix inversion using Gauss–Jordan elimination method.

C.4 Eigenvalue Problems

The nature of these problems is discussed in Section 1.3. Here, we are concerned with the so-called standard eigenproblems

$$[A - \lambda I][X] = 0 \tag{C.39}$$

or the generalized eigenproblem

$$[A - \lambda B][X] = 0 \tag{C.40}$$

To show that Equations C.39 and C.40 are solved in the same way, we premultiply Equation C.40 by B^{-1} to obtain

$$[B^{-1}A - \lambda I][X] = 0 \tag{C.41}$$

Assuming $C = B^{-1}A$ gives

$$[C - \lambda I][X] = 0 \tag{C.42}$$

showing that Equation C.39 is a special case of Equation C.40 in which $B = I$. Thus, the procedure for solving Equation C.39 applies to Equation C.40 or Equation C.42.

The eigenvalue problems of Equations C.39 and C.40 are solved by either direct or indirect methods. In direct methods, such as Jacobi's method, the relevant matrix elements are stored in the computer, and an explicit procedure is used to obtain some or all of the eigenvalues $\lambda_1, \lambda_2, ..., \lambda_n$ and eigenvectors $X_1, X_2, ..., X_n$. Indirect methods are basically iterative, and the matrix elements are usually generated rather than stored.

C.4.1 Iteration (or Power) Method

The most commonly used iterative method is this *power method*. This method is suitable in situations where either the greatest or the least eigenvalue is required. Suppose that one of the eigenvalues of A, say λ_1, satisfies the condition

$$|\lambda_1| > |\lambda_i|, \quad i \neq 1, \tag{C.43}$$

then $|\lambda_1|$ is said to be the *dominant* eigenvalue of A. In many applications, the dominant eigenvalue is the most important and is probably the only eigenvalue in which we are interested. The iteration method is specifically used for finding the dominant eigenvalues.

The iterative procedure is essentially based on the condition that should a trial vector $[X]_i$ be assumed, an approximate eigenvalue and a new trial eigenvector $[X]_{i+1}$ can be determined from Equation C.39 or Equation C.40. To find the largest eigenvalue $|\lambda_1|$, we rewrite Equation C.39 as

$$[A][X] = \lambda[X] \tag{C.44}$$

and follow these steps [2]:

1. Assume a trial vector $[X]_0 = (x_1, x_2, ..., x_n)$, for example, $[X]_0 = (1, 1, ..., 1)$. Substituting $[X]_0$ to the left-hand side of Equation C.44 gives the first approximation to the corresponding eigenvector, that is,

$$\lambda[X]_1 = (\lambda x_1, \lambda x_2, ..., \lambda x_n)$$

2. Normalize the new vector $\lambda[X]$ by dividing it by the magnitude of its first component or by dividing the vector $[X]$ by the magnitude of the first component.

3. Substitute the normalized vector into the left-hand side of Equation C.44 and obtain a new approximate eigenvector.

4. Repeat steps (2) and (3) until the components of the new and previous eigenvectors differ by some prescribed tolerance. The normalizing factor will be the *largest eigenvalue* λ_1 while $[X]$ is the associated eigenvector.

To find the smallest eigenvalue, we first premultiply Equation C.44 by the inverse of $[A]$ to obtain

$$[X] = \lambda[A]^{-1}[X]$$

or

$$[A]^{-1}[X] = \frac{1}{\lambda}[X] \tag{C.45}$$

Thus, the iteration formula becomes

$$[A]^{-1}[X]_i = \frac{1}{\lambda}[X]_{i+1} \tag{C.46}$$

In this form, the iteration converges to the largest value $1/\lambda$, which corresponds to the smallest eigenvalue λ of $[A]$.

Once the largest eigenvalue is found, the method can be used to obtain the next largest eigenvalue by transforming $[A]$ to another matrix possessing only the remaining eigenvalues [2]. This so-called *matrix deflation* procedure assumes that $[A]$ is symmetric. The matrix deflation is continued until all the eigenvalues have been extracted. Error propagation from one stage of the deflation to the next leads to inaccurate results, specially for large eigenproblems. Jacobi's method, to be discussed in the next section, is recommended for large eigenproblems.

The MATLAB program in Figure C.6 is useful for finding the largest eigenvalues of a matrix.

C.4.2 Jacobi's Method

Jacobi's method is perhaps the most reliable method for solving eigenvalue problems. Its major advantage is that it finds all eigenvalues and eigenvectors simultaneously with excellent accuracy.

```
% This example finds the largest Eigenvalue and the associated
% eigen vector of a matrix equation [A][X]=Lambda*[X] and
% compares this to the largest eigenvalue obtained by the
% MATLAB eig function.

clear;

%Define [A]
A = [10 -7 0;-3 6 1;2 -1 5];

%MATLAB eig function
eigval = eig(A);

%Begin Iterative Method
maxiter = 100;
epsi = 1e-3;
kk = 0;
X = ones(size(A,1),1);
valuefound = 0;

while ~valuefound
    D = A*X;
    Z = D/D(1);

    %Check if result is sufficient
    test = abs((X-Z)-epsi*abs(Z));
    if max(abs(X-Z))<epsi
        valuefound = 1;
    elseif maxiter<kk
        disp('Failed to find solution')
        valuefound = 1;
    end

    X = Z;
    disp(D')
    kk = kk + 1;
end

lambda = D(1);

thedifference = lambda - eigval(1)
```

FIGURE C.6
A program for finding the largest eigenvalue of equation [A][X] = Lambda [X].

The method transforms a symmetric matrix [A] into a diagonal matrix having the same eigenvalues as [A]. This is achieved by eliminating one pair of off-diagonal elements of [A] at a time. Given

$$[A][X] = \lambda[X], \tag{C.47}$$

let $\lambda_1, \lambda_2, ..., \lambda_n$ be the eigenvalues and $[V_1], [V_2], ..., [V_n]$ the corresponding eigenvectors. Then,

$$[A][V_1] = \lambda_1 [V_1]$$
$$[A][V_2] = \lambda_2 [V_2]$$
$$\vdots$$ (C.48)
$$[A][V_n] = \lambda_1 [V_n]$$

or simply

$$[A][V] = [V][\lambda] \qquad (C.49)$$

where

$$[V] = [[V_1],[V_2], \ldots, [V_n]] \qquad (C.50a)$$

$$[\lambda] = \begin{bmatrix} \lambda_1 & 0 & \ldots & 0 \\ 0 & \lambda_1 & \ldots & 0 \\ \vdots & \vdots & & \vdots \\ 0 & 0 & \ldots & \lambda_n \end{bmatrix} \qquad (C.50b)$$

From the theory of matrices, if $[A]$ is symmetric, $[V]$ is orthogonal, that is,

$$[V]^t = [V]^{-1} \qquad (C.51)$$

hence, premultiplying Equation C.49 by $[V]^t$ leads to

$$[V]^t[A][V] = [\lambda] \qquad (C.52)$$

signifying that the eigenvalues of $[V]^t[A][V]$, which is known as the orthogonal transformation of $[A]$, are the same as those of $[A]$. Thus, the problem of finding the eigenvalues is reduced to finding the $[V]$ matrix.

The $[V]$ matrix is constructed iteratively by using unitary matrix (or plane rotation matrix) $[R]$. If we let

$$[A_1] = [A]$$
$$[A_2] = [R_1]^t [A_1][R_1]$$
$$[A_3] = [R_2]^t [A_2][R_2] = [R_2]^t [R_1]^t [A][R_1][R_2] \qquad (C.53)$$
$$\vdots$$
$$[A_k] = [R_{k-1}]^t \ldots [R_1]^t [A][R_1] \ldots [R_{k-1}],$$

then as $k \to \infty$

$$[A_k] \to [\lambda]$$
$$[R_1][R_2] \ldots [R_{k-1}] \to [V] \qquad (C.54)$$

The unitary transformation matrix $[R]$ eliminates the pair of equal elements a_{pq} and a_{qp}. It is given by [1,2,7]

$$[R_k] = \begin{bmatrix} 1 & & & & & \\ & 1 & & & & \\ & & \cos\theta & & -\sin\theta & \\ & & & 1 & & \\ & & \sin\theta & & \cos\theta & \\ & & & & & 1 \end{bmatrix} \begin{matrix} \\ \\ p \\ \\ q \\ \\ \end{matrix}$$

(C.55a)

with column labels p and q.

that is,

$$R_{qq} = R_{pp} = \cos\theta$$

$$-R_{pq} = R_{qp} = \sin\theta$$

$$R_{ii} = 1, \quad i \neq p,q$$

$$R_{ij} = 0, \quad \text{elsewhere}$$

(C.55b)

The choice of θ in the transformation matrix must be such that new elements $a'_{pq} = a'_{qp} = 0$, that is,

$$a'_{pq} = (-a_{pp} + a_{qq})\cos\theta\sin\theta + a_{pq}(\cos^2\theta - \sin^2\theta) = 0$$

(C.56)

Hence,

$$\tan 2\theta = \frac{2a_{pq}}{a_{pp} - a_{qq}}, \quad -45° < \theta < 45°$$

(C.57)

An alternative manipulation of Equation C.56 gives

$$\cos\theta = \left[\frac{\sqrt{(a_{pp} - a_{qq})^2 + 4a_{pq}^2} + (a_{pp} - a_{qq})}{2\sqrt{(a_{pp} - a_{qq})^2 + 4a_{pq}^2}} \right]^{1/2}$$

(C.58a)

$$\sin\theta = \frac{a_{pq}}{\sqrt{(a_{pp} - a_{qq})^2 + 4a_{pq}^2} \cos\theta}$$

(C.58b)

Notice that Equation C.53 requires an infinite number of transformations because the elimination of elements a_{pq} and a_{qp} in one step will in general undo the elimination of previously treated elements in the same row or column. However, the transformation converges rapidly and ceases when all the off-diagonal elements become negligible in magnitude.

The program in Figure C.7 determines all the eigenvalues and eigenvectors of symmetric matrices employing Jacobi's method.

```
% The following program attempts to implement the Jacobi's method of
% finding eigenvalues and eigenvectors. Just for illustration we have
% taken a random 8x8 symmetric matrix. If the actual matrix is stored
% in a .mat file it has to be loaded using load command in the second
% line and the first line will be commented out. We have decided to have
% 10 iterations or fewer, when the determinant of the original matrix is
% close to that of the modified matrix. The difference E is treated as
% error. We wish that the error E is less than eps or # of iterations
% = 10. These parameters have to be appropriately chosen based on one's
% experience and the nature of the problem under consideration.
%
% *********** Description of the variables *********
% A = original matrix  saved in Ao to start with.
% R is the rotation matrix.
% Rp is the product of all R's.
% At the end of the while loop Rp has the eigen vectors
% in its columns and A has the eigenvalues along the main diagonal.
% In fact MATLAB has a built-in routine called eig that finds the eigen-
% values and the eigenvectors. So we wish to compare both the results;
% hence we have added a command  at the bottom of the code
% asking to dispaly both. Note that the eigenvalues are not sorted in
% this code.
%%%%%%%%%%%%%%%%%

clear all;close all;
A = 5*rand(8); A =  A+A';
%load Afile.mat
N = length(A(:,1)); % N = 8 in this case
Ao=A;
Rp=eye(N);
E=10; Niter=1;
while (E > eps)|(Niter<10)
for p = 1:N
    for q = p+1:N
        th = .5*atan(2*A(p,q)/(A(p,p)-A(q,q)));
        R = eye(N);
        R([p,q],[p,q]) = [cos(th) -sin(th);sin(th) cos(th)];
        A=R'*A*R;
        Rp=Rp*R;
    end
end
Niter = Niter+1;
E=det(A)-det(Ao);
end
Rp
A
[V,D]=eig(A)
E
```

FIGURE C.7
MATLAB code for finding all the eigenvalues and eigenvectors of equation $[A][X] =$ Lambda $[x]$.

References

1. A.W. Al-Khafaji and J.R. Tooley, *Numerical Methods in Engineering Practice*. New York: Rinehart and Winston, 1986, pp. 84–159, 203–270.
2. M.L. James et al. *Applied Numerical Methods for Digital Computation*, 3rd Edition. New York: Harper & Row, 1985, pp. 146–298.
3. S.A. Hovanessian and L.A. Pipes, *Digital Computer Methods in Engineering*. New York: McGraw-Hill, 1969, pp. 1–48.

4. W. Cheney and D. Kincaid, *Numerical Mathematics and Computing*, 2nd Edition. Monterey, CA: Brooks/Cole, 1985, pp. 201–257.

5. R.L. Ketter and S.P. Prawel, *Modern Methods of Engineering Computation*. New York: McGraw-Hill, 1969, pp. 66–117.

6. A. Ralston and H.S. Wilf (eds.), *Mathematical Methods for Digital Computers*. New York: John Wiley, 1960, pp. 62–72.

7. A. Jennings, *Matrix Computation for Engineers and Scientists*. New York: John Wiley, 1977, pp. 182–222, 250–254.

8. J.C. Nash, *Compact Numerical Methods for Computers: Linear Algebra and Function Minimization*. New York: John Wiley, 1979, pp. 195–199.

9. T.K. Sarkar et al., "A limited survey of various conjugate gradient methods for solving complex matrix equations arising in electromagnetic wave interaction," *Wave Motion*, vol. 10, no. 6, 1988, pp. 527–546.

10. A.F. Peterson and R. Mittra, "Method of conjugate gradients for the numerical solution of large-body electromagnetic scattering problems," *J. Opt. Soc. Am., Pt. A*, vol. 2, no. 6, June 1985, pp. 971–977.

11. T.K. Sarkar, "Application of the fast Fourier transform and the conjugate gradient method for efficient solution of electromagnetic scattering from both electrically large and small conducting bodies," *Electromagnetics*, vol. 5, 1985, pp. 99–122.

12. D.T. Borup and O.P. Gandhi, "Calculation of high-resolution SAR distributions in biological bodies using the FFT algorithm and conjugate gradient method," *IEEE Trans. Micro. Theo. Tech.*, vol. MTT-33, no. 5, May 1985, pp. 417–419.

13. R.W. Southworth and S.L. Deleeuw, *Digital Computation and Numerical Methods*. New York: McGraw-Hill, 1965, pp. 247–251.

14. S. Hovanessian, *Computational Mathematics in Engineering*. Lexington, MA: Lexington Books, 1976, p. 25.

Appendix D: Computational Electromagnetic Codes

Numerical modeling and simulation have revolutionized all aspects of engineering design to the extent that several software packages have been developed. In the 1970s, researchers developed their own computer programs to solve problems since very few commercial solvers existed. Nowadays, the situation is completely different. There are dozens of computational tools on the market. There is no longer motivation or a justifiable need to develop one's own software tools at universities and in industry. There is growing dependence on commercial software in designing complex electromagnetic problems both in industry and academia. While some of these software are commercial, some are free. They are powerful tools now available to engineers, not formally trained in CEM.

The widely used software packages for CEM include:

- *Numerical Electromagnetic Codes* (NEC): This is based on the method of moments (MoM) and was developed at Livermore National Laboratory. It is used for frequency domain antenna modeling code of wires and surface structures. NEC2 uses a text interface code and is a widely used 3D code. A free NEC code is 4nec2, which can be found at http://www.qsl.net/4nec2/ Online documentation can be obtained from http://www.nec2.org/

- *High-Frequency Structure Simulator* (HFSS): This is based on FEM and was developed by Ansoft, which was later acquired by Ansys. HFSS offers state-of the-art solver to solve a wide range of EM applications. More information about HFSS can be found in http://www.ansys.com/Products/Electronics/ANSYS-HFSS

- *Sonnet*: This software provides high-frequency 3D planar electromagnetic (EM) analysis of single and multilayer planar circuits. Free student version (Sonnet Lite) is available. For more information, visit their website: http://www.sonnetsoftware.com/

- *COMSOL*: This is based on the FEM. It is a powerful tool for various physics and engineering applications. More information about COMSOL is available at https://www.comsol.com/

- *FEKO*: This is an antenna simulation software based on the method of moments. It can be used to calculate the radiation pattern, impedance, and gain of an antenna. For more information, visit their website: http://www.feko.info

- *EMAP*: This is a family of three-dimensional electromagnetic modeling codes developed at the Missouri University of Science and Technology. EMAP3 is a vector FEM code, while EMAP5 is a vector FEM/MoM code.

- *MEEP*: This is a free, open-source finite-difference time-domain (FDTD) simulation software package developed at MIT. MEEP is an acronym for MIT Electromagnetic.

- Equation Propagation: It was first released in 2006 and it can be downloaded from http://ab-initio.mit.edu/meep

- *MaxFem*: This is an open software package for electromagnetic simulation based on the FEM. The package can solve problems in electrostatics, magnetostatics, and eddy-currents.

Each of these software has some limitations in their ability to solve EM problems. Several CEM software are available on the Web but these are the most popular ones.

Appendix E: Answers to Odd-Numbered Problems

Chapter 1

1.1   Proof

1.3   Proof

1.5   Proof, assume medium is free space

1.7   Proof

1.9   Proof

1.11 $\mathbf{H} = -\dfrac{D_o}{\beta}\cos(\omega t - \beta z)\mathbf{a}_y$

1.13 $\beta = \dfrac{\omega\mu}{\eta}$, $\eta = \sqrt{\dfrac{j\omega\mu}{\sigma + j\omega\varepsilon}}$

1.15 a. Yes if $\beta^2 = \omega^2\mu_o\varepsilon_o$

      b. $\beta = \omega\sqrt{\mu_o\varepsilon_o}$

      c. $\dfrac{\mu_o\beta}{\omega}\cos(\omega t - \beta z)\mathbf{a}_y$

1.17 Proof, $\mathbf{H}_s = -\dfrac{20}{\omega\mu_o}[k_y\sin(k_xx)\sin(k_yy)\mathbf{a}_x + k_x\cos(k_xx)\cos(k_yy)\mathbf{a}_y]$

1.19 a. $\mathbf{A} = \cos(\omega t - 2z)\mathbf{a}_x - \sin(\omega t - 2z)\mathbf{a}_y$

      b. $\mathbf{B} = -10\sin x\sin\omega t\mathbf{a}_x - 5\cos(\omega t - 2z + 45^o)\mathbf{a}_z$

      c. $C = 2\cos 2x\sin(\omega t - 3x) + e^{3x}\cos(\omega t - 4x)$

1.21 $\nabla^2 A_z - \mu\varepsilon\dfrac{\partial^2 A_z}{\partial t^2} = -\mu J_z$

1.23 Proof

1.25 (a) Parabolic, (b) elliptic, (c) elliptic

Chapter 2

2.1   If a and d are functions of x only; c and e are functions of y only; $b = 0$; and f is a sum of a function of x only and a function of y only.

2.2 a. $V(x,y) = \sum_{n=\text{odd}}^{\infty} A_n \sin(n\pi x/a) \sinh[n\pi(b-y)/a],$

where $A_n = \dfrac{2V_o(-1)^{n+1}}{n\pi \sinh(n\pi b/a)}$

 b. $V(x,y) = \dfrac{V_o \cos(\pi x/a) \cosh(\pi y/a)}{\cosh(\pi b/a)}$

2.5 a. $\Phi(\rho,\phi) = \dfrac{\sin\phi}{\rho}$

 b. $\Phi(\rho,\phi) = \dfrac{4}{\pi} \sum_{n=1,3,5}^{\infty} \dfrac{I_0(n\pi\rho/L)}{I_0(n\pi a/L)} \dfrac{\sin(n\pi z/L)}{n}$

 c. $\Phi(\rho,\phi,t) = 2\sum_{m=1}^{\infty} \dfrac{J_2(\rho X_m/a)}{X_m J_3(X_m)} \cos 2\phi \exp(-X_m^2 kt/a^2)$

where $J_2(X_m) = 0.$

2.7 $U(x,t) = \dfrac{8}{\pi^3} \sum_{n=\text{odd}}^{\infty} \dfrac{1}{n^3} e^{-n^2\pi^2 t} \sin(n\pi x)$

2.9 $U(x,y,t) = \dfrac{40}{\pi^2} \sum_{m=1}^{\infty} \sum_{n=1}^{\infty} \dfrac{\cos(m\pi)\cos(n\pi)}{mn} \sin(m\pi x) \sin(n\pi y) \exp(-\lambda_{mn}^2 t)$

where $\lambda_{mn}^2 = (m\pi)^2 + (n\pi)^2.$

2.11 $V(x,y) = \left[\sinh ny - \dfrac{\cosh ny}{\tanh n\pi}\right] \cos x$

2.13 $U(x,t) = \dfrac{8}{\pi^3} \sum_{n=\text{odd}}^{\infty} \dfrac{1}{n^3} \sin(n\pi x) \cos(an\pi t)$

2.15 $V(0.8a, 0.3L) = 0.26396V_o.$

2.17 $V(\rho,\phi) = \dfrac{4V_o}{\pi} \sum_{n=1,3,5}^{\infty} \dfrac{1}{n} \dfrac{\left[(\rho/a)^{3n} - (\rho/a)^{-3n}\right]}{\left[(b/a)^{3n} - (b/a)^{-3n}\right]} \sin 3n\phi$

2.19 $U(\rho,t) = 2T_o \sum_{n=1}^{\infty} \dfrac{J_0(\lambda_n \rho)}{\lambda_n J_1(\lambda_n)} \exp(-\lambda_n^2 t)$

2.21 Proof/Derivation

2.23 Proof

2.25 (a) $x^n J_n(x) + C,$ (b) $-x^{-n} J_{n+1}(x) + C$

2.27 The result is shown in the table below.

M	$J_0(x)$	$J_1(x)$	$J_2(x)$	$J_3(x)$	$J_4(x)$	$J_5(x)$
1	2.4048	3.8317	5.1356	6.3802	7.5883	8.7715
2	5.5201	7.0156	8.4127	9.7610	11.06747	12.3386
3	8.6537	10.1735	11.6198	13.0152	14.3725	15.7002
4	11.7915	13.3237	14.7960	16.2235	17.6160	18.9801
5	14.9309	16.4706	17.9598	19.4095	20.8769	22.2178

2.29 a. $f(x) = \dfrac{1}{2} + \dfrac{3}{4}P_1 - \dfrac{7}{16}P_3 + \cdots$

b. $f(x) = \dfrac{3}{5}P_1(x) + \dfrac{2}{5}P_3(x)$

c. $f(x) = \dfrac{1}{4}P_0(x) + \dfrac{1}{2}P_1(x) + \dfrac{5}{16}P_2(x) - \dfrac{3}{32}P_4(x) + \cdots$

d. $f(x) = \dfrac{1}{2} - \dfrac{5}{8}P_2(x) + \dfrac{3}{16}P_4(x) + \cdots$

2.31 a. $\dfrac{-x^3}{6} + 2x^2 - 6x + 4$

b. $4x^3 - 3x, \quad 8x^4 - 8x^2 + 1$

c. $8x^3 - 4x, \quad 16x^4 - 12x^2 + 1$

2.33 a. $8x^3 - 12x, \quad 16x^4 - 48x^2 + 12$

b. $\dfrac{1}{3!}(-x^3 + 9x^2 - 18x + 6), \quad \dfrac{1}{4!}(x^4 - 16x^3 + 72x^2 - 96x + 24)$

2.35 $1, \quad 2x, \quad 4x^2 - 2, \quad 8x^3 - 12x$

2.37 Inside the sphere $r < a$,

$$V(r,\theta) = \frac{V_o}{2}\left[1 + \frac{3r}{2a}P_1(\cos\theta) - \frac{7r^3}{8a^3}P_3(\cos\theta) + \cdots\right]$$

Outside the sphere $r > a$,

$$V(r,\theta) = \frac{V_o}{2r}\left[1 + \frac{3a}{2r}P_1(\cos\theta) - \frac{7a^3}{8r^3}P_3(\cos\theta) + \frac{11a^5}{16r^5}P_5(\cos\theta) + \cdots\right]$$

2.39 a. Proof

b. $P_6(x) = \dfrac{1}{16}(231x^6 - 315x^4 + 105x^2 - 5),$

$P_7(x) = \dfrac{1}{16}(429x^7 - 693x^5 + 315x^5 - 35x)$

2.41 Proof

2.43 $V(r,\theta) = 2(r/a)^2 P_2(\cos\theta) + 3(r/a)P_1(\cos\theta) + 2P_0(\cos\theta)$

2.45 $V_1(r,\theta) = -\dfrac{3E_o}{\varepsilon_r + 2}\cos\theta, \quad r \le a$

$V_2(r,\theta) = -E_o r\cos\theta + E_o\dfrac{a^3}{r^2}\dfrac{(\varepsilon_r - 1)}{\varepsilon_r + 2}\cos\theta, \quad r \ge a$

2.47 $V = \dfrac{1}{3}\cos 2\phi P_2^2(\cos\theta)$

2.49 a. $V(x,y,z) = \dfrac{32}{\pi^3} \displaystyle\sum_{m=1}^{\infty} \sum_{n=1}^{\infty} \sum_{p=1}^{\infty} \dfrac{m[1-(-1)^m e^{\pi}]}{(m^2+1)(2n-1)(2p-1)} \dfrac{\sin mx \sin(2n-1)y \sin(2p-1)z}{[m^3+(2n-1)^2+(2p-1)^2]}$

   b. $V(x,y,z) = -\dfrac{128}{\pi^3} \displaystyle\sum_{m=1}^{\infty} \sum_{n=1}^{\infty} \sum_{p=1}^{\infty} \dfrac{1}{(2m-3)(4m^2-1)(2n-1)(2p-1)}$

   $\times \dfrac{\sin(2m-1)x \sin(2n-1)y \sin(2p-1)z}{[(2m-1)^2+(2n-1)^2+(2p-1)^2]}$

2.51 $\Phi(\rho,t) = 2a^2 \Phi_o \displaystyle\sum_{n=1}^{\infty} \dfrac{J_0(\lambda_{0n}\rho/a)}{\lambda_{0n}^2 J_1(\lambda_{0n})} \dfrac{\left[\dfrac{\lambda_{0n}}{a}\sin \omega t - \omega \sin(\lambda_{0n}t/a)\right]}{\lambda_{0n}^2 - a^2 \omega^2}$

2.53 $E_g = a_\rho \rho_v \displaystyle\sum_{n=1}^{\infty} \dfrac{2J_1(\lambda_n\rho)\lambda_n}{C_n K_n}[\cosh(\lambda_n b)-1]\sinh[\lambda_n(b+c-z)]$

   $+ a_z \rho_v \displaystyle\sum_{n=1}^{\infty} \dfrac{2J_0(\lambda_n\rho)\lambda_n}{C_n K_n}[\cosh(\lambda_n b)-1]\cosh[\lambda_n(b+c-z)]$

   $E_\ell = a_\rho \rho_v \displaystyle\sum_{n=1}^{\infty} \dfrac{2J_1(\lambda_n\rho)\lambda_n}{\varepsilon_r C_n}\left[\dfrac{\sinh(\lambda_n z)}{K_n} T_n - \cosh(\lambda_n z)+1\right]$

   $- a_z \rho_v \displaystyle\sum_{n=1}^{\infty} \dfrac{2J_0(\lambda_n\rho)\lambda_n}{\varepsilon_r C_n}\left[\dfrac{\cosh(\lambda_n z)}{K_n} T_n - \sinh(\lambda_n z)\right]$

2.55 $V_g = \displaystyle\sum_{n=1}^{\infty} A_n \sin \beta x \sinh \beta(c-y)$

   $V_\ell = \displaystyle\sum_{n=1}^{\infty} \sin \beta x [C_n \sinh \beta y - F_p(\cosh \beta y - 1)]$

where

$$F_p = \begin{cases} 0, & n = \text{even} \\ \dfrac{4\rho_o a^2}{n^3 \pi^3 \varepsilon}, & n = \text{odd} \end{cases}$$

$$C_n = F_p \left[\dfrac{\varepsilon_r \sinh \beta b \sinh \beta(c-b)+\cosh \beta(c-b)[\cosh \beta b - 1]}{\varepsilon_r \cosh \beta b \sinh \beta(c-b)+\sinh \beta b \cosh \beta(c-b)}\right] \tag{12}$$

$$A_n = C_n \dfrac{\sinh \beta b}{\sinh \beta(c-b)} - F_p \dfrac{(\cosh \beta b - 1)}{\sinh \beta(c-b)}$$

2.57 Proof

2.59 Proof

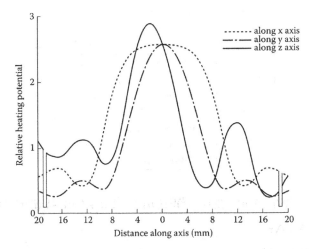

FIGURE E.1
For Problem 2.63.

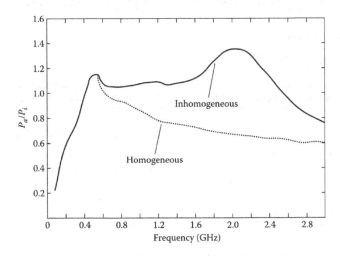

FIGURE E.2
For Problem 2.65.

2.63 See Figure E.1
2.65 See Figure E.2

Chapter 3

3.1 Proof
3.3 Results are shown in the table below.

	Φ			
t/x	0	0.25	0.5	0.75	1.0
0	0	0.7071	1.0	0.7971	0
0.03125	0	0.5	0.7071	0.5	0
0.0625	0	0.3526	0.5	0.3526	0
0.09375	0	0.25	0.3526	0.25	0
0.125	0	0.1768	0.25	0.1768	0
0.15625	0	0.1250	0.1768	0.125	0
0.1875	0	0.0884	0.125	0.0884	0

3.5 $V(i,n+1) = \alpha^2 V(i-1,n) + 2(1-\alpha^2)V(i,n) + \alpha^2 V(i+1,n) - V(i,n-1)$

   where $\alpha = \Delta t / \Delta x$.

3.7 $U(i,j,n) = \dfrac{(1+1/2i)}{4+\alpha} U(i+1,j,n) + \dfrac{(1-1/2i)}{4+\alpha} U(i-1,j,n)$

   $+ \dfrac{1}{4+\alpha}[U(i,j+1,n) + U(i,j-1,n)] + \dfrac{\alpha}{4+\alpha} U(i,j-1,n)$

   where $\alpha = h^2 / \Delta t$.

3.9   Results are tabulated below.

No. of Iterations	Φ(0)	Φ(0.25)	Φ(0.5)	Φ(0.75)	Φ(1.0)
0	0	0	0	0	1
1	0	−0.03916	−0.0664	0.4121	1
2	0	−0.07226	0.1232	0.5069	1
3	0	0.02254	0.2178	0.5542	1
4	0.06984	0.2651	0.5779	1	
5	0	0.09349	0.2888	0.5897	1

3.11   Proof

3.13 $V_A = 61.46$, $V_B = 21.96$, $V_C = 45.99$, $V_D = 21.96$, $V_E = 61.46$ V

3.15 $U_1 = U_2 = U_5 = U_6 = 4.56$, $U_3 = U_4 = 5.72$

3.17   Proof

3.19   Proof

3.21 $-j \leq jr \sin \beta\delta/2 \leq j$

3.23 $V_1 = 54.17 = V_5$, $V_2 = 58.33 = V_4$, $V_3 = 79.16$

3.25   a.   The exact solution gives $Z_o = 60.61\,\Omega$.

   b.   The exact solution gives $Z_o = 50\,\Omega$.

3.27 $k_c = 4.443$ (exact), $k_c = 3.5124$ (exact)

3.29   Proof

3.31   See Figure E.3

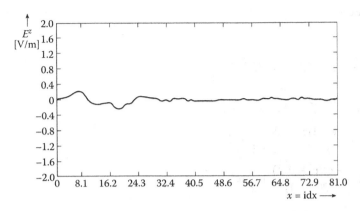

FIGURE E.3
For Problem 3.31.

3.33 a. Proof

b. $E_z^{n+1}(0,j,k+1/2) = E_z^n(1,j,k+1/2) + \dfrac{c_0\delta t - \delta}{c_0\delta t + \delta}\left[E_z^{n+1}(1,j,k+1/2) - E_z^{n+1}(0,j,k+1/2)\right]$

3.35 Proof

3.37 The results are tabulated below.

ρ	z	$V(\rho,z)$
5	18	65.852
5	10	23.325
5	2	6.3991
10	2	10.226
15	2	10.343

3.39 Proof

3.41 See table below.

t	Exact	FD
0.05	6.2475	6.3848
0.10	2.8564	2.9123
0.15	1.3059	1.2975
0.20	0.5971	0.5913
0.25	0.5971	0.27
0.30	0.1248	0.1233

3.43 (a) 3.66, (b) 3.67.

3.45 0.6321

3.47 $a_0 = a_3 = \dfrac{2}{9}$, $a_1 = a_2 = \dfrac{16}{9}$

3.49 Exact solution: $-0.4116 + j_0$

Chapter 4

4.1 (a) 1.333, (b) −4.667, (c) 157.08

4.3 0

4.5 a. $y(x) = e^x + (2-e)x - 1$

b. $y(x) = \dfrac{(e-1)e^x}{e+1} + \dfrac{2e^{-x+1}}{e+1} - 1$

c. $y(x) = \dfrac{-e^2(e^x - e^{-x})}{2(e^2-1)} + \dfrac{1}{2}xe^x$

4.7 a. $y = \dfrac{1}{8}(x^2 + 7x)$

b. $y = \cos x$

4.9 Proof

4.11 Proof

4.13 $\nabla \cdot J = 0$

4.15 Proof

4.17 $I(\Phi) = \displaystyle\int\limits_0^1 (\Phi'^2 + \Phi^2 - 8xe^x\Phi)\,dx - \Phi^2(1) + \Phi^2(0) + e\Phi(1) + \Phi(0)$

4.19 Compare your result with the following exact solutions.

a. $U(x) = \dfrac{\sin x + 2\sin(1-x)}{\sin 1} + x^2 - 2$

b. $U(x) = \dfrac{2\cos(1-x) - \sin x}{\sin 1} + x^2 - 2$

4.21 Compare your result with the following exact solutions.

a. $\Phi(x) = \dfrac{1}{\pi^2}(\cos \pi x + 2x - 1)$

b. $\Phi(x) = \dfrac{1}{\pi^2}(\cos \pi x - 1)$

4.23 a. For $m = 1$, $\tilde{\Phi} = a_1 u_1 = -\dfrac{1}{4}x(1-x)$

For $m = 2$, $\tilde{\Phi} = x(0.0184x^2 - 0.0263x - 0.1579)$

For $m = 3$, $\tilde{\Phi} = -0.223(x - x^2) - 0.506(x - x^3) - 0.157(x - x^4)$

b. For $m = 1$,

$$a_1 = -\frac{3}{\pi^3}$$

For $m = 2$,

$$\begin{bmatrix} a_1 \\ a_2 \end{bmatrix} = \begin{bmatrix} -\dfrac{2}{\pi^3} \\ \dfrac{1}{4\pi(\pi^2 - 1)} \end{bmatrix}$$

For $m = 3$,

$$\begin{bmatrix} a_1 \\ a_2 \\ a_3 \end{bmatrix} = \begin{bmatrix} -\dfrac{2}{\pi^3} \\ \dfrac{1}{\pi(\pi^2 - 1)} \\ -\dfrac{2}{27\pi^3} \end{bmatrix}$$

4.25 a. $a_1 = 0.1935,\ a_2 = 0.1843$

b. $a_1 = 0.1924,\ a_2 = 0.1707$

c. $a_1 = 1.8754,\ a_2 = 0.1695$

4.27 $\tilde{y} = \dfrac{100}{12} x(1 - x^3)$

4.29 $\tilde{\Phi} = (1 - x)(1 - 0.209x - 0.789x^2 + 0.209x^3)$

4.31 Exact: $\lambda_n = (n\pi)^2$, that is, $\lambda_1 = 9.8696,\ \lambda_2 = 39.48,\ \lambda_3 = 88.83$

4.33 0.1987

4.35 10.53

4.39 0.50032c/a

Chapter 5

5.1 Proof

5.3 a. Nonsingular, Fredholm integral of the second kind.

b. Nonsingular, Volterra integral of the second kind.

c. Fredholm integral of the second kind.

5.5 a. $\Phi = 5e^{x^2}$

b. $\Phi = \sin x$

5.7 Proof

5.9 $G(x \mid x_o) = \begin{cases} -\dfrac{\cos k(x_o - L)}{k \cos(2kL)} \sin k(x + L), & x < x_o \\[4mm] -\dfrac{\sin k(x_o + L)}{k \cos(2kL)} \cos k(x - L), & x > x_o \end{cases}$

5.11 $G(x, y, x', y') = -\dfrac{4}{\pi^2} \displaystyle\sum_{m=1}^{\infty} \sum_{n=1}^{\infty} \dfrac{\sin n\pi x \sin n\pi y \sin n\pi x' \sin n\pi y'}{m^2 + n^2}$

5.13 Proof

5.15 $G(x,y;x',h_1+h_2) = \displaystyle\sum_{n=1}^{\infty} \Phi_n(y)\sin(n\pi x'/a)\sin(n\pi x/a)$

where

$$\Phi_n(y) = \begin{cases} A_n \sinh n\pi y/a, & 0 \le y \le h_1 \\ B_n \sinh n\pi y/a + C_n \cosh n\pi y/a, & h_1 \le y \le h_1+h_2 \\ D_n \sinh[n\pi(b-y)/a], & h_1+h_2 \le y \le b \end{cases}$$

5.17 Proof/Derivation

5.19 $V(x,y) = \displaystyle\int_0^{\infty} f(x)\frac{\partial G(x,y;x',0)}{\partial n} dx' + \int_0^{\infty} G(x,y;0,y')h(y')dy'$

$$+ \int_0^{\infty}\int_0^{\infty} G(x,y,x',y')g(x',y')dx'\,dy'$$

where $G(x,y;x'y') = \dfrac{1}{4\pi}\ln\dfrac{[(x-x')^2+(y-y')^2][x+x')^2+(y-y')^2]}{[(x-x')^2+(y+y')^2][x+x')^2+(y+y')^2]}$

5.21 $\begin{bmatrix} Q_1 \\ Q_2 \\ Q_3 \end{bmatrix} = 4\pi\varepsilon_o d \begin{bmatrix} 0.1053 \\ 0 \\ -0.1053 \end{bmatrix}$

5.23 (a) 0.1019 pF, (b) 0.0679 pF.

5.25 The scattering patterns are shown in Figure E.4.

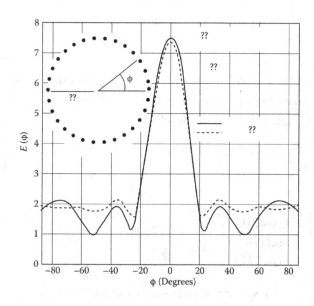

FIGURE E.4
For Problem 5.25.

5.27 See Figure E.5.

5.29 $[I] = \begin{bmatrix} 0.0099 + j0 \\ 0.0188 + j0.0001 \end{bmatrix}$

5.31 $B_1 = (0.0094, -0.00357)$, $B_2 = (0.00045, -0.00203)$. See Figure E.6 for the real and imaginary parts of $I(z)$.

5.33 Proof

5.35 The absorbed power density calculated at the center of each cell for the first and second layers is Figure E.7.

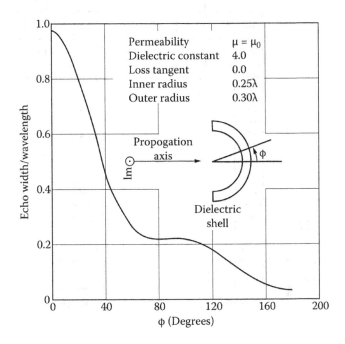

FIGURE E.5
For Problem 5.27.

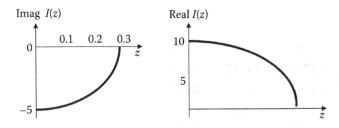

FIGURE E.6
For Problem 5.31.

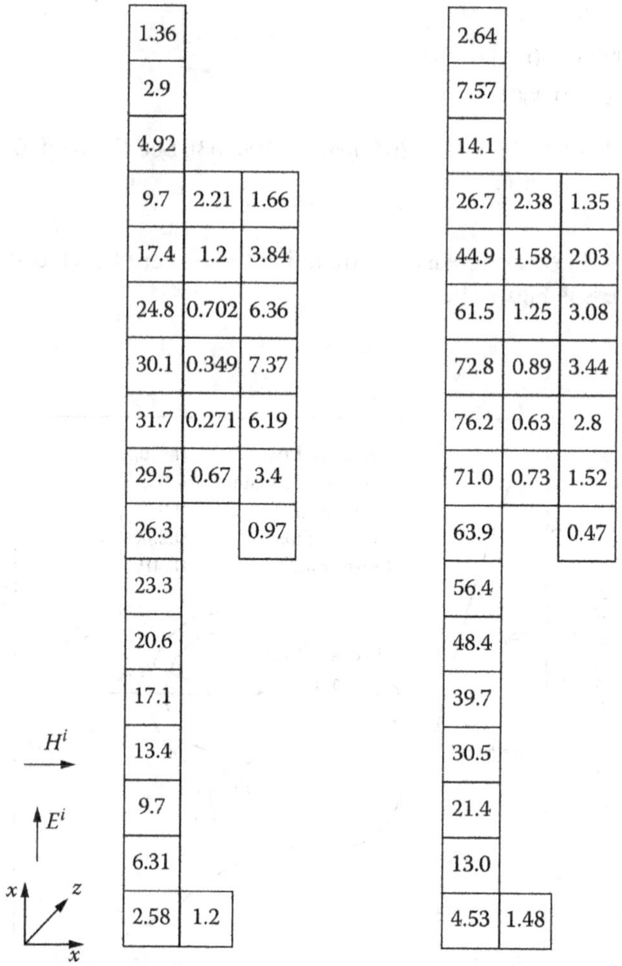

First layer		
1.36		
2.9		
4.92		
9.7	2.21	1.66
17.4	1.2	3.84
24.8	0.702	6.36
30.1	0.349	7.37
31.7	0.271	6.19
29.5	0.67	3.4
26.3		0.97
23.3		
20.6		
17.1		
13.4		
9.7		
6.31		
2.58	1.2	

Second layer		
2.64		
7.57		
14.1		
26.7	2.38	1.35
44.9	1.58	2.03
61.5	1.25	3.08
72.8	0.89	3.44
76.2	0.63	2.8
71.0	0.73	1.52
63.9		0.47
56.4		
48.4		
39.7		
30.5		
21.4		
13.0		
4.53	1.48	

$H^i \rightarrow$

$\uparrow E^i$

FIGURE E.7
For Problem 5.35.

Chapter 6

6.1 a. $C = \begin{bmatrix} 0.5909 & -0.1364 & -0.4545 \\ -0.1304 & 0.4545 & -0.3182 \\ -0.4545 & -0.3182 & 0.7727 \end{bmatrix}$

   b. $C = \begin{bmatrix} 0.6667 & -0.6667 & 0 \\ -0.6667 & 1.042 & -0.375 \\ 0 & -0.375 & 0.375 \end{bmatrix}$

6.3 (a) 10.667 V, (b) 10 V

6.5 $\quad \alpha_1 = \dfrac{1}{23}(4x + 3y - 24)$

$\alpha_2 = \dfrac{1}{23}(-5x + 2y + 30)$

$\alpha_3 = \dfrac{1}{23}(x - 5y + 17)$

6.7 Proof

6.9 $\quad W_e = \dfrac{\varepsilon}{4h_x h_y} \begin{bmatrix} h_y^2 V_1^2 - 2h_y^2 V_1 V_2 + h_x^2 V_3^2 - h_x^2 V_1 V_3 \\ + h_x^2 V_2^2 + h_y^2 V_2^2 - h_x^2 V_2 V_3 \end{bmatrix}$

6.13 Proof

6.15 **E** is uniform within each element and its value is calculated using the following formula:

$$E = -\frac{1}{2A}\sum_{i=3}^{3} P_i V_{ei} a_x - \frac{1}{2A}\sum_{i=3}^{3} Q_i V_{ei} a_y$$

6.17 See Figure E.8.

6.19 The exact value is given by

$$k_{mn}^2 = (\pi/a)^2[(m+n)^2 + n^2] = \left(\frac{2\pi}{\lambda_c}\right)^2$$

$$\lambda_{c,mn} = \frac{a}{2}\sqrt{(m+n)^2 + n^2}, \quad a = 1$$

6.23 $B = 14$. With the renumbered mesh, $B = 4$.

6.25 a. $\dfrac{A}{3}(x_1 + x_2 + x_3)$.

b. $\dfrac{A}{12}\left[x_1^2 + x_2^2 + x_3^2 + 9\hat{x}^2\right]$, where $\hat{x}$ is defined in part (c).

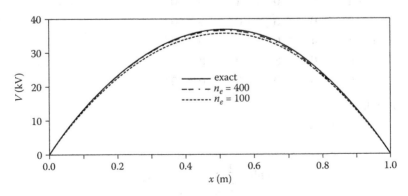

FIGURE E.8
For Problem 6.17.

c. $\dfrac{A}{12}\left[x_1y_1 + x_2y_2 + x_3y_3 + 9\hat{x}\hat{y}\right]$,

where

$$\hat{x} = \frac{1}{3}(x_1 + x_2 + x_3), \quad \hat{y} = \frac{1}{3}(y_1 + y_2 + y_3)$$

6.27 $\alpha_1 = \dfrac{y}{8}(y - 2)$

$\alpha_2 = \dfrac{y}{4}(4 - x - y)$

$\alpha_3 = \dfrac{1}{4}xy$

$\alpha_4 = \dfrac{1}{8}(4 - x - y)(2 - x - y)$

$\alpha_5 = \dfrac{1}{4}xy(4 - x - y)$

$\alpha_6 = \dfrac{x}{8}(x - 2)$

6.29 For $n = 1$,

$$Q^{(2)} = \frac{1}{2}\begin{bmatrix} 1 & 0 & -1 \\ 0 & 0 & 0 \\ -1 & 0 & 1 \end{bmatrix}, \quad Q^{(3)} = \frac{1}{2}\begin{bmatrix} 0 & 1 & -1 \\ 0 & -1 & 1 \\ 0 & 0 & 0 \end{bmatrix}$$

For $n = 2$,

$$Q^{(2)} = \frac{1}{6}\begin{bmatrix} 3 & 0 & -4 & 0 & 0 & 1 \\ 0 & 8 & 0 & 0 & -8 & 0 \\ -4 & 0 & 8 & 0 & 0 & -4 \\ 0 & 0 & 0 & 0 & 0 & 0 \\ 0 & -8 & 0 & 0 & 8 & 0 \\ 1 & 0 & -4 & 0 & 0 & 3 \end{bmatrix}$$

$$Q^{(3)} = \frac{1}{6}\begin{bmatrix} 3 & -4 & 0 & 1 & 0 & 0 \\ -4 & 8 & 0 & -4 & 0 & 0 \\ 0 & 0 & 8 & 0 & -8 & 0 \\ 1 & -4 & 0 & 3 & 0 & 0 \\ 0 & 0 & -8 & 0 & 8 & 0 \\ 0 & 0 & 0 & 0 & 0 & 0 \end{bmatrix}$$

6.31 a. Proof,

b. Proof. For $n = 1$,

$$Q^{(1)} = \frac{1}{2}\begin{bmatrix} 0 & 0 & 0 \\ 0 & 1 & -1 \\ 0 & -1 & -1 \end{bmatrix}$$

For $n = 2$,

$$Q^{(1)} = \frac{1}{6} \begin{bmatrix} 0 & 0 & 0 & 0 & 0 & 0 \\ 0 & 8 & -8 & 0 & 0 & 0 \\ 0 & -8 & 8 & 0 & 0 & 0 \\ 0 & 0 & 0 & 3 & -4 & -1 \\ 0 & 0 & 0 & -4 & 8 & -4 \\ 0 & 0 & 0 & 1 & -4 & 3 \end{bmatrix}$$

6.33 $\alpha_1 = \dfrac{1}{96}(96 - 24x - 8y - 8z), \quad \alpha_1 = \dfrac{1}{96}(24y)$

6.35 $T = \dfrac{\nu}{20} \begin{bmatrix} 2 & 1 & 1 & 1 \\ 1 & 2 & 1 & 1 \\ 1 & 1 & 2 & 1 \\ 1 & 1 & 1 & 2 \end{bmatrix}$

6.37 $B_1 = \dfrac{\partial}{\partial \rho} + jk + \dfrac{1}{2\rho}$

$B_2 = \dfrac{\partial}{\partial \rho} + jk + \dfrac{1}{2\rho} - \dfrac{1}{8\rho(1 + jk\rho)} - \dfrac{1}{2\rho(1 + jk\rho)} \dfrac{\partial^2}{\partial \phi^2}$

6.39 $40.587 \ \Omega$

6.41   See the table below.

w/h	$C_{11} = C_{22}$ (pF/m)	$C_{21} = C_{12}$ (pF/m)
2	64.46	−48.74
4	108.40	−88.42
8	195.999	−167.12

Chapter 7

7.1   Proof

7.3   Proof

7.5 $\Delta \ell / \lambda = 0.0501$

7.7   See Figure E.9

7.9   Proof

7.11   Proof

7.13   1/6 ns

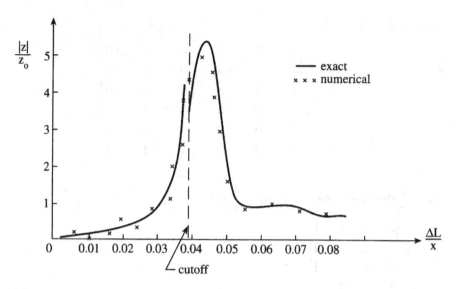

FIGURE E.9
For Problem 7.7.

7.15 See table below.

$\delta l / \lambda$	|Z|		arg (Z)	
	TLM	Exact	TLM	Exact
0.023	4.1961	6.1272	−0.2806	−0.0106
0.025	2.3822	2.4898	1.2546	1.0610
0.027	0.3281	0.3252	−0.7951	−0.8554
0.029	5.2724	5.1637	0.8459	0.8678
0.031	0.2963	0.3039	−1.1340	−1.1610
0.033	1.8117	1.8038	1.3408	1.3384
0.035	0.8505	0.8529	−1.3820	−1.4025
0.037	0.4912	0.4838	1.3914	1.3932
0.039	5.3772	5.4883	−1.1022	−1.1125
0.041	0.2115	0.2179	−1.2795	−1.3174

7.17 For $\varepsilon_r = 2$, $k_c a = 1.303$; for $\varepsilon_r = 8$, $k_c a = 0.968$
7.19 See Figure E.10

Chapter 8

8.3 a. 16, 187, 170, 429, 836, 47, 950, 369, 456, 307
    b. 997, 281, 13, 449, 277, 721, 133, 209, 757, 761

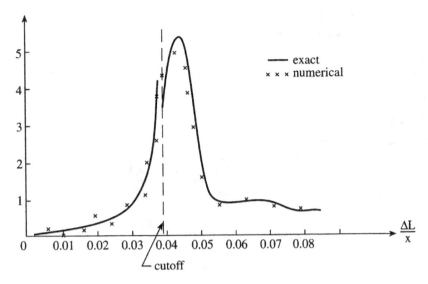

FIGURE E.10
For Problem 7.19.

8.7 $M = 5, a = 0, b = 1$. We generate the random variate as follows:
  1. Generate two uniform random variables U_1 and U_2 from (0,1)
  2. Check if $U_1 \le f_X(U_2)/M = U_2^2$.
  3. If the inequality holds, accept U_2 as the variate generated from $f_X(x)$.
  4. If the inequality is violated, reject U_1 and U_2 and repeat steps 1–3.

8.9 (a) 3.14156 (exact), (b) 0.4597 (exact), (c) 1.71828 (exact), (d) 2.0 (exact)

8.11 0.4053 (exact)

8.13 2.5 (exact)

8.15 See table below.

y	Exact	MCM (Floating)
1	0.408	0.402 ± 0.022
2	0.816	0.818 ± 0.026
3	1.224	1.226 ± 0.025
4	1.663	1.679 ± 0.039
5	2.041	2.045 ± 0.038
6	3.633	3.777 ± 0.046
7	5.224	5.262 ± 0.041
8	6.816	6.775 ± 0.048
9	8.408	8.395 ± 0.035

8.17 2.991 V

8.19 1.2 V (exact)

8.21 V(2,10) = 65.85, V(5,10) = 23.32, V(8,10) = 6.4, V(5,2) = 10.23, V(5,18) = 10.34

8.23 a. 0.33, 0.17, 0.17, 0.33

b. 0.455, 0.045, 0.045, 0.455, 0.455

8.25 12.11 V

8.27 10.44 V

8.29 35.55 V

8.31 23.41 V

8.33 See the table below.

t	Exact	MCM
0.05	1.491	1.534
0.10	0.563	0.6627
0.15	0.216	0.267
0.20	0.078	0.106
0.25	0.029	0.0419
0.30	0.0015	0.019

8.35 See the table below.

t	Exact	Exodus	FD
0.1	6.0191	6.2337	6.0573
0.2	3.3752	3.6034	3.3532
0.3	1.8933	2.1193	1.8770
0.4	1.0619	1.2488	1.0510
0.5	0.5955	0.736	0.589
1.0	0.0330	0.0266	0.0325

8.37 See the table below.

t	Exact	Random Walk	Exodus
0.05	1.491	1.534	1.5427
0.10	0.563	0.6627	0.6634
0.15	0.216	0.267	0.2727
0.20	0.078	0.106	0.1116
0.25	0.029	0.0419	0.0456
0.30	0.0015	0.019	0.0184

Chapter 9

9.1 Proof

9.3 $T_{ij} = \sqrt{\dfrac{2}{N+1/2}} \cos \dfrac{(i-0.5)(k-0.5)}{N+0.5}$

$\lambda_k = 2\sin\left(\dfrac{k-1/2}{N+1/2}\right)\pi$

9.7   See the table below.

(ρ, z)	Exact	MOL
(0,0.25)	5.375	5.370
(0,0.5)	7.123	7.196
(0,0.75)	5.375	5.370
(0.125,0.5)	7.433	7.415
(0.25,0.5)	8.065	8.046

9.9   Proof

Index

9781032339030